Management
tenth edition

Stephen P. Robbins
San Diego State University

Mary Coulter
Missouri State University

Neharika Vohra
Indian Institute of Management Ahmedabad

PEARSON

Authorized adaptation from the United States edition, entitled *Management, 10th Edition,* ISBN: 9780132090711 by Robbins, Stephen P.; Coulter, Mary, published by Pearson Education, Inc., publishing as Prentice Hall © 2009

Indian Subcontinent Adaptation

Published by Dorling Kindersley (India) Pvt. Ltd., licensees of Pearson Education in South Asia.

ISBN: 978-81-317-2720-1

First Impression, 2010
Seventh Impression, 2012
Eighth Impression, 2013

Head Office: 7th Floor, Knowledge Boulevard, A-8(A), Sector 62, Noida 201 309, India.
Registered Office: 14 Local Shopping Centre, Panchsheel Park, New Delhi 110 017, India.

Printed in India by Chennai Micro Print (P) Ltd., Chennai.

To my wife, Laura—SPR

To my husband, Ron—MC

About the Authors

Stephen P. Robbins (Ph.D., University of Arizona) is professor emeritus of management at San Diego State University and the world's best-selling textbook author in the areas of both management and organizational behavior. His books are used at more than 1,000 U.S. colleges and universities, have been translated into 19 languages, and have adapted editions for Canada, Australia, South Africa, and India. Dr. Robbins is also the author of the best-selling *The Truth About Managing People*, 2nd ed. (Financial Times/Prentice Hall, 2008) and *Decide & Conquer* (Financial Times/Prentice Hall, 2004).

In his "other life," Dr. Robbins actively participates in masters' track competitions. Since turning 50 in 1993, he's won 18 national championships, won 12 world titles, and set numerous U.S. and world age-group records at 60, 100, 200, and 400 meters. In 2005, Dr. Robbins was elected into the USA Masters' Track & Field Hall of Fame.

Mary Coulter received her Ph.D. in Management from the University of Arkansas in Fayetteville. Before completing her graduate work, she held different jobs, including high school teacher, legal assistant, and government program planner. She has taught at Drury University, the University of Arkansas, Trinity University, and, since 1983, Missouri State University. Dr. Coulter's research interests have focused on competitive strategies for not-for-profit arts organizations and the use of new media in the educational process. Her research on these and other topics has appeared in such journals as *International Journal of Business Disciplines, Journal of Business Strategies, Journal of Business Research, Journal of Nonprofit and Public Sector Marketing,* and *Case Research Journal.* In addition to *Management,* Dr. Coulter has published other books with Prentice Hall, including *Strategic Management in Action,* now in its fourth edition, and *Entrepreneurship in Action,* which is in its second edition. When she's not busy teaching or writing, she enjoys puttering around in her flower gardens, trying new recipes, reading all different types of books, and enjoying many different activities with Ron, Sarah and James, and Katie and Matt.

Neharika Vohra is a professor at IIM Ahmedabad in the Organizational Behavior Area. She completed her Ph.D. in Psychology from University of Manitoba, Canada. Her research interests are in the area of leadership, commitment, positive organization behavior and engagement/alienation, cross-cultural competence. She is the recipient of "Young Psychologist Award" at the International Congress of Psychology, and the "Best Teacher Award" at University of Manitoba, Arts Departement. She has published many papers in national and international journals and has been on the editorial board of several journals. She is currently the editor of "Vikalpa" the journal for decision makers, published by IIMA.

Brief Contents

Contents

Preface

You've made a good decision! You're taking a college course...maybe more than one. Although you may sometimes feel like you're wasting your time being in college, you're not. Yes, it's expensive. Yes, it's even sometimes hard. But what you're doing now will pay off in the long run. In a recent survey of job seekers, a whopping 92 percent said that a major disadvantage in competing for jobs was not having taken college courses. But that's not what you'll face because you *are* enrolled in a college course—the course for which you've purchased this book.

What This Course Is About and Why It's Important

This course and this book are about management and managers. Managers are the one thing that all organizations—no matter the size, kind, or location—need. And there's no doubt that the world that managers face has changed, is changing, and will continue to change. The dynamic nature of today's organizations means both rewards *and* challenges for the individuals who will be managing those organizations. Management is a dynamic subject, and a textbook on it should reflect those changes to help prepare you to manage under the current conditions. Thus, we've written this 10th edition of *Management* to provide you with the best possible understanding of what it means to be a manager confronting change.

Our Approach

Our approach to management is simple: Management is about people. Managers manage people. Thus, we introduce you to real managers, real people who manage people. We've talked with these real managers and asked them to share their experiences with you. You get to see what being a manager is all about—the problems these real managers have faced and how they have resolved those problems. Not only do you have the benefit of your professor's wisdom and knowledge, you also have access to your very own team of advisors and mentors.

What's Expected of You in This Course

It's simple. Come to class. Read the book. Do your assignments. And...study for your exams. If you want to get the most out of the money you've spent for this course and this textbook, that's what you need to do. In addition to writing this book, we teach. And that's what we expect of our students.

User's Guide

Your management course may be described as a "survey" course because a lot of topics are covered very quickly, and none of the topics are covered in great depth. It can be overwhelming at times! Your classroom professor is your primary source of information and will provide you with an outline of what you're expected to do during the course. That's also the person who will be evaluating your work and assigning you a grade, so pay attention to what is expected of you! View us, your textbook authors, as your supplementary professors. As your partners in this endeavor, we've provided you the best information possible both in the textbook and in the materials on mymanagementlab.com to help you succeed in this course. Now it's up to you to use them!

Getting the Most Out of Your Textbook: How to Get a Good Grade in This Course

Professors use a textbook because it provides a compact source of information that you need to know about the course's subject material. Professors like this particular textbook because it presents management from the perspective of the people who actually *do* management—real managers. So take advantage of that and read what these real managers have to say. See how they've handled managerial problems. Learn about their management styles and think about how you might manage.

In addition to what you can learn from these real managers, we provide several ways to help you get a good grade in this course. Use the *Quick Learning Reviews* scattered throughout the chapter. That's a great way to see if you understand the material you've just read. Also, use the materials on mymanagementlab.

Mymanagementlab is a powerful online tool that combines assessment, reporting, and personalized study to help you succeed. Mymanagementlab gives you the opportunity to test yourself on key concepts and skills, track your own progress through the course, and use the personalized study plan activities—all to help achieve success in the classroom.

Finally, we include a wide variety of useful learning experiences both in the textbook and on mymanagementlab. From ethical dilemmas and interactive skill-building modules to case analysis and hands-on management tasks, we've provided a lot of things to make your management course fun and worthwhile. Your professor will tell you what assignments you will be expected to do. But you don't need to limit your learning experiences to those. Try out some of the other activities, even if they aren't assigned. We know you won't be disappointed!

Student Supplements

CourseSmart is an exciting new choice for students looking to save money. As an alternative to purchasing the print textbook, students can purchase an electronic version of the same content and save up to 50 percent off the suggested list price of the print text. With a CourseSmart e-textbook, students can search the text, make notes online, print out reading assignments that incorporate lecture notes, and bookmark important passages for later review. For more information, or to purchase access to the CourseSmart e-textbook version of this text, visit www.coursesmart.com.

Acknowledgements

Every author relies on the comments of reviewers, and ours have been very helpful. We want to thank the following people for their insightful comments and suggestions for the 10th edition and previous editions of *Management*:

Suhail Abboushi, *Duquesne State*

Aline Arnold, *Eastern Illinois University*

Joseph Atallah, *DeVry Institute of Technology*

Robb Bay, *Community College of Southern Nevada*

Henry C. Bohleke, *San Juan College*

Ernest Bourgeois, *Castleton State College*

Jenell Bramlage, *University of Northwestern Ohio*

Jacqueline H. Bull, *Immaculata University*

James F. Cashman, *The University of Alabama*

Rick Castaldi, *San Francisco State University*

Bobbie Chan, *Open University of Hong Kong*

Jay Christensen-Szalanski, *University of Iowa*

Thomas Clark, *Xavier University*

Sharon Clinebell, *University of North Colorado*

Daniel Cochran, *Mississippi State University*

Augustus B. Colangelo, *Penn State*

Donald Conlon, *University of Delaware*

Roy Cook, *Fort Lewis College*

Anne C. Cowden, *California State University, Sacramento*

Claudia Daumer, *California State University, Chico*

Thomas Deckleman, *Owens Community College*

Mary Ann Edwards, *College of Mount St. Joseph*

Tan Eng, *Ngee Ann Polytechnic*

Allen D. Engle, Sr., *Eastern Kentucky University*

Judson C. Faurer, *Metro State College*

Dale M. Feinauer, *University of Wisconsin, Oshkosh*

Janice Feldbauer, *Austin Community College*

Diane L. Ferry, *University of Delaware*

Louis Firenze, *Northwood University*

Bruce Fischer, *Elmhurst College*

Phillip Flamm, *Angelo State University*

Barbara Foltz, *Clemson University*

June Freund, *Pittsburgh State University*

Michele Fritz, *DeAnza College*

Charles V. Goodman, *Texas A&M University*

H. Gregg Hamby, *University of Houston*

Frank Hamilton, *University of South Florida*

Robert W. Hanna, *California State University, Northridge*

James C. Hayton, *Utah State University*

Wei He, *Indiana State University*

Phyllis G. Holland, *Valdosta State College*

Henry Jackson, *Delaware County Community College*

Jim Jones, *University of Nebraska, Omaha*

Kathleen Jones, *University of North Dakota*

Marvin Karlins, *University of South Florida*

Andy Kein, *Keller Graduate School of Management*

David Kennedy, *Berkeley School of Business*

Russell Kent, *Georgia Southern University*

William H. Kirchman, *Fayetteville Technical Community College*

John L. Kmetz, *University of Delaware*

Gary Kohut, *University of North Carolina at Charlotte*

William Laing, *Anderson College*

Gary M. Lande, *Montana State University*

Ellis L. Langston, *Texas Tech University*

Les Ledger, *Central Texas College*

W. L. Loh, *Mohawk Valley Community College*

Susan D. Looney, *Delaware Technical and Community College*

James Mazza, *Middlesex Community College*

James McElroy, *Iowa State University*

Joseph F. Michlitsch, *Southern Illinois University–Edwardsville*

Sandy J. Miles, *Murray State University*

Lavelle Mills, *West Texas A&M University*

Corey Moore, *Angelo State University*

Rick Moron, *University of California, Berkeley*

Don C. Mosley, Jr., *University of South Alabama*

Anne M. O'Leary-Kelly, *Texas A&M University*

Rhonda Palladi, *Georgia State University*

Shelia Pechinski, *University of Maine*

Victor Preisser, *Golden Gate University*

Michelle Reavis, *University of Alabama, Huntsville*

Clint Relyea, *Arkansas State University, University of Arkansas*

James Robinson, *The College of New Jersey*

Patrick Rogers, *North Carolina A&T University*

James Salvucci, *Curry College*

Elliot M. Ser, *Barry University*

Tracy Huneycutt Sigler, *Western Washington University*

Eva Smith, *Spartanburg Technical College*

James Spee, *The Claremont Graduate School*

Roger R. Stanton, *California State University*

Dena M. Stephenson, *Calhoun Community College*

Charles Stubbart, *Southern Illinois University*

Ram Subramanian, *Grand Valley State University*

Thomas G. Thompson, *University of Maryland, University College*

Frank Tomassi, *Johnson & Wales University*

Isaiah O. Ugboro, *North Carolina A&T State University*

Philip M. VanAuken, *Baylor University*

Carolyn Waits, *Cincinnati State University*

Bill Walsh, *University of Illinois*

Emilia S. Westney, *Texas Tech University*

Gary L. Whaley, *Norfolk State University*

Bobbie Williams, *Georgia Southern University*

Wendy Wysocki, *Monroe Community College*

Our team at Prentice Hall has been amazing to work with! This team of editors, production experts, technology gurus, designers, marketing specialists, and sales representatives works hard to turn our digital files into a bound textbook and to see that it gets to faculty and students. We couldn't do this without all of you! Our thanks to the people who made this book "go" include Kim Norbuta, Nikki Jones, Judy Leale, Claudia Fernandes, Ana Jankowski, Ben Paris, Elisa Adams, Ann Courtney, Kitty Wilson, Steve Frim, Kristine Carney, Janet Slowik, Wendy Craven, and Sally Yagan.

A special thank you goes to Anita Looney, who worked very hard to coordinate all the materials and responses from our "real" managers. Anita, you were great! And simple words can't express how much we appreciate those "real" managers who so graciously gave their time to help us put together a textbook that is like no other on the market. Without their contributions, our belief in showing managers as real people would be hard to do. Thank you, thank you, thank you! Mary would also like to thank her college dean, Ron Bottin; her department head, Barry Wisdom; and her departmental secretary, Carole Hale, for all their support and encouragement.

Finally, Steve would like to thank his wife, Laura, for her encouragement and support. Mary would like to thank her husband, Ron, for being so supportive and understanding, and would like to tell Ron and her daughters and "new" sons-in-law—Sarah and James and Katie and Matt—that I love you all!

This book would not have been possible without the help of several people. I would like to acknowledge the help of various students, friends, colleagues, professionals who helped me put this information together to Indianize the book.

If anyone has tried to look for information specific to Indian industry they would know how difficult it is to find the information related to Indian organizations. Thus the help of many of my friends was really timely. I might forget some names but if I do please be sure that I still am grateful to you for your help and am sorry that I have missed you.

Rickden Lamba of Regalix helped me with many examples. Achal Raghavan was very helpful in providing many examples from the manufacturing sector. My students, currently professionals in their spheres, Subrat Beura, Sandeep Krishnan, Benny Joseph, Meenakshi Aggarwal, Nisha Nair, Sandeep, Patrali all graciously helped me with several examples and ideas. Sunil Jha also provided many examples in the service sector. Nimish Vohra, my brother also filled me in with several interesting experiences. My colleague, Professor Sriram, gave me several examples and material for the book. Sugatha Nair, my secretary, has helped in going through all the references and in managing the changes, copies of the book, couriering, etc.

All the managers who have been included in "meet the manager" section of the book have been extremely gracious in putting their thoughts together for the readers. They all responded to tight deadlines and delivered wonderful inputs for the book. I thank all of them specially.

The team from Pearson has also been very helpful and patient, Praveen Tiwari, Avnish Garg, and Ruchi Sachdev have been the people I have interacted with and they have been patient and great.

I would also like to acknowledge the support that my family members have provided while I worked on the book. My parents and my brother and sister, my nephews and nieces have all put up with my short phone calls, unavailability to them while I was working for several hours in a day for several months in putting this information together.

Part One

Defining the Manager's Terrain

▷ **Welcome to the world of management!** One thing is for certain: Organizations need managers. Good managers. No, not just good managers. They need *great* managers! They need people who can *set goals and plan* what needs to be done to achieve those goals. Organizations need people who can *organize and arrange* things so those goals can be met. They need people who can *lead and motivate* others in working toward those goals—who can pull up their sleeves and pitch in if needed. And they need people who can *evaluate* whether the goals were accomplished efficiently and effectively and who can change things when needed. Those "people" are managers. And great managers are essential for great organizations. We want to start you on your journey to being a *great* manager. What's it like to be a manager today?

In Part One, we look at the environment that managers face. Chapter 1 provides an introduction to management and organizations. Any time you face something new, you need to get a feel for it, and that's what we do in this chapter—give you a feel for management. In Chapter 2, we look at how the practice of management has changed over the years. In Chapter 3, we look at two constraints on how managers manage: the organization's culture and the external environment. Chapter 4 provides a look at the global aspects that today's organizations face. Finally, in Chapter 5, we look at the challenges managers face in being socially responsible and ethical.

Let's Get Real:
Meet the Manager

Shibaji Mandal
Director, Healthcare
CSR–Usha Martin Group

MY JOB: To manage and to be responsible for the healthcare initiatives of the group that includes running a chain of low-cost hospitals and clinics, and nursing schools.

BEST PART OF MY JOB: To be able to make decisions independently and to work in the rustic natural environment of rural India and the ability to make some difference.

WORST PART OF MY JOB: Dealing with nerve-wrecking bureaucracy, and difficulty in monitoring and assessing progress in the development sector.

BEST MANAGEMENT ADVICE EVER RECEIVED: Be honest and candid with your people and, practice differentiation in your workplace relating to performance and recognition.

You'll be hearing more from this real manager throughout the chapter.

Chapter 1

Introduction to Management and Organizations

In this chapter, we'll introduce you to who managers are and what they do. One thing you'll discover is that the work managers do is vitally important to organizations. But you'll also see that being a manager—a good manager—isn't easy. Focus on the following learning outcomes as you read and study this chapter.

LEARNING OUTCOMES

A Manager's Dilemma

Larsen & Toubro (L&T) Chairman A. P. Naik in a speech challenged the technocrats and asked them, "Who will build the nation?" L&T requires at least 2,000 engineers every year at its present growth rate but it is facing severe talent crunch. It is well known that in India there is a deep paucity of skilled labor to support the infrastructure, manufacturing, and construction needs. There is a growing trend in young trained engineers to opt for jobs in the Information Technology (IT) and IT-enabled services. M. S. Krishnamoorthy, Executive Vice President, of L&T, says, "we have been facing maximum attrition at the junior levels and have to resort to remunerative ways of retaining talent." Of course, pay is not the only reason for lack of applicants and attrition. Mr. Muthuraman, CMD TATA Steel, said that there are issues around the conditions of work that are less attractive to the youth today; he says, "like you can't have a blast furnace in an air-conditioned environment."[1] If you were an HR manager in L&T and mandated to hire entry-level engineers, what would be your strategy?

What would you do?

Today's successful managers need to possess leadership skills, the ability to build and manage teams, to motivate and keep employees engaged, and to manage the interface with the customers. This book is about managerial work similar to that of Mr. Naik, Mr. Krishnamoorthy, Mr. Muthuraman, and millions of other managers at various levels in the company do. The reality facing today's managers is that the world has changed. In workplaces of all types—offices, restaurants, retail stores, factories, and the like—managers must deal with new ways of organizing work. In this chapter, we introduce you to managers and management by looking at who managers are, what they do, and what an organization is. Finally, we wrap up the chapter by discussing why it's important to study management.

LEARNING **OUTCOME 1.1** ▷

WHO ARE MANAGERS?

Managers may not be who or what you might expect. They're under age 18 to over age 80. They run large corporations as well as entrepreneurial start-ups. They're found in government departments, hospitals, small businesses, not-for-profit agencies, museums, schools, and even such nontraditional organizations as political campaigns and consumer cooperatives. Managers can also be found doing managerial work in every country around the globe. In addition, some managers are top-level managers, while others are first-line managers. And today, managers are just as likely to be women as they are men, although the number of women who are top-level managers remains low (see Exhibit 1–1). According to the National Productivity Council's (NPC) State

Exhibit 1–1

Women in Managerial Positions Around the World

	Women in Management	Women in Top Manager's Job
Australia	41.9 percent	3.0 percent
Canada	36.3 percent	4.2 percent
Germany	35.6 percent	N/A
India	20 percent	1 percent
Japan	10.1 percent	N/A
Philippines	57.8 percent	N/A
United States	50.6 percent	2.6 percent

Sources: Metrics Pyramid Catalyst, www.catalyst.org/knowledge/metricspyramid, February 12, 2008; M. Fackler, "Career Women in Japan Find a Blocked Path," *New York Times* online, www.nytimes.com, August 6, 2007; "2006 Australian Census of Women in Leadership," Equal Opportunity for Women in the Workplace Agency, www.eowa.gov.au/Australian_Women_In_Leadership_Census.asp, http://www.csrindia.org/; http://www.ifj.org/en/articles/study-on-empowerment-of-women-in-the-workplace-in-india.

Gender Development Report 2005,[2] most Indian women entrepreneurs are still confined to marginal economic areas, such as micro and small enterprises. There are less than 10 professional Chief Executive Officers (CEOs) in top-renowned companies of India. In another more encouraging study by Grant Thornton,[3] Pallavi Bhakru, partner and director of international business at Grant Thornton, says that there are 14 percent women in the senior level management in India, and it has been mentioned that the number of women occupying management positions in leading companies both listed and privately held is increasing. In India, globalization has provided opportunities for educated, middle-class women to build their own dreams and excel in fields, which were hitherto perceived as male domains. The Confederation of Indian Industry released a study "Understanding the Levels of Empowerment of Women in the Workplace in India" covering 149 large- and medium-sized companies across regions. The report highlights that women comprise 16 percent at junior management level, 4 percent each at the middle and senior levels, and only 1 percent in the organizational leadership positions (CEOs). But no matter where managers are found or what gender they are, the fact is that managers have exciting and challenging jobs. And organizations need managers more than ever in these uncertain, complex, and chaotic times. *Managers do matter!* How do we know that? The Gallup Organization, which has polled millions of employees and tens of thousands of managers, has found that the single most important variable in employee productivity and loyalty isn't pay or benefits or workplace environment; it's the quality of the relationship between employees and their direct supervisors.[4] In addition, global consulting firm Watson Wyatt Worldwide found that the way a company manages its people can significantly affect its financial performance.[5] We can conclude from such reports that managers *do* matter!

It used to be fairly simple to define who managers were: They were the organizational members who told others what to do and how to do it. It was easy to differentiate *managers* from *nonmanagerial employees*. But it isn't quite so simple anymore. In many organizations, the changing nature of work has blurred the distinction between managers and nonmanagerial employees. Many nonmanagerial jobs now include managerial activities.[6] In India, there are two examples of companies being taken over by employees themselves, the Kamani Tubes Limited in 1998 and the India Coffee House. The India Coffee House is run by employee cooperatives and has more than 400 branches all over the country. Though some of them are ailing because of the advent of more swanky coffee shops such as Café Coffee Day and Barista, there are several patrons who still swear by the coffee and the ambience at the India Coffee House. In the India Coffee House, there isn't much distinction between the manager and the worker. In the CNBC Aawaaz Consumer Awards held recently, it secured the runners up position.[7]

The Chairman of Coal India, Partha Bhattacharjee, advocated the introduction of the multiskilled jobs to improve the efficiency of the coal miners. However, this experiment was not very successful, partly because of resistance from the unions and also

Exhibit 1–2

Levels of Management

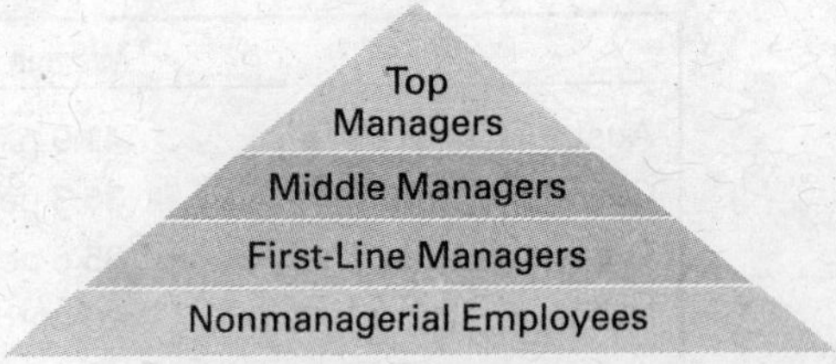

because of one or two accidents which occurred during that time. There are other manufacturing firms, such as Kirloskar and Sundaram Fasteners, where multiskilling was tried and has been successful.

So, how *do* we define who managers are? A **manager** is someone who coordinates and oversees the work of other people so that organizational goals can be accomplished. A manager's job is not about *personal* achievement—it's about helping *others* do their work. That may mean coordinating the work of a departmental group, or it might mean supervising a single person. It could involve coordinating the work activities of a team of people from different departments or even people outside the organization, such as temporary employees or employees who work for the organization's suppliers. Keep in mind, also, that managers may have work duties not related to coordinating and overseeing others' work. For example, an insurance claims supervisor may also process claims in addition to coordinating the work activities of other claims clerks.

Is there a way to classify managers in organizations? In traditionally structured organizations (which are said to be shaped like a pyramid because more employees are at lower organizational levels than at upper organizational levels), managers are often classified as first-line, middle, or top (see Exhibit 1–2). At the lowest level of management, **first-line managers** manage the work of nonmanagerial employees who typically are involved with producing the organization's products or servicing the organization's customers. First-line managers are often called *supervisors* but may also be called *shift managers, district managers, department managers,* or *office managers.* **Middle managers** are those found between the lowest and top levels of the organization. These managers manage the work of first-line managers and may have titles such as *regional manager, project leader, store manager,* or *division manager.* At the upper levels of the organization are the **top managers**, who are responsible for making organizationwide decisions and establishing the plans and goals that affect the entire organization. These individuals typically have titles such as *executive vice president, president, managing director, chief operating officer,* or *chief executive officer.* In our chapter-opening dilemma, Allyson Koteski is a middle manager. As the store manager, she's responsible for how her store performs, but she also is one of about 1,500 store managers companywide who report to someone at corporate headquarters.

Not all organizations get work done using this traditional pyramidal form, however. Some organizations, for example, are more loosely configured, with work being done by ever-changing teams of employees who move from one project to another as work demands arise. Although it's not as easy to tell who the managers are in these organizations, we do know that someone must fulfill that role—that is, there must be someone who coordinates and oversees the work of others, even if that "someone" changes as work tasks or projects change.

QUICK LEARNING REVIEW:

LEARNING OUTCOME 1.1

- Explain how managers differ from nonmanagerial employees.
- Describe how to classify managers in organizations.

Go to page 18 to see how well you know this material.

thinking critically about Ethics

How far should companies go to influence lives of the users through their services? In 2005, Google introduced a web-based Really Simple Syndication (RSS) reader, which permitted users to share their feeds with a selected group of friends. On December 14, 2007, Google permitted anyone in your Google contact book to see your "shared" list. People were very upset about the violation of privacy. There was no option to opt out of the "feature" all Google Reader users were opted in by default. Google changed both the rules, mid-game, and opted everyone into the change.[8] What do you think about this? Does technology provided by the companies also allow them to influence people's choices in this way?

LEARNING OUTCOME 1.2 ▷

WHAT IS MANAGEMENT?

Simply speaking, management is what managers do. But that simple statement doesn't tell us much, does it? A better explanation is that **management** involves coordinating and overseeing the work activities of others so that their activities are completed efficiently and effectively. We already know that coordinating and overseeing the work of others is what distinguishes a managerial position from a nonmanagerial one. However, this doesn't mean that managers can do what they want anytime, anywhere, or in any way. Instead, management involves ensuring that work activities are completed efficiently and effectively by the people responsible for doing them—or at least that's what managers aspire to do.

Let's Get Real: F2F

DO YOU THINK EFFICIENCY INITIATIVES THAT "WATCH" EMPLOYEES ARE ETHICAL?

There is a thin line between "watching" and "monitoring" employees that makes it difficult to demarcate between the two. Also, though unethical at times, it is a necessary evil that you need to practice before trust is built.

Efficiency refers to getting the most output from the least amount of inputs. Because managers deal with scarce inputs—including resources such as people, money, and equipment—they're concerned with the efficient use of those resources. It's often referred to as "doing things right"—that is, not wasting resources. For instance, the Hero Honda factory in Gurgaon has been using efficient manufacturing techniques such as utilization of CAD/CAM, flexible manufacturing systems, automatic material handling, robotics (for welding and sheet metal processing), giving importance to environmental issues, and the integrated use of information technology. The use of these manufacturing techniques has led to an increase in labor productivity, capacity utilization, on-time delivery, innovative product development, and has also enabled the companies to offer their products at competitive prices to their customers.[9]

It's not enough, however, just to be efficient. Management is also concerned with being effective, completing activities so that organizational goals are attained. **Effectiveness** is often described as "doing the right things"—that is, doing those work activities that will help the organization reach its goals. For instance, at the Hero Honda factory, goals include meeting customers' rigorous demands, executing world-class manufacturing strategies, and making employee jobs easier and safer. Through various work initiatives, these goals have been pursued *and* achieved. Whereas efficiency is concerned with the *means* of getting things done, effectiveness is concerned with the *ends*, or attainment of organizational goals (see Exhibit 1–3). In successful organizations, high efficiency and high effectiveness typically go hand in hand. Poor management (which leads to poor performance) usually involves being inefficient and ineffective or being effective but inefficient.

manager
Someone who coordinates and oversees the work of other people in order to accomplish organizational goals.

first-line managers
The lowest level of management who manage the work of nonmanagerial employees and typically are directly or indirectly involved with producing the organization's products or servicing the organization's customers.

middle managers
Managers between the lowest level and top levels of the organization who manage the work of first-line managers.

top managers
Managers at or near the upper levels of the organization structure who are responsible for making organizationwide decisions and establishing the goals and plans that affect the entire organization.

management
Coordination and oversight of the work activities of others so that their activities are completed efficiently and effectively.

efficiency
Doing things right, or getting the most output from the least amount of inputs.

effectiveness
Doing the right things, or completing activities so that organizational goals are attained.

Exhibit 1–3

Efficiency and Effectiveness in Management

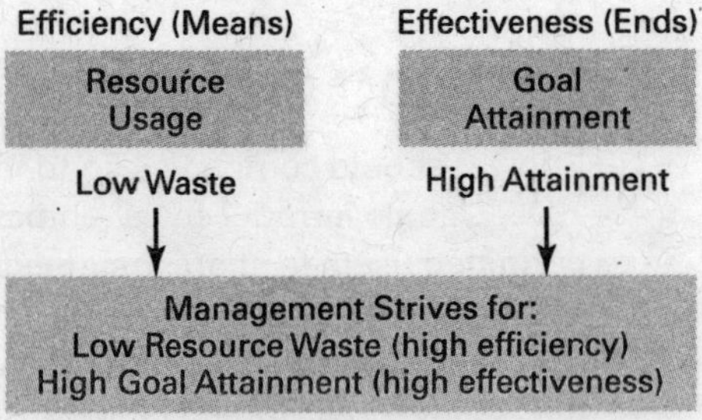

QUICK LEARNING REVIEW:

LEARNING OUTCOME 1.2

- **Define management.**
- **Explain why efficiency and effectiveness are important to management.**

Go to page 18 to see how well you know this material.

LEARNING **OUTCOME 1.3** ▷ WHAT DO MANAGERS DO?

Describing what managers do isn't easy. Just as no two organizations are alike, no two managers' jobs are alike. Despite this, management researchers have developed three approaches to describe what managers do: functions, roles, and skills. In this section, we'll examine each approach and take a look at how the manager's job is changing.

MANAGEMENT FUNCTIONS

According to the functions approach, managers perform certain activities or functions as they efficiently and effectively coordinate the work of others. What are these functions? Henri Fayol, a French businessman, first proposed in the early part of the twentieth century that all managers perform five functions: planning, organizing, commanding, coordinating, and controlling.[10] Today, these functions have been condensed to four: planning, organizing, leading, and controlling (see Exhibit 1–4). Let's briefly look at each function.

If you have no particular destination in mind, then any road will suffice. However, if you have someplace in particular you want to go, you need to plan the best way to get there. Because an organization exists to achieve some particular purpose, someone must define that purpose and the means for its achievement. Management is that someone. As managers engage in **planning**, they define goals, establish strategies for achieving those goals, and develop plans to integrate and coordinate activities.

Managers are also responsible for arranging and structuring work to accomplish the organization's goals. We call this function **organizing**. When managers organize, they determine what tasks are to be done, who is to do them, how the tasks are to be grouped, who reports to whom, and where decisions are to be made.

Every organization has people, and a manager's job is to work with and through people to accomplish goals. This is the **leading** function. When managers motivate

Exhibit 1–4

Management Functions

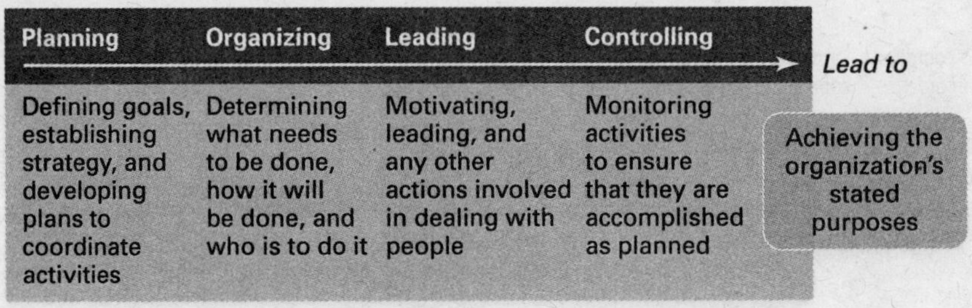

subordinates, help resolve work group conflicts, influence individuals or teams as they work, select the most effective communication channel, or deal in any way with employee behavior issues, they are leading.

The final management function is **controlling**. After goals and plans are set (planning), tasks and structural arrangements are put in place (organizing), and people are hired, trained, and motivated (leading), there has to be some evaluation of whether things are going as planned. To ensure that goals are being met and that work is being done as it should be, managers must monitor and evaluate performance. Actual performance must be compared with the set goals. If those goals aren't being achieved, it's management's job to get work back on track. This process of monitoring, comparing, and correcting is the controlling function.

Just how well does the functions approach describe what managers do? Do managers always plan, organize, lead, and then control? In reality, what a manager does may not always happen in this sequence. Regardless of the order in which these functions are performed, the fact is that managers do plan, organize, lead, and control as they manage. To illustrate, look back at the chapter-opening story. What examples of the management functions can you see? When Allyson is working to keep her employees motivated and engaged, that's leading. As she deals with unhappy customers, she has to control, lead, and maybe even plan.

Although the functions approach is popular for describing what managers do, some have argued that it isn't relevant.[11] So let's look at another perspective.

MANAGEMENT ROLES

Henry Mintzberg, a well-known management researcher, studied actual managers at work and concluded that what managers do can best be described by looking at the management roles they use at work.[12] The term **management roles** refers to specific actions or behaviors expected of a manager. (Think of the different roles you play—such as student, employee, student organization member, volunteer, sibling, and so forth—and the different things you're expected to do in these roles.) As shown in Exhibit 1–5, Mintzberg's 10 roles are grouped around interpersonal relationships, the transfer of information, and decision making.

The **interpersonal roles** are ones that involve people (subordinates and persons outside the organization) and other duties that are ceremonial and symbolic in nature. The three interpersonal roles are figurehead, leader, and liaison. The **informational roles** involve collecting, receiving, and disseminating information. The three informational roles are monitor, disseminator, and spokesperson. Finally, the **decisional roles** entail making decisions or choices. The four decisional roles are entrepreneur, disturbance handler, resource allocator, and negotiator.

As managers perform these roles, Mintzberg proposed that their activities included both reflection (thinking) and action (doing).[13] We can see both in our chapter opener. Reflection is shown as Allyson listens patiently to customers' problems, and action occurs when Allyson resolves those problems.

A number of follow-up studies have tested the validity of Mintzberg's role categories, and the evidence generally supports the idea that managers—regardless of the type of organization or level in the organization—perform similar roles.[14] However, the emphasis that managers give to the various roles seems to change with organizational level.[15] At higher levels of the organization, the roles of disseminator, figurehead, negotiator, liaison, and spokesperson are more important, while the leader role

planning
A management function that involves defining goals, establishing strategies for achieving those goals, and developing plans to integrate and coordinate activities.

organizing
A management function that involves arranging and structuring work to accomplish organizational goals.

leading
A management function that involves working with and through people to accomplish organizational goals.

controlling
A management function that involves monitoring, comparing, and correcting work performance.

management roles
Specific categories of managerial behavior.

interpersonal roles
Managerial roles that involve people and other duties that are ceremonial and symbolic in nature.

informational roles
Managerial roles that involve collecting, receiving, and disseminating information.

decisional roles
Managerial roles that revolve around making choices.

Exhibit 1–5

Mintzberg's Managerial Roles

Adapted from Mintzberg, Henry, The Nature of Managerial Work, 1st Edition, © 1980, pp. 93–94.

Mintzerg's Managerial Roles

Interpersonal Roles

- Figurehead
- Leader
- Liaison

Informational Roles

- Monitor
- Disseminator
- Spokesperson

Decisional Roles

- Entrepreneur
- Disturbance handler
- Resource allocator
- Negotiator

(as Mintzberg defined it) is more important for lower-level managers than it is for either middle- or top-level managers.

So which approach is better—functions or roles? Although each describes what managers do, the functions approach seems to be the best way of describing the manager's job. "The classical functions provide clear and discrete methods of classifying the thousands of activities that managers carry out and the techniques they use in terms of the functions they perform for the achievement of goals."[16] However, Mintzberg's role approach does offer another insight into managers' work.

According to Henry Mintzberg, managers fulfill several roles, one of which he called the interpersonal role. In this function, the manager relies heavily on his or her people skills to be both a leader who motivates subordinates, and a liaison who maintains a network of outside contacts that help provide a flow of information. The interpersonal role also calls on the manager to perform some symbolic duties such as signing legal and company documents.

Exhibit 1–6

Skills Needed at Different Managerial Levels

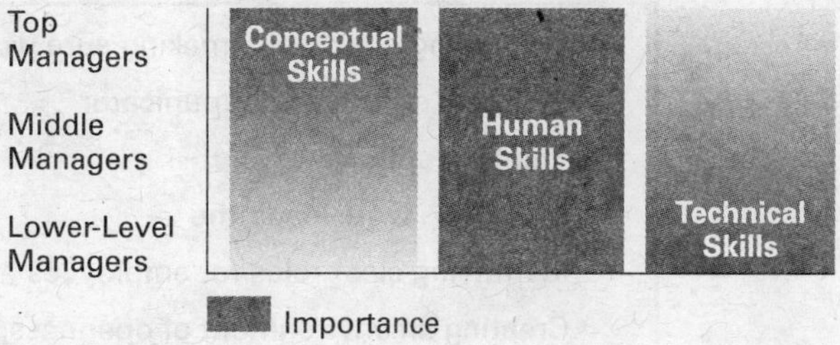

Let's Get Real: F2F

SKILLS THAT ARE IMPORTANT TO ME AS A MANAGER:

- Manage people, time, and task.
- Interpersonal skills and communication.
- Ability to withstand stress and navigate through chaos and contingencies.
- Ability to forecast and plan accordingly.

YOU CAN DEVELOP YOUR SKILLS BY…

- Judicious selection of a mentor at the workplace.
- Continuous updation of knowledge.
- Discussion and exchange of ideas among peers and friends.
- Having the courage to question.
- Challenging the status quo.
- Being open to failure and to learn from mistakes.

MANAGEMENT SKILLS

Dell Inc. is one company that understands the importance of management skills.[17] It started an intensive five-day offsite skills training program for first-line managers as a way to improve its operations. One of Dell's directors of learning and development thought this was the best way to develop "leaders who can build that strong relationship with their front-line employees." What have the supervisors learned from the skills training? Some things they have mentioned were how to communicate more effectively and how to refrain from jumping to conclusions when discussing a problem with a worker.

What types of skills do managers need? Robert L. Katz developed one approach to describing management skills; he concluded that managers need three essential skills: technical, human, and conceptual. Exhibit 1–6 shows the relationships of these skills and the levels of management.[18] **Technical skills** are the job-specific knowledge and techniques needed to proficiently perform work tasks. These skills tend to be more important for first-line managers because they typically manage employees who use tools and techniques to produce the organization's products or to service the organization's customers. Often, employees with excellent technical skills get promoted to first-line manager positions. However, success in managerial positions, merely relied on technical skills, can be fatal. This is understood by most of the top leaders; for example, Wipro's head Azim Premji in an interview with Ravi Aron, a professor at Wharton School of Business, mentioned, "our managers need to have a strong integration of managerial skills and technical understanding. One cannot substitute for the other. That is not an easy combination to get, especially if you want to sustain growth. Technical people tend to be more 'techie' and management people are more 'managerial.'"[19] To be able to strongly integrate managers who have a deep understanding of technology, Wipro provides leadership training at three levels: entry, supervisory, and senior/top management. Training inputs are linked with the evaluation of performance. Training is also supported by continuously exposing the managers to a blend of responsibility for people and technology. Several managers recognize the importance of **human skills**, which involve the ability to work well with other people both individually and in groups. Because all managers deal with people, these skills are equally important to all levels of management. Managers with good human skills get the best out of their people. They know how to communicate, motivate, lead, and inspire enthusiasm and trust. Finally, **conceptual skills** are the skills managers use to think and to conceptualize about abstract and complex situations. Using these skills, managers see the organization as a whole, understand the relationships among various subunits, and visualize how the organization fits into its broader environment. These skills are most important to top managers.

Some other important managerial skills that have been identified in various studies are listed in Exhibit 1–7. In today's demanding and dynamic workplace, employees who want to be valuable assets must constantly upgrade their skills and take on extra work outside their own specific job areas. We feel that understanding and developing management skills is so important that we've included a skills feature in our

technical skills
Job-specific knowledge and techniques needed to proficiently perform work tasks.

human skills
The ability to work well with other people individually and in a group.

conceptual skills
The ability to think and to conceptualize about abstract and complex situations.

Exhibit 1–7

Important Managerial Skills

- Delegating effectively (making sure work gets done right)
- Being an effective communicator
- Thinking critically
- Managing work load/time
- Identifying clear roles for employees
- Creating an environment of openness, trust, and challenge

Sources: Based on "Management Practices That Work," *McKinsey Quarterly Chart Focus Newsletter*, www.mckinsey quarterly.com/newsletters/chartfocus, September 2007; P. Korkki, "Young Workers: U Nd 2 Improve Ur Writing Skills," *New York Times* online, www.nytimes.com, August 26, 2007; and J. Jenkins, "The Bottom Line on Improperly Trained Leaders," *AMA Leader's EDGE*, www.amanet.org/LeadersEdge, December 2005.

mymanagementlab. There, you'll find material on skill building as well as several interactive skills exercises. As you study the four management functions throughout the rest of the book, you'll be able to practice some key management skills. Although a simple skill-building exercise won't make you an instant expert, it can provide an introductory understanding of some of the skills you'll need to master in order to be an effective manager.

HOW THE MANAGER'S JOB IS CHANGING

In India, in many companies even in the government, there has been a move to link pay to performance. People are not being paid for how much time they spend in the organization, rather, they are being paid for the quantity and quality of output they produce. Tanishq, the fashion jewellery maker, ABB India Ltd, and Home Appliances BPL Ltd have implemented schemes for pay to be linked to performance at all levels of the organization. In fact, Tanishq was among the first few manufacturing organizations to roll it out at the worker's level as well.[20] The railways and the postal department of the Government have also introduced measures of results rather than the time put in.[21] In many IT companies, top management is urging people to work smarter rather than harder and design their work such that they can spend quality time with their families.[22]

In today's world, managers are dealing with changing workplaces, security threats, ethical issues, global economic and political uncertainties, and technological advancements. For example, diamond cutter and polishers in Surat, Gujarat, have had to struggle to keep their operations open when faced with dwindling orders from Belgium and other countries. The uneducated cutters and polishers were suddenly left with no jobs. The Indian Express on December 22, 2008 reported that more than 1,200 people.working in the groundnut-processing units near Junagarh, Gujarat, had been rendered jobless because of fluctuations in the groundnuts and buyers returning shiploads of the crop from all over the world. Consider the management challenges faced by the managers of the Taj hotel and the Oberoi hotel in Mumbai, after the terror strike in November 2008 to induce their employees to come back to work. The fact that the Chattrapati Shivaji Terminus was able to resume functioning in less than 24 hours after the terror attack must have required great managerial talent. The challenges faced by the Delhi Metro to keep its employees motivated to perform despite the level of government salaries and opposition from the land-owners in crowded areas are not often discussed. Tata's decision to move Nano almost overnight from West Bengal to Gujarat must have posed serious managerial challenges with respect to relocating machinery and people, maintaining continuity, keeping the project on time, and keeping costs under control. Although most managers are not likely to have to manage under such demanding circumstances, the fact is that *how* managers manage is changing. Exhibit 1–8 shows some of the most important changes managers face. Throughout the rest of this book, we'll be discussing these and other changes and how they're affecting the way managers plan, organize, lead, and control.

Exhibit 1–8

Changes Affecting a Manager's Job

Change	Impact of Change
Changing Technology (Digitization)	Shifting organizational boundaries Virtual workplaces More mobile workforce Flexible work arrangements Empowered employees Work life-personal life balance
Changing Security Threats	Risk management Uncertainty over future energy sources/prices Restructured workplace Discrimination concerns Globalization concerns Employee assistance
Increased Emphasis on Organizational and Managerial Ethics	Redefined values Rebuilding trust Increased accountability
Increased Competitiveness	Customer service Innovation Globalization Efficiency/productivity

We want to highlight two of these changes: the increasing importance of customers and innovation.

Importance of Customers to the Manager's Job. John Chambers, CEO of Cisco Systems, likes to listen to voice mails forwarded to him from dissatisfied customers. He said, "E-mail would be more efficient, but I want to hear the emotion, I want to hear the frustration, I want to hear the caller's level of comfort with the strategy we're employing. I can't get that through e-mail."[23] This is a manager who understands the importance of customers. You need customers. Without them, most organizations would cease to exist. Yet, focusing on the customer has long been thought to be the responsibility of marketing types. "Let the marketers worry about the customers" is how many managers felt. We're discovering, however, that employee attitudes and behaviors play a big role in customer satisfaction. For a moment, imagine that you are a customer in a bank and were asked, "what are your essential needs vis-à-vis banking?"

HCL Infosystems, India's premier information-enabling ICT System Integrator and Distribution Company, is committed to providing excellent customer-care services. It has introduced "HCL TOUCH," a 24 × 7 round the clock service and support, for its Leaptop customers in over 4,000 towns in India. This is also backed by several other initiatives such as SMS request for support, pick and drop facility, and "On Site" direct company services anywhere in India at the customer's premises. The HCL TOUCH network consists of 14 remote support centers, 505 service offices, 390 warehouses, and 150 repair centers.[24]

Courtesy: HCL Technologies

Your list may have points such as quick service, friendly and efficient disposal of your needs, timely help, accurate information, safety, and security. If you were to reflect on this list, you would find that each of your needs is influenced by the actions of the employees in the bank. Today, the majority of employees in developed countries work in service jobs. For instance, some 79 percent of the U.S. labor force is employed in service industries. In Australia, 71 percent work in service industries. In the United Kingdom, Germany, and Japan, the percentages are 76, 70, and 73, respectively.[25] Examples of service jobs include technical support representatives, fast-food counter workers, sales clerks, teachers, food servers, nurses, computer repair technicians, front desk clerks, consultants, purchasing agents, credit representatives, financial planners, and bank tellers. Managers are recognizing that delivering consistent high-quality customer service is essential for survival and success in today's competitive environment and that employees are an important part of that equation.[26] The implication is clear: Managers must create a customer-responsive organization where employees are friendly and courteous, accessible, knowledgeable, prompt in responding to customer needs, and willing to do what's necessary to please customers.[27] We'll look at customer service management in several chapters.

Importance of Innovation to the Manager's Job. "Nothing is more risky than not innovating."[28] Innovation means doing things differently, exploring new territory, and taking risks. And innovation isn't just for high-tech or other technologically sophisticated organizations. Innovative efforts can be found in all types of organizations. For instance, the managers of HPCL, clearly understood the importance of getting their petrol pump personnel to be innovative, a task made particularly challenging because the average HPCL pump is often staffed by young adults in their first or second jobs. However, the managers tackled the problem by putting in place good training.[29] We'll study innovation in several chapters.

QUICK LEARNING REVIEW:

LEARNING OUTCOME 1.3

- Describe the four functions of management.
- Explain Mintzberg's managerial roles.
- Describe Katz's three essential managerial skills and how the importance of these skills changes depending on managerial level.
- Discuss the changes that are impacting managers' jobs.
- Explain why customer service and innovation are important to the manager's job.

Go to page 18 to see how well you know this material.

LEARNING OUTCOME 1.4 ▷ WHAT IS AN ORGANIZATION?

Managers work in organizations. But what is an **organization**? It's a deliberate arrangement of people to accomplish some specific purpose. Your college or university is an organization; so are fraternities and sororities, government departments, churches, Amazon.com, your neighborhood video rental store, the United Way, the Colorado Rockies baseball team, and the Mayo Clinic. All are considered organizations because all have the three common characteristics (see Exhibit 1–9).

Exhibit 1–9

Characteristics of Organizations

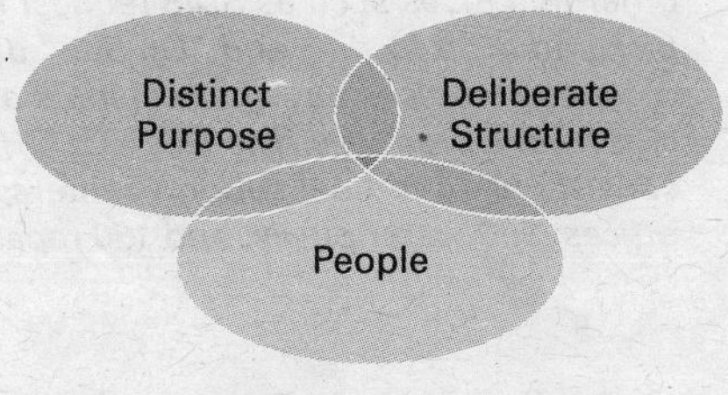

First, an organization has a distinct purpose. This purpose is typically expressed through goals that the organization hopes to accomplish. Second, each organization is composed of people. It takes people to perform the work that's necessary for the organization to achieve its goals. Third, all organizations develop some deliberate structure within which members do their work. That structure may be open and flexible, with no specific job duties or strict adherence to explicit job arrangements. For instance, at Google, most big projects, of which there are hundreds going on at the same time, are tackled by small focused employee teams that set up in an instant and complete work just as quickly.[30] Or the structure may be more traditional—like that of Procter & Gamble or General Motors—with clearly defined rules, regulations, job descriptions, and some members identified as "bosses" who have authority over other members.

Many of today's organizations are structured more like Google, with flexible work arrangements, employee work teams, open communication systems, and supplier alliances. In these organizations, work is defined in terms of tasks to be done. And workdays have no time boundaries because work can—and is—done anywhere, anytime. However, no matter what type of approach an organization uses, some deliberate structure is needed so work can get done efficiently and effectively.

QUICK LEARNING REVIEW:

LEARNING OUTCOME 1.4

- Explain the characteristics of an organization.
- Describe how today's organizations are structured.

Go to page 19 to see how well you know this material.

LEARNING **OUTCOME 1.5** ▷

WHY STUDY MANAGEMENT?

You may be wondering why you need to study management. If you're majoring in accounting or marketing or any field other than management, you may not understand how studying management is going to help you in your career. We can explain the value of studying management by looking at three things: the universality of management, the reality of work, and the rewards and challenges of being a manager.

THE UNIVERSALITY OF MANAGEMENT

Just how universal is the need for management in organizations? We can say with absolute certainty that management is needed in all types and sizes of organizations, at all organizational levels and in all organizational work areas, and in all organizations, no matter where they're located. This is known as the **universality of management** (see Exhibit 1–10). In all these organizations, managers must plan, organize, lead, and control. However, that's not to say that management is done the same way. What a supervisor in a software applications testing facility at Microsoft does versus what the CEO of Microsoft does is a matter of degree and emphasis, not of function. Because both are managers, both will plan, organize, lead, and control. How much and how they do so will differ, however.

Management is universally needed in all organizations, so we want to find ways to improve the way organizations are managed. Why? Because we interact with organizations every single day. Are you frustrated when you have to spend two hours in a governmental office to get your driver's license renewed? Do you get irritated when none of the salespeople in a retail store seems interested in helping

organization
A deliberate arrangement of people to accomplish some specific purpose.

universality of management
The reality that management is needed in all types and sizes of organizations, at all organizational levels, in all organizational areas, and in organizations no matter where located.

Exhibit 1–10

Universal Need for Management

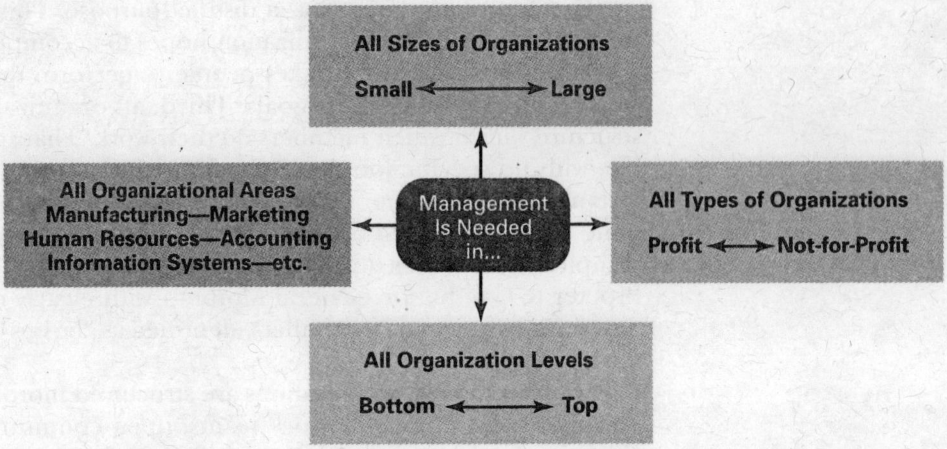

you? Is it annoying when you call an airline three times, and customer sales representatives quote you three different prices for the same trip? These are all examples of problems created by poor management. Organizations that are well managed—and we'll share many examples of these throughout the text—develop a loyal customer base, grow, and prosper. Those that are poorly managed find themselves losing customers and revenues. By studying management, you'll be able to recognize poor management and work to get it corrected. In addition, you'll be able to recognize and support good management, whether it's in an organization with which you're simply interacting or whether it's in an organization in which you're employed.

THE REALITY OF WORK

Besides the universality of management, another reason for studying management is the reality that for most of you, once you graduate from college and begin your career, you will either manage or be managed. For those who plan to be managers, an understanding of management forms the foundation on which to build your management skills. For those of you who don't see yourself managing, you're still likely to have to work with managers. Also, assuming that you will have to work for a living and recognizing that you are very likely to work in an organization, you'll probably have some managerial responsibilities even if you're not a manager. Our experience tells us that you can gain a great deal of insight into the way your boss (and fellow employees) behave and how organizations function by studying management. Our point is that you don't have to aspire to be a manager to gain something valuable from a course in management.

REWARDS AND CHALLENGES OF BEING A MANAGER

We can't leave our discussion of the value of studying management without looking at the rewards and challenges of being a manager (see Exhibit 1–11). What *does* it mean to be a manager in today's workplace?

First, there are many challenges. Management can be a tough and often thankless job. In addition, a portion of a manager's job (especially at lower organizational levels) may entail duties that are often more clerical (compiling and filing reports, dealing with bureaucratic procedures, or doing paperwork) than managerial.[31] Managers often have to deal with a variety of personalities and have to make do with limited resources. It can be a challenge to motivate workers in the face of uncertainty and chaos. And managers may find it difficult to successfully blend the knowledge, skills, ambitions, and experiences of a diverse work group. Finally, as a manager, you're not in full control of your destiny. Your success typically depends on others' work performance.

Yet, despite the challenges, being a manager *can be* rewarding. As a manager, you're responsible for creating a work environment in which organizational members

Let's Get Real: F2F

CHALLENGES I'VE FACED AS A MANAGER:

- Keeping pace with a continuously changing business environment.
- Energizing and motivating the team and managing ambitions and idiosyncrasies of star performers.
- Meeting deadlines and targets and walking the talk (read promises and commitments made) with the client at the same time.

REWARDS OF BEING A MANAGER:

- Satisfaction of creating a winning team.
- Making a difference in developing the careers of your junior reportees.
- Ability to turn around an organization through difficult situations.
- Create an impact among all stakeholders.
- Recognition and incentives (both monetary and nonmonetary)

Exhibit 1–11

Rewards and Challenges of Being a Manager

Rewards	Challenges
• Create a work environment in which organizational members can work to the best of their ability • Have opportunities to think creatively and use imagination • Help others find meaning and fulfillment in work • Support, coach, and nurture others • Work with a variety of people • Receive recognition and status in organization and community • Play a role in influencing organizational outcomes • Receive appropriate compensation in form of salaries, bonuses, and stock options • Good managers are needed by organizations	• Do hard work • May have duties that are more clerical than managerial • Have to deal with a variety of personalities • Often have to make do with limited resources • Motivate workers in chaotic and uncertain situations • Blend knowledge, skills, ambitions, and and experiences of a diverse work group • Success depends on others' work performance

can do their work to the best of their abilities and thus help the organization achieve its goals. You help others find meaning and fulfillment in their work. You get to support, coach, and nurture others and help them make good decisions. In addition, as a manager, you often have the opportunity to think creatively and use your imagination. You'll get to meet and work with a variety of people—both inside and outside the organization. Other rewards may include receiving recognition and status in your organization and in the community, playing a role in influencing organizational outcomes, and receiving attractive compensation in the form of salaries, bonuses, and stock options. Finally, organizations need good managers. It's through the combined efforts of motivated and passionate people working together that organizations accomplish their goals. As a manager, you can be assured that your efforts, skills, and abilities are needed.

QUICK LEARNING REVIEW:

LEARNING OUTCOME 1.5

- Discuss why it's important to study management.
- Explain the universality of management concept.
- Describe the rewards and challenges of being a manager.

Go to page 19 to see how well you know this material.

Let's Get Real:

My Turn

Shibaji Mandal

Director, Healthcare
CSR–Usha Martin Group

Essentially, I would think of a win–win strategy for the junior engineers and the company. If the company needs them next to the blast furnace rather than having them in the air-conditioned (AC) chambers then the compensatory mechanism for the trade off needs to be thought of as well. A well-defined career path needs to be chalked out for the new recruits so that they know what they can aspire for. A lot of "selling" needs to be done (I mean, a lot, really) on their importance of being hands on with the job on the floor and they should be convinced that the dividend of such an experience would eventually pay them off at a later stage. You need to brand the job well on the emotional appeal and attractiveness (say adjectives like "rugged," "resilient," and "tough," and think of the brand images of Levis or Marlboro or Victorinox). The job needs to be branded and positioned as "Cool" and not "Crude!"

LEARNING OUTCOMES SUMMARY

1.1 ▷ WHO ARE MANAGERS?

- Explain how managers differ from nonmanagerial employees.
- Describe how to classify managers in organizations.

Managers coordinate and oversee the work of other people so that organizational goals can be accomplished. Nonmanagerial employees work directly on a job or task and have no one reporting to them. Traditionally structured organizations have first-line, middle, and top managers. In other more loosely configured organizations, the managers may not be as readily identifiable, although someone must fulfill that role.

1.2 ▷ WHAT IS MANAGEMENT?

- Define management.
- Explain why efficiency and effectiveness are important to management.

Management involves coordinating and overseeing the efficient and effective completion of others' work activities. Efficiency means doing things right; effectiveness means doing the right things.

1.3 ▷ WHAT DO MANAGERS DO?

- Describe the four functions of management.
- Explain Mintzberg's managerial roles.

- Describe Katz's three essential managerial skills and how the importance of these skills changes depending on managerial level.
- Discuss the changes that are impacting managers' jobs.
- Explain why customer service and innovation are important to the manager's job.

The four functions of management are planning (defining goals, establishing strategies, and developing plans), organizing (arranging and structuring work), leading (working with and through people), and controlling (monitoring, comparing, and correcting work performance). Mintzberg's managerial roles include interpersonal, which involve people and other ceremonial/symbolic duties (figurehead, leader, and liaison); informational, which involve collecting, receiving, and disseminating information (monitor, disseminator, and spokesperson); and decisional, which involve making choices (entrepreneur, disturbance handler, resource allocator, and negotiator). Katz's managerial skills include technical (job-specific knowledge and techniques), human (ability to work well with people), and conceptual (ability to think and conceptualize). Technical skills are most important for lower-level managers, while conceptual skills are most important for top managers. Human skills are equally important for all managers. The changes affecting managers' jobs include technology (digitization), increased security threats, increased emphasis on ethics, and increased competitiveness. Managers must be concerned with customer service because employee attitudes and behaviors play a big role in customer satisfaction. Managers must also be concerned with innovation because it is important for organizations to be competitive.

1.4 ▷ WHAT IS AN ORGANIZATION?

- Explain the characteristics of an organization.
- Describe how today's organizations are structured.

An organization has three characteristics: a distinctive purpose, composed of people, and a deliberate structure. Many of today's organizations are more open, flexible, and responsive to changes than organizations once were.

1.5 ▷ WHY STUDY MANAGEMENT?

- Discuss why it's important to study management.
- Explain the universality of management concept.
- Describe the rewards and challenges of being a manager.

It's important to study management for three reasons: (1) the universality of management, which refers to the fact that managers are needed in all types and sizes of organizations, at all organizational levels and work areas, and in all global locations; (2) the reality of work—that is, you will either manage or be managed; and (3) the awareness that there are significant rewards (such as creating work environments to help people work to the best of their abilities, supporting and encouraging others, helping others find meaning and fulfillment in work, etc.) and challenges (such as it's hard work, may have more clerical than managerial duties, have to deal with a variety of personalities, etc.) in being a manager.

THINKING ABOUT MANAGEMENT ISSUES

1. Is your course instructor a manager? Discuss in terms of managerial functions, managerial roles, and skills.
2. "The manager's most basic responsibility is to focus people toward performance of work activities to achieve desired outcomes." What's your interpretation of this statement? Do you agree with it? Why or why not?
3. Is business management a profession? Why or why not? Do some external research in answering this question.
4. Is there one best "style" of management? Why or why not?
5. Does the way that contemporary organizations are structured appeal to you? Why or why not?
6. In today's environment, which is more important to organizations—efficiency or effectiveness? Explain your choice.
7. Researchers at Harvard Business School have found that the most important managerial behaviors involve two fundamental things: enabling people to move forward in their work and treating them decently as human beings. What do you think of these two managerial behaviors? What are the implications for someone, like yourself, who is studying management?
8. "Management was, is, and always will be the same thing: the art of getting things done." Do you agree? Why or why not?

YOUR TURN to be a Manager

- Use the most current *Occupational Outlook Handbook* (U.S. Department of Labor, Bureau of Labor Statistics) to research three different categories of managers. For each, prepare a bulleted list that describes the following: the nature of the work, training and other qualifications needed, earnings, and job outlook and projections data.
- Get in the habit of reading at least one current business periodical (*Wall Street Journal, BusinessWeek, Fortune, Fast Company, Forbes,* etc.). Keep a file of interesting information you find about managers or managing.
- Using current business periodicals, find 10 examples of managers you would describe as *master managers.* Write a paper describing these individuals as managers and why you feel they deserve this title.
- Steve's and Mary's suggested readings: Stephen P. Robbins, *The Truth About Managing People,* 2d ed. (Financial Times/Prentice Hall, 2007); Gary Hamel, *The Future of Management* (Harvard Business School, 2007); Rod Wagner and James K. Harter, *12 Elements of Great Managing* (Gallup Press, 2006); Marcus Buckingham, *First Break All the Rules: What the World's Greatest Managers Do Differently* (Simon & Schuster, 1999); and Peter F. Drucker, *The Executive in Action* (Harper Business, 1985 and 1964).
- Interview two different managers and ask them at least three of the questions listed in the *Let's Get Real: Meet the Managers* and *Let's Get Real: F2F* boxes in the chapter. Type up the questions and their answers to turn in to your professor.
- Accountants and other professionals have certification programs to verify their skills, knowledge, and professionalism. What about managers? Two certification programs for managers are the Certified Manager (Institute of Certified Professional Managers) and the Certified Business Manager (Association of Professional in Business Management). Research each of these programs. Prepare a bulleted list of what each involves.
- In your own words, write down three things you learned in this chapter about being a good manager.
- If you're involved in student organizations, volunteer for leadership roles or for projects where you can practice planning, organizing, leading, and controlling different projects and activities. You can also gain valuable experience by taking a leadership role in class team projects.
- Self-knowledge can be a powerful learning tool. Go to mymanagementlab and complete any of these self-assessment exercises: How Motivated Am I to Manage? How Well Do I Handle Ambiguity? How Confident Am I in My Abilities to Succeed? or What's My Attitude Toward Achievement? Using the results of your assessments, identify personal strengths and weaknesses. What will you do to reinforce your strengths and improve your weaknesses?

CASE APPLICATION

Managing the Virus Hunters

"Imagine what life would be like if your product were never finished, if your work were never done, if your market shifted 30 times a day. The computer-virus hunters at Symantec Corp. don't have to imagine." That's the reality of their daily work life. At the company's Response Lab in Santa Monica, California, described as the "dirtiest of all our networks at Symantec," software analysts collect viruses and other suspicious code and try to figure out how they work so the company can provide security updates to its customers. There's even a hazardous materials box by the door to the lab, marked DANGER, where they put all the discs, tapes, and hard drives with the nasty viruses that need to be completely disposed of. Symantec's situation may seem unique, but the company, which makes content and network security software for both consumers and businesses, reflects the realities facing many organizations today: quickly shifting customer expectations and continuously emerging global competitors that have drastically shortened product life cycles. Managing talented people in such an environment can be quite challenging as well.

Vincent Weafer, a native of Ireland, has been the leader of Symantec's virus hunting team since 1999. Back then, he said, "There were less than two dozen people, and . . . nothing really happened. We'd see maybe five new viruses a day, and they would spread in a matter of months, not minutes." Now, Symantec's virus hunters around the world deal with some 20,000 virus samples each month, not all of which are unique, stand-alone viruses. To make the hunters' jobs even more interesting, computer attacks are increasingly being spread by criminals wanting to steal information, whether corporate data or personal user account information that can be used in fraud. Dealing with these critical and time-sensitive issues requires special talents. The response center team is a diverse group whose members weren't easy to find. Said Weafer, "It's not as if colleges are creating thousands of anti-malware or security experts every year that we can hire. If you find them in any part of the world, you just go after them." The response center team's makeup reflects that. For instance, one senior researcher is from Hungary, another is from Iceland, and another works out of her home in Melbourne, Florida. But they all share something in common: They're all motivated by solving problems.

The launch of the Blaster-B worm changed the company's approach to dealing with viruses. The domino effect of Blaster-B and other viruses spawned by it meant that frontline software analysts were working around the clock for almost two weeks. The "employee burn-out" potential made the company realize that its virus-hunting team would now have to be much deeper, talent-wise. Now, the response center's team numbers in the hundreds, and managers can rotate people from the frontlines, where they're responsible for responding to new security threats that crop up, into groups where they can help with new-product development. Others write internal research papers. Still others are assigned to develop new tools that will help their colleagues battle the next wave of threats. There's even an individual who tries to figure out what makes the virus writers tick—and the day never ends for these virus hunters. When Santa Monica's team finishes its day, colleagues in Tokyo take over. When the Japanese team finishes its day, it hands off to Dublin, which then hands back to Santa Monica for the new day. It's a frenetic, chaotic, challenging work environment that spans the entire globe. But Weafer said his goals are to "try to take the chaos out, to make the exciting boring," to have a predictable and well-defined process for dealing with the virus threats, and to spread work evenly to the company's facilities around the world. It's a managerial challenge that Weafer has embraced.

Vincent Weafer, leader of Symantec's virus chasers, in the company's Response Lab.

Discussion Questions

1. Keeping professionals excited about work that is routine and standardized *and* chaotic is a major challenge for Vincent Weafer. How could he use technical, human, and conceptual skills to maintain an environment that encourages innovation and professionalism among the virus hunters?
2. What management roles would Vincent be playing as he (a) had weekly security briefing conference calls with coworkers around the globe, (b) assessed the feasibility of adding a new network security consulting service, and (c) kept employees focused on the company's commitments to customers?
3. Go to Symantec's Web site, www.symantec.com, and look up information about the company. What can you tell about its emphasis on customer service and innovation? In what ways does the organization support its employees in servicing customers and in being innovative?
4. What could other managers learn from Vincent Weafer and Symantec's approach?

Sources: Symantec Web site, www.symantec.com, February 23, 2008; J. Cox, "Cyber Threats Get Personal," *CNNMoney.com*, September 18, 2007; N. Rothbaum, "The Virtual Battlefield," *Smart Money*, January 2006, pp. 76–80; S. H. Wildstrom, "Viruses Get Smarter—And Greedy," *BusinessWeek* online, November 22, 2005; and S. Kirsner, "Sweating in the Hot Zone," *Fast Company*, October 2005, pp. 60–65.

Let's Get Real:

Meet The Manager

Ripu Lamba

General Manager, Training and Development
Datacom Solutions Pvt. Ltd

MY JOB: To enhance capabilities of employees, pan India.

BEST PART OF MY JOB: To plan and execute capability enhancement across hierarchies.

WORST PART OF MY JOB: When training is not given its due importance.

BEST MANAGEMENT ADVICE EVER RECEIVED: Ensure work-life balance.

You'll be hearing more from this real manager throughout the chapter.

Chapter

Management History

In this chapter, we're going to take a trip back in time to see how the field of study called management has evolved. What you're going to find out is that today's managers still use many elements of the historical approaches to management. Focus on the following learning outcomes as you read and study this chapter.

LEARNING OUTCOMES

- **2.1** Describe the historical background of management. page 24
- **2.2** Explain the various theories in the classical approach. page 26
- **2.3** Describe the quantitative approach. page 30
- **2.4** Discuss the development and uses of the behavioral approach. page 33
- **2.5** Explain the various theories in the contemporary approach. page 35

A Manager's Dilemma

Since January 2008, the managements' of call centers in the large metros and cities are worried. Following the rape of one woman employee traveling to her work in an HP Globalsoft call center in Bangalore and another woman employee meeting the same fate at the hands of a company-hired driver of a Wipro call center located in Pune, the Supreme Court has tried and found the HP CEO guilty. There are several other but less serious cases of harassment. Several attempts are being made at the industry level to place employee security at the top of the agenda. General packet radio service (GPRS) tracking of vehicles is being considered, and text-message alerts in addition to deploying more guards and working closely with the police to verify drivers' background are some of the measures taken. Sam Chopra, President of the Business Process Industry Association of India, says, "We are going to invest \$500,000 on a new security framework with a pilot phase in Gurgaon, where many call centers are located." However, ensuring safety requires full cooperation of its employees who may not want to be tracked all the time.[1] How can we make sure that the employees understand the necessity of safety procedures and also comply with them. Suppose you are a manager at the HP Globalsoft who has installed the GPRS tracking of vehicles, what would you do?

What would you do?

Attempts to find ways to make employees' jobs safer and better aren't unique. Many organizations are doing the same. Why? To help employees be efficient and effective, especially in today's global competitive environment. As you'll see in this chapter, searching for ways to help employees do their jobs better isn't new. Managers have always looked for ideas to help them do a better job of managing. In fact, the history of management is filled with evolutions and revolutions in ideas. And many seemingly "new" ideas do have a basis in the past.

LEARNING **OUTCOME 2.1** ▷

HISTORICAL BACKGROUND OF MANAGEMENT

Although studying history may not be at the top of your list of exciting things to do, it can help you understand today's management theories and practices and help see what has and has not worked. What is interesting, we think, is that management has been practiced a long time. Organized endeavors directed by people responsible for planning, organizing, leading, and controlling activities have existed for thousands of years. The Egyptian pyramids, the Great Wall of China, the Qutub Minar, and the Taj Mahal, for instance, are tangible evidence that projects of tremendous scope, employing tens of thousands of people, were completed in ancient times. The pyramids are a particularly remarkable example. The construction of a single pyramid occupied more than 100,000 workers for 20 years.[2] Who told each worker what to do? Who ensured that there would be enough stones at the site to keep workers busy? The answer is *managers.* Regardless of what these individuals were called, someone had to plan what was to be done, organize people and materials to do it, lead and direct the workers, and impose some controls to ensure that everything was done as planned.

Another example of early management can be seen during the 1400s in the city of Venice, a major economic and trade center. The Venetians developed an early form of business enterprise and engaged in many activities common to today's organizations. For instance, at the arsenal of Venice, warships were floated along the canals, and at each stop, materials and riggings were added to the ship. Doesn't that sound a lot like a car "floating" along an automobile assembly line and components being added to it? In addition, the Venetians used warehouse and inventory systems to keep track of materials, human resource management functions to manage the labor force, and an accounting system to keep track of revenues and costs.[3]

Two events are especially significant to management history. First, in 1776, Adam Smith published *The Wealth of Nations*, in which he argued the economic advantages that organizations and society would gain from the **division of labor** (or **job specialization**)—that is, the breakdown of jobs into narrow and repetitive tasks. Using the pin industry as an example, Smith claimed that 10 individuals, each doing a specialized task, could produce about 48,000 pins per day among them. However, if each person worked alone, performing each task separately, it would be quite an accomplishment to produce even 10 pins per day! Smith concluded that division of labor increased productivity by increasing each worker's skill and dexterity, saving time lost in changing tasks, and creating labor-saving inventions and machinery. Job specialization continues to be popular. For example, think of the specialized tasks performed by members of a hospital surgery team, meal preparation tasks done by workers in restaurant kitchens, or positions played by players on a football team.

The second important event is the **industrial revolution**. Starting in the late eighteenth century when machine power was substituted for human power, it became more economical to manufacture goods in factories than at home. These large, efficient factories needed someone to forecast demand, ensure that enough material was on hand to make products, assign tasks to people, direct daily activities, and so forth. That "someone" was managers, and these managers would need formal theories to guide them in running these large organizations. It wasn't until the early 1900s, however, that the first steps were taken toward developing such theories.

In this chapter, we'll look at four major approaches to management theory: classical, quantitative, behavioral, and contemporary (see Exhibit 2–1). Keep in mind that each approach is concerned with the same "animal"; the differences reflect the

Exhibit 2–1

Major Approaches to Management

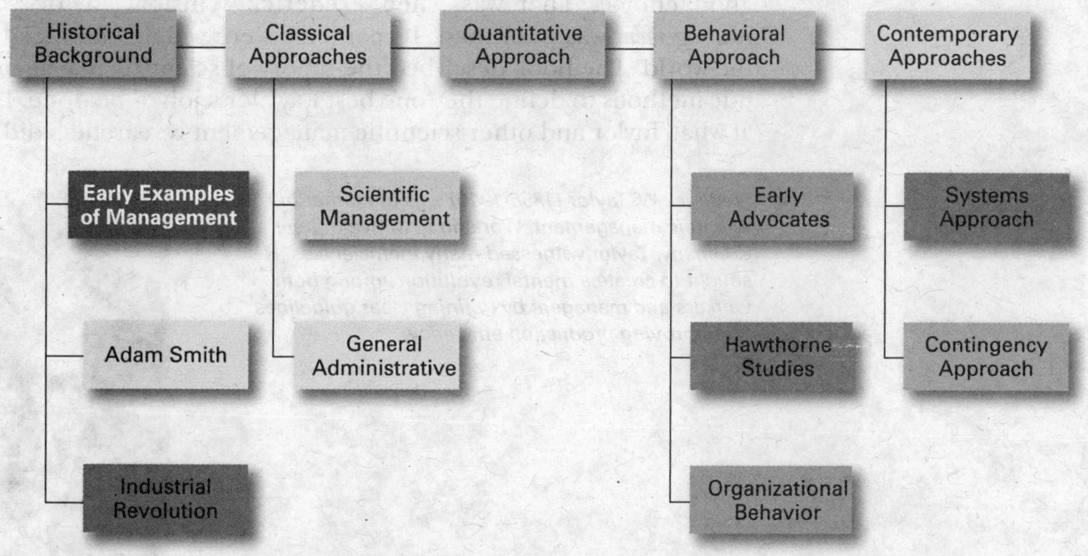

division of labor (or **job specialization**)
The breakdown of jobs into narrow and repetitive tasks.

industrial revolution
A period during the late eighteenth century when machine power was substituted for human power, making it more economical to manufacture goods in factories than at home.

backgrounds and interests of the writer. A relevant analogy is the classic story of the blindfolded men and the elephant, in which each declares the elephant to be like the part he is feeling: The first man, touching the side, declares that the elephant is like a wall; the second touches the trunk and says the elephant is like a snake; the third feels one of the elephant's tusks and believes the elephant to be like a spear; the fourth grabs a leg and says the elephant is like a tree; and the fifth touches the elephant's tail and concludes that the animal is like a rope. Each is encountering the same elephant, but what each observes depends on where he stands. Similarly, each of the four approaches contributes to our overall understanding of management, but each is also a limited view of a larger animal.

QUICK LEARNING REVIEW:
LEARNING OUTCOME 2.1

- **Explain why studying management history is important.**
- **Describe some early evidences of management practice.**
- **Describe two important historical events that are significant to the study of management.**

Go to page 38 to see how well you know this material.

LEARNING
OUTCOME 2.2 ▷

CLASSICAL APPROACH

Although we've seen how management has been used in organized efforts since early history, the formal study of management didn't begin until early in the twentieth century. These first studies of management, described as the **classical approach**, emphasized rationality and making organizations and workers as efficient as possible. Two major theories comprise the classical approach: scientific management and general administrative.

SCIENTIFIC MANAGEMENT

If you had to pinpoint when modern management theory was born, 1911 might be a good choice. That was when Frederick Winslow Taylor's *Principles of Scientific Management* was published. Its contents were widely embraced by managers around the world. The book described the theory of **scientific management**: the use of scientific methods to define the "one best way" for a job to be done. Let's look more closely at what Taylor and other scientific management researchers did.

Frederick W. Taylor (1856-1915) was the father of scientific management. Working at Midvale Steel Company, Taylor witnessed many inefficiencies. He sought to create a mental revolution among both workers and managers by defining clear guidelines for improving production efficiency.

Frederick Winslow Taylor Collection, Samuel C. Williams Library, Stevens Institute of Technology, Hoboken, NJ.

Frederick W. Taylor. Taylor worked at the Midvale and Bethlehem Steel Companies in Pennsylvania. As a mechanical engineer with a Quaker and Puritan background, he was continually appalled by workers' inefficiencies. Employees used vastly different techniques to do the same job. They often "took it easy" on the job, and Taylor believed that worker output was only about one-third of what was possible. Virtually no work standards existed. Workers were placed in jobs with little or no concern for matching their abilities and aptitudes with the tasks they were required to do. Taylor set out to remedy that by applying the scientific method to shop-floor jobs. He spent more than two decades passionately pursuing the "one best way" for such jobs to be done.

Taylor's experiences at Midvale led him to define clear guidelines for improving production efficiency. He argued that these four principles of management (see Exhibit 2–2) would result in prosperity for both workers and managers.[4] How did these scientific principles really work? Let's look at an example.

Probably the best known example of Taylor's scientific management efforts was the pig iron experiment. Workers loaded "pigs" of iron (each weighing 92 pounds) onto rail cars. Their daily average output was 12.5 tons. However, Taylor believed that by scientifically analyzing the job to determine the "one best way" to load pig iron, output could be increased to 47 or 48 tons per day. After scientifically applying different combinations of procedures, techniques, and tools, Taylor succeeded in getting that level of productivity. How? He put the right person on the job with the correct tools and equipment, had the worker follow his instructions exactly, and motivated the worker with an economic incentive of a significantly higher daily wage. Using similar approaches for other jobs, Taylor was able to define the "one best way" for doing each job. Overall, Taylor achieved consistent productivity improvements in the range of 200 percent or more. Based on his groundbreaking studies of manual work using scientific principles, Taylor became known as the "father" of scientific management. His ideas spread in the United States and to other countries and inspired others to study and develop methods of scientific management. His most prominent followers were Frank and Lillian Gilbreth.

Frank and Lillian Gilbreth. A construction contractor by trade, Frank Gilbreth gave up that career to study scientific management after hearing Taylor speak at a professional meeting. Frank and his wife, Lillian, a psychologist, studied work to eliminate inefficient hand-and-body motions. The Gilbreths also experimented with the design and use of the proper tools and equipment for optimizing work performance.[5]

Frank is probably best known for his bricklaying experiments. By carefully analyzing the bricklayer's job, he reduced the number of motions in laying exterior brick from 18 to about 5 and in laying interior brick from 18 to 2. Using Gilbreth's techniques, a bricklayer was more productive and less fatigued at the end of the day.

The Gilbreths were among the first researchers to use motion pictures to study hand-and-body motions. They invented a device called a microchronometer that recorded a worker's motions and the amount of time spent doing each motion.

Exhibit 2–2

Taylor's Scientific Management Principles

1. Develop a science for each element of an individual's work to replace the old rule-of-thumb method.
2. Scientifically select and then train, teach, and develop the worker.
3. Heartily cooperate with the workers so as to ensure that all work is done in accordance with the principles of the science that has been developed.
4. Divide work and responsibility almost equally between management and workers. Management does all work for which it is better suited than the workers.

classical approach
The first studies of management, which emphasized rationality and making organizations and workers as efficient as possible.

scientific management
An approach that involves using the scientific method to determine the "one best way" for a job to be done.

Frank and Lillian Gilbreth, parents of 12 children, ran their household using scientific management principles and techniques. Two of their children wrote a book, Cheaper by the Dozen, *which described life with the two masters of efficiency.*

Wasted motions missed by the naked eye could be identified and eliminated. The Gilbreths also devised a classification scheme to label 17 basic hand motions (such as search, grasp, hold), which they called **therbligs** (Gilbreth spelled backward, with the *th* transposed). This scheme gave the Gilbreths a more precise way of analyzing a worker's exact hand movements.

Let's Get Real: F2F

I HELP MY EMPLOYEES WORK MORE EFFICIENTLY BY:

- Assigning work that has sufficient challenges. When they encounter barriers, guide and help them without doing their work for them.

RULES FOR MANAGERS:

- Recognizing and appreciating good performance.
- Setting challenges.
- Planning/executing/monitoring for better results.

How Today's Managers Use Scientific Management. Many of the guidelines and techniques that Taylor and the Gilbreths devised for improving production efficiency are still used in organizations today. When managers analyze the basic work tasks that must be performed, use time-and-motion study to eliminate wasted motions, hire the best-qualified workers for a job, or design incentive systems based on output, they're using the principles of scientific management.

GENERAL ADMINISTRATIVE THEORY

A group of writers looked at the subject of management from the perspective of the entire organization. This approach, known as **general administrative theory**, focused more on what managers do and what constituted good management practice. The two most prominent individuals behind general administrative theory were Henri Fayol and Max Weber.

Henri Fayol. We introduced Fayol in Chapter 1 because he first identified five functions that managers perform: planning, organizing, commanding, coordinating, and controlling. Because his ideas were important, let's look closer at what he had to say.[6]

Fayol wrote during the same time period as Taylor. While Taylor was concerned with first-line managers and the scientific method, Fayol's attention was directed at the activities of *all* managers. He wrote from personal experience as he was the managing director of a large French coal-mining firm.

Fayol described the practice of management as something distinct from accounting, finance, production, distribution, and other typical business functions. His belief that management was an activity common to all business endeavors, government, and even the home led him to develop 14 **principles of management**—fundamental rules of management that could be applied to all organizational situations and taught in schools. These principles are shown in Exhibit 2–3.

Max Weber. Max Weber (pronounced "VAY-ber") was a German sociologist who studied organizations.[7] Writing in the early 1900s, he developed a theory of authority structures and relations based on an ideal type of organization he called a **bureaucracy**—a form of organization characterized by division of labor, a clearly defined hierarchy, detailed rules and regulations, and impersonal relationships (see Exhibit 2–4). Weber recognized that this "ideal bureaucracy" didn't exist in reality.

Exhibit 2–3

Fayol's 14 Principles of Management

1. *Division of Work.* Specialization increases output by making employees more efficient.
2. *Authority.* Managers must be able to give orders and authority gives them this right.
3. *Discipline.* Employees must obey and respect the rules that govern the organization.
4. *Unity of command.* Every employee should receive orders from only one superior.
5. *Unity of direction.* The organization should have a single plan of action to guide managers and workers.
6. *Subordination of individual interests to the general interest.* The interests of any one employee or group of employees should not take precedence over the interests of the organization as a whole.
7. *Remuneration.* Workers must be paid a fair wage for their services.
8. *Centralization.* This term refers to the degree to which subordinates are involved in decision making.
9. *Scalar chain.* The line of authority from top management to the lowest ranks is the scalar chain.
10. *Order.* People and materials should be in the right place at the right time.
11. *Equity.* Managers should be kind and fair to their subordinates.
12. *Stability of tenure of personnel.* Management should provide orderly personnel planning and ensure that replacements are available to fill vacancies.
13. *Initiative.* Employees who are allowed to originate and carry out plans will exert high levels of effort.
14. *Esprit de corps.* Promoting team spirit will build harmony and unity within the organization.

Exhibit 2–4

Weber's Bureaucracy

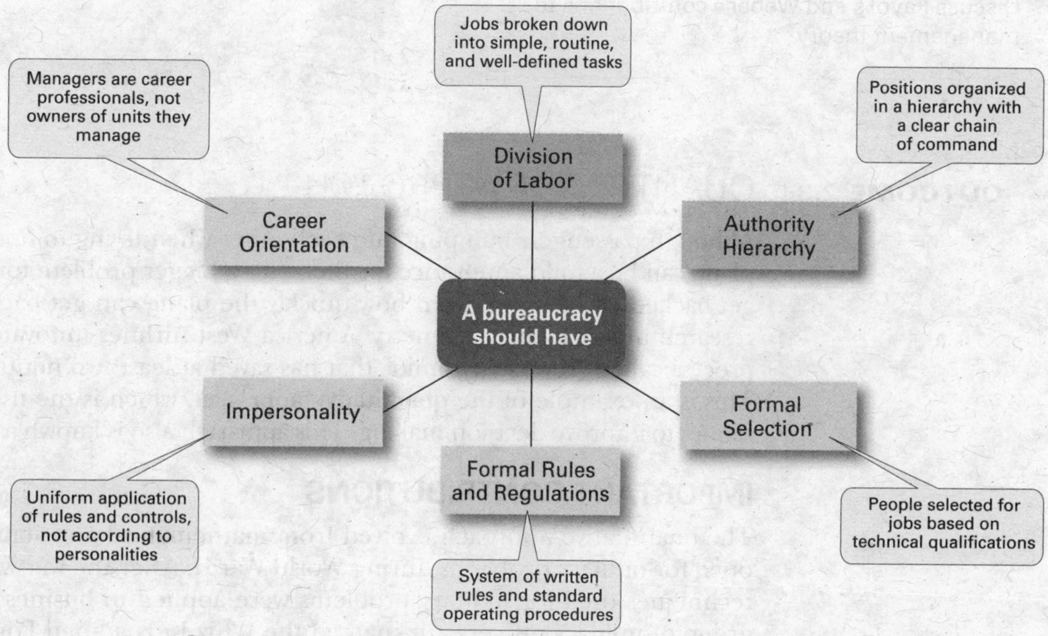

therbligs
A classification scheme for labeling 17 basic hand motions.

general administrative theory
An approach to management that focuses on describing what managers do and what constitutes good management practice.

principles of management
Fundamental rules of management that could be applied in all organizational situations and taught in schools.

bureaucracy
A form of organization characterized by division of labor, a clearly defined hierarchy, detailed rules and regulations, and impersonal relationships.

Instead, he intended it as a basis for theorizing about how work could be done in large groups. His theory became the structural design for many of today's large organizations.

Bureaucracy, as described by Weber, is a lot like scientific management in its ideology. Both emphasize rationality, predictability, impersonality, technical competence, and authoritarianism. Although Weber's ideas were less practical than Taylor's, the fact that his "ideal type" still describes many contemporary organizations attests to the importance of Weber's ideas.

How Today's Managers Use General Administrative Theories. Several of our current management ideas and practices can be directly traced to the contributions of general administrative theory. For instance, the functional view of the manager's job can be attributed to Fayol. In addition, his 14 principles serve as a frame of reference from which many current management concepts—such as managerial authority, centralized decision making, reporting to only one boss, and so forth—have evolved.

Weber's bureaucracy was an attempt to formulate an ideal prototype for organizations. Although many characteristics of Weber's bureaucracy are still evident in large organizations, his model isn't as popular today as it was in the twentieth century. Many managers feel that a bureaucratic structure hinders individual employees' creativity and limits an organization's ability to respond quickly to an increasingly dynamic environment. However, even in flexible organizations of creative professionals—such as Infosys, UTV, Cisco Systems, and Dr Reddy's Lab—some bureaucratic mechanisms are necessary to ensure that resources are used efficiently and effectively.

QUICK LEARNING REVIEW:

LEARNING OUTCOME 2.2

- Describe the important contributions made by Frederick W. Taylor and Frank and Lillian Gilbreth.
- Discuss Fayol's and Weber's contributions to management theory.
- Explain how today's managers use scientific management and general administrative theory.

Go to page 39 to see how well you know this material.

LEARNING **OUTCOME 2.3** ▷

QUANTITATIVE APPROACH

Although passengers bumping into each other when trying to find their seats on an airplane can be a mild annoyance for them, it's a bigger problem for airlines because lines get backed up, slowing down how quickly the plane can get back in the air. Based on research in space–time geometry, America West Airlines innovated a unique boarding process called "reverse pyramid" that has saved at least two minutes in boarding time.[8] This is an example of the **quantitative approach**, which is the use of quantitative techniques to improve decision making. This approach also is known as **management science**.

IMPORTANT CONTRIBUTIONS

The quantitative approach evolved from mathematical and statistical solutions developed for military problems during World War II. After the war was over, many of these techniques used for military problems were applied to businesses. For example, one group of military officers, nicknamed the Whiz Kids, joined Ford Motor Company in the mid-1940s and immediately began using statistical methods and quantitative models to improve decision making.

What exactly is the quantitative approach? It involves applying statistics, optimization models, information models, computer simulations, and other quantitative techniques to management activities. Linear programming, for instance, is a technique that managers use to improve resource allocation decisions. Work scheduling can be more efficient as a result of critical-path scheduling analysis. The economic order quantity model helps managers determine optimum inventory levels. Each of these is an example of quantitative techniques being applied to improve managerial decision

The Indian Coffee Board promoted the Indian Coffee House(s) during the British rule in the 1940s. In the 1950s, however, the business was not doing well and so the Indian Coffee Board decided to close down the Coffee Houses. Under the leadership of the communist leader A. K. Gopalan, the dismissed workers took over the place to run it without any management. The first Indian Coffee Workers Cooperative Society was founded in Bangalore on August 19, 1957. Today, there are around 400 Indian Coffee Houses managed by 13 cooperative societies. There is also a federation of the cooperative societies as the national umbrella organization to lead these societies.[9]

making. Another area where quantitative techniques are being used is known as total quality management.

TOTAL QUALITY MANAGEMENT

A quality revolution swept through both the business and public sectors in the 1980s and 1990s.[12] It was inspired by a small group of quality experts, the most famous being W. Edwards Deming and Joseph M. Juran. The ideas and techniques they advocated in the 1950s had few supporters in the United States but were enthusiastically embraced by Japanese organizations. As Japanese manufacturers began beating U.S. competitors in quality comparisons, however, Western managers soon took a more serious look at Deming's and Juran's ideas—ideas that became the basis for today's quality management programs.

Total quality management (TQM) is a management philosophy devoted to continual improvement and responding to customer needs and expectations (see Exhibit 2–5). The term *customer* includes anyone who interacts with the organization's product or

thinking critically about Ethics

When does technology overstep the bounds?[10] Since 2006, the Mahatma Gandhi Centennial Sindhu High School, the largest school in Nagpur, has 250 closed-circuit television (CCTV) cameras installed in their classrooms, corridors, staircases, labs, activity rooms, staff rooms, and everywhere. The students are watched every moment, and every move is captured.[11] An airport scanner that made its debut in a Phoenix airport early in 2007 uses "backscatter technology" to produce an image that sees through clothes and shows the contours of the body. In Japan, the National Police Agency uses a type of face recognition software to identify criminals whose images have been captured on surveillance video. A Canadian researcher has developed a portable system that can track eye movements of people viewing a billboard, an electronic screen, or a television. And casinos in Macau use face recognition software to identify casino cheats. What do you think about these situations? What ethical issues might arise? Are some okay and others not okay? Are there any circumstances when such technologies might be appropriate (and ethical) for organizations to use with employees? Explain.

quantitative approach (or **management science**)
The use of quantitative techniques to improve decision making.

total quality management (TQM)
A philosophy of management that is driven by continuous improvement and responsiveness to customer needs and expectations.

Exhibit 2–5

What Is Quality Management?

1. Intense focus on the *customer*. The customer includes outsiders who buy the organization's products or services and internal customers who interact with and serve others in the organization.
2. Concern for *continual improvement*. Quality management is a commitment to never being satisfied. "Very good" is not good enough. Quality can always be improved.
3. *Process focused*. Quality management focuses on work processes as the quality of goods and services is continually improved.
4. Improvement in the *quality of everything* the organization does. This relates to the final product, how the organization handles deliveries, how rapidly it responds to complaints, how politely the phones are answered, and the like.
5. *Accurate measurement*. Quality management uses statistical techniques to measure every critical variable in the organization's operations. These are compared against standards to identify problems, trace them to their roots, and eliminate their causes.
6. *Empowerment of employees*. Quality management involves the people on the line in the improvement process. Teams are widely used in quality management programs as empowerment vehicles for finding and solving problems.

services, internally or externally. It encompasses employees and suppliers as well as the people who purchase the organization's goods or services. *Continuous improvement* isn't possible without accurate measurements, which require statistical techniques that measure every critical variable in the organization's work processes. These measurements are compared against standards to identify and correct problems.

TQM was a departure from earlier management approaches that were based on the belief that keeping costs low was the only way to increase productivity. In 2006, Dell had to announce a worldwide recall of certain models of laptops that contained defective batteries. This recall was initiated after several reported instances of overheating of the battery and even in rare cases setting of fire. Even in October 2008, the recall was being re-announced because it was found that there were certain other laptop models that were also shipped with the affected battery. Fifteen thousand potentially affected units had been sold until October 2008 and had to be potentially replaced.[13] If you add the cost of rejects, repairing shoddy work, product recalls, and expensive controls to identify control problems, manufacturers are less productive than what they project.

Speed Tex was established in 1999 in Tiruppur, Tamil Nadu, and is one of the leading manufacturers in the worldwide knit apparel textile with a long tradition of quality and customer satisfaction. The company employs only 275 people in-house and has 2 full-time subcontractor factories (sewing facilities, 150 people) to produce approximately 20,000 pieces every month. It has been able to maintain its cost and quality advantage in the United Kingdom, Germany, Spain, and Switzerland by following total quality management (TQM) procedures. As a part of its TQM philosophy, it continuously searches for options to lowering the costs to give the best price with the best quality to its customers. One such innovation in reducing cost of manufacturing is to set up all production lines as "U"-line bundle system, including all finishing process.[14]

The Japanese demonstrated that it *was* possible for the highest-quality manufacturers to be among the lowest-cost producers. Manufacturers in many industries have now realized the importance of quality management and implemented many of its basic components.

HOW TODAY'S MANAGERS USE THE QUANTITATIVE APPROACH

No one likes to be in long queues. Many customers get very upset if they see a very long line and engage themselves in queue jumping based on their naïve theories of which queue is moving faster. This leads to irritation and indiscipline among others who are waiting. The Indian Railway reservation system, after trying out variations of how to manage the queue, uses a system where you join one queue that feeds into the next open counter. In most cases, customers find that they spend less time to get to the reservation window. The call centers also use the same logic to queue the receiver's call. In fact, tour operators, such as Grayline India, in their attempt to become one of the most preferred tour operators in India offer real time availability and confirmation to the domestic and inbound traveler in the City Tour, Sightseeing, and Incredible India holiday space by partnering with vendors who are able to provide services in the areas of inventory control, hosting, reservation, ticketing, internet booking engine, distribution, and departure control.[15] The science of keeping lines moving is known as *queue management*. And for Whole Foods, this quantitative technique has translated into strong sales at its Manhattan stores.

The quantitative approach contributes directly to management decision making in the areas of planning and control. For instance, when managers make budgeting, queuing, scheduling, quality control, and similar decisions, they typically rely on quantitative techniques. Specialized software has made the use of these techniques less intimidating for managers, although many still feel anxious about using them.

QUICK LEARNING REVIEW:
LEARNING OUTCOME 2.3

- Explain what the quantitative approach has contributed to the field of management.
- Describe total quality management.
- Discuss how today's managers use the quantitative approach.

Go to page 39 to see how well you know this material.

LEARNING OUTCOME 2.4 ▷ BEHAVIORAL APPROACH

As we know, managers get things done by working with people. This explains why some writers have chosen to look at management by focusing on an organization's people. The field of study that researches the actions (behavior) of people at work is called **organizational behavior (OB)**. Much of what managers do today when managing people—motivating, leading, building trust, working with a team, managing conflict, and so forth—has come out of OB research.

EARLY ADVOCATES OF ORGANIZATIONAL BEHAVIOR

Although a number of individuals in the early twentieth century recognized the importance of people to an organization's success, four stand out as early advocates of the OB approach: Robert Owen, Hugo Munsterberg, Mary Parker Follett, and Chester Barnard. Their contributions were varied and distinct, yet all believed that people were the most important asset of the organization and should be managed accordingly. Their ideas provided the foundation for such management practices as employee selection procedures, motivation programs, and work teams. Exhibit 2–6 summarizes their most important ideas.

THE HAWTHORNE STUDIES

Without question, the most important contribution to the OB field came out of the **Hawthorne Studies**, a series of studies conducted at the Western Electric Company Works in Cicero, Illinois. These studies, which started in 1924, were initially designed by Western Electric industrial engineers as a scientific management experiment. They wanted to examine the effect of various lighting levels on worker productivity. As with any other good scientific experiment, control and experimental groups were set up, with the experimental group being exposed to various lighting intensities, and the control group working under a constant intensity. If you were the industrial engineers in charge of this experiment, what would you have expected to happen? It's logical to think that individual output in the experimental group would be directly related to

organizational behavior (OB)
A field of study that researches the actions (behavior) of people at work.

Exhibit 2–6

Early OB Advocates

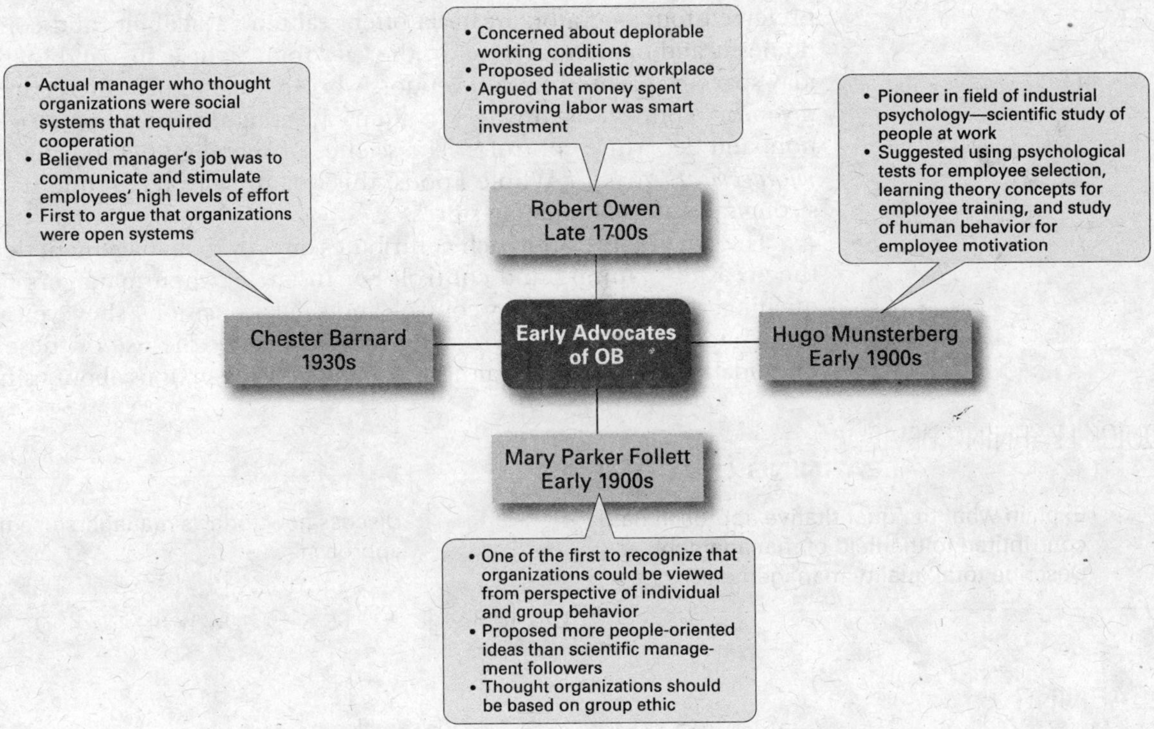

the intensity of the light. However, they found that as the level of light was increased in the experimental group, output for both groups increased. Then, much to the surprise of the engineers, as the light level was decreased in the experimental group, productivity continued to increase in both groups. In fact, a productivity decrease was observed in the experimental group *only* when the level of light was reduced to that of a moonlit night. What would explain these unexpected results? The engineers weren't sure but concluded that lighting intensity was not directly related to group productivity and that something else must have contributed to the results. They weren't able to pinpoint what that "something else" was, though.

Let's Get Real: F2F

CAN QUANTITATIVE TECHNIQUES HELP SOLVE A "PEOPLE" PROBLEM?

Only to a limited extent. I think it is useful as a fact-finding tool. The solution to the problem will require qualitative inputs as well.

In 1927, the Western Electric engineers asked Harvard professor Elton Mayo and his associates to join the study as consultants. Thus began a relationship that would last through 1932 and encompass numerous experiments in the redesign of jobs, changes in workday and workweek length, introduction of rest periods, and individual versus group wage plans.[16] For example, one experiment was designed to evaluate the effect of a group piecework incentive pay system on group productivity. The results indicated that the incentive plan had less effect on a worker's output than did group pressure, acceptance, and security. The researchers concluded that social norms or group standards were the key determinants of individual work behavior.

Scholars generally agree that the Hawthorne Studies had a dramatic impact on management beliefs about the role of people in organizations. Mayo concluded that people's behavior and attitudes are closely related, that group factors significantly affect individual behavior, that group standards establish individual worker output, and that money is less a factor in determining output than are group standards, group attitudes, and security. These conclusions led to a new emphasis on the human behavior factor in the management of organizations.

Although critics attacked the research procedures, analyses of findings, and conclusions, it's of little importance from a historical perspective whether the Hawthorne Studies were academically sound or their conclusions justified.[17] What *is* important is that they stimulated an interest in human behavior in organizations.

HOW TODAY'S MANAGERS USE THE BEHAVIORAL APPROACH

The behavioral approach has largely shaped how today's organizations are managed. From the way that managers design jobs to the way that they work with employee teams to the way that they communicate, we see elements of the behavioral approach. Much of what the early OB advocates proposed and the conclusions from the Hawthorne Studies have provided the foundation for our current theories of motivation, leadership, group behavior and development, and numerous other behavioral approaches.

QUICK LEARNING REVIEW:

LEARNING OUTCOME 2.4

- Describe the contributions of the early advocates of OB.
- Explain the contributions of the Hawthorne Studies to the field of management.
- Discuss how today's managers use the behavioral approach.

Go to page 39 and see how well you know this material.

LEARNING **OUTCOME 2.5** ▷ CONTEMPORARY APPROACH

As we've seen, many elements of the early approaches to management theory continue to influence how managers manage. Most of the early approaches focused on managers' concerns *inside* the organization. Starting in the 1960s, management researchers began to look at what was happening in the external environment *outside* the boundaries of the organization. Two contemporary management perspectives—systems and contingency—are part of this approach. Let's take a look at each.

SYSTEMS THEORY

Systems theory is a basic theory in the physical sciences but had never been applied to organized human efforts. In 1938, Chester Barnard, a telephone company executive, first wrote in his book, *The Functions of an Executive,* that an organization functioned as a cooperative system. However, it wasn't until the 1960s that management researchers began to look more carefully at systems theory and how it related to organizations.

A **system** is a set of interrelated and interdependent parts arranged in a manner that produces a unified whole. The two basic types of systems are closed and open. **Closed systems** are not influenced by and do not interact with their environment. In contrast, **open systems** are influenced by and do interact with their environment. Today, when we describe organizations as systems, we mean open systems. Exhibit 2–7 shows a diagram of an organization from an open systems perspective. As you can see, an organization takes in inputs (resources) from the environment and transforms or processes these resources into outputs that are distributed into the environment. The organization is "open" to and interacts with its environment.

The Systems Approach and Managers. How does the systems approach contribute to our understanding of management? Researchers envisioned an organization as being made up of "interdependent factors, including individuals, groups, attitudes, motives, formal structure, interactions, goals, status, and authority."[18] What this means is that as managers coordinate work activities in the various parts of the organization, they ensure that all these parts are working together so that the organization's goals can be achieved. For example, the systems approach recognizes that, no matter how efficient the production department might be, the marketing department better anticipate changes in customer tastes and work with the product development department in creating products customers want, or the organization's overall performance will suffer.

Hawthorne Studies
A series of studies during the 1920s and 1930s that provided new insights into individual and group behavior.

system
A set of interrelated and interdependent parts arranged in a manner that produces a unified whole.

closed systems
Systems that are not influenced by and do not interact with their environment.

open systems
Systems that interact with their environment.

Exhibit 2–7

Organization as an Open System

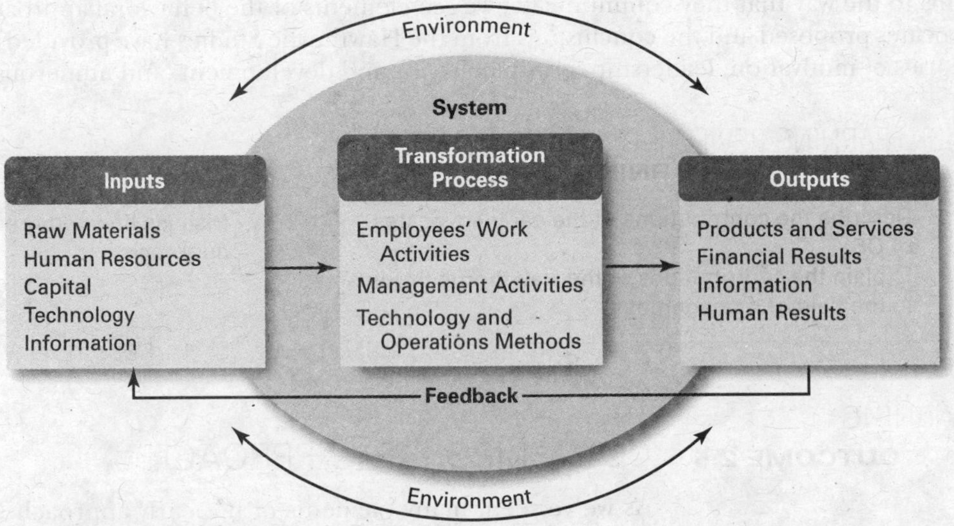

In addition, the systems approach implies that decisions and actions in one organizational area will affect other areas. For example, if the purchasing department doesn't acquire the right quantity and quality of inputs, the production department won't be able to do its job.

Finally, the systems approach recognizes that organizations are not self-contained. They rely on their environment for essential inputs and as outlets to absorb their outputs. No organization can survive for long if it ignores government regulations, supplier relations, or the varied external constituencies on which it depends. (We'll cover these external environmental forces in Chapter 3.)

How relevant is the systems approach to management? It's quite relevant. Consider, for example, a shift manager at a Kamat's restaurant (serving South Indian meals) must coordinate the work of the waiters filing in the customers' orders and the delivery and unloading of the supplies, manage the billing system, and address any customers' concerns that come up.

managing in a Virtual World

How IT Is Changing the Manager's Job

IT (an abbreviation for information technology) is changing many things in today's world.[19] On college campuses—the "world" you're most familiar with now—students are downloading professors' lectures to their iPods and other MP3 players. Laptops in classrooms are a common sight. Some experts say that this anytime/anywhere access for students has tangible benefits. Says one consultant, "To compete globally, we're going to have to produce a nation of problem-solvers and analytical thinkers, and we're going to have to do it using twenty-first-century tools." But IT isn't just drastically changing the educational world, it's having a major impact on businesses and how managers do their job. For instance, Zensar has moved from using a paper-based manual process of performance appraisal that involved sending documents as e-mail attachments to a web-based HR assessment portal. The use of appropriate technology has allowed employees to keep a track of their past performances and has greater control over their reviews. It has also shortened the review process by up to 50 percent. It has helped in creating transparency and enhancing employee productivity.

Although this may not mean much for many companies but at Zensar where the pressure on employees' time is immense, the manager's multitask at various levels to deliver projects on time; cutting the time for performance appraisal by 50 percent means a significant improvement for the company. Timely and quick appraisal helps to keep employees motivated.[20]

Electronic scorecards are just one example of a way that managers are using IT to manage their organizations more efficiently and effectively. As managers plan, organize, lead, and control, they can look to IT as a tool to help them collect and use information.

Exhibit 2–8

Popular Contingency Variables

Organization Size. As size increases, so do the problems of coordination. For instance, the type of organization structure appropriate for an organization of 50,000 employees is likely to be inefficient for an organization of 50 employees.
Routineness of Task Technology. To achieve its purpose, an organization uses technology. Routine technologies require organizational structures, leadership styles, and control systems that differ from those required by customized or nonroutine technologies.
Environmental Uncertainty. The degree of uncertainty caused by environmental changes influences the management process. What works best in a stable and predictable environment may be totally inappropriate in a rapidly changing and unpredictable environment.
Individual Differences. Individuals differ in terms of their desire for growth, autonomy, tolerance of ambiguity, and expectations. These and other individual differences are particularly important when managers select motivation techniques, leadership styles, and job designs.

THE CONTINGENCY APPROACH

The early management theorists came up with management principles that they generally assumed to be universally applicable. Later research found exceptions to many of these principles. For example, division of labor is valuable and widely used, but jobs can become *too* specialized. Bureaucracy is desirable in many situations, but in other circumstances, other structural designs are more effective. Management is not (and cannot be) based on simplistic principles to be applied in all situations. Different and changing situations require managers to use different approaches and techniques. The **contingency approach** (sometimes called the **situational approach**) says that organizations are different, face different situations (contingencies), and require different ways of managing.

The Contingency Approach and Managers. A good way to describe contingency is "if, then": *If* this is the way my situation is, *then* this is the best way for me to manage in this situation. This approach is intuitively logical because organizations and even units within the same organization differ—in terms of size, goals, work activities, and the like. It would be surprising to find universally applicable management rules that would work in *all* situations. But, of course, it's one thing to say that the way to manage "depends on the situation" and another to say what the situation is. Management researchers continue working to identify these situational variables. Exhibit 2–8 describes four popular contingency variables. Although the list is by no means comprehensive—more than 100 different variables have been identified—it represents those that are most widely used and gives you an idea of what we mean by the term *contingency variable.* The primary value of the contingency approach is that it stresses that there are no simplistic or universal rules for managers to follow.

QUICK LEARNING REVIEW:

LEARNING OUTCOME 2.5

- Describe an organization using the systems approach.
- Discuss how the systems approach helps us understand management.
- Explain how the contingency approach is appropriate for studying management.

Go to page 39 to see how well you know this material.

contingency approach (or **situational approach**)
A management approach which says that organizations are different, face different situations (contingencies), and require different ways of managing.

Let's Get Real:

My Turn

Ripu Lamba

General Manager, Training and Development
Datacom Solutions Pvt. Ltd

I would take the following measures:

"People have the feeling–"this is something that happens to others!" Safety cannot be taken for granted; hence, as a Manager of HP Globalsoft, I would organize workshops in the company, to sensitize employees, on their vulnerability, if enough caution is not taken. Without alarming, talk of occupational hazards in every sphere—medicine, where a surgeon could be exposed to human immunodeficiency virus (HIV) if they don't take precaution and hospitality (refer recent events in Mumbai). I would try to convince people that the onus of ensuring safety of self and of employees becomes a collective responsibility.

To implement compliance to the safety procedures, I would have workshops organized.

- These workshops would be facilitated in a manner that would bring out the need to partake in the "tracking system." Invite police or those who have first-hand interaction with the victims to come for brief interactions.
- The management will have to demonstrate that they are not infringing on the privacy of the employees but rather extending the assurance that tracking is only for SOS purpose. Never, to use the whereabouts for purpose other than intended (hence, supervisor training to sensitize on this is imperative).
- Have the security system be closely manned-inform alarms (without disclosing names), if any, as occurring in a similar or related industry.
- Work closely with the police.
- Invest in a resident counselor (someone with a psychology/OB background), in whom employees can confide/vent, and can have their fears allayed, which they would have doubted for the system.

LEARNING OUTCOMES

SUMMARY

2.1 ▷ HISTORICAL BACKGROUND OF MANAGEMENT

- Explain why studying management history is important.
- Describe some early evidences of management practice.
- Describe two important historical events that are significant to the study of management.

Studying history is important because it helps us see the origins of today's management practices and helps us see what has and has not worked. We can see early examples of management practice in the construction of the Egyptian pyramids and in the arsenal of Venice. One important historical event was the publication of Adam Smith's *Wealth of Nations*, in which he argued the benefits of division of labor (job specialization). The second was the industrial revolution, where it became more economical to manufacture in factories than at home. Managers were needed to manage these factories, and these managers needed formal management theories to guide them.

2.2 ▷ CLASSICAL APPROACH

- Describe the important contributions made by Frederick W. Taylor and Frank and Lillian Gilbreth.
- Discuss Fayol's and Weber's contributions to management theory.
- Explain how today's managers use scientific management and general administrative theory.

Taylor, known as the "father" of scientific management, studied manual work using scientific principles—that is, guidelines for improving production efficiency—to find the "one best way" to do those jobs. The Gilbreths' primary contribution was finding efficient hand-and-body motions and designing proper tools and equipment for optimizing work performance. Fayol believed that the functions of management were common to all business endeavors but also were distinct from other business functions. He developed 14 principles of management from which many current management concepts have evolved. Weber described an ideal type of organization he called a bureaucracy, which is a type still used by many of today's large organizations. Today's managers use the concepts of scientific management when they analyze basic work tasks to be performed, use time-and-motion study to eliminate wasted motions, hire the best-qualified workers for a job, and design incentive systems based on output. They use general administrative theory when they perform the functions of management and structure their organizations so that resources are used efficiently and effectively.

2.3 ▷ QUANTITATIVE APPROACH

- Explain what the quantitative approach has contributed to the field of management.
- Describe total quality management.
- Discuss how today's managers use the quantitative approach.

The quantitative approach involves applications of statistics, optimization models, information models, and computer simulations to management activities. Total quality management is a management philosophy devoted to continuous improvement and responsiveness to customer needs and expectations. Today's managers use the quantitative approach especially when making decisions as they plan and control work activities such as allocating resources, improving quality, scheduling work, or determining optimum inventory levels.

2.4 ▷ BEHAVIORAL APPROACH

- Describe the contributions of the early advocates of OB.
- Explain the contributions of the Hawthorne Studies to the field of management.
- Discuss how today's managers use the behavioral approach.

The early organizational behavior advocates (Robert Owen, Hugo Munsterberg, Mary Parker Follett, and Chester Barnard) contributed various ideas, but all believed that people were the most important asset of an organization and should be managed accordingly. The Hawthorne Studies dramatically affected management beliefs about the role of people in organizations, leading to a new emphasis on the human behavior factor in managing. The behavioral approach has largely shaped how today's organizations are managed. Many current theories of motivation, leadership, group behavior and development, and other behavioral issues can be traced to the early OB advocates and the conclusions from the Hawthorne Studies.

2.5 ▷ CONTEMPORARY APPROACH

- Describe an organization using the systems approach.
- Discuss how the systems approach helps us understand management.
- Explain how the contingency approach is appropriate for studying management.

The systems approach says that an organization takes in inputs (resources) from the environment and transforms or processes those resources into outputs that are distributed into the environment. It helps us understand management because managers must ensure that all the interdependent units are working together in order to achieve the organization's goals, helps managers realize that decisions and actions taken in one organizational area will affect others, and helps managers recognize that organizations are not self-contained but instead rely on their environment for essential inputs and as outlets to absorb their outputs. The contingency approach says that organizations are different, face different situations, and require different ways of managing. It helps us understand management because it stresses that there are no simplistic or universal rules for managers to follow. Instead, managers must look at their situation and determine that if this is the way my situation is, then this is the best way for me to manage.

THINKING ABOUT MANAGEMENT ISSUES

1. What kind of workplace would Henri Fayol create? How about Mary Parker Follett? How about Frederick W. Taylor?
2. Can a mathematical (quantitative) technique help a manager solve a "people" problem, such as how to motivate employees or how to distribute work equitably? Explain.
3. How do societal trends influence the practice of management? What are the implications for someone studying management?
4. Continual improvement is a cornerstone of TQM. Is continual improvement possible? What challenges do organizations face in searching for ways to continually improve? How can managers deal with those challenges?
5. How can an approach in which we say "it depends on the situation" be useful to managers? Discuss.

YOUR TURN to be a Manager

- Choose two non-management classes that you are currently enrolled in or have taken previously. Describe three ideas and concepts from those subject areas that might help you be a better manager.
- Read at least one current business article from any of the popular business periodicals each week for four weeks. Take one of those articles and describe what it is about and how it relates to any (or all) of the four approaches to management.
- Choose an organization with which you are familiar and describe the job specialization used there. Is it efficient and effective? Why or why not? How could it be improved?
- Can scientific management principles help you be more efficient? Choose a task that you do regularly (such as laundry, fixing dinner, grocery shopping, studying for exams, etc.). Analyze it by writing down the steps involved in completing that task. See if there are activities that could be combined or eliminated. Find the "one best way" to do this task. And the next time you have to do the task, try the scientifically managed way and see if you become more efficient (keeping in mind that changing habits isn't easy to do).
- How do business organizations survive for 100+ years? Obviously, they've seen a lot of historical events come and go! Choose one such company (for example, Coca-Cola, Procter & Gamble, Avon, General Electric) and research its history. How has the company changed over the years? From your research on this company, what did you learn that could help you be a better manager?
- Find the current top five best-selling management books. Read a review of each book, read the book covers (or even read the books). Write a short paragraph describing what each book is about. Also write about which of the historical management approaches you think the book fits with and how you think it fits into that approach.
- Choose one historical event from this century and do some research on it. Write a paper describing the impact that this event might be having or has had on how workplaces are managed.
- Steve's and Mary's suggested readings: Gary Hamel, *The Future of Management* (Harvard Business School Press, 2007); Malcolm Gladwell, *Blink* (Little, Brown and Company, 2005); James C. Collins, *Good to Great: Why Some Companies Make the Leap . . . and Others Don't* (Harper Business, 2001); Matthew J. Kiernan, *The Eleven Commandments of 21st Century Management* (Prentice Hall, 1996); and James C. Collins and Jerry I. Porras, *Built to Last: Successful Habits of Visionary Companies* (Harper Business, 1994).
- Come on, admit it. You multitask, don't you? And if not, you probably know people who do. Multitasking is common in the workplace. But does it make employees more efficient and effective? Pretend you're the manager in charge of a loan-processing department. Describe how you would research this issue, using each of the following management approaches or theories: scientific management, general administrative theory, quantitative approach, behavioral approach, systems theory, and contingency theory.

- In your own words, write down three things you learned in this chapter about being a good manager.
- Self-knowledge can be a powerful learning tool. Go to mymanagementlab and complete any of these self-assessment exercises: How Well Do I Respond to Turbulent Change? How Well Do I Handle Ambiguity? and What Do I Value? Using the results of your assessments, identify personal strengths and weaknesses. What will you do to reinforce your strengths and improve your weaknesses?

CASE APPLICATION

Fast-Forwarding Blockbuster

What name comes to mind when you think of renting movies on the weekend? Maybe it's Blockbuster. If so, you're not alone. Although Blockbuster is still the world's largest video rental company, with more than 5,000 stores in the United States, the past few years have not been good ones for this company. Blockbuster has posted losses in 9 of the past 11 years, closed hundreds of stores, and lost many customers to Netflix. The whole business of renting DVDs is changing, and Blockbuster is making changes in order to remain competitive.

One of the most important changes was hiring a new CEO, James W. Keyes. As the former CEO of 7-Eleven Stores, Keyes faced a similar situation as the convenience store industry went through a difficult transition in the 1980s and 1990s. By 1990, 7-Eleven was in bankruptcy. However, by 2004, his company had achieved 36 consecutive quarters of revenue growth and a profit of $106 million. How? What Keyes did at 7-Eleven was to rely on numbers. He implemented an approach in which quantitative data collected by each store dictated the product mix carried in that store. For instance, a store in one neighborhood might carry more Corona compared to another store across town that stocked more Coors Lite. An enthusiastic believer in the power of data, Keyes took into account all kinds of factors, calculating, for example, which doughnuts sold best in hot and cold weather.

Now, Keyes is doing the same for Blockbuster. Among other things, he hopes to customize titles at each store based on rental patterns. Although selling video rentals may be different than selling Slurpees, Keyes is unfazed. He says, "I think an Internet company, even if it is selling a service, is still another form of retail. It comes down to the ability to understand the needs of customers and satisfy those needs in a way that is better and different from the competitors."

Blockbuster CEO, James W. Keyes must meet competition from Netflix, video-on-demand services, and TiVo.

Discussion Questions

1. Mr. Keyes obviously is a big fan of the quantitative approach. How might principles of scientific management be useful to Blockbuster?
2. How might knowledge of organizational behavior help the company's frontline store supervisors manage their employees? Would Mr. Keyes and other top managers need to understand OB? Why or why not?
3. Using Exhibit 2–7, describe Blockbuster as a system.
4. Based on information from Blockbuster's Web site (www.blockbuster.com), what values does this company embrace that might be important for successful organizations in the twenty-first century? (Hint: Look at the About Us link and the Culture link found in Careers.)
5. What do you think Mr. Keyes's quote (last sentence in last paragraph) is saying? Do you agree or disagree with his statement? Explain.

Sources: Blockbuster Web site, www.blockbuster.com, February 28, 2008; A. Adam Newman, "New Boss Aims to Apply Some 7-Eleven Tactics to Blockbuster," *New York Times* online, www.nytimes.com, July 17, 2007; and M. Herbst, "Blockbuster Gives a Say on Pay," *BusinessWeek* online, www.businessweek.com, May 11, 2007.

Let's Get Real:

Meet The Manager

Atul Srivastava

Deputy General Manager, HR
Steel Authority of India Ltd (SAIL)

MY JOB: Establishment of matters (such as promotion, rotation, and development) of senior-level executives and championing new interventions such as performance management system and introduction of 360 DF.

BEST PART OF MY JOB: To interact with a cross-section of executives/senior management and to introduce new systems for the betterment of the HR.

WORST PART OF MY JOB: Some routines that do not add value to the organization.

BEST MANAGEMENT ADVICE EVER RECEIVED: To do something remarkable so that people remember you.

You'll be hearing more from this real manager throughout the chapter.

Chapter

Organizational Culture and Environment

Are managers free to do whatever they want? In this chapter, we're going to look at the factors that limit the discretion managers have in doing their jobs. These factors are both internal (the organization's culture) and external (the organizational environment). Focus on the following learning outcomes as you read and study this chapter.

LEARNING OUTCOMES

A Manager's Dilemma

LG's working environment is characterized by passion, empowerment, and engagement. To be able to address attrition, which is a common problem in all industries, LG started many initiatives to balance work-life and to stress career development. For example, LG banned work on Sundays. The HR department even made calls on home landlines of employees to find out if they were, in fact, not working. A family ambassadors program has been started to counsel employees and their families on personal problems. LG spent close to Rs 1 crore on designing a training program for 30 of its top employees to lead change and provide mentorship to selected 2,000 employees' careers. All employees spend at least 10 days on training in LG and every quarter work comes to standstill in LG worldwide so that all employees can take a test.[1] The high scorers are given attractive prizes. Excellence in performance is incentivized. As the company continues to grow and face competition, how can the managers at LG ensure that its culture continues?

What would you do?

LG's managers recognize that how important organizational culture is to their organization. The company has created a culture where employees are treated with care and respect and as part of a family. The company is driven by passion and challenge. But the company's managers also recognize the challenges in trying to manage the culture as the organization grows and economic environment changes. How much impact do managers have on an organization's success or on its failure? In the following section, we explore this important question.

LEARNING **OUTCOME 3.1** ▷

THE MANAGER: OMNIPOTENT OR SYMBOLIC?

How much difference does a manager make in how an organization performs? The dominant view in management theory and society in general is that managers are directly responsible for an organization's success or failure. We'll call this perspective the **omnipotent view of management**. In contrast, others have argued that much of an organization's success or failure is due to external forces outside managers' control. This perspective has been called the **symbolic view of management**. Let's look at each perspective so we can try to clarify just how much credit or blame managers should get for their organization's performance.

THE OMNIPOTENT VIEW

In Chapter 1, we stressed how important managers are to organizations. Differences in an organization's performance are assumed to be due to decisions and actions of its managers. Good managers anticipate change, exploit opportunities, correct poor performance, and lead their organizations. When profits are up, managers take the credit and reward themselves with bonuses, stock options, and the like. When profits are down, top managers are often fired, in the belief that "new blood" will bring improved results. For instance, the Royal Challengers of the Indian Premiere League (IPL) series on May 6 found their CEO Charu Sharma fired in the wake of its embarrassing defeats. Vijay Mallya, the owner of the team, of course later said that the Charu Sharma had left for personal reasons.[2] In the omnipotent view, someone has to be held accountable when organizations perform poorly, regardless of the reasons, and that "someone" is the manager. Of course, when things go well, managers also get the credit—even if they had little to do with achieving the positive outcomes.

The view of managers as omnipotent is consistent with the stereotypical picture of the take-charge business executive who overcomes any obstacle in seeing that the organization achieves its goals. And this view isn't limited to business organizations. It also explains the high turnover among college and professional sports coaches, who are considered the "managers" of their teams. Coaches who lose more games than they win are fired and replaced by new coaches who are expected to correct the poor performance.

THE SYMBOLIC VIEW

In the 1990s, Cisco Systems was the picture of success. Growing rapidly, it was widely praised by analysts for its "brilliant strategy, masterful management of acquisitions, and superb customer focus."[3] Then the tech bubble burst. As Cisco's performance declined, analysts said that its strategy was flawed, its acquisition approach was haphazard, and its customer service was poor. Was declining performance due to the managers' decisions and actions, or was it beyond their control? The symbolic view would suggest the latter.

The symbolic view says that a manager's ability to affect outcomes is influenced and constrained by external factors.[4] According to this view, it's unreasonable to expect managers to significantly affect an organization's performance. Instead, performance is influenced by factors managers don't control, such as the economy, customers, governmental policies, competitors' actions, industry conditions, and decisions made by previous managers.

This view is called "symbolic" because it's based on the belief that managers symbolize control and influence.[5] How? They develop plans, make decisions, and engage in other managerial activities to make sense out of random, confusing, and ambiguous situations. However, the part that managers actually play in organizational success or failure is limited according to this view.

REALITY SUGGESTS A SYNTHESIS

In reality, managers are neither all-powerful nor helpless. But their decision and action options are constrained. Internal constraints come from the organization's culture, and external constraints come from the organization's environment (see Exhibit 3–1). In the rest of this chapter, we explore both.

Exhibit 3–1

Constraints on Managerial Discretion

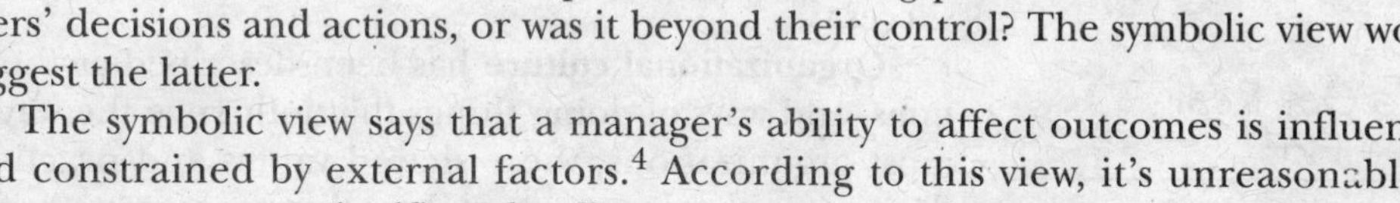

omnipotent view of management
The view that managers are directly responsible for an organization's success or failure.

symbolic view of management
The view that much of an organization's success or failure is due to external forces outside managers' control.

QUICK LEARNING REVIEW:

LEARNING OUTCOME 3.1

- **Contrast the actions of managers according to the omnipotent and symbolic views.**
- **Identify the two constraints on managers' discretion.**

Go to page 65 to see how well you know this material.

LEARNING **OUTCOME 3.2** ▷

ORGANIZATIONAL CULTURE

Each of us has a unique personality—traits and characteristics that influence the way we act and interact with others. When we describe someone as warm, open, relaxed, shy, or aggressive, we're describing personality traits. An organization, too, has a personality, which we call its *culture.*

WHAT IS ORGANIZATIONAL CULTURE?

MindTree is a global information technology (IT) and R&D Services company, co-headquartered in the United States and India. With a passion for customer satisfaction, MindTree partners with its clients to create a transparent, value-based relationship. MindTree promotes a diverse culture and possesses a clear vision focused on giving back to the community and cultivating a culture of empathy. Partnership, teamwork, and integrity form the cornerstone of MindTree's philosophy.[6] Puneet Jetli, General Manager of people function with MindTree Consulting, says, "our core values guide our day-to-day policies and actions. They are the cornerstone for everything we do."[7] MindTree Consulting believes that the human ability to create new knowledge and to lead with ideas is central to personal as well as for business success.

Organizational culture has been described as the shared values, principles, traditions, and ways of doing things that influence the way organizational members act. In most organizations, these shared values and practices have evolved over time and determine, to a large extent, how things are done in the organization.[8]

Our definition of culture implies three things. First, culture is a *perception.* It's not something that can be physically touched or seen, but employees perceive it on the basis of what they experience within the organization. Second, organizational culture is *descriptive.* It's concerned with how members perceive the culture, not with whether they like it. Finally, even though individuals may have different backgrounds or work at different organizational levels, they tend to describe the organization's culture in similar terms. That's the *shared* aspect of culture.

Research suggests that there are seven dimensions that describe an organization's culture.[9] Each of the seven dimensions (shown in Exhibit 3–2) ranges from low to high, meaning it's not very typical of the culture (low) or is very typical of the culture (high). Describing an organization using these seven dimensions gives a composite picture of the organization's culture. In many organizations, one cultural dimension often is emphasized more than the others and essentially shapes the organization's personality and the way organizational members work. For instance, at Sony Corporation, the focus is product innovation (innovation and risk taking). The company "lives and breathes" new product development, and employees' work behaviors support that goal. In contrast, Southwest Airlines has made its employees a central part of its culture (people orientation). Exhibit 3–3 describes how the dimensions can create significantly different cultures.

Let's Get Real: F2F

TO ME, ORGANIZATIONAL CULTURE IS:
The pulse of an organization. It is reflected in the way we act, perform, and interact with the individuals or the customers.

OUR CULTURE:
Is very rich. We believe in long-term relationships and in producing good quality steel.

STRONG CULTURES

All organizations have cultures, but not all cultures equally influence employees' behaviors and actions. **Strong cultures**—those in which the key values are deeply held and widely shared—have a greater influence on employees than do weaker

Exhibit 3–2

Dimensions of Organizational Culture

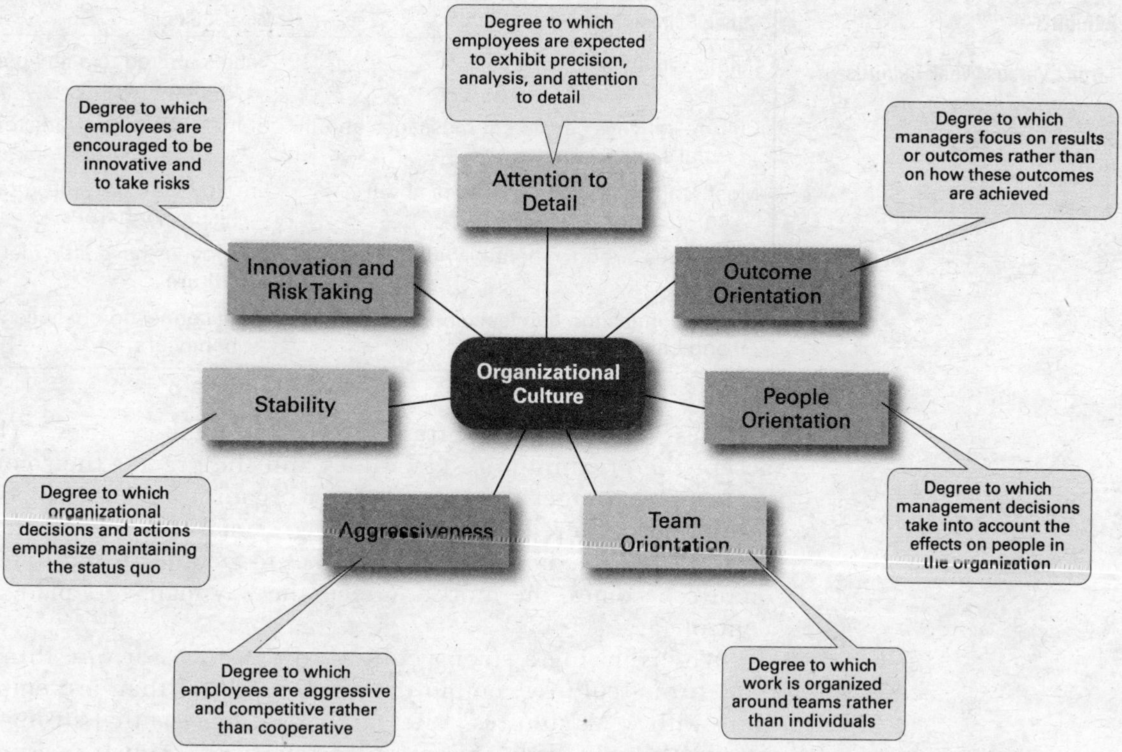

Exhibit 3–3 Contrasting Organizational Cultures

Organization A

This organization is a manufacturing firm. Managers are expected to fully document all decisions, and "good managers" are those who can provide detailed data to support their recommendations. Creative decisions that incur significant change or risk are not encouraged. Because managers of failed projects are openly criticized and penalized, managers try not to implement ideas that deviate much from the status quo. One lower-level manager quoted an often-used phrase in the company: "If it ain't broke, don't fix it."

Employees are required to follow extensive rules and regulations in this firm. Managers supervise employees closely to ensure that there are no deviations. Management is concerned with high productivity, regardless of the impact on employee morale or turnover.

Work activities are designed around individuals. There are distinct departments and lines of authority, and employees are expected to minimize formal contact with other employees outside their functional area or line of command. Performance evaluations and rewards emphasize individual effort, although seniority tends to be the primary factor in the determination of pay raises and promotions.

Organization B

This organization is also a manufacturing firm. Here, however, management encourages and rewards risk taking and change. Decisions based on intuition are valued as much as those that are well rationalized. Management prides itself on its history of experimenting with new technologies and its success in regularly introducing innovative products. Managers or employees who have a good idea are encouraged to "run with it," and failures are treated as "learning experiences." The company prides itself on being market driven and rapidly responsive to the changing needs of its customers.

There are few rules and regulations for employees to follow, and supervision is loose because management believes that its employees are hardworking and trustworthy. Management is concerned with high productivity but believes that this comes through treating its people right. The company is proud of its reputation as being a good place to work.

Job activities are designed around work teams, and team members are encouraged to interact with people across functions and authority levels. Employees talk positively about the competition between teams. Individuals and teams have goals, and bonuses are based on achievement of outcomes. Employees are given considerable autonomy in choosing the means by which the goals are attained.

organizational culture
The shared values, principles, traditions, and ways of doing things that influence the way organizational members act.

strong cultures
Organizational cultures in which the key values are intensely held and widely shared.

Exhibit 3–4

Strong Versus Weak Cultures

Strong Cultures	Weak Cultures
Values widely shared	Values limited to a few people—usually top management
Culture conveys consistent messages about what's important	Culture sends contradictory messages about what's important
Most employees can tell stories about company history/heroes	Employees have little knowledge of company history or heroes
Employees strongly identify with culture	Employees have little identification with culture
Strong connection between shared values and behaviors	Little connection between shared values and behaviors

cultures. (Exhibit 3–4 contrasts strong and weak cultures.) The more employees accept the organization's key values and the greater their commitment to those values, the stronger the culture. Most organizations have moderate to strong cultures; that is, there is relatively high agreement on what's important, what defines "good" employee behavior, what it takes to get ahead, and so forth. The stronger a culture becomes, the more it affects the way managers plan, organize, lead, and control.[10]

Why is having a strong culture important? For one thing, in organizations with strong cultures, employees are more loyal than are employees in organizations with weak cultures.[11] Research also suggests that strong cultures are associated with high organizational performance.[12] And it's easy to understand why. After all, if values are clear and widely accepted, employees know what they're supposed to do and what's expected of them, so they can act quickly to take care of problems. However, the drawback is that a strong culture also might prevent employees from trying new approaches, especially when conditions are changing rapidly.[13]

WHERE CULTURE COMES FROM AND HOW IT CONTINUES

Exhibit 3–5 illustrates how an organization's culture is established and maintained. The original source of the culture usually reflects the vision of the founders. For instance, as described earlier, in the early days, Godrej's culture reflected the values of its founder, Nadir Godrej. Company founders are not constrained by previous customs or approaches and can establish the early culture by articulating an image of what they want the organization to be. The small size of most new organizations makes it easier to instill that image with all organizational members.

When the culture is in place, certain organizational practices help maintain it. For instance, during the employee selection process, managers typically judge job candidates not only on the job requirements but also on how well they might fit into the organization. At the same time, job candidates find out information about the organization and determine whether they are comfortable with what they see.

Exhibit 3–5 Establishing and Maintaining Culture

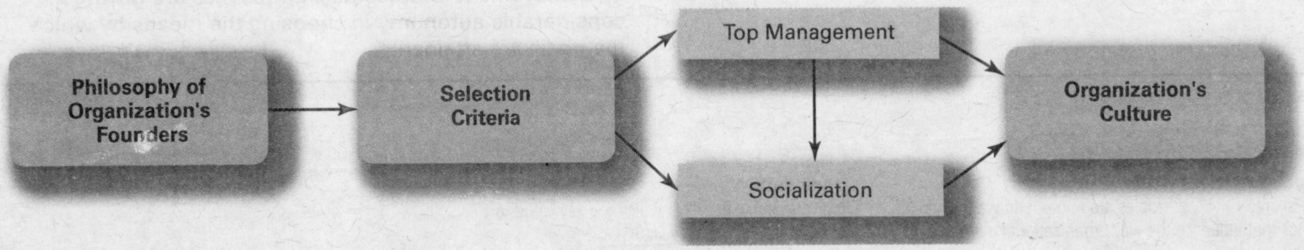

The actions of top managers also have a major impact on the organization's culture. The risk-taking appetite and irreverence to its large competitors of Sri Karsan Bhai Patel, the founder of Nirma has pervaded the growth and actions of Nirma to date. Similarly, Vijay Mallya's flamboyant lavish lifestyle influenced him to provide top of the line facilities to passengers in his airline Kingfisher. The dress of the flight attendants, the presentation of food, all reflect the personality of Vijay Mallya. The stepping out of Narayan Murthy from the CEO's role at the prime of his career has signaled to others, within the company, that they need to develop and mentor those below them and, constantly, rediscover their strength and own roles. Through what they say and how they behave, top managers establish norms that filter down through the organization and that can have a positive effect on employees' behaviors. For instance, IBM's CEO Sam Palmisano wanted employees to value teamwork, so he chose to take several million dollars from his yearly bonus and give it to his top executives, based on their teamwork. He said, "If you say you're about a team, you have to be a team. You've got to walk the talk, right?"[14] However, as we've seen in numerous corporate ethics scandals, the actions of top managers can also lead to undesirable outcomes.

Finally, organizations help employees adapt to the culture through **socialization**, a process that helps new employees learn the organization's way of doing things. For instance, new employees at HomeTown go through a 6-day intensive training that helps turn them into a home design consultants. They learn company philosophy, company jargon, and how to assist customers with decisions about beans, grind, and espresso machines. One benefit of this socialization is that it helps employees understand the culture and become enthusiastic and knowledgeable with customers.[15] Another benefit is that it minimizes the chance that new employees who are unfamiliar with the organization's culture might disrupt current beliefs and customs.

HOW EMPLOYEES LEARN CULTURE

Employees "learn" an organization's culture in a number of ways. The most common are stories, rituals, material symbols, and language.

Stories. Organizational "stories" typically contain a narrative of significant events or people, including such things as the organization's founders, rule breaking, and reactions to past mistakes.[16] For instance, managers at Nike feel that stories told about the company's past help shape the future. Whenever possible, corporate "storytellers" (senior executives) explain the company's heritage and tell stories that

One area where organizational culture influences employee behavior is how employees dress for work. Look at what each of these four individuals is wearing. At what kinds of organizations do you think each of these styles is appropriate work wear? What do you think the cultures might be like at the organizations where each person works?

socialization
The process that helps employees adapt to the organization's culture.

celebrate people getting things done. When they tell the story of how co-founder Bill Bowerman (now deceased) went to his workshop and poured rubber into his wife's waffle iron to create a better running shoe, they're celebrating and promoting Nike's spirit of innovation. These company stories provide examples that people can learn from.[17] At 3M Company, the product innovation stories are legendary. There's the story about the 3M scientist who spilled chemicals on her tennis shoe and came up with Scotchgard. Then there's the story about Art Fry, a 3M researcher, who wanted a better way to mark the pages of his church hymnal and invented the Post-it Note. These stories reflect what made 3M great and what it will take to continue that success.[18] To help employees learn the culture, organizational stories anchor the present in the past, provide explanations and legitimacy for current practices, exemplify what is important to the organization, and provide compelling pictures of an organization's goals.[19]

Rituals. The "Passing of the Pillars" is an important ritual at Boston Scientific's facility near Minneapolis. When someone has a challenging assignment, they're "awarded" a 2-foot-high plaster-of-Paris pillar to show that they've got support from all their colleagues. Corporate rituals are repetitive sequences of activities that express and reinforce the important values and goals of the organization.[20] At Aditya Birla, an Indian multinational company, there are three high-profile awards that are given by the company to recognize individual contributors every year, awards for superior teamwork, awards for internationally benchmarked outstanding achievement, and a monetary award for performance recognition in delivering excellence. The awards are presented by the Chairman in a very extravagant award ceremony. It conveys to the company's employees that delivering excellent performance is important.[21]

Material Symbols. When you walk into different businesses, do you get a feel for what type of work environment it is—formal, casual, fun, serious, and so forth? These reactions demonstrate the power of material symbols or artifacts in creating an organization's personality.[22] The layout of an organization's facilities, how employees dress, the types of automobiles provided to top executives, and the availability of corporate aircraft are examples of material symbols. Others include the size of offices, the elegance of furnishings, executive "perks" (extra benefits provided to managers, such as health club memberships, use of company-owned facilities, and so forth), employee fitness centers or on-site dining facilities, and reserved parking spaces for certain employees. At WorldNow, an important material symbol is an old dented drill that the founders purchased for $2 at a thrift store. The drill symbolizes the company's culture of "drilling down to solve problems." When an employee is presented with the drill in recognition of outstanding work, he or she is expected to personalize the drill in some way and devise a new rule for caring for it. One employee installed a Bart Simpson trigger; another made the drill wireless by adding an antenna. The company's "icon" carries on the culture even as the organization evolves and changes.[23]

Material symbols convey to employees who is important and the kinds of behavior (for example, risk taking, conservative, authoritarian, participative, individualistic) that are expected and appropriate.

Language. Many organizations and units within organizations use language as a way to identify and unite members of a culture. By learning this language, members attest to their acceptance of the culture and their willingness to help preserve it. At Cranium, a Seattle board game company, "chiff" is used to remind employees of the need to be incessantly innovative in everything they do. "Chiff" stands for "clever, high-quality, innovative, friendly, fun."[24] At another Seattle company, Microsoft, employees have their own unique vocabulary: *work judo*

Let's Get Real: F2F

STORIES, RITUALS, AND MATERIAL SYMBOLS:
People will follow you if they see that your intentions are good and you have your heart in the project.

HOW OUR CULTURE AFFECTS THE WAY I MANAGE:
Sincere actions with the intention of doing the best for the organization and the individual.

(deflecting a work assignment to someone else without making it appear that you're avoiding it), *eating your own dog food* (using your own software programs or products in the early stages as a way of testing it even if the process is disagreeable), *flat food*(goodies from the vending machine that can be slipped under the door to a colleague who's working on a tight deadline), *facemail* (talking to someone face-to-face), *death march* (countdown to shipping a new product), and so on.[25]

Over time, organizations often develop unique terms to describe equipment, key personnel, suppliers, customers, processes, or products related to their business. New employees are frequently overwhelmed with acronyms and jargon that, after a short period of time, become a natural part of their language. Once learned, this language acts as a common denominator that bonds members.

HOW CULTURE AFFECTS MANAGERS

Houston-based Apache Corp. has become one of the best performers in the independent oil drilling business because it has fashioned a culture that values risk taking and quick decision making. Potential hires are judged on how much initiative they've shown in getting projects done at other companies. And company employees are handsomely rewarded if they meet profit and production goals.[26]

Because an organization's culture constrains what they can and cannot do, it's particularly relevant to managers. Such constraints are rarely explicit. They're not written down. It's unlikely that they'll even be spoken. But they're there, and all managers quickly learn what to do and not do in their organization. For instance, you won't find the following values written down, but each comes from a real organization:

- Look busy even if you're not.
- If you take risks and fail around here, you'll pay dearly for it.
- Before you make a decision, run it by your boss so that he or she is never surprised.
- We make our product only as good as the competition forces us to.
- What made us successful in the past will make us successful in the future.
- If you want to get to the top here, you have to be a team player.

The link between values such as these and managerial behavior is fairly straightforward. Take, for example, a so-called "ready-aim-fire" culture. In such an organization, managers study and analyze proposed projects endlessly before committing to them. On the other hand, in a "ready-fire-aim" culture, managers take action and then analyze what has been done. Or, say that an organization's culture supports the belief that profits can be increased through cost cutting and that the company's best interests are served by achieving slow but steady increases in quarterly earnings. Managers are unlikely to pursue programs that are innovative, risky, long term, or expansionary. In an organization whose culture conveys a basic distrust of employees, managers are more likely to use an authoritarian leadership style than a democratic one. Why? The culture establishes for managers appropriate and expected behavior. At Mindtree, a software company, for example, a culture shaped by the value placed on freedom of expression, a lack of coercion and fear, and a determination to make work fun influences the way employees work and the way that managers plan, organize, lead, and control. The organization's culture is reinforced even by the office environment, which is open, versatile, and creative.[27]

As shown in Exhibit 3–6, a manager's decisions are influenced by the culture in which he or she operates. An organization's culture, especially a strong one, influences and constrains the way managers plan, organize, lead, and control.

Exhibit 3–6

Managerial Decisions Affected by Culture

Planning

- The degree of risk that plans should contain
- Whether plans should be developed by individuals or teams
- The degree of environmental scanning in which management will engage

Organizing

- How much autonomy should be designed into employees' jobs
- Whether tasks should be done by individuals or in teams
- The degree to which department managers interact with each other

Leading

- The degree to which managers are concerned with increasing employee job satisfaction
- What leadership styles are appropriate
- Whether all disagreements—even constructive ones—should be eliminated

Controlling

- Whether to impose external controls or to allow employees to control their own actions
- What criteria should be emphasized in employee performance evaluations
- What repercussions will occur from exceeding one's budget

QUICK LEARNING REVIEW:

LEARNING OUTCOME 3.2

- Identify the seven dimensions of organizational culture.
- Discuss the impact of a strong culture on organizations and managers.
- Explain how a culture is formed and maintained.
- Describe how culture affects managers.

Go to page 65 to see how well you know this material.

LEARNING **OUTCOME 3.3** ▷

CURRENT ORGANIZATIONAL CULTURE ISSUES

Nordstrom, the specialty retail chain, is renowned for its attention to customers. Nike's innovations in running shoe technology are legendary. Tom's of Maine is known for its commitment to doing things ethically and spiritually. Having rebounded from the largest racial discrimination lawsuit in U.S. history (with a

thinking critically about Ethics

When does success turn ugly? EMC Corporation prided itself on its hard-driving and immensely successful sales force.[28] The head of marketing bragged that he hired salespeople "with the passion to knock down walls and who would fit into a culture of doing whatever it takes." The sales team viewed itself as an elite force that made the company a market leader. However, behind the scenes was a much darker and offensive side: "a macho, frat-boy atmosphere that was intimidating and, at times, discriminatory to women." Interviews with former salespeople, both women and men, described "locker-room antics, company-paid visits to strip clubs, demeaning sexual remarks, or retaliation against women who complained about the atmosphere." Although such behaviors are illegal, the ethical issues are just as critical. What would you do at this point? Do you think this culture could be changed? What would you have to do to change it? What lessons are here for other organizations and managers?

$192.5 million settlement), the Coca-Cola Company has become one of the top companies for diversity. How have these organizations achieved such reputations? Their organizational cultures have played a crucial role. Let's look at five current cultural issues: creating an ethical culture, creating an innovative culture, creating a customer-responsive culture, creating a workplace culture that supports diversity, and nurturing workplace spirituality.

CREATING AN ETHICAL CULTURE

Andrew Fastow's name is one that forever will be linked to the Enron ethics scandal. The former chief financial officer of Enron Corporation (he was convicted and sentenced to six years in prison for wire and securities fraud) had a Lucite cube on his desk that laid out the company's values. It included the following inscription: "When Enron says it's going to rip your face off, it will rip your face off."[29] Other Enron employees described a culture in which personal ambition was valued over teamwork, youth over wisdom, and earnings growth at any cost.[30]

The content and strength of an organization's culture influences its ethical climate and the ethical behavior of its members.[31] If the culture is strong and supports high ethical standards, it should have a very powerful and positive influence on employee behavior. For example, Alfred P. West, founder and CEO of financial services firm SEI Investments Company, spends a lot of time emphasizing to employees his vision for the company—an open culture of integrity, ownership, and accountability. He says, "We tell our employees a lot about where the company is going. We over-communicate the vision and strategy and continually reinforce the culture."[32]

An organizational culture most likely to shape high ethical standards is one that's high in risk tolerance, low to moderate in aggressiveness, and focused on means as well as outcomes. Managers in such a culture are supported for taking risks and innovating, are discouraged from engaging in uncontrolled competition, and will pay attention to *how* goals are achieved as well as to *what* goals are achieved. Exhibit 3–7 outlines some suggestions for creating a more ethical culture.

CREATING AN INNOVATIVE CULTURE

You may not recognize IDEO's name, but you've probably used a number of its products. As a product design firm, it takes the ideas that corporations bring it and turns those ideas into reality. Some of its creations range from the first commercial mouse (for Apple Computer) to the first standup toothpaste tube (for Procter & Gamble) to the handheld personal organizer (for Palm). It's critical that IDEO's culture support creativity and innovation.[33] Another innovative organization is Cirque du Soleil, the Montreal-based creator of circus theatre. Its managers state that the culture is based on involvement, communication, creativity, and diversity, which they see as keys to innovation.[34] Although these two companies are in industries where innovation is important (product design and entertainment), the fact is that any successful organization needs a culture that supports innovation. How important is culture to innovation? In a recent survey of senior executives, more than half said that the most important driver of innovation for companies was a supportive corporate culture.[35]

Exhibit 3–7

Creating an Ethical Culture

- Be a *visible role model.*
- Communicate *ethical expectations.*
- Provide *ethics training.*
- Visibly *reward ethical acts and punish unethical ones.*
- Provide *protective mechanisms* so employees can discuss ethical dilemmas and report unethical behavior without fear.

Apple's risk-taking culture has helped it create such industry-changing innovations as the iPod and the iPhone and made it number-one on the World's 50 Most Innovative Companies list in 2007 and America's Most Admired Company in 2008. CEO Steve Jobs (shown here) has been the catalyst behind his company's risk-taking culture.

What does an innovative culture look like? According to Swedish researcher Goran Ekvall, it would be characterized by the following:

- **Challenge and involvement**—Are employees involved in, motivated by, and committed to the long-term goals and success of the organization?
- **Freedom**—Can employees independently define their work, exercise discretion, and take initiative in their day-to-day activities?
- **Trust and openness**—Are employees supportive and respectful to each other?
- **Idea time**—Do individuals have time to elaborate on new ideas before taking action?
- **Playfulness/humor**—Is the workplace spontaneous and fun?
- **Conflict resolution**—Do individuals make decisions and resolve issues based on the good of the organization versus personal interest?
- **Debates**—Are employees allowed to express opinions and put forth ideas for consideration and review?
- **Risk-taking**—Do managers tolerate uncertainty and ambiguity and are employees rewarded for taking risks?[36]

CREATING A CUSTOMER-RESPONSIVE CULTURE

Harrah's Entertainment, the world's largest gaming company, is fanatical about customer service—and for good reason. Company research showed that customers who were satisfied with the service they received at a Harrah's casino increased their gaming expenditures by 10 percent and those who were extremely satisfied increased their gaming expenditures by 24 percent. When customer service translates into these types of results, of course managers would want to create a customer-responsive culture![37]

What does a customer-responsive culture look like?[38] Exhibit 3–8 describes five characteristics of customer-responsive cultures and offers suggestions as to what managers can do to create that type of culture.

CREATING A CULTURE THAT SUPPORTS DIVERSITY

Today's organizations are characterized by **workforce diversity**, a workforce that's heterogeneous in terms of gender, race, ethnicity, age, and other characteristics that reflect differences. Managers must look long and hard at their culture to see whether the shared meaning and beliefs that were appropriate for a more homogeneous workforce will accept and promote diverse views. Organizations' diversity efforts are no longer driven only by federal mandate. Instead, organizations now recognize that diversity-supportive cultures are good for business. Among other things, diversity contributes to

Exhibit 3–8

Creating a Customer-Responsive Culture

Characteristics of Customer-Responsive Culture	Suggestions for Managers
Type of employee	Hire people with personalities and attitudes consistent with customer service: friendly, attentive, enthusiastic, patient, good listening skills
Type of job environment	Design jobs so employees have as much control as possible to satisfy customers, without rigid rules and procedures
Empowerment	Give service-contact employees the discretion to make day-to-day decisions on job-related activities
Role clarity	Reduce uncertainty about what service-contact employees can and cannot do by continual training on product knowledge, listening, and other behavioral skills
Consistent desire to satisfy and delight customers	Clarify organization's commitment to doing whatever it takes, even if it's outside an employee's normal job requirements

Let's Get Real: F2F

WE SUPPORT DIVERSITY BY:
Taking divergent views into account while formulating policies, etc.

more creative solutions and enhances employee moralc. But how can such a culture be encouraged? The Managing Workforce Diversity box discusses what managers can do.

SPIRITUALITY AND ORGANIZATIONAL CULTURE

What do Southwest Airlines, Timberland, and Hewlett-Packard have in common? They're among a growing number of organizations that have embraced workplace spirituality. What is **workplace spirituality**? It's a feature of a culture where organizational values promote a sense of purpose through meaningful work taking place in the context of community.[40] Organizations with a spiritual culture recognize that people have a mind and a spirit, seek to find meaning and purpose in their work, and desire to connect with other human beings and be part of a community. And such desires aren't limited to workplaces; a recent study showed that college students are also searching for meaning and purpose in life.[41]

Workplace spirituality seems to be important now for a number of reasons. Employees are looking for ways to cope with the stresses and pressures of a turbulent pace of life. Contemporary lifestyles—single-parent families, geographic mobility, temporary jobs, technologies that create distance between people—underscore the

managing workforce Diversity

Creating an Inclusive Workplace Culture

Creating a workplace culture that supports and encourages the inclusion of diverse individuals and views is a major organizational effort. How can managers create a culture that allows diversity to flourish?[39] There are two things they can do. First, managers must show that they value diversity through their decisions and actions. As they plan, organize, lead, and control, they need to recognize and embrace diverse perspectives. For instance, at the Marriott Marquis Hotel in New York's Times Square, managers take required diversity training classes, where they learn that the best way to cope with diversity-related conflict is by focusing on performance and not defining problems in terms of gender, culture, race, or disability. At Prudential, the annual planning process includes key diversity performance goals that are measured and tied to managers' compensation. The second thing managers can do is look for ways to reinforce employee behaviors that exemplify inclusiveness. Some suggestions include encouraging individuals to value and defend diverse views, creating traditions and ceremonies that celebrate diversity, rewarding "heroes" and "heroines" who accept and promote inclusiveness, and communicating formally and informally about employees who champion diversity issues.

workplace spirituality
A feature of a culture where organizational values promote a sense of purpose through meaningful work that takes place in the context of community.

lack of community that many people feel. As humans, we want involvement and connection. In addition, many managers as they reach midlife, look for something meaningful both at the job and outside the job. Others wish to integrate their personal life values with their professional lives. For others, formalized religion hasn't worked, and they continue to look for anchors to replace a lack of faith and to fill a growing sense of emptiness. What type of culture can do all these things? What differentiates spiritual organizations from their nonspiritual counterparts? Research shows that spiritual organizations tend to have five cultural characteristics.[42]

Strong Sense of Purpose. A spiritual organization builds its culture around a meaningful purpose. While profits are important, they're not the primary values of the organization. For instance, Timberland's slogan is "Boots, Brand, Belief," which embodies the company's intent to use its "resources, energy, and profits as a publicly traded footwear-and-apparel company to combat social ills, help the environment, and improve conditions for laborers around the globe...and to create a more productive, efficient, loyal, and committed employee base."[43]

Focus on Individual Development. Spiritual organizations recognize the worth and value of individuals. They aren't just providing jobs; they seek to create cultures in which employees can continually grow and learn.

Trust and Openness. Spiritual organizations are characterized by mutual trust, honesty, and openness. Managers aren't afraid to admit mistakes. And they tend to be upfront with employees, customers, and suppliers. The founder and CEO, Mr Ajit Singh, of Amar Coach Builders says that we believe in developing teamwork and relationship of truthful trust with the entire chain of customers, suppliers, and production people. Such a unique relationship has helped us in delivering the best value to our customers and for being the customer's preferred choice for all type of busses and coaches.[44]

Employee Empowerment. Managers trust employees to make thoughtful and conscientious decisions. For instance, at Southwest Airlines, employees—including flight attendants, baggage handlers, gate agents, and customer service representatives—are encouraged to take whatever action they deem necessary to meet customer needs or help fellow workers, even if it means going against company policies.

Toleration of Employee Expression. The final characteristic that differentiates spiritually based organizations from organizations that are not spiritual is that they don't stifle employee emotions. They allow people to be themselves—to express their moods and feelings without guilt or fear of reprimand.

Critics of the spirituality movement have focused on two issues: legitimacy (do organizations have the right to impose spiritual values on their employees?) and economics (are spirituality and profits compatible?).

An emphasis on spirituality clearly has the potential to make some employees uneasy. Critics might argue that secular institutions, especially businesses, should not impose spiritual values on employees. This criticism is probably valid when spirituality is defined as bringing religion into the workplace.[45] However, it's less valid when the goal is helping employees find meaning in their work. If concerns about today's lifestyles and pressures truly characterize a growing number of workers, then maybe it is time for organizations to help employees find meaning and purpose in their work and to use the workplace to create a sense of community.

The issue of whether spirituality and profits are compatible is certainly important. Limited evidence suggests that the two may be compatible. One study found that companies that introduced spiritually based techniques improved productivity and significantly reduced turnover.[46] Another found that organizations that provided their employees with opportunities for spiritual development outperformed those that didn't.[47] Others reported that spirituality in organizations was positively related to creativity, employee satisfaction, team performance, and organizational commitment.[48]

QUICK LEARNING REVIEW:
LEARNING OUTCOME 3.3

- Describe the characteristics of an ethical culture, an innovative culture, a customer-responsive culture, and a diversity-supportive culture.
- Explain why workplace spirituality seems to be an important concern.
- Describe the characteristics of a spiritual organization.

Go to page 65 to see how well you know this material.

LEARNING OUTCOME 3.4 ▷ THE ENVIRONMENT

Our discussion of an organization as an open system in Chapter 2 explained that an organization interacts with its environment as it takes in inputs and distributes outputs. Anyone who questions the impact the external environment has on managing should consider the following:

- The skyrocketing cost of food around the world has economists urgently studying the issue. Although many factors are contributing to the increase, the biggest factor is runaway demand.
- The Indian government aims to save 10,000 megawatts by 2012 through energy-saving measures. To this end, it has set a minimum target of 5 percent renewable energy to be procured by power grids on a competitive basis. Coal plants worth of 5,000 megawatts will be retired by 2012.[49]
- Although systems to automate tasks and processes have been in place for several years, these systems are now becoming interconnected through common standards for exchanging data and representing business processes in bits and bytes.[50]

As these examples show, environmental forces play a major role in shaping managers' actions. In this section, we'll identify some of these forces that affect managers and show how they constrain managerial discretion.

DEFINING THE EXTERNAL ENVIRONMENT

The term **external environment** refers to factors and forces outside the organization that affect the organization's performance. As shown in Exhibit 3–9, it includes two components: the specific environment and the general environment.

Exhibit 3–9

The External Environment

external environment
Factors and forces outside an organization that affect the organization's performance.

The Specific Environment. The **specific environment** includes external forces that directly impact managers' decisions and actions and are directly relevant to the achievement of the organization's goals. An organization's specific environment is unique to it. For instance, Timex and Rolex both make watches, but their specific environments differ because they operate in distinctly different market niches. The main forces that make up the specific environment are customers, suppliers, competitors, and pressure groups.

Customers An organization exists to meet the needs of customers who use its output. Customers represent potential uncertainty to an organization because their tastes can change or they can become dissatisfied with the organization's products or service. For example, shoppers are confused by different food rating systems—"Smart Choices," "Sensible Solution," "Best Life," etc. Grocery chains are taking steps to help them by creating simpler ways to assess the foods they're buying.[51]

Suppliers Managers seek to ensure a steady flow of needed inputs (supplies) at the lowest price possible. An organization's supplies being limited or delayed in delivery can constrain managers' decisions and actions. For instance, Pragati Maidan during trade fairs must make sure it has supplies of soft drinks, computers, food, flowers and other nursery stock, concrete, paper products, and so forth. Suppliers also provide financial and labor inputs. For example, a lack of qualified nurses continues to be a serious problem plaguing health care providers, affecting their ability to meet demand and keep service levels high.

Competitors All organizations—profit and not-for-profit—have competitors. Managers cannot afford to ignore the competition. For instance, Doordarshan used to control what you watched on television. Now, they face competition from digital cable, satellite, DVDs, and the Internet, all of which offer customers a much broader choice.

Pressure Groups Managers must recognize special-interest groups that attempt to influence the actions of organizations. For instance, pressure by PETA (People for the Ethical Treatment of Animals) on McDonald's over its handling of animals during the slaughter process led the company to stop buying beef from one supplier until it met higher standards for processing cattle. And it would be an unusual week if we didn't read about environmental or human rights activists picketing, boycotting, or threatening some organization in order to get managers to change some decision or action.

The General Environment. The **general environment** includes the broad economic, political/legal, sociocultural, demographic, technological, and global conditions that affect an organization. Although these external factors don't affect organizations to the extent that changes in their specific environment do, managers must consider them as they plan, organize, lead, and control.

Economic Conditions. Interest rates, inflation, changes in disposable income, stock market fluctuations, and the stage of the general business cycle are some economic factors that can affect management practices in an organization. For example, many specialty retailers, such as Crossword, Satya Paul, and Pantaloons, are acutely aware of the impact consumer disposable income has on their sales. When consumers' incomes fall or when their confidence about job security declines, they will postpone purchasing anything that isn't a necessity.

Political/Legal Conditions. Federal, state, and local laws, as well as global and other country laws and regulations, influence what organizations can and cannot do. Some federal legislation has significant implications. In February 1996, "The Persons with Disabilities (Equal Opportunities, Protection of Rights, and Full Participation) Act, 1995" became a law of the land under which it was promised to employers, both in public and private sectors, to ensure that at least 5 percent of their work force will be composed of persons with disabilities.[52] Exhibit 3–10 lists some of the significant legislation affecting businesses.

Although organizations spend a great deal of time and money to meet governmental regulations, the effects go beyond time and money; they reduce managerial discretion by limiting available choices. Consider the decision to dismiss an employee.[53] Historically,

Exhibit 3–10

Important Legislation

Year	Law of Ruling	Description
1923	Workmen's Compensation Act	This act is central legislation which provides for payment of compensation for injuries suffered by a workman in the course of and arising out of his employment.
1936	The Payment of Wages Act	An Act to regulate the payment of wages of certain classes of *employed persons.*
1946	Industrial Employment Act	An Act requiring employers in industrial establishments to formally define the conditions of employment under them. It applies to every industrial establishment wherein 100 or more workmen are employed.
1947	Industrial Disputes Act	An Act to make provision for the investigation and settlement of industrial disputes and for certain other purposes, such as to secure industrial peace.
1948	The Employees State Insurance Act	An Act to provide certain benefits to the employees in case of sickness, maternity, and employment injury and to make provision for certain other matters in relation thereto.
1948	Minimum Wages Act	An Act to prevent exploitation of labor through the payment of unduly low wages. The Act provides rules for fixing minimum rates of wages in respect of certain employments specified in the schedule thereto.
1948	The Factories Act	An Act to regulate health, safety, welfare, and other working conditions of workers in factories. It is enforced by the state governments through their factory inspectorates.
1955	The Protection of Civil Rights Act	An Act to prescribe punishment for the preaching and practice of "Untouchability" for the enforcement of any disability arising there from and for matters connected therewith.
1961	The Maternity Benefit Act	An Act to regulate the employment of women in certain establishments for certain period before and after child-birth and to provide maternity benefit and certain other benefits.
1963	The Personal Injuries (Compensation Insurance) Act	Seeks to impose on employers a liability to pay compensation to workers sustaining personal injuries and to provide for the insurance of employers against such responsibility.
1965	The Payment of Bonus Act	An Act to provide rule for the payment of bonuses to persons employed in certain establishments on the basis of profits or production or productivity and for the matters connected therewith.
1976	The Equal Remuneration Act	An Act to provide rule for the payment of equal remuneration to men and women workers and for the prevention of discrimination, on the ground of sex, against women in the matter of employment and for the matters connected therewith or incidental thereto.
1995	The Persons with Disabilities (Equal Opportunities, Protection of Rights, and Full Participation) Act	The Act is guided by the philosophy of empowering persons with disabilities and their associates. The endeavor of the Act has been to introduce an instrument for promoting equality and participation of persons with disability on the one hand, and eliminating discriminations of all kinds, on the other hand.
2007	The Protection of Women Against Sexual Harassement at Workplace Bill	A bill to provide rule for prevention and redressal of sexual harassment of women at workplace and for the matters connected therewith or incidental thereto.

Sources: http://www.erlaws.com/library/bills/PBAXVII-2002.pdf; http://www.erlaws.com/library/bills/POW2002.htm; http://www.nrcw.nic.in/shared/sublinkimages/173.htm; http://www.indiacode.nic.in/fullact1.asp?tfnm=197625; http://www.labour.nic.in/act/welcome.html; http://www.nrcw.nic.in/shared/sublinkimages/68.htm; http://www.socialwelfare.delhigovt.nic.in/disabilityact.htm.

specific environment
External forces that have a direct impact on managers' decisions and actions and are directly relevant to the achievement of an organization's goals.

general environment
Broad external conditions that may affect an organization.

India's youth has been described as ambitious, technology-oriented, and confident. Mats Lindgren, CEO and founder of Kairos Future Group, found in a survey of 17 countries that Indian youth were the most optimistic about themselves and their country. In another survey by ad agency McCann-Erickson Asia-Pacific, Asian youth voted Paris, London, and New York as the "coolest" cities except young Indians who voted for Bombay and New York. The youth of India confidently mix Indian values with Western packaging. The new generation of Indians entering the work force likes money; according to a survey conducted by Coca-Cola, the primary ambition of young Indians irrespective of where they live is to "'become rich." Young people hope to achieve this goal through enterprise and education.[55] How do you think these youth should be managed and motivated?

employees were free to leave an organization at any time, and employers had the right to fire an employee at any time, with or without cause. Laws and court decisions, however, have limited what employers can do. Employers are expected to deal with employees by following the principles of good faith and fair dealing. Employees who feel they've been wrongfully discharged often take their case to court, where juries decide what is "fair." This trend has made it more difficult for managers to fire poor performers or to dismiss employees for inappropriate off-duty conduct.

Other aspects of the political/legal sector are political conditions and stability of a country where an organization operates and the attitudes that elected governmental officials hold toward business. However, management is a global activity, and managers should be aware of political changes in countries in which they operate because these changes can affect decisions and actions.[54]

Sociocultural Conditions. Pepsi (and other food manufacturers) has responded to customers' changing attitudes about food by offering customers healthier versions of their favorite snacks. Managers must adapt their practices to the changing expectations of the society in which they operate. As these values, customs, and tastes change, managers must also change. For instance, as workers have begun seeking more balanced lives, organizations have had to adjust by offering family leave policies, flexible work hours, and even on-site child-care facilities. Sociocultural trends may pose a potential constraint to managers' decisions and actions.

Demographic Conditions. Demographic conditions encompass trends in population characteristics such as gender, age, level of education, geographic location, income, and family composition. Changes in these characteristics may constrain how managers plan, organize, lead, and control. For instance, population researchers have labeled specific age cohorts in the United States: the Depression group (born 1912–1921), the World War II group (born 1922–1927), the Postwar group (born 1928–1945), baby boomers (born 1946–1964), Generation X (born 1965–1977), and Generation Y (born 1978–1994). Although each group has its own unique characteristics, Gen Y is of particular interest because its members are learning, working, shopping, and playing in fundamentally different ways that are beginning to affect the way organizations are managed. We discuss some of the challenges in managing generational differences in Chapter 13.

Technological Conditions. In terms of the general environment, the most rapid changes have occurred in technology. For instance, the human genetic code has been cracked. Just think of the implications of such knowledge! Information gadgets are shrinking yet becoming more powerful. We have automated offices, electronic meetings, robotic manufacturing, lasers, integrated circuits, faster and more powerful microprocessors, synthetic fuels, and entirely new models of doing business in an electronic age. Companies that capitalize on technology, such as General Electric, eBay, and Google, prosper. In addition, many successful retailers, such as Wal-Mart, use sophisticated information systems to keep on top of sales trends. Technology has changed the fundamental ways that organizations are structured and the ways that managers manage.

Global Conditions. By the end of this decade, Nigeria will have a larger population than Russia, Ethiopia will have more people than Germany, and Morocco will be more populous than Canada.[56] Do these facts surprise you? They shouldn't. They simply reflect that globalization is one of the major factors affecting managers and organizations. Managers are challenged by an increasing number of global competitors and markets as part of the external environment. We cover this component of the external environment in detail in Chapter 4.

HOW THE ENVIRONMENT AFFECTS MANAGERS

Knowing *what* the various components of the environment are is important to managers. However, understanding *how* the environment affects managers is equally important. There are two ways the environment affects managers: first, through the degree of environmental uncertainty that is present and, second, through the various stakeholder relationships that exist between the organization and its external constituencies.

Assessing Environmental Uncertainty. Environments differ in terms of what we call **environmental uncertainty**, which is the degree of change and complexity in the organization's environment (see Exhibit 3–11).

The first dimension of uncertainty is the degree of change. If the components in an organization's environment change frequently, we call it a *dynamic* environment. If change is minimal, we call it a *stable* environment. A stable environment might be one in which there are no new competitors, few technological breakthroughs by current competitors, little activity by pressure groups to influence the organization, and so forth. For instance, Zippo Manufacturing, best known for its Zippo lighters, faces a relatively stable environment. There are few competitors and little technological change. Probably the main environmental concern for the company is the declining trend in tobacco smokers, although the company's lighters have other uses, and global markets remain attractive.

Exhibit 3–11

Environmental Uncertainty Matrix

Degree of Complexity	Degree of Change: Stable	Degree of Change: Dynamic
Simple	**Cell 1** Stable and predictable environment Few components in environment Components are somewhat similar and remain basically the same Minimal need for sophisticated knowledge of components	**Cell 2** Dynamic and unpredictable environment Few components in environment Components are somewhat similar but are continually changing Minimal need for sophisticated knowledge of components
Complex	**Cell 3** Stable and predictable environment Many components in environment Components are not similar to one another and remain basically the same High need for sophisticated knowledge of components	**Cell 4** Dynamic and unpredictable environment Many components in environment Components are not similar to one another and are continually changing High need for sophisticated knowledge of components

environmental uncertainty
The degree of change and complexity in an organization's environment.

In contrast, the recorded music industry faces a dynamic (highly uncertain and unpredictable) environment. Digital formats and music-downloading sites have turned the industry upside down and brought uncertainty. If change is predictable, is that considered dynamic? No. Think of department stores that typically make one-quarter to one-third of their sales in December. The drop-off from December to January is significant. But because the change is predictable, we don't consider the environment to be dynamic. When we talk about degree of change, we mean change that is unpredictable. If change can be accurately anticipated, it's not an uncertainty that managers must confront.

The other dimension of uncertainty describes the degree of **environmental complexity**. The degree of complexity refers to the number of components in an organization's environment and the extent of the knowledge that the organization has about those components. For example, Hasbro Toy Company, the second largest toy manufacturer (behind Mattel), simplified its environment by acquiring many of its competitors. When an organization has fewer competitors, customers, suppliers, government agencies, and so forth, the less complex and uncertain its environment is.

Complexity is also measured in terms of the knowledge an organization needs about its environment. For instance, managers at Regalix must know a great deal about their Internet service provider's operations if they want to ensure that their Web site is available, reliable, and secure for their customers. On the other hand, managers of college bookstores have a minimal need for sophisticated knowledge about their suppliers.

How does the concept of environmental uncertainty influence managers? Look again at Exhibit 3–11; each of the four cells represents different combinations of degree of complexity and degree of change. Cell 1 (stable and simple environment) represents the lowest level of environmental uncertainty, and cell 4 (dynamic and complex environment) the highest. Not surprisingly, managers have the greatest influence on organizational outcomes in cell 1 and the least in cell 4. Because uncertainty is a threat to an organization's effectiveness, managers try to minimize it. Given a choice, managers would prefer to operate in the least uncertain environments. However, they rarely control that choice. In addition, most industries today are facing more dynamic change, making their environments more uncertain.

Managing Stakeholder Relationships. What has made 9XM *the* cable channel for music-loving baby boomers? One reason is that it realizes the importance of building relationships with its various stakeholders: viewers, music celebrities, advertisers, affiliate TV stations, public service groups, and others. The nature of stakeholder relationships is another way in which the environment influences managers. The more obvious and secure these relationships, the more influence managers will have over organizational outcomes.

Exhibit 3–12

Organizational Stakeholders

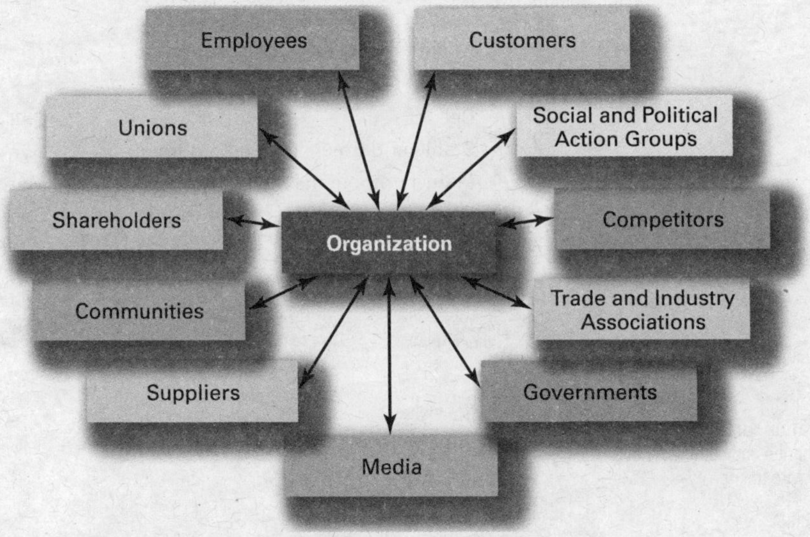

Stakeholders are any constituencies in an organization's environment that are affected by the organization's decisions and actions. These groups have a stake in or are significantly influenced by what the organization does. In turn, these groups can influence the organization. For example, think of the groups that might be affected by the decisions and actions of Reliance Fresh: farmers, employees, other retailers, local communities, and so forth. Some of these stakeholders also, in turn, may affect decisions and actions of Starbucks managers. The idea that organizations have stakeholders is now widely accepted by both management academics and practicing managers.[57]

Exhibit 3–12 identifies some of the most common stakeholders that an organization might have to deal with. Note that these stakeholders include both internal and external groups. Why? Because both can affect what an organization does and how it operates.

Why should managers even care about managing stakeholder relationships?[58] For one thing, doing so can lead to desirable organizational outcomes, such as improved predictability of environmental changes, more successful innovations, greater degree of trust among stakeholders, and greater organizational flexibility to reduce the impact of change. But does it affect organizational performance? The answer is yes! Management researchers who have looked at this issue are finding that managers of high-performing companies tend to consider the interests of all major stakeholder groups as they make decisions.[59]

Another reason for managing external stakeholder relationships is that it's the "right" thing to do. Because an organization depends on these external groups as sources of inputs (resources) and as outlets for outputs (goods and services), managers should consider their interests as they make decisions. We'll address this issue in more detail in Chapter 5, as we look at the concepts of managerial ethics and corporate social responsibility.

How can managers manage stakeholder relationships? First, they need to identify the organization's stakeholders. Groups that are likely to be influenced by and to influence organizational decisions are the organization's stakeholders. Second, they need to determine what particular interests or concerns the stakeholders might have—product quality, financial issues, safe working conditions, environmental protection, and so forth. Next, they need to decide how critical each stakeholder is to the organization's decisions and actions. In other words, how critical is it for the manager to consider a particular stakeholder's concerns as he or she plans, organizes, leads, and controls? The very idea of a stakeholder—a group that has a "stake" in what the organization does—means that it is important. But some stakeholders are more critical to the organization's decisions and actions than others. The final step is to determine how to manage the external stakeholder relationships. This decision depends on how critical an external stakeholder is to the organization and how uncertain the environment is.[60] The more critical the stakeholder and the more uncertain the environment, the more that managers need to rely on establishing explicit stakeholder partnerships rather than just acknowledging their existence.

QUICK LEARNING REVIEW:

LEARNING OUTCOME 3.4

- List the components of the specific and general environments.
- Explain the two dimensions of environmental uncertainty.
- Identify the most common organizational stakeholders.
- List the four steps in managing external stakeholder relationships.

Go to page 65 to see how well you know this material.

environmental complexity
The number of components in an organization's environment and the extent of the organization's knowledge about those components.

stakeholders
Any constituencies in an organization's environment that are affected by the organization's decisions and actions.

Let's Get Real: My Turn

Atul Srivastava

Deputy General Manager, HR
Steel Authority of India Ltd (SAIL)

In my opinion, stress cannot be reduced by these measures. Stress is not due to work requirement but it is mainly due to the lack of proper delegation, unrealistic targets, and inadequate competency to deal with the requirements of the work. I would suggest proper delegation, realistic target setting, finding out the competency required for the job, and providing specific solutions to the managers will help them in relieving undue stress.

LEARNING OUTCOMES **SUMMARY**

3.1 ▷ THE MANAGER: OMNIPOTENT OR SYMBOLIC?

- Contrast the actions of managers according to the omnipotent and symbolic views.
- Identify the two constraints on managers' discretion.

According to the omnipotent view, managers are directly responsible for an organization's success or failure. The symbolic view argues that much of an organization's success or failure is due to external forces outside managers' control. The two constraints on managers' discretion are the organization's culture (internal) and the environment (external). Managers aren't totally constrained by these two factors because they can and do influence their culture and environment.

3.2 ▷ ORGANIZATIONAL CULTURE

- Identify the seven dimensions of organizational culture.
- Discuss the impact of a strong culture on organizations and managers.
- Explain how a culture is formed and maintained.
- Describe how culture affects managers.

The seven dimensions of culture are attention to detail, outcome orientation, people orientation, team orientation, aggressiveness, stability, and innovation and risk taking. In organizations with strong cultures, employees are more loyal, and performance tends to be higher. The stronger a culture becomes, the more it affects the way managers plan, organize, lead, and control. The original source of a culture reflects the vision of organizational founders. A culture is maintained by employee selection practices, the actions of top managers, and socialization processes. Also, culture is transmitted to employees through stories, rituals, material symbols, and language. These elements help employees "learn" what values and behaviors are important as well as who exemplifies those values. The culture affects how managers plan, organize, lead, and control.

3.3 ▷ CURRENT ORGANIZATIONAL CULTURE ISSUES

- Describe the characteristics of an ethical culture, an innovative culture, a customer-responsive culture, and a diversity-supportive culture.
- Explain why workplace spirituality seems to be an important concern.
- Describe the characteristics of a spiritual organization.

An ethical culture is high in risk tolerance and low to moderate in aggressiveness, and it focuses on means as well as outcomes. An innovative culture features challenge and involvement, freedom, trust and openness, idea time, playfulness/humor, conflict resolution, debates, and risk taking. A customer-responsive culture has: outgoing and friendly employees; jobs with few rigid rules, procedures, and regulations; empowerment; clear roles and expectations; and employees who are conscientious in their desire to please the customer. A diversity-supportive culture has managers who show through their decisions and actions that they value diversity, and it reinforces employee behaviors that exemplify inclusiveness. Workplace spirituality is important because employees are looking for a counterbalance to the stresses and pressures of their turbulent lives, and they are also seeking involvement and connection that are often missing in contemporary lifestyles. Organized religion has failed to meet the needs of some people, and aging baby boomers in particular are looking for something meaningful in their lives. Spiritual organizations tend to have five characteristics: strong sense of purpose, focus on individual development, trust and openness, employee empowerment, and toleration of employee expression.

3.4 ▷ THE ENVIRONMENT

- List the components of the specific and general environments.
- Explain the two dimensions of environmental uncertainty.
- Identify the most common organizational stakeholders.
- List the four steps in managing external stakeholder relationships.

The specific environment includes customers, suppliers, competitors, and pressure groups. The general environment includes economic, political/legal, sociocultural, demographic, technological, and global conditions. The two dimensions of environmental uncertainty are degree of change (stable or dynamic) and degre of complexity (simple or complex). The most common stakeholders are customers, social and political action groups, competitors, trade and industry associations, governments, media, suppliers, communities, shareholders, unions, and employees. The four steps in managing stakeholder relationships are identifying the organization's stakeholders, determining the interests or concerns the stakeholders might have, deciding how critical each stakeholder is to the organization's decisions and actions, and determining how to manage the stakeholders.

THINKING ABOUT MANAGEMENT ISSUES

1. Refer to Exhibit 3–3. How would a first-line manager's job differ in these two organizations? How about a top-level manager's job?
2. Describe an effective culture for (a) a relatively stable environment and (b) a dynamic environment. Explain your choices.
3. Classrooms have cultures. Describe your classroom culture, using the seven dimensions of organizational culture. Does the culture constrain your instructor? How?
4. Can culture be a liability to an organization? Explain.
5. Why is it important for managers to understand the external forces that are acting on them and their organization?
6. "Businesses are built on relationships." What do you think this statement means? What are the implications for managing the external environment?
7. What would be the drawbacks to managing stakeholder relationships?

YOUR TURN to be a Manager

- Find two current examples in any of the popular business periodicals of the omnipotent and symbolic views of management. Write a paper describing what you found and how the two examples you found represent the views of management.
- Pick two organizations that you interact with frequently (as an employee or as a customer) and assess their culture by looking at the following aspects:

  Physical design (buildings, furnishings, parking lot, office or store design)—Where are they located and why? Where do customers and employees park? What does the office/store layout look like? What activities are encouraged or discouraged by the physical layout? What do these things say about what the organization values?

  Symbols (logos, dress codes, slogans, philosophy statements)—What values are highlighted? Where are logos displayed? Whose needs are emphasized? What concepts are emphasized? What actions are prohibited? What actions are encouraged? Are any artifacts prominently displayed? What do those artifacts symbolize? What do these things say about what the organization values?

  Words (stories, language, job titles)—What stories are repeated? How are employees addressed? What do job titles say about the organization? Are jokes/anecdotes used in conversation? What do these things say about what the organization values?

  Policies and activities (rituals, ceremonies, financial rewards, policies for how customers or employees are treated)—(Note: You may be able to assess this one only if you're an employee or know the organization well.) What activities are rewarded? Ignored? What kinds of people succeed? Fail? What rituals are important? Why? What events get commemorated? Why? What do these things say about what the organization values?
- If you belong to a student organization, evaluate its culture. How would you describe the culture? How do new members learn the culture? How is the culture maintained? If you don't belong to a student organization, talk to another student who does and evaluate it using the same questions.
- Steve's and Mary's suggested readings: Terrence E. Deal and Allan A. Kennedy, *Corporate Culture: The Rites and Rituals of Corporate Life* (Perseus Books Group, 2000); Edgar H. Schein, *The Corporate Culture Survival Guide* (Jossey-Bass, 1999); and Kim S. Cameron and Robert E. Quinn, *Diagnosing and Changing Organizational Culture* (Jossey-Bass, 2005).
- Pick three companies you're interested in. Identify the stakeholders that would be most important to these companies. Describe why these stakeholders are important to these companies.
- Find one example of a company that represents each of the current issues in organizational culture. Describe what the company is doing that reflects its commitment to this culture.

- Describe the general and specific environments of McDonald's and Anheuser-Busch. How are they similar? How are they different?
- In your own words, write down three things you learned in this chapter about being a good manager.
- Self-knowledge can be a powerful learning tool. Go to mymanagementlab and complete any of these self-assessment exercises: What's the Right Organizational Culture for Me? How Well Do I Respond to Turbulent Change? and Am I Experiencing Work/Family Conflict? Using the results of your assessments, identify personal strengths and weaknesses. What will you do to reinforce your strengths and improve your weaknesses?

CASE APPLICATION

Making You Say Wow

When you hear the name The Ritz-Carlton Hotels, what words come to mind? Luxurious? Elegant? Formal, or maybe even stodgy? Way beyond my budget constraints? Three words that the company hopes comes to mind are *exemplary customer service*. Ritz-Carlton is committed to treating its guests like royalty. It has one of the most distinctive corporate cultures in the lodging industry, and employees are referred to as "our ladies and gentlemen." Its motto is printed on a card that employees carry with them: "We are Ladies and Gentlemen serving Ladies and Gentlemen." And these ladies and gentlemen of Ritz have been trained in very precise standards and specifications for treating customers. These standards were established more than a century ago by founders Caesar Ritz and August Escoffier. Ritz employees are continually schooled in company lore and company values. Every day at 15-minute "lineup" sessions at each hotel property, managers reinforce company values and review service techniques. And these values are the basis for all employee training and rewards. Nothing is left to chance when it comes to providing exemplary customer service. Potential hires are tested both for cultural fit and for traits associated with an innate passion to serve. A company executive says, "The smile has to come naturally." Although staff members are expected to be warm and caring, their behavior toward guests had been extremely detailed and scripted. That's why a new customer service philosophy implemented in mid-2006 was such a radical departure from what the Ritz had been doing.

The company's new approach is almost the opposite from what the company had been doing: Don't tell employees how to make guests happy. Now they're expected to figure it out. Says Diana Oreck, vice president, "We moved away from that heavily prescriptive, scripted approach and toward managing to outcomes." The outcome didn't change, though. The goal is still a happy guest who's wowed by the service received. However, under the new approach, staff member interactions with guests are more natural, relaxed, and authentic rather than sounding like they're recited lines from a manual.

Getty Images, Inc.

A doorman stands at the main entrance of The Ritz-Carlton in Tokyo, Japan.

Discussion Questions

1. Using Exhibit 3–2 and the information from this case, describe the culture at The Ritz-Carlton. Why do you think this type of culture might be important to a luxury hotel? What might be the drawbacks of such a culture?
2. What challenges do you think the company faced in changing the culture? What is The Ritz-Carlton doing to maintain this new culture?
3. What kind of person do you think would be happiest and most successful in this culture? How do you think new employees "learn" the culture?
4. What could other organizations learn from The Ritz-Carlton about the importance of organizational culture?

Sources: The Ritz-Carlton, http://corporate.ritzcarlton.com, March 10, 2008; R. Reppa and E. Hirsh, "The Luxury Touch," *Strategy & Business*, Spring 2007, pp. 32–37; and J. Gordon, "Redefining Elegance," *Training*, March 2007, pp. 14–20.

Let's Get Real:

Meet The Manager

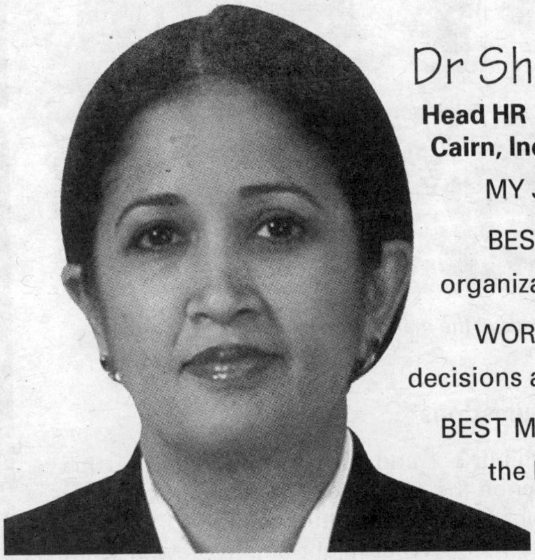

Dr Shalini Sarin

Head HR
Cairn, India

MY JOB: To manage people

BEST PART OF MY JOB: Contributing to individual and organizational growth.

WORST PART OF MY JOB: To implement tough people decisions as a management directive.

BEST MANAGEMENT ADVICE EVER RECEIVED: Understand the business and align the HR.

You'll be hearing more from this real manager throughout the chapter.

Chapter

Managing in a Global Environment

Every organization is affected in some way by the global environment. In this chapter, we're going to look at what managers need to know about managing globally, including regional trading alliances, how organizations go international, and cross-cultural differences. Focus on the following learning outcomes as you read and study this chapter.

LEARNING OUTCOMES

A Manager's Dilemma

Courtesy of W. R. Grace & Co.

The pink Eraser Man.[1] A cartoon mascot designed to convey a simple message to employees: "Erase" waste. However, as managers at W. R. Grace & Company soon discovered, its gimmick to promote being efficient (known as "lean" manufacturing) to employees throughout its global operations wasn't as harmless as they had thought it would be. When Eraser Man was introduced to the company's Asian staff at a focus group meeting in China, the attendees were perplexed and a little annoyed. The company's vice president of human resources recalls that they said, "Do you really want this program to be invisible?" Because that's what the concept of "erase" means in China . . . invisible. The other problem was the color of Eraser Man: pink. Although many employees wouldn't see anything odd about a pink eraser, it's "just not an acceptable color in China; it's feminine. No self-respecting man would want to be associated with a program that's marked by the color pink." This cross-cultural blunder was easily corrected. Eraser Man is now tan, and employees in China are encouraged to "simplify" or "reduce" rather than erase. But how could such problems be avoided in the future?

What would you do?

The Eraser Man story illustrates some of the challenges of managing in today's global environment. Even large, successful organizations with talented managers (such as W. R. Grace & Company) don't always do it right. Despite these challenges, going global is something that most organizations want to do. A study of U.S. manufacturing firms found that companies that operated in multiple countries had twice the sales growth of and significantly higher profitability than strictly domestic firms.[2] However, if managers don't closely monitor changes in their global environment or don't consider specific location characteristics as they plan, organize, lead, and control, they may find limited global success. In this chapter, we discuss the issues managers face as they manage in a global environment.

WHO OWNS WHAT?

One way to see how global the marketplace has become is to consider the country of ownership origin for some familiar products. You might be surprised to find that many products you thought were made by U.S. companies aren't! Take the following quiz[3] and then check your answers at the end of the chapter, on page 89.

1. Ben and Jerry's Ice Cream is owned by a company based in:
   a. Mexico b. Saudi Arabia c. United Kingdom d. United States
2. Lebedyansky juices are owned by a company based in:
   a. Japan b. United Kingdom c. United States d. Russia
3. Rajah spices are products of a company based in:
   a. Brazil b. Switzerland c. United States d. India

4. Tetley Tea is owned by a company located in:
   a. Great Britain b. India c. Japan d. Spain
5. Skippy brand peanut butter is a product of a company based in:
   a. United States b. Canada c. Venezuela d. United Kingdom
6. The 6,000-plus 7-Eleven stores are owned by a company based in:
   a. Japan b. United States c. Canada d. United Kingdom
7. The company that produces Boboli Pizza Crust is based in:
   a. United States b. Mexico c. Italy d. Spain
8. The parent company of Braun electric shavers is located in:
   a. Switzerland b. Germany c. United States d. Japan
9. The company that owns Sephora cosmetics retail stores is located in:
   a. Germany b. Canada c. France d. United States
10. Gerber Baby Products is owned by a company located in:
    a. United States b. Germany c. Japan d. Switzerland
11. Lean Cuisine frozen meals are products of a company based in:
    a. Germany b. United States c. Switzerland d. Brazil
12. Dr. Pepper and 7-Up are products of a company based in:
    a. United States b. Japan c. Canada d. United Kingdom
13. The company that markets Lipton Tea is based in:
    a. China b. United Kingdom c. Japan d. United States
14. Eight O'Clock Coffee is owned by a company located in:
    a. India b. Costa Rica c. United States d. Canada
15. Frédéric Fekkai hair care products are marketed by a company based in:
    a. United States b. Switzerland c. France d. Italy

How well did you do? Were you aware of how many products we use every day that are made by companies *not* based in the United States? Probably not! Most of us don't fully appreciate the truly global nature of today's marketplace.

LEARNING OUTCOME 4.1 ▷

WHAT'S YOUR GLOBAL PERSPECTIVE?

It's not unusual for Indians, Germans, Italians, or Indonesians to speak three or four languages. "More than half of all primary school children in China now learn English and the number of English speakers in India and China—500 million—now exceeds the total number of mother-tongue English speakers elsewhere in the world." On the other hand, most U.S. children study only English in school; only 24,000 are studying Chinese. And only 22 percent of the population in the United States speaks a language other than English.[4] Americans tend to think of English as the only international business language and don't see a need to study other languages. This could lead to future problems, as a major research report commissioned by the British Council said that the "competitiveness of both Britain and the United States is being undermined" by only speaking English.[5]

Monolingualism is one sign that a nation suffers from **parochialism**—viewing the world solely through one's own eyes and perspectives.[6] People with a parochial attitude do not recognize that others have different ways of living and working. They

parochialism
Viewing the world solely through one's own eyes and perspectives and not recognizing that others have different ways of living and working. Parochialism leads to an inability to recognize differences between people.

Reto Wittwer is the president and CEO of Kempinski Hotels. Born in Switzerland and educated in Catholic schools in France, he became a Buddhist when he fell in love with and married a Vietnamese woman after living in Asia for many years. Wittwer is a good example of a geocentric manager; in addition to German, French, Italian, and a Latin-based gypsy language called Rhaeto-Romanish, he speaks eight other languages.

ignore others' values and customs and rigidly apply an attitude of "ours is better than theirs" to foreign cultures. This type of narrow, restricted attitude is one approach that managers might take, but it isn't the only one.[7] In fact, there are three possible global attitudes. Let's look at each more closely.

First, an **ethnocentric attitude** is the parochialistic belief that the best work approaches and practices are those of the *home* country (the country in which the company's main offices are located). Managers with an ethnocentric attitude believe that people in foreign countries don't have the needed skills, expertise, knowledge, or experience to make the best business decisions, as people in the home country do. They don't trust foreign employees with key decisions or technology.

Next, a **polycentric attitude** is the view that employees in the *host* country (the foreign country in which the organization is doing business) know the best work approaches and practices for running their business. Managers with this attitude view every foreign operation as different and difficult to understand. Thus, they're likely to let employees in the host country figure out how best to do things.

The final type of global attitude managers might have is a **geocentric attitude**, a *world-oriented* view that focuses on using the best approaches and people from around the globe. Managers with this type of attitude have a global view and look for the best approaches and people, regardless of origin. For instance, the CEO of a fast-growing manufacturer of household accessories is a Chinese immigrant who describes the company's strategy as "combining Chinese costs with Japanese quality, European design, and American marketing."[8] A geocentric attitude requires eliminating parochial attitudes and developing an understanding of cross-cultural differences. That's the type of approach successful managers will need in today's global environment.

QUICK LEARNING REVIEW:

LEARNING OUTCOME 4.1

- **Define parochialism.**
- **Contrast ethnocentric, polycentric, and geocentric attitudes toward global business.**

Go to page 87 to see how well you know this material.

LEARNING OUTCOME 4.2 ▷ UNDERSTANDING THE GLOBAL ENVIRONMENT

One important feature of today's global environment is global trade, which, you may remember from history class, isn't new. Countries and organizations have been trading with each other for centuries.[9] And it continues today, as we saw in the chapter-opening quiz. Global trade today is shaped by two forces: regional trading alliances and trade agreements negotiated through the World Trade Organization.

REGIONAL TRADING ALLIANCES

Global competition was once considered country against country—the U.S. versus Japan, France versus Germany, Mexico versus Canada, etc. Now, global competition is shaped by regional trading agreements, including the South Asian Association for Regional Cooperation (SAARC), European Union (EU), the Association of Southeast Asian Nations (ASEAN), the North American Free Trade Agreement (NAFTA), and others.

South Asian Association for Regional Cooperation. The **South Asian Association for Regional Cooperation (SAARC)**[10] was established on 8 December 1985. It includes an alliance between Bangladesh, Bhutan, India, Maldives, Nepal, Pakistan, and Sri Lanka. There are seven countries with observer status (China, the EU, Iran, Japan, Korea, Mauritius, and the USA). The EU has observer status since 2006 and great values co-operation and regional integration in South Asia.

The Association provides a platform for the peoples of South Asia to work together with a spirit of friendship, trust, and understanding. Cooperation in the SAARC is based on respect for the principles of sovereign equality, territorial integrity, political independence, noninterference in internal affairs of the Member States, and mutual benefit. Decisions are taken on the basis of unanimity. Bilateral and contentious issues are excluded from the deliberations of SAARC.

SAARC members revisit the areas of cooperation often. The agreed areas of cooperation in 2006 were

1. Agriculture and Rural Development;
2. Health and Population Activities;
3. Women, Youth, and Children;
4. Environment and Forestry;
5. Science and Technology and Meteorology;
6. Human Resources Development; and
7. Transport.

In addition, high-level working groups have also been established to strengthen cooperation in the areas of Information and Communications Technology, Biotechnology, Intellectual Property Rights, Tourism, and Energy.

Also, the SAARC, composed of eight member states (India, Pakistan, Sri Lanka, Bangladesh, Bhutan, Nepal, the Maldives, and Afghanistan), began eliminating tariffs in 2006.[11] Its aim, like those of all the other regional trading alliances, is to allow free flow of goods and services.

The European Union. The **European Union (EU)** is an economic and political partnership of 27 democratic European countries. Three countries (Croatia, Macedonia, and Turkey) have applied for membership (see Exhibit 4–1). When the 12 original members formed the EU in 1992, the primary motivation was to reassert the region's economic position against the United States and Japan. Before then, each European nation had border controls, taxes, and subsidies; nationalistic policies; and protected industries. These barriers to travel, employment, investment, and trade prevented European companies from developing economic efficiencies. Now these barriers have been removed, and the economic power represented by the EU is

ethnocentric attitude
The parochialistic belief that the best work approaches and practices are those of the home country.

polycentric attitude
The view that the managers in the host country know the best work approaches and practices for running their business.

geocentric attitude
A world-oriented view that focuses on using the best approaches and people from around the globe.

South Asian Association for Regional Cooperation (SAARC)

European Union (EU)
An economic and political partnership of 27 democratic European countries created as a unified economic and trade entity. Three additional countries have applied for membership.

Exhibit 4–1

European Union

considerable. Its current membership covers a population base of nearly half a billion people, and it accounts for approximately 31 percent of the world's total economic output.[12]

Another step toward unification occurred when 15 EU member countries adopted the common currency, the **euro**. In addition, all new EU member countries must adopt the euro. Only Denmark, the United Kingdom, and Sweden have been allowed to opt out of using the euro.[13] Attempts to develop a unified European constitution have not been as successful. However, in December 2007, the Lisbon Treaty (or Reform Treaty), signed by the heads of state of all 27 member countries, appeared to be a first step.[14] This treaty provides the EU with a common legal framework and tools to meet the challenges of a rapidly changing world, including climatic and demographic changes, globalization, security, and energy. The EU will continue to evolve and assert its economic power in one of the world's richest markets, and European businesses will continue to play an important role in the global economy.

Association of Southeast Asian Nations (ASEAN). The **Association of Southeast Asian Nations (ASEAN)** is a trading alliance of 10 Southeast Asian nations (see Exhibit 4–2). The ASEAN region has a population of about 566 million, with a combined GDP of US$737 billion.[15] This fast-growing region means ASEAN will be an increasingly important regional economic and political alliance whose impact could eventually rival that of NAFTA and the EU.

North American Free Trade Agreement (NAFTA) and Other Latin American Agreements. When agreements in key issues covered by the **North American Free Trade Agreement (NAFTA)** were reached by the Mexican, Canadian, and U.S. governments in 1992, a vast economic bloc was created. As of 2008, it remains the largest trade bloc in the world in terms of combined GDP of its members. Between 1994, when NAFTA went into effect, and 2006 (the most recent year for complete statistics), overall trade in goods among the United States, Canada, and Mexico increased 198 percent.[16] Canada continues to be the top U.S. trading partner, with Mexico following at number three. (China is second.) Eliminating barriers to free trade (tariffs, import licensing requirements, customs user fees) has resulted in a strengthening of the economic power of all three countries.

Other Latin American nations are moving to become part of free-trade blocs. Colombia, Mexico, and Venezuela led the way when all three signed an economic pact

Exhibit 4–2

ASEAN Members

Source: Based on J. McClenahen and T. Clark, "ASEAN at Work," *IW*, May 19, 1997, p. 42.

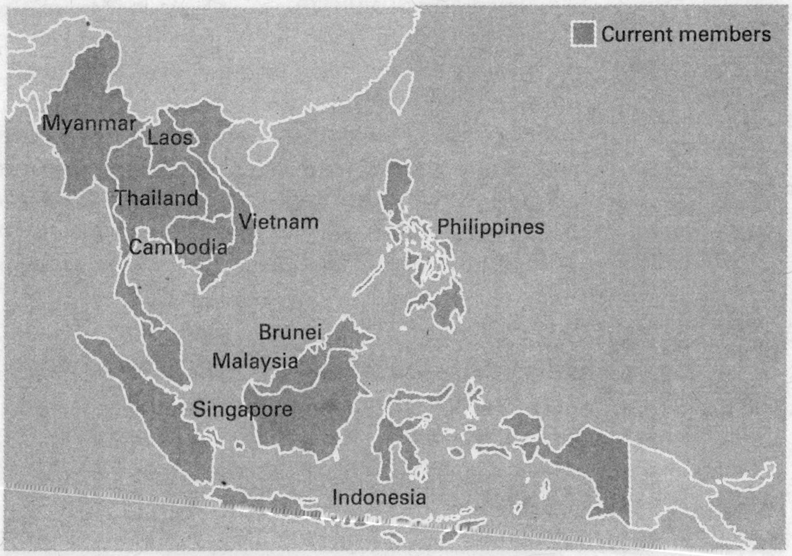

in 1994 to eliminate import duties and tariffs. Another agreement, the Central America Free Trade Agreement (CAFTA), promotes trade liberalization between the United States and five Central American countries: Costa Rica, El Salvador, Guatemala, Honduras, and Nicaragua. However, only El Salvador and Costa Rica have joined. The other countries have yet to change laws to be in line with the agreement.[17] The United States also signed a trade deal with Colombia that is said to be the "largest Washington has concluded with a Latin American country since signing" NAFTA.[18] Also, negotiators from 34 countries in the Western Hemisphere continue work on the Free Trade Area of the Americas (FTAA) agreement, which was to have been operational no later than 2005. However, leaders of these nations have yet to reach agreement, leaving the future of the FTAA still up in the air.[19] Already in existence is another free-trade bloc of 10 South American countries known as the Southern Common Market (Mercosur). Some South Americans see Mercosur as an effective way to combine resources to better compete against other global economic powers, especially the EU and NAFTA. And with the future of FTAA highly doubtful, this regional alliance could take on new importance.

Let's Get Real: F2F

HOW GLOBAL FACTORS AFFECT THE WAY I MANAGE:
Global factors have a great impact on oil price; hence on the business. They also cause an impact on the kind of talent that is available. Our retention is also largely impacted by global factors.

LANGUAGES I SPEAK:
English, Hindi, Punjabi, and Gujarati.

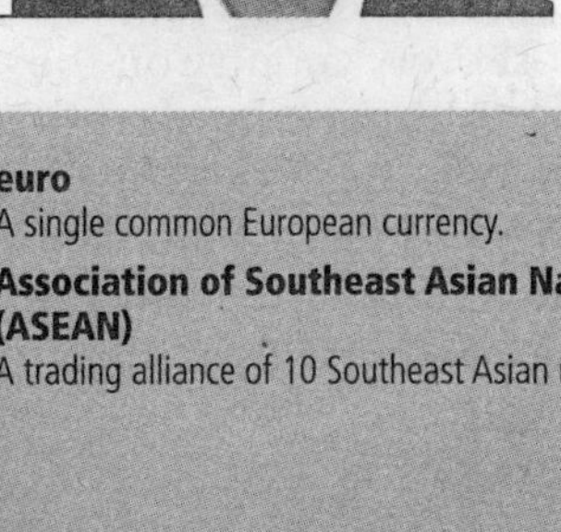

Other Trade Alliances. Other regions around the world continue to develop regional trading alliances. For instance, the 53-nation African Union (AU) came into existence in 2002, with the vision of "building an integrated, prosperous, and peaceful Africa."[20] Members of this alliance plan to create an economic development plan to achieve greater unity among Africa's nations. Like members of other trade alliances, these countries hope to gain economic, social, cultural, and trade benefits from their association.

THE WORLD TRADE ORGANIZATION

Global trade among nations doesn't just happen on its own. As trade issues arise, global trade systems ensure that trade continues efficiently and effectively. Indeed, one of the realities of globalization is the interdependence of countries—that is, what happens in one can affect others, good or bad. For example, the financial crisis in the United States in 2008 had the potential to disrupt economic growth around the globe, but the effect was minimal. Why? Because trade and financial mechanisms helped avert a potential crisis. One of these mechanisms is a multilateral trading system called the **World Trade Organization (WTO)**.[21]

euro
A single common European currency.

Association of Southeast Asian Nations (ASEAN)
A trading alliance of 10 Southeast Asian nations.

North American Free Trade Agreement (NAFTA)
An agreement among the Mexican, Canadian, and U.S. governments that has eliminated barriers to trade.

World Trade Organization (WTO)
A global organization of 153 countries that deals with the rules of trade among nations.

Formed in 1995, the WTO evolved from the General Agreement on Tariffs and Trade (GATT), a trade agreement in effect since the end of World War II. Today, the WTO is the only *global* organization that deals with trade rules among nations. Its membership encompasses 153 countries (as of July 23, 2008) and 30 observer governments (which have a specific time frame within which they must apply to become members). The goal of the WTO is to help countries conduct trade through a system of trade rules. Although critics have staged vocal protests against it, claiming that global trade destroys jobs and the natural environment, the WTO has an important role in monitoring and promoting global trade.

QUICK LEARNING REVIEW:

LEARNING OUTCOME 4.2

- **Describe the current status of the EU, NAFTA, ASEAN, and other regional trade alliances.**
- **Discuss the role of the WTO.**

Go to page 87 to see how well you know this material.

LEARNING **OUTCOME 4.3** ▷

DOING BUSINESS GLOBALLY

At 2 P.M. on a Saturday afternoon in Bucharest, Romania, a hypermarket operated by French retail group Auchan is jam-packed with shoppers. McDonald's says it's on track to continue expanding aggressively in China, even though it already has more than 1,000 locations there. Y. V. L. Pandit, head of SHL India, a subsidiary of a company, headquartered in UK, and an industry leader in assessment centers, says, "A lot of our business is serving large PSU's, Indian subsidiaries of multinational companies and the executives are from both India and other parts of the world." Infosys, Wipro, Patni, and all major software service providers have bought companies in different parts of the world. As these examples show, organizations in different industries and from different countries are pursuing global opportunities. But how do companies *do* business globally?

DIFFERENT TYPES OF INTERNATIONAL ORGANIZATIONS

Companies doing business globally aren't new. DuPont started doing business in China in 1863. H. J. Heinz Company was manufacturing food products in the United Kingdom in 1905. Ford Motor Company set up its first overseas sales branch in France in 1908. By the 1920s, other companies, including Fiat, Unilever, and Royal Dutch/Shell had gone international. But it wasn't until the mid-1960s that international companies became quite common. In 1969, the Aditya Birla group set up manufacturing in 17 countries in the then vibrant and free-market Southeast Asian countries. Today, few companies don't do business internationally. However, there's not a generally accepted approach to describing the different types of international companies; different authors call them by different names. We use the terms *multinational*, *multidomestic*, *global*, and *transnational*.[22] A **multinational corporation (MNC)** is any type of international company that maintains operations in multiple countries.

One type of MNC is a **multidomestic corporation**, which decentralizes management and other decisions to the local cou ntry. This type of globalization reflects the polycentric attitude. A multidomestic corporation doesn't attempt to replicate its domestic successes by managing foreign operations from its home country. Instead, local employees are typically hired to manage the business, and marketing strategies are tailored to that country's unique characteristics. For example, Switzerland-based Nestlé is a multidomestic corporation with operations in almost every country on the planet. Its managers match the company's products to its consumers; in parts of Europe, for example, Nestlé sells products that are not available in the United States or Latin America. Another example is Frito-Lay, a division of PepsiCo, which markets a Doritos chip in the British market that differs in both taste and texture from the U.S. and Canadian version. Many consumer product companies organize their global businesses using such an approach because they must adapt their products to meet the needs of local markets.

Another type of MNC is a **global company**, which centralizes its management and other decisions in the home country. This approach to globalization reflects the ethnocentric attitude. Global companies treat the world market as an integrated whole and focus on the need for global efficiency. Although these companies may have considerable global holdings, management decisions with companywide implications are made from headquarters in the home country. Some examples of global companies are Sony, Deutsche Bank AG, and Merrill Lynch.

Other companies use an arrangement that eliminates artificial geographic barriers. This type of MNC is often called a **transnational**, or **borderless, organization** and reflects a geocentric attitude.[23] For example, IBM dropped its organizational structure based on country and reorganized into industry groups. Ford Motor Company is pursuing what it calls the One Ford concept as it integrates its operations around the world. Another company, Thomson SA, which is legally based in France, has eight major locations around the globe. The CEO said, "we don't want people to think we're based anyplace."[24] Wilhelm Williamson, a Norwegian shipping major, has its recruitment arm registered as Barbers in India, its Maritime Training Wing also registered in India, and its shipping management arm registered and operated out of Dubai. Managers choose this approach to increase efficiency and effectiveness in a competitive global marketplace.[25]

HOW ORGANIZATIONS GO INTERNATIONAL

When organizations go international, they often use different approaches (see Exhibit 4–3). At first, managers may want to get into a global market with minimal investment. At this stage, they might start with **global sourcing** (also called **global outsourcing**), which is purchasing materials or labor from around the world based on lowest cost. The goal is to take advantage of lower costs in order to be more competitive. For instance, Massachusetts General Hospital uses radiologists in India to interpret CT scans.[26] Although global sourcing may be the first step to going international for many companies, they often continue to use this approach because of the competitive advantages it offers. Each successive stage of going international beyond global sourcing, however, requires more investment and thus entails more risk for the organization.

The next step in going international may involve **exporting** the organization's products to other countries—that is, making products domestically and selling them abroad. In addition, an organization might do **importing**, which involves acquiring

Exhibit 4–3

How Organizations Go Global

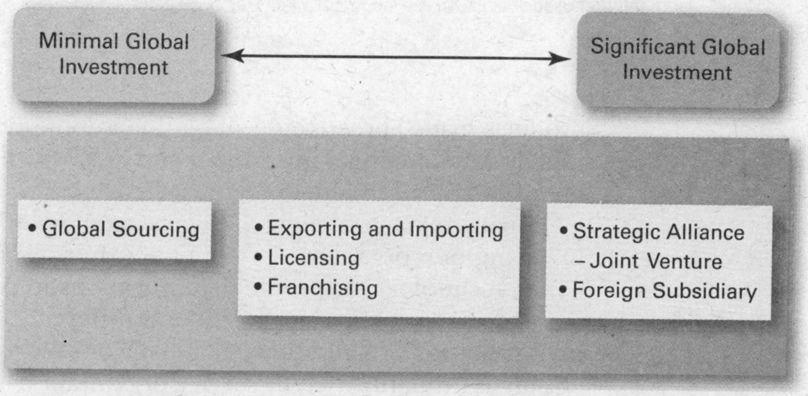

multinational corporations (MNCs)
A broad term that refers to any and all types of international companies that maintain operations in multiple countries.

multidomestic corporation
An international company that decentralizes management and other decisions to the local country.

global company
An international company that centralizes management and other decisions in the home country.

transnational (or borderless) organization
A type of international company in which artificial geographic barriers are eliminated.

global sourcing (or global outsourcing)
Purchasing materials or labor from around the world based on lowest cost.

exporting
Making products domestically and selling them abroad.

importing
Acquiring products made abroad and selling them domestically.

managing in a Virtual World

IT in a Global World

Managers around the globe are tapping into the power of information technology.[27] For example, managers at Dresdner Kleinwort Wasserstein, a London-based financial services firm, found that e-mail was extremely inefficient for employee collaboration. They started using **wikis** (server software that allows users to freely create and edit Web page content using any Web browser) and **blogs** (Web logs, or online diaries) as ways for its employees to create, edit, comment on, and revise projects in real time. The payoff: "e-mail volume down 75 percent, meeting times slashed, and team members more productive." South Korean company Samsung Electronics uses online links with suppliers and retailers to help improve demand forecasts, which has resulted in cutting inventory levels from 80 days to about 2 weeks. These examples—as well as many others—illustrate how important IT can be to global organizations.

As managers look to exploit the benefits of IT, might their geographic location have an impact? What do Internet usage statistics show? The following table summarizes these statistics for world regions. Are you surprised at what it shows? Although we can't conclude or even assume that organizations in the countries with the highest Internet usage rates will be more likely to use IT than others, we do need to consider some issues about where innovations and ideas on how best to manage IT efficiently and effectively may come from in the future.

World Internet Usage and Population Statistics

World Regions	Population (2007 Est.)	Population % of World	Internet Usage, Latest Data	% Population (Penetration)	Usage % of World	Usage Growth 2000–2007
Africa	941,249,130	14.2%	44,361,940	4.7%	3.4%	882.7%
Asia	3,733,783,474	56.5%	510,478,743	13.7%	38.7%	346.6%
Europe	801,821,187	12.1%	348,125,847	43.4%	26.4%	231.2%
Middle East	192,755,045	2.9%	33,510,500	17.4%	2.5%	920.2%
North America	334,659,631	5.1%	238,015,529	71.1%	18.0%	120.2%
Latin America/Caribbean	569,133,474	8.6%	126,203,714	22.2%	9.6%	598.5%
Oceania/Australia	33,569,718	0.5%	19,175,836	57.1%	1.5%	151.6%
World Total	6,606,971,659	100.0%	1,319,872,109	20.0%	100.0%	265.6%

Notes: (1) Internet usage and world population statistics are for December 31, 2007. (2) Demographic (population) numbers are based on data from the U.S. Census Bureau. (3) Internet usage information comes from data published by Nielsen/NetRatings, the International Telecommunications Union, local NIC, and other reliable sources.

Source: Internet World Stats, "Internet Usage Statistics: The Internet Big Picture," www.internetworldstats.com. Copyright © 2000–2008, Miniwatts Marketing Group.

products made abroad and selling them domestically. Both exporting and importing usually entail minimal investment and risk, which is why many small businesses often use these approaches to doing business globally.

Finally, managers might use **licensing** or **franchising**, which are similar approaches involving one organization giving another organization the right to use its brand name, technology, or product specifications in return for a lump-sum payment or a fee (usually based on sales). The only difference is that licensing is primarily used by manufacturing organizations that make or sell another company's products, and franchising is primarily used by service organizations that want to use another company's name and operating methods. For example, New Delhi consumers can enjoy Subway sandwiches, Hong Kong residents can dine on Shakey's Pizza, and Malaysians can consume Schlotzky's deli sandwiches—all because of *franchises* in these countries. On the other hand, Anheuser-Busch InBev has *licensed* the right to brew and market its Budweiser beer to brewers, such as Kirin in Japan and Crown Beers in India.

When an organization has been doing business internationally for a while and has gained experience in international markets, managers may decide to make more of a direct investment. One way to do this is through a **strategic alliance**, which is a partnership

Fast-food king McDonald's model for success is based on having more and more outlets in all countries of the world. McDonald's had to make definite departures in their menu to be able to serve the largely vegetarian and rich Gujaratis in Gujarat in India. The menu of McDonalds in Ahmedabad has many vegetarian options, such as McAloo Tikki and vegetable puffs. Keeping in line with India's culture, McDonald's does not offer any beef and pork items in India. The cheese and cold sauces used in India are 100 percent vegetarian. This separation of vegetarian and non-vegetarian food products is maintained throughout the various stages of procurement, cooking, and serving.

between an organization and a foreign company partner or partners in which both share resources and knowledge in developing new products or building production facilities. For example, Honda Motor Company and General Electric teamed up to produce a new jet engine. A specific type of strategic alliance in which the partners form a separate, independent organization for some business purpose is called a **joint venture**. For example, Hewlett-Packard has had numerous joint ventures with various suppliers around the globe to develop different components for its computer equipment. These partnerships provide a relatively easy way for companies to compete globally.

Finally, managers may choose to directly invest in a foreign country by setting up a **foreign subsidiary** as a separate and independent facility or office. This subsidiary can be managed as a multidomestic organization (local control) or as a global organization (centralized control). As you can probably guess, this arrangement involves the greatest commitment of resources and poses the greatest amount of risk. For instance, Orind set up a fully export-oriented unit to manufacture refractory bricks in a remote town in northern China. The owner of the company, Mr Jhunjhunwala, says that such investment was necessary to cater to the growing demand for refractory bricks in steel industries in central Asia and Eastern Europe at the best possible price and quality.

QUICK LEARNING REVIEW:

LEARNING OUTCOME 4.3

- Contrast multinational, multidomestic, global, and transnational organizations.
- Describe the different ways organizations can go international.

Go to page 87 and see how well you know this material.

LEARNING OUTCOME 4.4 ▷ MANAGING IN A GLOBAL ENVIRONMENT

Assume for a moment that you're a manager going to work for a branch of a global organization in a foreign country. You know that your environment will differ from the one at home, but how? What should you look for?

wikis
Server software that allows users to freely create and edit Web page content using any Web browser.

blogs
Web logs or online diaries.

licensing
An agreement in which an organization gives another organization the right to make or sell its products, using its technology or product specifications.

franchising
An agreement in which an organization gives another organization the right to use its name and operating methods.

strategic alliance
A partnership between an organization and a foreign company partner(s) in which both share resources and knowledge in developing new products or building production facilities.

joint venture
A specific type of strategic alliance in which the partners agree to form a separate, independent organization for some business purpose.

foreign subsidiary
A direct investment in a foreign country that involves setting up a separate and independent facility or office.

thinking critically about Ethics

A survey released in November 2007 reported that more than 80 percent of Russian companies use pirated software. The study, commissioned by Microsoft, indicated that many Russian managers believe that buying software legally is not cost-effective. What do you think of this situation? Who gets harmed by this action? Have *you* ever used copies of illegally duplicated software or given friends copies of software, music, or video games? Is that any different? Discuss.

Any manager who finds himself or herself in a new country faces challenges. In this section, we look at some of these challenges. Although our discussion is presented through the eyes of a U.S. manager, this framework could be used by any manager, regardless of national origin, who manages in a foreign environment.

Let's Get Real: F2F

ADVICE FOR SOMEONE WHO HAS LITTLE GLOBAL EXPERIENCE: Understand context and details before you assume anything. It is never foolish to ask questions as long as they give you better understanding.

THE POLITICAL/LEGAL ENVIRONMENT

U.S. managers are accustomed to a stable legal and political system. Changes are slow, and legal and political procedures are well established. Elections are held at regular intervals, and even when the political party in power changes after an election, it's unlikely that anything radical will happen. The stability of laws allows for accurate predictions. However, this certainly isn't true for all countries. Managers must stay informed of the specific laws in countries where they do business.

Also, some countries have risky political climates. Chicago-based Aon Corporation's political risk assessment found that businesses faced the highest level of risk in Iran, Nigeria, and Venezuela. A level of elevated risk was present in 13 other countries: China, India, Mexico, Turkey, Saudi Arabia, South Africa, Argentina, Thailand, Colombia, Pakistan, the Philippines, Algeria, and Egypt.[28] Managers of businesses in countries with high risk levels face dramatic uncertainty. In addition, political interference is a fact of life in some regions, especially in some Asian countries. For instance, many companies have cautiously approached doing business in China because of government controls. However, as Chinese consumers gain more power, that's likely to change.

A country's political/legal environment doesn't have to be risky or unstable to be a concern to managers. Just the fact that it differs from that of the United States is important. Managers must recognize these differences if they hope to understand the constraints and opportunities that exist.

THE ECONOMIC ENVIRONMENT

A global manager must be aware of economic issues when doing business in other countries. First, it's important to understand a country's type of economic system. The two major types are a free market economy and a planned economy. A **free market economy** is one in which resources are primarily owned and controlled by the private sector. A **planned economy** is one in which economic decisions are planned by a central government. Actually, no economy is purely free market or planned. For instance, the United States and United Kingdom are at the free market end of the spectrum but do have some governmental controls. The economies of India, Vietnam and North Korea are more planned. China is also a more planned economy, but it is moving toward being more free market. Why would managers need to know about a country's economic system? Because it has the potential to constrain decisions. Other economic issues managers need to understand include currency exchange rates, inflation rates, and diverse tax policies.

An MNC's profits can vary dramatically, depending on the strength of its home currency and the currencies of the countries in which it operates. For instance, the rising value of rupee against the dollar in July and August of 2008 affected the profit margins of several export-oriented IT companies that received their revenue in dollars. Currency exchange revaluations can affect managers' decisions and the level of a company's profits.

Inflation means that prices for products and services are increasing. But it also affects interest rates, exchange rates, the cost of living, and the general confidence in a country's political and economic system. Country inflation rates can, and do, vary widely.[29] The *World Factbook* shows rates ranging from a negative 3.6 percent in Nauru to a whopping positive 26,470 percent in Zimbabwe.[30] Managers need to monitor inflation trends so they can anticipate possible changes in a country's monetary policies and make good decisions.

Finally, tax policies can be a major economic worry. Some countries' tax laws are more restrictive than those in an MNC's home country. Others are more lenient. About the only certainty is that they differ from country to country. Managers need exact information on tax rules in countries in which they operate to minimize a business's overall tax obligation.

THE CULTURAL ENVIRONMENT

Managing today's talented global workforce isn't easy.[31] A large multinational oil company found that employee productivity in one of its Mexican plants was off 20 percent and sent a U.S. manager to find out why. After talking to several employees, the manager discovered that the company used to have a monthly fiesta in the parking lot for all the employees and their families. Another U.S. manager had canceled the fiestas, saying they were a waste of time and money. The message employees were getting was that the company didn't care about their families anymore. When the fiestas were reinstated, productivity and employee morale soared. At Hewlett-Packard, a cross-global team of U.S. and French engineers were assigned to work together on a software project. The U.S. engineers sent long, detailed e-mails to their counterparts in France. The French engineers viewed the lengthy e-mails as patronizing and replied with quick, concise e-mails. This made the U.S. engineers think that the French were hiding something from them. The situation spiraled out of control and negatively affected output until team members went through cultural training.[32]

As we know from Chapter 3, organizations have different cultures. Countries have cultures, too. **National culture** is the values and attitudes shared by individuals from a specific country that shape their behavior and their beliefs about what is important.[33]

Which is more important to a manager—national culture or organizational culture? For example, would the Indorama factory in Indonesia be likely to reflect Indonesian work values or Indorama's organizational culture? Or, is an IBM facility in India more likely to reflect Indian culture or IBM's corporate culture? Research indicates that national culture has a greater effect on employees than does their organization's culture.[34] German employees at an IBM facility in Munich will be influenced more by German culture than by IBM's culture. This means that as influential as organizational culture may be on managerial practice, national culture is even more influential.

Legal, political, and economic differences among countries are fairly obvious. An Indian manager who works in the Poland or his or her Polish counterpart who works in India can get information about laws or tax policies without too much difficulty. Getting information about cultural differences isn't quite that easy. The primary reason? It's difficult for natives to explain their country's unique cultural characteristics to someone else. For instance, if you were born and raised in the India, how would you describe Indian culture? In other words, what are Indians like? Think about it for a moment and see which characteristics in Exhibit 4–4 you identified.

Hofstede's Framework for Assessing Cultures. Geert Hofstede developed one of the most widely referenced approaches to helping managers better understand differences between national cultures. His research found that countries vary on five dimensions of national culture. These dimensions are described in Exhibit 4–5, which also shows some of the countries characterized by those dimensions.

free market economy
An economic system in which resources are primarily owned and controlled by the private sector.

planned economy
An economic system in which all economic decisions are planned by a central government.

national culture
The values and attitudes shared by individuals from a specific country that shape their behavior and beliefs about what is important.

Exhibit 4–4

What Are Indians Like?

- Indians are *informal*. They do not say thank you, please, etc. as much as people say from other countries. They do not think that it is important to make an appointment before going to a friends place. If two people are talking to each other, a third will not think twice before joining the conversation. They will not be shy of asking a stranger about his or her personal information.
- Indians are *indirect*. Indians will find ways of saying unpleasant things so that they do not sound rude.
- Indians are both *collectivistic and individualistic*. Indians will depend on others to get their work done and will also help others in their needs. They do not believe that people control their own destiny and at the same time would not mind to be on their own to achieve their goals.
- Indians view time as *stretchable*. Indians are not very upset if a meeting does not start on time. Time is defined by how long it takes to get something done.
- Indians have mostly *external locus of control*. A typical Indian will not blame oneself for all their failures. It is very common for them to say that this was because of their destiny.
- Indians *value relationships*. Indians will go to great extent to build relationships and their view for relationships is long term. Family is a very important source of relationships for the Indians.
- Indians are *paternalistic*. The head of the family is revered and respected. A leader or a boss is looked upon as the caretaker and father figure. A caring boss can get a lot of work done from the Indian subordinates. Whenever you go out to meet a family or for a formal business meeting, it is common to first greet the eldest or the senior-most person in the setting.
- Indians value *"keeping of word."* Once trust is established and a word is given, Indians will try hard to honor it. They might be late or a little slow to honor their word but will eventually come back to you and will keep their promise.

Exhibit 4–5

Hofstede's Five Dimensions of National Culture

(1) ***Individualistic***—people look after their own and family interests
Collectivistic—people expect group to look after and protect them

Individualistic ←→ *Collectivistic*
United States, Canada, Australia — Japan — Mexico, Thailand

(2) ***High power distance***—Accepts wide differences in power, great deal of respect for those in authority
Low power distance—Plays down inequalities: employees are not afraid to approach nor are in awe of the boss

High power distance ←→ *Low power distance*
Mexico, Singapore, France — Italy, Japan — United States, Sweden

(3) ***High uncertainty avoidance***—Threatened with ambiguity and experience high levels of anxiety
Low uncertainty avoidance—Comfortable with risks; tolerant of different behavior and opinions

High uncertainty avoidance ←→ *Low uncertainty avoidance*
Italy, Mexico, France — United Kingdom — Canada, United States, Singapore

(4) ***Achievement***—Values such as assertiveness, acquiring money and goods, and competition prevail
Nurturing—Values such as relationships and concern for others prevail

Achievement ←→ *Nurturing*
United States, Japan, Mexico — Canada, Greece — France, Sweden

(5) ***Long-term orientation***—People look to the future and value thrift and persistence
Short-term orientation—People value tradition and the past

Short-term thinking ←→ *Long-term thinking*
Germany, Australia, United States, Canada — China, Taiwan, Japan

Let's Get Real: F2F

KNOWING ABOUT GLOBAL BUSINESS IS IMPORTANT BECAUSE:

The geographies are shrinking and we are largely impacted by what's happening in the world.

MINIMIZE CROSS-CULTURAL PROBLEMS BY:

Sensitizing people through communication, workshops, training, etc.

The GLOBE Framework for Assessing Cultures. The GLOBE (Global Leadership and Organizational Behavior Effectiveness) research program extended Hofstede's work by investigating cross-cultural leadership behaviors. It gives managers additional information to help them identify and manage cultural differences. Using data from more than 18,000 managers in 62 countries, the GLOBE research team (led by Robert House) has identified nine dimensions on which national cultures differ.[35] Two dimensions (power distance and uncertainty avoidance) fit directly with Hofstede's. Four are similar to Hofstede's (assertiveness, which is similar to achievement–nurturing; humane orientation, which is similar to the nurturing dimension; future orientation, which is similar to long-term and short-term orientation; and institutional collectivism, which is similar to individualism–collectivism). The remaining three (gender differentiation, in-group collectivism, and performance orientation) offer additional insights into a country's culture:

- **Power distance**—The degree to which members of a society expect power to be unequally shared.
- **Uncertainty avoidance**—A society's reliance on social norms and procedures to alleviate the unpredictability of future events.
- **Assertiveness**—The extent to which a society encourages people to be tough, confrontational, assertive, and competitive rather than modest and tender.
- **Humane orientation**—The degree to which a society encourages and rewards individuals for being fair, altruistic, generous, caring, and kind to others.
- **Future orientation**—The extent to which a society encourages and rewards future-oriented behaviors such as planning, investing in the future, and delaying gratification.
- **Institutional collectivism**—The degree to which individuals are encouraged by societal institutions to be integrated into groups in organizations and society.
- **Gender differentiation**—The extent to which a society maximizes gender role differences, as measured by how much status and decision-making responsibilities women have.
- **In-group collectivism**—The extent to which members of a society take pride in membership in small groups, such as their families, their circles of close friends, and the organizations in which they're employed.
- **Performance orientation**—The degree to which a society encourages and rewards group members for performance improvement and excellence.

Exhibit 4–6 provides information on how different countries rank on these nine dimensions.

GLOBAL MANAGEMENT IN TODAY'S WORLD

IRKUT, a Russian jet maker widely admired for its Sukhoi military fighter jets, is building passenger planes for both domestic and international markets, but it may find the demands of the commercial market to be a lot different from what it's used to. Nissan Motor Company is exporting a limited number of its U.S.-made Quest minivans to China, a move that reflects how deeply Japan's automakers are integrated into the U.S. economy. Because of high costs, Danish toy maker Lego Group transferred production from a factory in Switzerland to the Czech Republic. In Bangalore, India, General Electric invested more than $80 million into creating its largest research center outside the United States, a risky move considering the fragile nature of relations between India and Pakistan. The global electronics industry must comply with European rules called RoHS (Restriction of Hazardous Substances), which for the most part ban the use of lead, cadmium, mercury, and certain flame retardants in most electrical and electronic products.[36] Doing business globally today isn't easy! Managers face serious

Globe
The Global Leadership and Organizational Behavior Effectiveness research program, a program that studies cross-cultural leadership behaviors.

Exhibit 4–6

GLOBE Highlights

Dimension	Countries Rating Low	Countries Rating Moderate	Countries Rating High
Assertiveness	Sweden	Egypt	Spain
	New Zealand	Ireland	United States
	Switzerland	Philippines	Greece
Future orientation	Russia	Slovenia	Denmark
	Argentina	Egypt	Canada
	Poland	Ireland	Netherlands
Gender differentiation	Sweden	Italy	South Korea
	Denmark	Brazil	Egypt
	Slovenia	Argentina	Morocco
Uncertainty avoidance	Russia	Israel	Austria
	Hungary	United States	Denmark
	Bolivia	Mexico	Germany
Power distance	Denmark	England	Russia
	Netherlands	France	Spain
	South Africa	Brazil	Thailand
Individualism/collectivism*	Denmark	Hong Kong	Greece
	Singapore	United States	Hungary
	Japan	Egypt	Germany
In-group collectivism	Denmark	Japan	Egypt
	Sweden	Israel	China
	New Zealand	Qatar	Morocco
Performance orientation	Russia	Sweden	United States
	Argentina	Israel	Taiwan
	Greece	Spain	New Zealand
Humane orientation	Germany	Hong Kong	Indonesia
	Spain	Sweden	Egypt
	France	Taiwan	Malaysia

*A low score is synonymous with collectivism.

Source: M. Javidan and R. J. House, "Cultural Acumen for the Global Manager: Lessons from Project GLOBE," *Organizational Dynamics*, Spring 2001, pp. 289–305.

challenges—challenges arising from the openness associated with globalization and from significant cultural differences.

The push to go global has been widespread. Advocates praise the economic and social benefits that come with globalization. Yet globalization has created challenges because of the openness that's necessary for it to work. One challenge is the increased threat of terrorism by a truly global terror network. Globalization is meant to open up trade and to break down the geographic barriers separating countries. Yet opening up means being open to the bad as well as the good. From the Philippines and the United Kingdom to Israel and Pakistan, organizations and employees face the risk of terrorist attacks. Another challenge from openness is the economic interdependence of trading countries. If one country's economy falters, it could potentially have a domino effect on other countries with which it does business. So far, however, the world economy has proven to be quite resilient. And there are structures in place, such as the World Trade Organization and the International Monetary Fund, to isolate and address potential problems.

But it's not simply the challenges from openness that managers must be prepared for. The far more serious challenges for managers reflect intense underlying and fundamental cultural differences—differences that encompass traditions, history, religious beliefs, and deep-seated values. Managing in such an environment can be extremely complicated. Although globalization has long been praised for its economic benefits, some people think that globalization is simply a euphemism for "Westernization"—that is, the way Western cultural values and Westerns business philosophy are said to be slowly taking over the world.[37] Proponents of Westernization hope others will see how progressive, efficient, industrious, and free Western society and businesses are and want to emulate that way of doing things. However, critics claim that this attitude of the "almighty American dollar wanting to spread the American way to every single country"[38] has created many problems. Although history is filled with clashes between civilizations, what's unique about the current time period is the speed and ease with which misunderstandings and disagreements can erupt and escalate. The Internet, television and other media, and global air travel have brought the good and the bad of entertainment, products, and behaviors to every part of the globe. For those who don't like what people from other countries do, say, or believe, this can lead to resentment, dislike, distrust, and even outright hatred.

Successfully managing in today's global environment will require incredible sensitivity and understanding. Managers from any country will need to be aware of how their decisions and actions will be viewed not only by those who may agree but, more importantly, by those who may disagree. They will need to adjust their leadership styles and management approaches to accommodate these diverse views. Yet, as always, they will need to do this while still being as efficient and effective as possible in reaching the organization's goals.

QUICK LEARNING REVIEW:

LEARNING OUTCOME 4.4

- **Explain how the global political/legal and economic environments affect managers.**
- **Discuss Hofstede's dimensions and the GLOBE dimensions for assessing country cultures.**
- **Describe the challenges of doing business globally in today's world.**

Go to page 87 to see how well you know this material.

Let's Get Real: My Turn

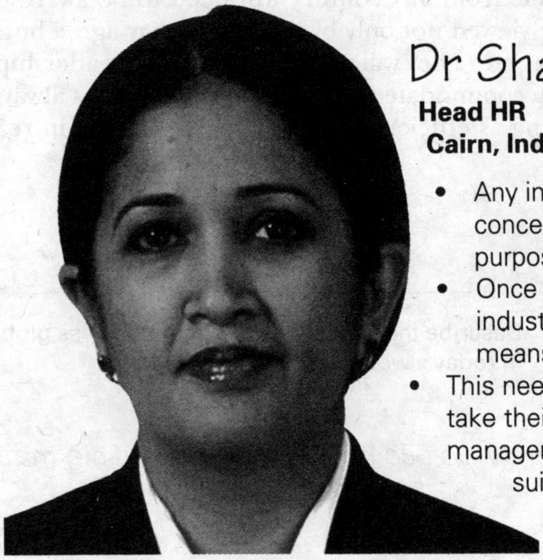

Dr Shalini Sarin

Head HR
Cairn, India

- Any initiative to be rolled out needs to be conceptualized with clear definitions of the objective or purpose.
- Once the purpose or the objective is defined, a quick industry study needs to be collated to establish the means to roll it out.
- This needs to be then presented to the local HR team to take their inputs and bounced off with a few line managers before it is suitably modified or customized to suit the local environment and the objective for which it was designed.
- It then needs to be rolled out by local HR in partnership with the line managers.

LEARNING OUTCOMES **SUMMARY**

4.1 ▷ WHAT'S YOUR GLOBAL PERSPECTIVE?

- Define parochialism.
- Contrast ethnocentric, polycentric, and geocentric attitudes toward global business.

Parochialism is viewing the world solely through one's own eyes and perspectives and not recognizing that others have different ways of living and working. An ethnocentric attitude is the parochialistic belief that the best work approaches and practices are those of the home country. A polycentric attitude is the view that the managers in the host country know the best work approaches and practices for running their business. A geocentric attitude is a world-oriented view that focuses on using the best approaches and people from around the globe.

4.2 ▷ UNDERSTANDING THE GLOBAL ENVIRONMENT

- Describe the current status of the EU, NAFTA, ASEAN, and other regional trade alliances.
- Discuss the role of the WTO

The European Union consists of 27 democratic countries, with 3 countries having applied for membership. Fifteen countries have adopted the euro, and all new member countries must adopt it. The recently signed Lisbon Treaty now gives the EU a common legal framework. NAFTA continues to help Canada, Mexico, and the United States strengthen their global economic power. CAFTA is still trying to get off the ground, as is the proposed FTAA. Because of the delays for CAFTA and FTAA, Mercosur (Southern Common Market) will likely take on new importance. ASEAN is a trading alliance of 10 nations in Southeast Asia, a region that remains important in the global economy. The African Union and SAARC are relatively new but will continue to see benefits from their alliances. To counteract some of the risks in global trade, the World Trade Organization (WTO) plays an important role in monitoring and promoting trade relationships.

4.3 ▷ DOING BUSINESS GLOBALLY

- Contrast multinational, multidomestic, global, and transnational organizations.
- Describe the different ways organizations can go international.

A multinational corporation is an international company that maintains operations in multiple countries. A multidomestic organization is an MNC that decentralizes management and other decisions to the local country (the polycentric attitude). A global organization is an MNC that centralizes management and other decisions in the home country (the ethnocentric attitude). A transnational organization (the geocentric attitude) is an MNC that has eliminated artificial geographic barriers and uses the best work practices and approaches, regardless of where they're from. Global sourcing is purchasing materials or labor from around the world based on lowest cost. Exporting is making products domestically and selling them abroad. Importing is acquiring products made abroad and selling them domestically. Licensing is used by manufacturing organizations that make or sell another company's products and gives that organization the right to use its brand name, technology, or product specifications. Franchising is similar to licensing but is usually used by service organizations that want to use another company's name and operating methods. A global strategic alliance is a partnership between an organization and foreign company partners in which they share resources and knowledge to develop new products or build facilities. A joint venture is a specific type of strategic alliance in which the partners agree to form a separate, independent organization for some business purpose. A foreign subsidiary is a direct investment in a foreign country that a company creates by establishing a separate and independent facility or office.

4.4 ▷ MANAGING IN A GLOBAL ENVIRONMENT

- Explain how the global political/legal and economic environments affect managers.
- Discuss Hofstede's dimensions and the GLOBE dimensions for assessing country cultures.
- Describe the challenges of doing business globally in today's world.

The laws and political stability of a country are issues in the global political/legal environment with which managers must be familiar. Likewise, managers must be aware of a country's economic issues, such as currency exchange rates, inflation rates, and tax policies. Geert Hofstede identified five dimensions for assessing a country's culture: individualism–collectivism, power distance, uncertainty avoidance, achievement–nurturing, and long-term/short-term orientation. The GLOBE studies identified nine dimensions for assessing country cultures: power distance, uncertainty avoidance, assertiveness, humane orientation, future orientation, institutional collectivism, gender differentiation, in-group collectivism, and performance orientation. The main challenges of doing business globally in today's world involve the openness associated with globalization and the significant cultural differences between countries.

THINKING ABOUT MANAGEMENT ISSUES

1. What are the managerial implications of a borderless organization?
2. Can the GLOBE framework presented in this chapter be used to guide managers in a Thai hospital or a government agency in Venezuela? Explain.
3. Compare the advantages and drawbacks of the various approaches to going global.
4. What challenges might confront a Mexican manager transferred to the United States to manage a manufacturing plant in Tucson, Arizona? Will these be the same for a U.S. manager transferred to Guadalajara? Explain.
5. In what ways do you think global factors have changed the way organizations select and train managers? What impact might the Internet have on this? Explain.
6. How might a continued war on terrorism affect U.S. managers and companies doing business globally?
7. How might the cultural differences in Hofstede's dimensions affect how managers (a) use work groups, (b) develop goals/plans, (c) reward outstanding employee performance, and (d) deal with employee conflict?

YOUR TURN to be a Manager

- Find two current examples of each of the ways that organizations go international. Write a short paper describing what these companies are doing.
- The U.K.-based company Kwintessential has several cultural awareness "quizzes" on its Web site, www.kwintessential.co.uk/resources/culture-tests.html. Go to the Web site and try two or three of them. Were you surprised at your score? What does your score tell you about your cultural awareness?
- On the Kwintessential Web site, you'll also find Country Etiquette Guides. Pick two countries to study (from different regions) and compare them. How are they the same? Different? How would this information help a manager?
- Interview two or three professors or students at your school who are from other countries. Ask them to describe what the business world is like in their country. Write a short paper describing what you found out.
- Create a timeline illustrating the history of the European Union and a timeline illustrating the history of NAFTA.
- Take advantage of opportunities that you might have to travel to other countries, either on personal trips or on school-sponsored trips.
- Suppose that you were being sent on an overseas assignment to another country (you decide which one). Research that country's economic, political/legal, and cultural environments. Write a report summarizing your findings.
- If you don't have your passport yet, go through the process to get one. (The current fee in the USA is $100.)
- Steve's and Mary's suggested readings: Nancy J. Adler, *International Dimensions of Organizational Behavior*, 5th edition (South-Western Publishing, 2008); Kenichi Ohmae, *The Next Global Stage* (Wharton School Publishing, 2005); John Hooker, *Working Across Cultures* (Stanford Business Books, 2003); and Thomas L. Friedman, *The Lexus and the Olive Tree* (Anchor Books, 2000).

- If you want to better prepare yourself for working in an international setting, take additional classes in international management and international business.
- You've been put in charge of designing a program to prepare your company's managers to go on an overseas assignment. What should (and would) this program include? Be specific. Be thorough. Be creative.
- In your own words, write down three things you learned in this chapter about being a good manager.
- Self-knowledge can be a powerful learning tool. Go to mymanagementlab and complete these self-assessment exercises: Am I Well-Suited for a Career as a Global Manager? and What Are My Attitudes Toward Workplace Diversity? Using the results of your assessments, identify personal strengths and weaknesses. What will you do to reinforce your strengths and improve your weaknesses?

CASE APPLICATION

Learning to Love Globalization

Léo Apotheker is what you would call a true global manager. He speaks German, French, and English fluently and can converse in Hebrew. As the second-in-command at German software giant SAP, Léo spent 19 years in the company's Paris office and helped turn SAP into a dominant global player. However, his apparent ease and familiarity with different global settings isn't shared elsewhere throughout the company, especially at company headquarters in Walldorf, Germany. There, a global perspective hasn't been as easily embraced.

In 2002, SAP executives decided they needed to make the company more global and did several things to make that happen. First, they hired thousands of programmers in various foreign locations, such as the United States, China, and India. In fact, by 2005, the company had eight global software labs, each with different areas of expertise. Large software projects were split up and sent to units around the world. Employees in India specialized in analytical tools; those in Palo Alto, California, worked on the products' look and feel; and Walldorf programmers continued to manage the specific software coding. In addition, English became the official language for corporate meetings, even at headquarters. Finally, SAP began recruiting hundreds of foreign managers, and the non-Germans now make up half of the company's top ranks. The changes haven't been easy. "The newcomers sought to inject a faster pace and open SAP's insular culture to more outside influences. The resulting tensions show how the challenge of globalization goes far beyond navigating different languages and time zones."

Exterior of SAP building, (Firmensitz der SAP AG), a German Software Company.

In Walldorf, long-time employees felt the company was changing too quickly and too much. Veteran software developers objected to the loss of autonomy and the perceived Americanization of the company. Said one developer who had spent his entire 25-year career at SAP, "We used to be kings here." However, employees at the recently added locations worried that the company wasn't changing enough. Top executives at SAP state, "What we're doing to globalize R&D is the right thing and we will continue."

Discussion Questions

1. What global attitude do you think characterized SAP prior to 2002? How do you know? What global attitude do you think most characterizes it now? Explain.
2. Do some cultural research on Germany, the United States, and India. Compare the cultural characteristics of Germany and the United States. What similarities and differences exist? How about with Germany and India? How might these cultural differences be affecting the situation at SAP?

3. What could SAP managers do to support, promote, and encourage cultural awareness among the various global locations? Explain.
4. What might other managers learn from SAP's experiences in going global?

Sources: L. Abboud, "SAP Appoints Co-CEO, Paving Way for Succession," *Wall Street Journal,* April 3, 2008, p. B3; M. Schiessl, "SAP's Very Big Small Biz Challenge," *BusinessWeek* online, www.businessweek.com, September 14, 2007; A. Ricadela, "SAP Reassures Silicon Valley, Post-Agassi," *BusinessWeek* online, www.businessweek.com, June 8, 2007; J. Ewing, "SAP's Tough Guy Ready to Rumble," *BusinessWeek* online, www.businessweek.com, May 16, 2007; and P. Dvorak and L. Abboud, "SAP's Plan to Globalize Hits Cultural Barriers," *Wall Street Journal,* May 11, 2007, pp. A1+.

Answers to "Who Owns What" Quiz

1. c. United Kingdom
   Ben & Jerry's Ice Cream was purchased by Unilever, PLC, in April 2000.
2. c. United States
   The maker of Lebedyansky juices was acquired by PepsiCo Inc. and Pepsi Bottling Group Inc. in March 2008.
3. c. United States
   Rajah Spices are products of the Lea & Perrins sauce division, which the H. J. Heinz Company acquired in June 2005.
4. b. India
   Tetley Tea is owned by the Tata Tea Group, a subsidiary of Indian conglomerate Tata Group.
5. d. United Kingdom
   Skippy is a product of BestFoods, which Unilever PLC purchased in 2000.
6. a. Japan
   The 7-Eleven stores are owned by Japanese conglomerate Seven & I Holdings.
7. b. Mexico
   Grupo Bimbo, one of the world's largest bakers, bought the rights to make and distribute Boboli pizza crusts in 2002.
8. c. United States
   Braun electric shavers are a part of Global Gillette, which was purchased by the Procter & Gamble Company in October 2005.
9. c. France
   LMVH Moët Hennessy Louis Vuitton SA, the world's largest luxury-goods group, owns Sephora.
10. d. Switzerland
    Nestlé SA, the food and drink company, acquired the Gerber Products Company in 2008.
11. c. Switzerland
    Nestlé SA purchased the maker of Lean Cuisine frozen meals in 2002.
12. d. United Kingdom
    Cadbury Schweppes PLC owns the Dr. Pepper/7-Up businesses.
13. b. United Kingdom
    Lipton Tea is a product of BestFoods, which Unilever PLC purchased in 2000.
14. a. India
    Tata Coffee, a division of Indian conglomerate Tata Group, purchased Eight O'Clock Coffee in 2006.
15. a. United States
    Consumer products giant Procter & Gamble purchased the luxury hair-care brand from a private equity firm in 2008.

Let's Get Real:

Meet the Manager

Dr G. S. Rao

Vice President
Lavasa Corporation Ltd

MY JOB: To implement GIS as a technology in the organization.

BEST PART OF MY JOB: Technically challenging job with new things to learn.

WORST PART OF MY JOB: Delays in decision making at all levels in the organization.

BEST MANAGEMENT ADVICE EVER RECEIVED: Work sincerely and take care of your team.

You'll be hearing more from this real manager throughout the chapter.

Chapter 5

Social Responsibility and Managerial Ethics

How important is it for organizations and managers to be socially responsible and ethical? In this chapter, we're going to look at what it means to be socially responsible and ethical and what role managers play in both. Focus on the following learning outcomes as you read and study this chapter.

LEARNING OUTCOMES

A Manager's Dilemma

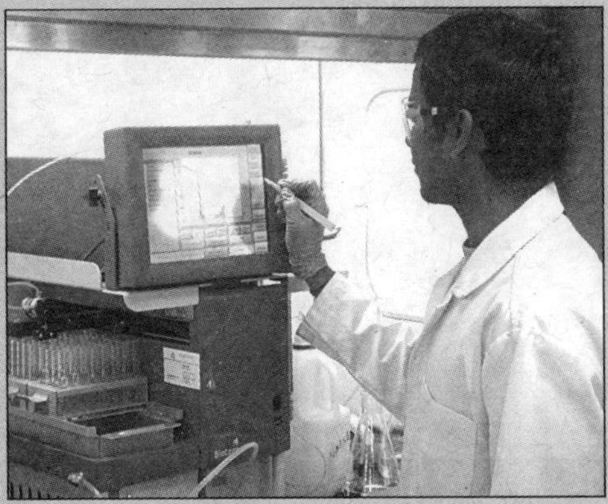

Courtesy: Jubilant Organosys

Jubilant Organosys offers products and services to global life science companies. The company works with a concept of triple bottomline—economic, environmental, and social. It recognizes that the manufacturing of the industrial chemicals for providing them to the life science companies is polluting. Thus, it has introduced environmentally sensitive practices in all its processes. Jubilant has been involved with the World Health Organization in Western UP (in Gajraula, near its manufacturing locations) to help in the eradication of tuberculosis (TB). Jubilant Organosys, which exports to different countries, such as the United Kingdom, United States, and Japan, has acquired global certifications, such as ISO 14000 and OHSAS 18000. It is one of the five companies in India who publish an annual Sustainability Report (SR) that displays its commitment toward being transparent and accountable to its stakeholders.[1] If you were a manager at Jubilant Organosys, how would you instill the importance of being socially responsible as well as profitable among new recruits?

What would you do?

Deciding how socially responsible an organization needs to be is just one example of the complicated types of ethical and social responsibility issues that managers may have to cope with as they plan, organize, lead, and control. As managers manage, these issues can and do influence their actions.

LEARNING OUTCOME 5.1 ▷

WHAT IS SOCIAL RESPONSIBILITY?

By using digital technology and file sharing Web sites, music and video users all over the world obtain and share many of their favorite recordings for free. Large global corporations lower their costs by outsourcing to countries where human rights are not a high priority and justify it by saying they're bringing in jobs and helping strengthen the local economies. Businesses facing a drastically changed industry environment offer employees early retirement and buyout packages. Are these companies being socially responsible? Managers regularly face decisions that have a dimension of social responsibility, such as those involving employee relations, philanthropy, pricing, resource conservation, product quality and safety, and doing business in countries that devalue human rights. What does it mean to be socially responsible?

FROM OBLIGATIONS TO RESPONSIVENESS TO RESPONSIBILITY

The concept of *social responsibility* has been described in different ways. For instance, it has been called "profit making only," "going beyond profit making," "any discretionary corporate activity intended to further social welfare," and "improving social or environmental conditions."[2] We can understand it better if we first compare it to two similar concepts: social obligation and social responsiveness.[3] **Social obligation** is a firm's engaging in social actions because of its obligation to meet certain economic and legal responsibilities. The organization does what it's obligated to do and nothing more.

This idea reflects the **classical view** of social responsibility, which says that management's only social responsibility is to maximize profits. The most outspoken advocate of this approach is economist and Nobel laureate Milton Friedman. He argued that managers' primary responsibility is to operate the business in the best interests of the stockholders, whose primary concerns are financial.[4] He also argued that when managers decide to spend the organization's resources for "social good," they add to the costs of doing business, which have to be passed on to consumers through higher prices or absorbed by stockholders through smaller dividends. You need to understand that Friedman doesn't say that organizations shouldn't be socially responsible. But his interpretation of social responsibility is to maximize profits for stockholders.

The other two concepts—social responsiveness and social responsibility—reflect the **socioeconomic view**, which says that managers' social responsibilities go beyond making profits to include protecting and improving society's welfare. This view is based on the belief that corporations are *not* independent entities responsible only to stockholders but have an obligation to the larger society. Organizations around the world have embraced this view, as shown by a recent survey of global executives in which 84 percent said that companies must balance obligations to shareholders with obligations to the public good.[5] But how do these two concepts differ?

Social responsiveness means that a company engages in social actions in response to some popular social need. Managers in these companies are guided by social norms and values and make practical, market-oriented decisions about their actions.[6] For instance, managers at American Express Company identified three themes—community service, cultural heritage, and leaders for tomorrow—to guide it in deciding which worldwide projects and organizations to support. By making these choices, managers "responded" to what they felt were important social needs.[7]

A socially *responsible* organization views things differently. It goes beyond what it's obligated to do or chooses to do because of some popular social need and does what it can to help improve society because it's the right thing to do. We define **social responsibility** as a business's intention, beyond its legal and economic obligations, to do the right things and act in ways that are good for society.[8] Our definition assumes that a business obeys the law and cares for its stockholders, and it adds an ethical imperative to do those things that make society better and not to do those that make it worse. As Exhibit 5–1 shows, a socially responsible organization does what is right because it feels it has an ethical responsibility to do so. For example, Abt Electronics in Glenview, Illinois, would be described as socially responsible according to our definition. As one of the largest single-store electronics retailers in the United States, it responded to soaring energy costs and environmental concerns by shutting off lights

thinking critically about Ethics

In an effort to be (or at least appear to be) socially responsible, many organizations donate money to philanthropic and charitable causes. In addition, many organizations ask their employees to make individual donations to these causes. Suppose you're the manager of a work team, and you know that several of your employees can't afford to pledge money right now because of personal or financial problems. You've also been told by your supervisor that the CEO has been known to check the list of individual contributors to see who is and is not "supporting these very important causes." What would you do? What ethical guidelines might you suggest for individual and organizational contributions in such a situation?

social obligation
A firm's engaging in social actions because of its obligation to meet certain economic and legal responsibilities.

classical view
The view that management's only social responsibility is to maximize profits.

socioeconomic view
The view that management's social responsibility goes beyond making profits and includes protecting and improving society's welfare.

social responsiveness
A firm's engaging in social actions in response to some popular social need.

social responsibility
A business's intention, beyond its legal and economic obligations, to do the right things and act in ways that are good for society.

Exhibit 5–1

Social Responsibility Versus Social Responsiveness

	Social Responsibility	Social Responsiveness
Major consideration	Ethical	Pragmatic
Focus	Ends	Means
Emphasis	Obligation	Responses
Decision framework	Long term	Medium and short term

Source: Adapted from S. L. Wartick and P. L. Cochran, "The Evolution of the Corporate Social Performance Model," *Academy of Management Review*, October 1985, p. 766.

more frequently and reducing air conditioning and heating. However, an Abt family member said, "These actions weren't just about costs, but about doing the right thing. We don't do everything just because of money."[9]

So, how should we view an organization's social actions? A manufacturer that meets pollution control standards or hires scheduled caste applicants to meet its constitutional quota does it because laws mandate these actions. However, when it provides childcare facilities for employees or runs a program for helping widows in the community, it's being socially responsive. Why? Working parents and environmentalists have voiced these social concerns and demanded such actions.

For many businesses, their social actions are better viewed as being socially responsive than socially responsible (at least according to our definition). However, such actions are still good for society. For example, Times of India started an initiative to address the issues of illiteracy. Educated readers of the paper were urged to donate two hours of their time for children who needed help for reading and writing. As a part of this program, the paper ran advertisements in all its publications reading "TEACH INDIA: Lets learn to teach" and "WE'VE JUST IDENTIFIED THE PERFECT SOLUTION TO ILLITERACY - YOU."[10]

Let's Get Real: F2F

BEING SOCIALLY RESPONSIBLE MEANS:

Using the resources of the organization to address, in a positive way, the issues faced by the society.

SHOULD ORGANIZATIONS BE SOCIALLY INVOLVED?

Other than meeting their social obligations (which they *must* do), should organizations be socially involved? One way to look at this is by examining arguments for and against social involvement. Several points are outlined in Exhibit 5–2.[11]

Another way to look at this is whether social involvement affects a company's economic performance, which numerous studies have done.[12] Although most have found a small positive relationship, no generalizable conclusions can be made because the studies haven't use standardized measures of social responsibility and economic performance.[13] Another concern in these studies has been causation: If a study showed that social involvement and economic performance were positively related, this didn't necessarily mean that social involvement *caused* higher economic performance. It could simply mean that high profits afforded companies the "luxury" of being socially involved.[14] Such methodological concerns can't be taken lightly. In fact, one study found that if the flawed empirical analyses in these studies were "corrected," social responsibility had a neutral impact on a company's financial performance.[15] Another found that participating in social issues not related to the organization's primary stakeholders was negatively associated with shareholder value.[16] A recent re-analysis of several studies concluded that managers can afford to be (and should be) socially responsible.[17]

Another way to view social involvement and economic performance is by looking at socially responsible investing (SRI) funds, which provide a way for individual investors to support socially responsible companies. (You can find a list of SRI funds at www.socialfunds.com.) Typically, these funds use some type of **social screening**; that is, they apply social and environmental criteria to investment decisions (Exhibit 5–3). For instance, SRI funds usually do not invest in companies which are involved in liquor, gambling, tobacco, nuclear power, weapons, price fixing, or fraud or in companies that have poor product safety, employee relations, or environmental track records. In March 2008, it was reported that ABN AMRO is the only company which floats a fund that invests in the sustainable development.[18] The AMRO Sustainable Development Fund invests in companies with high-disclosure policies on environmental, social, and corporate governance yardsticks. ABN AMRO's mutual fund invests 65 percent of its capital in companies that deliver well on ethical parameters.[19]

Exhibit 5–2

Arguments For and Against Social Responsibility

For	Against
Public expectations Public opinion now supports businesses pursuing economic and social goals.	**Violation of profit maximization** Business is being socially responsible only when it pursues its economic interests.
Long-run profits Socially responsible companies tend to have more secure long-run profits.	**Dilution of purpose** Pursuing social goals dilutes business's primary purpose—economic productivity.
Ethical obligation Businesses should be socially responsible because responsible actions are the right thing to do.	**Costs** Many socially responsible actions do not cover their costs and someone must pay those costs.
Public image Businesses can create a favorable public image by pursuing social goals.	**Too much power** Businesses have a lot of power already and if they pursue social goals they will have even more.
Better environment Business involvement can help solve difficult social problems.	**Lack of skills** Business leaders lack the necessary skills to address social issues.
Discouragement of further governmental regulation By becoming socially responsible, businesses can expect less government regulation.	**Lack of accountability** There are no direct lines of accountability for social actions.
Balance of responsibility and power Businesses have a lot of power and an equally large amount of responsibility is needed to balance against that power.	
Stockholder interests Social responsibility will improve a business's stock price in the long run.	
Possession of resources Businesses have the resources to support public and charitable projects that need assistance.	
Superiority of prevention over cures Businesses should address social problems before they become serious and costly to correct.	

Exhibit 5–3

Trends in SRI in US

Socially Responsible Investing in the US • 1995–2007

(In billions)	1995	1997	1999	2001	2003	2005	2007
Social Screening	$162	$529	$1,497	$2,010	$2,143	$1,685	$2,098
Shareholder Advocacy	$473	$736	$922	$897	$448	$703	$739
Screening and Shareholder	N/A	($84)	($265)	($592)	($441)	($117)	($151)
Community Investing	$4	$4	$5	$8	$14	$20	$26
Total	$639	$1,185	$2,159	$2,323	$2,164	$2,290	$2,711

Source: Social Investment Forum Foundation.

Notes: Social Screening includes socially and environmentally screened funds and separate account assets. Overlapping assets involved in Screening and Shareholder Advocacy are subtracted to avoid potential double-counting. Tracking Screening and Shareholder Advocacy together only began in 1997, so there is no datum for 1995. There are also potentially overlapping assets in the relatively small screened funds categories of Alternative Investments and Other Pooled Products; therefore these categories are also excluded from the SRI universe aggregated in this Report. See Chapter II for details.

social screening
Applying social criteria (screens) to investment decisions.

In helping Indians invest in socially responsible firms, CRISIL launched a 50-scrip ESG (environment, social, and governance) Index which tracks the top Indian companies on the basis of corporate governance, ethics, and social responsibility. The top 10 companies by weight in the index are Infosys, ITC, Aditya Birla Nuvo, Dr Reddy Labs, Wipro, Jubilant Organosys, Axis Bank, GTL, RIL, and HUL. But, more important than the total amount invested in these funds is that the Social Investment Forum reports that the performance of SRI funds is comparable to the performance of non-SRI funds.[20]

So, what can we conclude about social involvement and economic performance? It appears that a company's social actions *don't hurt* its economic performance. Given political and societal pressures to be socially involved, managers probably need to take into consideration social issues and goals as they plan, organize, lead, and control.

QUICK LEARNING REVIEW:

LEARNING OUTCOME 5.1

- Differentiate between social obligation, social responsiveness, and social responsibility.
- Discuss whether organizations should be socially involved.
- Describe what conclusion can be reached regarding social involvement and economic performance.

Go to page 115 to see how well you know this material.

LEARNING **OUTCOME 5.2** ▷

GREEN MANAGEMENT

The plastic shopping bag. An ugly symbol of development. Some 110 billions are used each year in the United States (numbers not available for India) and only an estimated 2 percent of those bags are recycled. Plastic shopping bags can last 1,000 years in landfills. These bags are made from oil, and our "bag habit" costs 1.6 billion gallons each year.[21] But the good news is that things are changing. Being green is in! For instance, IKEA encourages customers to use fewer bags by charging a nickel (which it donates to American Forests) for each bag used. It also cut the price of its large reusable totes from 99 cents to 59 cents. Many stores such as Fab India and Colour Plus only give paper bags. Many shops encourage people to bring their own bags. The largest solar–steam cooking system for 15,000 persons per day has been set up at Tirupati Tirumala Devasthanam and 3,600 remote villages or hamlets, including those in Sunderbans, Bastar, Ladakh, and the North East, have been electrified through solar energy.[22] India's largest tyre company, MRF, has recently launched a new tubeless, eco-friendly rubber tyre that reduces rolling resistance and results in lower fuel consumption. Bajaj Auto has installed wind power generation units in three factories that not only conserve energy but also save Rs. 25 crore in power costs every year. The Orchid Ecotel Hotel—Asia's first eco-friendly five-star hotel chain-is conserving natural resources without compromising quality of service.

Until the late 1960s, few people (and organizations) paid attention to the environmental consequences of their decisions and actions. Although some groups were concerned with conserving natural resources, about the only reference to saving the environment was the ubiquitous printed request "Please Don't Litter." However, a number of environmental disasters brought a new spirit of environmentalism to individuals, groups, and organizations. Increasingly, managers have begun to consider the impact of their organizations on the natural environment, which we call **green management**. What do managers need to know about going green?

HOW ORGANIZATIONS GO GREEN

Managers and organizations can do many things to protect and preserve the natural environment.[23] Some do no more than what is required by law—that is, they fulfill their social obligation. However, others have radically changed their products and production processes. For instance, Fiji Water is using renewable energy sources, preserving forests, and conserving water. Carpet-maker Shaw Industries transforms its carpet and wood manufacturing waste into energy. Google and Intel initiated an effort to

Exhibit 5–4

Green Approaches

Source: Based on R. E. Freeman, J. Pierce, and R. Dodd, *Shades of Green: Business Ethics and the Environment* (New York: Oxford University Press, 1995).

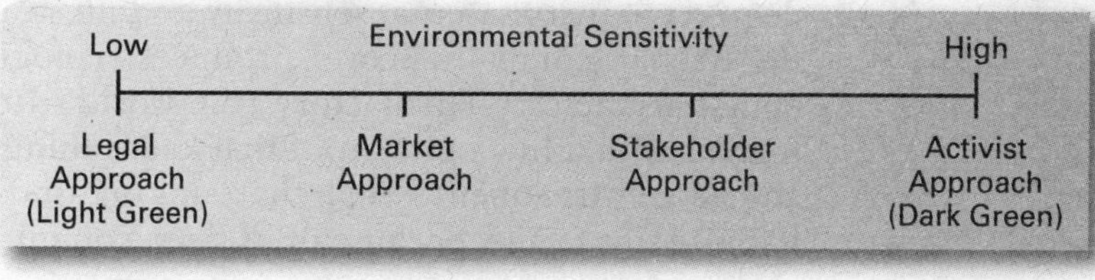

get computer makers and customers to adopt technologies that reduce energy consumption. Paris-based TOTAL, SA, one of the world's largest integrated oil companies, is going green by implementing tough new rules on oil tanker safety and working with groups such as Global Witness and Greenpeace. UPS, the world's largest package delivery company, has done several things—from retrofitting its aircraft with advanced technology and fuel-efficient engines to developing a computer network that efficiently dispatches its fleet of brown trucks to using alternative fuel to run those trucks. Although interesting, these examples don't tell us much about how organizations go green. One model uses the terms *shades of green* to describe the different environmental approaches that organizations may take (see Exhibit 5–4).[24]

The first approach, the *legal* (or *light green*) *approach*, is simply doing what is required legally. In this approach, which illustrates social obligation, organizations exhibit little environmental sensitivity. They obey laws, rules, and regulations without legal challenge, and that's the extent of their being green.

As an organization becomes more sensitive to environmental issues, it may adopt the *market approach* and respond to environmental preferences of customers. Whatever customers demand in terms of environmentally friendly products will be what the organization provides. For example, DuPont developed a new type of herbicide that helped farmers around the world reduce their annual use of chemicals by more than 45 million pounds. By developing this product, the company was responding to the demands of its customers (farmers) who wanted to minimize the use of chemicals on their crops. This is a good example of social responsiveness, as is the next approach.

In the *stakeholder approach*, an organization works to meet the environmental demands of multiple stakeholders, such as employees, suppliers, or community. For instance, Hewlett-Packard has several corporate environmental programs in place for its supply-chain (suppliers), product design and product recycling (customers and society), and work operations (employees and community).

Finally, if an organization pursues an *activist* (or *dark green*) *approach*, it looks for ways to protect the earth's natural resources. The activist approach reflects the highest

Subway is helping to lead the way among retail food stores when it comes to "going green." The sandwich chain is testing ways to reduce its use of paper by serving its famous subs with less wrapping and using baskets made partly of recycled material to hold food instead. It has also switched to 100% recycled paper napkins, which the store estimates is saving 147,000 trees each year.

green management
A form of management in which managers consider the impact of their organization on the natural environment.

degree of environmental sensitivity and illustrates social responsibility. For example, the Belgian company Ecover produces ecological cleaning products in a near-zero-emissions factory. This factory (the world's first ecological one) is an engineering marvel, with a huge grass roof that keeps things cool in summer and warm in winter and a water treatment system that runs on wind and solar energy. The company chose to build this facility because of its deep commitment to the environment.

EVALUATING GREEN MANAGEMENT ACTIONS

As businesses become "greener," they often release detailed reports on their environmental performance. Some 1,500 companies around the globe are voluntarily reporting their efforts in promoting environmental sustainability, using the guidelines developed by the Global Reporting Initiative (GRI). These reports, which can be found on the GRI Web site (www.globalreporting.org), describe the numerous green actions of these organizations.

Another way that organizations show their commitment to being green is through pursuing standards developed by the nongovernmental International Organization for Standardization (ISO). Although the ISO has developed more than 17,000 international standards, it's probably best known for its ISO 9000 (quality management) and ISO 14000 (environmental management) standards. An organization that wants to become ISO 14000 compliant must develop a total management system for meeting environmental challenges. This means it must minimize the effects of its activities on the environment and continually improve its environmental performance. If an organization can meet these standards, it can state that it's ISO 14000 compliant, which organizations in 138 countries have achieved. In addition to its environmental management standards, ISO is developing standards for social responsibility and for energy management. The one for social responsibility (known as ISO 26000) will be published in 2010 and will be voluntary, which means that organizations won't be able to obtain any type of certification for meeting the standards. And no date has been announced for the energy management standards because the committee developing those standards was only recently created.[25]

The final way to evaluate a company's green actions is to use the Global 100 list of the most sustainable corporations in the world (www.global100.org).[26] To be named to this list, which is announced each year at the renowned World Economic Forum in Davos, Switzerland, a company must have displayed a superior ability to effectively manage environmental and social factors. In 2008, the United Kingdom led the list with 23 Global 100 companies. The United States followed with 19 and Japan with 13. Some companies on the 2008 list included BASF (Germany), Diageo PLC (United Kingdom), Mitsubishi (Japan), and Nike (United States).

QUICK LEARNING REVIEW:

LEARNING OUTCOME 5.2

- Define green management.
- Describe how organizations can go green.
- Explain how green management actions can be evaluated.

Go to page 115 to see how well you know this material.

LEARNING

OUTCOME 5.3 ▷

MANAGERS AND ETHICAL BEHAVIOR

Two weeks after firing seven top managers for failing to meet company standards, Wal-Mart issued an extensive ethics policy for employees. Takafumi Horie, founder of the Tokyo-based Internet company Livedoor, was sentenced to 2.5 years in jail for securities violations. Former WorldCom CEO Bernie Ebbers is serving a 25-year prison sentence for financial fraud, conspiracy, and false filings. The Gemological Institute of America, which grades diamonds for independent dealers and large retailers, fired four employees and made changes to top management after an internal investigation showed that lab workers took bribes to inflate the quality of diamonds in grading reports.[27] When you hear about such behaviors—especially after the high-profile

Let's Get Real: F2F

BEING ETHICAL MEANS:
Being sensitive to other peoples' constraints and dealing with people without compromising on principles.

financial misconduct at Enron, WorldCom, and other companies—you might conclude that businesses aren't ethical. Although that's not the case, managers—at all levels, in all areas, in all sizes, and in all kinds of organizations—do face ethical issues and dilemmas. For instance, is it ethical for a sales representative to bribe a purchasing agent as an inducement to buy? Would it make a difference if the bribe came out of the sales rep's commission? Is it ethical for someone to use a company car for private use? How about using company e-mail for personal correspondence or using a company phone to make personal phone calls? What if you managed an employee who worked all weekend on an emergency situation, and you told him to take off two days sometime later and mark it down as "sick days" because your company had a clear policy that overtime would not be compensated for any reason?[28] Would that be okay? How will you handle such situations? As managers plan, organize, lead, and control, they must consider ethical dimensions.

What do we mean by **ethics**? We define it as the principles, values, and beliefs that define right and wrong decisions and behavior.[29] Many decisions that managers make require them to consider both the process and who's affected by the result.[30] To better understand the ethical issues involved in such decisions, let's look at the factors that determine whether a person acts ethically or unethically.

FACTORS THAT DETERMINE ETHICAL AND UNETHICAL BEHAVIOR

Whether someone behaves ethically or unethically when faced with an ethical dilemma is influenced by several things: his or her stage of moral development and other moderating variables, including individual characteristics, the organization's structural design, the organization's culture, and the intensity of the ethical issue (see Exhibit 5–5). People who lack a strong moral sense are much less likely to do the wrong things if they're constrained by rules, policies, job descriptions, or strong cultural norms that disapprove of such behaviors. Conversely, intensely moral individuals can be corrupted by an organizational structure and culture that permits or encourages unethical practices. Let's look more closely at these factors.

Stage of Moral Development. Research confirms there are three levels of moral development, each having two stages.[31] At each successive stage, an individual's moral judgment becomes less dependent on outside influences and more internalized.

At the first level, the *preconventional* level, a person's choice between right and wrong is based on personal consequences from outside sources, such as physical punishment, reward, or exchange of favors. At the second level, the *conventional* level, ethical decisions rely on maintaining expected standards and living up to the expectations of others. At the *principled* level, individuals define moral values apart from the authority of the groups to which they belong or society in general. The three levels and six stages are described in Exhibit 5–6.

Exhibit 5–5

Factors that Determine Ethical and Unethical Behavior

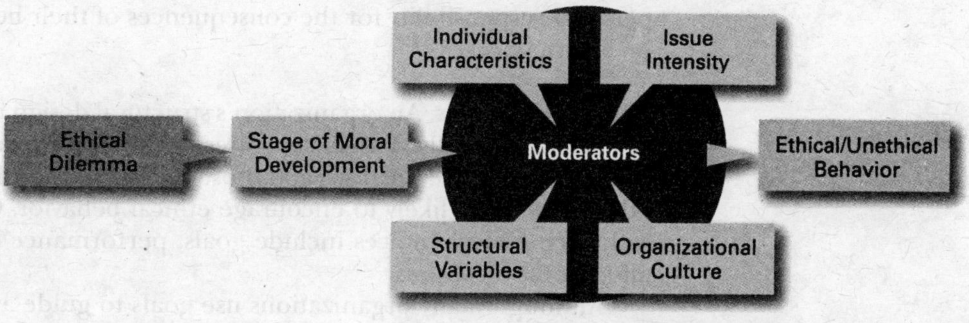

ethics
Principles, values, and beliefs that define what is right and what is wrong behavior.

Exhibit 5–6

Stages of Moral Development

Source: Based on L. Kohlberg, "Moral Stages and Moralization: The Cognitive-Development Approach," in T. Lickona (ed.), *Moral Development and Behavior: Theory, Research, and Social Issues* (New York: Holt, Rinehart & Winston, 1976), pp. 34–35.

Level	Description of Stage
Principled	6. Following self-chosen ethical principles even if they violate the law 5. Valuing rights of others and upholding absolute values and rights regardless of the majority's opinion
Conventional	4. Maintaining conventional order by fulfilling obligations to which you have agreed 3. Living up to what is expected by people close to you
Preconventional	2. Following rules only when doing so is in your immediate interest 1. Sticking to rules to avoid physical punishment

What can we conclude about moral development?[32] First, people proceed through the six stages sequentially. Second, there is no guarantee of continued moral development. Third, the majority of adults are at stage 4: They're limited to obeying the rules and will be inclined to behave ethically, although for different reasons. A manager at stage 3 is likely to make decisions based on peer approval; a manager at stage 4 will try to be a "good corporate citizen" by making decisions that respect the organization's rules and procedures; and a stage 5 manager is likely to challenge organizational practices that he or she believes to be wrong.

Individual Characteristics. Two individual characteristics—values and personality—play a role in determining whether a person behaves ethically. Each person comes to an organization with a relatively entrenched set of personal **values**, which represent basic convictions about what is right and wrong. Our values develop from a young age, based on what we see and hear from parents, teachers, friends, and others. Thus, employees in the same organization often possess very different values.[33] Although *values* and *stage of moral development* may seem similar, they're not. Values are broad and cover a wide range of issues; the stage of moral development is a measure of independence from outside influences.

Two personality variables have been found to influence an individual's actions according to his or her beliefs about what is right or wrong: ego strength and locus of control. **Ego strength** measures the strength of a person's convictions. People with high ego strength are likely to resist impulses to act unethically and instead follow their convictions. That is, individuals high in ego strength are more likely to do what they think is right and be more consistent in their moral judgments and actions than those with low ego strength.

Locus of control is the degree to which people believe they control their own fate. People with an *internal* locus of control believe they control their own destinies. They're more likely to take responsibility for consequences and rely on their own internal standards of right and wrong to guide their behavior. They're also more likely to be consistent in their moral judgments and actions. People with an *external* locus believe that what happens to them is due to luck or chance. They're less likely to take personal responsibility for the consequences of their behavior and more likely to rely on external forces.[34]

Structural Variables. An organization's structural design can influence whether employees behave ethically. Those structures that minimize ambiguity and uncertainty with formal rules and regulations and those that continuously remind employees of what is ethical are more likely to encourage ethical behavior. Other structural variables that influence ethical choices include goals, performance appraisal systems, and reward allocation procedures.

Although many organizations use goals to guide and motivate employees, those goals can create some unexpected problems. One study found that people who don't reach set goals are more likely to engage in unethical behavior, regardless of whether there are economic incentives to do so. The researchers concluded that "goal setting can lead to unethical behavior."[35] Examples of such behaviors abound—from companies shipping unfinished products just to reach sales goals or

Let's Get Real: F2F

MY COMPANY'S VALUES: Integrity, natural harmony, diligence, and honesty to oneself.

"managing earnings" to meet financial analysts' expectations, to schools excluding certain groups of students when reporting standardized test scores to make their "pass" rate look better.[36]

An organization's performance appraisal system can also influence ethical behavior. Some systems focus exclusively on outcomes, while others evaluate means as well as ends. When employees are evaluated only on outcomes, they may be pressured to do whatever is necessary to look good on the outcomes and not be concerned with how they got those results. Research suggests that "success may serve to excuse unethical behaviors."[37] The danger of such thinking is that if managers are more lenient in correcting unethical behaviors of successful employees, other employees will model their behavior on what they see.

Closely related to the organization's appraisal system is how rewards are allocated. The more that rewards or punishment depend on specific goal outcomes, the more employees are pressured to do whatever they must to reach those goals, perhaps to the point of compromising their ethical standards.

Organization's Culture. As Exhibit 5–6 shows, the content and strength of an organization's culture influence ethical behavior.[38] We learned in Chapter 3 that an organization's culture consists of the shared organizational values. These values reflect what the organization stands for and what it believes in, and they create an environment that influences employee behavior ethically or unethically. When it comes to ethical behavior, a culture most likely to encourage high ethical standards is one that's high in risk tolerance, control, and conflict tolerance. Employees in such a culture are encouraged to be aggressive and innovative, are aware that unethical practices will be discovered, and feel free to openly challenge expectations they consider to be unrealistic or personally undesirable.

Because shared values can be powerful influences, many organizations are using **values-based management**, in which the organization's values guide employees in the way they do their jobs. For instance, Timberland is an example of a company that uses values-based management. Based on the simple statement "Make It Better," employees at Timberland know what's expected and valued: They know they need to find ways to "make it better"—whether it's creating quality products for customers, performing community service activities, designing employee training programs, or figuring out ways to make the company's packaging more environmentally friendly. As CEO Jeffrey Swartz says on the company's Web site, "Everything we do at Timberland grows out of our relentless pursuit to find a way to make it better." And Timberland isn't alone in its use of values-based management. A survey of global companies found that a large number—more than 89 percent—said they had written corporate values statements.[39] This survey also found that most of the companies believed that their values influenced relationships and reputation, the top-performing companies consciously connected values with the way employees did their work, and top managers were important to reinforcing the importance of the values throughout the organization.

Thus, an organization's managers do play an important role in ethics. They're responsible for creating an environment that encourages employees to embrace the culture and the desired values as they do their jobs. In fact, research shows that the behavior of managers is the single most important influence on an individual's decision to act ethically or unethically.[40] People look to see what those in authority are doing and use that as a benchmark for acceptable practices and expectations.

Finally, as we discussed in Chapter 3, a strong culture exerts more influence on employees than a weak one. A culture that is strong and supports high ethical standards has a very powerful and positive influence on the decision to act ethically or

values
Basic convictions about what is right and what is wrong.

ego strength
A personality measure of the strength of a person's convictions.

locus of control
A personality attribute that measures the degree to which people believe they control their own fate.

values-based management
A form of management in which an organization's values guide employees in the way they do their jobs.

unethically. For example, IBM has a strong culture that has long stressed ethical dealings with customers, employees, business partners, and communities.[41] To reinforce the importance of ethical behaviors, the company developed an explicitly detailed set of guidelines for business conduct and ethics. And the penalty for violating the guidelines: disciplinary actions, including dismissal. IBM's managers continually reinforce the importance of ethical behavior and reinforce the fact that a person's actions and decisions are important to the way the organization is viewed.

Issue Intensity. A student who would never consider breaking into an instructor's office to steal an accounting exam may not think twice about asking a friend who took the same course from the same instructor last semester what questions were on an exam. Similarly, a manager might think nothing about taking home a few office supplies yet be highly concerned about the possible embezzlement of company funds. These examples illustrate the final factor that influences ethical behavior: the intensity of the ethical issue itself.[42]

As Exhibit 5–7 shows, six characteristics determine issue intensity or how important an ethical issue is to an individual: greatness of harm, consensus of wrong, probability of harm, immediacy of consequences, proximity to victim(s), and concentration of effect. These factors suggest that the greater the number of people harmed, the more agreement that the action is wrong; the greater the likelihood that the action will cause harm, the more immediately that the consequences of the action will be felt; and the closer the person feels to the victim(s) and the more concentrated the effect of the action on the victim(s), the greater the issue intensity or importance. When an ethical issue is important, employees are more likely to behave ethically.

ETHICS IN AN INTERNATIONAL CONTEXT

Are ethical standards universal? Although some common moral beliefs exist, social and cultural differences between countries are important factors that determine ethical and unethical behavior.[43] For example, say that a manager in a Mexican firm bribes several high-ranking government officials in Mexico City to secure a profitable government contract. Although this business practice is unethical (and illegal) in the United States, it's acceptable in Mexico.

Should Coca-Cola employees in Saudi Arabia adhere to U.S. ethical standards, or should they follow local standards of acceptable behavior? If Airbus (a European company) pays a "broker's fee" to a middleman to get a major contract with a

Exhibit 5–7

Issue Intensity

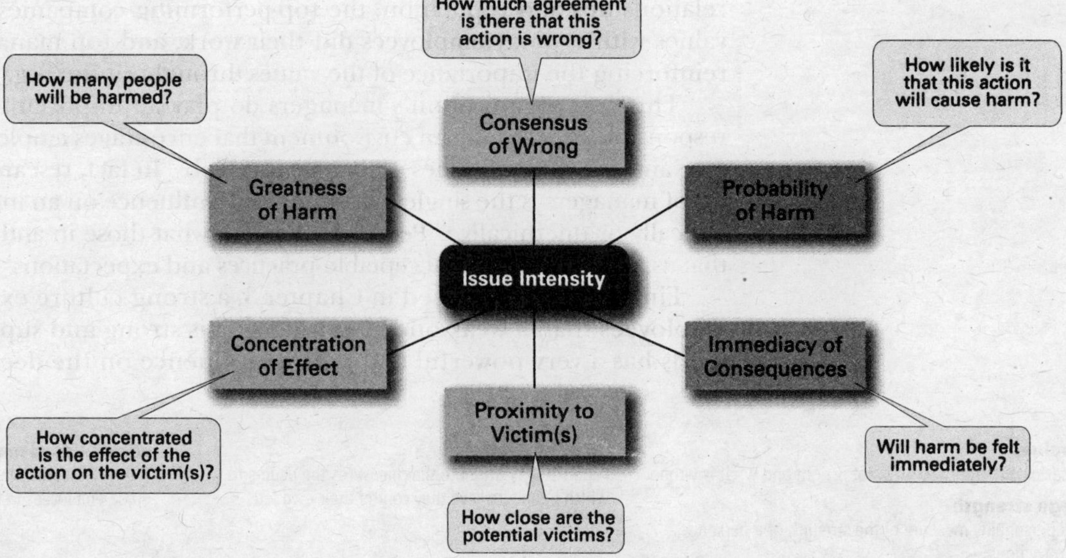

Middle Eastern airline, should Boeing be restricted from doing the same because such practices are considered improper in the United States? (Note that in the United Kingdom, the Law Commission, a governmental advisory body, has said that bribing officials in foreign countries should be a criminal offense. It said that claims of "it's local custom" should not be a reason for allowing it.)[44]

In the case of payments to influence foreign officials or politicians, laws of the host country can make it illegal to knowingly bribe an official. However, even when there is a law to help make the decision that the ethical dilemma is not black or white. In some cases, government bureaucrats' salaries are low and they receive small payments from whom they serve. Payoffs to these officers "grease the machinery" and ensure that things get done. No law or rule expressly prohibits small payoffs to government employees as gifts or money to get the work done if it is a part of the corporate ethos.

It's important for individual managers working in foreign cultures to recognize the social, cultural, and political/legal influences on what is appropriate and acceptable behavior.[45] And international businesses must clarify their ethical guidelines so that employees know what's expected of them while working in a foreign location, which adds another dimension to making ethical judgments.

Another guide to being ethical in international business is the Global Compact, which is a document created by the United Nations that outlines principles for doing business globally in the areas of human rights, labor, and the environment and anti-corruption (see Exhibit 5–8). "More than 3,000 CEOs have signed the Compact, making it the world's largest voluntary corporate citizenship initiative."[46] The goal of the Compact is a more sustainable and inclusive global economy. Organizations making this commitment do so because they believe that the world business community plays a significant role in improving economic and social conditions. In addition, the Organisation for Economic Co-operation and Development (OECD) has made fighting bribery and corruption in international business a high priority. The centerpiece of its efforts is the Anti-Bribery Convention (or set of rules and guidelines), which was the first global instrument to combat corruption in cross-border business deals. To date, significant gains have been made in fighting corruption in the 37 countries that have ratified it.[47]

Exhibit 5–8

Ten Principles of the UN

Human Rights

Principle 1:	Support and respect the protection of international human rights within their sphere of influence.
Principle 2:	Make sure business corporations are not complicit in human rights abuses.
Labor Standards	
Principle 3:	Freedom of association and the effective recognition of the right to collective bargaining.
Principle 4:	The elimination of all forms of forced and compulsory labor.
Principle 5:	The effective abolition of child labor.
Principle 6:	The elimination of discrimination in respect of employment and occupation.
Environment	
Principle 7:	Support a precautionary approach to environmental challenges.
Principle 8:	Undertake initiatives to promote greater environmental responsibility.
Principle 9:	Encourage the development and diffusion of environmentally friendly technologies.
Principle 10:	Businesses should work against corruption in all its forms, including extortion and bribery.

Source: Courtesy of UN Global Compact.

QUICK LEARNING REVIEW:
LEARNING OUTCOME 5.3

- **Define ethics.**
- **Discuss the factors that influence whether a person behaves ethically or unethically.**
- **Describe what managers need to know about international ethics.**

Go to page 115 and see how well you know this material.

LEARNING **OUTCOME 5.4** ▷

ENCOURAGING ETHICAL BEHAVIOR

Managers can do a number of things if they're serious about encouraging ethical behaviors—hire employees with high ethical standards, establish codes of ethics, lead by example, and so forth. By themselves, such actions won't have much of an impact. But having a comprehensive ethics program in place can potentially improve an organization's ethical climate. The key variable, however, is *potentially.* There are no guarantees that a well-designed ethics program will lead to the desired outcome. Sometimes corporate ethics programs are little more than public relations gestures that do little to influence managers and employees. For instance, Infosys has been very vocal about its ethical practices. However, its ethical ethos could not stop managers of certain locations to curry favors to the bureaucrats who gave them permissions to set up business or prying into the personal lives of the employees, such as those who were having an affair, and then selectively firing employees based on such information or treating service providers differentially based on the service providers' size and reputation. Even Enron, often referred to as the "poster child" of corporate wrongdoing, outlined values in its 2000 Annual Report that most would consider ethical—communication, respect, integrity, and excellence. Yet the way top managers behaved didn't reflect those values at all.[48] Let's look at some specific ways that managers can encourage ethical behavior and create a comprehensive ethics program.

EMPLOYEE SELECTION

The selection process (interviews, tests, background checks, and so forth) should be viewed as an opportunity to learn about an individual's level of moral development, personal values, ego strength, and locus of control.[49] But even a carefully designed selection process isn't foolproof! Even under the best circumstances, individuals with questionable standards of right and wrong may be hired. However, this shouldn't be a problem if other ethics controls are in place.

CODES OF ETHICS AND DECISION RULES

Since 1999, the "Tata Code of Conduct," which is simple, easy to understand, and easy to follow, has been communicated to all employees of the organization.[50] The content of the code covers areas such as commitment toward national interest, maintaining harmonious relations with the employees, abhorrence of bribery and corruption, avoidance of conflicts of interest, and emphasis on corporate social responsibility. However, that's not the way it is in all organizations.

Uncertainty about what is and is not ethical can be a problem for employees. A **code of ethics**, a formal statement of an organization's values and the ethical rules it expects employees to follow, is a popular choice for reducing that ambiguity. Research shows that 97 percent of organizations with more than 10,000 employees have written codes of ethics. Even in smaller organizations, nearly 93 percent have them.[51] And codes of ethics are becoming more popular globally. Research by the Institute for Global Ethics says that shared values such as honesty, fairness, respect, responsibility, and caring are pretty much universally embraced worldwide.[52] In addition, a survey of businesses in 22 countries found that 78 percent have formally stated ethics standards and codes of ethics.[53]

What should a code of ethics look like? It should be specific enough to show employees the spirit in which they're supposed to do things yet loose enough to allow

Exhibit 5–9

Codes of Ethics

Cluster 1. Be a Dependable Organizational Citizen

1. Comply with safety, health, and security regulations.
2. Demonstrate courtesy, respect, honesty, and fairness.
3. Illegal drugs and alcohol at work are prohibited.
4. Manage personal finances well.
5. Exhibit good attendance and punctuality.
6. Follow directives of supervisors.
7. Do not use abusive language.
8. Dress in business attire.
9. Firearms at work are prohibited.

Cluster 2. Do Not Do Anything Unlawful or Improper That Will Harm the Organization

1. Conduct business in compliance with all laws.
2. Payments for unlawful purposes are prohibited.
3. Bribes are prohibited.
4. Avoid outside activities that impair duties.
5. Maintain confidentiality of records.
6. Comply with all antitrust and trade regulations.
7. Comply with all accounting rules and controls.
8. Do not use company property for personal benefit.
9. Employees are personally accountable for company funds.
10. Do not propagate false or misleading information.
11. Make decisions without regard for personal gain.

Cluster 3. Be Good to Customers

1. Convey true claims in product advertisements.
2. Perform assigned duties to the best of your ability.
3. Provide products and services of the highest quality.

Source: F. R. David, "An Empirical Study of Codes of Business Ethics: A Strategic Perspective," paper presented at the 48th Annual Academy of Management Conference, Anaheim, California, August 1988.

for freedom of judgment. A survey of companies' codes of ethics found that their content tended to fall into three categories, as shown in Exhibit 5–9.[54]

Unfortunately, codes of ethics don't appear to work very well. A survey of employees in U.S. businesses found that 56 percent of those surveyed had observed ethical or legal violations in the previous 12 months, including such things as conflicts of interest, abusive or intimidating behavior, and lying to employees. And 42 percent of those employees don't report observed misconduct. Even in companies with comprehensive ethics and compliance programs, 29 percent failed to report misconduct.[55] Does this mean that codes of ethics shouldn't be developed? No. However, in doing so, managers should use these suggestions:[56]

1. Organizational leaders should model appropriate behavior and reward those who act ethically.
2. All managers should continually reaffirm the importance of the ethics code and consistently discipline those who break it.

code of ethics
A formal statement of an organization's primary values and the ethical rules it expects its employees to follow.

3. The organization's stakeholders (employees, customers, and so forth) should be considered as an ethics code is developed or improved.
4. Managers should communicate and reinforce the ethics code regularly.
5. Managers should use the 12 questions approach (see Exhibit 5–10) to guide employees when faced with ethical dilemmas.[57]

TOP MANAGEMENT'S LEADERSHIP

Doing business ethically requires a commitment from top managers. Why? Because they're the ones who uphold the shared values and set the cultural tone. They're role models in terms of both words and actions, though what they *do* is far more important than what they *say*. If top managers, for example, take company resources for their personal use, inflate their expense accounts, or give favored treatment to friends, they imply that such behavior is acceptable for all employees.

Top managers also set the tone with their reward and punishment practices. The choices of whom and what are rewarded with pay increases and promotions send a strong signal to employees. As we said earlier, when an employee is rewarded for achieving impressive results in an ethically questionable manner, it indicates to others that those ways are acceptable. When an employee does something unethical, managers must punish the offender and publicize the fact by making the outcome visible to everyone in the organization. This practice sends a message that doing wrong has a price, and it's not in employees' best interests to act unethically.

JOB GOALS AND PERFORMANCE APPRAISAL

Employees in three Internal Revenue Service offices were found in the bathrooms flushing tax returns and other related documents down the toilets. When questioned, they openly admitted doing it but offered an interesting explanation for their behavior. The employees' supervisors had been pressuring them to complete more work in less time. If the piles of tax returns weren't processed and moved off their desks more quickly, they were told their performance reviews and salary raises would be adversely affected. Frustrated because they had few resources and an overworked computer system, the employees decided to "flush away" the paperwork on their desks, even though they knew what they were doing was wrong. This story illustrates how powerful unrealistic goals and performance appraisals can be.[58]

Exhibit 5–10

12 Questions Approach

1. Have you defined the problem accurately?
2. How would you define the problem if you stood on the other side of the fence?
3. How did this situation occur in the first place?
4. To whom and to what do you give your loyalty as a person and as a member of the corporation?
5. What is your intention in making this decision?
6. How does this intention compare with the probable results?
7. Whom could your decision or action injure?
8. Can you discuss the problem with the affected parties before you make the decision?
9. Are you confident that your position will be as valid over a long period of time as it seems now?
10. Could you disclose without qualm your decision or action to your boss, your chief executive officer, the board of directors, your family, society as a whole?
11. What is the symbolic potential of your action if understood? If misunderstood?
12. Under what conditions would you allow exceptions to your stand?

Source: Reprinted by permission of *Harvard Business Review*. An exhibit from "Ethics Without the Sermon," by L. L. Nash. November–December 1981, p. 81.

Under the stress of unrealistic goals, otherwise ethical employees may feel that they have no choice but to do whatever is necessary to meet those goals. Also, goal achievement is usually a key issue in performance appraisal. If performance appraisals focus only on economic goals, ends will begin to justify means. To encourage ethical behavior, both ends and means should be evaluated. For example, a manager's annual review of employees might include a point-by-point evaluation of how their decisions measured up against the company's code of ethics as well as how well goals were met.

ETHICS TRAINING

Organizations are increasingly setting up seminars, workshops, and similar ethics training programs to encourage ethical behavior. Such training programs aren't without controversy; the primary concern is whether ethics can be taught. Critics stress that the effort is pointless because people establish their individual value systems when they're young. Proponents note, however, that several studies have shown that values can be learned after early childhood. In addition, they cite evidence which shows that teaching ethical problem solving can make a difference in ethical behaviors;[59] that training has increased individuals' level of moral development;[60] and that, if nothing else, ethics training increases awareness of ethical issues in business.[61]

How can ethics be taught? Let's look at an example involving global defense contractor Lockheed Martin, one of the pioneers in the case-based approach to ethics training.[62] Lockheed Martin's employees take annual ethics training courses delivered by their managers. The main focus of these short courses is Lockheed Martin–specific case situations "chosen for their relevance to department or job-specific issues." In each department, employee teams review and discuss the cases and then apply an "Ethics Meter" to "rate whether the real-life decisions were ethical, unethical, or somewhere in between." For example, one of the possible ratings on the Ethics Meter, "On Thin Ice," is explained as "bordering on unethical and should raise a red flag." After the teams have applied their ratings, managers lead discussions about the ratings and examine "which of Lockheed Martin's core ethics principles were applied or ignored in the cases." In addition to its ethics training, Lockheed Martin has a widely used written code of ethics, an ethics helpline that employees can call for guidance on ethical issues, and ethics officers based in the company's various business units.

INDEPENDENT SOCIAL AUDITS

The fear of being caught can be an important deterrent to unethical behavior. Independent social audits, which evaluate decisions and management practices in terms of the organization's code of ethics, increase that likelihood. Such audits can

For the last 30 years, the Barefoot College, located in Tilonia, has been working to improve the quality of life of the rural poor. Those who work at the Barefoot College follow a code of conduct which includes living and working in close proximity with the rural community; not discriminating on the basis of caste, religion, or political thinking; ensuring gender equality within the organization; having respect for collective, traditional knowledge, beliefs, wisdom, and practices of the community; being committed to the preservation of natural resources and not endorse processes that destroy, exploit, or abuse natural resources; and using appropriate technologies that sustain the community and not encouraging technologies that deprive people of their livelihoods. All people living in the campus model the code of conduct. The entire campus is lighted with solar power and harvests 700,000 liter of rainwater every year.

be regular evaluations, or they can occur randomly, with no prior announcement. An effective ethics program probably needs both. To maintain integrity, auditors should be responsible to the company's board of directors and present their findings directly to the board. This arrangement gives the auditors clout and reduces the opportunity for retaliation from those being audited. Because the Sarbanes-Oxley Act holds businesses to more rigorous standards of financial disclosure and corporate governance, more organizations are finding the idea of independent social audits appealing. As the publisher of *Business Ethics* magazine stated, "The debate has shifted from *whether* to be ethical to *how* to be ethical."[63]

PROTECTIVE MECHANISMS

Employees who face ethical dilemmas need protective mechanisms so they can do what's right without fear of reprimand. An organization might designate ethical counselors for employees facing an ethics dilemma. These advisors also might advocate the ethically "right" alternatives. Other organizations have appointed ethics officers who design, direct, and modify the organization's ethics programs as needed.[64] The Ethics and Compliance Officer Association reports its total membership at over 1,300 (including more than half of the *Fortune* 100 companies) and covering several countries, including the United States, Germany, India, Japan, and Canada.[65]

QUICK LEARNING REVIEW:

LEARNING OUTCOME 5.4

- **Describe managers' important role in encouraging ethical behavior.**
- **Discuss specific ways managers can encourage ethical behavior.**

Go to page 116 and see how well you know this material.

LEARNING **OUTCOME 5.5** ▷

SOCIAL RESPONSIBILITY AND ETHICS ISSUES IN TODAY'S WORLD

Today's managers continue to face challenges in being socially responsible and ethical. Next we examine three current issues: managing ethical lapses and social irresponsibility, encouraging social entrepreneurship, and promoting positive social change.

MANAGING ETHICAL LAPSES AND SOCIAL IRRESPONSIBILITY

Even after public outrage over the Enron-era misdeeds, irresponsible and unethical practices by managers in all kinds of organizations haven't gone away. What's more alarming is what's going on "in the trenches" in offices, warehouses, and stores. One survey reported that of more than 5,000 employees, 45 percent admitted having fallen asleep at work; 22 percent said they had spread a rumor about a coworker; 18 percent said they had snooped after hours; and 2 percent said that they had taken credit for someone else's work.[66]

Unfortunately, it's not just at work that we see such behaviors. They're prevalent throughout society. Studies conducted by the Center for Academic Integrity showed that 26 percent of college and university business majors admitted to "serious cheating" on exams, and 54 percent admitted to cheating on written assignments. But business students weren't the worst cheaters; that distinction belonged to journalism majors, of whom 27 percent said they had cheated.[67] And a survey by Students in Free Enterprise (SIFE) found that only 19 percent of students would report a classmate who cheated.[68] But even more frightening is what today's teenagers say is "acceptable." In a survey, 23 percent said they thought violence toward another person is acceptable on some level.[69] What do such statistics say about what managers may have to deal with in the future? It's not too far-fetched to say that organizations may have

Let's Get Real: F2F

AN ETHICAL LEADER IS:
An ethical leader sets examples in the best way to deal with the moral dilemmas. He or she clearly demonstrates the balancing act between difficult choices and organizational goals. At every opportunity, the leader promotes the ethical code of the organization and makes all efforts to ensure that it is internalized by every employee.

difficulty upholding high ethical standards when their future employees so readily accept unethical behavior.

What can managers do? Two actions seem to be particularly important: ethical leadership and protection of those who report wrongdoing.

Ethical Leadership. Not long after Herb Baum took over as CEO of Dial Corporation, he got a call from Reuben Mark, the CEO of competitor Colgate-Palmolive, who told him he had a copy of Dial's strategic marketing plan that had come from a former Dial salesperson who had recently joined Colgate-Palmolive. Mark told Baum that he had not looked at it, didn't intend to look at it, and was returning it. In addition, he himself was going to deal appropriately with the new salesperson.[70] As this example illustrates, managers must provide ethical leadership. As we said earlier, what managers *do* has a strong influence on employees' decisions to behave ethically or not. When managers cheat, lie, steal, manipulate, take advantage of situations or people, or treat others unfairly, what kind of signal are they sending to employees (or other stakeholders)? Probably not the one they want to send. Exhibit 5–11 gives some suggestions on how managers can provide ethical leadership.

Protection of Employees Who Raise Ethical Issues. What would you do if you saw other employees doing something illegal, immoral, or unethical? Would you step forward? Many of us wouldn't because of the perceived risks. That's why it's important for managers to assure employees who raise ethical concerns or issues that they will face no personal or career risks. These individuals, often called **whistle-blowers**, can be a key part of any company's ethics program. One well-known whistle-blower in recent memory is Sherron Watkins, a vice president at Enron who clearly outlined her concerns about the company's accounting practices in a letter to chairman Ken Lay. Her statement "I am incredibly nervous that we will implode in a wave of accounting scandals" couldn't have been more prophetic.[71] However, surveys show that most observers of wrongdoing don't report it, and that's the attitude managers have to address.[72] How can they protect employees so they're willing to step up if they see unethical or illegal things occurring?

One way is to set up toll-free ethics hotlines. For instance, Dell has an ethics hotline that employees can call anonymously to report infractions that the company will then investigate.[73] In addition, managers need to create a culture where bad news can be heard and acted on before it's too late. Michael Josephson, founder of the Josephson Institute of ethics (www.josephsoninstitute.org), said, "It is absolutely and unequivocally important to establish a culture where it is possible for employees to complain and protest and to get heard."[74] Even if some whistle-blowers have a personal agenda they're pursuing, it's important to take them seriously. Finally, the Sarbanes-Oxley Act offers some legal protection. Any manager who retaliates against an employee for reporting violations faces a stiff penalty: a 10-year jail sentence.[75] Unfortunately, despite this protection, hundreds of employees who have stepped forward and revealed wrongdoings at their companies have been fired or let go from their jobs.[76] So at the present time, it's not a perfect solution, but it is a step in the right direction.

Exhibit 5–11
Being an Ethical Leader

- Be a good role model by being ethical and honest.
  - Tell the truth always.
  - Don't hide or manipulate information
  - Be willing to admit your failures.
- Share your personal values by regularly communicating them to employees.
- Stress the organization's or team's important shared values.
- Use the reward system to hold everyone accountable to the values.

whistle-blower
An individual who raises ethical concerns or issues to others.

ENCOURAGING SOCIAL ENTREPRENEURSHIP

The world's social problems are many, and viable solutions are few. But numerous people and organizations are trying to do something. For instance, Teresa Fritschi, James Potemkin, and Kirit Shroff share a common bond even though they don't know each other. Each sells unique handmade items made by artisans from different parts of the globe—Scotland, Mexico, Guatemala, India, Pakistan, Peru. But they also share a passionate belief in fair trade and act on this belief by paying their supplier artisans more than the going rate for their works. Fair trade proponents seek to "give businesses or solo artists in poor or marginalized parts of the world a higher price for what they create and a more direct route into lucrative markets in America, Europe, and Asia."[77] Each of these individuals is also an example of a **social entrepreneur**, an individual or organization who seeks out opportunities to improve society by using practical, innovative, and sustainable approaches.[78] "What business entrepreneurs are to the economy, social entrepreneurs are to social change."[79] Social entrepreneurs want to make the world a better place and have a driving passion to make that happen. For instance, Devi Prasad Shetty strives to make sophisticated healthcare available to all, irrespective of their economic situation or geographic location. He founded the Narayana Hrudayalaya Hospital in Bangalore in 2001. He has built a network of 39 telemedicine centers to reach out to patients in remote rural areas. These hospitals perform 32 heart surgeries a day. Almost half the patients are children and babies. Sixty percent of the treatments are provided below cost or for free.[80,81] Aravind Eye Care Hospital in Madurai set out to eliminate unnecessary blindness provides free, high-quality eye care to 66 percent of the patients. It manufactures interocular lenses to be able to make eye care affordable to the rural poor. Dr Venkataswamy, who started these hospitals at the age of 56, was determined to provide the best eye care to the rural poor.[82] Also, social entrepreneurs use creativity and ingenuity to solve problems. For example, Give India foundation was set up to match those who wished to give money to credible organizations. Give India has reviewed over 1,000 nonprofit organizations from all over India to identify nearly 100 organizations that work for causes ranging from child welfare and education to disability, poverty, and women's empowerment that have met with the Credibility Alliance's norms.[83] Also, social entrepreneur use creativity and ingenuity to solve problems. For example, Nidan, started by Arbind Singh, has helped to develop sustainable businesses, cooperatives, trade unions, and "people's institutions" led by the most excluded categories of the poor in Bihar. It has promoted and built 20 independent profit-making ventures governed and owned by the urban poor including waste workers, ragpickers, vegetable vendors, construction laborers, domestic helpers, microfarmers, street traders, and other marginalized occupational groups. Another social enterprise run by Prema Gopalan, Swayam Shikshan Prayog (SSP), is building networks of rural "social businesses" that are cocreated by private corporations and women survivors of disasters, such as the 2004 Asian Tsunami and the Latur and Gujarat earthquakes (of 1993 and 2001, respectively). SSP has facilitated networks of rural women entrepreneurs to launch retail businesses in renewable home energy products, home groceries, and health funds in partnership with BP (previously known as British Petroleum), LIC, and others.[84]

What can we learn from social entrepreneurs? Although many organizations have committed to doing business ethically and responsibly, perhaps there is more they can do, as social entrepreneurs show. Maybe, as in the case of Give India Foundation, it's simply a matter of business organizations collaborating with public groups or nonprofit organizations to address a social issue. Or maybe, as in the case of the Aravind Eye Care Hospital, it's providing expertise where needed. Or, it may involve nurturing individuals who passionately and unwaveringly believe they have an idea that could make the world a better place and simply need the organizational support to pursue it.

BUSINESSES PROMOTING POSITIVE SOCIAL CHANGE

Since the last three decades, HDFC contributes 7 percent of its income to support community needs.[85] And it's not alone in those efforts. "Over the past two decades, a growing number of corporations, both within and beyond India, have been engaging in activities that promote positive social change."[86] Businesses can do this in a couple ways: through corporate philanthropy and through employee volunteering.

Corporate Philanthropy. Corporate philanthropy can be an effective way for companies to address societal problems.[87] ICICI donated to Pratham, an initiative to help municipal school children living in slums with their education, Mahindra Tech employees donated one-day salary to help victims of Bihar floods, and Infosys donates to the mid-day meal provided by ISCON in Bangalore. ACC undertook social volunteering practices almost from its inception; the company's earliest initiatives in community development date backs to the 1940s in a village on the outskirts of Mumbai while the first formal Village Welfare Scheme was launched in 1952.[88] Others have funded their own foundations through to support various social issues. For example, Wipro has set up a foundation, named Azim Premji Foundation, to help improve education of the elementary schools in rural India. Google has set up a foundation called DotOrg to support five areas: developing systems to help predict and prevent disease pandemics, empowering the poor with information about public services, creating jobs by investing in small and midsize businesses in developing countries, accelerating the commercialization of plug-in cars, and making renewable energy cheaper than coal.[89]

Employee Volunteering Efforts. Employee volunteering is a popular way for businesses to be involved in promoting social change. For example, Satyam encourages its employees to volunteer its time for its social projects such as research behind the Emergency 108 number. Also in the quarter ended September 30, 2008, Infosys announced that those employees who have worked for Infosys for the last 2 years could opt for a 1-year sabbatical to engage themselves in philanthropic activities. They would continue to draw 50 percent of their salary during the period while the other half will be given by the respective NGOs they work with.[90] HSBC launched Employee Volunteering in India in 2007 in a new format to enhance volunteering. The program has been named "Catalyst" and is being deployed across HSBC branches and Group Companies in India.[91] Many businesses have found that such efforts not only benefit communities but enhance employees' work efforts and motivation.

QUICK LEARNING REVIEW:

LEARNING OUTCOME 5.5

- Describe how managers can manage ethical lapses and social irresponsibility.
- Explain the role of social entrepreneurs.
- Discuss how businesses can promote positive social change.

Go to page 116 and see how well you know this material.

social entrepreneur
An individual or organization who seeks out opportunities to improve society by using practical, innovative, and sustainable approaches.

Let's Get Real: My Turn

Dr G. S. Rao

Vice President
Lavasa Corporation Ltd

The most important thing to realize is that profits come in different forms. While the organization wishes to maximize its economic returns, there are commitments also toward environmental and social objectives. The desire to make profits should not supersede the environmental and ecological objectives. (For example, restoring the glory of natural vegetation, which may have been damaged due to company activities, is a case in point. Although this involves additional expenditure, it is beneficial in the long run.) An often-neglected aspect is that of intangible benefits. The manager must emphasize that the company values intangible benefits as much as the monetary economic benefits. Stories of good practices can be told to the new managers and awards can be instituted for those young managers who help the organization to think green.

LEARNING OUTCOMES SUMMARY

5.1 ▷ WHAT IS SOCIAL RESPONSIBILITY?

- Differentiate between social obligation, social responsiveness, and social responsibility.
- Discuss whether organizations should be socially involved.
- Describe what conclusion can be reached regarding social involvement and economic performance.

Social obligation, which reflects the classical view of social responsibility, involves a firm engaging in social actions because of its obligation to meet certain economic and legal responsibilities. Social responsiveness involves a firm engaging in social actions in response to some popular social need. Social responsibility is a business's intention, beyond its economic and legal obligations, to pursue long-term goals that are good for society. Both of these reflect the socioeconomic view of social responsibility. Determining whether organizations should be socially involved can be done by looking at arguments for and against it. Other ways are to assess the impact of social involvement on a company's economic performance and evaluate the performance of SRI funds versus non-SRI funds. Based on such information, we can conclude that a company's being socially responsible doesn't appear to hurt its economic performance.

5.2 ▷ GREEN MANAGEMENT

- Define *green management*.
- Describe how organizations can go green.
- Explain how green management actions can be evaluated.

With green management, managers consider the impact of their organization on the natural environment. Organizations can "go green" in different ways. The light green approach involves doing what is required legally, which is social obligation. Using the market approach, organizations respond to the environmental preferences of their customers. Using the stakeholder approach, organizations respond to the environmental demands of multiple stakeholders. Both the market and stakeholder approaches can be viewed as social responsiveness. The activist, or dark green, approach involves an organization looking for ways to respect and preserve the earth and its natural resources, which can be viewed as social responsibility.

Green actions can be evaluated by examining reports that companies compile about their environmental performance, by looking for compliance with global standards for environmental management (ISO 14000), and by using the Global 100 list of the most sustainable corporations in the world.

5.3 ▷ MANAGERS AND ETHICAL BEHAVIOR

- Define *ethics*.
- Discuss the factors that influence whether a person behaves ethically or unethically.
- Describe what managers need to know about international ethics.

Ethics refers to the principles, values, and beliefs that define right and wrong decisions and behavior. The factors that affect ethical and unethical behavior include an individual's level of moral development (preconventional, conventional, or principled); individual characteristics (values and personality variables—ego strength and locus of control); structural variables (structural design, use of goals, performance appraisal systems, and reward allocation procedures); organizational culture (shared values and cultural strength); and issue intensity (greatness of harm, consensus of wrong, probability of harm, immediacy of consequences, proximity to victims, and concentration of effect).

Because ethical standards aren't universal, managers should know what they can and cannot do legally, according to the Foreign Corrupt Practices Act. It's also important to recognize any cultural differences and to clarify ethical guidelines for employees working in different global locations. Finally, managers should know about the principles of the Global Compact and the Anti-Bribery Convention.

5.4 ▷ ENCOURAGING ETHICAL BEHAVIOR

- Describe managers' important role in encouraging ethical behavior.
- Discuss specific ways managers can encourage ethical behavior.

The behavior of managers is the single most important influence on an individual's decision to act ethically or unethically. Some specific ways managers can encourage ethical behavior include paying attention to employee selection, having and using a code of ethics, recognizing the important ethical leadership role they play and how what they do is far more important than what they say, making sure that goals and the performance appraisal process don't reward goal achievement without taking into account how goals were achieved, using ethics training and independent social audits, and establishing protective mechanisms.

5.5 ▷ SOCIAL RESPONSIBILITY AND ETHICS ISSUES IN TODAY'S WORLD

- Describe how managers can manage ethical lapses and social irresponsibility.
- Explain the role of social entrepreneurs.
- Discuss how businesses can promote positive social change.

Managers can manage ethical lapses and social irresponsibility by being strong ethical leaders and by protecting employees who raise ethical issues. The example set by managers has a strong influence on whether employees behave ethically. Ethical leaders also are honest, share their values, stress important shared values, and use the reward system appropriately. Managers can protect whistle-blowers (employees who raise ethical issues or concerns) by encouraging them to come forward, by setting up toll-free ethics hotlines, and by establishing a culture where employees can complain and get heard without fear of reprisal. Social entrepreneurs play an important role in solving social problems by seeking out opportunities to improve society by using practical, innovative, and sustainable approaches. Social entrepreneurs want to make the world a better place and have a driving passion to make that happen. Businesses can promote positive social change through corporate philanthropy and employee volunteering efforts.

THINKING ABOUT MANAGEMENT ISSUES

1. What does social responsibility mean to you personally? Do *you* think business organizations should be socially responsible? Explain.
2. Do you think values-based management is just a "do-gooder" ploy? Explain your answer.
3. Internet file sharing programs are popular among college students. These programs work by allowing non-organizational users to access any local network where desired files are located. These types of file sharing programs tend to clog bandwidth and reduce local users' ability to access and use a local network. What ethical and social responsibilities does a university have in such a situation? To whom does it have a responsibility? What guidelines might you suggest for university decision makers?
4. What are some problems that could be associated with employee whistle-blowing for (a) the whistle-blower and (b) the organization?
5. Describe the characteristics and behaviors of someone you consider to be an ethical person. How could the types of decisions and actions this person engages in be encouraged in a workplace?
6. This question was posed in an article in the October 10, 2006, issue of *USA Today*: "Is capitalism going to be the salvation of the world or the cause of its demise?" Discuss.

YOUR TURN to be a Manager

- **Find five different examples of organizational codes of ethics. Using Exhibit 5–9, describe what each contains. Compare and contrast the examples.**
- **Using the examples of codes of ethics you found, create what you feel would be an appropriate and effective organizational code of ethics. In addition, create your own *personal code of ethics* that you can use as a guide to ethical dilemmas.**

- Start a portfolio that contains each of the "Thinking Critically About Ethics" dilemmas found in each chapter. Write a response to each of the dilemmas and include these responses in your portfolio.
- Take advantage of volunteer opportunities and be sure to include them on your résumé. If possible, try to do things in volunteer positions that will improve your managerial skills in planning, organizing, leading, or controlling.
- Go to the Global Reporting Initiative Web site, www.globalreporting.org, and choose three businesses from the list of organizations that have filed reports. Look at those reports and describe/evaluate what's in them. In addition, identify the stakeholders that might be affected and how they might be affected by the company's actions.
- Find out what green management activities your school or employer is doing and write up a list of them. Do some research on being green. Are there additional things your school or employer could be doing? Write a report to your school or employer, describing any suggestions. (Also look for ways that you can be green in your personal life.)
- Over the course of two weeks, see what ethical "dilemmas" you observe. These could be ones that you personally face, or they could be ones that others (friends, colleagues, other students talking in the hallway or before class starts, and so forth) face. List these dilemmas and think about what you might do if faced with each one.
- Interview two different managers about how they encourage their employees to be ethical. Write down their comments and discuss how these ideas might help you be a better manager.
- Steve's and Mary's suggested readings: Bethany McLean and Peter Elkind, *The Smartest Guys in the Room: The Amazing Rise and Scandalous Fall of Enron* (Portfolio, 2003); Barbara Ley Toffler, *Final Accounting: Ambition, Greed, and the Fall of Arthur Andersen* (Broadway Books, 2003); Joseph L. Badaracco, Jr., *Leading Quietly: An Unorthodox Guide to Doing the Right Thing* (Harvard Business School Press, 2002); and Kenneth Blanchard and Norman Vincent Peale, *The Power of Ethical Management* (Morrow, 1988).
- If you have the opportunity, take a class on ethics (business or management) or on social responsibility—often called business and society—or both. Not only will this look good on your résumé, it could help you personally grapple with some of the tough issues managers face in being ethical and responsible.
- In your own words, write down three things you learned in this chapter about being a good manager.
- Self-knowledge can be a powerful learning tool. Go to mymanagementlab and complete these self-assessment exercises: What Do I Value? How Do My Ethics Rate? Do I Trust Others? and Do Others See Me as Trusting? Using the results of your assessments, identify personal strengths and weaknesses. What will you do to reinforce your strengths and improve your weaknesses?

CASE APPLICATION

Not Just Another Outdoor Company

We opened the chapter with a story about an outdoor company, and we end it with a story about another outdoor company. The company we're discussing this time, based in Portland, Oregon, was the brainchild of a small group of executives who left big-time jobs at Patagonia, Nike, and Adidas. These individuals shared a belief that "in addition to generating a profit, companies have an equal responsibility to create positive social and environmental change." Putting their beliefs into action, the group formed Nau (which is Maori for "Welcome! Come in"). And Nau is not just another outdoor company!

When deciding what Nau was going to be like and how it was going to do business, the founders knew they didn't want to do things the way they'd always been done by traditional businesses. CEO Chris Van Dyke said, "We started with a clean whiteboard. We believed every single operational element in our business was an opportunity to turn traditional business notions inside out, integrating environmental, social, and economic factors." From design to sales to finances, Nau is driven by these factors. Everything in Nau's operation has been approached with a sustainability and social justice "filter."

Smith Rock State Park, Oregon.

In the design area, the company, in partnership with its suppliers, developed 24 of its 32 fabrics to be more sustainable and to combine performance and visual appeal. Each supplier, manufacturer, and even Nau itself is bound by a code of conduct. To ensure that all parties are living up to the standards, their actions are overseen by an independent, nonprofit auditing and research firm. In the sales area, the way the company retails its product is also unique. Using a concept it calls a "Web-front," Nau has combined the efficiency of the Web with the intimacy of a gallery-like boutique. In the "store," customers can try on clothes, but they use self-serve kiosks to purchase from the Web. Because in-store inventory is greatly reduced, the stores are small (2,400 square feet compared to the standard 4,000-plus-square-foot outdoor retail store). This approach saves operating expenses because less energy and fewer materials are used. Good for the planet ... good for the business. Finally, Nau has a unique financial approach it calls "aggressive altruism." The company has pledged 5 percent of sales to charitable organizations dedicated to solving crucial environmental and humanitarian problems. The "philanthropic gold standard" is 1 percent of sales, and the average among all corporations is .047 percent. But although the amount it gives is unusual, what happens with Nau's dollars is really exceptional: Nau puts the giving decision in the hands of its customers. They're asked to indicate which "Partners for Change" they'd like their 5 percent to go to. Using this "conscious choice" process, Nau is "calling its customers out, daring them to connect the dots."

Discussion Questions

1. What do you think of Nau's approach to doing business? Is it being ethical and responsible? Discuss.
2. Will Nau's approach have a limited appeal, or do you think it has staying power? What drawbacks might there be to what Nau is doing?
3. Is it a business's responsibility to get customers to "connect the dots" and make choices about social issues?
4. Are there lessons here for other businesses? Discuss.

Sources: Nau Web site, www.nau.com; and P. LaBarre, "Leap of Faith," *Fast Company*, June 2007, pp. 96–103.

Part Two

Planning

Many have observed the importance of plans and planning, as demonstrated by the abundance of familiar quotes about it. From the noted Chinese philosopher Confucius, who said, "A man who does not plan long ahead will find trouble at his door" to the legendary Crimson Tide football coach Paul "Bear" Bryant, who said, "Have a plan. Follow the plan, and you'll be surprised how successful you can be. Most people don't have a plan. That's why it's easy to beat most folks," we see how important planning is. *Great* managers need to learn how to plan and then do it. As novelist Richard Cushing said, "Always plan ahead. It wasn't raining when Noah built the ark."

In Part Two, we look at the first management function—planning. Chapter 6 introduces how managers make decisions. In Chapter 7, we look at the fundamental elements of planning—goals and plans. In Chapter 8, we look at an important aspect of planning—strategic management.

We can't resist one more quote about planning, this one from Benjamin Franklin: "By failing to prepare, you are preparing to fail."

Let's Get Real:

Meet The Manager

Rajesh Kumar Duggal

General Manager
Maruti Suzuki India Ltd

MY JOB: Service network development

BEST PART OF MY JOB: Devising new ways and processes and having the opportunities to implement them.

WORST PART OF MY JOB: Limited scope to work across many functions in the company.

BEST MANAGEMENT ADVICE EVER RECEIVED: Yearly review gives us one opportunity in a year to review and correct the situation; monthly review provides 12 opportunities, while daily review gives us 365 opportunities.

You'll be hearing more from this real manager throughout the chapter.

Chapter 6

Managers as Decision Makers

Managers make decisions. And they want those decisions to be good decisions. In this chapter, we're going to study the steps in the decision-making process. We'll also look at the various things that influence a manager as he or she makes decisions. Focus on the following learning outcomes as you read and study this chapter.

LEARNING OUTCOMES

A Manager's Dilemma

The head of operations, Mr Vibhor, of a pharmaceutical plant of a pharmacy mainly located in Ankleshwar, Gujarat, realized that the quantity of output was less than the raw materials being input.[1] He called a meeting of the production engineers, supervisors, and workers to solve the problem. At first, everyone was asked to define the problem and then brainstorm as a group on the possible causes. All ideas were jotted on a board and lists of causes were agreed upon. Small groups of engineers, supervisors, and workers were formed to evaluate the suitability of the listed causes. Based on the deliberations of the groups, it was decided that the calibration of the raw material was needed to be checked and leakages in the formulation area were needed to be reduced. Mr Vibhor was intuitively happy with the outcome of the session. What decision criteria might Mr Vibhor use to evaluate the effectiveness of these decisions?

What would you do?

Like other managers everywhere, Vibhor needs to make decisions as he manages. Decision making is the essence of management. It's what managers do (or try to avoid). And all managers would like to make good decisions since they're judged on the outcomes of those decisions. In this chapter, we examine the concept of decision making and how managers make decisions.

LEARNING **OUTCOME 6.1** ▷

THE DECISION-MAKING PROCESS

On December 2006, when the Tsunami struck the coasts of Tamil Nadu unexpectedly in a matter of hours, 73 habitations were affected along 187.9 km of coastline. It was required that doctors and paramedical staff, the deputy collector, electricity board officials, water and drainage board officials, Public Work Department officials, the block development officer, municipal commissioner, highways officials, fisheries inspector, deputy superintendent of police, and fire and rescue officials all work together. Mr Radhakrishnan, an IAS officer, the collector of Nagapattinam at that time led the team of 1,000 officers and staff from Nagapattinam and 380 officers and staff from other districts who worked round the clock. The decisions that Mr Radhakrishnan had to take were about the removal and safe disposal of dead bodies, restoration of water, electricity, roads, communication, transport, distribution of relief material by government and NGOs, reduction of panic and building public confidence, building temporary shelters and getting people back, restoration of important services such as child-welfare centers and health posts, and opening of schools. He had to continuously motivate the officials to manage the stress.[2] Although most decisions managers make don't involve the weather, you can see that decisions play an important role in what an organization has to do or is able to do.

Managers at all levels and in all areas of organizations make **decisions**. That is, they make choices. For instance, top-level managers make decisions about their organization's goals, where to locate manufacturing facilities, or what new markets to move into.

Middle- and lower-level managers make decisions about production schedules, product quality problems, pay raises, and employee discipline. Making decisions isn't something that just managers do; all organizational members make decisions that affect their jobs and the organization they work for. But our focus is on how *managers* make decisions.

Although decision making is typically described as choosing among alternatives, that view is too simplistic. Why? Because decision making is a process, not just a simple act of choosing among alternatives.[3] Even for something as straightforward as deciding where to go for lunch, you do more than just choose chinese or *tandoori*. Granted, you may not spend a lot of time contemplating your lunch decision, but you still go through the process when making that decision. Exhibit 6–1 shows the eight steps in the decision-making process. This process is as relevant to personal decisions as it is to corporate decisions. Let's use an example—a manager deciding what laptop computers to purchase—to illustrate the steps in the process.

STEP 1: IDENTIFYING A PROBLEM

Every decision starts with a **problem**, a discrepancy between an existing and a desired condition.[4] Neeta is a sales manager whose reps need new laptops because their old

Exhibit 6–1

Decision-Making Process

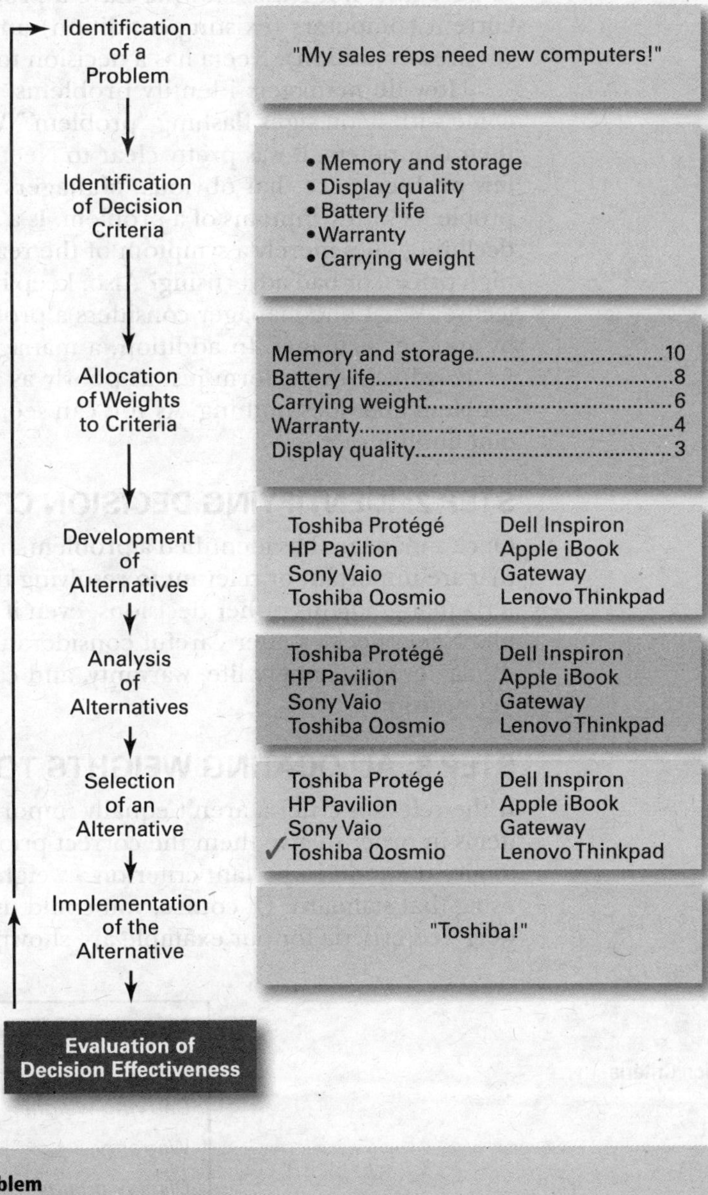

decision
A choice from two or more alternatives.

problem
An obstacle that makes achieving a desired goal or purpose difficult.

The choice of a new laptop computer relies on specific decision criteria like price, convenience, memory and storage capacity, display quality, battery life, and even carrying weight.

ones are outdated and inadequate for doing their job. To make it simple, assume that it's not economical to add memory to the old computers and it's the company's policy to purchase, not lease. Now we have a problem—a disparity between the sales reps' current computers (existing condition) and their need to have more efficient ones (desired condition). Neeta has a decision to make.

How do managers identify problems? In the real world, most problems don't come with neon signs flashing "problem." When her reps started complaining about their computers, it was pretty clear to Neeta that something needed to be done, but few problems are that obvious. Managers also have to be cautious not to confuse problems with symptoms of a problem. Is a 5 percent drop in sales a problem? Or are declining sales merely a symptom of the real problem, such as poor-quality products, high prices, or bad advertising? Also, keep in mind that problem identification is subjective. What one manager considers a problem might not be considered a problem by another manager. In addition, a manager who resolves the wrong problem perfectly is likely to perform just as poorly as a manager who doesn't even recognize a problem and does nothing. As you can see, effectively identifying problems is important but not easy.[5]

STEP 2: IDENTIFYING DECISION CRITERIA

Once a manager has identified a problem, he or she must identify the **decision criteria** that are important or relevant to resolving the problem. Every decision maker has criteria that guide his or her decisions, even if they're not explicitly stated. In our example, Neeta decides after careful consideration that memory and storage capabilities, display quality, battery life, warranty, and carrying weight are the relevant criteria in her decision.

STEP 3: ALLOCATING WEIGHTS TO THE CRITERIA

If the relevant criteria aren't equally important, the decision maker must weight the items in order to give them the correct priority in the decision. How? A simple way is to give the most important criterion a weight of 10 and then assign weights to the rest using that standard. Of course, you could use any number as the highest weight. The weighted criteria for our example are shown in Exhibit 6–2.

Exhibit 6–2

Important Decision Criteria

Memory and storage	*10*
Battery life	*8*
Carrying weight	*6*
Warranty	*4*
Display quality	*3*

Exhibit 6–3
Possible Alternatives

	Memory and Storage	Battery Life	Carrying Weight	Warranty	Display Quality
Toshiba Protégé	10	3	10	8	5
Dell Inspiron	8	7	7	8	7
HP Pavilion	8	5	7	10	10
Apple iBook	8	7	7	8	7
Sony Vaio	7	8	7	8	7
Gateway	8	3	6	10	8
Toshiba Qosmio	10	7	8	6	7
Lenovo Thinkpad	4	10	4	8	10

STEP 4: DEVELOPING ALTERNATIVES

The fourth step in the decision-making process requires the decision maker to list viable alternatives that could resolve the problem. This is the step where a decision maker needs to be creative. At this point, the alternatives are only listed, not evaluated. Our sales manager, Neeta, identifies eight laptops as possible choices (see Exhibit 6–3).

STEP 5: ANALYZING ALTERNATIVES

Once alternatives have been identified, a decision maker must evaluate each one. How? By using the criteria established in step 2. Exhibit 6–3 shows the assessed values that Neeta gave each alternative after doing some research on them. Keep in mind that these data represent an assessment of the eight alternatives using the decision criteria but *not* the weighting. When you multiply each alternative by the assigned weight, you get the weighted alternatives, as shown in Exhibit 6–4. The total score for each alternative, then, is the sum of its weighted criteria.

There are times when a decision maker might not have to do this step. If one alternative scored highest on every criterion, you wouldn't need to consider the weights because that alternative would already be the top choice. Or, if the weights were all equal, you could evaluate an alternative merely by summing up the assessed values for them all (see Exhibit 6–3). For example, the score for the Toshiba Protégé would be 36, and the score for the Gateway would be 35.

STEP 6: SELECTING AN ALTERNATIVE

The sixth step in the decision-making process is choosing the best alternative or the one that generated the highest total in step 5. In our example (see Exhibit 6–4), Neeta would choose the Toshiba Qosmio because it scored highest (249 total).

STEP 7: IMPLEMENTING THE ALTERNATIVE

In step 7 in the decision-making process, you put the decision into action by conveying it to those affected and getting their commitment to it. We know that if the people who must implement a decision participate in the process, they're more likely to support it

Exhibit 6–4
Evaluation of Alternatives

	Memory and Storage	Battery Life	Carrying Weight	Warranty	Display Quality	Total
Toshiba Protégé	100	24	60	32	15	231
Dell Inspiron	80	56	42	32	21	231
HP Pavilion	80	40	42	40	30	232
Apple iBook	80	56	42	32	21	231
Sony Vaio	70	64	42	32	21	229
Gateway	80	24	36	40	24	204
Toshiba Qosmio	100	56	48	24	21	249
Lenovo Thinkpad	40	80	24	32	30	206

decision criteria
Criteria that define what's important or relevant in resolving a problem.

than if you just tell them what to do. Another thing managers may need to do during implementation is reassess the environment for any changes, especially with a long-term decision. Are the criteria, alternatives, and choices still the best ones, or has the environment changed in such a way that you need to reevaluate?

STEP 8: EVALUATING DECISION EFFECTIVENESS

The last step in the decision-making process involves evaluating the outcome or result of the decision to see if the problem was resolved. If the evaluation shows that the problem still exists, then the manager needs to assess what went wrong. Was the problem incorrectly defined? Were errors made when evaluating alternatives? Was the right alternative selected but poorly implemented? The answers might lead you to redo an earlier step or might even require starting the whole process over.

QUICK LEARNING REVIEW:

LEARNING OUTCOME 6.1

- **Define decision.**
- **Describe the eight steps in the decision-making process.**

Go to page 140 and see how well you know this material.

LEARNING **OUTCOME 6.2** ▷

MANAGERS MAKING DECISIONS

Although everyone in an organization makes decisions, decision making is particularly important to managers. As Exhibit 6–5 shows, it's part of all four managerial functions. In fact, that's why we say that decision making is the essence of management.[6] And that's why managers—when they plan, organize, lead, and control—are called *decision makers.*

The fact that almost everything a manager does involves making decisions doesn't mean that decisions are always time-consuming, complex, or evident to an outside observer. Most decision making is routine. Every day of the year, you make a decision about what to eat for dinner. It's no big deal. You've made the decision thousands of times before. It's a pretty simple decision and can usually be handled quickly. It's the type of decision you almost forget *is* a decision. And managers also make dozens of these routine decisions every day—for example, which employee will work what shift next week, what information should be included in a report, or how to resolve a customer's complaint. Keep in mind that even though a decision seems easy or has been faced by a manager a number of times before, it still is a decision. Let's look at three perspectives on how managers make decisions.

Let's Get Real: F2F

I MAKE THE BEST DECISIONS BY:

- **Thinking hard, listening to others, and when self-convinced.**

MAKING DECISIONS: RATIONALITY

When Hewlett-Packard (HP) acquired Compaq, the company did no research on how customers viewed Compaq products until "months after then-CEO Carly Fiorina publicly announced the deal and privately warned her top management team that she didn't want to hear any dissent pertaining to the acquisition."[7] By the time the company discovered that customers perceived Compaq products as inferior—just the opposite of what customers felt about HP products—it was too late. HP's performance suffered, and Fiorina lost her job.

We assume that managers' decision making will be **rational**; that is, we assume that they'll make logical and consistent choices to maximize value.[8] After all, managers have all sorts of tools and techniques to help them be rational decision makers. (See "Managing in a Virtual World" on p. 133.) But as the HP example illustrates, managers aren't always rational. What does it mean to be a rational decision maker?

Assumptions of Rationality. A rational decision maker would be fully objective and logical. The problem faced would be clear and unambiguous, and the decision maker would have a clear and specific goal and know all possible alternatives and consequences. Finally, making decisions rationally would consistently lead to selecting the

Exhibit 6–5

Decisions Managers May Make

Planning

- What are the organization's long-term objectives?
- What strategies will best achieve those objectives?
- What should the organization's short-term objectives be?
- How difficult should individual goals be?

Organizing

- How many employees should I have report directly to me?
- How much centralization should there be in the organization?
- How should jobs be designed?
- When should the organization implement a different structure?

Leading

- How do I handle employees who appear to be low in motivation?
- What is the most effective leadership style in a given situation?
- How will a specific change affect worker productivity?
- When is the right time to stimulate conflict?

Controlling

- What activities in the organization need to be controlled?
- How should those activities be controlled?
- When is a performance deviation significant?
- What type of management information system should the organization have?

alternative that maximizes the likelihood of achieving that goal. These assumptions apply to any decision—personal or managerial. However, for managerial decision making, we need to add one additional assumption: Decisions are made in the best interests of the organization. These assumptions of rationality aren't very realistic, but the next concept can help explain how most decisions get made in organizations.

MAKING DECISIONS: BOUNDED RATIONALITY

Despite the unrealistic assumptions, managers are expected to be rational when making decisions.[9] They understand that "good" decision makers are supposed to do certain things and exhibit good decision-making behaviors as they identify problems, consider alternatives, gather information, and act decisively but prudently. When they do so, they show others that they're competent and that their decisions are the result of intelligent deliberation. However, a more realistic approach to describing how managers make decisions is the concept of **bounded rationality**, which says that managers make decisions rationally but are limited (bounded) by their ability to process information.[10] Because they can't possibly analyze all information on all alternatives, managers **satisfice** rather than maximize. That is, they accept solutions that are "good enough." They're being rational within the limits (bounds) of their ability to process information. Let's look at an example.

Suppose that you're a finance major, and upon graduation you want a job, preferably as a personal financial planner, with a minimum salary of Rs. 70,000 and within 100 kilometers of your hometown. You accept a job offer as a business credit analyst—not exactly a personal financial planner, but still in the finance field—at a bank 50 kilometers from home at a starting salary of Rs. 85,000. If you had done a more

rational decision making
A type of decision making in which choices are logical and consistent and maximize value.

bounded rationality
Decision making that's rational but limited (bounded) by an individual's ability to process information.

satisfice
To accept solutions that are "good enough."

Let's Get Real:
F2F

MAKING GOOD DECISIONS IS HARD BECAUSE:
It involves serious and good understanding. Decision has to be not only correct but also implementable.

comprehensive job search, you would have discovered a job in personal financial planning at a trust company only 25 kilometers from your hometown and starting at a salary of Rs. 95,000. You weren't a perfectly rational decision maker because you didn't maximize your decision by searching all possible alternatives and then choosing the best. But because the first job offer was satisfactory (or "good enough"), you behaved in a bounded rationality manner by accepting it.

Most decisions don't fit the assumptions of perfect rationality, so managers satisfice. However, keep in mind that managers' decision making is also likely influenced by the organization's culture, internal politics, power considerations, and a phenomenon called **escalation of commitment**, which is an increased commitment to a previous decision despite evidence that it may have been wrong.[11] The *Challenger* space shuttle disaster is often used as an example of escalation of commitment. Decision makers chose to launch the shuttle that day even though the decision was questioned by several individuals who believed that it was a bad one. Why would decision makers escalate commitment to a bad decision? Because they don't want to admit that their initial decision may have been flawed. Rather than search for new alternatives, they simply increase their commitment to the original solution.

MAKING DECISIONS: THE ROLE OF INTUITION

When managers at stapler-maker Swingline saw the company's market share declining, they used a logical scientific approach to address the issue. For three years, they exhaustively researched stapler users before deciding what new products to develop. However, at Accentra, Inc., founder Todd Moses used a more intuitive decision approach to come up with his line of unique PaperPro staplers.[12]

Like Todd Moses, other managers often use their intuition to help their decision making. **Intuitive decision making** is making decisions on the basis of experience, feelings, and accumulated judgment. Researchers studying managers' use of intuitive decision making have identified five different aspects of intuition, which are described in Exhibit 6–6.[13] How common is intuitive decision making? One survey found that almost half of the executives surveyed "used intuition more often than formal analysis to run their companies."[14]

Intuitive decision making can complement both rational and boundedly rational decision making.[15] First of all, a manager who has had experience with a similar type of problem or situation often can act quickly with what appears to be limited information because of that past experience. In addition, a recent study found that individuals who experienced intense feelings and emotions when making decisions actually

Exhibit 6–6 What Is Intuition?

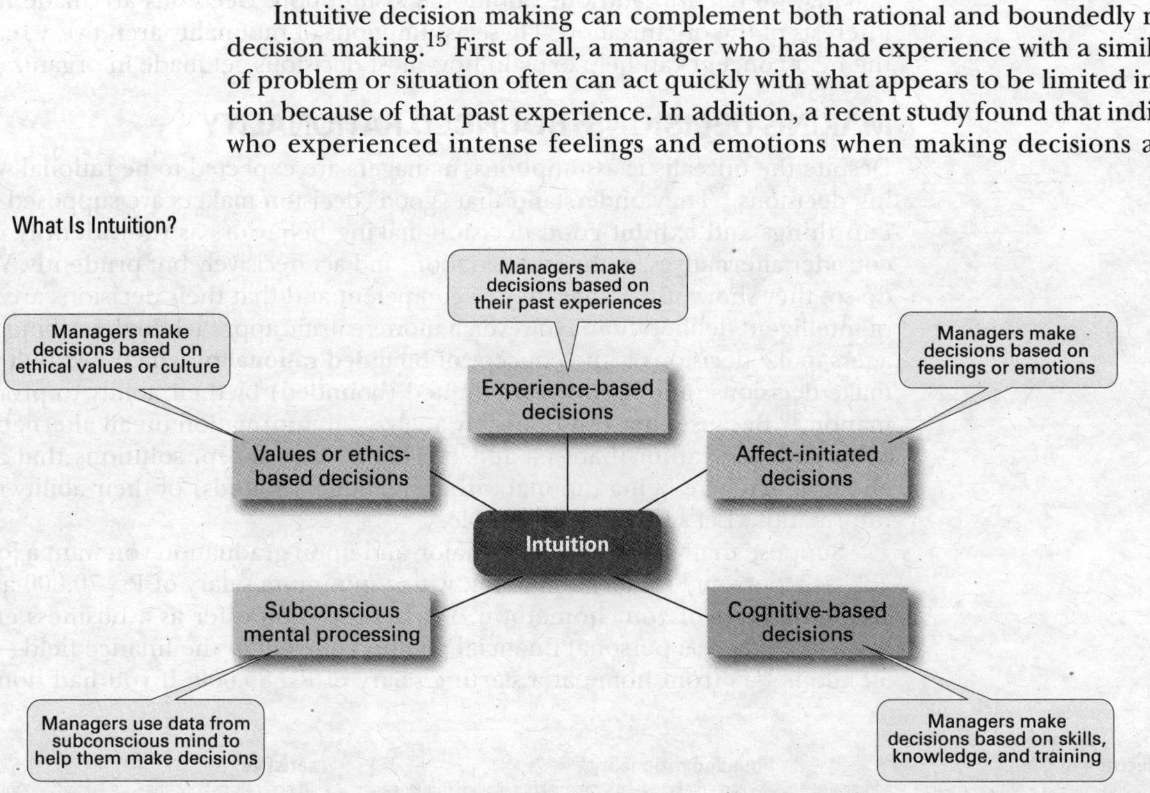

Source: Based on L. A. Burke and M. K. Miller, "Taking the Mystery Out of Intuitive Decision Making," *Academy of Management Executive*, October 1999, pp. 91–99.

achieved higher decision-making performance, especially when they understood their feelings as they were making decisions. The old belief that managers should ignore emotions when making decisions may not be the best advice.[16]

QUICK LEARNING REVIEW:

LEARNING OUTCOME 6.2

- Discuss the assumptions of rational decision making.
- Describe the concepts of bounded rationality, satisficing, and escalation of commitment.
- Explain intuitive decision making.

Go to page 140 and see how well you know this material.

LEARNING OUTCOME 6.3 ▷

TYPES OF DECISIONS AND DECISION-MAKING CONDITIONS

Restaurant managers in Illinois make routine decisions weekly about purchasing food supplies and scheduling employee work shifts. It's something they've done numerous times. But now they're facing a different kind of decision—one they've never encountered: how to adapt to the new statewide smoking ban.

TYPES OF DECISIONS

Managers in all kinds of organizations face different types of problems and decisions as they do their jobs. Depending on the nature of the problem, a manager can make one of two different types of decisions.

Structured Problems and Programmed Decisions. Some problems are straightforward. The decision maker's goal is clear, the problem is familiar, and information about the problem is easily defined and complete. Examples might include when a customer returns a purchase to a store, when a supplier is late with an important delivery, a news team's response to a fast-breaking event, or a college's handling of a student wanting to drop a class. Such situations are called **structured problems** because they're straightforward, familiar, and easily defined. For instance, a server spills a drink on a customer's coat. The customer is upset, and the manager needs to do something. Because it's not an unusual occurrence, there's probably some standardized routine for handling it. For example, the manager offers to have the coat cleaned at the restaurant's expense. This is what we call a **programmed decision**, a repetitive decision that can be handled using a routine approach. Because the problem is structured, the manager doesn't have to go to the trouble and expense of going through an involved decision-making process. With this type of decision, the "develop-the-alternatives" stage of the decision-making process either doesn't exist or is given little attention. Why? Because once the structured problem is defined, the solution is usually self-evident or at least reduced to a few alternatives that are familiar and have proved successful in the past. The spilled drink on the customer's coat doesn't require the restaurant manager to identify and weight decision criteria or to develop a long list of possible solutions. Instead, the manager relies on one of three types of programmed decisions: procedure, rule, or policy.

A **procedure** is a series of sequential steps a manager uses to respond to a structured problem. The only difficulty is identifying the problem. Once the problem is

escalation of commitment
An increased commitment to a previous decision despite evidence that it may have been a poor decision.

intuitive decision making
Making decisions on the basis of experience, feelings, and accumulated judgment.

structured problem
A straightforward, familiar, and easily defined problem.

programmed decision
A repetitive decision that can be handled using a routine approach.

procedure
A series of sequential steps used to respond to a well-structured problem.

clear, so is the procedure. For instance, say that a purchasing manager receives a request from a warehouse manager for 15 handheld computers for the inventory clerks. The purchasing manager knows how to make this decision by following the established purchasing procedure.

A **rule** is an explicit statement that tells a manager what can or cannot be done. Rules are frequently used because they're simple to follow and ensure consistency. For example, rules about lateness and absenteeism permit supervisors to make disciplinary decisions rapidly and fairly.

The third type of programmed decisions is a **policy**, which is a guideline for making a decision. In contrast to a rule, a policy establishes general parameters for the decision maker rather than specifically stating what should or should not be done. Policies typically contain an ambiguous term that leaves interpretation up to the decision maker. Here are some sample policy statements:

- The customer always comes first and should always be *satisfied.*
- We promote from within *whenever possible.*
- Employee wages shall be *competitive* within community standards.

Notice that the terms *satisfied, whenever possible,* and *competitive* require interpretation. For instance, the policy of paying competitive wages doesn't tell a company's human resources manager the exact amount he or she should pay, but it does guide the person in making the decision.

Unstructured Problems and Nonprogrammed Decisions. Not all the problems managers face can be solved using programmed decisions. Many organizational situations involve **unstructured problems**, which are problems that are new or unusual and for which information is ambiguous or incomplete. Whether to build a new manufacturing facility in China is an example of an unstructured problem. So, too, is the problem facing restaurant managers in Illinois who must decide how to modify their businesses to comply with the new smoking ban. When problems are unstructured, managers must rely on nonprogrammed decision making in order to develop unique solutions. **Nonprogrammed decisions** are unique and nonrecurring and involve custom-made solutions.

Exhibit 6–7 describes the differences between programmed and nonprogrammed decisions. Lower-level managers mostly rely on programmed decisions (procedures, rules, and policies) because they confront familiar and repetitive problems. As managers move up the organizational hierarchy, the problems they confront become more unstructured. Why? Because lower-level managers handle the routine decisions and let upper-level managers deal with the unusual or difficult decisions. Also, upper-level managers delegate routine decisions to their subordinates so they can deal with more

Many people believe that China is the next big market for powerful brand-name products, and Zong Qinghou, founder of China's Wahaha beverage group, is ready. But brand names are a new concept in Chinese markets, and Zong prefers his own firsthand information to market research. He will face many nonprogrammed decisions as he tries to keep his brand strong at home and abroad.

Exhibit 6–7
Programmed Versus Nonprogrammed Decisions

Characteristic	Programmed Decisions	Nonprogrammed Decisions
Type of problem	Structured	Unstructured
Managerial level	Lower levels	Upper levels
Frequency	Repetitive, routine	New, unusual
Information	Readily available	Ambiguous or incomplete
Goals	Clear, specific	Vague
Time frame for solution	Short	Relatively long
Solution relies on	Procedures, rules, policies	Judgment and creativity

difficult issues.[17] Thus, few managerial decisions in the real world are either fully programmed or nonprogrammed. Most fall somewhere in between.

DECISION-MAKING CONDITIONS

When making decisions, managers may face three different conditions: certainty, risk, and uncertainty. Let's look at the characteristics of each.

Certainty. The ideal situation for making decisions is one of **certainty**, which is a situation in which a manager can make accurate decisions because the outcome of every alternative is known. For example, when Finance Controller decides where to deposit the Provident Fund monies, he knows exactly the interest rate being offered by each bank and the amount that will be earned on the funds. He is certain about the outcomes of each of the alternatives. As you might expect, most managerial decisions aren't like this.

Risk A far more common situation than decision making under certainty is one of **risk**, conditions in which the decision maker is able to estimate the likelihood of certain outcomes. Under risk, managers have historical data from past personal experiences or secondary information that lets them assign probabilities to different alternatives. Let's look at an example.

Suppose that you manage a hotel in Mount Abu and you're thinking about adding more rooms. Obviously, your decision will be influenced by the additional revenue the new rooms would generate, which depends on the number of expected tourists. It is believed that the warmer it is in the plains around Mount Abu the more number of

thinking critically about Ethics

Decisions can and do create ethical dilemmas. School board members at a high school in Minnesota learned of a threat of violence at the school on a Sunday afternoon.[18] After investigating, they found it to be baseless. However, on Sunday evening, another threat of violence reached the administration. By this time, it was 10:30 P.M. The high school principal, school superintendent, and board chair decided to close the school on Monday and say that it was due to a broken water main. To get the word out quickly, they contacted the media. When the second rumor also turned out to be false, the public was informed that school had been canceled because of a rumored threat of violence, not because of a broken water main. Not only did the media get upset, but some parents demanded the superintendent's resignation. What do you think about this situation? Was the "right" decision made? Could the situation have been handled better? Discuss.

rule
An explicit statement that tells managers what can or cannot be done.

policy
A guideline for making decisions.

unstructured problem
A problem that is new or unusual and for which information is ambiguous or incomplete.

nonprogrammed decision
A unique and nonrecurring decision that requires a custom-made solution.

certainty
A situation in which a decision maker can make accurate decisions because all outcomes are known.

risk
A situation in which the decision maker is able to estimate the likelihood of certain outcomes.

Exhibit 6–8

Expected Value

Event	Expected Revenues in INR	×	Probability	=	Expected Value of Each Alternative in INR
Extreme summers	6,00,000		0.3		1,80,000
Normal summers	4,50,000		0.5		2,25,000
Mild summers	3,50,000		0.2		70,000
					4,75,000

tourists you would receive there. You have fairly reliable data on the maximum temperature reached in Ahmedabad (one of the large cities around Mt. Abu) in the last 10 years—3 years of extremely warm summers, 5 years of normal heat, and 2 years of fairly mild summers. You can use this information to help make your decision by calculating expected value—the expected return from each possible outcome—by multiplying expected revenues by warm temperature in planes probabilities. The result is the average revenue you can expect over time if the given probabilities hold. As Exhibit 6–8 shows, the expected revenue from adding a new room is INR 4,75,000. Of course, whether that's enough to justify a decision to build depends on the costs involved in generating that revenue.

Uncertainty. What happens if you face a decision and you're not certain about the outcomes and can't even make reasonable probability estimates? We call this condition **uncertainty**. Managers do face decision-making situations of uncertainty. Under these conditions, the choice of alternative is influenced by the limited amount of available information and by the psychological orientation of the decision maker. An optimistic manager will follow a *maximax* choice (maximizing the maximum possible payoff), a pessimist will follow a *maximin* choice (maximizing the minimum possible payoff), and a manager who desires to minimize his maximum "regret" will opt for a *minimax* choice. Let's look at these different choice approaches using an example.

A marketing manager at Visa has determined four possible strategies (S1, S2, S3, and S4) for promoting the Visa card throughout the Western India. The marketing manager also knows that major competitor MasterCard has three competitive actions (CA1, CA2, CA3) it's using to promote its card in the same region. For this example, we'll assume that the Visa manager has no previous knowledge that allows her to determine probabilities of success of any of the four strategies. She formulates the matrix shown in Exhibit 6–9 to show the various Visa strategies and the resulting profit, depending on the competitive action used by MasterCard.

In this example, if the Visa manager is an optimist, she'll choose strategy 4 (S4) because that could produce the largest possible gain: INR 56 lakh. Note that this choice maximizes the maximum possible gain (maximax choice).

If the manager is a pessimist, she'll assume that only the worst can occur. The worst outcome for each strategy is as follows: S1 = INR 22 lakh, S2 = INR 18 lakh, S3 = INR 30 lakh, and S4 = INR 28 lakh. These are the most pessimistic outcomes from each strategy. Following the *maximin* choice, she would maximize the minimum payoff; in other words, she'd select S3 (INR 30 lakh is the largest of the minimum payoffs).

In the third approach, managers recognize that once a decision is made, it will not necessarily result in the most profitable payoff. There may be a "regret" of profits given up—*regret* referring to the amount of money that could have been made had a different

Exhibit 6–9

Payoff Matrix

(in lakh of rupees)	MasterCard's Response		
Visa Marketing Strategy	CA1	CA2	CA3
S1	26	28	22
S2	18	30	36
S3	48	42	30
S4	36	28	56

Exhibit 6–10

Regret Matrix

(in lakh of rupees) Visa Marketing Strategy	MasterCard's Response CA1	CA2	CA3
S1	22	14	34
S2	30	12	20
S3	0	0	26

strategy been used. Managers calculate regret by subtracting all possible payoffs in each category from the maximum possible payoff for each given event—in this case, for each competitive action. For the Visa manager, the highest payoff, given that MasterCard engages in CA1, CA2, or CA3, is INR 48 lakh, INR 42 lakh, or INR 56 lakh, respectively (the highest number in each column). Subtracting the payoffs in Exhibit 6–9 from those figures produces the results shown in Exhibit 6–10.

The maximum regrets are S1 = INR 34 lakh, S2 = INR 30 lakh, S3 = INR 26 lakh, and S4 = INR 14 lakh. The *minimax* choice minimizes the maximum regret, so the Visa manager would choose S4. By making this choice, she'll never have a regret of profits given up of more than INR 14 lakh. This result contrasts, for example, with a regret of INR 30 lakh had she chosen S2 and MasterCard had taken CA1.

Although managers try to quantify a decision when possible by using payoff and regret matrices, uncertainty often forces them to rely more on intuition, creativity, hunches, and "gut feel."

managing in a Virtual World

Making Better Decisions with IT

BudNet is the "crown jewel of the King of Beers." What is it? It's Anheuser-Busch's (A-B's) powerful and sophisticated information system. Every night, data are collected from A-B distributors' computer servers. Each morning, managers can see what brands are selling in which packages, using which promotional materials and pricing discounts. According to "dozens of analysts, beer-industry veterans, and distributor executives ... Anheuser has made a deadly accurate science out of finding out what beer lovers are buying, as well as when, where, and why." All this information allows A-B managers to continually adjust production and fine-tune marketing campaigns.

Most companies are "drowning in data" and don't know how to make sense out of it.[19] As this example shows, however, one of the primary uses for IT can be to help managers—and other employees—make better decisions by sorting through tons of data, looking for trends, patterns, and other insights. As we saw in our discussion of bounded rationality, a person's ability to process such a massive amount of information would be severely limited. So managers use IT to help make sense of all this information so they can make better decisions.

Another way that IT can help managers make better decisions is by using software tools that help them analyze data. Consultants estimate that some 75 percent of individual managers rely on personal productivity tools, such as spreadsheets, which can be used to gather and report information to help them make decisions in their own local area of responsibility. However, when you have each manager using his or her own data collection tools, there are no linkages or collaboration. Thus, on the organizationwide level, there's the more sophisticated **business performance management (BPM) software**, also sometimes called corporate performance management software, to help make decisions. BPM, which provides key performance indicators that help companies monitor efficiency of projects and employees, was initially believed to be the "silver bullet that had the potential to help corporate managers control their organization's performance in an increasingly volatile world." Although BPM software hasn't quite lived up to those lofty expectations, as it improves, it will increasingly be a tool managers use to help make better decisions.

uncertainty
A situation in which a decision maker has neither certainty nor reasonable probability estimates available.

business performance management (BPM) software
IT software that provides key performance indicators to help managers monitor efficiency of projects and employees. Also known as corporate performance management software.

QUICK LEARNING REVIEW:

LEARNING OUTCOME 6.3

- Explain the two types of problems and decisions.
- Contrast the three decision-making conditions.
- Describe maximax, maximin, and minimax decision approaches.

Go to page 140 and see how well you know this material.

LEARNING **OUTCOME 6.4** ▷

DECISION-MAKING STYLES

William D. Perez's tenure as Nike's CEO lasted a short and turbulent 13 months. Analysts attributed his abrupt dismissal to a difference in decision-making approaches between him and Nike co-founder Phil Knight. Perez tended to rely more on data and facts when making decisions, while Knight highly valued, and had always used, his judgment and feelings to make decisions.[20] As this example clearly shows, managers have different styles when it comes to making decisions.

LINEAR–NONLINEAR THINKING STYLE PROFILE

Suppose that you're a new manager. How will you make decisions? Recent research done with four distinct groups of people says that the way a person approaches decision making is likely affected by his or her thinking style.[21] Your thinking style reflects two things: (1) the source of information you tend to use (external data and facts or internal sources, such as feelings and intuition) and (2) how you process that information (linear—rational, logical, analytical; or nonlinear—intuitive, creative, insightful). These four dimensions are collapsed into two styles. The first, the **linear thinking style**, is characterized by a person's preference for using external data and facts and processing this information through rational, logical thinking to guide decisions and actions. The second, the **nonlinear thinking style**, is characterized by a preference for internal sources of information (feelings and intuition) and processing this information with internal insights, feelings, and hunches to guide decisions and actions. Look back at the earlier Nike example, and you'll see both styles described.

managing workforce Diversity

The Value of Diversity in Decision Making

Have you decided what your major is going to be? How did you decide? Do you feel that your decision is a good one? Is there anything you could have done differently to make sure that your decision was the best one?[22]

Making good decisions is tough! Managers continuously make decisions—for instance, developing new products, establishing weekly or monthly goals, implementing an advertising campaign, reassigning an employee to a different work group, resolving a customer's complaint, or purchasing new laptops for sales reps. One important suggestion for making better decisions is to tap into the diversity of the work group. Drawing on diverse employees can prove valuable to a manager's decision making. Why? Diverse employees can provide fresh perspectives on issues. They can offer different interpretations of how a problem is defined and may be more open to trying new ideas. Diverse employees can be more creative in generating alternatives and more flexible in resolving issues. And getting input from diverse sources increases the likelihood that creative and unique solutions will be generated.

Even though diversity in decision making can be valuable, there are drawbacks. The lack of a common perspective usually means that more time is spent discussing the issues. Communication may be a problem particularly if language barriers are present. In addition, seeking out diverse opinions can make the decision-making process more complex, confusing, and ambiguous. And with multiple perspectives on the decision, it may be difficult to reach agreement or to agree on specific actions. Although these drawbacks are valid concerns, the value of diversity in decision making outweighs the potential disadvantages.

Now, about that decision on a major. Did you ask others for their opinions? Did you seek out advice from professors, family members, friends, or coworkers? Getting diverse perspectives on an important decision like this could help you make the best decision. Managers should also consider the value to be gained from diversity in decision making.

Let's Get Real:
F2F

GOOD DECISION-MAKING SKILLS ARE:

Most important for the role of a leader. Decisions have to be not only correct but also timely with the involvement of stakeholders benefiting all.

Managers need to recognize that their employees may use different decision-making styles. Some employees may take their time weighing alternatives and relying on how they feel about it, while others may rely on external data before logically making a decision. This doesn't make one person's approach better than the other. It just means that their decision-making styles are different. The "Managing Workforce Diversity" box addresses some of the issues associated with valuing diversity in decision making.

DECISION-MAKING BIASES AND ERRORS

When managers make decisions, they not only use their own particular style, they may use "rules of thumb," or **heuristics**, to simplify their decision making. Heuristics can be useful because they help make sense of complex, uncertain, and ambiguous information.[23] Even though managers may use rules of thumb, that doesn't mean those rules are reliable. Why? Because they may lead to errors and biases in processing and evaluating information. Exhibit 6–11 identifies 12 common decision errors and biases that managers make. Let's look at each.[24]

When decision makers tend to think they know more than they do or hold unrealistically positive views of themselves and their performance, they're exhibiting the *overconfidence bias.* The *immediate gratification bias* describes decision makers who tend to want immediate rewards and to avoid immediate costs. For these individuals, decision choices that provide quick payoffs are more appealing than those that may provide payoffs in the future. The *anchoring effect* describes the situation when decision makers fixate on initial information as a starting point and then, once set, fail to adequately adjust for subsequent information. First impressions, ideas, prices, and estimates carry unwarranted weight relative to information received later. When decision makers selectively organize and interpret events based on their biased perceptions, they're using the *selective perception bias.* This influences the information they pay attention to, the problems they identify, and the alternatives they develop. Decision makers who seek out information that reaffirms their past choices and discount information that contradicts past judgments exhibit the *confirmation bias.* These people tend to

Exhibit 6–11

Common Decision-Making Errors and Biases

linear thinking style
A decision style characterized by a person's preference for using external data and facts and processing this information through rational, logical thinking.

nonlinear thinking style
A decision style characterized by a person's preference for internal sources of information and processing this information with internal insights, feelings, and hunches.

heuristics
Rules of thumb that managers use to simplify decision making.

The choice of an advertising agency is often made by those with a nonlinear approach to decision making. Marketing executives from Virgin Atlantic Airways saw presentations from five ad agencies before choosing Crispin Porter & Bogusky, a small firm whose innovative proposal showed how efficiently the airline's $15 million ad budget could be spent. The marketing team from Virgin allowed 10 weeks to make a decision; it took 4 days. The winning team is pictured here with the paper airplanes that played a part in their pitch.

accept at face value information that confirms their preconceived views and are critical and skeptical of information that challenges these views. With the *framing bias*, decision makers select and highlight certain aspects of a situation while excluding others. By drawing attention to specific aspects of a situation and highlighting them, while at the same time downplaying or omitting other aspects, they distort what they see and create incorrect reference points. The *availability bias* causes decision makers to tend to remember events that are the most recent and vivid in their memory. This bias distorts their ability to recall events in an objective manner and results in distorted judgments and probability estimates. When decision makers assess the likelihood of an event based on how closely it resembles other events or sets of events, that's the *representation bias.* Managers exhibiting this bias draw analogies and see identical situations where they don't exist. The *randomness bias* occurs when decision makers try to create meaning out of random events. They do this because most decision makers have difficulty dealing with chance, even though random events happen to everyone, and there's nothing that can be done to predict them. With the *sunk costs error*, decision makers forget that current choices can't correct the past. They incorrectly fixate on past expenditures of time, money, or effort in assessing choices rather than on future consequences. Instead of ignoring sunk costs, they can't forget them. Decision makers who are quick to take credit for their successes and to blame failure on outside factors exhibit the *self-serving bias.* Finally, the *hindsight bias* is the tendency for decision makers to falsely believe, after that outcome is actually known, that they could have accurately predicted the outcome of an event.

Managers avoid the negative effects of these decision errors and biases by being aware of them and then not using them. Managers should also pay attention to how they make decisions and try to identify the heuristics they typically use and critically evaluate how appropriate those are. Finally, managers might want to ask those around them to help identify weaknesses in their decision-making style and try to improve on those weaknesses.

OVERVIEW OF MANAGERIAL DECISION MAKING

Exhibit 6–12 provides an overview of managerial decision making. Because it's in their best interests to do so, managers *want* to make good decisions—that is, choose the "best" alternative, implement it, and determine whether it takes care of the problem, which is the reason the decision was needed in the first place. Their decision-making process is affected by four factors: the decision-making approach, the type of problem, decision-making conditions, and their decision-making style. In addition, certain decision-making errors and biases may affect the process. Each factor plays a role in determining how the manager makes a decision. So whether that decision involves addressing an employee's habitual tardiness, resolving a product quality problem, or determining whether to enter a new market, remember that it has been shaped by a number of factors.

Exhibit 6–12 Overview of Managerial Decision Making

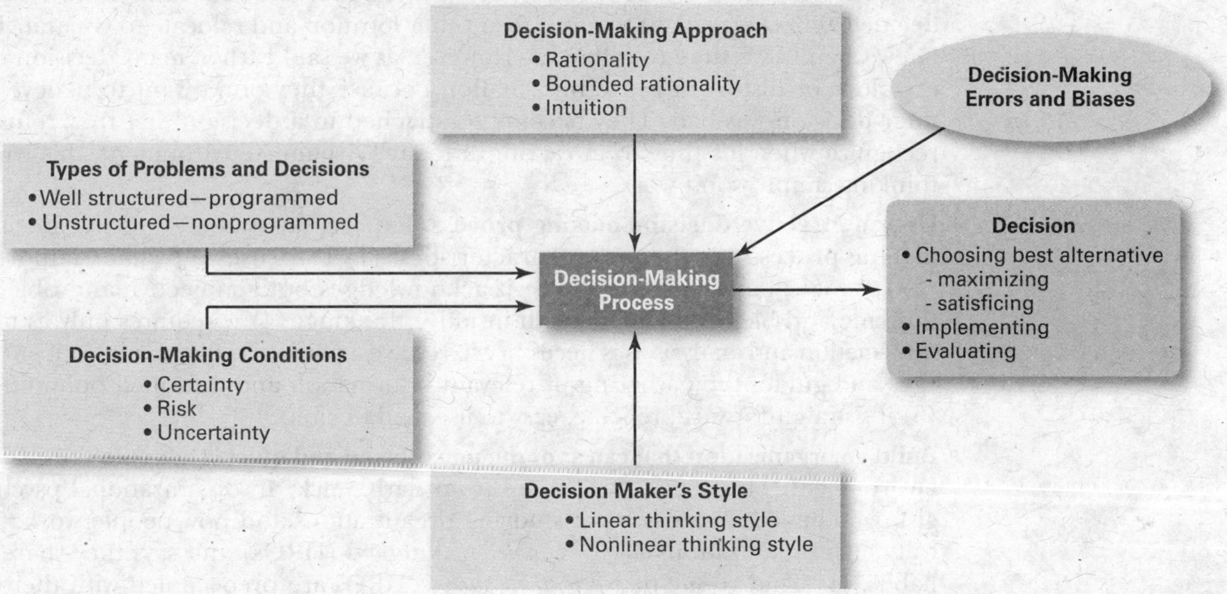

QUICK LEARNING REVIEW:

LEARNING OUTCOME 6.4

- Describe two decision-making styles.
- Discuss the 12 decision-making biases.
- Explain the managerial decision-making model.

Go to page 140 and see how well you know this material.

LEARNING OUTCOME 6.5 ▷ EFFECTIVE DECISION MAKING IN TODAY'S WORLD

Per Carlsson, a product development manager at IKEA, "spends his days creating Volvo-style kitchens at Yugo prices." His job is to take the "problems" identified by the company's product-strategy council (a group of globe-trotting senior managers that monitors consumer trends and establishes product priorities) and turn them into furniture that customers around the world want to buy. One "problem" recently identified by the council: The kitchen has replaced the living room as the social and entertaining center in the home. Customers are looking for kitchens that convey comfort and cleanliness while still allowing them to pursue their gourmet aspirations. Carlsson must use this information to make things happen. There are a lot of decisions to make—both programmed and nonprogrammed—and the fact that IKEA is a global company makes it even more challenging. Comfort in Asia means small, cozy appliances and spaces, while North American customers want oversized glassware and giant refrigerators. His ability to make good decisions quickly has significant implications for IKEA's success.[25]

Today's business world revolves around making decisions, often risky ones, usually with incomplete or inadequate information, and under intense time pressure. Most managers make one decision after another; and as if that weren't challenging enough, more is at stake than ever before. Bad decisions can cost millions. What do managers need to do to make effective decisions in today's fast-moving world? Here are some guidelines:

- **Understand cultural differences.** Managers everywhere want to make good decisions. However, is there only one "best" way worldwide to make decisions? Or does the "best way depend on the values, beliefs, attitudes, and behavioral patterns of the people involved?"[26]
- **Know when it's time to call it quits.** When it's evident that a decision isn't working, don't be afraid to pull the plug. For instance, the CEO of Tata Motors and the Tata Group

pulled the plug on locating the Nano plant in Singur, West Bengal; despite trying hard to satisfy the different stakeholders, farmers, political parties, and unions, he found that he was not able to satisfy any of them enough to start the work. Thus, he decided almost overnight to move to another location and relocate to Gujarat. They knew when it was time to call quits. However, as we said earlier, many decision makers block or distort negative information because they don't want to believe that their decision was bad. They become so attached to a decision that they refuse to recognize when it's time to move on. In today's dynamic environment, this type of thinking simply won't work.

- **Use an effective decision-making process.** Experts say that an effective decision-making process has these six characteristics: (1) It focuses on what's important; (2) it's logical and consistent; (3) it acknowledges both subjective and objective thinking and blends analytical with intuitive thinking; (4) it requires only as much information and analysis as is necessary to resolve a particular dilemma; (5) it encourages and guides the gathering of relevant information and informed opinion; and (6) it's straightforward, reliable, easy to use, and flexible.[27]
- **Build an organization that can spot the unexpected and quickly adapt to the changed environment.** This suggestion comes from Karl Weick, an organizational psychologist, who has made a career of studying organizations and how people work.[28] He calls such organizations *highly reliable organizations* (HROs) and says they share five habits. (1) They're *not tricked by their success.* HROs are preoccupied with their failures. They're alert to the smallest deviations and react quickly to anything that doesn't fit with their expectations. Weick talks about Navy aviators who describe "leemers—a gut feeling that something isn't right." Typically, these leemers turn out to be accurate. Something, in fact, is wrong. Organizations need to create climates where people feel safe trusting their leemers. (2) They *defer to the experts on the frontline.* Frontline workers—those who interact day in and day out with customers, products, suppliers, and so forth—have firsthand knowledge of what can and cannot be done, what will and will not work. Get their input. Let them make decisions. (3) They *let unexpected circumstances provide the solution.* One of Weick's best-known works is his study of the Mann Gulch fire in Montana that killed 13 smoke jumpers in 1949. The event was a massive, tragic organizational failure. However, the reaction of the foreman illustrates how effective decision makers respond to unexpected circumstances: When the fire was nearly on top of his men, he invented the escape fire—a small fire that consumed all the brush around the team, leaving an area where the larger fire couldn't burn. His action was contrary to everything firefighters are taught (that is, you don't start fires—you extinguish them), but at the time it was the best decision. (4) They *embrace complexity.* Because business is complex, these organizations recognize that it "takes complexity to sense complexity." Rather than simplifying data, which we instinctively try to do when faced with complexity, these organizations aim for deeper understanding of the situation. They ask "why" and keep asking why as they probe more deeply into the causes of the problem and possible solutions. (5) Finally, they *anticipate but also recognize their limits.* These organizations do try to anticipate as much as possible, but they recognize that they can't anticipate everything. As Weick says, they don't "think, then act. They think by acting. By actually doing things, you'll find out what works and what doesn't."

Making decisions in today's fast-moving world isn't easy. Successful managers need good decision-making skills to plan, organize, lead, and control.

QUICK LEARNING REVIEW:

LEARNING OUTCOME 6.5

- Explain how managers can make effective decisions in today's world.
- List the six characteristics of an effective decision-making process.
- List the five habits of highly reliable organizations.

Go to page 141 and see how well you know this material.

Let's Get Real:

My Turn

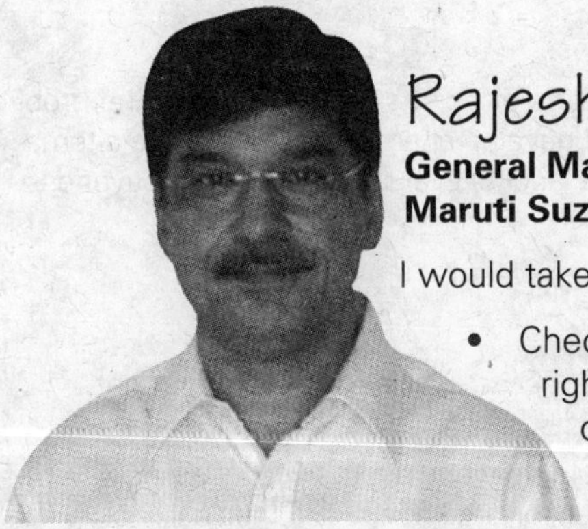

Rajesh Kumar Duggal

General Manager
Maruti Suzuki India Ltd

I would take action in many areas as listed below:

- Check accuracy of calibration of the raw material and set up right procedures and processes for continued correct calibration.
- Identification and measurement of leakages in the formulation area.
- Remove or control leakages.
- If leakage is still happening, study the trend of leakages over a period of time, whether the same is reducing or not.
- Establish ratio between output and input and study the trend over a period of time.
- Continuous review with teams.
- Benchmark against other pharmaceutical plants.

LEARNING OUTCOMES SUMMARY

6.1 ▷ THE DECISION-MAKING PROCESS

- Define decision.
- Describe the eight steps in the decision-making process.

A decision is a choice. The decision-making process consists of eight steps: (1) identify the problem, (2) identify decision criteria, (3) weight the criteria, (4) develop alternatives, (5) analyze alternatives, (6) select an alternative, (7) implement the alternative, and (8) evaluate decision effectiveness.

6.2 ▷ MANAGERS MAKING DECISIONS

- Discuss the assumptions of rational decision making.
- Describe the concepts of bounded rationality, satisficing, and escalation of commitment.
- Explain intuitive decision making.

The assumptions of rationality are as follows: The problem is clear and unambiguous; a single, well-defined goal is to be achieved; all alternatives and consequences are known; and the final choice will maximize the payoff. Bounded rationality says that managers make rational decisions but are bounded (limited) by their ability to process information. Decision makers satisfice when they accept solutions that are good enough. Escalation of commitment is when managers increase commitment to a decision even when they have evidence that it may have been a wrong decision. Intuitive decision making is making decisions on the basis of experience, feelings, and accumulated judgment.

6.3 ▷ TYPES OF DECISIONS AND DECISION-MAKING CONDITIONS

- Explain the two types of problems and decisions.
- Contrast the three decision-making conditions.
- Describe maximax, maximin, and minimax decision approaches.

Programmed decisions are repetitive decisions that can be handled by a routine approach and are used when the problem being resolved is straightforward, familiar, and easily defined (structured). Nonprogrammed decisions are unique decisions that require a custom-made solution and are used when the problems are new or unusual (unstructured) and when the information for the problems is ambiguous or incomplete. Certainty is a situation in which a manager can make accurate decisions because all outcomes are known. Risk is a situation in which a manager can estimate the likelihood of certain outcomes. Uncertainty is a situation in which a manager is not certain about the outcomes and can't even make reasonable probability estimates. When decision makers face uncertainty, their psychological orientation determines whether they follow a maximax choice (maximizing the maximum possible payoff), a maximin choice (maximizing the minimum possible payoff), or a mimimax choice (minimizing the maximum regret—the amount of money that could have been made if a different decision had been made).

6.4 ▷ DECISION-MAKING STYLES

- Describe two decision-making styles.
- Discuss the 12 decision-making biases.
- Explain the managerial decision-making model.

A person's thinking style reflects two things: the source of information the person tends to use (external or internal) and how the person processes that information (linear or nonlinear). These four dimensions were collapsed into two styles. The linear thinking style is characterized by a person's preference for using external data and processing this information through rational, logical thinking. The nonlinear thinking style is characterized by a preference for internal sources of information and processing this information with internal insights, feelings, and hunches. The 12 common decision-making errors and biases are overconfidence, immediate gratification, anchoring, selective perception, confirmation, framing, availability, representation, randomness, sunk costs, self-serving bias, and hindsight. The managerial decision-making model helps explain how the decision-making process is used to choose the best alternative(s) either through maximizing or satisficing and then implement and evaluate the decision. It also helps explain what factors affect the decision-making process, including the decision-making approach (rationality, bounded rationality, intuition), the types of problems and decisions (structured and programmed or unstructured and nonprogrammed), the decision-making conditions (certainty, risk, uncertainty), and the decision maker's style (linear or nonlinear).

6.5 ▷ EFFECTIVE DECISION MAKING IN TODAY'S WORLD

- Explain how managers can make effective decisions in today's world.
- List the six characteristics of an effective decision-making process.
- List the five habits of highly reliable organizations.

Managers can make effective decisions by understanding cultural differences in decision making, knowing when it's time to call it quits, using an effective decision-making process, and building an organization that can spot the unexpected and quickly adapt to the changed environment. The six characteristics of an effective decision-making process are (1) focuses on what's important, (2) is logical and consistent, (3) acknowledges both subjective and objective thinking and blends both analytical and intuitive approaches, (4) requires only "enough" information to resolve a problem, (5) encourages and guides gathering relevant information and informed opinions, and (6) is straightforward, reliable, easy to use, and flexible. The five habits of highly reliable organizations are (1) not being tricked by their successes, (2) deferring to experts on the frontline, (3) letting unexpected circumstances provide a solution, (4) embracing complexity, and (5) anticipating but also recognizing limits.

THINKING ABOUT MANAGEMENT ISSUES

1. Why is decision making often described as the essence of a manager's job?
2. How might an organization's culture influence the way managers make decisions?
3. All of us bring biases to the decisions we make. What would be the drawbacks of having biases? Could there be any advantages to having biases? Explain. What are the implications for managerial decision making?
4. Would you call yourself a linear or nonlinear thinker? What are the implications for choosing the type of organization where you want to work?
5. "As managers use computers and software tools more often, they'll be able to make more rational decisions." Do you agree or disagree with this statement? Why?
6. How can managers blend the guidelines for making effective decisions in today's world with the rationality and bounded rationality models of decision making? Or can managers not do this? Explain.
7. Is there a difference between wrong decisions and bad decisions? Why do good managers sometimes make wrong decisions? Bad decisions? How can managers improve their decision-making skills?

YOUR TURN to be a Manager

- For one week, pay close attention to the decisions you make and how you make them. Write a description of five of those decisions, using the steps in the decision-making process as your guide. Also, describe whether you relied on external or internal sources of information to help you make each decision and whether you think you were more linear or nonlinear in how you processed that information.
- When you feel you haven't made a good decision, assess how you could have made a better decision.
- Find two examples each of procedures, rules, and policies. Bring your examples to class and be prepared to share them.
- Write a procedure, a rule, and a policy for your instructor to use in your class. Be sure that each one is clear and understandable. And be sure to explain how it fits the characteristics of being a procedure, a rule, or a policy.

- Find three examples of managerial decisions described in any of the popular business periodicals (*Wall Street Journal, BusinessWeek, Fortune,* etc.). Write a paper describing each decision and any related information, such as what led to the decision, what happened as a result of the decision, etc. What did you learn about decision making from these examples?
- Interview two managers and ask them for suggestions on what it takes to be a good decision maker. Write down their suggestions and be prepared to present them in class.
- Steve's and Mary's suggested readings: Noel M. Tichy and Warren G. Bennis, *Judgment: How Winning Leaders Make Great Calls* (Portfolio, 2007); Gerd Gigerenzer, *Gut Feelings: The Intelligence of the Unconscious* (Viking, 2007); Stephen P. Robbins, *Decide & Conquer: Make Winning Decisions and Take Control of Your Life* (Financial Times Press, 2004); and John S. Hammond, Ralph L. Keeney, and Howard Raiffa, *Smart Choices: A Practical Guide to Making Better Decisions* (Harvard Business School Press, 1999).
- Do a Web search on the phrase "101 dumbest moments in business." Get the most current version of this end-of-year list. Choose three of the examples and describe what happened. What's your reaction to each example? How could the managers have made better decisions?
- In your own words, write down three things you learned in this chapter about being a good manager.
- Self-knowledge can be a powerful learning tool. Go to mymanagementlab and complete these self-assessment exercises: How Well Do I Handle Ambiguity? How Well Do I Respond to Turbulent Change? and What's My Decision-Making Style? Using the results of your assessments, identify personal strengths and weaknesses. What will you do to reinforce your strengths and improve your weaknesses?

CASE APPLICATION

Designing for Dollars

Great product design is absolutely critical for most consumer products companies. But how do these companies know when a design feature will pay off, especially when every dollar counts? How do they make those tough decisions? That's the challenge that faced Whirlpool's chief designer, Chuck Jones. He knew he had to come up with a better way.

Chuck's realization that the whole process of making design decisions needed to be improved came after a meeting with Whirlpool's resource allocation team. Chuck wanted to add some ornamentation to a KitchenAid refrigerator that was being redesigned, but it would have added about $5 in extra cost. When the team asked him to estimate the return on investment (that is, would it pay off financially to add this cost?), he couldn't give them any data. His "trust me, I'm a designer" argument didn't sway them either. Chuck resolved to improve the approach to investing in design.

His first step was to survey other "design-centric" companies, including BMW, Nike, and Nokia. Surprisingly, only a few had a system for forecasting return on design. Most of them simply based future investments on past performance. Chuck said, "No one had really figured this stuff out." With so many smart, talented people in the field, why had no one been able to come up with a good way to make those decisions? According to two accounting professors, one reason is that it's incredibly difficult to discern design's contribution from all the other business functions (marketing, manufacturing, distribution, etc.). And even the design profession couldn't

KitchenAid mixers are shown at the International Home and Housewares show at McCormick Place in Chicago.

agree on how to approach this problem. Despite the obstacles, Chuck continued his quest to find a way to objectively measure the benefits of design.

What he eventually concluded was that a focus on customer preferences would work better than a focus on bottom-line returns. If his team could objectively measure what customers want in a product and then meet those needs, the company could realize financial returns. Chuck's design team created a standardized companywide process that puts design prototypes in front of customer focus groups and then gets detailed measurements of their preferences about aesthetics, craftsmanship, technical performance, ergonomics, and usability. They chart the results against competing products and the company's own products. This metrics-based approach gives decision makers a baseline of objective evidence from which to make investment decisions. Design investment decisions are now based on fact, not opinion. The "new" decision-making approach has transformed the company's culture and led to bolder designs because the designers can now make a strong case for making those investments.

Discussion Questions

1. Would you characterize product design decisions as structured or unstructured problems?
2. Describe and evaluate the process Chuck went through to change the way design decisions were made. Describe and evaluate the company's new design decision process.
3. What criteria does Whirlpool's design team use in design decisions? What do you think each of these criteria involves?

Sources: B. Breen, "No Accounting for Design," *Fast Company*, February 2007, pp. 38–39; and R. Siegel, "Meet the Whirlwind of Whirlpool," *BusinessWeek* online, www.businessweek.com, April 11, 2006.

Let's Get Real:

Meet The Managers

Sasmita Biswal

Senior Manager
Accenture Services

MY JOB: Managing delivery of application-outsourcing projects from India to overseas clients and business development.

BEST PARTS OF MY JOB: Challenges to meet the stringent targets.

WORST PART OF MY JOB: Meeting deadly targets sometimes all alone.

BEST MANAGEMENT ADVICE EVER RECEIVED: Always keep parallel channels open.

Uma Ramaseshan

Vice President (Compliance)
Axis Bank Ltd

MY JOB: Working as Money Laundering Reporting Officer, Compliance Officer for the BCSBI Codes on Customer Service, and grievance handling.

BEST PART OF MY JOB: The ability to generate acceptable levels of work efficiency and productivity from my teams.

WORST PART OF MY JOB: The feeling that I could have done better.

BEST MANAGEMENT ADVICE EVER RECEIVED: Management is getting work done by people but good management is developing people through good working standards and work culture.

You'll be hearing more from these real managers throughout the chapter.

Chapter 7

Foundations of Planning

In this chapter, we begin our study of the first of the management functions: planning. Planning is important because it establishes what an organization is doing. We'll look at how managers set goals as well as how they establish plans. Focus on the following learning outcomes as you read and study this chapter.

LEARNING OUTCOMES

A Manager's Dilemma

Courtesy of Alibaba.com.

Just as in the fictional story *Ali Baba and the 40 Thieves*, in which Ali Baba gains access to a treasure of gold, Hong Kong–based Alibaba.com provides access to an important treasure for many Chinese businesses.[1] This treasure—Web access—is extremely vital in today's environment. And Alibaba offers businesses simple and efficient Internet solutions through its three online marketplaces. For example, soon after prototypes of a new dirt bike were posted on its Alibaba Web site, scooter manufacturer Zhejiang Bifei had German buyers clamoring for shipments. And Guangdong Gemacki Appliance had nearly 200 buyers in Europe and the Middle East ... all because customers found them on its Alibaba Web site. And these are just a few of the customer success stories. Now, Alibaba's top managers have set some ambitious global goals. They're eyeing markets in places such as India, South Korea, and Taiwan. As a vice president of Alibaba.com, Trudy Dai is responsible for managing sales and customer service. What types of plans might she need to ensure that the sales and customer service areas remain strong as the company expands?

What would you do?

You may think that planning isn't something that's relevant to you right now. But when you figure out your class schedule for the next term or when you decide what you need to do to finish a class project on time, you're planning. And planning is something that all managers, like Trudy Dai, need to do. Although what they plan and how they plan may differ, it's still important that they do plan. In this chapter we present the basics: what planning is, why managers plan, and how they plan.

LEARNING OUTCOME 7.1 ▷

THE WHAT AND WHY OF PLANNING

Boeing called its new 787 aircraft the Dreamliner, but the project turned into a nightmare for managers. The new plane has been the company's most popular product ever, mostly because of its innovations, especially in fuel efficiency. However, one month before its original delivery date of May 2008, the company announced another delivery delay (the third one), which pushed back its debut another 15 months. The company admitted that the project's timeline was way too ambitious, even though every detail had been meticulously planned.[2] What happens if Boeing's customers (the airlines that ordered the jets) get tired of waiting and cancel their orders? Could the managers have planned better?

WHAT IS PLANNING?

As we stated in Chapter 1, **planning** involves defining the organization's goals, establishing strategies for achieving those goals, and developing plans to integrate and coordinate work activities. It's concerned with both ends (what) and means (how).

When we use the term *planning*, we mean *formal* planning. In formal planning, specific goals covering a specific time period are defined. These goals are written and shared with organizational members to reduce ambiguity and create a common understanding about what needs to be done. Finally, specific plans exist for achieving these goals.

WHY DO MANAGERS PLAN?

Planning seems to take a lot of effort. So why should managers plan? We can give you at least four reasons. First, planning *provides direction* to managers and nonmanagers alike. When employees know what their organization or work unit is trying to accomplish and what they must contribute in order to reach goals, they can coordinate their activities, cooperate with each other, and do what it takes to accomplish those goals. Without planning, departments and individuals might work at cross-purposes and prevent the organization from efficiently achieving its goals.

Next, planning *reduces uncertainty* by forcing managers to look ahead, anticipate change, consider the impact of change, and develop appropriate responses. Although planning won't eliminate uncertainty, managers plan so they can respond effectively.

In addition, planning *minimizes waste and redundancy*. When work activities are coordinated around plans, inefficiencies become obvious and can be corrected or eliminated.

Finally, planning *establishes the goals or standards used in controlling*. When managers plan, they develop goals and plans. When they control, they see whether the plans have been carried out and the goals met. Without planning, there would be no goals against which to measure or evaluate work effort.

PLANNING AND PERFORMANCE

Is planning worthwhile? Numerous studies have looked at the relationship between planning and performance.[3] Although most have shown generally positive relationships, we can't say that organizations that formally plan *always* outperform those that don't plan. What *can* we conclude?

First, generally speaking, formal planning is associated with positive financial results—higher profits, higher return on assets, and so forth. Second, it seems that doing a good job planning and implementing those plans plays a bigger part in high performance than does how much planning is done. Next, in studies where formal planning didn't lead to higher performance, the external environment often was the culprit. When external forces—such as governmental regulations or powerful labor unions—constrain managers' options, they reduce the impact planning has on an organization's performance. Finally, the planning–performance relationship seems to be influenced by the planning time frame. It seems that at least four years of formal planning is required before it begins to affect performance.

Nirma started in 1969 as a one-man operation. Dr Karsanbhai Patel used to make detergent in the backyard of his home and sell it at Rs. 3 per kg, while the cheapest detergent in the market was sold at Rs. 13 per kg. In 1980s, Nirma surpassed Surf, a well-established brand in detergents, with the help of indigenous process, packaging, and low-profiled marketing. Since then, Nirma has continued to grow on its strength. It has invested in setting up a deemed university. Today, Nirma has 14,000 employees and an annual turnover of more than 25,000 crores. It has taken a lot of hardwork, risk-taking, and planning to grow to this extent.[4]

Courtesy: Nirma University

planning
Defining an organization's goals, establishing an overall strategy for achieving those goals, and developing plans to integrate and coordinate work activities.

QUICK LEARNING REVIEW:

LEARNING OUTCOME 7.1

- Define planning.
- Describe the purposes of planning.
- Explain what studies have shown about the relationship between planning and performance.

Go to page 159 and see how well you know this material.

LEARNING OUTCOME 7.2 ▷

GOALS AND PLANS

Planning is often called the primary management function because it establishes the basis for all the other things managers do as they organize, lead, and control. It involves two important aspects: goals and plans.

Goals (objectives) are desired outcomes or targets.[5] They guide management decisions and form the criteria against which work results are measured. That's why they're often called the foundation of planning. You have to know the desired target or outcome before you can establish plans for reaching it. **Plans** are documents that outline how goals are going to be met. They usually include resource allocations, schedules, and other necessary actions to accomplish the goals. As managers plan, they develop both goals and plans.

TYPES OF GOALS

It might seem that organizations have a single goal: for businesses, to make a profit, and for not-for-profit organizations, to meet the needs of some constituent group(s). However, an organization's success can't be determined by a single goal. And if managers emphasize only one goal, other goals necessary for long-term success are ignored. Also, as we discussed in Chapter 5, having a single goal such as profit may result in unethical behaviors because managers and employees will ignore other aspects of their jobs in order to look good on that one measure.[6] In reality, all organizations have multiple goals. For instance, businesses may want to increase market share, keep employees enthused about working for the organization, and work toward more environmentally sustainable practices. And a church may provide a place for religious practices and also act as a social gathering place for church members and assist economically disadvantaged individuals in its community.

Most companies' goals can be classified as either strategic or financial. Financial goals are related to the financial performance of the organization, while strategic goals are related to all other areas of an organization's performance. In January 2008, BSNL announced that the company has planned to double the revenues to Rs. 80,000 crores (Rs. 800 billion) in 3 years.[7] An example of a strategic goal would be the plans of Kochi-based JRG, a leading stock, commodity, and insurance-broking company that planned to recruit 500 new employees and open 30 new regional offices across the country in the next 1 year so that the company emerges with the largest trading network in the country with 1,500 terminals.[8]

The goals described here are **stated goals**: official statements of what an organization says—and what it wants its stakeholders to believe—its goals are. However, stated goals—which can be found in an organization's charter, annual report, or public relations announcements or in public statements made by managers—are often conflicting and influenced by what various stakeholders think an organization should do. For instance, Nike's goal is to "bring inspiration and innovation to every athlete." Canadian company EnCana's vision is "to be the world's high performance benchmark independent oil and gas company." Infosys' goal is "to be a global transformation partner", International Public relations Agency Network, a public relation (PR) company's goals is to constantly innovate and to be the force behind PR being used as a strategic tool by corporates and organizations to achieve their strategic goals, and Deutsche Bank's goal is "to be the leading global provider of financial solutions for demanding clients creating exceptional value for our shareholders and people."[9] Such statements are vague and probably better represent management's public relations skills than being meaningful guides to what the organization is actually trying to accomplish. It shouldn't be surprising, then, to find that an organization's stated goals are often irrelevant to what actually goes on.[10]

If you want to know an organization's **real goals**—the goals an organization actually pursues—you should observe what organizational members are doing. Actions define priorities. For example, a number of universities say their goal is limiting class sizes, facilitating close student–faculty relationships, and actively involving students in the learning process, but they commonly hold 300+-student lecture classes! Knowing that real and stated goals may differ is important for recognizing what you might otherwise think are inconsistencies.

TYPES OF PLANS

The most popular ways to describe organizational plans are in terms of breadth (strategic versus operational), time frame (short term versus long term), specificity (directional versus specific), and frequency of use (single use versus standing). As Exhibit 7–1 shows, these types of plans aren't independent. That is, strategic plans are usually long term, directional, and single use whereas operational plans are usually short term, specific, and standing. What does each include?

Strategic plans are plans that apply to an entire organization and establish the organization's overall goals. Plans that encompass a particular operational area of the organization are called **operational plans**. These two types of plans differ in that strategic plans are broad, while operational plans are narrow.

The number of years used to define short-term and long-term plans has declined considerably because of environmental uncertainty. *Long-term* used to mean anything over seven years. Try to imagine what you're likely to be doing in seven years, and you can begin to appreciate how difficult it would be for managers to establish plans that far in the future. We define **long-term plans** as those with a time frame beyond three years.[11] **Short-term plans** are those covering one year or less. Any time period in between would be an intermediate plan. Although these time classifications are fairly common, an organization can use any planning time frame it wants.

Intuitively, it would seem that specific plans would be preferable to directional, or loosely guided, plans. **Specific plans** are plans that are clearly defined and leave no

Exhibit 7–1 Types of Plans

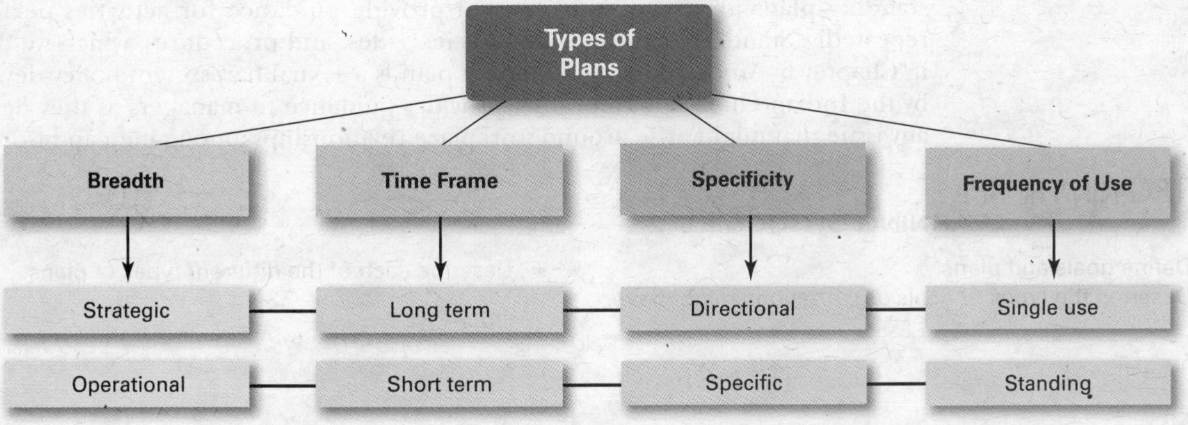

goals (objectives)
Desired outcomes or targets.

plans
Documents that outline how goals are going to be met.

stated goals
Official statements of what an organization says—and what it wants its various stakeholders to believe—its goals are.

real goals
Goals that an organization actually pursues, as defined by the actions of its members.

strategic plans
Plans that apply to an entire organization and establish the organization's overall goals.

operational plans
Plans that encompass a particular operational area of an organization.

long-term plans
Plans with a time frame beyond three years.

short-term plans
Plans covering one year or less.

specific plans
Plans that are clearly defined and that leave no room for interpretation.

When the founders of Blue Man Group, a 3-man band/performance troupe, decided they were ready to expand by hiring performers to do Blue Man shows around the country, they were eager to keep the acts consistent with their original vision. They finally realized they needed a specific plan to guide the new performers they were bringing on board, so they locked themselves in an apartment and talked through their creative vision in great detail. The result? A 132-page operating manual that tells the story of the Blue Man show and allows it to be produced by others. Ironically, by writing the plan, although it is a somewhat unorthodox one, the founders were able to express artistic ideals that had been understood among them but never stated.

room for interpretation. They have clearly defined objectives, so there's no ambiguity and no problem with misunderstanding. For example, a manager who seeks to increase his or her unit's work output by 8 percent over a given 12-month period might establish specific procedures, budget allocations, and schedules of activities to reach that goal.

However, when uncertainty is high and managers must be flexible in order to respond to unexpected changes, directional plans are preferable. **Directional plans** are flexible plans that set out general guidelines. They provide focus but don't lock managers into specific goals or courses of action. For example, Sylvia Rhone, President of Motown Records, said she has a simple goal: to "sign great artists."[12] So instead of creating a specific plan to produce and market 10 albums from new artists this year, she might formulate a directional plan to use a network of people around the world to alert her to new and promising talent so she can increase the number of new artists she has under contract. Keep in mind, however, that the flexibility of directional plans must be weighed against the lack of clarity of specific plans.

Some plans that managers develop are ongoing, while others are used only once. A **single-use plan** is a one-time plan specifically designed to meet the needs of a unique situation. For instance, when Big Bazaar decided to expand the number of its stores in India, top-level executives formulated a single-use plan as a guide. In contrast, **standing plans** are ongoing plans that provide guidance for activities performed repeatedly. Standing plans include policies, rules, and procedures, which we defined in Chapter 6. An example of a standing plan is a sexual harassment policy developed by the Indian Oil Corporation that provides guidance to managers as they deal with any issue that might arise around workplace relationships among men and women.

QUICK LEARNING REVIEW:

LEARNING OUTCOME 7.2

- Define goals and plans.
- Describe the types of goals organizations might have.
- Describe each of the different types of plans.

Go to page 159 and see how well you know this material.

LEARNING OUTCOME 7.3 ▷ SETTING GOALS AND DEVELOPING PLANS

Subrat Das has been elected the secretary of his business school's alumni association. He wants the association to be more actively involved in the business school than it has been. Monica Khan graduated from XIM Bhubaneshwar with specialization in Marketing and information technology 5 years ago and went to work for a national level consulting services firm. She was recently promoted to manage a 10-person e-business development team and hopes to strengthen the team's financial contribution to the firm. What should Subrat and Monica do to achieve their goals?

APPROACHES TO SETTING GOALS

As we stated earlier, goals provide the direction for all management decisions and actions and form the criterion against which actual accomplishments are measured. Everything organizational members do should be oriented toward achieving goals. Goals can be set either through a process of traditional goal setting or by using management by objectives.

In **traditional goal setting**, goals set by top managers flow down through the organization and become subgoals for each organizational area. This traditional perspective assumes that top managers know what's best because they see the "big picture." And the goals passed down to each succeeding level to guide individual employees as they work to achieve those assigned goals. If Subrat Das were to use this approach, she would see what goals the dean or director of the school of business had set and develop goals for her group that would contribute to achieving those goals. Or, take a manufacturing business, for example. The president tells the vice president of production what he expects manufacturing costs to be for the coming year and tells the marketing vice president what level he expects sales to reach for the year. These goals are passed to the next organizational level and written to reflect the responsibilities of that level, passed to the next level, and so forth. Then, at some later time, performance is evaluated to determine whether the assigned goals have been achieved. Or that's the way it's supposed to happen. But in reality, it doesn't always do so. Turning broad strategic goals into departmental, team, and individual goals can be a difficult and frustrating process.

Another problem with traditional goal setting is that when top managers define the organization's goals in broad terms—such as achieving "sufficient" profits or increasing "market leadership"—these ambiguous goals have to be made more specific as they flow down through the organization. Managers at each level define the goals and apply their own interpretations and biases as they make them more specific. However, what often happens is that clarity is lost as the goals make their way down from the top of the organization to lower levels. Exhibit 7–2 illustrates what can happen. But it doesn't have to be that way. For example, at Tijuana-based dj Orthopedics de Mexico, employee teams see the impact of their daily work output on company goals. Says human resource manager Joaquin Samaniego, "When people get a close connection with the result of their work, when they everyday know what they are supposed to do and how they achieved the goals, that makes a strong connection with the company and their job."[13]

Exhibit 7–2

The Downside of Traditional Goal Setting

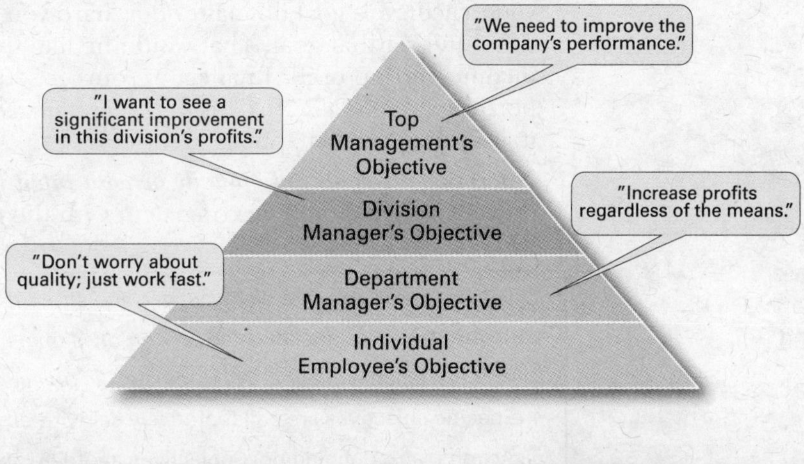

directional plans
Plans that are flexible and that set out general guidelines.

single-use plan
A one-time plan specifically designed to meet the needs of a unique situation.

standing plans
Ongoing plans that provide guidance for activities performed repeatedly.

traditional goal setting
An approach to setting goals in which top managers set goals that flow down through the organization and become subgoals for each organizational area.

Let's Get Real: F2F

HOW GOALS ARE SET: Goals are set based on the objectives and deliverables. They are also set in consideration with the resource availability and capability.

When the hierarchy of organizational goals *is* clearly defined, as it is at dj Orthopedics, it forms an integrated network of goals, or a **means–ends chain.** Higher-level goals (or ends) are linked to lower-level goals, which serve as the means for their accomplishment. In other words, the goals achieved at lower levels become the means to reach the goals (ends) at the next level. And the accomplishment of goals at that level becomes the means to achieve the goals (ends) at the next level and on up through the different organizational levels. That's how traditional goal setting is supposed to work.

Instead of using traditional goal setting, many organizations use **management by objectives (MBO)**, a process of setting mutually agreed upon goals and using those goals to evaluate employee performance. If Monica Khan were to use this approach, he would sit down with each member of his team and set goals and periodically review whether progress was being made toward achieving those goals. MBO programs have four elements: goal specificity, participative decision making, an explicit time period, and performance feedback.[14] Instead of using goals to make sure employees are doing what they're supposed to be doing, MBO uses goals to motivate employees as well. The appeal is that MBO focuses on employees working to accomplish goals they've had a hand in setting. Exhibit 7–3 lists the steps in a typical MBO program.

Does MBO work? Studies have shown that it can increase employee performance and organizational productivity. For example, one review of MBO programs found productivity gains in almost all of them.[15] But is MBO relevant for today's organizations? If it's viewed as a way of setting goals, then yes, it is relevant; research shows that goal setting can be an effective approach to motivating employees.[16]

Let's Get Real: F2F

HOW GOALS ARE SET: Goals are set and re-evaluated mainly with a purpose to ensure that the overall objective of the organization is met.

Characteristics of Well-Written Goals. Goals aren't all written the same way. Some are better than others at making clear what the desired outcomes are. Managers should be able to write well-written goals. Exhibit 7–4 lists the characteristics of a "well-written" goal.[17]

Steps in Goal Setting. Managers should follow five steps when setting goals:

1. ***Review the organization's* mission, *or purpose.*** A mission is a broad statement that provides an overall guide to what organizational members think is important. Managers should review the mission before writing goals because goals should reflect that mission.
2. ***Evaluate available resources.*** You don't want to set goals that are impossible to achieve, given your available resources. Even though goals should be challenging, they should be realistic. After all, if the resources you have to work with won't allow you to achieve a goal no matter how hard you try or how much effort is exerted, you shouldn't set that goal. That would be like the person with a INR 1,50,000 annual income and no other financial resources setting a goal of building an investment portfolio worth INR 50 lakhs in three years. No matter how hard he or she works at it, it's not going to happen.
3. ***Determine the goals individually or with input from others.*** The goals reflect desired outcomes and should be congruent with the organizational mission and with goals

Exhibit 7–3

Steps in MBO

1. The organization's *overall objectives and strategies* are formulated.
2. Major objectives are allocated among *divisional and departmental units.*
3. Unit managers *collaboratively set specific objectives* for their units with their managers.
4. Specific objectives are collaboratively set with *all department members.*
5. *Action plans,* defining how objectives are to be achieved, are specified and agreed upon by managers and employees.
6. The action plans are *implemented.*
7. Progress toward objectives is *periodically reviewed,* and *feedback is provided.*
8. Successful achievement of objectives is reinforced by *performance-based rewards.*

Exhibit 7–4

Well-Written Goals

- Written in terms of outcomes rather than actions
- Measurable and quantifiable
- Clear as to a time frame
- Challenging yet attainable
- Written down
- Communicated to all necessary organizational members

in other organizational areas. These goals should be measurable and specific, and they should include a time frame for accomplishment.

4. ***Write down the goals and communicate them to all who need to know.*** Writing down and communicating goals forces people to think them through. Written goals also become visible evidence of the importance of working toward something.
5. ***Review results and whether goals are being met.*** If goals aren't being met, change them as needed.

Once goals have been established, written down, and communicated, a manager is ready to develop plans for pursuing the goals.

DEVELOPING PLANS

The process of developing plans is influenced by three contingency factors and by the planning approach followed.

Contingency Factors in Planning. Look back at our chapter-opening "A Manager's Dilemma." How will Trudy Dai know what types of plans to develop for guiding Alibaba? Three contingency factors affect the choice of plans: organizational level, degree of environmental uncertainty, and length of future commitments.[18]

Exhibit 7–5 shows the relationship between a manager's level in the organization and the type of planning done. For the most part, lower-level managers do operational planning, while upper-level managers do strategic planning.

The second contingency factor is environmental uncertainty. When uncertainty is high, plans should be specific but flexible. Managers must be prepared to change or amend plans as they're implemented. At times, managers may even have to abandon the plans.[19] For example, at Continental Airlines, the former CEO and his management team established a specific goal of focusing on what customers wanted most—on-time flights—to help the company become more competitive in the highly uncertain airline industry. Because of the high level of uncertainty, the management team identified a "destination but not a flight plan" and changed plans as necessary to achieve the goal of on-time service.

Exhibit 7–5

Planning and Organizational Level

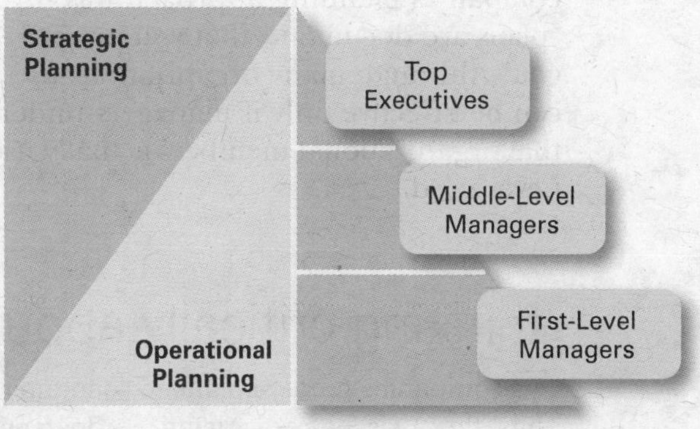

means–ends chain
An integrated network of goals in which the accomplishment of goals at one level serves as the means for achieving the goals, or ends, at the next level.

management by objectives (MBO)
A process of setting mutually agreed upon goals and using those goals to evaluate employee performance.

mission
A statement of the purpose of an organization.

Let's Get Real: F2F

PLANNING I DO:
I do short-, medium-, and long-term planning.

The last contingency factor is also related to the time frame of plans. The **commitment concept** says that plans should extend far enough to meet the commitments made when the plans were developed. Planning for too long or too short a time period is inefficient and ineffective. An example of the importance of the commitment concept can be seen at data centers where companies' computers are housed. Many have found that their "power-hungry computers" created challenges. For instance, at Pomona Valley Medical Center, the expansion from 30 servers to 70 servers generated so much heat that the medical center's air-conditioning system was overwhelmed. At San Antonio–based Rackspace Ltd., which manages servers for clients, utility power needs have swelled from 3 to 8 megawatts, resulting in sky-high electric bills. And at the Department of Energy's National Energy Research Computing Center, providing adequate power for its supercomputer meant digging up a parking lot in order to knock a hole in a basement wall to install locomotive-sized power sources and air-conditioning units.[20] How does this illustrate the commitment concept? As organizations expand their computing technology, they're "committed" to whatever future expenses are generated by that plan. They have to live with the decision and its consequences, good and bad.

APPROACHES TO PLANNING

Federal, state, and local government officials are working together on a plan to boost populations of wild salmon in the northwestern United States. Managers in the Global Fleet Graphics division of 3M Company are developing detailed plans to satisfy increasingly demanding customers and to battle more aggressive competitors. Emilio Azcárraga Jean, chairman, president, and CEO of Grupo Televisa, gets input from many different people before setting company goals and then turns over the planning for achieving the goals to various executives. In each of these situations, planning is done a little differently. *How* an organization plans can best be understood by looking at *who* does the planning.

Let's Get Real: F2F

PLANNING I DO:
Predetermine and prioritize the deadlines to be met and targets to be achieved on a daily or monthly and quarterly basis.

In the traditional approach, planning is done entirely by top-level managers who often are assisted by a **formal planning department**, a group of planning specialists whose sole responsibility is to help write the various organizational plans. Under this approach, plans developed by top-level managers flow down through other organizational levels, much like the traditional approach to goal setting. As they flow down through the organization, the plans are tailored to the particular needs of each level. Although this approach makes managerial planning thorough, systematic, and coordinated, all too often, the focus is on developing "the plan," a thick binder (or binders) full of meaningless information, that's stuck away on a shelf and never used by anyone for guiding or coordinating work efforts. In fact, in a survey of managers about formal top-down organizational planning processes, more than 75 percent said that their company's planning approach was unsatisfactory.[22] A common complaint was that, "plans are documents that you prepare for the corporate planning staff and later forget." Although many organizations use traditional top-down planning, this approach can be effective only if managers understand the importance of creating documents that organizational members actually use, not documents that look impressive but are never used.

thinking critically about Ethics

As companies prepare plans to keep their businesses operating if a bird-flu pandemic hits, thorny issues are arising. For instance, Procter & Gamble "asked its company doctors whether it should try to secure a private stash of the avian-influenza drug Tamiflu for its staff in Asia." The company's medical leader in southern Asia said, "How ethical would it be if we were holding supplies that the general public didn't have access to but badly needed?"[21] What do you think? Would it be unethical for a company to protect its own employees? What other alternatives might there be?

Another approach to planning is to involve more organizational members in the process. In this approach, plans aren't handed down from one level to the next but instead are developed by organizational members at the various levels and in the various work units to meet their specific needs. For instance, at Dell, employees from production, supply management, and channel management meet weekly to make plans based on current product demand and supply. In addition, work teams set their own daily schedules and track their progress against those schedules. If a team falls behind, team members develop "recovery" plans to try to get back on schedule.[23] When organizational members are more actively involved in planning, they see that the plans are more than just something written down on paper. They can actually see that the plans are used in directing and coordinating work.

QUICK LEARNING REVIEW:

LEARNING OUTCOME 7.3

- Discuss how traditional goal setting and MBO work.
- Describe well-written goals and explain how to set them.
- Discuss the contingency factors that affect planning.
- Describe the approaches to planning.

Go to page 159 and see how well you know this material.

LEARNING **OUTCOME 7.4** ▷

CONTEMPORARY ISSUES IN PLANNING

We conclude this chapter by addressing two contemporary issues in planning. Specifically, we look at criticisms of planning and at how managers can plan effectively in dynamic environments.

CRITICISMS OF PLANNING

Formalized organizational planning became popular in the 1960s, and it still is today, for the most part. It makes sense for an organization to establish targets and some direction. But critics have challenged some of the basic assumptions of planning:

1. ***Planning may create rigidity.***[24] Formal plans can lock an organization into specific goals to be achieved within specific time frames. When these goals were set, the assumption may have been that the environment wouldn't change. If that assumption is faulty, managers who follow a plan may face trouble. Rather than remain flexible—maybe even discard the plan—managers who continue to pursue the original goals may not be able to cope with the changed environment. Staying "on course" when the environment is changing can be a recipe for disaster.
2. ***Plans can't be developed for a dynamic environment.***[25] If a basic assumption of planning—that the environment won't change—is faulty, then how can you plan? Today's business environment is often random and unpredictable. Managing under those conditions requires flexibility, and that may mean not being tied to formal plans.
3. ***Formal plans can't replace intuition and creativity.***[26] Organizations often succeed because of someone's innovative vision, and routine planning efforts may impede such a vision. For example, the rapid growth of Apple Computer in the 1970s and 1980s was attributed, in part, to the innovative and creative approaches of co-founder Steve Jobs. As the company grew, Jobs felt it needed more formalized management—something he was uncomfortable providing. He hired a CEO who ultimately ousted Jobs from his own company. With Jobs's departure came

commitment concept
A concept which says that plans should extend far enough to meet the commitments made when the plans were developed.

formal planning department
A group of planning specialists whose sole responsibility is helping to write organizational plans.

Let's Get Real: F2F

BE A BETTER PLANNER BY: Linking all the components of resources such as manpower, capital, space, and, of course, business.

Let's Get Real: F2F

BE A BETTER PLANNER BY: Work out a plan based on a realistic view of ones ability to meet with the plan considering the time constraints and other factors involved in achieving the same.

increased organizational formality, including detailed planning—the same things Jobs despised so much because he felt they hampered creativity. By the mid-1990s, Apple, once an industry leader, was struggling for survival. The situation became so bad that the CEO was ousted, and Jobs was brought back as CEO to get Apple back on track. The company's renewed focus on creativity led to the iMac in 1998, the iPod in 2001, a radically new look for the iMac in 2002, the online iTunes music store in 2003, a video iPod in 2005, and the iPhone in 2007.

4. ***Planning focuses managers' attention on today's competition, not on tomorrow's survival.***[27] Formal planning has a tendency to focus on how to capitalize on existing business opportunities within an industry but may not allow managers to consider creating or reinventing an industry. Consequently, formal plans may result in costly blunders when other competitors take the lead. On the other hand, companies such as Intel, General Electric, Nokia, and Sony have found success forging into new industries.
5. ***Formal planning reinforces success, which may lead to failure.***[28] Success breeds success. That's an American tradition. If something's not broken, don't fix it, right? Well, maybe not! Success may, in fact, breed failure in an uncertain environment. It's difficult to change or discard previously successful plans—to leave the comfort of what works for the anxiety of the unknown. Successful plans may provide a false sense of security, generating more confidence in the formal plans than is warranted. Many managers will not face the unknown until they're forced to do so by environmental changes. By then, it may be too late!
6. ***Just planning isn't enough.*** It's not enough for managers just to plan. They have to start doing![29] When executives at the *Wall Street Journal* needed to respond to a prolonged slump in advertising, they developed a plan for how best to accomplish that goal. And then they set about doing it. One of the first things they did was change the newspaper's design by adding more color to its pages, redesigning the typeface, and making other format changes. Another thing they did was launch a Saturday edition. Next, they decreased the physical size of the newspaper.[30] As this example shows, just planning to do something doesn't get it done. Planning to make enough money so you can retire at age 35 isn't enough. You have to put that plan into motion and do it. Managers need to plan, and they also need to see that the plan is carried out.

How valid are these criticisms? Should managers forget about planning? No! Although rigid and inflexible planning may lead to such problems, today's managers can effectively plan if they understand planning in dynamic uncertain environments.

EFFECTIVE PLANNING IN DYNAMIC ENVIRONMENTS

The external environment is continually changing. For instance, Wi-Fi has revolutionized all kinds of industries, from airlines to automobile manufacturing to supermarkets. Companies are using the Internet for customer-driven product design. Amounts spent on eating out instead of cooking at home are predicted to decline. Prices for crude oil have reached all-time highs. And experts believe that China and India are transforming the twenty-first century global economy.

How can managers effectively plan when the external environment is continually changing? We already discussed uncertain environments as one of the contingency factors that affect the types of plans managers develop. Because dynamic environments are more the norm than the exception, let's look at managers they can effectively plan in such environments.

In an uncertain environment, managers should develop plans that are specific but flexible. Although this may seem contradictory, it's not. To be useful, plans need some specificity, but the plans should not be set in stone. Managers need to recognize that planning is an ongoing process. Plans serve as a road map, although the destination may change due to dynamic market conditions. Managers should be ready to change directions if environmental conditions warrant. This flexibility is particularly important as plans are implemented. Managers need to stay alert to environmental changes that may affect implementation and respond as needed. Keep in mind, also, that even when the environment is highly uncertain, it's important to continue formal planning in order to see any effect on organizational performance. Persistence in planning contributes to significant performance improvement. Why? It seems that, as with most other activities, managers learn to plan, and the quality of their planning improves when they continue to do it.[31]

Finally, making the organizational hierarchy flatter helps make planning more effective in dynamic environments. This means allowing lower organizational levels to set goals and develop plans because there's little time for goals and plans to flow down from the top. Managers should teach their employees how to set goals and to plan and then trust them to do it. We can look to Bangalore, India, to find a company that effectively understands this. Just a decade ago, Wipro Limited was "an anonymous conglomerate selling cooking oil and personal computers, mostly in India." Today, it is a $3.4 billion-per-year global company with most of its business (some 90 percent) coming from information technology services.[32] Accenture, EDS, IBM, and the big U.S. accounting firms know all too well the competitive threat Wipro represents. Not only are Wipro's employees economical, they're knowledgeable and skilled. And they play an important role in the company's planning. Because the information services industry is continually changing, employees are taught to analyze situations and to define the scale and scope of a client's problems in order to offer the best solutions. These employees are the ones on the frontline with the clients, and it's their responsibility to establish what to do and how to do it. It's an approach that positions Wipro for success no matter how the industry changes.

QUICK LEARNING REVIEW:

LEARNING OUTCOME 7.4

- **Explain the criticisms of planning.**
- **Describe how managers can effectively plan in today's dynamic environment.**

Go to page 159 and see how well you know this material.

Let's Get Real: Our Turn

Sasmita Biswal

Senior Manager
Accenture Services

As the name Alibaba suggests the clientele will be on the look out for new and innovative cost-effective products, such as the treasures on the Web. Sales, customer services, and after-sales operations are the key focus areas as the company expands to new geographies. Planning for market survey for ascertaining the expectations of the customers from Alibaba.com and a promotion strategy in international media will improve the visibility of the site.

Uma Ramaseshan

Vice President (Compliance)
Axis Bank Ltd

- Identify sales controllers to monitor sales results (Web-hits) on weekly basis to ensure that there is sales generation, customer enlistment, and retention.
- Ensure that all the sales channels required for delivery of a service or product is either procured or acquired or recruited to meet with the growing demands in the new Asian markets as well as the growing existing markets. Time schedules for acquisition of such delivery channels are to be predetermined.
- Promotional activities to be undertaken on Web as well as through alternate audio-visual media, such as television and FM channels, keeping in view the sensitivities of the target audience.
- Encourage the repeat customers that frequently visit the Website to write clients' testimonials or product reviews, which would be available for global viewing.
- Implement an effective customer feedback system to ensure quality control and efficient post-sales customer service.
- Customers may also be offered customized replacement of faulty products or buyback of old products in lieu of new ones as marketing cum customer service strategy.

LEARNING OUTCOMES **SUMMARY**

7.1 ▷ THE WHAT AND WHY OF PLANNING

- Define planning.
- Describe the purposes of planning.
- Explain what studies have shown about the relationship between planning and performance.

Planning involves defining an organization's goals, establishing an overall strategy for achieving those goals, and developing plans for organizational work activities. The four purposes of planning are to provide direction, reduce uncertainty, minimize waste and redundancy, and establish the goals or standards used in controlling. Studies of the planning–performance relationship have concluded that formal planning is associated with positive financial performance, for the most part; it's more important to do a good job of planning and implementing the plans than to do more extensive planning; the external environment is usually the reason companies that plan don't achieve high levels of performance; and the planning–performance relationship seems to be influenced by the planning time frame.

7.2 ▷ GOALS AND PLANS

- Define goals and plans.
- Describe the types of goals organizations might have.
- Describe each of the different types of plans.

Goals are desired outcomes. Plans are documents that outline how goals are going to be met. Goals can be strategic or financial, and they can be stated or real. Strategic plans apply to an entire organization, while operational plans encompass a particular functional area. Long-term plans are those with a time frame beyond three years. Short-term plans are those covering one year or less. Specific plans are clearly defined and leave no room for interpretation. Directional plans are flexible and set out general guidelines. A single-use plan is a one-time plan designed to meet the needs of a unique situation. Standing plans are ongoing plans that provide guidance for activities performed repeatedly.

7.3 ▷ SETTING GOALS AND DEVELOPING PLANS

- Discuss how traditional goal setting and MBO work.
- Describe well-written goals and explain how to set them.
- Discuss the contingency factors that affect planning.
- Describe the approaches to planning.

In traditional goal setting, goals are set at the top of the organization and then become subgoals for each organizational area. Management by objectives (MBO) is a process of setting mutually agreed upon goals and using those goals to evaluate employee performance. Well-written goals have six characteristics: (1) written in terms of outcomes, (2) measurable and quantifiable, (3) clear as to time frame, (4) challenging but attainable, (5) written down, and (6) communicated to all organizational members who need to know them. Goal setting involves these steps: review the organization's mission, evaluate available resources, determine the goals individually or with input from others, write down the goals and communicate them to all who need to know them, and review results and change goals as needed. The contingency factors that affect planning include the manager's level in the organization, the degree of environmental uncertainty, and the length of future commitments. The two main approaches to planning are the traditional approach and MBO. In the traditional approach, top managers develop plans that flow down through other organizational levels; this approach may use a formal planning department. MBO involves more organizational members in the planning process.

7.4 ▷ CONTEMPORARY ISSUES IN PLANNING

- Explain the criticisms of planning.
- Describe how managers can effectively plan in today's dynamic environment.

The main criticisms of planning are that (1) planning may create rigidity; (2) plans can't be developed for a dynamic environment; (3) formal plans can't replace intuition and creativity; (4) planning focuses managers' attention on today's competition, not tomorrow's; (5) formal planning reinforces success, which may lead to failure; and (6) just planning isn't enough. These criticisms are valid if planning is rigid and inflexible. Managers can effectively plan in today's dynamic environment by using plans that are specific but flexible. It's also important to push responsibility for establishing goals and developing plans to lower organizational levels.

THINKING ABOUT MANAGEMENT ISSUES

1. Will planning become more or less important to managers in the future? Why?
2. If planning is so crucial, why do some managers choose not to do it? What would you tell these managers?
3. Explain how planning involves making decisions today that will have an impact later.
4. How might planning in a not-for-profit organization such as the American Cancer Society differ from planning in a for-profit organization such as Coca-Cola?
5. What types of planning do you do in your personal life? Describe these plans in terms of being (a) strategic or operational, (b) short term or long term, and (c) specific or directional.
6. The late Peter Drucker, an eminent management author, identified the SMART format for setting goals back in 1954: S (specific), M (measurable), A (attainable), R (relevant), and T (time bound). Is this format still relevant today? Discuss.
7. Many companies have a goal of becoming more environmentally sustainable. One of the most important steps they can take is controlling paper waste. Choose a company—any type, any size. Imagine that you've been put in charge of creating a program to control paper waste for the company. Set goals and develop plans. Prepare a report for your boss (that is, your professor), outlining these goals and plans.

YOUR TURN to be a Manager

- Practice setting goals for various aspects of your personal life, such as academics, career preparation, family, hobbies, and so forth. Set at least two short-term goals and at least two long-term goals for each area.
- For the goals that you have set, write out plans for achieving those goals. Think in terms of what you will have to do to accomplish each one. For instance, if one of your academic goals is to improve your grade-point average, what will you have to do to reach that goal?
- Write a personal mission statement. Although this may sound simple to do, it's not going to be simple or easy. Our hope is that it will be something that you'll want to keep, use, and revise when necessary... that it will be something that helps you be the you you'd like to be and helps you live the life you'd like to live. Start by doing some research on personal mission statements. There are some wonderful Web resources that can guide you. Good luck!
- Interview three managers about the types of planning they do. Ask them for suggestions on how to be a better planner. Write a report that describes and compares your findings.
- Choose two companies, preferably in different industries. Research the companies' Web sites and find examples of goals that they have stated. (Hint: A company's annual report is often a good place to start.) Evaluate these goals. Are they well written? Rewrite those that don't exhibit the characteristics of well-written goals so that they do.
- Steve's and Mary's suggested readings: Peter F. Drucker, *Management: Tasks, Responsibilities, Practices* (Harper Business, 1974); Peter F. Drucker, *The Executive in Action: Managing for Results* (Harper Business, 1967); and Peter F. Drucker, *The Practice of Management* (HarperCollins, 1954).
- What does it take to be a good planner? Do some research on this. As part of your research, talk to professors and other professionals. Make a bulleted list of suggestions. Be sure to cite your sources.
- In your own words, write down three things you learned in this chapter about being a good manager.
- Self-knowledge can be a powerful learning tool. Go to mymanagementlab and complete these self-assessment exercises: What's My Attitude Toward Achievement? What Are My Course Performance Goals? What Time of Day Am I Most Productive? and How Good Am I at Personal Planning? Using the results of your assessments, identify personal strengths and weaknesses. What will you do to reinforce your strengths and improve your weaknesses?

CASE APPLICATION

Hologram on MasterCard credit card.

Mastering the Plan . . . Priceless

When MasterCard became a public corporation in May 2006, the momentous occasion signified the start of a new way of doing things for the company's 4,600 employees around the world. Company executives wanted to ensure that every employee understood what such a change meant and how MasterCard would be different after the initial public offering (IPO). To do that, they decided to hold the "largest single learning event in the company's history."

This learning event, dubbed the RoadMap to the Future, was to be a series of intensive, 4.5-hour seminars conducted in 110 workshops in 36 cities over a three-week time frame. Rebecca Ray, the company's senior vice president for global learning, was put in charge of the event. She recognized that pulling it off effectively and efficiently would require some serious and detailed planning. Dozens of company human resource (HR) specialists and hundreds of local managers around the world would be serving as trainers and facilitators. They would be the ones face-to-face with employees, teaching them about what being a publicly traded company meant and what changes they could expect. Preparing these individuals for such an important task would require significant planning. The goal of the training was to ensure that every employee understood the business strategy and how MasterCard would be different after the IPO.

The training program was to be anchored by three "learning maps" or topics. The first, called "Universe of Opportunity," would describe the company's competitive landscape and industry challenges/opportunities. The second, titled "How We Make Money," was to focus on MasterCard's financial models and how it fit into the industry. The final one, titled "New Climate, New Culture, New Company," would be very detailed about the company's strategy as a public company and what it would take to successfully pursue that strategy.

Discussion Questions

1. What role do you think goals would play in planning for this training event? List some goals you think might be important. (Make sure these goals have the characteristics of well-written goals.)
2. What types of plans would be needed for actually doing the event? (For instance, strategic or operational or both? Short term, long term, or both?) Explain why you think these plans would be important.
3. What challenges might there be in doing such an event? How about doing such an event in different global locations over a short time frame? How could they best prepare for those challenges?
4. What did this case story teach you about planning?

Sources: J. Gordon, "MasterCard's Master Plan," *Training*, October 2007, pp. 58–62.

Let's Get Real:

Meet The Manager

Rupinder S. Anand

Director, Sales and Marketing
Diebold Systems Pvt. Ltd

MY JOB: Responsible for driving sales for Diebold through addressing and, sometimes, creating business opportunities for key accounts.

BEST PART OF MY JOB: People management and customer interactions.

WORST PART OF MY JOB: Firing people.

BEST MANAGEMENT ADVICE EVER RECEIVED: People are definitely a company's greatest asset. It doesn't make any difference whether the product is cars or cosmetics. A company is only as good as the people it keeps.

You'll be hearing more from this real manager throughout the chapter.

Chapter 8

Strategic Management

In this chapter, we look at an important part of the planning that managers do: developing organizational strategies. Every organization has strategies for doing what it's in business to do. And managers must manage those strategies effectively. Focus on the following learning outcomes as you read and study this chapter.

LEARNING OUTCOMES

A Manager's Dilemma

As chairman of the Arab world's leading real estate development company, Mohamed Ali Alabbar has a goal of meeting the demand for quality housing and other properties in his region and, increasingly, in growing markets such as India and Egypt.[1] Villas and apartments built by his company, Emaar Properties PJSC, have sold quickly. The sweet combination of inexpensive land and strong demand have led to sky-high profit margins. Sky-high would also describe Emaar's biggest project to date—the Downtown Burj project, a US$20 billion development housing the world's tallest building (2,300 feet tall) and the world's biggest shopping mall (1,500 stores). This development also includes an area called Old Town, with buildings that have buff-colored domes resembling a traditional Arab city. As the company continues to rapidly expand—it's now operating in 16 countries—Mohamed is finding that his greatest challenge is strategy implementation. One of his top managers says, "We have the vision and the strategy. The question is, do we have the skills across the organization to execute?" Put yourself in Mohamed's shoes. How could SWOT analysis help him as his company continues to grow?

What would you do?

The importance of having good strategies can be seen in what Mohamed Ali Alabbar has accomplished with Emaar Properties. By recognizing real estate opportunities and formulating effective strategies to exploit these opportunities, his company has become the leader in the Arab world. As he continues to expand, strategic management will continue to play an important role. An underlying theme in this chapter is that effective strategies result in high organizational performance.

LEARNING OUTCOME 8.1 ▷

STRATEGIC MANAGEMENT

Suzlon Energy Ltd, a leading wind turbine manufacturer, decides to sell 10 percent stake in the gearbox company Hansen Transmissions International. Funskool is capitalizing on its partnership with the Industrial Design Centre (IDC) of IIT Mumbai to sell six Indian theme board games all over India. RBI cut its key rates—repo and reverse repo—to ensure availability of credit at lower rates.[2] These are just some of the business stories from a Business Daily in a single day, and each one is about a company or an organization's strategies.

WHAT IS STRATEGIC MANAGEMENT?

The discount retail industry is a good place to see what strategic management is all about. Wal-Mart and Kmart Corporation (now part of Sears Holdings) have battled for market dominance since 1962, the year both companies were founded. The two chains have other striking similarities: store atmosphere, names, markets served, and organizational purpose. Yet, Wal-Mart's performance (financial and otherwise) has far surpassed that of Kmart. Wal-Mart is the world's largest retailer, and Kmart was the largest retailer ever to seek Chapter 11 bankruptcy protection. Why the difference in performance? Because of different strategies and competitive abilities.[3] Wal-Mart has excelled by using strategic management, while Kmart has struggled.

Strategic management is what managers do to develop an organization's strategies. It's an important task that involves all the basic management functions—planning, organizing, leading, and controlling. What are an organization's **strategies**? They're the plans for how the organization will do whatever it's in business to do, how it will compete successfully, and how it will attract and satisfy its customers in order to achieve its goals.[4]

One term often used in strategic management is **business model**, which simply is how a company is going to make money. The business model focuses on two factors: (1) whether customers will value what the company is providing and (2) whether the company can make any money doing that.[5] For instance, Dell pioneered a new business model for selling computers to consumers directly from the Internet instead of selling through computer retailers, like all the other manufacturers. Did customers "value" that? Absolutely! Did Dell make money doing it that way? Absolutely! As managers think about strategies, they need to think about the economic viability of their company's business model.

WHY IS STRATEGIC MANAGEMENT IMPORTANT?

In 2000, Star Plus aired Saas Bhi Kabhi Bahu Thi that became one of the most popular shows in India. Until November 2008 (this was when the program ended), a large number of viewers tuned in every evening to watch the story unfold. Similarly, Ramayan, the first mythological serial, was aired in 1987–1988. It became quickly popular and earned a place in the Limca Book of Records; Ramayan caused a standstill in India whenever it aired on Doordarshan. It was aired again on NDTV in 2007. The producers of the IPL 20-20 match seem to understand the importance of strategic management because they are developing and exploiting every aspect of the cricket match, talk shows, cheer leaders, brand managers, merchandize, etc.

Let's Get Real: F2F

WHY MANAGERS NEED TO KNOW ABOUT STRATEGIC MANAGEMENT:

- Managers need to be sensitive to strategic management as markets are changing. Strategies are built on assumptions, which in the absence of perfect knowledge will never be perfectly correct.

Why is strategic management so important? There are three reasons. The most significant one is that it can make a difference in how well an organization performs. Why do some businesses succeed and others fail, even when faced with the same environmental conditions? (Remember our Wal-Mart and Kmart example.) Research has found a generally positive relationship between strategic planning and performance.[6] In other words, it appears that organizations that use strategic management do have higher levels of performance. And that makes it pretty important for managers!

Another reason it's important has to do with the fact that managers in organizations of all types and sizes face continually changing situations. They cope with this uncertainty by using the strategic management process to examine relevant factors and decide what actions to take.

Finally, strategic management is important because organizations are complex and diverse. Each part needs to work toward achieving the organization's goals; strategic management helps do this. For example, with 12,540 employees Reliance Industries use strategic management to help coordinate and focus employees' efforts on what's important.

Today, both business organizations and not-for-profit organizations use strategic management. For instance, the India Post found itself in competitive battles with email, couriers, etc. The Indian Post and the Telegraph Department used strategic management to come up with a response. For example, to help eliminate dependence on budgetary support and to become a profit-making unit, India Post planned to spend more than Rs. 6,000 crores on technology as part of a 5-year national restructuring exercise in 2008. Although strategic management in not-for-profits hasn't been as well researched as that in for-profit organizations, we know it's important for those organizations as well.

strategic management
What managers do to develop an organization's strategies.

strategies
Plans for how an organization will do what it's in business to do, how it will compete successfully, and how it will attract and satisfy its customers in order to achieve its goals.

business model
A design for how a company is going to make money.

QUICK LEARNING REVIEW:

LEARNING OUTCOME 8.1

- Define strategic management, strategy, and business model.
- Give three reasons why strategic management is important.

Go to page 180 and see how well you know this material.

LEARNING **OUTCOME 8.2** ▷

THE STRATEGIC MANAGEMENT PROCESS

The **strategic management process** (see Exhibit 8–1) is a six-step process that encompasses strategy planning, implementation, and evaluation. Although the first four steps describe the planning that must take place, implementation and evaluation are just as important! Even the best strategies can fail if management doesn't implement or evaluate them properly.

STEP 1: IDENTIFYING THE ORGANIZATION'S CURRENT MISSION, GOALS, AND STRATEGIES

Every organization needs a **mission**—a statement of its purpose. Defining a mission forces managers to identify what the organization is in business to do. For instance, the mission of GMR group is "To build entrepreneurial organizations that make a difference to society through creation of value." The mission of Infosys is "To achieve our objectives in an environment of fairness, honesty, and courtesy toward our clients, employees, vendors and society at large." The mission of L. V. Prasad Eye Institute is to be a center of excellence in eye care services, basic and clinical research into eye diseases and vision-threatening conditions, training, product development, and rehabilitation for those with incurable visual disability, with a focus on extending equitable and efficient eye care to underserved populations in the developing world. These statements provide clues to what these organizations see as their purpose. What should a mission statement include? Exhibit 8–2 describes some typical components.

It's also important for managers to identify the current goals and strategies. Why? So managers have a basis for assessing whether they need to be changed.

STEP 2: DOING AN EXTERNAL ANALYSIS

What impact might the following trends have for businesses?

- Denmark, Sweden, and Switzerland lead the rankings of the world's most technologically advanced economies.
- Researchers studying the role of colleges and universities said that, "Traditional colleges and universities are doomed. Technology, lethargy, and astronomical costs will destroy the current model of higher education to create a \$100-billion-a-year opportunity for businesses and investors."

Exhibit 8–1 The Strategic Management Process

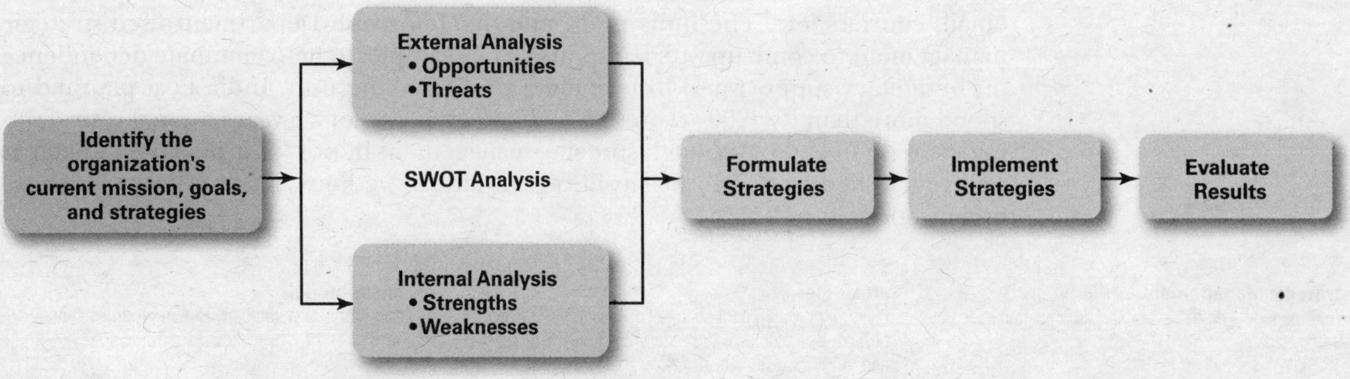

Exhbit 8–2

Components of a Mission Statement

Customers: Who are the firm's customers?

Markets: Where does the firm compete geographically?

Concern for survival, growth, and profitability: Is the firm committed to growth and financial stability?

Philosophy: What are the firm's basic beliefs, values, and ethical priorities?

Concern for public image: How responsive is the firm to societal and environmental concerns?

Products or services: What are the firm's major products or services?

Technology: Is the firm technologically current?

Self-concept: What are the firm's major competitive advantage and core competencies?

Concern for employees: Are employees a valuable asset of the firm?

Source: Based on F. David, *Strategic Management*, 11 ed. (Upper Saddle River, NJ: Prentice Hall, 2007), p. 70.

- Web-enabled computers and other digital tools have accelerated the amount of electronic multitasking going on, particularly among "Gen M"—the term being used to describe young teens and young adults who seem to be simultaneously talking on their cell phones, carrying on multiple IM (instant messaging) chats, watching TV, and doing homework.[7]

We described the external environment in Chapter 3 as an important constraint on a manager's actions. Analyzing that environment is a critical step in the strategic management process. Managers do an external analysis so they know, for instance, what the competition is doing, what pending legislation might affect the organization, or what the labor supply is like in locations where it operates. In an external analysis, managers should examine both the specific and general environments to see the trends and changes.

Once they've analyzed the environment, managers need to pinpoint opportunities that the organization can exploit and threats that it must counteract or buffer against. **Opportunities** are positive trends in the external environment; **threats** are negative trends.

STEP 3: DOING AN INTERNAL ANALYSIS

Now we move to the internal analysis, which provides important information about an organization's specific resources and capabilities. An organization's **resources** are its assets—financial, physical, human, and intangible—that it uses to develop, manufacture, and deliver products to its customers. They're "what" the organization has. On the other hand, its **capabilities** are its skills and abilities in doing the work activities needed in its business—"how" it does its work. The major value-creating capabilities of the organization are known as its **core competencies.**[8] Both resources and core competencies determine the organization's competitive weapons.

After completing an internal analysis, managers should be able to identify organizational strengths and weaknesses. Any activities the organization does well or any unique resources it has are called **strengths**. **Weaknesses** are activities the organization doesn't do well or resources it needs but doesn't possess.

The combined external and internal analyses are called a **SWOT analysis**, which is an analysis of the organization's strengths, weaknesses, opportunities, and threats.

strategic management process
A six-step process that encompasses strategic planning, implementation, and evaluation.

mission
A statement of the purpose of an organization.

opportunities
Positive trends in external environmental factors.

threats
Negative trends in external environmental factors.

resources
An organization's assets that are used to develop, manufacture, and deliver products to its customers.

capabilities
An organization's skills and abilities in doing the work activities needed in its business.

core competencies
The organization's major value-creating capabilities that determine its competitive weapons.

strengths
Any activities an organization does well or any unique resources that it has.

weaknesses
Any Activities an organization does not do well or resources it needs but does not possess.

SWOT analysis
An analysis of an organization's strengths, weaknesses, opportunities, and threats.

thinking critically about Ethics

How much profit is too much profit? Is it okay that the aggregate net profit of Mukesh Ambani group has increased 10.2 percent in the first half of 2008–2009 while investors in their company have been losing money over the last year? Is it okay that the net profit for the Anil Dhirubhai Ambani Group (ADAG)-promoted Reliance Capital Ltd (RCL) announced a consolidated net profit of Rs. 3,43,000 crores for the first quarter of 2008 while the budget for the entire state of Orissa for 2007–2008 was Rs. 22,000 crores?[9] How can and should a company ethically earn? As a business student, what do you think? How would you explain this situation to your friends who are not business students?

Let's Get Real: F2F

WHY DO SWOT ANALYSIS? SWOT essentially gives us an opportunity to see the market that we are in from a balanced perspective. SWOT helps to clear misconceptions, prompts timely actions and responses, and corrects the course.

After completing a SWOT analysis, managers are ready to formulate appropriate strategies—that is, strategies that (1) exploit an organization's strengths and external opportunities, (2) buffer or protect the organization from external threats, or (3) correct critical weaknesses.

STEP 4: FORMULATING STRATEGIES

As managers formulate strategies, they should consider the realities of the external environment and their available resources and capabilities and design strategies that will help the organization achieve its goals. There are three main types of strategies managers formulate: corporate, business, and functional. We'll describe each shortly.

STEP 5: IMPLEMENTING STRATEGIES

Once strategies are formulated, they must be implemented. No matter how effectively an organization has planned its strategies, performance will suffer if the strategies aren't implemented properly.

STEP 6: EVALUATING RESULTS

The final step in the strategic management process is evaluating results. How effective have the strategies been at helping the organization reach its goals? What adjustments are necessary? After assessing the results of the previous strategies and determining that changes were needed in them, Mr Gupta of Lupin Pharmaceutical made strategic adjustments to regain market share and improve the company's bottom line. The company improved production, revamped one manufacturing plant, and reorganized management.

managing in a Virtual World

IT's Role in Company Strategy

How important is information technology (IT) to a company's strategy? We'll look at two examples that illustrate just how important it can be. Harrah's Entertainment, the world's largest gaming company, is fanatical about customer service—and for good reason. Company research showed that customers who were satisfied with the service they received at a Harrah's casino increased their gaming expenditures by 10 percent, and those who were extremely satisfied increased their gaming expenditures by 24 percent. Harrah's discovered this important customer service–expenditures connection because of its incredibly sophisticated information system. But an organization's IT doesn't always have such a positive payoff, as the next example shows. At Prada's Manhattan flagship store, store designers were hoping for a "radically new shopping experience" that combined "cutting-edge architecture and twenty-first-century customer service." Or at least that was the strategy. Prada invested almost one-fourth of the store's budget into IT, including wireless networks linked to an inventory database. The store envisioned sales staff roaming the store armed with PDAs, checking whether items were in stock. Even the dressing rooms would have touch screens so customers could check on items. But the strategy didn't work as planned. The equipment malfunctioned, and the staff was overwhelmed with trying to cope with crowds and equipment that didn't work. In this case, the multi-million-dollar investment might not have been the best strategy.

So how important is IT to a company's strategy? Undoubtedly, when an IT system is working as it's supposed to, it's a wonderful asset and tool, as the Harrah's example shows. However, as Prada so painfully discovered, when an IT system isn't working as it's supposed to, serious problems can arise.[10]

QUICK LEARNING REVIEW:
LEARNING OUTCOME 8.2

- **Describe the six steps in the strategic management process.**
- **Define SWOT (strengths, weaknesses, opportunities, and threats).**

Go to page 180 and see how well you know this material.

LEARNING
OUTCOME 8.3 ▷

CORPORATE STRATEGIES

As we said earlier, organizations use three types of strategies: corporate, business, and functional (see Exhibit 8–3). Top-level managers typically are responsible for corporate strategies, middle-level managers for competitive strategies, and lower-level managers for functional strategies. In this section, we'll look at corporate strategies.

WHAT IS A CORPORATE STRATEGY?

A **corporate strategy** is one that specifies what businesses a company is in or wants to be in and what it wants to do with those businesses. It's based on the mission and goals of the organization and the roles that each business unit of the organization will play. We can see both of these aspects with PepsiCo, for instance. Its mission is "to be the world's premier consumer products company focused on convenient foods and beverages." PepsiCo pursues its mission with a corporate strategy that has put it in different businesses, including PepsiCo International, Frito-Lay North America, PepsiCo Beverages North America, and Quaker Foods North America. The other part of corporate strategy is top managers deciding what to do with those businesses.

WHAT ARE THE TYPES OF CORPORATE STRATEGY?

The three main types of corporate strategy are growth, stability, and renewal. Let's look at each type.

Growth Strategies. Wal-Mart, the world's largest retailer, continues to grow internationally and in the United States. With a **growth strategy**, an organization expands the number of markets served or products offered, either through its current business(es) or through new business(es). Because of its growth strategy, an organization may increase its revenues, number of employees, or market share. Organizations grow by using concentration, vertical integration, horizontal integration, or diversification.

An organization that grows using *concentration* focuses on its primary line of business and increases the number of products offered or markets served in that primary business. For instance, CRI pumps, based out of Coimbatore, Tamil Nadu, a 250-crore company

Exhibit 8–3 Types of Organizational Strategies

corporate strategy
An organizational strategy that specifies what businesses a company is in or wants to be in and what it wants to do with those businesses.

growth strategy
A corporate strategy that's used when an organization wants to expand the number of markets served or products offered, either through its current business(es) or through new business(es).

with an annual production of 750,000 pump that exports to 40 countries, has used concentration as its strategy. CRI pumps started in 1961 out of a small nonferrous foundry set up by K. Gopal. Another example is Bose Corporation of Framingham, Massachusetts, which focuses on developing innovative audio products and has become one of the world's leading manufacturers of speakers for home entertainment, automotive, and pro audio markets, with annual sales of more than $2 billion.

A company also might choose to grow by using *vertical integration*, either backward, forward, or both. In backward vertical integration, an organization becomes its own supplier so it can control its inputs. For instance, eBay owns an online payment business that helps it provide more secure transactions and control one of its most critical processes. In forward vertical integration, an organization becomes its own distributor and is able to control its outputs. For example, Apple has more than 80 retail stores to distribute its product.

In *horizontal integration*, a company grows by combining with competitors. For instance, French cosmetics giant L'Oréal acquired The Body Shop. Horizontal integration has been used in a number of industries in the past few years—financial services, consumer products, airlines, department stores, and software, among others. The U.S. Federal Trade Commission usually scrutinizes these combinations closely to see if consumers might be harmed by decreased competition. Till January 2005, there was a Press Note 18 mandating any foreign company to first get no objection from its local partner in the existing venture, which means, if the Honda group wants to set up an independent unit and break away from Hero Honda, it would have to get a no objection from the Hero group with which they had the joint venture, Hero Honda.[11] Other countries may have similar restrictions. For instance, managers at Oracle Corporation had to get approval from the European Commission, the "watchdog" for the European Union, before it could acquire rival business-software maker PeopleSoft.

Finally, an organization can grow through *diversification*, either related or unrelated. With related diversification, a company combines with other companies in different, but related, industries. For example, Godrej is in a variety of businesses, including locks and storage (original business), steel furniture, office equipment, consumer durables, personal and household care products, pest control, agricultural products, international trading, real estate, and software. The strategic fit is the basic value proposition on delivering quality products and services that the customer can trust and all those are at competitive costs.[12] With unrelated diversification, a company combines with firms in different and unrelated industries. For instance, the Tata Group of India has businesses in chemicals, communications and IT, consumer products, energy, engineering, materials, and services. This is an odd mix, and in this case, there's no strategic fit among the businesses.

Stability Strategies. As sales of candy and chocolate slow down, Cadbury Schweppes—with almost half of its confectionery sales coming from chocolate—is maintaining things as they are. A **stability strategy** is a corporate strategy in which an organization continues

Steven Shore (on left in photo) and Barry Prevor (on right, in checkered shirt) of retailer Steve & Barry's fueled their company's growth through low-cost operations. The two CEOs saved money on everything from the low rent they paid in midsize malls hungry for tenants to hefty allowances they earned for building the interiors of their stores. Buying direct from overseas factories also cut expenses as did word-of-mouth advertising. However, despite their low-cost approach, the tough economic climate in 2008 led to the company's filing for bankruptcy. A turnaround specialist bought the company and plans to continue operating the 276 stores.

to do what it is currently doing. Examples of this strategy include continuing to serve the same clients by offering the same product or service, maintaining market share, and sustaining an organization's current business operations. With this type of strategy, the organization doesn't grow, but it doesn't fall behind, either.

Renewal Strategies In 2008, many real estate-related companies, such as DLF and Unitech, faced serious financial issues. When an organization is in trouble, something needs to be done. Managers need to develop strategies, called **renewal strategies**, that address declining performance. There are two main types of renewal strategies: retrenchment and turnaround strategies. A *retrenchment strategy* is a short-run renewal strategy used for minor performance problems. This type of strategy helps an organization stabilize operations, revitalize organizational resources and capabilities, and prepare to compete once again. When an organization's problems are more serious, more drastic action—a *turnaround strategy*—is needed. Managers do two things for both retrenchment and turnaround strategies: cut costs and restructure organizational operations. However, in a turnaround strategy, these measures are more extensive than in a retrenchment strategy.

HOW ARE CORPORATE STRATEGIES MANAGED?

When an organization's corporate strategy encompasses a number of businesses, managers can manage this collection, or portfolio, of businesses using a tool called a corporate portfolio matrix. This matrix provides a framework for understanding diverse businesses and helps managers establish priorities for allocating resources.[13] The first portfolio matrix—the **BCG matrix**—was developed by Boston Consulting Group and introduced the idea that an organization's various businesses could be evaluated using a 2 × 2 matrix (see Exhibit 8–4) to identify which ones offered high potential and which were draining organizational resources.[14] The horizontal axis represents market share (low or high), and the vertical axis indicates anticipated market growth (low or high). A business unit is evaluated using a SWOT analysis and placed in one of the four categories.

What are the strategic implications of using the BCG matrix? The dogs should be sold off or liquidated as they have low market share in markets with low growth potential. Managers should "milk" cash cows for as much as they can, limit any new investment in them, and use the large amounts of cash generated to invest in stars and question marks with strong potential to improve market share. Heavy investment in stars will help take advantage of the market's growth and help maintain high market share. The stars, of course, will eventually develop into cash cows as their markets mature and sales growth slows. The hardest decisions for managers relate to the question marks. After careful analysis, some will be sold off and others turned into stars.

Exhibit 8–4

BCG Matrix

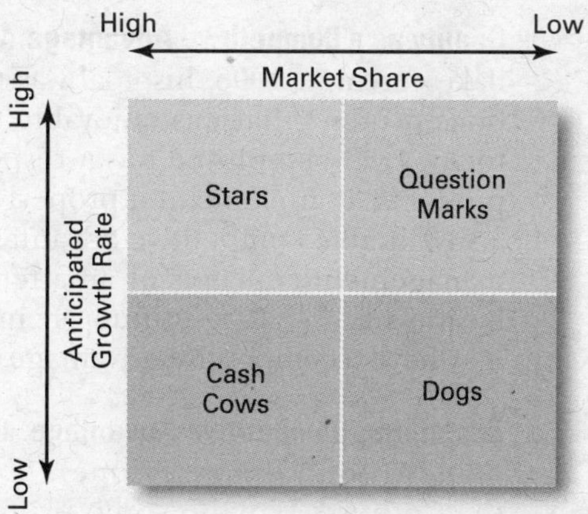

stability strategy
A corporate strategy in which an organization continues to do what it is currently doing.

renewal strategy
A corporate strategy designed to address declining performance.

BCG matrix
A strategy tool that guides resource allocation decisions on the basis of market share and growth rate of SBUs.

QUICK LEARNING REVIEW:

LEARNING OUTCOME 8.3

- Describe the three major types of corporate strategies.
- Explain how the BCG matrix is used to manage corporate strategies.

Go to page 180 and see how well you know this material.

LEARNING OUTCOME 8.4 ▷ COMPETITIVE STRATEGIES

A **competitive strategy** is a strategy for how an organization will compete in its business(es). For a small organization in only one line of business or a large organization that has not diversified into different products or markets, the competitive strategy describes how it will compete in its primary or main market. For organizations in multiple businesses, however, each business has its own competitive strategy that defines its competitive advantage, the products or services it will offer, the customers it wants to reach, and the like[15]. For example, the French company LVMH-Moët Hennessy Louis Vuitton SA has different competitive strategies for its businesses, which include Donna Karan fashions, Louis Vuitton leather goods, Guerlain perfume, TAG Heuer watches, Dom Perignon champagne, and other luxury products. When an organization is in several different businesses, those single businesses that are independent and formulate their own competitive strategies are often called **strategic business units (SBUs)**.

THE ROLE OF COMPETITIVE ADVANTAGE

Michelin has mastered a complex technological process for making superior radial tires. Coca-Cola has created the world's most powerful brand, using specialized marketing and merchandising capabilities. The Ritz-Carlton hotels have a unique ability to deliver personalized customer service. Each of these companies has created a competitive advantage.

Developing an effective competitive strategy requires an understanding of **competitive advantage**, which is what sets an organization apart—that is, its distinctive edge. That distinctive edge comes from the organization's core competencies because the organization does something that others cannot do or does it better than others can do it. For example, Southwest Airlines has a competitive advantage because of its skills at giving passengers what they want—convenient and inexpensive service. Or competitive advantage can come from a company's resources—the organization has something that its competitors do not have. For instance, Wal-Mart's state-of-the-art information system allows it to monitor and control inventories and supplier relations more efficiently than its competitors, which Wal-Mart has turned into a cost advantage.

Quality as a Competitive Advantage. When W. K. Kellogg started manufacturing his cornflake cereal in 1906, his goal was to provide his customers with a high-quality, nutritious product that was enjoyable to eat. That emphasis on quality is still important today. Every employee has a responsibility to maintain the high quality of Kellogg products. If implemented properly, quality can be a way for an organization to create a sustainable competitive advantage.[16] That's why many organizations apply quality management concepts in an attempt to set themselves apart from competitors. If a business is able to continuously improve the quality and reliability of its products, it may have a competitive advantage that can't be taken away.[17]

Sustaining Competitive Advantage. Every organization has resources (assets) and capabilities (how work gets done). So what makes some organizations more successful than others? Why do some professional baseball teams consistently win championships or draw large crowds? Why do some organizations have consistent and continuous growth in revenues and profits? Why do some colleges, universities, or departments experience continually increasing enrollments? Why do some companies consistently appear at the top of lists ranking the "best," or the "most admired,"

At Luxottica's factory in Agordo, Italy, designer-brand eyeglass frames are carefully inspected for scratches or other imperfections before being shipped to stores all over the world. Quality has remained a top priority at the company even as it has grown dramatically in the last several years. The focus begins in the design phase and extends all the way through production, 80 percent of which is still done in Italy in the heart of the mountainous areas where artisans have specialized in eyeglasses for generations.

or the "most profitable"? The answer is that not every organization is able to effectively exploit its resources and to develop the core competencies that can provide it with a competitive advantage. And it's not enough simply to create a competitive advantage. An organization must be able to sustain that advantage; that is, it must be able to keep its edge, despite competitors' actions or evolutionary changes in the industry. But that's not easy to do! Market instabilities, new technology, and other changes can challenge managers' attempts at creating a long-term, sustainable competitive advantage. However, by using strategic management, managers can better position their organizations to get a sustainable competitive advantage. Many important ideas in strategic management have come from the work of Michael Porter.[18] One of his major contributions was explaining how managers can create a sustainable competitive advantage. An important part of doing this is an industry analysis, which is done using the five forces model.

Five Forces Model. In any industry, five competitive forces dictate the rules of competition. Together, these five forces (see Exhibit 8–5) determine industry attractiveness and profitability, which managers assess using these five factors:

1. ***Threat of new entrants***—How likely is it that new competitors will come into the industry?
2. ***Threat of substitutes***—How likely is it that other industries' products can be substituted for our industry's products?
3. ***Bargaining power of buyers***—How much bargaining power do buyers (customers) have?
4. ***Bargaining power of suppliers***—How much bargaining power do suppliers have?
5. ***Current rivalry***—How intense is the rivalry among current industry competitors?

CHOOSING A COMPETITIVE STRATEGY

Once managers have assessed the five forces and done a SWOT analysis, they're ready to select an appropriate competitive strategy—that is, one that fits the competitive strengths (resources and capabilities) of the organization and its industry. According to Porter, no firm can be successful by trying to be all things to all people. He proposed that managers select a strategy that will give the organization a competitive advantage, either from having lower costs than all other industry competitors or by being significantly different from competitors.

competitive strategy
An organizational strategy for how an organization will compete in its business(es).

strategic business units (SBUs)
The single businesses of an organization that are independent and formulate their own competitive strategies.

competitive advantage
The factor that sets an organization apart; its distinctive edge.

Exhibit 8–5

Five Forces Model

Source: Based on M. E. Porter, *Competitive Strategy: Techniques for Analyzing Industries and Competitors* (New York: The Free Press, 1980).

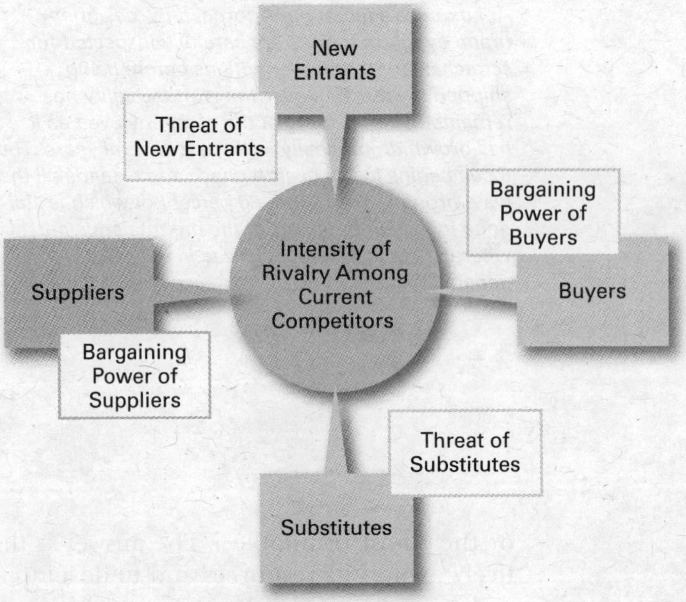

When an organization competes on the basis of having the lowest costs in its industry, it's following a *cost leadership strategy*. A low-cost leader is highly efficient. Overhead is kept to a minimum, and the firm does everything it can to cut costs. You won't find expensive art or interior décor at offices of low-cost leaders. For example, at Wal-Mart's headquarters in Bentonville, Arkansas, office furnishings are functional, not elaborate—maybe not what you'd expect for the world's largest retailer. Although a low-cost leader doesn't place a lot of emphasis on "frills," its products must be perceived as comparable in quality to those offered by rivals, or they must at least be acceptable to buyers.

A company that competes by offering unique products that are widely valued by customers is following a *differentiation strategy*. Product differences might come from exceptionally high quality, extraordinary service, innovative design, technological capability, or an unusually positive brand image. Practically any successful consumer product or service can be identified as an example of the differentiation strategy; for instance, Erstwhile, Air Deccan (renamed as Kingfisher Red after its takeover by Kingfisher) focused on customer service, Ritu Berry Apparels on design and brand image, Liberty on comfort and durability of its shoes, and Apple on product design.

Whereas the cost leadership strategy and differentiation strategy are aimed at the broad market, the final type of competitive strategy—the *focus strategy*—involves a cost advantage (cost focus) or a differentiation advantage (differentiation focus) in a narrow segment or niche. Segments can be based on product variety, customer type, distribution channel, or geographic location. For example, Denmark's Bang & Olufsen, whose revenues are over $671 million, focuses on high-end audio equipment sales. Whether a focus strategy is feasible depends on the size of the segment and whether the organization can make money serving that segment.

What happens if an organization can't develop a cost or differentiation advantage? Porter called that being *stuck in the middle* and warned that it's not a good place to be.

Although Porter said that an organization had to pursue either the low-cost or the differentiation advantage to prevent being stuck in the middle, more recent research has shown that organizations *can* successfully pursue both a low-cost and a differentiation advantage and achieve high performance.[19] Needless to say, it's not easy to pull off! A company has to keep costs low *and* be truly differentiated. But companies such as FedEx, Intel, and Coca-Cola have been able to do it.

Before we leave this section, we want to point out the final type of organizational strategy, the **functional strategies**, which are the strategies used by an organization's various functional departments to support the competitive strategy. For example, when R. R. Donnelley & Sons Company, a Chicago-based printer, wanted to become more competitive and invested in high-tech digital printing methods, its marketing department had to develop new sales plans and promotional pieces, the production department had to incorporate the digital equipment in the printing plants, and the human resources department had to update its employee selection and training programs. We don't cover specific functional strategies in this book, as you'll cover them in other business courses you take.

QUICK LEARNING REVIEW:

LEARNING OUTCOME 8.4

- Describe the role of competitive advantage.
- Explain Porter's five forces model.
- Describe Porter's three competitive strategies.

Go to page 180 and see how well you know this material.

LEARNING **OUTCOME 8.5** ▷

CURRENT STRATEGIC MANAGEMENT ISSUES

There's no better example of the strategic challenges faced by managers in today's market environment than the recorded music industry. Overall sales of recorded music declined again in 2007—about 10 percent. However, one bright spot is that digital music sales rose slightly, but not enough to offset the sales decline of compact discs. In addition, rampant global piracy (according to the IFPI—an organization representing the worldwide recording industry—one in three music discs sold worldwide is an illegal copy), economic uncertainty, and intense competition from other forms of entertainment have devastated the industry. The industry continues to change, and managers are struggling to find strategies that will help their organizations succeed in such an environment.[20] But it isn't just the music industry that's dealing with strategic challenges. Managers everywhere face increasingly intense global competition and high performance expectations of investors and customers. How have they responded to these new realities? In this section we look at some current strategy issues, including the need for strategic flexibility and how managers are designing strategies to emphasize e-business, customer service, and innovation.

THE NEED FOR STRATEGIC FLEXIBILITY

Jürgen Schrempp, former CEO of Daimler AG, stated, "My principle always was ... move as fast as you can and [if] you indeed make mistakes, you have to correct them. . . . It's much better to move fast, and make mistakes occasionally, than move too slowly."[21] You wouldn't think that smart individuals who are paid lots of money to manage organizations would make mistakes when it comes to strategic decisions. But even when managers use the strategic management process, there's no guarantee that the chosen strategies will lead to positive outcomes. And reading any of the current business periodicals would certainly support this assertion! But the key is responding quickly when it's obvious that a strategy isn't working. In other words, managers need **strategic flexibility**—that is, the ability to recognize major external changes, to quickly commit resources, and to recognize when a strategic decision isn't working. Given the highly uncertain environment that managers face today, strategic flexibility seems absolutely necessary. Exhibit 8–6 provides suggestions for developing strategic flexibility.

NEW DIRECTIONS IN ORGANIZATIONAL STRATEGIES

ESPN.com gets more than 16 million unique users per month. 16 million! That's almost twice the population of New York City. And its popular online business is just one of many businesses that ESPN is in. Originally founded as a television channel,

functional strategies
The strategies used by an organization's various functional departments to support the organization's competitive strategy.

strategic flexibility
The ability to recognize major external changes, to quickly commit resources, and to recognize when a strategic decision was a mistake.

Exhibit 8–6

Strategic Flexibility

- Know what's happening with strategies currently being used by *monitoring and measuring results.*
- Encourage employees to *be open about disclosing and sharing negative information.*
- *Get new ideas and perspectives from outside* the organization.
- Have *multiple alternatives* when making strategic decisions.
- *Learn from mistakes.*

Source: Based on K. Shimizu and M. A. Hitt, "Strategic Flexibility: Organizational Preparedness to Reverse Ineffective Strategic Decisions," *Academy of Management Executive,* November 2004, pp. 44–59.

ESPN is now into original programming, radio, online, publishing, gaming, X Games, ESPY Awards, ESPN Zones, and global.[22] The company president, George Bodenheimer, "runs one of the most successful and envied franchises in entertainment," and he obviously understands how to successfully manage its various strategies in today's environment! We think three strategies are important in today's environment: e-business, customer service, and innovation.

E-Business Strategies. Managers use e-business strategies to develop a sustainable competitive advantage.[23] A cost leader can use e-business to lower costs in a variety of ways. For instance, it might use online bidding and order processing to eliminate the need for sales calls and to decrease sales force expenses, it could use Web-based inventory control systems that reduce storage costs, or it might use online testing and evaluation of job applicants.

A differentiator needs to offer products or services that customers perceive and value as being unique. For instance, a business might use Internet-based knowledge systems to shorten customer response times, provide rapid online responses to service requests, or automate purchasing and payment systems so that customers have detailed status reports and purchasing histories.

Finally, since the focuser targets a narrow market segment with customized products, it might provide chat rooms or discussion boards for customers to interact with others who have common interests, design niche Web sites that target specific groups with specific interests, or use Web sites to perform standardized office functions such as payroll or budgeting.

Research has shown that an important e-business strategy might be a clicks-and-bricks strategy. A *clicks-and-bricks firm* is one that uses both online (clicks) and traditional stand-alone locations (bricks).[24] For example, ICICI established a site for online banking to reduce the number of customers coming to the bank. However, many customers still prefer to visit the bank at least once in 4–6 months. So, it is important to have a clicks and bricks strategy to provide appropriate service.

Let'l Get Real: F2F

SKILLS NEEDED BY A GOOD STRATEGIC LEADER:

- Ability to think and take difficult decisions.

Customer Service Strategies. Companies emphasizing excellent customer service need strategies that cultivate that atmosphere from top to bottom. Such strategies involve giving customers what they want, communicating effectively with them, and providing employees with customer service training. Let's look first at the strategy of giving customers what they want.

It shouldn't surprise you that an important customer service strategy is giving customers what they want, which is a major aspect of an organization's overall marketing strategy. For instance, New Balance, maker of athletic shoes, gives customers a truly unique product: shoes in varying widths. No other athletic shoe manufacturer has shoes for narrow or wide feet and in practically any size.[25]

Having an effective customer communication system is an important customer service strategy. Managers should know what's going on with customers. They need to find out what customers liked and didn't like about their purchase encounter—from their interactions with employees to their experience with an actual product or service. It's also important to let customers know what's going on with the company that might affect future purchase decisions. For example, Bikki Oberoi, Vice Chairman and MD of East India Hotels, is fanatical about customer feedbacks. At all Oberoi Hotels, it is mandatory that all negative feedback should be forwarded to Bikki Oberoi, who personally responds to all the unhappy customers. Even the front-desk clerk is empowered to go beyond the call of his or her duty to satisfy a customer.

Finally, an organization's culture is important to providing excellent customer service. This typically requires that employees be trained to provide exceptional customer service. For example, Singapore Airlines is well known for its customer treatment. "On everything facing the customer, they do not scrimp," says an analyst based in Singapore.[26] Employees are expected to "get service right," leaving employees with no doubt about the expectations about how to treat customers.

Innovation Strategies. When Procter & Gamble purchased the Iams pet food business, it did what it always does—used its renowned research division to look for ways to transfer technology from its other divisions to make new products.[27] One outcome of this cross-divisional combination: a new tartar-fighting ingredient from toothpaste that's included in all its dry adult pet foods. As this example shows, innovation strategies aren't necessarily focused on just the radical, breakthrough products. They can include applying existing technology to new uses. And organizations have successfully used both approaches.

What types of innovation strategies do organizations need in today's environment? The strategies used should reflect an organization's innovation philosophy, which is shaped by two strategic decisions: innovation emphasis and innovation timing. Managers must first decide where the emphasis of their innovation efforts will be. Is the organization going to focus on basic scientific research, product development, or process improvement? Basic scientific research requires the most resource commitment because it involves the nuts-and-bolts work of scientific research. In numerous industries (for instance, genetic engineering, pharmaceuticals, information technology, or cosmetics), an organization's expertise in basic research is the key to a sustainable competitive advantage. However, not every organization requires this extensive commitment to scientific research to achieve high performance levels. Instead, many depend on product development strategies. Although such a strategy also requires a significant resource investment, it's not in areas associated with scientific research. Instead, the organization takes existing technology and improves on it or applies it in new ways, just as Procter & Gamble did when it applied tartar-fighting knowledge to pet food products. Both basic scientific research and product development can help an organization achieve high levels of differentiation, which can be a significant source of competitive advantage.

Finally, the last strategic approach to innovation emphasis is a focus on process development. Using this strategy, an organization looks for ways to improve and enhance its work processes. The organization innovates new and improved ways for employees to do their work in all organizational areas. This innovation strategy can lead to lower costs, which can, of course, also be a significant source of competitive advantage.

Once managers have determined the focus of their innovation efforts, they must decide on an innovation timing strategy. Some organizations want to be the first with innovations, whereas others are content to follow or mimic the innovations. An organization that's first to bring a product innovation to market or to use a new process innovation is called a **first mover**. Being a first mover has certain strategic advantages and disadvantages, as shown in Exhibit 8–7. Some organizations pursue this route, hoping to develop a sustainable competitive advantage. Others have successfully developed a sustainable competitive advantage by being followers in the industry. They let the first movers pioneer the innovations and then mimic their products or processes. Which approach managers choose depends on their organization's innovation philosophy and specific resources and capabilities.

first mover
An organization that's first to bring a product innovation to market or to use a new process innovation.

Exhibit 8–7

First-Mover Advantages and Disadvantages

Advantages	Disadvantages
• Reputation for being innovative and industry leader	• Uncertainty over exact direction technology and market will go
• Cost and learning benefits	• Risk of competitors imitating innovations
• Control over scarce resources and keeping competitors from having access to them	• Financial and strategic risks
• Opportunity to begin building customer relationships and customer loyalty	• High development costs

QUICK LEARNING REVIEW:

LEARNING OUTCOME 8.5

- Explain why strategic flexibility is important.
- Describe e-business strategies.
- Discuss what strategies organizations might use to become more customer oriented and to be more innovative.

Go to page 180 and see how well you know this material.

Let's Get Real:
My Turn

Rupinder S. Anand
Director, Sales and Marketing
Diebold Systems Pvt. Ltd

Start with a SWOT for each country and answer the questions given below:

Strengths (internal): What does the company do well? What are its assets? What advantages does the company have over its competitors?

Weaknesses (internal): What is done badly? Does it have people who understand the local market or customer in these other countries? How might a change in leadership at these locations affect the staff?

Opportunities (external): What has the competition missed in these countries? What are the emerging needs of the customer? Do customers have financial resources in these countries to afford the Emaar quality projects? What should Emaar be doing better?

Threats (external): Are the company's competitors getting stronger? Is the land affordable in these countries? Does the company have cash to fund development? Does the economic and political climate support such projects in that country?

Proposed Action:

Obviously Emaar has a great understanding of the customer needs and resources in UAE/Dubai and it is leveraging on this to build world-class quality infrastructure in a way that will also give it global recognition. It also has great execution capabilities, which probably beat that of competition.

1. Build a competency or skills requirement for the leadership positions for each of the geographies.
2. Do a competency assessment for the existing leadership or management in each of the target countries identified as per the SWOT above to determine if Emaar has the right people and, if yes, what are their development needs.
3. Put together a development plan to address the gaps.
4. Parallely, Mohammad Ali will need to consistently communicate his strategy to the core team who, in turn, need to do the same to their respective teams.
5. The execution plan along with the resource requirement, timelines, checkpoints, and employee engagement initiatives to be formulated for the identified countries.
6. Ali will also need to consistently monitor the execution in each country through the concerned leadership.
7. Perform a SWOT for the country from time to time to determine if the strategy needs a course correction in its execution.

LEARNING OUTCOMES **SUMMARY**

8.1 ▷ STRATEGIC MANAGEMENT

- Define *strategic management, strategy,* and *business model.*
- Give three reasons why strategic management is important.

Strategic management is what managers do to develop an organization's strategies. Strategies are the plans for how the organization will do whatever it's in business to do, how it will compete successfully, and how it will attract and satisfy its customers in order to achieve its goals. A business model is how a company is going to make money. Strategic management is important for three reasons. First, it makes a difference in how well organizations perform. Second, it's important for helping managers cope with continually changing situations. Finally, strategic management helps coordinate and focus employee efforts on what's important.

8.2 ▷ THE STRATEGIC MANAGEMENT PROCESS

- Describe the six steps in the strategic management process.
- Define *SWOT (strengths, weaknesses, opportunities, and threats).*

The six steps in the strategic management process are (1) identify the current mission, goals, and strategies; (2) do an external analysis; (3) do an internal analysis; (4) formulate strategies; (5) implement strategies; and (6) evaluate strategies. Steps 2 and 3 are collectively known as SWOT analysis. Strengths are any activities the organization does well or unique resources that it has. Weaknesses are activities the organization doesn't do well or resources it needs but doesn't have. Opportunities are positive trends in the external environment. Threats are negative trends.

8.3 ▷ CORPORATE STRATEGIES

- Describe the three major types of corporate strategies.
- Explain how the BCG matrix is used to manage corporate strategies.

With a growth strategy, an organization expands the number of markets served or products offered, either through current or new businesses. The types of growth strategies include concentration, vertical integration (backward and forward), horizontal integration, and diversification (related and unrelated). With a stability strategy, an organization makes no significant changes in what it's doing. Both renewal strategies—retrenchment and turnaround—address organizational weaknesses that are leading to performance declines. Using the BCG matrix is a way to analyze a company's portfolio of businesses by looking at a business's market share and its industry's anticipated growth rate. The four categories of the BCG matrix are cash cows, stars, question marks, and dogs.

8.4 ▷ COMPETITIVE STRATEGIES

- Describe the role of competitive advantage.
- Explain Porter's five forces model.
- Describe Porter's three competitive strategies.

An organization's competitive advantage is what sets it apart, its distinctive edge. A company's competitive advantage becomes the basis for choosing an appropriate business or competitive strategy. Porter's five forces model assesses the five competitive forces that dictate the rules of competition in an industry: threat of new entrants, threat of substitutes, bargaining power of buyers, bargaining power of suppliers, and current rivalry. Porter's three competitive strategies are cost leadership (competing on the basis of having the lowest costs in the industry), differentiation (competing on the basis of having unique products that are widely valued by customers), and focus (competing in a narrow segment, with either a cost advantage or a differentiation advantage).

8.5 ▷ CURRENT STRATEGIC MANAGEMENT ISSUES

- Explain why strategic flexibility is important.
- Describe e-business strategies.
- Explain what strategies organizations might use to become more customer oriented and to be more innovative.

Strategic flexibility—that is, the ability to recognize major external environmental changes, to quickly commit resources, and to recognize when a strategic decision isn't working—is important because managers often face highly uncertain environments. Managers can use e-business strategies to reduce costs, to differentiate their firm's products and services, or to target (focus on) specific customer groups or to lower costs by standardizing certain office functions. Another important e-business strategy is the clicks-and-bricks strategy, which combines online and traditional stand-alone locations. Strategies managers can use to become more customer oriented include giving customers what they want, communicating effectively with customers, and cultivating a culture that emphasizes customer service. Strategies managers can use to become more innovative include deciding their organization's innovation emphasis (basic scientific research, product development, or process development) and its innovation timing (first mover or follower).

THINKING ABOUT MANAGEMENT ISSUES

1. Perform a SWOT analysis on a local business you know well. What, if any, competitive advantage does this organization have?
2. How might the process of strategy formulation, implementation, and evaluation differ for (a) large businesses, (b) small businesses, (c) not-for-profit organizations, and (d) global businesses?
3. "The concept of competitive advantage is as important for not-for-profit organizations as it is for for-profit organizations." Do you agree or disagree with this statement? Explain, using examples to make your case.
4. Should ethical considerations be included in analyses of an organization's internal and external environments? Why or why not?
5. How could the Internet be helpful to managers as they follow the steps in the strategic management process?
6. Find examples of five different organizational mission statements. Using the mission statements, describe what types of corporate-level and business-level strategies each organization might use to fulfill its mission statement. Explain your rationale for choosing each strategy.

YOUR TURN to be a Manager

- Do a personal SWOT analysis. Assess your personal strengths and weaknesses (skills, talents, abilities). What are you good at? What are you not so good at? What do you enjoy doing? What don't you enjoy doing? Then, identify career opportunities and threats by researching job prospects in the industry you're interested in. Look at trends and projections. You might want to check out the information the Bureau of Labor Statistics provides on job prospects. Once you have all this information, write a specific career action plan. Outline five-year career goals and what you need to do to achieve those goals.
- Using current business periodicals, find two examples of each of the corporate and competitive strategies. Write a description of what these businesses are doing and how each represents a particular strategy.
- Pick five companies from the latest version of *Fortune*'s "Most Admired Companies" list. Research these companies and identify, for each, its (a) mission statement, (b) strategic goals, and (c) strategies being used.
- Steve's and Mary's suggested readings: Adrian Slywotzky and Richard Wise, *How to Grow When Markets Don't* (Warner Business Books, 2003); Jim Collins, *Good to Great: Why Some Companies Make the Leap... and Others Don't* (Harper Business, 2001); Michael E. Porter, *On Competition* (Harvard Business School Press, 1998); James C. Collins and Jerry I. Porras, *Built to Last: Successful Habits of Visionary Companies* (Harper Business, 1994); and Gary Hamel and C. K. Prahalad, *Competing for the Future* (Harvard Business School Press, 1994).
- Customer service, e-business, and innovation strategies are particularly important to managers today. We described in the chapter specific ways that companies can pursue these strategies. Your task is to pick either customer service, e-business, or innovation and find one example for each of the specific approaches in that category. For instance, if you choose customer service, find an example of (a) giving customers what they want, (b) communicating effectively with customers, and (c) providing employees with customer service training. Write a report describing your examples.
- In your own words, write down three things you learned in this chapter about being a good manager.
- Self-knowledge can be a powerful learning tool. Go to mymanagementlab and complete these self-assessment exercises: How Well Do I Handle Ambiguity? How Creative Am I? How Well Do I Respond to Turbulent Change? Using the results of your assessments, identify personal strengths and weaknesses. What will you do to reinforce your strengths and improve your weaknesses?

CASE APPLICATION

Living Large

Although the music business is struggling, Live Nation is sitting pretty. It's the world's largest events and live music promoter, with more than 64 million people attending some 28,000 of those events each year. The company also owns the House of Blues chain of venues, where customers can enjoy different genres of live music. CEO Michael Rapino has guided the company since it was spun off as a separate business in 2005 from radio giant Clear Channel Communications.

On its Web site, Live Nation describes itself as the "future of the music business." Through live concerts, music venues and festivals, and the most comprehensive concert search engine on the Web, Live Nation is revolutionizing the music industry both onstage and online. Its strategy is to connect the artists to the fans. And Rapino isn't satisfied with dominating only the concert business. Although Live Nation will continue to focus on its live music assets, Rapino is going after the record labels' most important assets—the music stars. He's offering them a one-stop operation that handles their every musical need. That offer is: "We already operate your tours. Why not let us make your albums, sell your merchandise, run your web site, and produce your videos and a range of other products you haven't yet thought of." In October 2007, Rapino landed a big name when he signed a first-of-its-kind deal with Madonna, who left her longtime label Warner Records and signed a 10-year contract estimated at $120 million to let Live Nation handle every part of her business except publishing. Madonna's manager said, "The labels are in a jam. For a company to do well in music now, it's got to be in all aspects of the business. And Live Nation is the risk-taker. It's leading the charge." LiveNation has signed Shakira, Jay-Z, and Nickelback to similar deals and hopes to add more superstars to its roster.

Michael Rapino, CEO of Live Nation.

The key to Live Nation's growth strategy is the ability to connect to those millions of people who attend shows every year. The company's valuable database containing contact information for those fans gives it an efficient way to offer them additional music-related products and services. Will Rapino's strategy live or die?

Discussion Questions

1. What growth strategy does Live Nation appear to be using? What competitive advantage do you think Live Nation has?
2. How might SWOT analysis be useful to Mike Rapino?
3. Find Live Nation's most current annual report. What goals is the company pursuing? What strategies is it using? Do its strategies appear to be helping it reach these goals?
4. What do you think of Rapino's strategic direction for Live Nation?

Sources: Live Nation Web site, www.livenation.com, April 28, 2008; E. Smith, "Live Nation Sings a New Tune," *Wall Street Journal*, July 11, 2008, pp. B1+; B. Sisario, "Nickelback Signs Up with Live Nation," *New York Times* online, www.nytimes.com, July 9, 2008; E. Smith, "Live Nation's Leaders Battle Over Strategy," *Wall Street Journal*, June 12, 2008, p. B1; C. Robertson, "Live Nation Finds a Buyer for Its Theater Business," *New York Times* online, www.nytimes.com, January 25, 2008; P. Sloan, "Keep on Rocking in the Free World," *Fortune*, December 10, 2007, pp. 156–160; and E. Smith, "Live Nation's New Act," *Wall Street Journal*, November 30, 2007, pp. B1+.

Part Three

Organizing

Organizing is an important task of managers—one that's not always understood or appreciated. However, when the organization's goals and plans are in place, the organizing function sets in motion the process of seeing that those goals and plans are pursued. When managers organize, they're defining what work needs to get done and creating a structure that enables those work activities to be completed efficiently and effectively.

In Part Three, we look at the management function of organizing. Chapter 9 introduces the concepts of organizational structure and organizational design. When the organizational structure is in place, it's time to find people to fill the jobs that have been created. In Chapter 10, we discuss the human resource management function. In Chapter 11, we focus on work teams and the influence they've had on how work gets done and on organizational structures. Finally, in Chapter 12, we look at how organizational change and innovation affect an organization's structure and design.

Let's Get Real:

Meet the Managers

Sharad Srivastava

Staffing Consultant
Microsoft India

MY JOB: Encompasses managing staffing programs and learning and development programs for a set of internal business group. My work is a combination of managing routine programs (rhythm of business) and at the same time strategic and change impacting.

BEST PART OF MY JOB: Are those where I get to bring about crucial changes, which will significantly alter the quality of output and make programs more scalable and flexible.

WORST PART OF MY JOB: Are those where I have multiple stakeholders and where it takes time to reach decisions and approvals.

BEST MANAGEMENT ADVICE EVER RECEIVED: It is not the strongest of the species that survive nor the most intelligent that survive; it is the one that is the most adaptable to change that survives.

Manju Saini

Vice President
WNS

MY JOB: Strategy head and responsible for organizational development.

BEST PART OF MY JOB: To get buy-in from operations for new plans and work.

WORST PART OF MY JOB: To manage the politics between two co-CEOs.

BEST MANAGEMENT ADVICE EVER RECEIVED: Mentor your people in the right direction. You grow when you grow your people.

You'll be hearing more from these real managers throughout the chapter.

Chapter 9

Organizational Structure and Design

Once managers are done planning, then what? This is when managers need to begin to "work the plan." And the first step in doing that involves designing an appropriate organizational structure. This chapter covers the decisions involved with designing this structure. Focus on the following learning outcomes as you read and study this chapter.

LEARNING OUTCOMES

A Manager's Dilemma

Danish company Bang & Olufsen (B&O) is known globally for its high-end audio and video equipment.[1] Many of its incredibly beautiful and artistic products—most of which are made in Denmark—are part of the collection at New York's Museum of Modern Art. Needless to say, product design is critically important to B&O. What's even more unique than its futuristic products, however, is CEO Torben Ballegaard Sørensen's approach to the design process. Unlike the conventional design approach used by most companies, in which employees conduct consumer market research and then decide design direction, Sørensen uses contract designers to create the company's products. And he has empowered these designers to veto any product they don't like. Giving such power to individuals who weren't employees would frighten most managers. However, it works well for Sørensen. This "business-by-genius model depends on the instincts of a handful of quirky and creative individuals and the ability of executives to manage them." How could Sørensen use organizational design to ensure that such an approach will continue to work well?

What would you do?

Torben Ballegaard Sørensen understands the importance of organizational structure and design. His approach to empowering the contract designers his company uses to create innovative products is risky. However, it seems to have worked well. Although this structural approach might not be right for others, it illustrates the importance of designing an organizational structure that helps accomplish organizational goals. In this chapter, we'll look at what's involved with that.

LEARNING **OUTCOME 9.1** ▷

DESIGNING ORGANIZATIONAL STRUCTURE

The TVS Group is India's leading supplier of automotive components and one of the country's most respected business conglomerates. It employs more than 40,000 people worldwide and has a turnover in excess of USD 2.2 billion. Underlying the success of the group is an ethos of commitment to the values of Quality, Service, and Reliability. The factories are well organized and work processes are well defined. The work environment is mundane, structured, and disciplined coupled with high levels of care and commitment.[2] Work also gets done efficiently and effectively at Cisco Systems, although not in a structured and formal way. At Cisco, some 70 percent of the employees work from home at least 20 percent of the time.[3] Both of these organizations get needed work done, although each does so using a different structure.

Few topics in management have undergone as much change in the past few years as organizing and organizational structure. Managers are reevaluating traditional approaches to find new structural designs that best support and facilitate employees' doing the organization's work—designs that can achieve efficiency but are also flexible. In Chapter 1 we defined **organizing** as arranging and structuring work to accomplish organizational goals. It's an important process during which managers

Exhibit 9–1

Purposes of Organizing

- Divides work to be done into specific jobs and departments.
- Assigns tasks and responsibilities associated with individual jobs.
- Coordinates diverse organizational tasks.
- Clusters jobs into units.
- Establishes relationships among individuals, groups, and departments.
- Establishes formal lines of authority.
- Allocates and deploys organizational resources.

design an organization's structure. **Organizational structure** is the formal arrangement of jobs within an organization. This structure, which can be shown visually in an **organizational chart**, also serves many purposes (see Exhibit 9–1). When managers create or change the structure, they're engaged in **organizational design**, a process that involves decisions about six key elements: work specialization, departmentalization, chain of command, span of control, centralization and decentralization, and formalization.[4]

WORK SPECIALIZATION

An ophthalmologist at the Aravind Eye Care Hospital in Madurai is able to perform 2,500 cataract surgeries in a year compared to 300 similar surgeries in other hospitals in India. To be able to do this, the work has been carefully analyzed. The trained nurses prepare the patient, counsel them, and provide them post-operative care, while the doctor spends only 10–12 minutes that are needed to do the cataract surgery.[5] This is an example of **work specialization**, which is dividing work activities into separate job tasks. Individual employees specialize in doing part of an activity rather than the entire activity in order to increase work output. It's also known as *division of labor*, a concept we introduced in Chapter 2.

Today's View Most managers today see work specialization as an important organizing mechanism because it helps employees be more efficient. For example, McDonald's uses high work specialization to get its products made and delivered to customers efficiently. However, when it's carried to extremes, work specialization can lead to problems, including boredom, fatigue, stress, poor quality, increased absenteeism, reduced performance, and increased turnover.[6] That's why companies such as Cairn, Tata Steel, Hallmark, and American Express use minimal work specialization and instead give employees a broad range of tasks to do.

DEPARTMENTALIZATION

Does your college have a department of student services or a financial aid department? Are you taking this course through a management department? After deciding what job tasks will be done by whom, common work activities need to be grouped back together so work gets done in a coordinated and integrated way. How jobs are grouped together is called **departmentalization**. There are five common forms of departmentalization (see Exhibit 9–2), although an organization may use its own unique classification. Large organizations often combine most or all of these forms of departmentalization. For example, a major Japanese electronics firm organizes its divisions along functional lines, its manufacturing units around processes, its sales units around seven geographic regions, and its sales regions into four customer groupings.

organizing
Arranging and structuring work to accomplish an organization's goals.

organizational structure
The formal arrangement of jobs within an organization.

organizational chart
The visual representation of an organization's structure.

organizational design
Creating or changing an organization's structure.

work specialization
Dividing work activities into separate job tasks.

departmentalization
The basis on which jobs are grouped together.

Exhibit 9–2 The Five Common Forms of Departmentalization

Functional Departmentalization—Groups Jobs According to Function

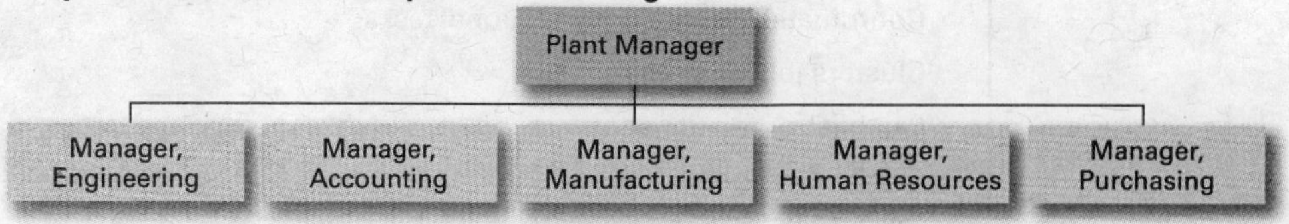

- \+ Efficiencies from putting together similar specialties and people with common skills, knowledge, and orientations
- \+ Coordination within functional area
- \+ In-depth specialization
- – Poor communication across functional areas
- – Limited view of organizational goals

Geographical Departmentalization—Groups Jobs According to Geographic Region

- \+ More effective and efficient handling of specific regional issues that arise
- \+ Serve needs of unique geographic markets better
- – Duplication of functions
- – Can feel isolated from other organizational areas

Product Departmentalization—Groups Jobs by Product Line

Source: Bombardier Annual Report.

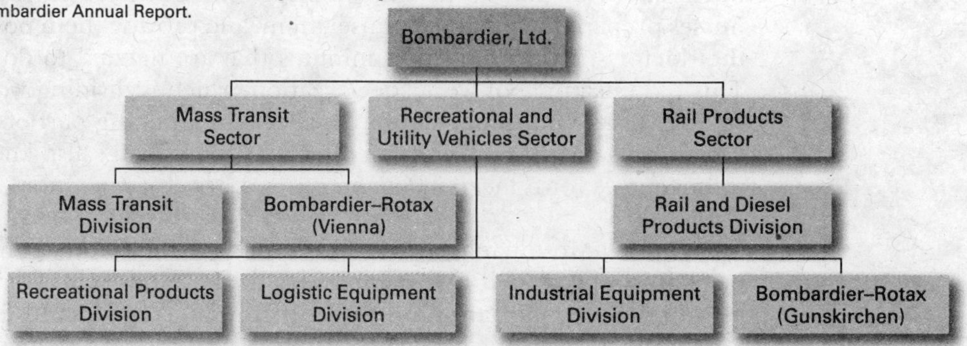

- \+ Allows specialization in particular products and services
- \+ Managers can become experts in their industry
- \+ Closer to customers
- – Duplication of functions
- – Limited view of organizational goals

Process Departmentalization—Groups Jobs on the Basis of Product or Customer Flow

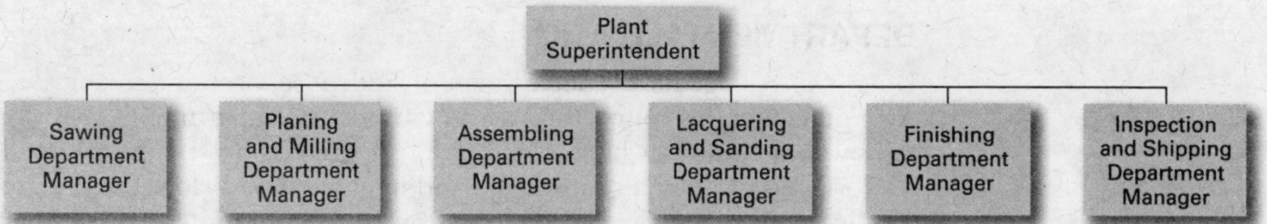

- \+ More efficient flow of work activities
- – Can only be used with certain types of products

Customer Departmentalization—Groups Jobs on the Basis of Specific and Unique Customers Who Have Common Needs

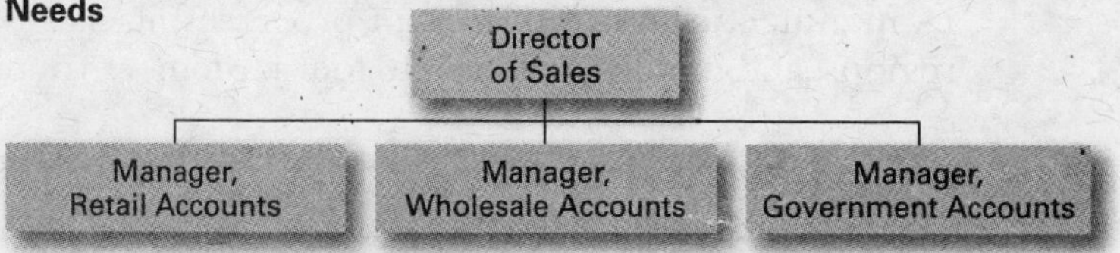

- \+ Customers' needs and problems can be met by specialists
- – Duplication of functions
- – Limited view of organizational goals

Today's View One popular departmentalization trend is the increasing use of customer departmentalization. Because getting and keeping customers is essential for success, this approach works well because it emphasizes monitoring and responding to changes in customers' needs. Another popular trend is **cross-functional teams**, which are work teams composed of individuals from various functional specialties. For instance, at ITC's Saharnpur plant, a 77-year-old establishment, which manufactures processed leaf tobacco and packed cigarettes, has set up cross-functional teams to devise ways to save energy. The teams hold regular meetings to deliberate on issues related to energy conservation and implement them of their own. The multiple options of monitoring and reporting have lead to several improvement ideas.[7] We'll discuss cross-functional teams more fully in Chapter 11.

Let's Get Real: F2F

MY SPAN OF CONTROL:
Five people

CHAIN OF COMMAND

The **chain of command** is the line of authority extending from upper organizational levels to lower levels, which clarifies who reports to whom. Managers need to consider it when organizing work because it helps employees with questions such as "Who do I report to?" and "Who do I go to if I have a problem?"

To understand the chain of command, you have to understand three other concepts: authority, responsibility, and unity of command. **Authority** refers to the rights inherent in a managerial position to tell people what to do and to expect them to do it.[8] Managers in the chain of command have authority to do their job of coordinating and overseeing the work of others. As managers assign work to employees, those employees assume an obligation to perform any assigned duties. This obligation or expectation to perform is known as **responsibility**. Finally, the **unity of command** principle (1 of Fayol's 14 management principles) states that a person should report to only one manager. Without unity of command, conflicting demands from multiple bosses may create problems, as it did for Smitashree Menon, 39, Head (Talent Acquisition, India and Emerging Markets), Motorola, where she reports to two bosses and has deliverables spread across five countries. "I report to Bangalore and Beijing. But my deliverables are quite diverse as I also study markets and provide inputs for Vietnam, Indonesia, The Philippines, and Thailand," she says.[9] The same is the case for a marketing manager at Sun Microsystems in India. The manager was reporting to two bosses: one in charge of the loyalty program of top 10 key accounts and the other in charge of providing technical solutions to all the clients. The manager tried to minimize the conflict by marking all copies of his communication to both managers and sending them his understanding of what he thought he needed to do for the clients he was servicing.

Let's Get Real: F2F

MY SPAN OF CONTROL:
20 people

Today's View Although early management theorists (Fayol, Weber, Taylor, and others) believed that chain of command, authority, responsibility, and unity of command were essential, times have changed,[10] and these concepts are far less important today. For example, at the Michelin plant in Tours, France, managers have replaced the top-down chain of command with "birdhouse" meetings, in which employees meet for five minutes at regular intervals throughout the day at a column on the shop floor and study simple tables and charts to identify production bottlenecks. Instead of being bosses, shop managers are enablers.[11] In addition, information technology has made such concepts less relevant today. In a matter of a few seconds, employees can access information that used to be available only to managers. And employees can communicate with anyone else in the organization without going through the chain of command.

cross-functional teams
Work teams composed of individuals from various functional specialties.

chain of command
The line of authority extending from upper organizational levels to the lowest levels, which clarifies who reports to whom.

authority
The rights inherent in a managerial position to tell people what to do and to expect them to do it.

responsibility
The obligation or expectation to perform any assigned duties.

unity of command
The management principle that each person should report to only one manager.

Exhibit 9–3

Contrasting Spans of Control

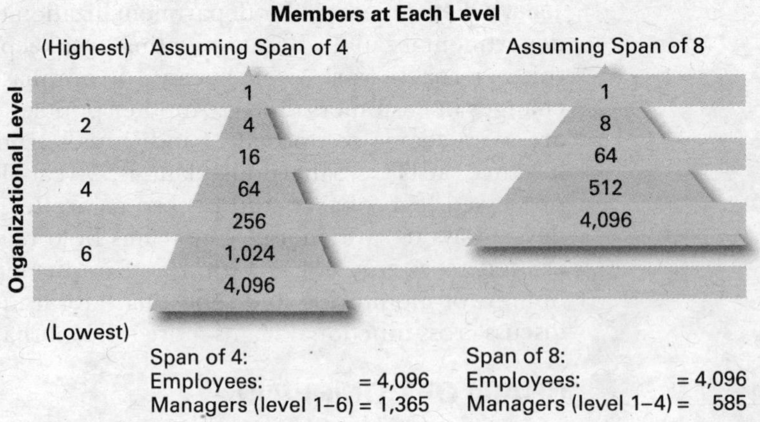

SPAN OF CONTROL

How many employees can a manager efficiently and effectively manage? That's what **span of control** is all about. The traditional view was that managers could not—and should not—directly supervise more than five or six subordinates. Determining the span of control is important because to a large degree, it determines the number of levels and managers in an organization—an important consideration in how efficient an organization will be. All other things being equal, the wider or larger the span, the more efficient it is. Here's why: Assume that two organizations each have approximately 4,100 employees. As Exhibit 9–3 shows, if one organization has a span of four and the other a span of eight, the organization with the wider span will have two fewer levels and approximately 800 fewer managers. At an average manager's salary of INR 10,00,000 per year, the organization with the wider span would save over INR 800 crore million per year! Obviously, wider spans are more efficient in terms of cost. However, at some point, wider spans may reduce effectiveness if employee performance worsens because managers no longer have the time to lead.

Today's View The contemporary view of span of control recognizes that there is no magic number. Many factors influence the number of employees that a manager can efficiently and effectively manage. These factors include the skills and abilities of the manager and the employees, as well as the characteristics of the work being done. For instance, managers with well-trained and experienced employees can function well with a wider span. Other contingency variables that determine the appropriate span include similarity and complexity of employee tasks, the physical proximity of subordinates, the degree to which standardized procedures are in place, the sophistication of the organization's information system, the strength of the organization's culture, and the preferred style of the manager.[12]

The trend in recent years has been toward larger spans of control, which is consistent with managers' efforts to speed up decision making, increase flexibility, get closer to customers, empower employees, and reduce costs. Managers are beginning to recognize that they can handle a wider span when employees know their jobs well and when those employees understand organizational processes. For instance, at PepsiCo's Gemasa cookie plant in Mexico, 56 employees now report to each manager. However, to ensure that performance doesn't suffer because of these wider spans, employees are thoroughly briefed on company goals and processes. Also, new pay systems reward quality, service, productivity, and teamwork.[13]

CENTRALIZATION AND DECENTRALIZATION

Centralization is the degree to which decision making takes place at upper levels of the organization. If top managers make key decisions with little input from below, then the organization is more centralized. On the other hand, the more that lower-level employees provide input or actually make decisions, the more **decentralization** there is. Keep in mind that centralization–decentralization is relative, not absolute—that is, an

Exhibit 9–4

Centralization or Decentralization

More Centralization	More Decentralization
• Environment is stable. • Lower-level managers are not as capable or experienced at making decisions as upper-level managers. • Lower-level managers do not want a say in decisions. • Decisions are relatively minor. • Organization is facing a crisis or the risk of company failure. • Company is large. • Effective implementation of company strategies depends on managers retaining say over what happens.	• Environment is complex, uncertain. • Lower-level managers are capable and experienced at making decisions. • Lower-level managers want a voice in decisions. • Decisions are significant. • Corporate culture is open to allowing managers a say in what happens. • Company is geographically dispersed. • Effective implementation of company strategies depends on managers having involvement and flexibility to make decisions.

organization is never completely centralized or decentralized. Exhibit 9–4 lists some of the factors that affect an organization's use of centralization or decentralization.[14]

Today's View As organizations have become more flexible and responsive to environmental trends, there's been a distinct shift toward decentralized decision making. This is also known as **employee empowerment**, which is giving employees more authority (power) to make decisions. (We'll address this concept more thoroughly in our discussion of leadership in Chapter 16.) In large companies especially, lower-level managers are "closer to the action" and typically have more detailed knowledge about problems and how best to solve them than do top managers. For instance, MindTree covers 100 percent of its employees under its Employee Stock Option Plan for which highly subsidized grant price is offered. Sasken prides itself in giving no perquisites to its employees based on position. For example, no one—including those at the top management positions—is provided with perks, such as company car, chauffeur, telephone, and housing. All employees including the founder are eligible to travel by economy-class air travel or 2-tier AC class train. Irrespective of the level, employees are eligible for staying in a 3-star hotel. All employees are eligible for the same daily allowance while on travel.[15] As another example, at Ternary Software Inc., managers run the company as a democracy, and decisions have to be unanimous. At a recent meeting about a new incentive plan for employees, one of the programmers criticized what the CEO was proposing. After extensive discussion, meeting attendees voted to handle the incentives differently.[16]

FORMALIZATION

Formalization refers to how standardized an organization's jobs are and the extent to which employee behavior is guided by rules and procedures. In highly formalized organizations, there are explicit job descriptions, numerous organizational rules, and clearly defined procedures covering work processes. Employees have little discretion over what's done, when it's done, and how it's done. However, where formalization is low, employees have more discretion in how they do their work.

span of control
The number of employees a manager can efficiently and effectively manage.

centralization
The degree to which decision making is concentrated at upper levels of the organization.

decentralization
The degree to which lower-level employees provide input or actually make decisions.

employee empowerment
Giving employees more authority (power) to make decisions.

formalization
How standardized an organization's jobs are and the extent to which employee behavior is guided by rules and procedures.

Steve Sanghi took over Microchip Technology Inc., a failing spin-off of General Instrument, in 1990 and turned it around to make it a highly profitable company. At Microchip, the organizational chart is viewed upside down, where the customer is at the top and the CEO is at the bottom, with the engineers, managers, and divisional heads in descending order between them. Steve believes that he gets the performance of his people because of the structure and "Aggregate System" culture that is prevalent in the company.[17]

Courtesy: Microchip Technology Inc.

Today's View Although some formalization is necessary for consistency and control, many organizations today rely less on strict rules and standardization to guide and regulate employee behavior. For instance, consider the following situation:

> It is 2:30 P.M. and a customer at a branch of Scientific Diagnostic (a pathology lab) wants to get a random blood test done for diabetes and results to be made available at 5 P.M. The lab policy is that same-day test results are only promised for those who give their samples by 1:30 P.M. The front-office executive knows the rule. At the same time he wants to be accommodating given the pleading look on the customers face. He knows that the test can be done and the results can be given on the same day. He decides to accept the customers request, and, in doing so, to violate the policy. He just hopes that the manager does not find it out?

Has this employee done something wrong? He did "break" the rule. But by "breaking" the rule, he actually brought in revenue and provided good customer service.

Considering that there are numerous situations in which rules may be too restrictive, many organizations have allowed employees some latitude, giving them sufficient autonomy to make decisions that they feel are best under the circumstances. This doesn't mean throwing out all organizational rules because there *will* be rules that are important for employees to follow—and those rules should be explained so employees understand why it's important to adhere to them. But for other rules, employees may be given some leeway.[18]

QUICK LEARNING REVIEW:

LEARNING OUTCOME 9.1

- Discuss the traditional and contemporary views of work specialization, chain of command, and span of control.
- Describe each of the five forms of departmentalization.
- Differentiate authority, responsibility, and unity of command.
- Explain how centralization–decentralization and formalization are used in organizational design.

Go to page 202 and see how well you know this material.

LEARNING **OUTCOME 9.2** ▷

MECHANISTIC AND ORGANIC STRUCTURES

Organizations don't, and won't, have identical structures. A company with 30 employees isn't going to look like one with 30,000 employees. But even organizations of comparable size don't necessarily have similar structures. What works for one organization may not work for another. How do managers decide what organizational design to use? That decision often depends on certain contingency factors. In this section, we look at two generic models of organizational design and then at the contingency factors that favor each.

Exhibit 9–5

Mechanistic Versus Organic Organizations

Mechanistic	Organic
• High specialization	• Cross-functional teams
• Rigid departmentalization	• Cross-hierarchical teams
• Clear chain of command	• Free flow of information
• Narrow spans of control	• Wide spans of control
• Centralization	• Decentralization
• High formalization	• Low formalization

TWO MODELS OF ORGANIZATIONAL DESIGN

Exhibit 9–5 describes two organizational models.[19] A **mechanistic organization** is a rigid and tightly controlled structure characterized by high specialization, rigid departmentalization, narrow spans of control, high formalization, a limited information network (mostly downward communication), and little participation in decision making by lower-level employees.

Mechanistic organizational structures strive for efficiency and rely heavily on rules, regulations, standardized tasks, and similar controls. This design tries to minimize the impact of differing personalities, judgments, and ambiguity because these human traits are seen as inefficient and inconsistent. Although no organization is totally mechanistic, almost all large ones have some of these mechanistic characteristics.

The other organizational design model is an **organic organization**, which is a structure that's highly adaptive and flexible. Organic organizations may have specialized jobs, but those jobs are not standardized and can change as needs require. Work is frequently organized around employee teams. Employees are highly trained and empowered to handle diverse activities and problems, and they require minimal formal rules and little direct supervision. For example, an organizational redesign at GlaxoSmithKline, a London-based pharmaceutical company, made the company more of an organic structure. Before the restructuring, product research was hampered by slow-moving bureaucracy. Decisions about which drugs to fund were made by a committee of research and development executives far removed from the research labs—a time-consuming process not at all appropriate for a company dependent on scientific breakthroughs. Now, lab scientists set the priorities and allocate the resources. The change has "helped produce an entrepreneurial environment akin to a smaller, biotechnology outfit."[20]

When is a mechanistic structure preferable and when is an organic one more appropriate? Let's look at the main contingency factors that influence this decision.

Let's Get Real: F2F

MECHANISTIC OR ORGANIC?

My organization is organic.

Let's Get Real: F2F

MECHANISTIC OR ORGANIC?

My organization is organic.

CONTINGENCY FACTORS

ICICI Prudential, a joint venture between ICICI Bank and Prudential UK, has a portfolio of over 6.5 million policies. India's biggest private sector life insurer has not merely held on to its share but grown it; at the end of January 2008, the firm commanded 29 percent of the share owned by private sector players. Shikha Sharma, Managing Director, ICICI Prudential, has been consistently working on the design and selling of insurance to the customer for the last 7 years.[21] Top managers typically put a great deal of thought into designing an appropriate structure. What that appropriate structure is depends on four contingency variables: the organization's strategy, size, technology, and degree of environmental uncertainty.

Strategy and Structure An organization's structure should facilitate goal achievement. Because goals are an important part of the organization's strategies, it's only logical that strategy and structure are closely linked. Alfred Chandler initially researched this relationship.[22] He studied several large U.S. companies and concluded that changes in corporate strategy led to changes in an organization's structure that support the strategy.

Research has shown that certain structural designs work best with different organizational strategies.[23] For instance, the flexibility and free-flowing information of the

mechanistic organization
An organizational design that's rigid and tightly controlled.

organic organization
An organizational design that's highly adaptive and flexible.

organic structure works well when an organization is pursuing meaningful and unique innovations. The mechanistic organization, with its efficiency, stability, and tight controls, works best for companies that want to tightly control costs.

Size and Structure There's considerable evidence that an organization's size affects its structure.[24] Large organizations—typically considered to be those with more than 2,000 employees—tend to have more specialization, departmentalization, centralization, and rules and regulations than do small organizations. However, once an organization grows past a certain size, size has less influence on structure. Why? Essentially, once there are around 2,000 employees, the organization is already fairly mechanistic. Adding another 500 employees won't impact the structure much. On the other hand, adding 500 employees to an organization that has only 300 employees is likely to make it more mechanistic.

Technology and Structure Every organization uses some form of technology to convert its inputs into outputs. For instance, workers at Whirlpool's Manaus, Brazil, facility build microwave ovens and air-conditioners on a standardized assembly line. Employees at FedEx Kinko's produce custom design and print jobs for individual customers. And employees at Bayer's facility in Karachi, Pakistan, make pharmaceutical products using a continuous-flow production line.

The initial research on technology's effect on structure can be traced to Joan Woodward, who studied small manufacturing firms in southern England to determine the extent to which structural design elements were related to organizational success.[25] Woodward couldn't find any consistent pattern until she divided the firms into three distinct technology categories that had increasing levels of complexity and sophistication. The first category, **unit production**, described the production of items in units or small batches. The second category, **mass production**, described large-batch manufacturing. Finally, the third and most technically complex group, **process production**, included continuous-process production. A summary of Woodward's findings is shown in Exhibit 9–6.

Other studies have also shown that organizations adapt their structures to their technology depending on how routine their technology is for transforming inputs into outputs.[26] In general, the more routine the technology, the more mechanistic the structure can be, and organizations with more nonroutine technology are more likely to have organic structures.[27]

Environmental Uncertainty and Structure Some organizations face stable and simple environments with little uncertainty; others face dynamic and complex environments with a lot of uncertainty. Managers try to minimize environmental uncertainty by adjusting the organization's structure.[28] In stable and simple environments, mechanistic designs can be more effective. On the other hand, the greater the uncertainty, the more an organization needs the flexibility of an organic design. For example, the uncertain nature of the oil industry means that oil companies need to be flexible. Soon after being named CEO of Royal Dutch Shell PLC, Jeroen van der Veer streamlined the corporate structure to counteract some of the industry volatility. One thing he did was eliminate the company's cumbersome, overly analytical process of making deals with OPEC countries and other major oil producers.[29]

Exhibit 9–6

Woodward's Findings on Technology and Structure

	Unit Production	Mass Production	Process Production
Structural characteristics:	Low vertical differentiation Low horizontal differentiation Low formalization	Moderate vertical differentiation High horizontal differentiation High formalization	High vertical differentiation Low horizontal differentiation Low formalization
Most effective structure:	Organic	Mechanistic	Organic

Today's View The evidence on the environment–structure relationship helps explain why many managers today are restructuring their organizations to be lean, fast, and flexible. Global competition, accelerated product innovation by competitors, and increased demands from customers for high quality and faster deliveries are examples of dynamic environmental forces. Mechanistic organizations are not equipped to respond to rapid environmental change and environmental uncertainty. As a result, many organizations are becoming more organic.

QUICK LEARNING REVIEW:

LEARNING OUTCOME 9.2

- Contrast mechanistic and organic organizations.
- Explain the contingency factors that affect organizational design.

Go to page 202 and see how well you know this material.

LEARNING **OUTCOME 9.3** ▷

COMMON ORGANIZATIONAL DESIGNS

In making structural decisions, managers have some common designs from which to choose: traditional ones and more contemporary ones.

TRADITIONAL ORGANIZATIONAL DESIGNS

When designing a structure, managers may choose one of the traditional organizational designs. These structures tend to be mechanistic in nature. (See Exhibit 9–7 for a summary of the strengths and weaknesses of each.)

Exhibit 9–7

Traditional Organizational Designs

Simple Structure

- Strengths: Fast; flexible; inexpensive to maintain; clear accountability.
- Weaknesses: Not appropriate as organization grows; reliance on one person is risky.

Functional Structure

- Strengths: Cost-saving advantages from specialization (economies of scale, minimal duplication of people and equipment); employees are grouped with others who have similar tasks.
- Weaknesses: Pursuit of functional goals can cause managers to lose sight of what's best for the overall organization; functional specialists become insulated and have little understanding of what other units are doing.

Divisional Structure

- Strengths: Focuses on results—division managers are responsible for what happens to their products and services.
- Weaknesses: Duplication of activities and resources increases costs and reduces efficiency.

unit production
The production of items in units or small batches.

mass production
The production of items in large batches.

process production
The production of items in continuous processes.

thinking critically about **Ethics**

Responding to customer demand and the lure of extra sales, many fast-food restaurants are opening earlier and closing later. However, this decision may be having unintended consequences, as employee deaths have increased as restaurants stay open longer. One professor of criminal justice said, "Some fast-food chains have come up with special food menus after midnight, but what they really need are special late-night security menus." What do you think? Are corporations putting the lives of employees—often young kids working the late shift—at risk? Is this an ethical issue?

Source: B. Horovitz, "Late Shift Proves Deadly to More Fast-Food Workers," USA Today, December 13, 2007, p. 1B+.

Simple Structure Most companies start as entrepreneurial ventures using a **simple structure**, which is an organizational design with low departmentalization, wide spans of control, authority centralized in a single person, and little formalization.[30] As employees are added, however, most companies don't remain as simple structures. The structure tends to become more specialized and formalized. Rules and regulations are introduced, work becomes specialized, departments are created, levels of management are added, and the organization becomes increasingly bureaucratic. At this point, managers might choose a functional structure or a divisional structure.

Functional Structure A **functional structure** is an organizational design that groups similar or related occupational specialties together. You can think of this structure as functional departmentalization applied to an entire organization.

Divisional Structure The **divisional structure** is an organizational structure made up of separate business units or divisions.[31] In this structure, each division has limited autonomy, with a division manager who has authority over his or her unit and is responsible for performance. In divisional structures, however, the parent corporation typically acts as an external overseer to coordinate and control the various divisions, and it often provides support services such as financial and legal services. Wal-Mart, for example, has two divisions: retail (Wal-Mart Stores, International, Sam's Clubs, and others) and support (distribution centers).

CONTEMPORARY ORGANIZATIONAL DESIGNS

Managers are finding that traditional designs often aren't appropriate for today's increasingly dynamic and complex environment. Instead, organizations need to be lean, flexible, and innovative; that is, they need to be more organic. So managers are finding creative ways to structure and organize work.[32] (See Exhibit 9–8 for a summary of these designs.)

Team Structures Larry Page and Sergey Brin, co-founders of Google, have created a corporate structure that "tackles most big projects in small, tightly focused teams."[33] A **team structure** is one in which the entire organization is made up of work teams that do the organization's work.[34] In this structure, employee empowerment is crucial because there is no line of managerial authority from top to bottom. Rather, employee teams design and do work in the way they think is best, but they are also held responsible for all work performance results in their respective areas.

In large organizations, the team structure complements what is typically a functional or divisional structure. This allows the organization to have the efficiency of a bureaucracy while providing the flexibility that teams provide. For instance, companies such as Amazon, Boeing, Hewlett-Packard, Louis Vuitton, Motorola, and Xerox extensively use employee teams to improve productivity.

Matrix and Project Structures In addition to the team structure, other popular contemporary designs are the matrix and project structures. In the **matrix structure**, specialists

Exhibit 9–8

Contemporary Organizational Designs

Team Structure	
• What it is:	A structure in which the entire organization is made up of work groups or teams.
• Advantages:	Employees are more involved and empowered. Reduced barriers among functional areas.
• Disadvantages:	No clear chain of command. Pressure on teams to perform.
Matrix-Project Structure	
• What it is	Matrix is structure that assigns specialists from different functional areas to work on projects but who return to their areas when the project is completed. Project is a structure in which employees continuously work on projects. As one project is completed, employees move on to the next project.
• Advantages:	Fluid and flexible design that can respond to environmental changes. Faster decision making.
• Disadvantages:	Complexity of assigning people to projects. Task and personality conflicts.
Boundaryless Structure	
• What it is:	A structure that is not defined by or limited to artificial horizontal, vertical, or external boundaries; includes *virtual* and *network* types of organizations.
• Advantages:	Highly flexible and responsive. Utilizes talent wherever it's found.
• Disadvantages:	Lack of control. Communication difficulties.

from different functional departments work on projects that are led by a project manager. One unique aspect of this design is that it creates a *dual chain of command* in which employees have two managers—their functional area manager and their product or project manager—who share authority. The project manager has authority over the functional members who are part of his or her project team in areas related to the project's goals. However, any decisions about promotions, salary recommendations, and annual reviews typically remain the functional manager's responsibility. To work effectively, both managers have to communicate regularly, coordinate work demands on employees, and resolve conflicts together.

Many organizations are using a **project structure**, in which employees continuously work on projects. Unlike a matrix structure, a project structure has no formal departments where employees return at the completion of a project. Instead, employees take their specific skills, abilities, and experiences to other projects. Also, all work in project structures is performed by teams of employees. For instance, at design firm IDEO, project teams form, disband, and form again as the work requires. Employees "join" a project team because they bring needed skills and abilities to that project. Once a project is completed, however, they move on to the next one.[35]

Project structures tend to be flexible organizational designs. There's no departmentalization or rigid organizational hierarchy to slow down making decisions or taking action. In this structure, managers serve as facilitators, mentors, and coaches. They eliminate or minimize organizational obstacles and ensure that teams have the resources they need to effectively and efficiently complete their work.

simple structure
An organizational design with low departmentalization, wide spans of control, centralized authority, and little formalization.

functional structure
An organizational design that groups together similar or related occupational specialties.

divisional structure
An organizational structure made up of separate, semiautonomous units or divisions.

team structure
An organizational structure in which the entire organization is made up of work groups or teams.

matrix structure
An organizational structure that assigns specialists from different functional departments to work on one or more projects.

project structure
An organizational structure in which employees continuously work on projects.

Bill Green, shown here at a press conference in Mumbai, India, is the CEO of Accenture, Ltd., the international consulting firm that's also a virtual organization. Green doesn't maintain a permanent office, and the company has no operational headquarters or branch offices. Its top-level executives are scattered around the world, and many of its employees spend their days traveling to or working with clients in their clients' offices. The company's culture is one of constant motion and its managers thrive on personal contact with clients. "We don't get to go down the hall to the coffee pot, ask someone how their weekend was, and then ask a business question," says Green, who logs hundreds of thousands of air miles in a typical year. "We spend time together in the countries where our clients are, which is more important if you're running a global company."

The Boundaryless Organization. Another contemporary organizational design is the **boundaryless organization**, which is an organization whose design is not defined by, or limited to, the horizontal, vertical, or external boundaries imposed by a predefined structure.[36] Former GE Chairman Jack Welch coined the term because he wanted to eliminate vertical and horizontal boundaries within GE and break down external barriers between the company and its customers and suppliers. Although the idea of eliminating boundaries may seem odd, many of today's most successful organizations are finding that they can operate most effectively by remaining flexible and *un*structured—that the ideal structure for them is *not* having a rigid, bounded, and predefined structure.[37]

What do we mean by *boundaries*? There are two types: (1) *internal*—the horizontal boundaries imposed by work specialization and departmentalization and the vertical ones that separate employees into organizational levels and hierarchies; and (2) *external*—the boundaries that separate the organization from its customers, suppliers, and other stakeholders. To minimize or eliminate these boundaries, managers might use virtual or network structural designs.

A **virtual organization** consists of a small core of full-time employees and outside specialists temporarily hired as needed to work on projects.[38] An example is StrawberryFrog, a global advertising agency with offices in Amsterdam and New York. It does its work with a small administrative staff but has a global network of freelancers who are assigned client work. By relying on these freelancers, the company enjoys a network of talent without all the unnecessary overhead and structural complexity.[39] The inspiration for this structural approach comes from the film industry. There, people are essentially "free agents" who move from project to project, applying their skills—directing, talent casting, costuming, makeup, set design, and so forth—as needed.

Another structural option for managers who want to minimize or eliminate organizational boundaries is a **network organization**, in which a company uses its own employees to do some work activities and uses networks of outside suppliers to provide other needed product components or work processes.[40] This organizational form is sometimes called a *modular organization* among manufacturing firms.[41] This structural approach allows organizations to concentrate on what they do best by contracting out other activities to companies that do those activities best. Many companies are using such an approach for certain organizational work activities. For instance, the head of development for Boeing's 787 airplane manages thousands of employees and some 100 suppliers at more than 100 sites in different countries.[42] Sweden's Ericsson contracts its manufacturing and even some of its research and development to more cost-effective contractors in New Delhi, Singapore, California, and other global locations.[43] And at Penske Truck Leasing, dozens of business processes, such as securing permits and titles, entering data from drivers' logs, and processing data for tax filings and accounting, have been outsourced to Mexico and India.[44]

Let's Get Real: F2F

MANAGING AT A DISTANCE: When I was working onsite with Infosys in Australia, I managed my team by communicating regularly, keeping the work specifications clear and keeping the group on the same page at all times.

TODAY'S ORGANIZATIONAL DESIGN CHALLENGES

As managers look for organizational designs that will best support and facilitate employees doing their work efficiently and effectively, they must contend with certain challenges. These include keeping employees connected, building a learning organization, and managing global structural issues.

Keeping Employees Connected Many organizational design concepts were developed during the twentieth century, when work tasks were fairly predictable and constant, most jobs were full time and continued indefinitely, and work was done at an employer's place of business, under a manager's supervision.[45] That's not what it's like in many organizations today, as you saw in our preceding discussion of virtual and network organizations. A major structural design challenge for managers is finding a way to keep widely dispersed and mobile employees connected to the organization. The "Managing in a Virtual World" box describes ways that information technology can help.

Building a Learning Organization. Doing business in an intensely competitive global environment, British retailer Tesco realizes how important it is for its stores to run well behind the scenes. And it does so by using a proven "tool" called Tesco in a Box, which promotes consistency in operations and acts as a way to share innovations. Tesco is an example of a **learning organization**, an organization that has developed the capacity to

managing in a Virtual World

IT's Impact on Organizational Design

It's fair to say that the world of work will never be like it was 10 years ago. IT has opened up new possibilities for employees to do their work in locations as remote as Patagonia or in the middle of downtown Seattle. Although organizations have always had employees who traveled to distant corporate locations to take care of business, these employees no longer have to find the nearest pay phone or wait to get back to "the office" to see what problems have cropped up. Instead, mobile computing and communication have given organizations and employees ways to stay connected and to be more productive.[46] Let's look at some of the technologies that are changing the way work is done:

- Handheld devices that enable a worker to access e-mail, calendars, and contacts can be used anywhere there's a wireless network. And these devices can be used to log in to corporate databases and company intranets.
- Employees can videoconference using broadband networks and Webcams.
- Many companies are giving employees key fobs with constantly changing encryption codes that allow them to log on to the corporate network to access e-mail and company data from any computer that is hooked up to the Internet.
- Cell phones switch seamlessly between cellular networks and corporate Wi-Fi connections.

The biggest issue in doing work anywhere, anytime is security. Companies must protect their important and sensitive information. However, software and other disabling devices have minimized security issues considerably. Even insurance providers are more comfortable giving their mobile employees access to information. For instance, Health Net Inc. gave BlackBerrys to many of its managers so they can tap into customer records from anywhere. As one tech company CEO said, "Companies now can start thinking about innovative [applications] they can create and deliver to their workers anywhere."

boundaryless organization
An organization whose design is not defined by or limited to the horizontal, vertical, and external boundaries imposed by a predefined structure.

virtual organization
An organization that consists of a small core of full-time employees and outside specialists temporarily hired as needed to work on projects.

network organization
An organization that uses its own employees to do some work activities and networks of outside suppliers to provide other needed product components or work processes.

learning organization
An organization that has developed the capacity to continuously learn, adapt, and change.

Let's Get Real: F2F

MANAGING AT A DISTANCE?

I have to frequently get involved in mentoring the team for managing their deliverables.

continuously learn, adapt, and change.[47] In a learning organization, employees continually acquire and share new knowledge and apply that knowledge in making decisions or doing their work. Some organizational theorists even go so far as to say that an organization's ability to do this—that is, to learn and to apply that learning—may be the only sustainable source of competitive advantage.[48] What structural characteristics does a learning organization need?

In a learning organization, employees throughout the entire organization—across different functional specialties and even at different organizational levels—must share information and collaborate on work activities. This requires minimal structural and physical barriers. In such a boundaryless environment, employees work together and collaborate in doing the organization's work the best way they can, and they learn from each other. Finally, because of the need to collaborate, empowered work teams tend to be an important feature of a learning organization's structural design. These teams make decisions about doing whatever work needs to be done or resolving issues. With empowered employees and teams, there's little need for "bosses" to direct and control. Instead, managers serve as facilitators, supporters, and advocates.

Managing Global Structural Issues Are there global differences in organizational structures? Are Australian organizations structured like those in the United States? Are German organizations structured like those in France or Mexico? Given the global nature of today's business environment, this is an issue with which managers need to be familiar. Researchers have concluded that the structures and strategies of organizations worldwide are similar, "while the behavior within them is maintaining its cultural uniqueness."[49] What does this mean for designing effective and efficient structures? When designing or changing structure, managers may need to think about the cultural implications of certain design elements. For instance, one study showed that formalization—rules and bureaucratic mechanisms—may be more important in less economically developed countries and less important in more economically developed countries, where employees may have higher levels of professional education and skills.[50] Other structural design elements may be affected by cultural differences as well.

A Final Thought No matter what structural design managers choose for their organizations, the design should help employees do their work in the best—most efficient and effective—way they can. The structure should support and facilitate organizational members as they carry out the organization's work. After all, the structure is simply a means to an end.

QUICK LEARNING REVIEW:

LEARNING OUTCOME 9.3

- Contrast the three traditional organizational designs.
- Describe the contemporary organizational designs.
- Discuss the organizational design challenges that managers face today.

Go to page 202 and see how well you know this material.

Let's Get Real:

Our Turn

Sharad Srivastava

Staffing Consultant
Microsoft India

I would advise the CEO to use principles of natural system design. These principles are also useful in crafting characteristics of a learning organization. By using this design, it will be possible to decrease formalization, have horizontal structure, decentralize decision-making, and increase empowerment. It is important to manage by information sharing, have greater personal communication, and have higher levels of work iteration so that work is done well.

Manju Saini

Vice President
WNS

Sorensen's methodology of getting new ideas, i.e. business-by-genius, is a short-term model because it might increase the feeling of incompetence in his employees. Instead, I will try to identify the gaps in the existing team and try to provide appropriate training and environment so as to get the results which Sorenson is getting from people outside the organization on need basis. Growing your people will give long-lasting results. Rather than opting for a short-term solution, it is better to coach employees who are aware of a performance problem or skill gap and who are motivated to address the situation.

LEARNING OUTCOMES SUMMARY

9.1 ▷ DESIGNING ORGANIZATIONAL STRUCTURE

- Discuss the traditional and contemporary views of work specialization, chain of command, and span of control.
- Describe each of the five forms of departmentalization.
- Differentiate authority, responsibility, and unity of command.
- Explain how centralization–decentralization and formalization are used in organizational design.

Traditionally, work specialization was viewed as a way to divide work activities into separate job tasks. Today's view is that work specialization is an important organizing mechanism, but it can lead to problems. The chain of command and its companion concepts—authority, responsibility, and unity of command—were viewed as important ways of maintaining control in organizations. The contemporary view is that they are less relevant in today's organizations. The traditional view of span of control was that managers should directly supervise no more than five or six individuals. The contemporary view is that the span of control depends on the skills and abilities of the manager and the employees and on the characteristics of the situation.

The various forms of departmentalization are as follows: Functional groups jobs by functions performed; product groups jobs by product lines; geographical groups jobs by geographical region; process groups jobs on product or customer flow; and customer groups jobs on specific and unique customer groups.

Authority refers to the rights inherent in a managerial position to tell people what to do and to expect them to do it. Responsibility is the obligation or expectation to perform assigned duties. Unity of command states that a person should report to only one manager.

Centralization–decentralization is a structural decision about who makes decisions—upper-level managers or lower-level employees. Formalization concerns an organization's use of standardization and strict rules to provide consistency and control.

9.2 ▷ MECHANISTIC AND ORGANIC STRUCTURES

- Contrast mechanistic and organic organizations.
- Explain the contingency factors that affect organizational design.

A mechanistic organization is a rigid and tightly controlled structure. An organic organization is highly adaptive and flexible.

An organization's structure should support its strategy. If the strategy changes, the structure should also change. An organization's size can affect its structure up to a certain point. Once an organization reaches a certain size (usually around 2,000 employees), it's fairly mechanistic. An organization's technology can also affect its structure. An organic structure is most effective with unit production and process production technology. A mechanistic structure is most effective with mass production technology. The more uncertain an organization's environment, the more it needs the flexibility of an organic design.

9.3 ▷ COMMON ORGANIZATIONAL DESIGNS

- Contrast the three traditional organizational designs.
- Describe the contemporary organizational designs.
- Discuss the organizational design challenges that managers face today.

A simple structure is one with low departmentalization, wide spans of control, authority centralized in a single person, and little formalization. A functional structure groups similar or related occupational specialties together. A divisional structure is made up of separate business units or divisions. In a team structure, the entire organization is made up of work teams. In the matrix structure, specialists from different functional departments work on one or more projects that are led by project managers. A project structure is one in which employees continuously work on projects. A virtual organization consists of a small core of full-time employees and outside specialists temporarily hired as needed to work on projects. A network organization is an organization that uses its own employees to do some work activities and networks of outside suppliers to provide other needed product components or work processes.

Three organizational design challenges face today's managers: keeping employees connected, building a learning organization, and managing global structural issues.

THINKING ABOUT MANAGEMENT ISSUES

1. Can an organization's structure be changed quickly? Why or why not? Should it be changed quickly? Explain.
2. Would you rather work in a mechanistic organization or an organic organization? Why?
3. What types of skills would a manager need to effectively work in a project structure? In a boundaryless organization? In a learning organization?
4. The boundaryless organization has the potential to create a major shift in the way we work. Do you agree or disagree? Explain.
5. With the availability of advanced information technology that allows an organization's work to be done anywhere, anytime, is organizing still an important managerial function? Why or why not?
6. Researchers are now saying that efforts to simplify work tasks actually have negative results for both companies and their employees. Do you agree? Why or why not?

YOUR TURN to be a Manager

- Find three different examples of organizational charts. In a report, describe each of them. Try to decipher the organization's use of organizational design elements, especially departmentalization, chain of command, centralization–decentralization, and formalization.
- Survey at least 10 different managers about how many employees they supervise. Also ask them whether they feel they could supervise more employees or whether they feel the number they supervise is too many. Graph your survey results and write a report describing what you found. Draw some conclusions about span of control.
- Draw an organizational chart of an organization with which you're familiar (where you work, a student organization to which you belong, your college or university, etc.). Be very careful in showing the departments (or groups) and especially be careful to get the chain of command correct. Be prepared to share your chart with the class.
- Using the organizational chart you just created, redesign the organization's structure. What structural changes might make this organization more efficient and effective? Write a report describing what you would do and why. Be sure to include an example of the original organizational chart as well as a chart of your proposed revision of the organizational structure.
- Steve's and Mary's suggested readings: Gary Hamel, *The Future of Management* (Harvard Business School Press, 2007); Thomas Friedman, *The World Is Flat 3.0* (Picador, 2007); Harold J. Leavitt, *Top Down: Why Hierarchies Are Here to Stay and How to Manage Them More Effectively* (Harvard Business School Press, 2005); and Thomas W. Malone, *The Future of Work* (Harvard Business School Press, 2004).
- Choose one of the three topics discussed in the section on today's organizational design challenges. Research this topic and write a paper about it. Focus on finding current information and current examples of companies dealing with these issues.
- In your own words, write down three things you learned in this chapter about being a good manager.

- Self-knowledge can be a powerful learning tool. Go to mymanagementlab and complete these self-assessment exercises: How Well Do I Handle Ambiguity? What Type of Organizational Structure Do I Prefer? How Good Am I at Playing Politics? How Willing Am I to Delegate? Using the results of your assessments, identify personal strengths and weaknesses. What will you do to reinforce your strengths and improve your weaknesses?

CASE APPLICATION

A New Kind of Structure

Admit it. Sometimes the projects you're working on (school, work, or both) can get pretty boring and monotonous. Wouldn't it be nice to have a magic button you could push to get someone else to do the boring, time-consuming stuff for you? At Pfizer, such a button is a reality for a large number of employees.

As a global pharmaceutical company, Pfizer is continually looking for ways to be more efficient and effective. The company's senior director of organizational effectiveness, Jordan Cohen, found that the "Harvard MBA staff we hired to develop strategies and innovate were instead Googling and making PowerPoints." Indeed, internal studies conducted to find out just how much time its valuable talent was spending on menial tasks was startling. The average Pfizer employee was spending 20 percent to 40 percent of his or her time on support work (creating documents, typing notes, doing research, manipulating data, scheduling meetings) and only 60 percent to 80 percent on knowledge work (strategy, innovation, networking, collaborating, critical thinking). And the problem wasn't just at lower levels. Even the highest-level employees were affected. That's when Cohen began looking for solutions. The solution he chose turned out to be the numerous knowledge-process outsourcing companies based in India.

Initial tests of outsourcing the support tasks didn't go well at all. However, Cohen continued to tweak the process until everything worked. Now Pfizer employees can click the OOF (Office of the Future) button in Microsoft Outlook, and they are connected to an outsourcing company where a single worker in India receives the request and assigns it to a team. The team leader calls the employee to clarify the request. The team leader then e-mails back a cost specification for the requested work. At this point, the Pfizer employee can say yes or no. Cohen says that the benefits of OOF are unexpected. Time spent on analysis of data has been cut—sometimes in half. The financial benefits are also impressive. And Pfizer employees love it. Cohen says, "It's kind of amazing. I wonder what they used to do."

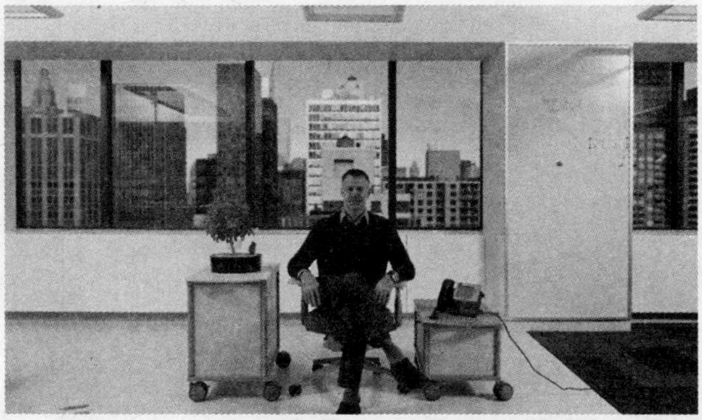

Courtesy of Mark Mahaney.

Jordan Cohen, senior director of organizational effectiveness at Pfizer.

Discussion Questions

1. Describe and evaluate what Pfizer is doing.
2. What structural implications—good and bad—does this approach have? (Think in terms of the six organizational design elements.)
3. Do you think this arrangement would work for other types of organizations? Why or why not?
4. What role do you think organizational structure plays in an organization's efficiency and effectiveness? Explain.

Sources: FAST COMPANY by A. Cohen. Copyright 2008 by Mansueto Ventures LLC. Reproduced with permission of Mansueto Ventures LLC in the format Textbook via Copyright Clearance Center.

Let's Get Real:

Meet The Manager

Avinash Deolekar

Vice President, HR

Pepsico India Holdings Pvt. Ltd

MY JOB: To support the organization with all human related matters, from recruitment to relieving and partner the business in strategic decision making and attract and nurture talent.

BEST PART OF MY JOB: To sustain high performance of the company by building and nurturing talents.

WORST PART OF MY JOB: Closure of a business unit by way of restructuring.

BEST MANAGEMENT ADVICE EVER RECEIVED: People do their business very well, if you take care of people properly.

You'll be hearing more from this real manager Throughout the chapter.

Chapter

Managing Human Resources

Once an organization's structure is in place, it's time to find the people to fill the jobs that have been created. That's where human resource management comes in. It's an important task that involves getting the right number of the right people in the right place at the right time. Focus on the following learning outcomes as you read and study this chapter.

LEARNING OUTCOMES

A Manager's Dilemma

Indian Railways has a great legacy as a nation builder. Of the 27 million people employed in the organized sector, Indian Railways account for 6 percent directly and an additional 2.5 percent indirectly. On one hand, the Indian Railways runs 11,000 goods and passenger trains daily over a massive stretch of nearly 63,000 route kilometers efficiently. On the other hand given the pressure on the trains, it is very common to encounter rude railwaymen across the length and breadth of the country. The enquiry telephones are often unanswered, ticket window clerks are impatient and careless at times, ticket collectors are loud and stressed, station masters are not helpful and unable to help harried passengers. It is a constant struggle to find ways to instill customer orientation among railway employees who interface with passengers.[1] If you are an HR manager at Indian Railways and have been charged with the service skill upgradation of its 1.5 lakh employees what strategies would you employ?

What would you do?

It is clearly understood by our HR Manager in Indian Railways quality of service depends a great deal on the quality of training it imparts and the culture it sets up. And like the other managers today, they are facing a major human resource management (HRM) challenge: ensuring that their company has a high-quality workforce. Getting and keeping competent employees is critical to the success of every organization, whether an organization is just starting or has been in business for years. If an organization doesn't take its HRM responsibilities seriously, performance may suffer. Therefore, part of every manager's job when organizing is HRM. All managers engage in some HRM activities, such as interviewing job candidates, orienting new employees, and evaluating their employees' work performance, even if there is a separate department dedicated to HRM.

LEARNING OUTCOME 10.1 ▷

THE HUMAN RESOURCE MANAGEMENT PROCESS

People are at the heart of success at Godrej. "It's not the machinery, the money or the structure that is important. It is the people. The company's (Godrej's) main asset is its people. And the main asset of people is not competency, hard work, time, it is their mind and their heart", says Marc Fourcade a consultant to Godrej.[2] Like Godrej, many other organizations profess that their people are their most important asset and acknowledge the important role that employees play in organizational success.

WHY IS HRM IMPORTANT?

HRM is important for three reasons. First, it can be a significant source of competitive advantage, as various studies have concluded.[3] And that's true for organizations around the world. The Human Capital Index, a comprehensive study of more than

2,000 global firms, concluded that people-oriented HR gives an organization an edge by creating superior shareholder value.[4]

Second, HRM is an important part of organizational strategies. Achieving competitive success through people means managers must change how they think about their employees and how they view the work relationship. They must work with people and treat them as partners, not just as costs to be minimized or avoided. That's what people-oriented organizations such as Southwest Airlines and W. L. Gore do.

Finally, the way organizations treat their people has been found to significantly affect organizational performance.[5] For instance, one study reported that significantly improving those work practices could increase market value by as much as 30 percent.[6] Another study that tracked average annual shareholder returns of companies on *Fortune's* annual "100 Best Companies to Work For" list found that these companies significantly beat the S&P 500 over 10-year, 5-year, 3-year, and 1-year periods.[7] Work practices that lead to both high individual and high organizational performance are known as **high-performance work practices**. (See some examples in Exhibit 10–1.) The common thread among these practices seems to be a commitment to improving the knowledge, skills, and abilities of an organization's employees; increasing their motivation; reducing loafing on the job; and enhancing the retention of quality employees while encouraging low performers to leave.

Even if an organization doesn't use high-performance work practices, there are specific HRM activities that must be completed in order to ensure that the organization has qualified people to perform the work that needs to be done—activities that comprise the HRM process. Exhibit 10–2 shows the eight activities in this process. The first three activities ensure that competent employees are identified and selected, the next two involve providing employees with up-to-date knowledge and skills, and the final three ensure that the organization retains competent and high-performing employees. Before we discuss those specific activities, we need to look at external factors that affect the HRM process.

EXTERNAL FACTORS THAT AFFECT THE HRM PROCESS

The entire HRM process is influenced by the external environment. The factors that most directly influence it include employee labor unions, governmental laws and regulations, and demographic trends.

Employee Labor Unions. A **labor union** is an organization that represents workers and seeks to protect their interests through collective bargaining. In unionized organizations, many HRM decisions are dictated by collective bargaining agreements, which usually define things such as recruitment sources; criteria for hiring, promotions, and layoffs; training eligibility; and disciplinary practices. About 12.1 percent of the U.S. workforce is unionized, but the percentage is higher in other countries except France, where only 9.6 percent of workers are unionized. For instance in India,

Exhibit 10–1

High-Performance Work Practices

- Self-managed teams
- Decentralized decision making
- Training programs to develop knowledge, skills, and abilities
- Flexible job assignments
- Open communication
- Performance-based compensation
- Staffing based on person–job and person–organization fit

Source: Based on W. R. Evans and W. D. Davis, "High-Performance Work Systems and Organizational Performance: The Mediating Role of Internal Social Structure," *Journal of Management*, October 2005, p. 760.

high-performance work practices
Work practices that lead to both high individual performance and high organizational performance.

labor union
An organization that represents workers and seeks to protect their interests through collective bargaining.

Exhibit 10–2 HRM Process

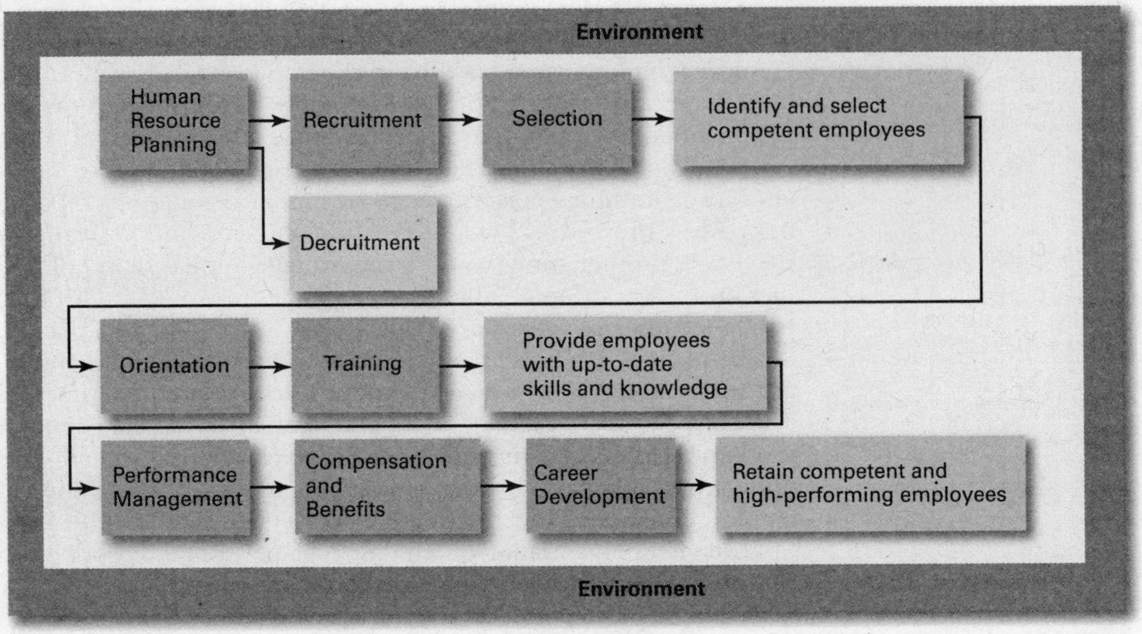

approximately 31 percent of the organized labor force belong to a union, in Japan, some 19.6 percent of the labor force belongs to a union; in Germany, 27 percent; in Denmark, 75 percent; in Australia, 22.7 percent; in Canada, 30 percent; and in Mexico, 19 percent.[8] The average membership per unions decreased from 900 in 2001 to 893 in 2002 in India.[9] Although labor unions can affect an organization's HRM practices, the most significant environmental constraint is governmental laws, especially in North America.

Governmental Laws. $33.5 million. That's the amount Wal-Mart Stores agreed to pay in back wages plus interest to settle a federal lawsuit that accused the company of violating overtime laws.[10] An organization's HRM practices are affected by a country's laws. (See Exhibit 10–3 for some of the important work-related laws for women in India and, in addition, also refer to Exhibit 3–10 for work-related legislation that affect HRM processes). For example, decisions regarding who will be hired or which employees will be chosen for a training program or what an employee's compensation will be must be made without regard to race, sex, religion, age, color, national origin, or disability. Exceptions can occur only in special circumstances. For instance, a community fire department can deny employment to a firefighter applicant who is confined to a wheelchair; but if that same individual is applying for a desk job, such as a dispatcher position, the disability cannot be used as a reason to deny employment. The issues, however, are rarely so clear-cut. For example, employment laws protect most employees whose religious beliefs require a specific style of dress—robes, long shirts, long hair, and the like. However, if the specific style of dress may be hazardous or unsafe in the work setting (such as when operating machinery), a company could refuse to hire a person who won't adopt a safer dress code.[11]

Trying to balance the "shoulds" and "should-nots" of many laws often falls within the realm of **affirmative action**. Through affirmative action programs, an organization actively seeks to enhance the status of members of protected groups. However, managers are not completely free to choose whom they hire, promote, or fire, nor are they free to treat employees any way they want. Although laws have helped reduce employment discrimination and unfair work practices, they have, at the same time, reduced managers' discretion over HRM decisions. Because workplace lawsuits are increasingly targeting supervisors, as well as their organizations, managers must know what they

Exhibit 10–3 Employment of Women–Protective Legal Provisions

S.No.	Name of Enactment	Protective Provisions
1.	The *Beedi* & Cigar Workers (Condition of Employment Act, 1966)	Provision of creches for the benefit of women workers
2.	The Plantation Labour Act, 1951	Women workers to be provided time off for feeding children.
3.	The Contract Labour (Regulation & Abolition) Act, 1970	Not to be required to work beyond 9 hours between 6 a.m. and 7 p.m.– with the exception of mid-wives and nurses in plantations.
4.	The Inter State Migrant Workmen (Regulation of Employment and Condition of Service) Act, 1979	Separate toilets and washing facilities to be provided.
5.	The Factories Act, 1948	In factories, women not to be engaged for cleaning, lubricating, or adjusting an part of prime or transmission machinery; maternity leave upto 12 weeks with wages to be provided.
6.	The Mines Act, 1952	Employment in mines below ground prohibited.
7.	Maternity Benefit Act, 1961	Maternity benefits to be provided on completion of 80 days working. Not required to work during six weeks immediately following the day of delivery or miscarriage. No work of arduous nature; long hours of standing likely to interfere with pregnancy/normal development of foetus; or which may cause miscarriage or is likely to affect health to be given for a period of one month immediately preceeding the period of six weeks before delivery. On medical certificate, advance maternity benefit to be allowed. Rs. 250/- as medical bonus to be given when no pre-natal confinement and post natal care is provided free of charge.
8.	Equal Remuneration Act, 1976	Payment of equal remuneration to men and women workers for same or similar nature of work protected under the act. No discrimination permissible in recruitment and service conditions except where employment of women is prohibited or restricted by or under any law.
9.	Employees' State Insurance (General) Regulation, 1950	Claim for maternity benefit becomes due on the date the medical certificate is issued for miscarriage, sickness arriving out of pregnancy, confinement, or premature birth of child.
10.	Lime Stone and Dolomite Mines Labour Welfare Fund Act,1972.	Appointment of a women member in Advisory Committee and Central Advisory Committee is mandatory under these acts.
11.	The Protection Of Women Against Sexual Harassement At Workplace Bill, 2007	To provide for prevention and redressal of sexual harassment of women at workplace and for matters connected therewith or incidental thereto.
12.	*Beedi* Workers Welfare Fund Act, 1976	
13.	Iron Ore Mines, Manganese Ore Mines and Chrome Ore Mines Labour Welfare Fund Act 1976.	
14	Mica Mines Labour Welfare Fund Act, 1946.	

can and cannot do according to the law.[12] Also, it's important that managers in other countries be familiar with the specific laws that apply there.

Demographic Trends. The statistics are clear: By 2020 more than half of the working population will be more than 40 in India. The number of less than 15 years of age is 35 percent, which means in the next five years these children will enter the workforce much before the existing employees are willing to or are asked to make way for new employees. More women are entering the workplace. It is required by the state that the disabled should be given opportunity to work more regularly. The directive principles of the constitution require that the scheduled caste and scheduled tribe should be given special treatment while being educated and when asked to join the

affirmative action
Organizational programs that enhance the status of members of protected groups.

organization.[13] These and other demographic trends are important because of the impact they're having on current and future HRM practices.

QUICK LEARNING REVIEW:

LEARNING OUTCOME 10.1

- **Explain why the HRM process is important.**
- **Discuss the environmental factors that most directly affect the HRM process.**

Go to page 229 and see how well you know this material.

LEARNING **OUTCOME 10.2** ▷

IDENTIFYING AND SELECTING COMPETENT EMPLOYEES

Every organization needs people to do whatever work is necessary for doing what the organization is in business to do. How do organizations get those people? And more importantly, what can they do to ensure that they get competent, talented people? This first phase of the HRM process involves three tasks: human resource planning, recruitment and decruitment, and selection.

HUMAN RESOURCE PLANNING

Human resource planning is the process by which managers ensure that they have the right number and kinds of capable people in the right places and at the right times. Through planning, organizations avoid sudden people shortages and surpluses.[14] HR planning entails two steps: (1) assessing current human resources and (2) meeting future HR needs.

Current Assessment. Managers begin HR planning by taking inventory of the current employees. This inventory usually includes information on employees such as name, education, training, prior employment, languages spoken, special capabilities, and specialized skills. Sophisticated databases make getting and keeping this information quite easy. For example, Stephanie Cox, Schlumberger's director of personnel for North and South America, uses a company planning program called PeopleMatch to help pinpoint managerial talent. Suppose that she needs a manager for Brazil. She types in the qualifications: someone who can relocate, speak Portuguese, and is a high-potential employee. Within a minute, 31 names of possible candidates pop up.[15] That's what good HR planning should do—help managers identify the people they need.

An important part of a current assessment is **job analysis**, an assessment that defines a job and the behaviors necessary to perform it. For instance, what are the duties of a level 3 accountant who works for Kodak? What minimal knowledge, skills, and abilities are necessary to adequately perform this job? How do these requirements compare with those for a level 2 accountant or for an accounting manager? Information for a job analysis is gathered by directly observing individuals on the job, interviewing employees individually or in a group, having employees complete a

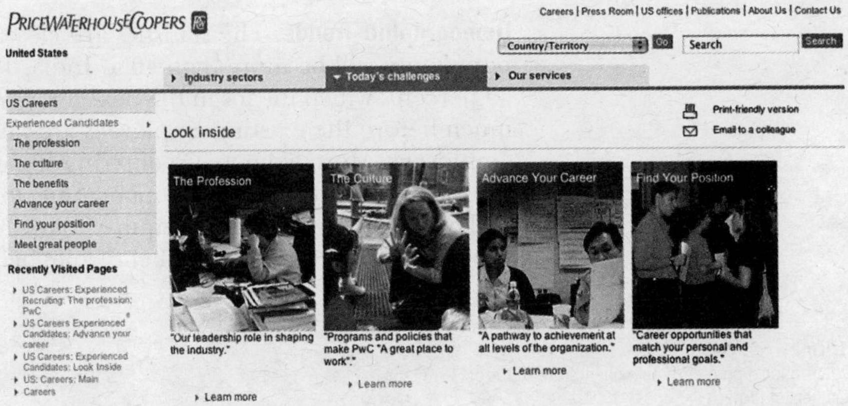

PricewaterhouseCoopers, one of the world's largest accounting firms, has developed a Web site designed for recruiting a very specific group of people—the under-22 college market. Found at www.pwc.tv, the interactive site includes videos of interns and associates talking about their jobs and offers career advice and tips. The firm, voted an "ideal employer" by more than 9,000 business undergraduates, is committed to recruiting in the college market and chose the Web as its medium for the campaign because it is "radio, TV, newspaper, entertainment, and information all rolled up into one."

questionnaire or record daily activities in a diary, or having job "experts" (usually managers) identify a job's specific characteristics.

Using this information from the job analysis, managers develop or revise job descriptions and job specifications. A **job description** is a written statement that describes a job—typically job content, environment, and conditions of employment. A **job specification** states the minimum qualifications that a person must possess to successfully perform a given job. It identifies the knowledge, skills, and attitudes needed to do the job effectively. Both the job description and job specification are important documents when managers recruit and select employees.

Let's Get Real: F2F

BEST WAYS TO FIND THE BEST JOB CANDIDATES:
Blend of young highly talented people from premier institutes and those with EQ levels matching the competencies required in the organization.

Meeting Future HR Needs. Future HR needs are determined by the organization's mission, goals, and strategies. Demand for employees results from demand for the organization's products or services. For instance, Huawei's expansion into India was slowed by a lack of qualified employees. To continue its growth strategy, it had to plan how to find qualified employees.[16]

After assessing both current capabilities and future needs, managers can estimate areas in which the organization will be understaffed or overstaffed. Then they're ready to proceed to the next step in the HRM process

RECRUITMENT AND DECRUITMENT

If employee vacancies exist, managers should use the information gathered through job analysis to guide them in **recruitment**—that is, locating, identifying, and attracting capable applicants.[17] On the other hand, if HR planning shows a surplus of employees, managers may want to reduce the organization's workforce through **decruitment**.[18]

Recruitment. Some organizations have interesting approaches to finding employees. For instance, a Southwest Airlines recruitment team distributed airsickness bags printed with the slogan "Sick of your job?" at job fairs.[19] To find qualified computer science and engineering candidates, Google holds Google Games, a day devoted to student team competitions on the company's campus.[20] Accounting firm Deloitte & Touche created its Deloitte Film Festival to solicit employee team-produced films about "life" at Deloitte to use in college recruiting.[21] Exhibit 10–4 explains different recruitment approaches managers can use to find potential job candidates.[22]

Although online recruiting is popular and allows organizations to identify applicants inexpensively and quickly, applicant quality may not be as good as that found

Exhibit 10–4
Recruiting Sources

Source	Advantages	Disadvantages
Internet	Reaches large numbers of people; can get immediate feedback	Generates many unqualified candidates
Employee referrals	Knowledge about the organization provided by current employee; can generate strong candidates because a good referral reflects on the recommender	May not increase the diversity and mix of employees
Company Web site	Wide distribution; can be targeted to specific groups	Generates many unqualified candidates
College recruiting	Large centralized body of candidates	Limited to entry-level positions
Professional recruiting organizations	Good knowledge of industry challenges and requirements	Little commitment to specific organization

human resource planning
A method of planning to ensure that the organization has the right number and kinds of capable people in the right places and at the right times.

job analysis
An assessment that defines jobs and the behaviors necessary to perform them.

job description
A written statement that describes a job.

job specification
A written statement of the minimum qualifications that a person must possess to perform a given job successfully.

recruitment
Locating, identifying, and attracting capable applicants.

decruitment
Reducing an organization's workforce.

Exhibit 10–5

Decruitment Options

Option	Description
Firing	Permanent involuntary termination
Layoffs	Temporary involuntary termination; may last only a few days or extend to years
Attrition	Not filling openings created by voluntary resignations or normal retirements
Transfers	Moving employees either laterally or downward; usually does not reduce costs but can reduce intraorganizational supply–demand imbalances
Reduced workweeks	Having employees work fewer hours per week, share jobs, or perform their jobs on a part-time basis
Early retirements	Providing incentives to older and more senior employees for retiring before their normal retirement date
Job sharing	Having employees share one full-time position

using other sources. Research has shown that employee referrals generally produce the best candidates.[23] Why? Because current employees know both the job and the person being recommended, they tend to refer applicants who are well qualified. Also, current employees often feel that their reputation is at stake and refer someone only when they're confident that the person will not make them look bad.

Decruitment. In addition to recruitment, another approach to controlling labor supply is decruitment, which is not a pleasant task for any manager. Decruitment options are shown in Exhibit 10–5. Although employees can be fired, other choices may be better. However, no matter how you do it, it's never easy to reduce an organization's workforce.

SELECTION

Once you have a pool of candidates, the next step in the HRM process is **selection**, screening job applicants to determine who is the best qualified for the job. Managers need to select carefully because hiring errors can have significant implications. For example, instances of police torture of hapless citizens in many cases are because the concerned policeman have difficulty in handling power. In special services such as the CID, Defence etc. the selection process is such that odds of hiring people who will serve the goals of the organization are higher. Similarly, Infosys is credited with thinking of testing for learning orientation among applicants given that the nature of work requires employees who are able to learn rather than know what they have been taught in their respective schools well.

What Is Selection? Selection involves predicting which applicants will be successful if hired. For example, in hiring for a sales position, the selection process should predict which applicants will generate a high volume of sales. As shown in Exhibit 10–6, any selection decision can result in four possible outcomes—two correct outcomes and two errors.

A decision is correct when the applicant was predicted to be successful and proved to be successful on the job, or when the applicant was predicted to be unsuccessful

Exhibit 10–6

Selection Decision Outcomes

Later Job Performance	Selection Decision: Accept	Selection Decision: Reject
Successful	Correct Decision	Reject Error
Unsuccessful	Accept Error	Correct Decision

and was not hired. In the first instance, the organization has successfully accepted; in the second, it has successfully rejected.

Problems arise when errors are made in rejecting candidates who would have performed successfully on the job (reject errors) or accepting those who ultimately perform poorly (accept errors). These problems can be significant. Given today's HR laws and regulations, reject errors can cost more than the additional screening needed to find acceptable candidates. Why? Because they can expose the organization to discrimination charges, especially if applicants from protected groups are disproportionately rejected. On the other hand, the costs of accept errors include the cost of training the employee, the profits lost because of the employee's incompetence, the cost of severance, and the subsequent costs of further recruiting and screening. The major emphasis of any selection activity should be reducing the probability of reject errors or accept errors while increasing the probability of making correct decisions. Managers do this by using selection procedures that are both valid and reliable.

Validity and Reliability. For a selection device to be valid, there must be a proven relationship between the selection device and some relevant criterion. Federal employment laws prohibit managers from using a test score to select employees unless there is clear evidence that, once on the job, individuals with high scores on this test outperform individuals with low test scores. The burden is on managers to show that any selection device they use to differentiate applicants is validly related to job performance.

A reliable selection device measures the same thing consistently. On a test that's reliable, any single individual's score should remain fairly consistent over time, assuming that the characteristics being measured are also stable. No selection device can be effective if it's not reliable. Using an unreliable selection device would be like weighing yourself every day on an erratic scale. If the scale is unreliable—randomly fluctuating, say, 5 to 10 pounds every time you step on it—the results don't mean much.

Types of Selection Tools. The best-known selection tools include application forms, written and performance-simulation tests, interviews, background investigations, and, in some cases, physical exams. Exhibit 10–7 lists the strengths and weaknesses of each.[24] Because many selection tools have limited value for making selection decisions, managers should use tools that effectively predict performance for a given job.

Realistic Job Previews. One thing managers need to carefully watch is how they portray their organization and the work that an applicant will be doing. If they tell applicants only the good aspects, they're likely to have a workforce that's dissatisfied and prone to high turnover.[25] Negative things can happen when the information an applicant receives is excessively inflated. First, mismatched applicants probably won't withdraw from the selection process. Second, inflated information builds unrealistic expectations, so new employees may quickly become dissatisfied and leave the organization. Third, new hires become disillusioned and less committed to the organization when they face the unexpected harsh realities of the job. In addition, these individuals may feel that they were misled during the hiring process and then become problem employees.

To increase employee job satisfaction and reduce turnover, managers should consider a **realistic job preview (RJP)**, which includes both positive and negative information about the job and the company. For instance, in addition to the positive comments typically expressed during an interview, the job applicant might be told that there are limited opportunities to talk to coworkers during work hours, that promotional advancement is unlikely, or that work hours are erratic and employees may have to work weekends.

selection
Screening job applicants to ensure that the most appropriate candidates are hired.

realistic job preview (RJP)
A preview of a job that provides both positive and negative information about the job and the company.

Exhibit 10–7

Selection Tools

Application Forms
- Almost universally used
- Most useful for gathering information
- Can predict job performance but not easy to create one that does

Written Tests
- Must be job related
- Include intelligence, aptitude, ability, personality, and interest tests
- Are popular (e.g., personality tests; aptitude tests)
- Relatively good predictor for supervisory positions

Performance-Simulation Tests
- Use actual job behaviors
- Work sampling—test applicants on tasks associated with that job; appropriate for routine or standardized work
- Assessment center—simulate jobs; appropriate for evaluating managerial potential

Interviews
- Almost universally used
- Must know what can and cannot be asked
- Can be useful for managerial positions

Background Investigations
- Used for verifying application data—valuable source of information
- Used for verifying reference checks—not a valuable source of information

Physical Examinations
- Are for jobs that have certain physical requirements
- Mostly used for insurance purposes

Research indicates that applicants who receive an RJP have more realistic expectations about the jobs they'll be performing and are better able to cope with the frustrating elements than are applicants who receive only inflated information.

thinking critically about Ethics

What you say online *can* come back to haunt you. Organizations are using Google, MySpace, and Facebook to check out applicants and current employees. A survey by Vault.com found that 44 percent of employers use social networking sites to examine the profiles of job candidates, and 39 percent have looked up the profile of a current employee. In fact, some organizations see Google as a way to get "around discrimination laws, inasmuch as employers can find out all manner of information—some of it for a nominal fee-that is legally off-limits in interviews: your age, your marital status, fraternity pranks, stuff you wrote in college, political affiliations and so forth." And for those individuals who like to rant and rave about employers, there might be consequences at some point. Some individuals are removing their Facebook profiles because employers are looking at these things.[26] What do you think of what these employers are doing? What positives and negatives are there to such behavior? What are the ethical implications? What guidelines might you suggest for an organization's selection process?

QUICK LEARNING REVIEW:

LEARNING OUTCOME 10.2

- Define *job analysis, job description,* and *job specification.*
- Discuss the major sources of potential job candidates.
- Describe the different selection devices and which work best for different jobs.
- Explain why a realistic job preview is important.

Go to page 229 and see how well you know this material.

LEARNING **OUTCOME 10.3** ▷

PROVIDING EMPLOYEES WITH NEEDED SKILLS AND KNOWLEDGE

Did you participate in some type of organized "introduction to college life" when you started school? If so, you may have been told about your school's rules and the procedures for activities such as applying for financial aid, cashing a check, or registering for classes, you were probably also introduced to some of the college administrators. A person starting a new job needs the same type of introduction to his or her job and the organization. This introduction is called **orientation**.

There are two types of orientation. *Work unit orientation* familiarizes an employee with the goals of the work unit, clarifies how his or her job contributes to the unit's goals, and includes an introduction to his or her new coworkers. *Organization orientation* informs a new employee about the company's goals, history, philosophy, procedures, and rules. It should also include relevant HR policies and maybe even a tour of the facilities.

Many organizations have formal orientation programs, and others use a more informal approach, in which a manager assigns a new employee to a senior member of the work group, who introduces the new hire to immediate coworkers and shows him or her where important things are located. And then there are intense orientation programs like that at Tata Consultancy Servicers (TCS). The company has an intense two month program in Trivandrum for all new recruits. The program covers both technical skills and life skills program. The life skills program includes communication, grooming and personal etiquette, business orientation, and systems thinking. "The paradigm used is that of kindling a flame rather than filling a bucket. The motto is to teach "How to learn." With the emphasis on facilitation to learn and self-discovery, it fits into the classical adult learning paradigm of andragogy" says Narayanan Vice-President (Education & Training).[27] And managers do have an obligation to effectively and efficiently integrate any new employee into the organization. They should openly discuss mutual obligations of the organization and the employee.[28] It's in the best interests of both the organization and the new employee to get the person up and running in the job as soon as possible. Successful orientation results in an outsider-insider transition that makes the new employee feel comfortable and fairly well adjusted, lowers the likelihood of poor work performance, and reduces the probability of a surprise resignation only a week or two into the job.

EMPLOYEE TRAINING

In order to meet international standards, in construction Larsen & Toubro has entered into a Memorandum of Understanding with Henry Boot Training Ltd and the Construction Industry Training Board of UK for the development of Modular Training. At present, basic training is imparted in the following trades: formwork carpentry; masonry (brick work and block work); bar-bending & steel fixing; plumbing & sanitary; electrical wireman; trade assistants; welding; and Pre-stressing.

orientation
Education that introduces a new employee to his or her job and the organization.

Infosys Technologies, India's fast-growing software company, has created Infosys University, one of the largest training centers in the world, to train the new employees the company hires each year. The $120 million campus-like facility bans alcohol and offers only single-sex dorms, but there are three movie theaters, a pool and gym, and dozens of instructors as well as online courses to teach recruits everything from technical skills and team building to interpersonal communication and corporate etiquette. Says CEO Nandan Nilekani, "Companies haven't been investing enough in people. Rather than train them, they let them go. Our people are our capital. The more we invest in them, the more they can be effective."

Thermax is among the few companies who has been able to beat the industry average of 20–25 percent attrition by continuous training, movement, and challenge; their attrition level is less than 10 percent on an annualized basis. Thermax recruits over 100 graduate engineers every year who undergo rigorous training for 12 months. Mr Sohoni Executive Vice-President, HR, IR & Administration, of Thermax Ltd says "We regularly offer challenging jobs ahead of the expectation curve of our people and hence they are encouraged to stretch both their mental and physical limits. Apart from this, we offer job rotation and movement between locations and businesses, which prove useful as the employee never falls prey to monotony."[29]

Employee training is an important HRM activity. As job demands change, employee skills have to change. On an average, major IT companies spend $1 billion annually on training.[30] Managers, of course, are responsible for deciding what type of training employees need, when they need it, and what form that training should take.

Types of Training. Exhibit 10–8 describes the major types of training that organizations provide. Some of the most popular types include profession-/industry-specific training, management/supervisory skills, mandatory/compliance information (such as information about sexual harassment, safety, etc.), and customer service training.[31] For many organizations, employee interpersonal skills training—communication, conflict resolution, team building, customer service, and so forth—is a high priority. For example, at MphasiS the goal is to have at least 1 percent of its work force comprised of people with disabilities. Thus, they have invested in setting up training programs with NGOs to train people with disability in effective communication, analytical skills, computer skills and personality development, content and methodology of successfully training people to join the corporate world. Jeya Kumar, CEO, of MphasiS said that in his experience it is the soft skills that are as important as other skills to help integrate people with disability in corporations.

Exhibit 10–8

Types of Training

Type	Includes
General	Communication skills, computer systems application and programming, customer service, executive development, management skills and development, personal growth, sales, supervisory skills, and technological skills and knowledge
Specific	Basic life/work skills, creativity, customer education, diversity/cultural awareness, remedial writing, managing change, leadership, product knowledge, public speaking/presentation skills, safety, ethics, sexual harassment, team building, wellness, and others

Source: Based on "2005 Industry Report—Types of Training." *Training*, December 2005, p. 22.

Exhibit 10–9

Training Methods

Traditional Training Methods
On-the-job—Employees learn how to do tasks simply by performing them, usually after an initial introduction to the task.
Job rotation—Employees work at different jobs in a particular area, getting exposure to a variety of tasks.
Mentoring and coaching—Employees work with an experienced worker who provides information, support, and encouragement; also called apprenticeships in certain industries.
Experiential exercises—Employees participate in role playing, simulations, or other face-to-face types of training.
Workbooks/manuals—Employees refer to training workbooks and manuals for information.
Classroom lectures—Employees attend lectures designed to convey specific information.
Technology-Based Training Methods
CD-ROM/DVD/videotapes/audiotapes/podcasts—Employees listen to or watch selected media that convey information or demonstrate certain techniques.
Videoconferencing/teleconferencing/satellite TV—Employees listen to or participate as information is conveyed or techniques demonstrated.
E-learning—Internet-based learning where employees participate in multimedia simulations or other interactive modules

Training Methods. Although employee training can be done in traditional ways, many organizations are increasingly relying on technology-based training methods because of their accessibility, cost, and ability to deliver information. Exhibit 10–9 provides descriptions of the various traditional and technology-based training methods that managers might use. Of all these training methods, experts believe that organizations will increasingly rely on e-learning applications to deliver important information and to develop employees' skills.

QUICK LEARNING REVIEW:

LEARNING OUTCOME 10.3

- Explain why orientation is so important.
- Describe the different types of training and how each type of training can be provided.

Go to page 229 and see how well you know this material.

LEARNING **OUTCOME 10.4** ▷

RETAINING COMPETENT, HIGH-PERFORMING EMPLOYEES

When an organization has invested significant money in recruiting, selecting, orienting, and training employees, it wants to keep those employees, especially the competent, high-performing ones! Two HRM activities that play a role in doing this are managing employee performance and developing an appropriate compensation and benefits program.

EMPLOYEE PERFORMANCE MANAGEMENT

Managers need to know whether their employees are performing their jobs efficiently and effectively or whether there is need for improvement. That's what a **performance management system** does: It establishes performance standards that are used to evaluate employee performance. How do managers evaluate employees' performance? That's where the different performance appraisal methods come in.

performance management system
A system that establishes performance standards that are used to evaluate employee performance.

Let's Get Real: F2F

MOST CHALLENGING HR PROBLEM: Changing the mind sets of senior management.

Performance Appraisal Methods. More than 70 percent of managers admit that they have trouble giving a critical performance review to an underachieving employee.[32] Appraising someone's performance is never easy, especially with employees who aren't doing their jobs well, but managers can be better at it by using any of the seven different performance appraisal methods. A description of each of these methods, including advantages and disadvantages, is shown in Exhibit 10–10.

COMPENSATION AND BENEFITS

In most public sector organizations in India, high performing committed employees find themselves a bit demoralized. The compensation program does not allow for differentiating between paying high and low performers differently. Thus, many public sector companies have devised other means to reward and recognize their top performers. For instance in Steel Authority of India Ltd, top performers are given awards, special incentive schemes, comparatively faster promotions, foreign visits, or training.

Most of us expect to receive appropriate compensation from our employers. Developing an effective and appropriate compensation system is an important part of the HRM process.[33] It can help attract and retain competent and talented individuals who help the organization accomplish its mission and goals. In addition, an organization's compensation system has been shown to have an impact on its strategic performance.[34]

Managers must develop a compensation system that reflects the changing nature of work and the workplace in order to keep people motivated. Organizational compensation can include many different types of rewards and benefits, such as base wages and salaries, wage and salary add-ons, incentive payments, and other benefits and services. Some organizations offer employees some unusual, but popular, benefits.

Exhibit 10–10

Performance Appraisal Methods

Written Essay

Evaluator writes a description of employee's strengths and weaknesses, past performance, and potential; provides suggestions for improvement

\+ Simple to use

– May be better measure of evaluator's writing ability than of employee's actual performance

Critical Incident

Evaluator focuses on critical behaviors that separate effective and ineffective performance

\+ Rich examples, behaviorally based

– Time-consuming, lacks quantification

Graphic Rating Scale

Popular method that lists a set of performance factors and an incremental scale; evaluator goes down the list and rates employee on each factor

\+ Provides quantitative data; not time-consuming

– Doesn't provide in-depth information on job behavior

BARS (Behaviorally Anchored Rating Scale)

Popular approach that combines elements from critical incident and graphic rating scale; evaluator uses a rating scale, but items are examples of actual job behaviors

\+ Focuses on specific and measurable job behaviors

– Time-consuming; difficult to develop

Multiperson Comparison

Employees are rated in comparison to others in work group

\+ Compares employees with one another

– Difficult with large number of employees; legal concerns

MBO

Employees are evaluated on how well they accomplish specific goals

\+ Focuses on goals; results oriented

– Time-consuming

360-Degree Appraisal

Utilizes feedback from supervisors, employees, and coworkers

\+ Thorough

– Time-consuming

Exhibit 10–11

What Determines Pay and Benefits?

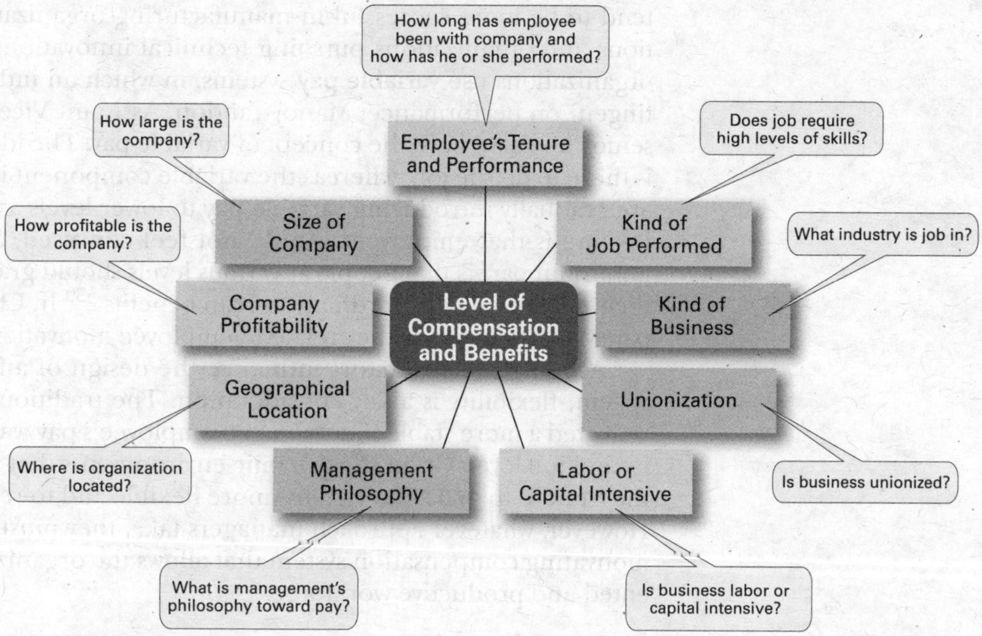

Let's Get Real: F2F

SHOULD HR PROCESSES BE AUTOMATED?

Yes. But a balance has to be maintained in the technological/personal touch and also the core HR processes to be retained with HR function.

Financial Technologies give its senior employees a meal of their choice, a blackberry and a laptop, and medical insurance for dependent parents.[35] Tata Steel assures a job to the child of the employee as per his/her qualifications. Advances and bonus are given on salary to clerical staff in Indian Oil Corporation. Meals are provided at subsidized rates to all employees of Hindustan Petroleum Corporation Limited. Every time HCL Comnet employee travels abroad for work, the company gives free telephone coupons to the family members to help them keep in touch.[36]

How do managers determine who gets paid what? Several factors influence the compensation and benefits packages that different employees receive. Exhibit 10–11 summarizes these factors, which are job based and business or industry based. Many organizations, however, are using alternative approaches to determining compensation, including skill-based pay and variable pay.

Skill-based pay systems reward employees for the job skills and competencies they demonstrate. Under this type of pay system, an employee's job title doesn't define his or her pay category; skills do.[37] Research has shown that these types of pay systems

managing in a Virtual World

HR and IT

HR has gone digital.[40] Using software that automates many basic HR processes associated with recruiting, selecting, orienting, training, appraising performance, and storing and retrieving employee information, HR departments have cut costs and optimized service. The main area where IT has had a significant impact is in training.

In a survey by the American Society for Training and Development, 95 percent of the responding companies reported using some form of e-learning. Using technology to deliver needed knowledge, skills, and abilities has had many benefits. As one researcher said, "The ultimate purpose of e-learning is not to reduce the cost of training, but to improve the way your organization does business." And in many instances, it seems to do that! For example, when Hewlett-Packard looked at how its customer service was affected by a blend of e-learning and other instructional methods, rather than just classroom training, it found that "sales representatives were able to answer questions more quickly and accurately, enhancing customer-service provider relations." And Unilever found that after it provided e-learning training for sales employees, sales increased by several million dollars.

skill-based pay
A pay system that rewards employees for the job skills they demonstrate.

tend to be more successful in manufacturing organizations than in service organizations and organizations pursuing technical innovations.[38] On the other hand, many organizations use **variable pay** systems, in which an individual's compensation is contingent on performance; Manoj Tandon, Assistant Vice-president, CSC India said "At senior levels we have the concept of variable pay. The idea is that the fixed component is there to do the job, whereas the variable component is for going beyond the job. We are gradually introducing variable pay to lower levels as well. There are two aspects to it—one is that employees should not feel confused, and the second and the most important part is that people at various levels should gradually adapt to it and enhance their performance accordingly to reap benefits."[39] In Chapter 15, we'll discuss variable pay systems further as they relate to employee motivation.

Although many factors influence the design of an organization's compensation system, flexibility is a key consideration. The traditional approach to paying people reflected a more stable time when an employee's pay was largely determined by seniority and job level. Given the dynamic environments that many organizations face today, the trend is to make pay systems more flexible and to reduce the number of pay levels. However, whatever approach managers take, they must establish a fair, equitable, and motivating compensation system that allows the organization to recruit and keep a talented and productive workforce.

QUICK LEARNING REVIEW:
LEARNING OUTCOME 10.4

- **Describe the different performance appraisal methods.**
- **List the factors that influence employee compensation and benefits.**
- **Describe skill-based and variable pay systems.**

Go to page 229 and see how well you know this material.

LEARNING **OUTCOME 10.5** ▷

CONTEMPORARY ISSUES IN MANAGING HUMAN RESOURCES

We'll conclude this chapter by looking at some contemporary HR issues that today's managers face: managing downsizing, managing workforce diversity, managing sexual harassment, managing work–life balance, and controlling HR costs.

MANAGING DOWNSIZING

Downsizing is the planned elimination of jobs in an organization. When an organization has too many employees—which can happen when an organization is faced with declining market share, has grown too aggressively, or has been poorly managed—one option for improving profits is to eliminate some of those excess workers. Many well-known companies have downsized or eased out employees because of poor performance, including, Jet Airways, Kingfisher, GoAir, HLL, Infosys, IBM, ICICI, and Accenture. How can managers best manage a downsized workplace? Disruptions in the workplace and in employees' personal lives should be expected. Stress, frustration, anxiety, and anger are typical reactions of both individuals being laid off and job survivors. Exhibit 10–12 lists some ways that managers can reduce the trauma.[41]

MANAGING WORKFORCE DIVERSITY

We've discussed the changing makeup of the workforce in several places throughout the book and provided insights in "Managing Workforce Diversity" boxes in several chapters. In this section, we discuss how workforce diversity is directly affected by HRM activities.

Recruitment. To improve workforce diversity, managers need to widen their recruiting net. For example, the popular practice of relying on employee referrals as a source of job applicants tends to produce candidates who are similar to present employees. The

Exhibit 10–12

Tips for Managing Downsizing

- Communicate openly and honestly:
  - Inform those being let go as soon as possible
  - Tell surviving employees the new goals and expectations
  - Explain impact of layoffs
- Follow any laws regulating severance pay or benefits
- Provide support/counseling for surviving employees
- Reassign roles according to individuals' talents and backgrounds
- Focus on boosting morale:
  - Offer individualized reassurance
  - Continue to communicate, especially one-on-one
  - Remain involved and available
- Have a plan for the empty office spaces/cubicles so it isn't so depressing for surviving employees

Tata Group has used this to its advantage in recruiting diverse employees. The company's positive experience with a hearing-impaired employee led to hiring other hearing-impaired employees through employee referrals. But not every organization has this option. So managers may have to look for diverse job applicants in places they might not have looked before: women's job networks, over-50 clubs, urban job banks, training centers for disabled individuals, ethnic newspapers, and gay rights organizations, for example. Such nontraditional recruiting should enable an organization to broaden its pool of diverse applicants.

Selection. When a diverse set of applicants exists, managers must ensure that the selection process does not discriminate. Moreover, applicants need to be made comfortable with the organization's culture and be made aware of management's desire to accommodate their needs. For instance, Microsoft strives to hire a high percentage of female applicants, especially for technical jobs, and to make sure those women have a successful experience once they're on the job.[42]

Orientation and Training. The outsider–insider transition is often more challenging for women and minorities than for white males. Many organizations provide special workshops to raise diversity awareness issues. For example, at a Kraft manufacturing facility, managers developed an ambitious diversity program that reflected the increased value the organization had placed on incorporating diverse perspectives. Among other things, they trained more than half of the plant's employees in diversity issues.[43]

MANAGING SEXUAL HARASSMENT

In India, it has been 11 years since sexual harassment was for the first time recognized by the Supreme Court as human rights violation and gender based systemic discrimination that affects women's right to life and livelihood. The Court has given mandatory guidelines, known as *Vishaka* Guidelines, for resolution and prevention of sexual harassment. The Sexual Harassment of women at Work Place (Prevention, Prohibition And Redressal) Bill was passed in 2006. 92 percent women said sexual harassment had a detrimental effect on their work (*Sanhita* study), and studies by the International Labor Organization also document the high costs of sexual harassment as poor organizational image, low work productivity, poor employee morale, high legal costs, mounting medical bills, and high employee turnover.[44] In a study of working women

variable pay
A pay system in which an individual's compensation is contingent on performance.

downsizing
The planned elimination of jobs in an organization.

in Maharashtra, the incidence of sexual harassment of women at work place was reported to be 17 percent in organized sector and 20 percent in unorganized sector.[45]

Sexual harassment is not a local problem. It's a global issue. For instance, data collected by the European Commission found that between 30 and 50 percent of female employees in European Union countries had experienced some form of sexual harassment.[46] And sexual harassment charges have been filed against employers in other countries, such as Japan, Australia, New Zealand, and Mexico.[47]

Even though discussions of sexual harassment cases often focus on the large awards granted by courts, there are other concerns for employers. Sexual harassment creates an unpleasant work environment and undermines workers' ability to perform their job.

Sexual harassment is defined as any unwanted action or activity of a sexual nature that explicitly or implicitly affects an individual's employment, performance, or work environment. It can occur between members of the opposite sex or of the same sex.

Many problems associated with sexual harassment involve determining exactly what constitutes this illegal behavior. The Bill defines "Sexual Harassment" as such unwelcome sexually determined behavior such as physical contact, advances, sexually colored remarks, showing pornography, or making sexual demands, whether verbal, textual, graphic or electronic, or by any other actions, which may contain: (a) implied or overt promise of preferential treatment, (b) an implied or overt threat of detrimental effects the present or future employment status of that employee and includes the creation of a hostile working environment, and (c) the conduct interferes with an employee's work or creates an intimidating, hostile, or offensive work environment or be humiliating and may constitute a health and safety problem.[48] For many organizations, it's the offensive or hostile environment issue that is problematic. Managers must be aware of what constitutes such an environment. Another thing that managers must understand is that the victim doesn't necessarily have to be the person harassed but could be anyone affected by the offensive conduct.[49] The key is being attuned to what makes other employees uncomfortable—and if we don't know, we should ask![50]

Let's Get Real: F2F

SEXUAL HARASSMENT TRAINING WORKS BEST WHEN:

Women are part of the discussion/workshop. Opportunities are made available through confidential processes to address this important and necessary value system in the organization.

What can an organization do to protect itself against sexual harassment claims?[51] The courts want to know two things: First, did the organization know about, or should it have known about, the alleged behavior? And second, what did managers do to stop it? It's vital that all employees should be educated on sexual harassment matters. In addition, organizations need to ensure that no retaliatory actions—such as cutting back hours or assigning back-to-back work shifts without a rest break—are taken against a person who has filed harassment charges. One final area of interest we want to discuss in terms of sexual harassment is workplace romances.

Workplace Romances. Have you ever dated someone at work? If not, have you ever been attracted to someone in your workplace and thought about pursuing a relationship? Such situations are more common than you might think: 16 percent of employees surveyed in India by the *Team Lease Services* said that they have had an office romance.[52] And another survey in US found that 43 percent of single men and 28 percent of single women said that they would be open to dating a coworker.[53] The environment in today's organizations, with mixed-gender work teams and long work hours, has likely contributed to this situation. "People realize they're going to be at work such long hours, it's almost inevitable that this takes place," said one survey director.[54] But workplace romances can potentially become big problems for organizations.[55] In addition to the potential conflicts and retaliation between coworkers who decide to stop dating or to end a romantic relationship, more serious problems stem from the potential for sexual harassment accusations, especially with relationships between a supervisor and a subordinate. The standard used by judicial courts has been that workplace sexual conduct is prohibited sexual harassment *if* it is unwelcome. If it's welcome, it still may be inappropriate, but it usually is not unlawful.

What should managers do about workplace romances? It's important to have some type of policy regarding workplace dating, particularly in terms of educating employees about the potential for sexual harassment. However, because the potential liability is especially serious when it comes to supervisor–subordinate relationships, a more proactive approach is needed in terms of discouraging such relationships and perhaps even requiring supervisors to report any such relationships to the HR department. At some point, the organization may even want to consider banning such relationships, although an outright ban may be difficult to enforce.

MANAGING WORK-LIFE BALANCE

Smart managers recognize that employees don't leave their families and personal lives behind when they come to work. Although managers can't be sympathetic with every detail of an employee's family life, organizations are becoming more attuned to the fact that employees have sick children, elderly parents who need special care, and other family issues that may require special arrangements. In response, many organizations are offering **family-friendly benefits**, which accommodate employees' needs for work–life balance. They have introduced programs such as on-site child care, summer day camps, flextime, job sharing, time off for school functions, telecommuting, and part-time employment. Work–life conflicts are as relevant to male workers with children and women without children as they are for female employees with children. Heavy work loads and increased travel demands have made it difficult for many employees to satisfactorily juggle both work and personal responsibilities. A *Fortune* survey found that 84 percent of male executives surveyed said that "they'd like job options that let them realize their professional aspirations while having more time for things outside work."[56] Also, 87 percent of these executives believed that any company that restructured top-level management jobs in ways that would both increase productivity and make more time available for life outside the office would have a competitive advantage in attracting talented employees. Younger employees, particularly, put a higher priority on family and a lower priority on jobs and are looking for organizations that give them more work flexibility.[57]

Today's progressive workplaces must accommodate the varied needs of a diverse workforce. How? By providing a wide range of scheduling options and benefits that allow employees more flexibility at work and that allow them to better balance or integrate their work and personal lives. Despite these organizational efforts, work–life programs certainly have room for improvement. One survey showed that more than 31 percent of college-educated male workers spend 50 or more hours per week at work (up from 22 percent in 1980) and that about 40 percent of U.S. adults get less than 7 hours of sleep on weekdays (up from 34 percent in 2001).[58] What about women? Another survey showed that the percentage of American women working 40 hours or more per week had increased. By the way, this same survey showed that the percentage of

While some companies are finding more ways to free employees from their physical offices by letting them keep in touch digitally from home, about 80 other firms around the country are allowing parents to bring their babies into the office. Allie Hewlett is a benefits administrator for an advertising agency in Austin, Texas, whose daughter Scout is 7 months old. Her company has decided that although there may be occasional work disruptions with babies present, the benefits gained from helping employees achieve work-life balance make it worthwhile.

© Joel Salcido/USA Today, March 31.

sexual harassment
Any unwanted action or activity of a sexual nature that explicitly or implicitly affects an individual's employment, performance, or work environment.

family-friendly benefits
Benefits that accommodate employees' needs for work–life balance.

European women working 40 hours or more had actually declined.[59] According to the findings of a survey conducted by TeamLease in February 2008[60] on "India's changing world of work," 87 percent of the respondents were satisfied with their work-life balance and about 70 percent felt that Indian organizations have flexible policies for employees that allow them to maintain work-life balance. This survey had only 401 respondents from 500 top companies of India. However, 54 percent feel that they have become "weekend parents," 34 percent feel that two careers substantially increase the chances of a divorce, and 32 percent feel that it is not possible for women to pursue their careers after a child break. Other workplace surveys still show high levels of employee stress stemming from work–life conflicts. And large groups of women and minority workers remain unemployed or underemployed because of family responsibilities and bias in the workplace.[61] So what can managers do?

Research on work–life balance has shown that there are positive outcomes when individuals are able to combine work and family roles.[62] As one study participant noted, "I think being a mother and having patience and watching someone else grow has made me a better manager. I am better able to be patient with other people and let them grow and develop in a way that is good for them."[63] In addition, individuals who have family-friendly workplace support appear to be more satisfied on the job.[64] This seems to strengthen the notion that organizations benefit by creating a workplace in which employee work–life balance is possible. And the benefits are financial as well. Research has shown a significant, positive relationship between work–life initiatives and an organization's stock price.[65] However, managers need to understand that people differ in their preferences for work–life scheduling options and benefits.[66] Some prefer organizational initiatives that better *segment* work from their personal lives, and others prefer programs that facilitate *integration*. For instance, flextime schedules segment because they allow employees to schedule work hours that are less likely to conflict with personal responsibilities. On the other hand, on-site child care integrates the boundaries between work and family responsibilities. People who prefer segmentation are more likely to be satisfied and committed to their jobs when offered options such as flextime, job sharing, and part-time hours. People who prefer integration are more likely to respond positively to options such as on-site child care, gym facilities, and company-sponsored family picnics.

CONTROLLING HR COSTS

HR costs are skyrocketing, especially the costs of employee health care and employee pensions. Organizations are looking for ways to control these costs.

Employee Health Care Costs. HCL Comnet has set up Employee First Councils in five areas of which wellness is one of them. The Wellness council conducts health checkups for employees.[67] Johnson and Johnson reports saving $8.5 million per annum in the form of reduced employee medical claims and administrative savings because of focusing on a healthy employee and healthy workplace.[68] Ritu Anand, at TCS, says that they strive to make work a joyful experience through "Maitree," a unique platform for all employee engagement activities. Different clubs like Theatre Club, Adventure and Trekking Club, Fitness Club, Sanctuary Club, among others facilitate fun at work. These clubs organize various events for TCS employees.[69] All these examples illustrate how companies are trying to control skyrocketing employee health care costs. The Employer Health and Wellness Program study estimates that China will lose $558 billion, India $237 billion, and Russia $303 billion in national income from 2005 to 2015 due to only three chronic diseases: heart disease, stroke, and diabetes.[70] And smokers raise the rates for companies even more: They cost about 25 percent more for health care than do nonsmokers.[71] Another health care cost for companies is obesity, which costs organizations an estimated $45 billion per year in medical expenditures and absenteeism.[72] A study of manufacturing organizations found that presenteeism, which is defined as employees showing up at work but not performing at full capacity, was 1.8 percent higher for workers with moderate to severe obesity than for all other employees.[73] The reason for the lost productivity is likely the result of reduced mobility because of body size or pain problems such as arthritis. Is it any wonder that organizations are looking for ways to control their health care costs?

How can organizations control their health care costs? First, many organizations are providing opportunities for employees to lead healthy lifestyles. From financial incentives to company-sponsored health and wellness programs, the goal is to limit rising health care costs. About 41 percent of companies are using some type of positive incentives aimed at encouraging healthy behavior, up from 34 percent in 1996.[74] Another recent study indicated that nearly 90 percent of companies surveyed planned to aggressively promote healthy lifestyles to their employees during the next three to five years.[75] Many are starting sooner: Google, Akruti City of Mumbai, Patni, and others are putting health food in company break rooms, cafeterias, and vending machines; providing deliveries of fresh organic fruit; and imposing "calorie taxes" on fatty foods.[76] In the case of smokers, some companies have taken a more aggressive stance by increasing the amount smokers pay for health insurance or by firing them if they refuse to stop smoking.

Employee Pension Plan Costs. In addition to health care, another area where organizations are looking to control costs is employee pension plans. Corporate pensions have been around since the nineteenth century.[77] But the days when companies could afford to give each employee a broad-based pension that provided a guaranteed retirement income have changed. Pension commitments have become such an enormous burden that companies can no longer afford them. In fact, the corporate pension system has been described as "fundamentally broken."[78] It's not just struggling companies that have eliminated employee pension plans. Lots of reasonably sound companies—for instance, NCR, FedEx, Lockheed Martin, and Motorola—no longer provide pensions. Even IBM, which closed its pension plan to new hires in December 2004, told employees that their pension benefits would be frozen.[79] Obviously, the pension issue is one that directly affects HR decisions. On the one hand, organizations want to attract talented, capable employees by offering them desirable benefits such as pensions. On the other hand, organizations have to balance that with the costs of providing such benefits.

QUICK LEARNING REVIEW:

LEARNING OUTCOME 10.5

- Explain how managers can manage downsizing.
- Discuss how managers can manage workforce diversity.
- Explain what sexual harassment is and what managers need to know about it.
- Describe how organizations are dealing with work-life balance issues.
- Discuss how organizations are controlling HR costs.

Go to page 230 and see how well you know this material.

Let's Get Real:

My Turn

Avinash Deolekar

Vice President, HR
Pepsico India Holdings Pvt. Ltd

More than skill up-gradation issue, this is an issue of providing the right channel to the energies of the employees. The most crucial and first aspect of the problem is whether these employees know what is expected out of them? The steps that I would follow would be the following:

- Defining the consumer orientation with senior management.
- Rolling the same down the line through some innovative and connecting medium – like movies, self-experience.
- Putting in place reward mechanisms for employees who show consumer orientation as per definition.
- Putting the review mechanism in place to review the progress and suggest changes if any.

Once this is done, other HR systems would have direct/indirect linkages with the consumer orientation matrix, like compensation/rewards/communication etc.

LEARNING OUTCOMES **SUMMARY**

10.1 ▷ THE HUMAN RESOURCE MANAGEMENT PROCESS

- Explain why the HRM process is important.
- Discuss the environmental factors that most directly affect the HRM process.

HRM is important for three reasons. First, it can be a significant source of competitive advantage. Second, it's an important part of organizational strategies. Finally, the way organizations treat their employees has been found to significantly affect organizational performance.

The external factors that most directly affect the HRM process are labor unions, governmental laws, and demographic trends.

10.2 ▷ IDENTIFYING AND SELECTING COMPETENT EMPLOYEES

- Define *job analysis*, *job description*, and *job specification*.
- Discuss the major sources of potential job candidates.
- Describe the different selection devices and which work best for different jobs.
- Explain why a realistic job preview is important.

A job analysis is an assessment that defines a job and the behaviors necessary to perform it. A job description is a written statement that describes a job; it typically includes job content, environment, and conditions of employment. A job specification is a written statement that specifies the minimum qualifications that a person must possess to successfully perform a given job.

The major sources of potential job candidates include the Internet, employee referrals, company Web sites, college recruiting, and professional recruiting organizations.

The different selection devices include application forms (best used for gathering employee information), written tests (must be job related), work sampling (appropriate for complex nonmanagerial and routine work), assessment centers (most appropriate for top-level managers), interviews (widely used, but most appropriate for managerial positions, especially top-level managers), background investigations (useful for verifying application data, but reference checks are essentially worthless), and physical exams (useful for work that involves certain physical requirements and for insurance purposes).

A realistic job preview is important because it gives an applicant realistic expectations about the job. This in turn should increase employee job satisfaction and reduce turnover.

10.3 ▷ PROVIDING EMPLOYEES WITH NEEDED SKILLS AND KNOWLEDGE

- Explain why orientation is so important.
- Describe the different types of training and how each type of training can be provided.

Orientation is important because it eases the outsider–insider transition and makes the new employee feel comfortable and fairly well adjusted, reduces the likelihood of poor work performance, and reduces the probability of an early surprise resignation.

The two types of training are general (includes communication skills, computer skills, customer service, personal growth, etc.) and specific (includes basic life/work skills, customer education, diversity/cultural awareness, managing change, etc.). This training can be provided using traditional training methods (on-the-job, job rotation, mentoring and coaching, experiential exercises, workbooks/manuals, and classroom lectures) or using technology-based methods (CDs/DVDs/videotapes/audiotapes, videoconferencing or teleconferencing, or e-learning).

10.4 ▷ RETAINING COMPETENT, HIGH-PERFORMING EMPLOYEES

- Describe the different performance appraisal methods.
- List the factors that influence employee compensation and benefits.
- Describe skill-based and variable pay systems.

The different performance appraisal methods are written essays, critical incidents, graphic rating scales, BARS, multiperson comparisons, MBO, and 360-degree appraisals.

The factors that influence employee compensation and benefits include the employee's tenure and performance, kind of job performed, kind of business/industry, unionization, labor or capital intensity, management philosophy, geographic location, company profitability, and size of company.

Skill-based pay systems reward employees for the job skills and competencies they demonstrate. In a variable pay system, an employee's compensation is contingent on performance.

10.5 ▷ CONTEMPORARY ISSUES IN MANAGING HUMAN RESOURCES

- Explain how managers can manage downsizing.
- Discuss how managers can manage workforce diversity.
- Explain what sexual harassment is and what managers need to know about it.
- Describe how organizations are dealing with work–life balance issues.
- Discuss how organizations are controlling HR costs.

Managers can manage downsizing by communicating openly and honestly, following appropriate laws regarding severance pay or benefits, providing support/counseling for surviving employees, reassigning roles according to individuals' talents and backgrounds, focusing on boosting morale, and having a plan for empty office spaces.

Managers should manage workforce diversity by focusing on recruitment of diverse employees, making sure the selection process doesn't discriminate, and then making sure the outsider–insider transition process is solid.

Sexual harassment is any unwanted action or activity of a sexual nature that explicitly or implicitly affects an individual's employment, performance, or work environment. Managers need to be aware of what constitutes an offensive or hostile work environment, educate employees on sexual harassment, and ensure that no retaliatory actions are taken against any person who files harassment charges. Also, they may need to have a policy in place for workplace romances.

Organizations are dealing with work–life balance issues by offering family-friendly benefits such as on-site child care, flextime, and telecommuting. Managers need to understand that people may prefer programs that segment work and personal lives, while others prefer programs that integrate their work and personal lives.

Organizations are controlling HR costs by controlling employee health care costs through employee health initiatives (encouraging healthy behaviors and penalizing unhealthy behaviors) and controlling employee pension plans by eliminating or severely limiting them.

THINKING ABOUT MANAGEMENT ISSUES

1. How does HRM affect all managers?
2. Should an employer have the right to choose employees without governmental interference? Support your position.
3. Some critics claim that corporate HR departments have outlived their usefulness and are not there to help employees but to shield the organization from legal problems. What do you think? What benefits are there to having a formal HRM process? What drawbacks?
4. Studies show that women's salaries still lag behind men's, and even with equal opportunity laws and regulations, women are paid about 76 percent of what men are paid. How would you design a compensation system that would address this issue?
5. What drawbacks, if any, do you see in implementing flexible benefits? (Consider this question from the perspective of the organization and the perspective of an employee.)
6. What are the benefits and drawbacks of realistic job previews? (Consider this question from the perspective of the organization and the perspective of an employee.)
7. What, in your view, constitutes sexual harassment? Describe how companies can minimize sexual harassment in the workplace.
8. Go to the Society for Human Resource Management Web site (www.shrm.org) and find the HR News section. Pick a news story and read it. (Note: Some of these are available only to SHRM members, but others are available to any visitor.) Write a summary of the information. At the end of your summary, discuss the implications of the topic for managers.

YOUR TURN to be a Manager

- Use the Internet to research five different companies that interest you and check out what they say about careers or their employees. Put this information in a bulleted-format report. Be prepared to make a presentation to the class about your findings.
- Work on your résumé. If you don't have one already, research what a good résumé should include. If you have one already, make sure it provides specific information that explicitly describes your work skills and experience rather than meaningless phrases such as *results-oriented.*
- If you're working, note what types of HRM activities your managers do (such as interview and appraise performance). Ask them what they've found to be effective in getting and keeping good employees. What hasn't been effective? What can you learn from this? If you're not working, interview three different managers about what HRM activities they do and which they've found to be effective and not effective.
- Research your chosen career by finding out what it's going to take to be successful in that career in terms of education, skills, experience, and so forth. Write a personal career guide that details this information.
- Complete the skill building modules Interviewing and Valuing Diversity found on mymanagementlab. Your professor will tell you what you need to do with this information.
- Steve's and Mary's suggested readings: Thomas W. Malone, *The Future of Work* (Harvard Business School Press, 2004); Charles A. O'Reilly III and Jeffrey Pfeffer, *Hidden Value* (Harvard Business School Press, 2000); Jeffrey Pfeffer, *The Human Equation* (Harvard Business School Press, 1998); Richard W. Judy and Carol D'Amico, *Workforce 2020* (Hudson Institute, 1997); and Robert Johansen and Rob Swigart, *Upsizing the Individual in the Downsized Organization* (Addison-Wesley, 1996).
- Pick one of the five topics in the section "Contemporary Issues in Managing Human Resources." Research this topic and write a paper about it. Focus on finding current information and current examples of companies dealing with these issues.
- In your own words, write down three things you learned in this chapter about being a good manager.
- Self-knowledge can be a powerful learning tool. Go to mymanagementlab and complete these self-assessment exercises: How Good Am I at Giving Feedback? How Satisfied Am I with My Job? Am I Experiencing Work–Family Conflict? And What Are My Attitudes Toward Workplace Diversity? Using the results of your assessments, identify personal strengths and weaknesses. What will you do to reinforce your strengths and improve your weaknesses?

CASE APPLICATION

Busted

Like many other companies today, Scott's Miracle-Gro is facing the dilemma of persuading employees to take better care of themselves without diminishing employee morale or getting hit with employee lawsuits. It's on the leading edge of companies looking to monitor and change employee behavior. But sometimes that edge can be razor sharp.

Scott's Miracle-Gro's CEO, Jim Hagedorn (in photo), acknowledges that his company's wellness program is controversial. In 2000, he, like many other CEOs, watched as his company's health care costs skyrocketed.

Jim Hagedorn, Chief Executive of Scott's Miracle-Gro Co.

No help was in sight from either the government or the health insurance industry, and the company's employees were, he said, "bingeing on health care." By February 2003, workers' health care insurance premiums had doubled, and employee morale had plummeted. Following his usual tell-it-like-it-is style, Hagedorn confronted the issue head-on with employees. He wanted them to know what they were up against: 20 percent of the company's net profits were going to health care. The company's health-risk assessment showed that half of the 6,000 employees were overweight or morbidly obese and one-quarter of the employees smoked. After seeing a CNN program late one night in which a doctor was arguing that employers needed to get serious about employee obesity, smoking, and diabetes, Hagedorn knew what he had to do. Despite the late hour, he immediately called his HR chief and told her that he wanted to ban smoking and tackle obesity.

Getting that done wasn't so easy. The legal department worried that the plan might violate federal laws. Other advisors told Hagedorn not to do it or that he was moving too fast. But he wasn't easily dissuaded. He found a law firm that helped determine that in 21 states (including the company's home base in Ohio), it wasn't illegal to hire and fire people based on their smoking habits. Hagedorn also implemented a companywide wellness program but realized that he needed a third party to run it so managers couldn't discriminate against employees based on their health.

Today, Scott's Miracle-Gro employees are encouraged to take exhaustive health-risk assessments. Those who don't pay $40 per month more in premiums. Each employee found to be at moderate to high risk is assigned to and works closely with a health coach. Those who don't comply pay an additional $67 per month on top of the $40. Many employees find the policy intrusive, but Hagedorn hasn't budged. He's adamant about bringing down health care costs and getting employees all the help they need to be healthy and lead healthier lives. One employee who was fired on his 30th birthday because he failed a drug test for nicotine use is suing the company. (The lawsuit is still proceeding through the court system.) However, another employee who was prodded by his health coach to see a doctor had his life saved when surgeons found a 95 percent blockage in his heart that would have killed him within five days without the stents that were inserted.

Discussion Questions

1. What do you think about Hagedorn's approach to controlling employee health care costs? Do you agree with it? Why or why not?
2. What benefits and drawbacks are there to this type of wellness program for (a) employees and (b) the company?
3. Research company wellness programs. What types of things are companies doing to encourage employee wellness? Are there any things that you found that you might recommend that Hagedorn implement? Describe.

Sources: M. Freudenheim, "Seeking Savings, Employers Help Smokers Quit," *New York Times* online, www.nytimes.com, October 26, 2007; and M. Conlin, "Get Healthy or Else," *BusinessWeek*, February 26, 2007, pp. 58–69.

Let's Get Real:

Meet The Manager

Pradeep Waychal

Vice President
Patni Computer Systems Ltd

MY JOB: Head of the Delivery, Innovation and Quality Department. Drive various innovation initiatives to assure productivity, quality, and cycle time. Incubate various customer-facing practices or CoEs.

BEST PART OF MY JOB: Catalyzing success of team members and executing innovative initiatives.

WORST PART OF MY JOB: Compromising on correct views to accommodate various stakeholders.

BEST MANAGEMENT ADVICE EVER RECEIVED: Maintaining one-to-one relationship between an initiative and owner. For every initiative, there should be one fully accountable person. If a person becomes accountable for more than one initiative, chances of success of initiatives start diminishing rapidly.

You'll be hearing more from this real manager throughout the chapter.

Chapter

Managing Teams

Few trends have influenced how work gets done in organizations as much as the use of work teams. Organizations are increasingly structuring work around teams rather than individuals. Managers need to understand what influences team performance and satisfaction. Focus on the following learning outcomes as you read and study this chapter.

LEARNING OUTCOMES

A Manager's Dilemma

Management theory suggests that compared to an individual, a diverse group of people will be more creative because team members will bring a variety of ideas, perspectives, and approaches to the group.[1] This information might be important to a company such as Google; not only is innovation critical to the company's success but teams are a way of life at Google. As the company's Web site states, "Googlers thrive in small, focused teams and high-energy environments." If management theory about teams is correct, then Google's R&D center in India should excel at innovation. Why? Because there, you find broad diversity even though all employees are from India. These Googlers include Indians, Sikhs, Hindus, Muslims, Buddhists, Christians, and Jains. And they speak English, Hindi, Tamil, Bengali, and more of India's 22 officially recognized languages. One skill that Google looks for in potential hires is the ability to work as a team member. Suppose that you're a manager at Google's Hyderabad facility. What would you do to maintain your team's ability to be innovative when new designers join the group?

What Would You Do?

You've probably had a lot of experience working in groups—class project teams, maybe an athletic team, a fundraising committee, or even a sales team at work. Work teams are one of the realities—and challenges—of managing in today's dynamic global environment. Many organizations have made the move to restructure work around teams rather than individuals. Why? What do these teams look like? And, as with the challenge Google faces in India, how can managers build effective teams? These are some of the questions we'll be answering in this chapter. Before we can understand teams, however, we first need to understand some basics about groups and group behavior.

LEARNING **OUTCOME 11.1** ▷

GROUPS AND GROUP DEVELOPMENT

Each person in the group had his or her assigned role: The Spotter, The Back Spotter, The Gorilla, and the Big Player. For over 10 years, this group—former MIT students who were members of the secret Black Jack Club—used their extraordinary mathematical abilities, expert training, teamwork, and interpersonal skills to take millions of dollars from some of the major casinos in the United States.[2] Although most groups aren't formed for such dishonest purposes, the success of this group at its task was impressive. Managers would like their groups to be successful at their tasks also. The first step is understanding what a group is and how groups develop.

Exhibit 11–1

Examples of Formal Work Groups

- *Command groups*—Groups that are determined by the organization chart and composed of individuals who report directly to a given manager.
- *Task groups*—Groups composed of individuals brought together to complete a specific job task; their existence is often temporary because when the task is completed, the group disbands.
- *Cross-functional teams*—Groups that bring together the knowledge and skills of individuals various work areas or groups whose members have been trained to do each others' jobs.
- *Self-managed teams*—Groups that are essentially independent and that, in addition to their own tasks, take on traditional managerial responsibilities, such as hiring, planning and scheduling, and evaluating performance.

WHAT IS A GROUP?

A **group** is defined as two or more interacting and interdependent individuals who come together to achieve specific goals. *Formal groups* are work groups that are defined by an organization's structure and have designated work assignments and specific tasks directed at accomplishing organizational goals. Exhibit 11–1 provides some examples. *Informal groups* are social groups. These groups occur naturally in the workplace and tend to form around friendships and common interests. For example, five employees from different departments who regularly eat lunch together are an informal group.

STAGES OF GROUP DEVELOPMENT

Research has shown that groups develop through five stages.[3] As shown in Exhibit 11–2, these five stages are *forming, storming, norming, performing,* and *adjourning.*

The **forming** stage has two phases. The first occurs as people join the group. In a formal group, people join because of some work assignment. After they've joined, the second phase begins: defining the group's purpose, structure, and leadership. This phase involves a great deal of uncertainty as members "test the waters" to determine what types of behavior are acceptable. This stage is complete when members begin to think of themselves as part of a group.

Exhibit 11–2

Stages of Group Development

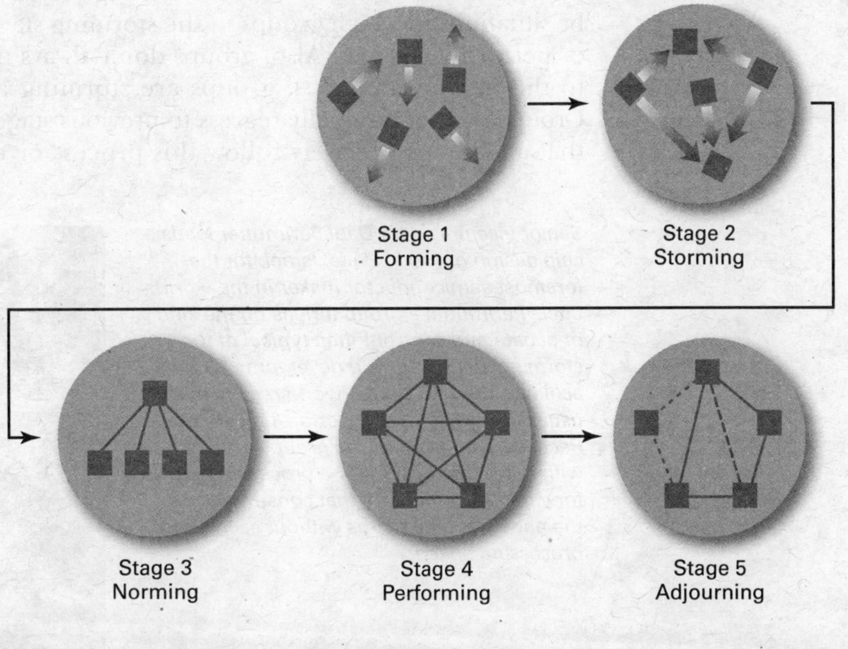

group
Two or more interacting and interdependent individuals who come together to achieve specific goals.

forming stage
The first stage of group development, in which people join the group and then define the group's purpose, structure, and leadership.

The **storming** stage is named because of the intragroup conflict that occurs over who will control the group and what the group needs to be doing. When this stage is complete, there is a relatively clear hierarchy of leadership and agreement on the group's direction.

The **norming** stage is one in which close relationships develop and the group becomes cohesive. There's now a strong sense of group identity and camaraderie. This stage is complete when the group structure solidifies and the group has assimilated a common set of expectations (or norms) regarding member behavior.

The fourth stage of group development is **performing**. The group structure is in place and accepted by group members. Their energies have moved from getting to know and understand each other to working on the group's task. This is the last stage of development for permanent work groups. However, for temporary groups—project teams, task forces, or similar groups that have a limited task to do—the final stage is **adjourning**. In this stage, the group prepares to disband. Attention is focused on wrapping up activities instead of task performance. Group members react in different ways. Some are upbeat, thrilled about the group's accomplishments. Others may be sad over the loss of camaraderie and friendships.

Many of you have probably experienced these stages as you've worked on a group project for a class. Group members are selected or assigned and then meet for the first time. There's a "feeling out" period to assess what the group is going to do and how it's going to be done. This is usually followed by a battle for control: Who's going to be in charge? When this issue is resolved and a "hierarchy" is agreed on, the group identifies specific work that needs to be done, who's going to do each part, and dates by which the assigned work needs to be completed. General expectations are established. These decisions form the foundation for what you hope will be a coordinated group effort culminating in a project that's been done well. When the project is complete and turned in, the group breaks up. Of course, some groups don't get much beyond the forming or storming stages. These groups may have serious interpersonal conflicts, turn in disappointing work, and get low grades.

Does a group become more effective as it progresses through the first four stages? Some researchers say yes, but it's not that simple.[4] That assumption may be generally true, but what makes a group effective is a complex issue. Under some conditions, high levels of conflict are conducive to high levels of group performance. There might be situations in which groups in the storming stage outperform those in the norming or performing stages. Also, groups don't always proceed sequentially from one stage to the next. Sometimes, groups are storming and performing at the same time. Groups even occasionally regress to previous stages. Therefore, it's not safe to assume that all groups precisely follow this process or that performing is always the most

Senior vice president Dadi Perlmutter leads a chip design group in Haifa, Israel, for the foremost semiconductor maker in the world, Intel. Perlmutter's group thrives on the kind of debate and confrontation typical of the storming stage of group development, but it achieves the kind of real-world results that usually characterize the performing stage. Recently, for instance, the group came up with a winning design for a processor chip for wireless computers that consumes half the power of other chips without sacrificing processing speed.

preferable stage. It's better to think of this model as a general framework that underscores the fact that groups are dynamic entities and managers need to know the stage a group is in so they can understand the problems and issues that are most likely to surface.

QUICK LEARNING REVIEW:

LEARNING OUTCOME 11.1

- **Describe the different types of groups.**
- **Describe the five stages of group development.**

Go to page 256 and see how well you know this material.

LEARNING **OUTCOME 11.2** ▷

WORK GROUP PERFORMANCE AND SATISFACTION

Let's Get Real: F2F

HOW MUCH WORK IS DONE BY GROUPS?

Almost all work is done by groups.

Many people consider them the most successful "group" of our times. Who? The Beatles. "The Beatles were great artists and entertainers, but in many respects they were four ordinary guys who, as a group, found a way to achieve extraordinary artistic and financial success and have a great time together while doing it. Every business team can learn from their story."[5]

Why *are* some groups more successful than others? Why do some groups achieve high levels of performance and high levels of member satisfaction and others do not? The answers are complex but include variables such as the abilities of the group's members, the size of the group, the level of conflict, and the internal pressures on members to conform to the group's norms. Exhibit 11–3 presents the major factors that determine group performance and satisfaction.[6] Let's look at each of them.

EXTERNAL CONDITIONS IMPOSED ON THE GROUP

A work group is affected by the external conditions imposed on it. These include the organization's strategy, authority relationships, formal rules and regulations, the availability of resources, employee selection criteria, the performance management system and culture, and the general physical layout of the group's work space. For instance, some groups have modern, high-quality tools and equipment to do their jobs, while other groups don't. Or the organization might be pursuing a strategy of lowering costs or improving quality, which will affect what a group does and how it does it.

GROUP MEMBER RESOURCES

A group's performance potential depends to a large extent on the resources each individual brings to the group. These resources include knowledge, skills, abilities, and personality traits, and they determine what members can do and how effectively they

Exhibit 11–3

Group Performance Satisfaction Model

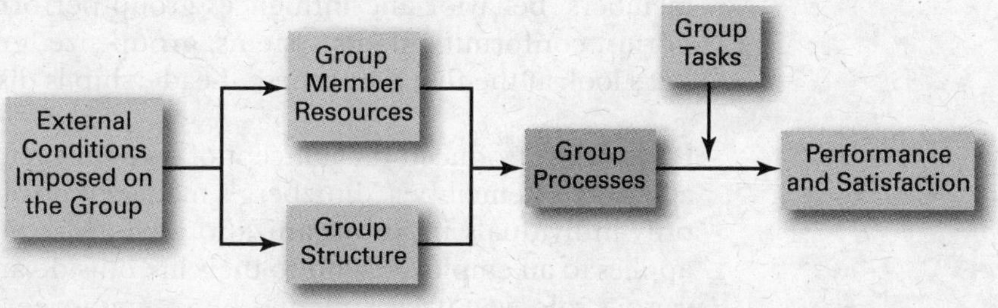

storming stage
The second stage of group development, which is characterized by intragroup conflict.

norming stage
The third stage of group development, which is characterized by close relationships and cohesiveness.

performing stage
The fourth stage of group development, when the group is fully functional and works on the group task.

adjourning stage
The final stage of group development for temporary groups, during which group members are concerned with wrapping up activities rather than task performance.

managing workforce Diversity

The Challenge of Managing Diverse Teams

Managing teams composed of people who are similar isn't always easy. But add in diverse members, and it can be even more challenging! However, the benefits from the diverse perspectives, skills, and abilities of diverse members are worth it.[9] Four interpersonal factors are important for meeting the challenge of coordinating a diverse work team: understanding, empathy, tolerance, and communication.

You know that people aren't the same, yet they need to be treated fairly and equitably. And differences (cultural, physical, or other) can cause people to behave in different ways. You need to understand and accept these differences and encourage each team member to do the same.

Empathy is closely related to understanding. As a team leader, you should try to understand others' perspectives. Put yourself in their place and encourage team members to empathize as well. For instance, suppose an Asian woman joins a team of Caucasian and Hispanic men. They can make her feel more welcome and comfortable by identifying with how she might feel. Is she excited or disappointed about her new work assignment? What were her previous work experiences? How can they help her feel more comfortable? By empathizing with her, existing team members can work together better as an effective group.

Tolerance is another important interpersonal consideration. Just because you understand that people are different and you empathize with them doesn't mean that it's any easier to accept different perspectives or behaviors. But it's important to be tolerant and open-minded about different values, attitudes, and behaviors.

Finally, open and two-way communication is important to managing a diverse team. Diversity problems may intensify if people are afraid or unwilling to openly discuss issues that concern them. If a person wants to know whether a certain behavior is offensive to someone else, it's best to ask. Likewise, a person who is offended by another's behavior should explain his or her concerns and ask that person to stop. Such communication exchanges can be positive when they're handled in a nonthreatening, low-key, and friendly manner.

will perform in a group. Interpersonal skills—especially conflict management and resolution, collaborative problem solving, and communication—consistently emerge as important for work groups to perform well.[7]

Personality traits also affect group performance because they strongly influence how an individual will interact with other group members. Research has shown that traits that are viewed as positive in our culture (such as sociability, self-reliance, and independence) tend to be positively related to group productivity and morale. In contrast, negative personality characteristics (such as authoritarianism, dominance, and unconventionality) tend to be negatively related to group productivity and morale.[8]

GROUP STRUCTURE

Work groups aren't unorganized crowds. They have an internal structure that shapes members' behavior and influences group performance. The structure defines roles, norms, conformity, status systems, group size, group cohesiveness, and leadership. Let's look at the first six of these. Leadership is discussed in Chapter 16.

Roles. We introduced the concept of roles in Chapter 1 when we discussed what managers do. (Remember Mintzberg's managerial roles.) Of course, managers aren't the only individuals in an organization who play various roles. The concept of roles applies to all employees and to their life outside an organization as well. (Think of the various roles you play: student, sibling, employee, spouse or significant other, etc.)

A **role** refers to behavior patterns expected of someone occupying a given position in a social unit. In a group, individuals are expected to do certain things because of their position (role) in the group. These roles are generally oriented toward either getting work done or keeping group members happy.[10] Think about groups that you've been in and the roles that you played. Were you continually trying to keep the group focused on getting its work done? If so, you were in a task-accomplishment role. Or were you more concerned that group members had the opportunity to offer ideas and that they were satisfied with the experience? If so, you were performing a group member satisfaction role. Both roles are important to the group's ability to function effectively and efficiently.

A problem that arises is that individuals play multiple roles and adjust their roles to the group to which they belong at the time. Because of the different expectations of these roles, employees face *role conflicts.*

Norms. All groups have **norms**—standards or expectations that are accepted and shared by a group's members. Norms dictate things such as work output levels, absenteeism, promptness, and the amount of socializing on the job.

For example, norms dictate in the secretariat in Bhubaneshwar that most people come to their departments between 10 A.M. and 10:15 A.M. (official time is 10:00 A.M.) and then after they have kept their bag, pen on the file, and glasses on the table then they go out to have a "paan" or tea. Anyone who starts work as soon as they come would be frowned upon and told indirectly to join them.

Although a group has its own unique set of norms, common organizational norms focus on effort and performance, dress, and loyalty. The most widespread norms are those related to work effort and performance. Work groups typically provide their members with explicit cues on how hard to work, level of output, when to look busy, when it's acceptable to goof off, and the like. These norms are very powerful influences on an individual employee's performance. They're so powerful that you can't predict someone's performance based solely on his or her ability and personal motivation. Dress norms frequently dictate what's acceptable to wear to work. If the norm is more formal dress, anyone who dresses casually may face subtle pressure to conform. Finally, loyalty norms will influence whether individuals work late, work on weekends, or move to locations they might not prefer to live.

One negative thing about group norms is that being part of a group can increase an individual's antisocial actions. If the norms of the group include tolerating deviant behavior, someone who normally wouldn't engage in such behavior might be more likely to do so. For instance, one study found that those working in a group were more likely to lie, cheat, and steal than were individuals working alone.[11] Why? Because groups provide anonymity, thus giving individuals—who might otherwise be afraid of getting caught—a false sense of security.

Conformity. Because individuals want to be accepted by groups to which they belong, they're susceptible to pressures to conform. Early experiments done by Solomon Asch demonstrated the impact that conformity has on an individual's judgment and attitudes.[12] In these experiments, groups of seven or eight people were asked to compare two cards held up by the experimenter. One card had three lines of different lengths, and the other had one line that was equal in length to one of the three lines on the other card (see Exhibit 11–4). Each group member was to announce aloud

Exhibit 11–4

Examples of Asch's Cards

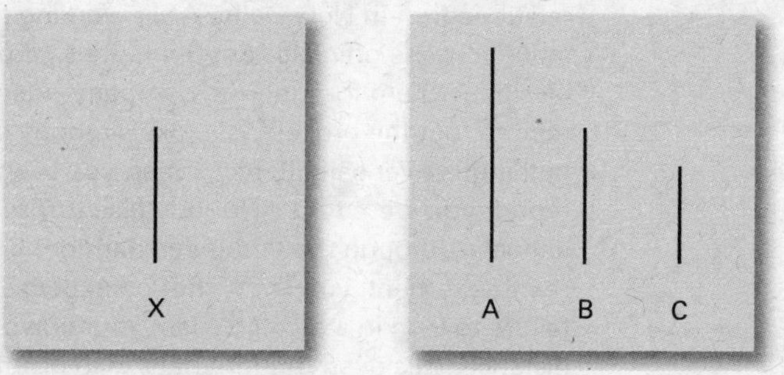

role
Behavior patterns expected of someone occupying a given position in a social unit.

norms
Standards or expectations that are accepted and shared by a group's members.

which of the three lines matched the single line. Asch wanted to see what would happen if members began to give incorrect answers. Would pressures to conform cause individuals to give wrong answers just to be consistent with the others? The experiment was "fixed" so that all but one of the members (the unsuspecting subject) were told ahead of time to start giving obviously incorrect answers after one or two rounds. Over many experiments and trials, the unsuspecting subject conformed over one-third of the time.

Are these conclusions still valid? Research suggests that conformity levels have declined since Asch's studies. However, managers can't ignore conformity because it can still be a powerful force in groups.[13] Group members often want to be seen as one of the group and avoid being visibly different. People tend to find it more pleasant to agree than to be disruptive, even if being disruptive may improve the group's effectiveness. So we conform. But conformity can go too far, especially when an individual's opinion differs significantly from that of others in the group. When this happens, the group often exerts intense pressure on the individual to align his or her opinion to conform to others' opinions, a phenomenon known as **groupthink**. Groupthink seems to occur when there is a clear group identity, members hold a positive group image that they want to protect, and the group perceives a collective threat to this positive image.[14]

Status Systems. Status systems are an important factor in understanding groups. **Status** is a prestige grading, position, or rank within a group. As far back as researchers have been able to trace groups, they have found status hierarchies. Status can be a significant motivator with behavioral consequences, especially when individuals see a disparity between what they perceive their status to be and what others perceive it to be.

Status may be informally conferred by characteristics such as education, age, skill, or experience. Anything can have status value if others in the group evaluate it that way. Of course, just because status is informal doesn't mean that it's unimportant or that it's hard to determine who has it or who does not. Group members have no problem placing people into status categories and usually agree about who has high or low status.

Status is also formally conferred, and it's important for employees to believe that the organization's formal status system is congruent—that is, that there's consistency between the perceived ranking of an individual and the status symbols he or she receives from the organization. For instance, status incongruence would occur when a supervisor earns less than his or her subordinates, a desirable office is occupied by a person in a low-ranking position, or paid country club memberships are provided to

thinking critically about **Ethics**

You've been hired as a summer intern in the events planning department of a public relations firm in New Delhi. After working there about a month, you conclude that the attitude in the office is "anything goes." Employees know that supervisors won't discipline them for ignoring company rules. For example, employees turn in expense reports, but the process is a joke. Nobody submits receipts to verify reimbursement, and nothing is ever said. In fact, when you tried to turn in your receipts with your expense report, you were told, "Nobody else turns in receipts, and you don't really need to either." Although the employee handbook says that receipts are required for reimbursement, you know that no expense check has ever been denied because of failure to turn in a receipt. Also, your coworkers use company phones for personal long-distance calls even though that's also prohibited in the employee handbook. And one of the permanent employees told you to "help yourself" to any paper, pens, or pencils you might need here or at home. What are the norms of this group? Suppose that you were the supervisor in this area. How would you go about changing the norms?

division managers but not to vice presidents. Employees expect the "things" an individual receives to be congruent with his or her status. When they're not, employees may question the authority of their managers and may not be motivated by job promotion opportunities.

Group Size. What's an appropriate size for a group? At Amazon, work teams have considerable autonomy to innovate and to investigate their ideas. And Jeff Bezos, founder and CEO, uses a "two-pizza" philosophy; that is, a team should be small enough that it can be fed with two pizzas. This "two-pizza" philosophy usually limits groups to five to seven people, depending, of course, on team member appetites.[15]

Group size affects performance and satisfaction, but the effect depends on what the group is supposed to accomplish.[16] Research indicates, for instance, that small groups are faster at completing tasks than are larger ones. However, for groups engaged in problem solving, large groups consistently get better results than smaller ones. What does this mean in terms of specific numbers? Large groups—those with a dozen or more members—are good for getting diverse input. Thus, if the goal of the group is to find facts, a larger group should be more effective. On the other hand, smaller groups—those with five to seven members—are better at doing something productive with those facts.

One important research finding related to group size concerns **social loafing**, which is the tendency for an individual to expend less effort when working collectively than when working individually.[17] Social loafing may occur because people believe that others in the group aren't doing their fair share. Thus, they reduce their work efforts in an attempt to make the work load fairer. Also, the relationship between an individual's input and the group's output is often unclear. Thus, individuals may become "free riders" and coast on the group's efforts because individuals believe that their contribution can't be measured.

The implications of social loafing are significant. When managers use groups, they must find a way to identify individual efforts. If they do not, group productivity and individual satisfaction may decline.[18]

Let's Get Real: F2F

DO YOU USE ONLINE COLLABORATIVE TOOLS?

We use tools like Messenger, WebEx, and E-mail.

Group Cohesiveness. Cohesiveness is important because it has been found to be related to a group's productivity. Groups in which there's a lot of internal disagreement and lack of cooperation are less effective in completing their tasks than are groups in which members generally agree, cooperate, and like each other. Research in this area has focused on **group cohesiveness**, or the degree to which members are attracted to a group and share the group's goals.[19]

Research has generally shown that highly cohesive groups are more effective than less cohesive ones.[20] However, the relationship between cohesiveness and effectiveness is complex. A key moderating variable is the degree to which the group's attitude aligns with its goals or with the goals of the organization[21] (see Exhibit 11–5). The more cohesive the group, the more its members will follow its goals. If the goals are desirable (for instance, high output, quality work, cooperation with individuals outside the group), a cohesive group is more productive than a less cohesive group. But if cohesiveness is high and attitudes are unfavorable, productivity decreases. If cohesiveness is low but goals are supported, productivity increases, but not as much as when both cohesiveness and support are high. When cohesiveness is low and goals are not supported, there's no significant effect on productivity.

groupthink
A phenomenon in which a group exerts extensive pressure on an individual to align his or her opinion with others' opinions.

status
A prestige grading, position, or rank within a group.

social loafing
The tendency for individuals to expend less effort when working collectively than when working individually.

group cohesiveness
The degree to which group members are attracted to one another and share the group's goals.

Exhibit 11–5

Group Cohesiveness and Productivity

Alignment of Group and Organizational Goals	Cohesiveness: High	Cohesiveness: Low
High	Strong Increase in Productivity	Moderate Increase in Productivity
Low	Decrease in Productivity	No Significant Effect on Productivity

GROUP PROCESSES

In addition to group member resources and structure, another factor that determines group performance and satisfaction concerns the processes that go on within a work group, such as communication, decision making, and conflict management. These processes are important to understanding work groups because they influence group performance and satisfaction positively or negatively. An example of a positive process factor is the synergy of four people on a marketing research team who are able to generate far more ideas as a group than the members could produce individually. However, the group may also have negative process factors, such as social loafing, high levels of conflict, or poor communication, that may hinder group effectiveness. We'll look at two important group processes: group decision making and conflict management.

Group Decision Making. It's a rare organization that doesn't use committees, task forces, review panels, study teams, or other similar groups to make decisions. Studies show that managers may spend up to 30 hours per week in group meetings.[23] Undoubtedly, a large portion of that time is spent formulating problems, developing solutions, and determining how to implement the solutions. It's possible, in fact, for groups to be assigned any of the eight steps in the decision-making process. (Refer to Chapter 6 to review these steps.)

managing in a Virtual World

IT and Groups

Work groups need information to do their work. With work groups often being not just steps away but continents away from each other, it's important to have a way for group members to communicate and collaborate. That's where IT comes in. Technology has enabled greater online communication and collaboration within groups of all types.[22]

The idea of technologically aided collaboration originated with online search engines. The Internet itself was initially intended as a way for groups of scientists and researchers to share information. Then, as more and more information was put on the Web, users relied on a variety of search engines to help them find that information. Now, we see many examples of collaborative technologies, such as wiki pages, blogs, and multiplayer virtual reality games.

Today, online collaborative tools have given work groups more efficient and effective ways to get work done. For instance, engineers at Toyota use collaborative communication tools to share process improvements and innovations. They have developed a "widely disseminated, collectively owned pool of common knowledge, which drives innovation at a speed few other corporate systems can match." And there's no disputing the successes Toyota has achieved. Managers everywhere should look to the power of IT to help work groups improve the way work gets done.

What advantages do group decisions have over individual decisions? One is that groups generate more complete information and knowledge. They bring a diversity of experience and perspectives to the decision process that an individual cannot. In addition, groups generate more diverse alternatives because they have a greater amount and diversity of information. Next, groups increase acceptance of a solution. Group members are reluctant to fight or undermine a decision that they helped develop. Finally, groups increase legitimacy. Decisions made by groups may be perceived as more legitimate than decisions made by one person.

Group decisions also have disadvantages. One is that groups almost always take more time to reach a solution than would an individual. Another is that a dominant and vocal minority can heavily influence a group's final decision. In addition, groupthink can undermine critical thinking in a group and harm the quality of the final decision.[24] Finally, in a group, members share responsibility, but the responsibility of any single member is ambiguous.

Let's Get Real: F2F

DECISIONS MADE BY A GROUP OR BY INDIVIDUALS?

Decisions are made by groups with a person leading the group and taking full responsibility for that.

Determining whether groups are effective at making decisions depends on the criteria used to assess effectiveness.[25] If accuracy, creativity, and degree of acceptance are important, then a group decision may work best. However, if speed and efficiency are important, then an individual decision may be the best. In addition, decision effectiveness is influenced by group size. Although a larger group provides more diverse representation, it also requires more coordination and time for members to contribute their ideas. Evidence indicates that groups of five, and to a lesser extent seven, are the most effective for making decisions.[26] Having an odd number in the group helps avoid decision deadlocks. Also, these groups are large enough for members to shift roles and withdraw from unfavorable positions but still small enough for quieter members to participate actively in discussions.

What techniques can managers use to help groups make more creative decisions? Exhibit 11–6 describes three possibilities.

Conflict Management. In addition to decision making, another important group process is how a group manages conflict. As a group performs its assigned tasks, disagreements inevitably arise. **Conflict** is *perceived* incompatible differences resulting in some form of interference or opposition. Whether the differences are real is irrelevant. If people in a group perceive that differences exist, then there is conflict.

Three different views have evolved regarding conflict.[27] The **traditional view of conflict** argues that conflict must be avoided—that it indicates a problem within the group. Another view, the **human relations view of conflict**, argues that conflict is a

Exhibit 11–6

Creative Group Decision Making

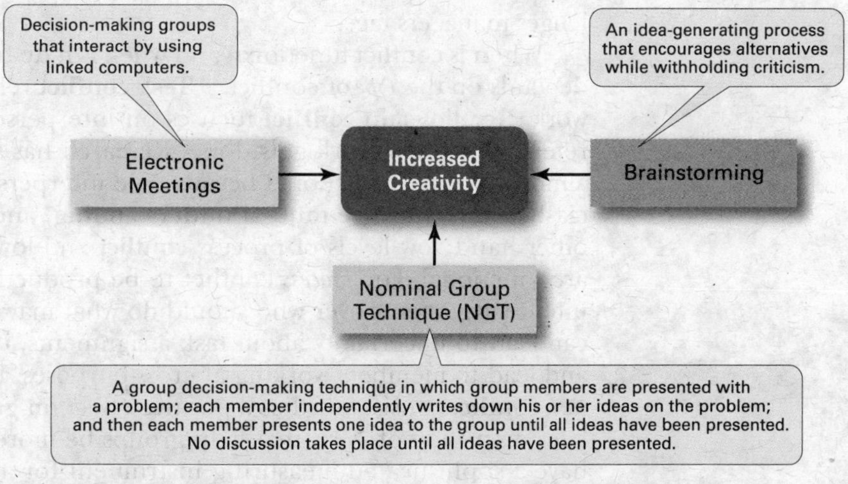

conflict
Perceived incompatible differences that result in interference or opposition.

traditional view of conflict
The view that all conflict is bad and must be avoided.

human relations view of conflict
The view that conflict is a natural and inevitable outcome in any group.

Exhibit 11–7

Conflict and Group Performance

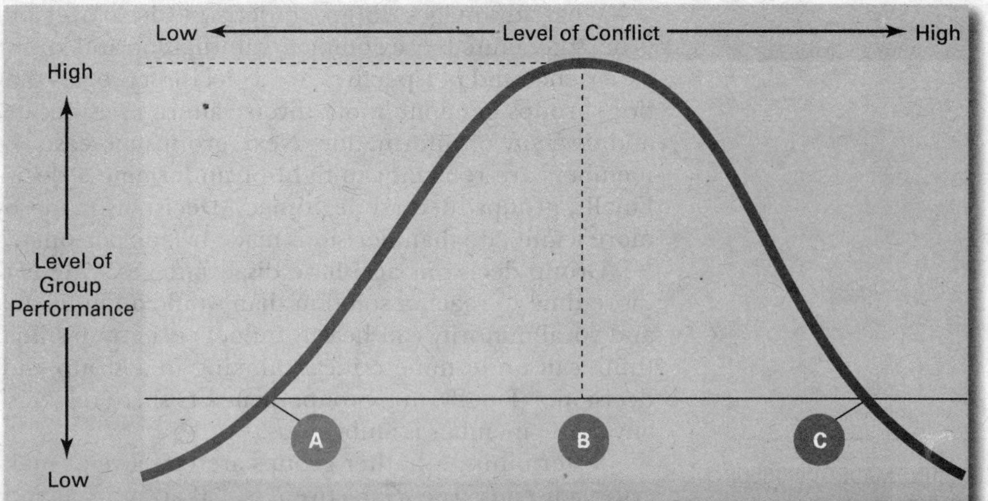

Situation	A	B	C
Level of Conflict	Low or none	Optimal	High
Type of Conflict	Dysfunctional	Functional	Dysfunctional
Group's Internal Characteristics	Apathetic Stagnant Unresponsive to Change Lack of New Ideas	Viable Self-Critical Innovative	Disruptive Chaotic Uncooperative
Level of Group Performance	Low	High	Low

natural and inevitable outcome in any group and need not be negative but has potential to be a positive force in contributing to a group's performance. The third and most recent view, the **interactionist view of conflict**, proposes that not only can conflict be a positive force in a group, but some conflict is *absolutely necessary* for a group to perform effectively.

The interactionist view doesn't suggest that all conflicts are good. Some conflicts—**functional conflicts**—are constructive and support the goals of the work group and improve its performance. Other conflicts—**dysfunctional conflicts**—are destructive and prevent a group from achieving its goals. Exhibit 11–7 illustrates the conflict challenges managers face.

When is conflict functional and when is it dysfunctional? Research indicates that it depends on the *type* of conflict.[28] **Task conflict** relates to the content and goals of the work. **Relationship conflict** focuses on interpersonal relationships. **Process conflict** refers to how the work gets done. Research has shown that *relationship* conflicts are almost always dysfunctional because the interpersonal hostilities increase personality clashes and decrease mutual understanding, and the tasks don't get done. On the other hand, low levels of process conflict and low to moderate levels of task conflict are functional. For *process* conflict to be productive, it must be minimal. Otherwise, intense arguments over who should do what may become dysfunctional because they can lead to uncertainty about task assignments, increase the time to complete tasks, and lead to members working at cross-purposes. However, a low to moderate level of *task* conflict consistently has a positive effect on group performance because it stimulates discussion of ideas that help groups be more innovative.[29] Because we don't yet have a sophisticated measuring instrument for assessing whether conflict levels are optimal, too high, or too low, a manager must try to judge that intelligently.

When group conflict levels are too high, managers can select from five conflict management options: avoiding, accommodating, forcing, compromising, and

Exhibit 11–8

Conflict-Management Techniques

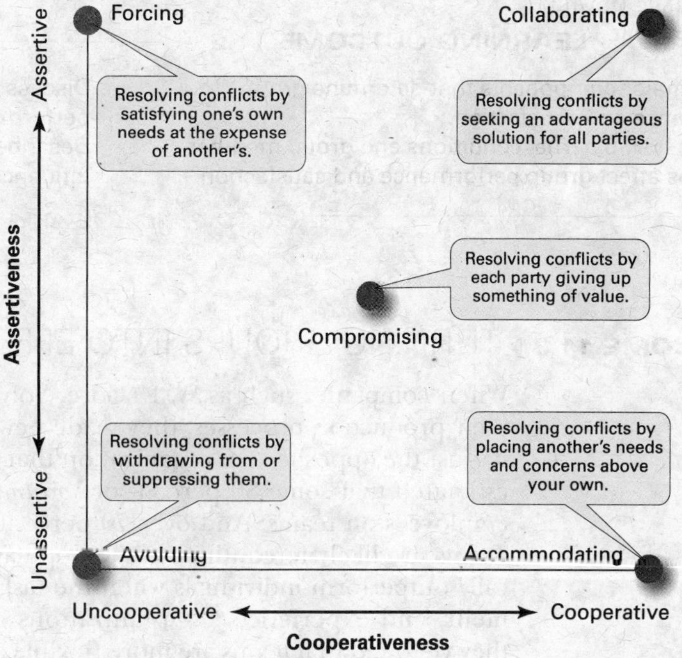

collaborating.[30] (See Exhibit 11–8 for descriptions of these techniques.) Keep in mind that no one option is ideal for every situation. Which approach to use depends on the circumstances.

GROUP TASKS

At Indian Express (or any newspaper as a matter of fact), the bringing out of an edition daily is the product of teamwork; the reporter, editor, placement, and the production are mutually dependent on each other. At Chinmaya Hospital, Bangalore, daily reviews of each patient are conducted in multidisciplinary rounds by teams of nurses, consultant doctors, and in-hospital doctors and registrar. Advertising teams at O&M comprised of the creative team, the production team, manager of the account, and the brand manager of the product work together to deliver a campaign for a client. At D-Link, one of the few successful manufacturers of hardware in India, assemblers work in teams to manufacture defect-free motherboards.

As the group performance/satisfaction model (Exhibit 11–3) shows, the impact that group processes have on group performance and member satisfaction is modified by the task the group is doing. More specifically, the *complexity* and *interdependence* of tasks influence a group's effectiveness.[31]

Tasks are either simple or complex. Simple tasks are routine and standardized. Complex tasks tend to be novel or nonroutine. It appears that the more complex the task, the more a group benefits from group discussion about alternative work methods. Group members don't need to discuss such alternatives for a simple task but can rely on standard operating procedures. Similarly, if there's a high degree of interdependence among the tasks that group members must perform, they'll need to interact more. Thus, effective communication and controlled conflict are most relevant to group performance when tasks are complex and interdependent.

interactionist view of conflict
The view that some conflict is necessary for a group to perform effectively.

functional conflicts
Conflicts that support a group's goals and improve its performance.

dysfunctional conflicts
Conflicts that prevent a group from achieving its goals.

task conflict
Conflicts over content and goals of work.

relationship conflict
Conflict based on interpersonal relationships.

process conflict
Conflict over how work gets done.

QUICK LEARNING REVIEW:

LEARNING OUTCOME 11.2

- List the major components that determine group performance and satisfaction.
- Describe how external conditions and group member resources affect group performance and satisfaction.
- Discuss how group structure influences group performance and satisfaction.
- Describe how group processes and group tasks influence group performance and satisfaction.

Go to page 256 and see how well you know this material.

LEARNING OUTCOME 11.3 ▷ TURNING GROUPS INTO EFFECTIVE TEAMS

When companies such as W. L. Gore, Volvo, and Kraft Foods introduced teams into their production processes, they made news because no one else was doing it. Today, it's just the opposite—an organization that *doesn't* use teams would be newsworthy. It's estimated that some 80 percent of *Fortune* 500 companies have at least half of their employees on teams. And over 70 percent of U.S. manufacturers use work teams.[32] Teams are likely to continue to be popular. Why? Research suggests that teams typically outperform individuals when the tasks being done require multiple skills, judgment, and experience.[33] Organizations are using team-based structures because they've found that teams are more flexible and responsive to changing events than are traditional departments or other permanent work groups. Teams have the ability to quickly assemble, deploy, refocus, and disband. In this section, we'll discuss what a work team is, the different types of teams that organizations might use, and how to develop and manage work teams.

WHAT IS A WORK TEAM?

Most of you are probably familiar with teams, especially if you've watched or participated in organized sports. Work *teams* differ from work *groups* and have their own unique traits (see Exhibit 11–9). Work groups interact primarily to share information and to make decisions to help each member do his or her job more efficiently and effectively. There's no need or opportunity for work groups to engage in collective work that requires joint effort. On the other hand, **work teams** are groups whose members work intensely on a specific, common goal, using their positive synergy, individual and mutual accountability, and complementary skills.

TYPES OF WORK TEAMS

Teams can do a variety of things. They can design products, provide services, negotiate deals, coordinate projects, offer advice, and make decisions.[34] For instance, at TVS Lakshmi Senior Secondary Matriculation School in Madurai, a team of teachers

Exhibit 11–9

Groups versus Teams

Work Teams	Work Groups
• Leadership role is shared	• One leader clearly in charge
• Accountable to self and team	• Accountable only to self
• Team creates specific purpose	• Purpose is same as broader organizational purpose
• Work is done collectively	• Work is done individually
• Meetings characterized by open-ended discussion and collaborative problem-solving	• Meetings characterized by efficiency; no collaboration or open-ended discussion
• Peformance is measured directly by evaluating collective work output	• Performance is measured indirectly according to its influence on others
• Work is decided upon and done together	• Work is decided upon by group leader and delegated to individual group members

Source: J. R. Katzenbach and D. K. Smith, "The Wisdom of Teams," *Harvard Business Review*, July–August 2005, p. 161.

studied, planned, and implemented a major change initiative in all aspects of the running of school. At JK Tyre National Racing Championships held each year, you can see teams work in action during drivers' pit stops. The four most common types of work teams are problem-solving teams, self-managedwork teams, cross-functional teams, and virtual teams.

When work teams first became popular, most were **problem-solving teams**, which are teams from the same department or functional area involved in efforts to improve work activities or to solve specific problems. Members share ideas or offer suggestions on how work processes and methods can be improved. However, these teams are rarely given the authority to implement any of their suggested actions.

Although problem-solving teams were helpful, they didn't go far enough in getting employees involved in work-related decisions and processes. This led to another type of team, the **self-managed work team**, which is a formal group of employees who operate without a manager and are responsible for a complete work process or segment. A self-managed team is responsible for getting the work done *and* for managing itself. This usually includes planning and scheduling work, assigning tasks to members, collectively controlling the pace of work, making operating decisions, and taking action on problems. For instance, teams at Corning have no shift supervisors and work closely with other manufacturing divisions to solve production-line problems and coordinate deadlines and deliveries. The teams have the authority to make and implement decisions, finish projects, and address problems.[35] Other organizations, such as Xerox, Boeing, PepsiCo, and Hewlett-Packard, also use self-managed teams. It's estimated that about 30 percent of U.S. employers now use this form of team; among large firms, the number is probably closer to 50 percent.[36] Most organizations that use self-managed team find them to be effective.[37]

Let's Get Real: F2F

TOP THREE CHARACTERISTICS OF EFFECTIVE TEAMS:

- A capable leader who understands the problem and solution domain.
- The leader focusing on the goal and a "mother" who is managing relationship between the team members and the leaders.
- Respect for each other.

The third type of team is the **cross-functional team**, which we introduced in Chapter 9 and defined as a work team composed of individuals from various specialties. Many organizations use cross-functional teams. For example, ArcelorMittal, the world's biggest steel company, uses cross-functional teams of scientists, plant managers, and salespeople to review and monitor product innovations.[38] The concept of cross-functional teams is being applied in all industries. In 2001, in an attempt to cut direct material cost, Tata Motors (formerly known as Telco) formed cross-functional teams to find ways to drastically or non-linearly cut costs across the organization. The cross-functional teams comprise of high-potential employees working in different divisions of Telco. Four value-engineering teams were leverage of similarity of design principles and tap synergies across vehicle families. An outcome of such an effort has been that the Indica alternator is being used on its multiutility vehicle Sumo. Another team is working on how to cut production costs.[39]

The final type of team is the **virtual team**, which is a team that uses technology to link physically dispersed members in order to achieve a common goal. For instance, a virtual team at Boeing-Rocketdyne played a pivotal role in developing a radically new product.[40] Another company, Decision Lens, uses a virtual team environment to generate and evaluate creative ideas.[41] In a virtual team, members collaborate online with tools such as wide-area networks, videoconferencing, fax, e-mail, or Web sites where the team can hold online conferences.[42] Virtual teams can do all the things that other teams can do, including share information, make decisions, and complete

work teams
Groups whose members work intensely on a specific, common goal, using their positive synergy, individual and mutual accountability, and complementary skills.

problem-solving team
A team from the same department or functional area that's involved in efforts to improve work activities or to solve specific problems.

self-managed work team
A type of work team that operates without a manager and is responsible for a complete work process or segment.

cross-functional team
A work team composed of individuals from various specialties.

virtual team
A type of work team that uses technology to link physically dispersed members in order to achieve a common goal.

tasks; however, they lack the normal give-and-take of face-to-face discussions. That's why virtual teams tend to be more task oriented, especially if the team members have never met in person.

CREATING EFFECTIVE WORK TEAMS

Teams are not always effective. They don't always achieve high levels of performance. However, research on teams provides insights into the characteristics typically associated with effective teams.[43] These characteristics are listed in Exhibit 11–10.

Clear Goals A high-performance team has a clear understanding of the goal to be achieved. Members are committed to the team's goals, know what they're expected to accomplish, and understand how they will work together to achieve these goals.

Relevant Skills. Effective teams are composed of competent individuals who have the necessary technical and interpersonal skills to achieve the desired goals while working well together. This last point is important because not everyone who is technically competent has the interpersonal skills to work well as a team member.

Mutual Trust. Effective teams are characterized by high mutual trust among members. That is, members believe in each other's ability, character, and integrity. But as you probably know from personal relationships, trust is fragile. Maintaining trust requires careful attention by managers.

Unified Commitment. Unified commitment is characterized by dedication to a team's goals and a willingness to expend extraordinary amounts of energy to achieve them. Members of an effective team exhibit intense loyalty and dedication to the team and are willing to do whatever it takes to help their team succeed.

Good Communication. Not surprisingly, effective teams are characterized by good communication. Members convey messages, verbally and nonverbally, between each other in ways that are readily and clearly understood. Also, feedback helps guide team members and correct misunderstandings. Like a couple who has been together for many years, members of high-performing teams are able to quickly and efficiently share ideas and feelings.

Exhibit 11–10

Characteristics of Effective Teams

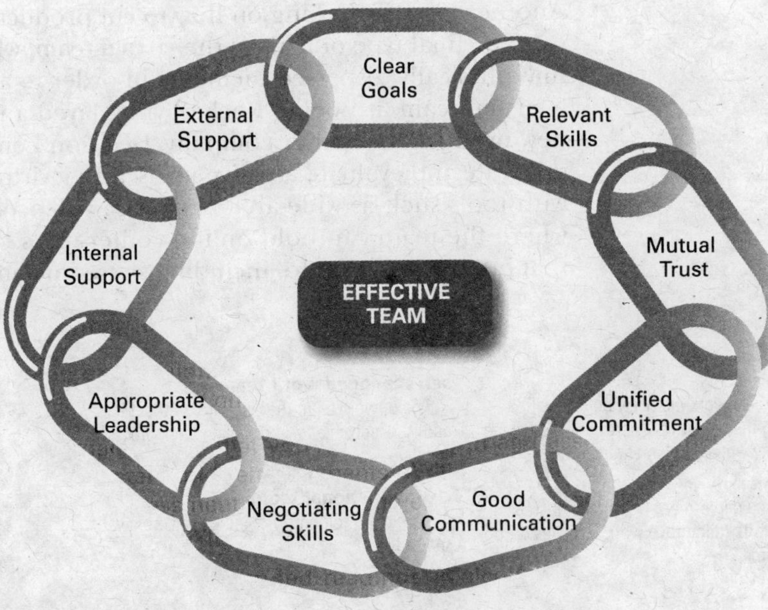

All the work that employees do at Whole Foods Market is based around teamwork. Characteristics of effective teams like job skills, commitment, trust, communication, and effective training and support are important for making this kind of structure successful and contributing to the rapid growth of the organic-food retailer.

Negotiating Skills. Effective teams are continually making adjustments to who does what. This flexibility requires team members to possess negotiating skills. Because problems and relationships are regularly changing in teams, members need to be able to confront and reconcile differences.

Appropriate Leadership. Effective leaders can motivate a team to follow them through the most difficult situations. How? By clarifying goals, demonstrating that change is possible by overcoming inertia, increasing the self-confidence of team members, and helping members to more fully realize their potential. Increasingly, effective team leaders are acting as coaches and facilitators. They help guide and support a team but don't control it.

Internal and External Support. The final condition necessary for an effective team is a supportive climate. Internally, the team should have a sound infrastructure, which means proper training, a clear and reasonable measurement system that team members can use to evaluate their overall performance, an incentive program that recognizes and rewards team activities, and a supportive human resource system. The right infrastructure should support members and reinforce behaviors that lead to high levels of performance. Externally, managers should provide the team with the resources needed to get the job done.

QUICK LEARNING REVIEW:

LEARNING OUTCOME 11.3

- Compare groups and teams.
- Describe the four most common types of teams.
- List the characteristics of effective teams.

Go to page 256 and see how well you know this material.

LEARNING OUTCOME 11.4 ▷ CURRENT CHALLENGES IN MANAGING TEAMS

Few trends have influenced how work gets done in organizations as much as the use of work teams. The shift from working alone to working on teams requires employees to cooperate with others, share information, confront differences, and sublimate personal interests for the greater good of the team. Managers can build effective teams by understanding what influences performance and satisfaction. However, managers also face some current challenges in managing teams, primarily those associated with managing global teams and with understanding organizational social networks.

MANAGING GLOBAL TEAMS

Two characteristics of today's organizations are obvious: They're global, and work is increasingly done by teams. This means that any manager is likely to have to manage a global team. What do we know about managing global teams? We know there are both drawbacks and benefits in using global teams (see Exhibit 11–11). Using our group model as a framework, we can see some of the issues associated with managing global teams.

Group Member Resources in Global Teams. In global organizations, understanding the relationship between group performance and group member resources is especially challenging because of the unique cultural characteristics represented by members of a global team. In addition to recognizing team members' knowledge, skills, abilities, and personality, managers need to be familiar with and clearly understand the cultural characteristics of the groups and the group members they manage.[44] For instance, is the global team from a culture in which uncertainty avoidance is high? If so, members will not be comfortable dealing with unpredictable and ambiguous tasks. Also, as managers work with global teams, they need to be aware of the potential for stereotyping, which can lead to problems.

Group Structure. Some of the structural areas where we see differences in managing global teams include conformity, status, social loafing, and cohesiveness.

Are conformity findings generalizable across cultures? Research suggests that Asch's findings are culture bound.[45] For instance, as might be expected, conformity to social norms tends to be higher in collectivistic cultures than in individualistic cultures. Despite this, however, groupthink tends to be less of a problem in global teams because members are less likely to feel pressured to conform to the ideas, conclusions, and decisions of the group.[46]

Also, the importance of status varies between cultures. The Indian, for example, are extremely status conscious. Also, countries differ on the criteria that confer status. For instance, in Latin America and Asia, status tends to come from family position and formal roles held in organizations. In contrast, while status is important in countries such as the United States and Australia, it tends to be less "in your face." And it tends to be given based on accomplishments rather than on titles and family history. Managers must understand who and what holds status when interacting with people from a culture different from their own. An Indian manager who doesn't understand that office size isn't a measure of a Japanese executive's position or who fails to grasp the importance the British place on family genealogy and social class is likely to unintentionally offend others and reduce his or her interpersonal effectiveness.

Social loafing has a Western bias. It's consistent with individualistic cultures, such as the United States and Canada, which are dominated by self-interest. It's not consistent with collectivistic societies, in which individuals are motivated by group goals. For

Exhibit 11–11

Global Teams

Drawbacks	Benefits
• Disliking team members	• Greater diversity of ideas
• Mistrusting team members	• Limited groupthink
• Stereotyping	• Increased attention on understanding others' ideas, perspectives, etc.
• Communication problems	
• Stress and tension	

Source: Based on N. Adler, *International Dimensions in Organizational Behavior,* 4th ed. (Cincinnati, OH: South-western Publishing, 2002), pp. 141–147

instance, in studies comparing employees from the United States with employees from the People's Republic of China and Israel (both collectivistic societies), the Chinese and Israelis showed no propensity to engage in social loafing. In fact, they actually performed better in a group than when working alone.[47]

Cohesiveness is another group structural element with which managers may face special challenges. In a cohesive group, members are unified and "act as one." There's a great deal of camaraderie, and group identity is high. In global teams, however, cohesiveness is often more difficult to achieve because of higher levels of "mistrust, miscommunication, and stress."[48]

Group Processes. The processes that global teams use to do their work can be particularly challenging for managers. For one thing, communication issues often arise because not all team members may be fluent in the team's working language. This can lead to inaccuracies, misunderstandings, and inefficiencies.[49] However, research has also shown that a multicultural global team is better able to capitalize on the diversity of ideas represented if a wide range of information is used.[50]

Managing conflict in global teams isn't easy, especially when those teams are virtual teams. Conflict can interfere with how a team uses information. However, research has shown that in collectivistic cultures, a collaborative conflict management style can be most effective.[51]

The Manager's Role. Despite the challenges associated with managing global teams, there are things managers can do to provide a group with an environment in which efficiency and effectiveness are enhanced.[52] First, because communication skills are vital, managers should focus on developing those skills. Also, as mentioned earlier, managers must consider cultural differences when deciding what type of global team to use. For instance, evidence suggests that self-managed teams have not fared well in Mexico largely due to that culture's low tolerance of ambiguity and uncertainty and employees' strong respect for hierarchical authority.[53] Finally, it's vital that managers be sensitive to the unique differences of each member of a global team. But it's also important that team members be sensitive to each other.

UNDERSTANDING SOCIAL NETWORKS

We can't leave this chapter on managing teams without looking at the patterns of informal connections among individuals within groups—that is, at the **social network structure**.[54] What actually happens *within* groups? How *do* group members relate to each other, and how does work get done?

Managers need to understand the social networks and social relationships of work groups. Why? Because a group's informal social relationships can help or hinder its effectiveness. For instance, research on social networks has shown that when people need help getting a job done, they'll choose a friendly colleague over someone who may be more capable.[55] Another recent review of team studies showed that teams with high levels of interpersonal interconnectedness actually attained their goals better and were more committed to staying together.[56] Organizations are recognizing the practical benefits of knowing the social networks within teams. For instance, when Ken Loughridge, an IT manager with MWH Global, was transferred from Cheshire, England, to New Zealand, he had a "map" of the informal relationships and connections among company IT employees. This map had been created a few months earlier, using the results of a survey that asked employees who they "consulted most frequently, who they turned to for expertise, and who either boosted or drained their energy levels." Not only did this map help him identify well-connected

social network structure
The patterns of informal connections among individuals within a group.

technical experts, it helped him minimize potential problems when a key manager in the Asia region left the company because Loughridge knew who this person's closest contacts were. Loughridge said, "It's as if you took the top off an ant hill and could see where there's a hive of activity. It really helped me understand who the players were."[57]

QUICK LEARNING REVIEW:

LEARNING OUTCOME 11.4

- **Discuss the challenges of managing global teams.**
- **Explain the role of informal (social) networks in managing teams.**

Go to page 257 and see how well you know this information.

Let's Get Real:

My Turn

Pradeep Waychal

Vice President
Patni Computer Systems Ltd

- I will involve, if not all, good number of people in the selection process of a new designer.
- I will understand strengths and improvement areas of the new designer and based on them allocate him or her to the suitable groups. I will publicize his or her strengths to the entire group.
- I will have more team awards rather than individual awards.

LEARNING OUTCOMES SUMMARY

11.1 ▷ GROUPS AND GROUP DEVELOPMENT

- Describe the different types of groups.
- Describe the five stages of group development.

A group is two or more interacting and interdependent individuals who come together to achieve specific goals. Formal groups are work groups that are defined by an organization's structure and have designated work assignments and specific tasks directed at accomplishing organizational goals. Informal groups are social groups.

The forming stage of group development consists of two phases: joining the group and defining the group's purpose, structure, and leadership. The storming stage involves intragroup conflict over who will control the group and what the group will be doing. In the norming stage, close relationships and cohesiveness develop as norms are determined. During the performing stage, group members begin to work on the group's task. At the adjourning stage, the group prepares to disband.

11.2 ▷ WORK GROUP PERFORMANCE AND SATISFACTION

- List the major components that determine group performance and satisfaction.
- Describe how external conditions and group member resources affect group performance and satisfaction.
- Discuss how group structure influences group performance and satisfaction.
- Describe how group processes and group tasks influence group performance and satisfaction.

The major components that determine group performance and satisfaction are external conditions, group member resources, group structure, group processes, and group tasks.

External conditions such as availability of resources and organizational goals affect work groups. Group member resources (knowledge, skills, abilities, and personality traits) can influence what members can do and how effectively they will perform in a group.

Group roles generally involve getting group work done or keeping group members happy. Group norms are powerful influences on a person's performance and dictate factors such as work output levels, absenteeism, and promptness. Pressures to conform can heavily influence a person's judgment and attitudes. If carried to extremes, groupthink can be a problem. Status systems can be a significant motivator with individual behavioral consequences, especially if there's incongruence in status. What size group is most effective and efficient depends on the task the group is supposed to accomplish. Cohesiveness can affect a group's productivity positively or negatively.

Group decision making and conflict management are important group processes that play a role in performance and satisfaction. If accuracy, creativity, and degree of acceptance are important, a group decision may work best. Relationship conflicts are almost always dysfunctional. Low levels of process conflict and low to moderate levels of task conflict are functional. Effective communication and controlled conflict are most relevant to group performance when tasks are complex and interdependent.

11.3 ▷ TURNING GROUPS INTO EFFECTIVE TEAMS

- Compare groups and teams.
- Describe the four most common types of teams.
- List the characteristics of effective teams.

Work groups have the following characteristics: a strong, clearly focused leader; individual accountability; a purpose that's the same as the broader organizational mission; an individual work product; efficient meetings; effectiveness measured by influence on others; and the ability to discuss, decide, and delegate together. Teams have the following characteristics: shared leadership roles; individual and mutual accountability; specific team purpose; collective work products; meetings with open-ended discussion and active problem solving; performance measured directly on collective work products; and the ability to discuss, decide, and do real work.

A problem-solving team is a team that's focused on improving work activities or solving specific problems. A self-managed work team is responsible for a complete work process or segment and manages itself. A cross-functional team is composed of individuals from various specialties. A virtual team uses technology to link physically dispersed members in order to achieve a common goal.

The characteristics of an effective team include clear goals, relevant skills, mutual trust, unified commitment, good communication, negotiating skills, appropriate leadership, and internal and external support.

11.4 ▷ CURRENT CHALLENGES IN MANAGING TEAMS

- Discuss the challenges of managing global teams.
- Explain the role of informal (social) networks in managing teams.

The challenges of managing global teams can be seen in the group member resources, especially the diverse cultural characteristics; group structure, especially conformity, status, social loafing, and cohesiveness; group processes, especially with communication and managing conflict; and the manager's role in making it all work.

Managers need to understand the patterns of informal connections among individuals within groups because those informal social relationships can help or hinder the group's effectiveness.

THINKING ABOUT MANAGEMENT ISSUES

1. Think of a group to which you belong (or have belonged). Trace its development through the stages of group development shown in Exhibit 11–2. How closely did its development parallel the group development model? How might the group development model be used to improve this group's effectiveness?
2. How do you think scientific management theorists would react to the increased reliance on teams in organizations? How would behavioral science theorists react?
3. How do you explain the popularity of work teams in the United States when U.S. culture places such high value on individualism and individual effort?
4. Why might a manager want to stimulate conflict in a group or team? How could conflict be stimulated?
5. A 20-year study done at Stanford University found that one quality fast-track executives had was the ability to function well as a member of a team. Do you think that everyone should be expected to be a team player, given the trends we're seeing in the use of teams? Discuss.
6. "To have a successful team, first find a great leader." What do you think of this statement? Do you agree? Why or why not?

YOUR TURN to be a Manager

- **What traits do you think good team players have? Do some research to answer this question and write up a report detailing your findings using a bulleted list format.**
- **Select two of the characteristics of effective teams listed in Exhibit 11–10 and develop a team-building exercise for each characteristic that will help a group improve that characteristic. Be creative. Write a report describing your exercises and be sure to explain how your exercise will help a group improve or develop that characteristic.**
- **Complete the skill building modules Developing Trust, Creating Effective Teams, and Running Productive Meetings found on mymanagementlab. Your professor will tell you what you need to do with this information.**
- **Select one group to which you belong. Write a report describing the following things about this group: stage of group development, types of roles played by group members, group norms, group conformity issues, status system, size of group and how effective/efficient it is, and group cohesiveness.**
- **Using the same group, describe how decisions are made. Is the process effective? Efficient? Describe what types of conflicts seem to arise most often (relationship, process, or task) and how those conflicts are handled. Add this information to your report on the group's structure.**
- **When working in a group (any group to which you're assigned or to which you belong), pay careful attention to what happens in the group as tasks are completed. How does the group's structure affect how successful the group is at completing its task? How about its processes?**

- Steve's and Mary's suggested readings: Tom Rath, *Vital Friends* (Gallup Press, 2006); Jon R. Katzenbach and Douglas K. Smith, *The Wisdom of Teams: Creating the High Performance Organization* (McGraw-Hill, 2005); Patrick Lencioni, *Overcoming the 5 Dysfunctions of a Team* (Jossey-Bass, 2005); Ben Mezrich, *Bringing Down the House: The Inside Story of Six MIT Students Who Took Vegas for Millions* (The Free Press, 2002); Jon R. Katzenbach and Douglas K. Smith, *The Discipline of Teams* (Wiley, 2001); and Jean Lipman-Blumen and Harold J. Leavitt, *Hot Groups* (Oxford, 1999).
- Research brainstorming and write a report to your professor explaining what it is and listing suggestions for making it an effective group decision-making tool.
- In your own words, write down three things you learned in this chapter about being a good manager.
- Self-knowledge can be a powerful learning tool. Go to mymanagementlab and complete these self-assessment exercises: Do I Trust Others? Do Others See Me As Trusting? How Good Am I at Building and Leading a Team? What's My Preferred Conflict-Handling Style? Using the results of your assessments, identify personal strengths and weaknesses. What will you do to reinforce your strengths and improve your weaknesses?

CASE APPLICATION

Mixing It Up

How do you combine two packaged-food companies, both with very well-known household brand names, and make the new company work? That's the challenge managers at General Mills faced when it acquired Pillsbury. The company's chief learning officer Kevin Wilde (standing at left in photo) said, "Let's get the best out of both of our marketing organizations. And let's not stop there." So they decided to identify, share, and integrate the best practices from both companies. And employee teams played a major role in how the company proceeded.

An intensive training program called Brand Champions was created and launched. The program was designed not just for marketing specialists but for all employees from different functional areas who worked on particular brands. These cross-functional teams attended the in-house training together as a unified group. According to one of the program developers (Beth Gunderson, seated in photo), specific benefits of including these teams soon became evident. "A person from human resources, for instance, would ask a provocative question precisely because she wasn't a marketer. And you'd see the look on the marketers' faces: Whoa, I never thought of that." It helped employees understand and appreciate different perspectives.

Kevin Wilde, Ami Anderson, and Beth Gunderson from General Mills.

Another benefit of including people from different functions was improved communication throughout the company. People were no longer griping about what other functional areas were doing. Employees began to understand how the other functional areas worked and how each area's contribution was important to the overall success of the company.

The training program has been so successful that now General Mills's production plants have asked for a mini-version of the course. "They want to understand the language marketers speak and why things are done as they are." Oh . . . and there's one other example of how successful the program has been. Betty Crocker is well known for packaged cake mixes and less so for cookie mixes. Inspired by input from the group, the cookie-mix team decided to go after scratch bakers (people who bake from scratch rather than from a boxed mix). As one person said, they were "taking on grandma." The cookie mixes were reformulated, and now the brand owns 90 percent of the dry cookie mix category.

Discussion Questions

1. What benefits did the cross-functional teams bring to General Mills?
2. What challenges would there be in creating an effective cross-functional team? How could managers deal with these challenges?
3. Discuss how each component of the group performance/satisfaction model (see Exhibit 11–3) might affect these teams.
4. Explain how each of the characteristics of effective teams (see Exhibit 11–10) would be important for an effective cross-functional team.

Sources: L. Gratton and T. J. Erickson, "8 Ways to Build Collaborative Teams," *Harvard Business Review*, November 2007, pp. 100–109; and J. Gordon, "Building Brand Champions," *Training*, January/February 2007, pp. 14–17.

Let's Get Real:

Meet The Manager

Sunil Jha

Senior Vice President
Essar Telecom Business Group

MY JOB: To handle strategic HR for the entire telecom business.

BEST PART OF MY JOB: Dealing with people, emotions, and opportunity to contribute to a fellow professional's well-being.

WORST PART OF MY JOB: None.

BEST MANAGEMENT ADVICE EVER RECEIVED: As HR leaders, one must be highly visible, credible, and value-added business partners to be effective in the world of business.

You'll be hearing more from this real manager throughout the chapter.

Chapter 12

Managing Change and Innovation

Change is a constant for organizations and thus for managers. Because change can't be eliminated, managers must learn how to manage it successfully. Because innovation is often closely tied to an organization's change efforts, managers must know how to manage it as well. Focus on the following learning outcomes as you read and study this chapter.

LEARNING OUTCOMES

A Manager's Dilemma

Over the last 3 years, Marico Company achieved a top line compound annual growth rate (CAGR) of 29 percent, making it among the fastest growing FMCG companies in India, and has achieved a turnover of Rs. 1,907 crore in 2007–2008. All three businesses, those of consumer products in India, consumer products in the international markets, and the skin care solutions business of Kaya have contributed to the growth. Marico has perfected the art of observing the customer. On the basis of what they hear and conceptualize, they develop insights into future directions. They use what is called the "insighting process." For example, an insighting team found that men are most comfortable to talk about the well-being of their hair with their barbers. Thus, in launching a hair care product for men, they tailored their advertising, marketing, and programs and ran a separate program to influence the barbers. The same insighting process was also used to launch Kaya Clinic in India, "Sundari" brand in developed world and boxed mixes for preparing roti to manage cholesterol and diabetes.[1] How can Harish Mariwala, CEO of Marico, ensure that innovation continues to thrive in his organization?

What would you do?

The managerial challenges Harish Marewala faces in encouraging creativity and innovation among all his employees are certainly not unique. Big companies and small businesses, universities and colleges, state and city governments, and even the military are forced to be innovative. Although innovation has always been a part of the manager's job, it has become especially important in recent years. In this chapter, we'll look at why innovation is important and how managers can manage innovation. Because innovation is often closely tied to an organization's change efforts, we'll start by looking at change and how managers manage change.

LEARNING **OUTCOME 12.1** ▷

THE CHANGE PROCESS

Ajay Piramal, Chairman of Nicholas Piramal India Ltd, a pharmaceutical company, is like many other managers today who are taking steps to make their workplaces more efficient and flexible and globally competitive. Why? In Piramal's case, it's the threat of the new World Trade Organization rules and competition from other Indian pharmaceutical majors. In many manufacturing firms, it is because of losing market share because of the cheap goods from China. Piramal is doing what managers everywhere must do—change!

If it weren't for change, a manager's job would be relatively easy. Planning would be simple because tomorrow would be no different from today. The issue of effective organizational design would also be resolved because the environment would not be

Exhibit 12–1

External and Internal Forces for Change

External

- Changing consumer needs and wants
- New governmental laws
- Changing technology
- Economic changes

Internal

- New organizational strategy
- Change in composition of workforce
- New equipment
- Changing employee attitudes

uncertain, and there would be no need to redesign the structure. Similarly, decision making would be dramatically streamlined because the outcome of each alternative could be predicted with almost certain accuracy. But that's not the way it is. Change is an organizational reality.[2] Organizations face change because external and internal factors create the need for change (see Exhibit 12–1). When managers recognize that change is needed, then what? How do they respond?

TWO VIEWS OF THE CHANGE PROCESS

Two very different metaphors can be used to describe the change process.[3] One metaphor envisions the organization as a large ship crossing a calm sea. The ship's captain and crew know exactly where they're going because they've made the trip many times before. Change comes in the form of an occasional storm, a brief distraction in an otherwise calm and predictable trip. In the calm waters metaphor, change is seen as an occasional disruption in the normal flow of events. In the other metaphor, the organization is seen as a small raft navigating a raging river with uninterrupted white-water rapids. Aboard the raft are half a dozen people who have never worked together before, who are totally unfamiliar with the river, who are unsure of their eventual destination, and who (as if things weren't already bad enough) are traveling at night. In the white-water rapids metaphor, change is expected, and managing it is a continual process. These two metaphors present very different approaches to understanding and responding to change. Let's take a closer look at each one.

The Calm Waters Metaphor Until the late 1980s, the calm-waters metaphor was fairly descriptive of the situation that managers faced. It's best illustrated by Kurt Lewin's three-step change process (see Exhibit 12–2).[4]

According to Lewin, successful change can be planned and requires *unfreezing* the status quo, *changing* to a new state, and *refreezing* to make the change permanent. The status quo is considered equilibrium. To move away from this equilibrium, unfreezing is necessary. Unfreezing can be thought of as preparing for the needed change. It can be done by increasing the *driving forces*, which are forces pushing for change; by decreasing the *restraining forces*, which are forces that resist change; or by combining the two approaches.

Exhibit 12–2

The Three-Step Change Process

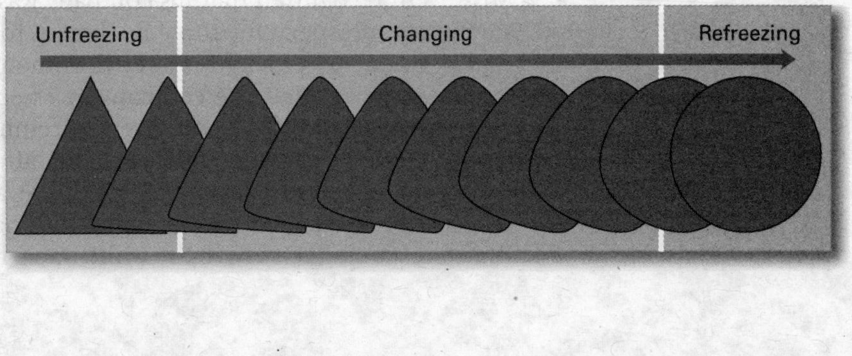

Once unfreezing is done, the change can be implemented. However, merely introducing change doesn't ensure that it will take hold. The new situation needs to be *refrozen* so that it can be sustained over time. Unless this last step is taken, there's a strong chance that employees will revert to the old equilibrium state—that is, the old ways of doing things. The objective of refreezing, then, is to stabilize the new situation by reinforcing the new behaviors.

Lewin's three-step process treats change as a move away from the organization's current equilibrium state. It's a calm waters scenario where an occasional disruption (a "storm") means changing to deal with the disruption. Once the disruption has been dealt with, however, things can continue on. This type of environment isn't what most managers face today.

White-Water Rapids Metaphor The ways of advertising and marketing have changed with the advent of internet, the world of media research business is changing with the introduction of technology (now you can tell the ratings of various shows in a matter of hours after the shows as done by Amap, a company based in MICA, Ahmedabad), and the world of how music is sold has changed with MP3 players and IPods. Nimish Vohra, Country Head of Regalix, a digital marketing company based in Bangalore, says, "My work involves managing the business that is constantly evolving in an industry and technology which is rapidly changing." That's a pretty accurate description of what change is like in the white-water rapids change metaphor. It's also consistent with a world that's increasingly dominated by information, ideas, and knowledge.[5]

To get a feeling of what managing change might be like in a white-water rapids environment, imagine that you attend a college that has the following environment: Courses vary in length. When you sign up, you don't know how long a course will run. It might go for 2 weeks or 30 weeks. Furthermore, the faculty can end a course at any time, with no prior warning. If that isn't challenging enough, the length of the class changes each time it meets: Sometimes the class lasts 20 minutes; other times it runs for 3 hours. And the time of the next class meeting is set by the faculty during this class. There's one more thing: All exams are unannounced, so you have to be ready for a test at any time. To succeed in this type of environment, you'd have to respond quickly to changing conditions. Students who are overly structured or uncomfortable with change wouldn't succeed.

Let's Get Real: F2F

CALM WATERS OR WHITE-WATER RAPIDS?

I keep questioning all paradigms in life and it helps pushing my own boundaries. My company is definitely white-water rapids.

Increasingly, managers are realizing that their job is much like what a student would face in such a college. The stability and predictability of the calm waters metaphor don't exist. Disruptions in the status quo are not occasional and temporary, and they are not followed by a return to calm waters. Many managers never get out of the rapids. Like Nimish Vohra, they face constant change.

Is the white-water rapids metaphor an exaggeration? Probably not! Although you'd expect a chaotic and dynamic environment in high-tech industries, even organizations in non-high-tech industries face constant change. Take the case of Blue Star. You might think that an air conditioning/home appliances industry wouldn't be all that difficult—after all, most households need the products, which are fairly uncomplicated—but that impression would be wrong. Blue Star's Chairman Ashok Advani confronted several challenges.[6] First, there's the challenge of developing products that will appeal to a wide range of global customers. Then there's the challenge of less expensive alternatives flooding the market. Because workforce in India belongs to a labor union, companies face expectations in terms of how they treat their employees. However, Mr Advani recognized that his company was going to have to change if it was going to survive and prosper. One thing he did was to set up R&D facilities to bring out products that were energy efficient and environmentally friendly. Then, to deliver world-class customer experience, the company invested in 3P's, people, processes, and products. Training of all employees was given a premium. Products were paid attention from the design phase to warranty. Processes that affect the customer—the sales, process order execution, and after-sales service—were revamped. Finally, to make the company nimble, changes were made in the organizational structure.[7] The changes were essential if Blue Star wanted to survive the white-water rapids environment in which it operated.

Today, any organization that treats change as an occasional disturbance in an otherwise calm and stable world runs a great risk. Too much is changing too fast for an organization or its managers to be complacent. It's no longer business as usual. And managers must be ready to efficiently and effectively manage the changes facing their organization or their work area.

QUICK LEARNING REVIEW:

LEARNING OUTCOME 12.1

- Explain Lewin's three-step model of the change process.
- Contrast the calm waters and white-water rapids metaphors of change.

Go to page 279 and see how well you know this material.

LEARNING **OUTCOME 12.2** ▷

TYPES OF ORGANIZATIONAL CHANGE

Managers at Hallmark, the world's largest greeting card company, know that as cultural values and people's lifestyles change, so do the types of greeting cards customers are looking for. Some of the emerging consumer trends they identified included "from me to we" (the choices we make inevitably affect someone else), "great expectations—not" (what began as the desire for better and better and more and more will lead to unrealistic expectations), and "it *is* My Movie!" (a dramatic increase in acute self-interest).[8] To accommodate these trends, Hallmark's managers may want to change the company's products, advertising, and perhaps even their HR practices.

WHAT IS ORGANIZATIONAL CHANGE?

Most managers, at one point or another, will have to change some things in their workplace. We classify these changes as **organizational change**, which is any alteration of people, structure, or technology. Organizational changes often need someone to act as a catalyst and assume the responsibility for managing the change process—that is, a **change agent**. A change agent could be a manager within the organization or could be a nonmanager—for example, a change specialist from the HR department or even an outside consultant. For major changes, an organization often hires outside consultants to provide advice and assistance. Because consultants are from the outside, they have an objective perspective that insiders may lack. But outside consultants have a limited understanding of the organization's history, culture, operating procedures, and people. They're also more likely than insiders to initiate drastic change because they don't have to live with the repercussions after the change is implemented. In contrast, internal managers may be more thoughtful but possibly overcautious because they must live with the consequences of their decisions.

TYPES OF CHANGE

Managers might make three main types of change: changes in structure, technology, and people (see Exhibit 12–3). Changing *structure* includes any changes in structural variables such as reporting relationships, coordination mechanisms, employee empowerment, or job redesign. Changing *technology* encompasses modifications in the way work is performed or the methods and equipment used. Changing *people* refers to changes in attitudes, expectations, perceptions, and behavior of individuals or groups.

organizational change
Any alteration of people, structure, or technology in an organization.

change agent
Someone who acts as a catalyst and assumes the responsibility for managing the change process.

Exhibit 12–3

Three Types of Change

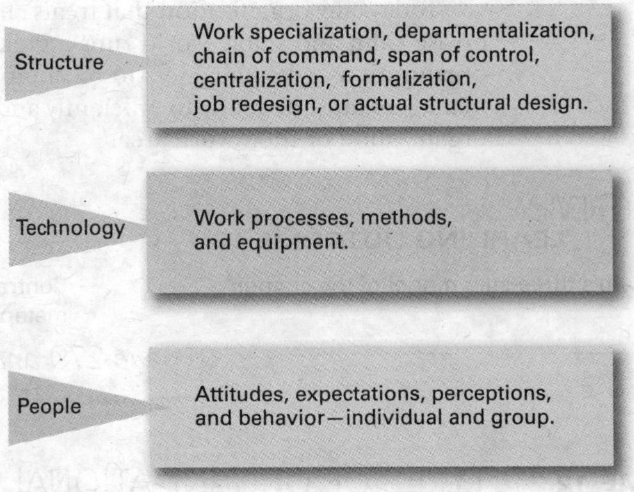

Changing Structure Changing conditions or changing strategies often lead to changes in the organizational structure. An organization's structure is defined by its work specialization, departmentalization, chain of command, span of control, centralization and decentralization, and formalization, and managers can alter one or more of these *structural components.* For instance, departmental responsibilities could be combined, organizational levels eliminated, or spans of control widened to make the organization flatter and less bureaucratic. More rules and procedures could be implemented to increase standardization. Or decentralization could be increased to make decision making faster.

Another option would be to make major changes in the actual *structural design.* For instance, when Hewlett Packard acquired Compaq Computer, product divisions were dropped, merged, and expanded. Structural design changes might also include a shift from a functional structure to a product structure or the creation of a project structural design. Karnataka State Road Corporation, for example, undertook many organizational innovations around information technology, workforce training, passenger service, etc.

Changing Technology Managers can change the technology used to convert inputs into outputs. Most early management studies dealt with changing technology. Scientific management implemented changes that would increase production efficiency. Today, technological changes usually involve the introduction of new equipment, tools, or methods; automation; or computerization.

Let's Get Real: F2F

HARDEST THING ABOUT CHANGE: People are threatened with anxiety leading to denial as a defense mechanism.

Competitive factors or new innovations in an industry often require managers to introduce *new equipment, tools,* or *operating methods.* For example, coal mining companies have updated operational methods, installed more efficient coal handling equipment, and made changes in work practices to be more productive.

Automation is a technological change that replaces certain tasks done by people with tasks done by machines. Automation has been introduced in organizations such as the Indian Postal Service, where electronic weighing and stamping machines are used, and in automobile assembly lines, where robots are programmed to do jobs that workers used to perform.

The most visible technological changes have come from *computerization.* Most organizations have sophisticated information systems. For instance, supermarkets and other retailers use scanners linked to computers that provide instant inventory information. Also, most offices are computerized. At Indian Oil Corporation, employees had to learn how to deal with the personal visibility and accountability brought about by the implementation of an enterprisewide information system. The integrative nature of this system meant that what any employee did on his or her computer automatically affected other computer systems on the internal network.[9] Benetton Group SpA uses computers to link together its manufacturing

Four bikers covered 18,780 kilometers across India in the most difficult terrains over 108 days, starting from Delhi in August 2006, and kept their several fans, online community of readers of blogs, and the audiences of TV and Radio informed on real-time basis about their trip. They used technologies such as the CDMA data card, SMS, cameras, and mobile phones as cameras to connect as many Indians as possible to deliver maximum possible content to them.

plants outside Treviso, Italy, with the company's various sales outlets and a highly automated warehouse.[10]

Changing People. Changing people involves changing attitudes, expectations, perceptions, and behaviors, but it's not easy to do. **Organizational development (OD)** is the term used to describe change methods that focus on people and the nature and quality of interpersonal work relationships.[11] The most popular OD techniques are described in Exhibit 12–4. Each seeks to bring about changes in the organization's people and make them work together better. For example, executives at State Bank of India, India's largest bank, knew that the success of a new customer sales and service strategy depended on changing employee attitudes and behaviors. Managers used different OD techniques during the strategic change, including team building, survey feedback, and intergroup development. One indicator of how well these techniques worked in getting people to change was that every branch of State Bank of India has been automated.

Exhibit 12–4

Popular OD Techniques

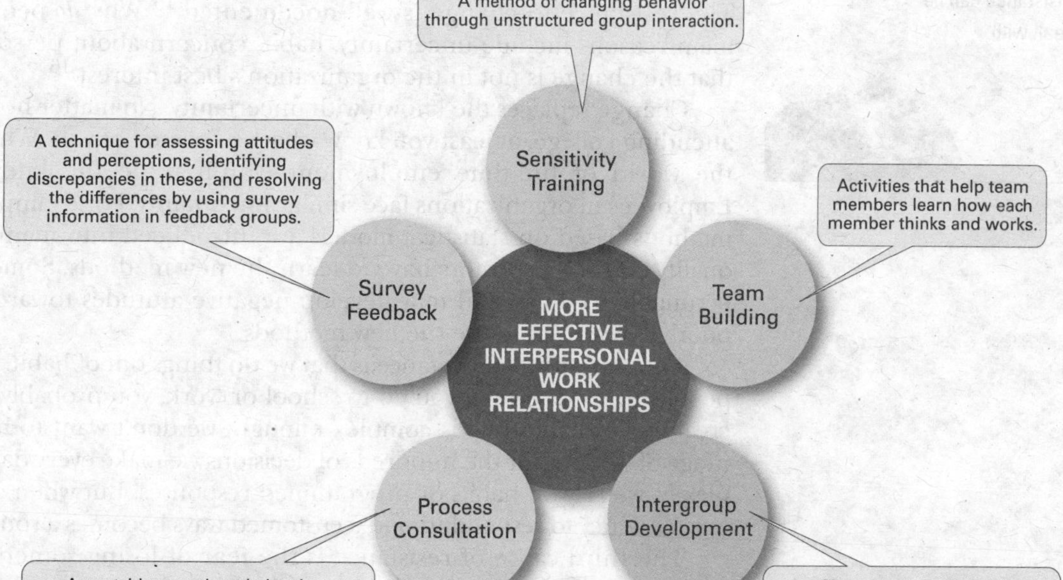

organizational development (OD)
Change methods that focus on people and the nature and quality of interpersonal work relationships.

Much of what we know about OD practices has come from North American research. However, managers need to recognize that although there may be some similarities in the types of OD techniques used, some techniques that work for U.S. organizations may not be appropriate for organizations or organizational divisions based in other countries.[12] For instance, a study of OD interventions showed that "multirater [survey] feedback as practiced in the United States is not embraced in Taiwan" because the cultural value of "saving face is simply more powerful than the value of receiving feedback from subordinates."[13] What's the lesson for managers? Before using the techniques they usually use to implement behavioral changes, especially across different countries, managers need to be sure that they've taken into account cultural characteristics and whether the techniques "make sense for the local culture."

QUICK LEARNING REVIEW:

LEARNING OUTCOME 12.2

- Define *organizational change.*
- Describe how managers might change structure, technology, and people.

Go to page 279 and see how well you know this material.

LEARNING **OUTCOME 12.3** ▷

MANAGING RESISTANCE TO CHANGE

Let's Get Real: F2F

MANAGE RESISTANCE TO CHANGE BY:
First and foremost, lead it by an example. If you are there personally into it, people's resistance can be dealt with.

Whenever the chief people officer at SAS Institute gave a speech about how to win employee loyalty, it didn't take long for someone in the audience to raise his or her hand and offer excuses as to why it couldn't be done.[14] Change can be a threat to people in an organization. Organizations can build up inertia that motivates people to resist changing their status quo, even though change might be beneficial. Why do people resist change, and what can be done to minimize their resistance?

WHY DO PEOPLE RESIST CHANGE?

It's often said that most people hate any change that doesn't jingle in their pockets. The resistance to change is well documented.[15] Why *do* people resist change? The main reasons include uncertainty, habit, concern about personal loss, and the belief that the change is not in the organization's best interest.[16]

Change replaces the known with uncertainty. No matter how much you may dislike attending college, at least you know what's expected of you. When you leave college for the world of full-time employment, you'll trade the known for the unknown. Employees in organizations face similar uncertainty. For example, when quality control methods based on statistical models are introduced into manufacturing plants, many quality control inspectors have to learn the new methods. Some may fear that they will be unable to do so and may develop negative attitudes toward the change or behave poorly if required to use the new methods.

Another cause of resistance is that we do things out of habit. If you're like most other people, every day when you go to school or work, you probably go the same way. We're creatures of habit. Life is complex enough; we don't want to have to consider the full range of options for the hundreds of decisions we make every day. To cope with this complexity, we rely on habits or programmed responses. But when confronted with change, our tendency to respond in our accustomed ways becomes a source of resistance.

The third cause of resistance is the fear of losing something already possessed. Change threatens the investment you've already made in the status quo. The more that people have invested in the current system, the more they resist change. Why? They fear the loss of status, money, authority, friendships, personal convenience, or other economic benefits that they value. This helps explain why older workers tend to resist change more than younger workers. Older employees generally have more invested in the current system and thus have more to lose by changing.

A final cause of resistance is a person's belief that a change is incompatible with the goals and interests of the organization. For instance, an employee who believes

Exhibit 12–5 Reducing Resistance to Change

(1) Education and Communication

- Communicate with employees to help them see the logic of change.
- Educate employees through one-on-one discussions, memos, group meetings, or reports.
- Appropriate if source of resistance is either poor communication or misinformation.
- Must be mutual trust and credibility between managers and employees.

(2) Participation

- Allows those who oppose a change to participate in the decision.
- Assumes that they have expertise to make meaningful contributions.
- Involvement can reduce resistance, obtain commitment to seeing change succeed, and increase quality of change decision.

(3) Facilitation and Support

- Provide supportive efforts such as employee counseling or therapy, new skills training, or short, paid leave of absence.
- Can be time-consuming and expensive.

(4 and 5) Manipulation and Co-optation

- Manipulation is covert attempts to influence such as twisting or distorting facts, withholding damaging information, or creating false rumors.
- Co-optation is a form of manipulation and participation.
- Inexpensive and easy ways to gain support of resisters.
- Can fail miserably if targets feel they've been tricked.

(6) Selecting People Who Accept Change

- Ability to easily accept and adapt to change is related to personality.
- Select people who are open to experience, take a positive attitude toward change, are willing to take risks, and are flexible in their behavior.

(7) Coercion

- Using direct threats or force.
- Inexpensive and easy way to get support.
- May be illegal. Even legal coercion can be perceived as bullying.

that a proposed new job procedure will reduce product quality can be expected to resist the change. This type of resistance can actually be beneficial to the organization if it's expressed in a positive way.

TECHNIQUES FOR REDUCING RESISTANCE TO CHANGE

When managers see resistance to change as dysfunctional, they can use any of seven actions to deal with the resistance.[17] These seven actions are described in Exhibit 12–5. Managers should view these techniques as tools and, depending on the type and source of the resistance, use the most appropriate one.

QUICK LEARNING REVIEW:

LEARNING OUTCOME 12.3

- Explain why people resist change.
- Describe the techniques for reducing resistance to change.

Go to page 279 and see how well you know this material.

LEARNING **OUTCOME 12.4** ▷

CONTEMPORARY ISSUES IN MANAGING CHANGE

Today's main change issues—changing organizational culture, handling employee stress, and making change happen successfully—are critical concerns for managers. Let's look more closely at these topics.

CHANGING ORGANIZATIONAL CULTURE

When Mr N. G. Dhal joined as Executive Director of a large engineering college in eastern India after quitting his job from being a service engineer in a large multinational airline, he tried to implement the processes and procedures from his previous company. But he soon discovered that what was the routine at his previous job was not heard of in the college. For example, he showed up at a meeting of the directors right on time and was surprised that others came in late, gave precedence to phone calls received, and spent most of their time talking about things never mentioned in the agenda. When he pulled up an office assistant for making a mistake very bluntly, he was surprised that though she did not say anything to him she sent her resignation letter within an hour. Mr Dhal realized very soon that the culture of the engineering college was very informal and not so task-focused. He realized that he would have to learn to understand the culture of the college for him to be able to function and possibly change it in the future. The fact that an organization's culture is made up of relatively stable and permanent characteristics tends to make it very resistant to change.[18] A culture takes a long time to form, and once established, it tends to become entrenched. Strong cultures are particularly resistant to change because employees have become so committed to them. For instance, it didn't take long for Lou Gerstner, who was CEO of IBM from 1993 to 2002, to discover the power of a strong culture. Gerstner, the first outsider to lead IBM, needed to overhaul the ailing, tradition-bound company if it was to regain its role as the dominant player in the computer industry. However, accomplishing that in an organization that prided itself on its long-standing culture was Gerstner's biggest challenge. He said, "I came to see in my decade at IBM that culture isn't just one aspect of the game—it *is* the game."[19] Over time, if a certain culture becomes a handicap, there might be little a manager can do to change it, especially in the short run. Even under the most favorable conditions, cultural changes have to be viewed in terms of years, not weeks or even months.

Understanding the Situational Factors What "favorable conditions" facilitate cultural change? One is that *a dramatic crisis occurs,* such as an unexpected financial setback, the loss of a major customer, or a dramatic technological innovation by a competitor. Such a shock can weaken the status quo and make people start thinking about the relevance of the current culture. Or, *leadership changes hands.* New top leadership can provide an alternative set of key values and may be perceived as more capable of responding to the crisis than the old leaders were. Another "favorable condition" is that *the organization is young and small.* The younger the organization, the less entrenched its culture. And it's easier for managers to communicate new values in a small organization than in a large one. Finally, the *culture is weak.* Weak cultures are more receptive to change than are strong ones.[20]

Making Changes in Culture If conditions are right, how do managers change culture? No single action is likely to have the impact necessary to change something that's ingrained and highly valued. Managers need a strategy for managing cultural change, as described in Exhibit 12–6. These suggestions focus on specific actions that managers can take. Following them, however, doesn't guarantee that the cultural change efforts will succeed. Organizational members don't quickly let go of values that they understand and that have worked well for them in the past. Change, if it comes, will be slow. Also, managers must stay alert to protect against any return to old, familiar traditions.

Exhibit 12–6

Changing Culture

- *Set the tone through management behavior*; top managers, particularly, need to be positive role models.
- Create *new stories, symbols, and rituals* to replace those currently in use.
- Select, promote, and support employees who *adopt the new values.*
- *Redesign socialization processes* to align with the new values.
- To encourage acceptance of the new values, *change the reward system.*
- Replace unwritten norms with *clearly specified expectations.*
- *Shake up current subcultures* through job transfers, job rotation, and/or terminations.
- Work to get consensus through *employee participation* and creating a climate with a high level of trust.

managing workforce Diversity

The Paradox of Diversity

When organizations bring in diverse individuals and socialize them into the culture, a paradox is created.[21] Managers want the new employees to accept the organization's core cultural values so they don't have a difficult time fitting in or being accepted. At the same time, managers want to openly acknowledge, embrace, and support the diverse perspectives and ideas that the new employees bring to the workplace.

Strong organizational cultures pressure employees to conform, and the range of acceptable values and behaviors is limited—hence the paradox. Organizations hire diverse individuals because of their unique strengths, yet their diverse behaviors and strengths are likely to diminish in strong cultures as people attempt to fit in.

A manager's challenge is to balance two conflicting goals: to encourage employees to accept the organization's dominant values and to encourage employees to accept differences. When changes are made in the organization's culture, managers need to remember the importance of keeping diversity alive.

EMPLOYEE STRESS

As a student, you've probably experienced stress—class projects, exams, even juggling a job and school. Then, there's the stress associated with getting a decent job after graduation. But even after you've landed that job, stress isn't likely to stop. For many employees, organizational change creates stress. Due to the uncertain environment characterized by time pressures, increasing work loads, mergers, and restructuring, a large number of employees are overworked and stressed.[22] In fact, depending on which survey you look at, the number of employees experiencing job stress in the United States ranges anywhere from 40 percent to 80 percent.[23] However, workplace stress isn't just an American problem. Global studies indicate that some 50 percent of workers surveyed in 16 European countries reported that stress and job responsibility have risen significantly over a five-year period; 35 percent of Canadian workers surveyed said they are under high job stress; in Australia, cases of occupational stress jumped 21 percent in a one-year period; more than 57 percent of Japanese employees suffer from work-related stress; some 83 percent of call-center workers in India suffer from sleeping disorders; and a study of stress in China showed that managers are experiencing more stress.[24] Another interesting study found that stress was the leading cause of people quitting their jobs. Surprisingly, however, employers were clueless. They said that stress wasn't even among the top five reasons people leave and instead wrongly believed that insufficient pay was the main reason.[25]

What Is Stress? Stress is the adverse reaction people have to excessive pressure placed on them due to extraordinary demands, constraints, or opportunities.[26] Stress isn't always bad. Although it's often discussed in a negative context, stress can be positive, especially when it offers a potential gain. For instance, functional stress helps an athlete, a stage performer, or an employee to perform at his or her highest level at crucial times.

However, stress is more often associated with constraints and demands. A constraint prevents you from doing what you desire; demands refer to the loss of something desired. When you take a test at school or have your annual performance review at work, you feel stress because you confront opportunity, constraints, and demands. A good performance review may lead to a promotion, greater responsibilities, and a higher salary. But a poor review may keep you from getting the promotion, and an extremely poor review might lead to your being fired.

stress
The adverse reaction people have to excessive pressure placed on them from extraordinary demands, constraints, or opportunities.

Exhibit 12–7

Symptoms of Stress

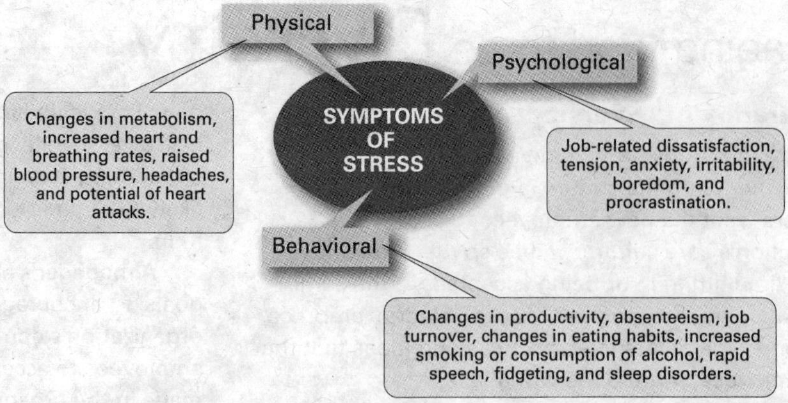

One other thing to understand about stress is that just because the conditions are right for stress to surface doesn't always mean it will. Two conditions are necessary for *potential* stress to become *actual* stress.[27] First, there must be uncertainty about the outcome, and second, the outcome must be important.

What Causes Stress? Stress can be caused by personal factors and by job-related factors. Clearly, change of any kind—personal or job related—has the potential to cause stress as it can involve demands, constraints, or opportunities. Because organizational changes frequently create a climate of uncertainty around issues that are important to employees, it's not surprising that change is a major stressor.

What Are the Symptoms of Stress? We see stress in a number of ways. For instance, an employee who is experiencing high stress may become depressed, accident prone, or argumentative; may have difficulty making routine decisions; may be easily distracted; and so on. As Exhibit 12–7 shows, stress symptoms can be grouped under three general categories: physical, psychological, and behavioral. All these can significantly affect an employee's work.

Let's Get Real: F2F

WE EASE STRESS BY:

Our own ability and belief to push boundaries.

Japan has seen a stress phenomenon called karoshi (pronounced "kah-roe-she"), which is translated literally as "death from overwork." During the late 1980s, "several high-ranking Japanese executives still in their prime years suddenly died without any previous sign of illness."[28] As public concern increased, even the Japanese Ministry of Labour got involved, and it now publishes statistics on the number of karoshi deaths. As Japanese multinational companies expand operations to China, Korea, and Taiwan, it's feared that the karoshi culture may follow.

How Can Stress Be Reduced? As mentioned earlier, not all stress is dysfunctional. Because stress can never be totally eliminated from a person's life, managers want to reduce the stress that leads to dysfunctional work behavior. How? They can do so by controlling certain organizational factors to reduce job-related stress, and, to a more limited extent, offering help for personal stress.

Things that managers can do in terms of job-related factors begin with employee selection. Managers need to make sure that an employee's abilities match the job requirements. When employees are in over their heads, their stress levels are typically high. A realistic job preview during the selection process can minimize stress by reducing ambiguity about job expectations. Improved organizational communications will keep ambiguity-induced stress to a minimum. Similarly, a performance planning program such as MBO can clarify job responsibilities, provide clear performance goals, and reduce ambiguity through feedback. Job redesign can also help reduce stress. If stress can be traced to boredom or to work overload, jobs should be redesigned to increase challenge or to reduce the work load. Redesigns that increase opportunities for employees to participate in decisions and to gain social support have also been found to reduce stress.[29] For instance, at UK pharmaceutical maker GlaxoSmithKline, a team-resilience program in which employees can shift assignments, depending on people's work load and deadlines, has helped reduce work-related stress by 60 percent.[30]

thinking critically about **Ethics**

One in five companies offers some form of stress management program.[31] Although these programs are available, many employees may choose not to participate in them. Why? Many employees are reluctant to ask for help, especially if a major source of that stress is job insecurity. After all, there's still a stigma associated with stress. Employees don't want to be perceived as being unable to handle the demands of their jobs. Although they may need stress management now more than ever, few employees want to admit that they're stressed. What can be done about this paradox? Do organizations even *have* an ethical responsibility to help employees deal with stress?

Stress from an employee's personal life raises two problems. First, it's difficult for a manager to control directly. Second, there are ethical considerations. Specifically, does a manager have the right to intrude—even in the most subtle ways—in an employee's personal life? If a manager believes it's ethical and that the employee is receptive, there are a few approaches the manager can consider. Employee *counseling* can provide stress relief. Employees often want to talk to someone about their problems, and an organization—through its managers, in-house human resource counselors, or free or low-cost outside professional help—can meet that need. Companies such as Citicorp, AT&T, and Johnson & Johnson provide extensive counseling services for their employees. A *time management program* can help employees whose personal lives suffer from a lack of planning to sort out their priorities.[32] Still another approach is organizationally sponsored *wellness programs.* For example, Infosys offers employees an on-site health and fitness facility that is open 7 days a week. Employees of Cairn at oil rigs are provided a wellness program tailored to the unique demands of their environment. And at the training center of customs and excise officials in Faridabad, there are regular yoga classes and art of living courses. EID Parry has a well-formulated program to help employees help themselves.

MAKING CHANGE HAPPEN SUCCESSFULLY

Organizational change is an ongoing daily challenge facing managers in the United States *and* around the globe. In a global study of organizational changes in more than 2,000 organizations in Europe, Japan, the United States, and the United Kingdom, 82 percent of the respondents had implemented major information systems changes, 74 percent had created horizontal sharing of services and information, 65 percent had implemented flexible human resource practices, and 62 percent had decentralized operational decisions.[33] Each of these major changes entailed numerous other changes in structure, technology, and people. When changes are needed, who makes them happen? Who manages them? Although you may think that it's just top-level managers, actually managers at *all* organizational levels are involved in the change process.

Even with the involvement of all levels of managers, change efforts don't always work the way they should. In fact, a global study of organizational change concluded that "Hundreds of managers from scores of U.S. and European companies [are] satisfied with their operating prowess… [but] dissatisfied with their ability to implement change."[34] How can managers make change happen successfully? They can (1) make the organization change capable, (2) understand their own role in the process, and (3) give individual employees a role in the change process. Let's look at each of these suggestions.

In an industry where growth is slowing and competitors are becoming stronger, United Parcel Service (UPS) prospers. How? By embracing change! Managers spent a decade creating new worldwide logistics businesses because they anticipated slowing domestic shipping demand. They continue change efforts in order to exploit new opportunities.[35] UPS is what we call a *change-capable organization.* What does it take to be a change-capable organization? Exhibit 12–8 summarizes the characteristics.

Exhibit 12–8

Change-Capable Organizations

- *Link the present and the future.* Think of work as more than an extension of the past; think about future opportunities and issues and factor them into today's decisions.
- *Make learning a way of life.* Change-friendly organizations excel at knowledge sharing and management.
- *Actively support and encourage day-to-day improvements and changes.* Successful change can come from the small changes as well as the big ones.
- *Ensure diverse teams.* Diversity ensures that things won't be done like they've always been done.
- *Encourage mavericks.* Since their ideas and approaches are outside the mainstream, mavericks can help bring about radical change.
- *Shelter breakthroughs.* Change-friendly organizations have found ways to protect those breakthrough ideas.
- *Integrate technology.* Use technology to implement changes.
- *Build and deepen trust.* People are more likely to support changes when the organization's culture is trusting and managers have credibility and integrity.

Source: Based on P. A. McLagan, "The Change-Capable Organization," T&D, January 2003, pp. 50–59.

The second component of making change happen successfully is for managers to recognize their own important role in the process. Managers can, and do, act as change agents. But their role in the change process includes more than being catalysts for change; they must also be change leaders. When organizational members resist change, it's the manager's responsibility to lead the change effort. But even when there's no resistance to the change, someone has to assume leadership. That someone is managers.

The final aspect of making change happen successfully revolves around getting all organizational members involved. Successful organizational change is not a one-person job. Individual employees are a powerful resource in identifying and addressing change issues. "If you develop a program for change and simply hand it to your people, saying, 'Here, implement this,' it's unlikely to work. But when people help to build something, they will support it and make it work."[36] Managers need to encourage employees to be change agents—to look for day-to-day improvements and changes that individuals and teams can make. For instance, a study of organizational change found that 77 percent of changes at the work group level were reactions to specific, current problems or to suggestions from people outside the work group; and 68 percent of those changes occurred in the course of employees' day-to-day work.[37]

QUICK LEARNING REVIEW:

LEARNING OUTCOME 12.4

- Explain why changing organizational culture is so difficult and how managers can do it.
- Describe employee stress and how managers can help employees deal with stress.
- Discuss what it takes to make change happen successfully.

Go to page 279 and see how well you know this material.

LEARNING OUTCOME 12.5 ▷ STIMULATING INNOVATION

"'Innovate or stagnate' could well be a warning sign on the road to economic growth in this century," said Kiran Karnik, President of NASSCOM India, in a talk on innovation's importance: Powering Economic Growth.[38] And how true it is! Success in business today demands innovation. Such is the stark reality facing today's managers. In the dynamic,

Exhibit 12–9 World's Most innovative Companies

Apple and Google reign worldwide. But respondents from different regions often favored local companies.*

Rank	ASIA	EUROPE	NORTH AMERICA
1	Apple	Apple	Apple
2	Google	Google	Google
3	Toyota Motor	Toyota Motor	Toyota Motor
4	Tata Group	Microsoft	General Electric
5	Nintendo	Nintendo	Procter & Gamble
6	Nokia	General Electric	Microsoft Corp
7	General Electric	Amazon.com	Nintendo
8	Reliance Industries	Tata Group	Research in Motion
9	Microsoft	Nokia	Disney
10	Sony	Procter & Gamble	IBM
11	Samsung Electronics	BMW	General Motors
12	IBM	Sony	Amazon.com
13	Hewlett-Packard	IBM	Tata Group
14	Procter & Gamble	Audi	Honda Motor
15	3M	Hewlett-Packard	Sony
16	Goldman Sachs Group	Boeing	Target
17	Honda Motor	Goldman Sachs Group	Hewlett-Packard
18	McDonald's	Fiat	Boeing
19	BMW	Facebook	BMW
20	Facebook	3M	Wal-Mart Stores

Source: "The World's Most Innovative Companies by Region," *BusinessWeek*, **BusinessWeekOnline**, April 15, 2008, businessweek.com.

chaotic world of global competition, organizations must create new products and services and adopt state-of-the-art technology if they're going to compete successfully.[39]

What companies come to mind when you think of successful innovators? Maybe Sony Corporation, with its MiniDisks, PlayStations, Aibo robot pets, Cyber-Shot digital cameras, and MiniDV Handycam camcorders. Maybe Toyota, with its continual advancements in product and manufacturing process designs. (See Exhibit 12–9 for a list of companies that executives from around the world named as most innovative in a *BusinessWeek* survey.) What's the secret to the success of these innovator champions? What can other managers do to make their organizations more innovative? In the following pages, we'll try to answer those questions as we discuss the factors behind innovation.

CREATIVITY VERSUS INNOVATION

Creativity refers to the ability to combine ideas in a unique way or to make unusual associations between ideas.[40] A creative organization develops unique ways of working or novel solutions to problems. But creativity by itself isn't enough. The outcomes of the creative process need to be turned into useful products or work methods, which is defined as **innovation**. Thus, an innovative organization is characterized by its ability to channel creativity into useful outcomes. When managers talk about changing an organization to make it more creative, they usually mean they want to stimulate and nurture innovation.

STIMULATING AND NURTURING INNOVATION

The systems model can help us understand how organizations become more innovative.[41] As you can see in Exhibit 12–10, getting the desired outputs (innovative products and work methods) involves transforming inputs. These inputs include creative people and groups within the organization. But having creative people isn't enough. The right environment is needed to help transform inputs into innovative products or work methods. This "right" environment—that is, an environment that stimulates innovation—includes three variables: the organization's structure, culture, and human resource practices (see Exhibit 12-11).

Structural Variables Research into the effect of structural variables on innovation shows five things.[42] First, an organic-type structure positively influences innovation. Because this structure is low in formalization, centralization, and work specialization,

creativity
The ability to combine ideas in a unique way or to make unusual associations between ideas.

innovation
The process of turning creative ideas into useful products or work methods.

Exhibit 12–10

Systems View of Innovation

Source: Adapted from R. W. Woodman, J. E. Sawyer, and R. W. Griffin, "Toward a Theory of Organizational Creativity," *Academy of Management Review*, April 1993, p. 309.

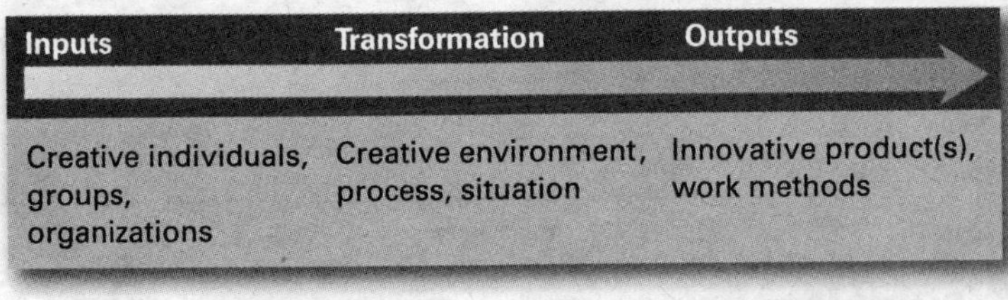

it facilitates the flexibility and sharing of ideas that are critical to innovation. Second, the availability of plentiful resources provides a key building block for innovation. With an abundance of resources, managers can afford to purchase innovations, can afford the cost of instituting innovations, and can absorb failures. Third, frequent communication between organizational units helps break down barriers to innovation.[43] Cross-functional teams, task forces, and other such organizational designs facilitate interaction across departmental lines and are widely used in innovative organizations. Fourth, innovative organizations try to minimize extreme time pressures on creative activities, despite the demands of white-water rapids–type environments. Although time pressures may spur people to work harder and may make them feel more creative, studies show that these pressures actually cause people to be less creative.[44] Finally, studies have shown that an employee's creative performance is enhanced when an organization's structure explicitly supports creativity. Beneficial kinds of support include things like encouragement, open communication, readiness to listen, and useful feedback.[45]

Cultural Variables "Throw the bunny" is part of the lingo used by a product development team at toy company Mattel. It refers to a juggling lesson in which team members learn to juggle two balls and a stuffed bunny. Most people easily learn to juggle two balls but can't let go of that third object. Creativity, like juggling, is learning to let go—that is, to "throw the bunny." And for Mattel, having a culture where people are encouraged to "throw the bunny" is important to its continued product innovations.[46]

Exhibit 12–11

Innovation Variables

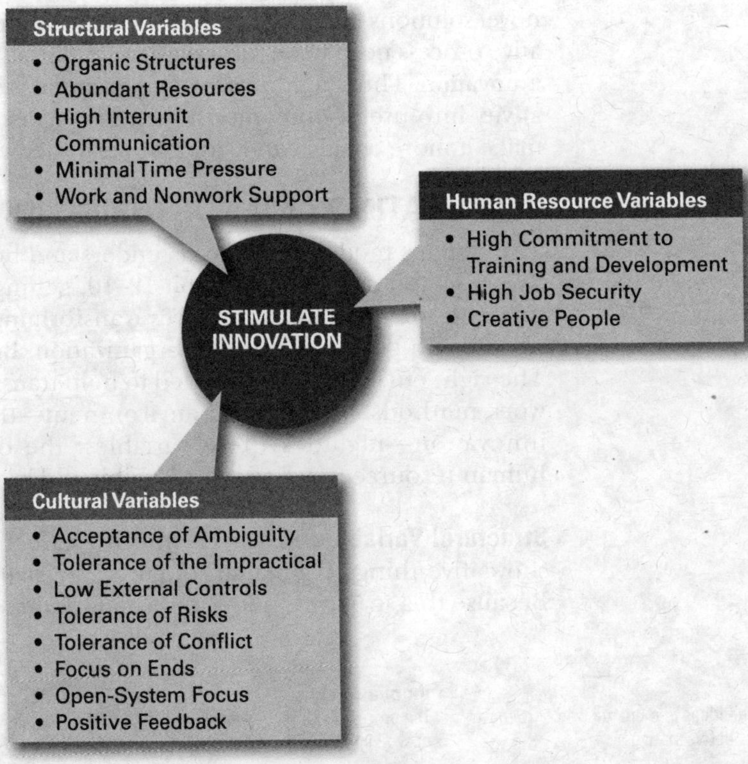

Innovative organizations tend to have similar cultures.[47] They encourage experimentation, reward both successes and failures, and celebrate mistakes. An innovative organization is likely to do the following:

- ***Accept ambiguity***—Too much emphasis on objectivity and specificity constrains creativity.
- ***Tolerate the impractical***—Individuals who offer impractical, even foolish, answers to what-if questions are not stifled. What at first seems impractical might lead to innovative solutions.
- ***Keep external controls minimal***—Rules, regulations, policies, and similar organizational controls are kept to a minimum.
- ***Tolerate risk***—Employees are encouraged to experiment without fear of consequences should they fail. Mistakes are treated as learning opportunities.
- ***Tolerate conflict***—Diversity of opinions is encouraged. Harmony and agreement between individuals or units are *not* assumed to be evidence of high performance.
- ***Focus on ends rather than means***—Goals are made clear, and individuals are encouraged to consider alternative routes toward meeting the goals. Focusing on ends suggests that there might be several right answers to any given problem.
- ***Use an open-system focus***—Managers closely monitor the environment and respond to changes as they occur. For example, at Starbucks, product development depends on "inspiration field trips to view customers and trends." Michelle Gass, now the company's senior vice president of global strategy, "took her team to Paris, Düsseldorf, and London to visit local Starbucks and other restaurants to get a better sense of local cultures, behaviors, and fashions." She says, "You come back just full of different ideas and different ways to think about things than you would had you read about it in a magazine or e-mail."[48] Taj Hotels also organizes trips for the general managers of their luxury hotels to travel and stay in luxury hotels in various parts of the world so that they can learn from the experience.
- ***Provide positive feedback***—Managers provide positive feedback, encouragement, and support so employees feel that their creative ideas receive attention. For instance, at Research in Motion, Mike Lazaridis, president and co-CEO, says, "I think we have a culture of innovation here, and [engineers] have absolute access to me. I live a life that tries to promote innovation."[49]

Let's Get Real: F2F

INNOVATION IS IMPORTANT BECAUSE:
It helps to channelize creativity.

Human Resource Variables Innovative organizations actively promote the training and development of their members so their knowledge remains current; offer their employees high job security to reduce the fear of getting fired for making mistakes; and encourage individuals to become **idea champions**, actively and enthusiastically supporting new ideas, building support, overcoming resistance, and ensuring that innovations are implemented. Research has found that idea champions have common personality characteristics: extremely high self-confidence, persistence, energy, and a tendency toward risk taking. They also display characteristics associated with dynamic leadership: They inspire and energize others with their vision of the potential of an innovation and through their strong personal conviction in their mission. They're also good at gaining the commitment of others to support their mission. In addition, idea champions have jobs that provide considerable decision-making discretion. This autonomy helps them introduce and implement innovations in organizations.[50]

QUICK LEARNING REVIEW:

LEARNING OUTCOME 12.5

- Explain how creativity and innovation differ from one another.
- Describe the structural, cultural, and human resource variables that are necessary for innovation.

Go to page 280 and see how well you know this material.

idea champion
An individual who actively and enthusiastically supports new ideas, builds support, overcomes resistance, and ensures that innovations are implemented.

Let's Get Real: My Turn

Sunil Jha

Senior Vice President
Essar Telecom Business Group

The system for sustaining innovation is the knowledge-brokering cycle. The first step is to capture good ideas from a wide variety of sources. The second step is to keep those ideas alive by playing with them, discussing them, and using them. Imagining new uses for old ideas is the third part. The fourth is turning promising concepts into real services, products, processes, or business models. One can use all or part of the cycle. Large companies, such as Marico, may like to move ideas from one place to another or one may want to build full-fledged consulting groups dedicated to internal knowledge brokering.

LEARNING OUTCOMES **SUMMARY**

12.1 ▷ THE CHANGE PROCESS

- Explain Lewin's three-step model of the change process.
- Contrast the calm waters and white-water rapids metaphors of change.

The calm waters metaphor suggests that change is an occasional disruption in the normal flow of events and can be planned and managed as it happens. In the white-water rapids metaphor, change is ongoing, and managing it is a continual process.

Lewin's three-step model says change can be managed by unfreezing the status quo (old behaviors), changing to a new state, and refreezing the new behaviors.

12.2 ▷ TYPES OF ORGANIZATIONAL CHANGE

- Define *organizational change*.
- Describe how managers might change structure, technology, and people.

Organizational change is any alteration of people, structure, or technology. Making changes often requires a change agent to act as a catalyst and guide the change process.

Changing structure involves any changes in structural components or structural design. Changing technology involves introducing new equipment, tools, or methods; automation; or computerization. Changing people involves changing attitudes, expectations, perceptions, and behaviors.

12.3 ▷ MANAGING RESISTANCE TO CHANGE

- Explain why people resist change.
- Describe the techniques for reducing resistance to change.

People resist change because of uncertainty, habit, concern about personal loss, and the belief that a change is not in the organization's best interest.

The techniques for reducing resistance to change include education and communication (educating employees about and communicating to them the need for the change), participation (allowing employees to participate in the change process), facilitation and support (giving employees the support they need to implement the change), negotiation (exchanging something of value to reduce resistance), manipulation and co-optation (using negative actions to influence), and coercion (using direct threats or force).

12.4 ▷ CONTEMPORARY ISSUES IN MANAGING CHANGE

- Explain why changing organizational culture is so difficult and how managers can do it.
- Describe employee stress and how managers can help employees deal with stress.
- Discuss what it takes to make change happen successfully.

The shared values that comprise an organization's culture are relatively stable, which makes change difficult. Managers can encourage change by being positive role models; creating new stories, symbols, and rituals; selecting, promoting, and supporting employees who adopt new values; redesigning socialization processes; changing the reward system, clearly specifying expectations; shaking up current subcultures; and getting employees to participate in change.

Stress is the adverse reaction people have to excessive pressure placed on them from extraordinary demands, constraints, or opportunities. To help employees deal with stress, managers can address job-related factors by making sure an employee's abilities match the job requirements, improving organizational communications, using a performance planning program, or redesigning jobs. Addressing personal stress factors is trickier, but managers could offer employee counseling, time management programs, and wellness programs.

Making change happen successfully involves focusing on making the organization change capable, making sure managers understand their own role in the process, and giving individual employees a role in the process.

12.5 ▷ STIMULATING INNOVATION

- Explain how creativity and innovation differ from one another.
- Describe the structural, cultural, and human resource variables that are necessary for innovation.

Creativity is the ability to combine ideas in a unique way or to make unusual associations between ideas. Innovation is turning the outcomes of the creative process into useful products or work methods.

Important structural variables include an organic-type structure, abundant resources, frequent communication between organizational units, minimal time pressure, and support. Important cultural variables include accepting ambiguity, tolerating the impractical, keeping external controls minimal, tolerating risk, tolerating conflict, focusing on ends rather than means, using an open-system focus, and providing positive feedback. Important human resource variables include having high commitment to training and development, having high job security, and encouraging individuals to be idea champions.

THINKING ABOUT MANAGEMENT ISSUES

1. Can a low-level employee be a change agent? Explain your answer.
2. Innovation requires allowing people to make mistakes. However, being wrong too many times can be disastrous to your career. Do you agree? Why or why not? What are the implications for nurturing innovation?
3. How are opportunities, constraints, and demands related to stress? Give an example of each.
4. Planned change is often thought to be the best approach to take in organizations. Can unplanned change ever be effective? Explain.
5. Organizations typically have limits to how much change they can absorb. As a manager, what signs would you look for that might suggest that your organization has exceeded its capacity to change?

YOUR TURN to be a Manager

- Take responsibility for your own future career path. Don't depend on your employer to provide you with career development and training opportunities. Right now, sign up for things that will help you enhance your skills—workshops, seminars, continuing education courses, and so on.
- Pay attention to how you handle change. Try to figure out why you resist certain changes and not others.
- Pay attention to how others around you handle change. When friends or family members resist change, practice using different approaches to managing this resistance to change.
- When you find yourself experiencing dysfunctional stress, write down what's causing the stress, what stress symptoms you're exhibiting, and how you're dealing with the stress. Keep this information in a journal and evaluate how well your stress reducers are working and how you could handle stress better. Your goal is to get to a point where you recognize that you're stressed and can take positive actions to deal with the stress.
- Research information on how to be a more creative person. Write down suggestions in a bulleted list format and be prepared to present your information in class.
- Complete the skill building modules Managing Resistance to Change and Solving Problems Creatively found on mymanagementlab.
- Choose two organizations that you're familiar with and assess whether they face calm waters or white-water rapids environments. Write a short report, describing these organizations and your assessment of the change environment each faces. Be sure to explain your choice of change environment.
- Steve's and Mary's recommended readings: Malcolm Gladwell, *Blink* (Little, Brown, 2005); Peter Senge and others, *Presence* (Doubleday, 2004); Tom Peters, *Re-Imagine!*

(Dorling Kindersely, 2003); John P. Kotter and Dan S. Cohen, *The Heart of Change* (Harvard Business School Press, 2002); Malcolm Gladwell, *The Tipping Point* (Back Bay Books, 2002); Tom Kelley, *The Art of Innovation* (Doubleday, 2001); and Ian Morrison, *The Second Curve* (Ballantine Books, 1996).

- Choose an organization with which you're familiar (employer, student organization, family business, etc.). Describe its culture (shared values and beliefs). Select two of its values/beliefs and describe how you would go about changing them. Put this information in a report.
- In your own words, write down three things you learned in this chapter about being a good manager.
- Self-knowledge can be a powerful learning tool. Go to mymanagementlab and complete these self-assessment exercises: How Well Do I Handle Ambiguity? How Creative Am I? How Well Do I Respond to Turbulent Change? How Stressful Is My Life? Am I Burned Out? Using the results of your assessments, identify personal strengths and weaknesses. What will you do to reinforce your strengths and improve your weaknesses?

CASE APPLICATION

That Special Touch

Starwood is one of the world's largest hotel and leisure companies. Its well-known luxury and upscale brands include Four Points, Sheraton, Westin, St. Regis, and W Hotels. In an industry that's highly dependent on attracting and keeping customers, Starwood knows it has to be innovative. In 2006, the company's Westin chain spotted a market opportunity after a survey found that 34 percent of frequent travelers feel lonely away from home. The challenge was to be innovative in finding ways to exploit that opportunity. Again, showing its willingness to be different, the company took an unusual approach

Instead of hiring the usual consultants, Starwood's management team turned to Six Sigma, a quality management process best known for increasing efficiency and reducing product defects. Many companies have had great success with Six Sigma, but it has often been described as a creativity killer. "Combining creativity and efficiency is a delicate managerial maneuver that few service companies can pull off." But that reputation didn't deter Starwood's managers. One thing they had in their favor was that the company already had a strong culture of creativity. In the 1990s, it designed its popular W Hotels as a blend of high-fashion and high energy and captured a lot of attention and customers looking for a trendy experience.

Today, Starwood uses Six Sigma to dream up projects for its properties around the world. Since the program launched in 2001, 150 employees have been trained as "black belts" and 2,700 as "green belts" in the art of Six Sigma. These employees are based mostly at the hotels, with the black belts overseeing the projects and the green belts taking care of the details. The Six

Starwood Hotels

Sigma specialists help hotel employees find ways to meet their goals. In fact, almost 100 percent of the creative ideas have come from in-house staff.

Starwood's innovation process starts with hotel teams who "pitch" the Six Sigma group on a new idea. A Six Sigma Council (composed of 13 people) evaluates the idea's merit, based on the goals and expected payoffs. If the council approves a project, the Six Sigma specialists are deployed to help the local hotel teams carry it out.

Discussion Questions

1. Do some research on Six Sigma. Explain what it is. Do you think its reputation as a "creativity killer" is justified? Why or why not?
2. What does this say about what Starwood has accomplished with its Six Sigma program? What could other companies learn from Starwood's experience?
3. How important do you think it is to have a culture of creativity? Explain. If a company doesn't have such a culture, what could managers do?
4. What else might Starwood do to promote creativity and innovation?
5. Do you think the hotel/resort industry environment is more calm waters or white-water rapids? Explain your choice.

Sources: S. A. Ante, "Rubbing Customers the Right Way," *BusinessWeek*, October 8, 2007, pp. 88–89; S. A. Ante, "Six Sigma Kick-Starts Starwood," *BusinessWeek* online, www.businessweek.com, August 30, 2007; and R. Jana, "Starwood Hotels Explore Second Life First," *BusinessWeek* online, www.businessweek.com, August 23, 2006.

Part Four

Leading

After people are hired and brought into an organization, managers must oversee and coordinate their work so that organizational goals can be pursued and achieved. This is the leading function of management. Because this function involves an organization's people, it's an important one. However, precisely because it involves people means it can also be quite challenging. Managing people successfully involves understanding their attitudes, behaviors, personalities, motivation, and so forth. In addition, it requires effective and efficient lines of communication. Understanding how people behave and why they do the things they do is downright difficult at times.

In Part Four, we look at the management function of leading. Chapter 13 provides important information to help you begin to understand the complex nature of individual behavior in organizations. In Chapter 14, we look at the ways that managers communicate with employees and at some of the issues that can arise. Every manager strives to get employees to put forth high levels of effort, so Chapter 15 provides an all-encompassing look at how and why people are motivated. Finally, in Chapter 16, we look at a manager from the leadership perspective. After all, managers need to be great leaders in addition to being great managers.

Let's Get Real:

Meet The Manager

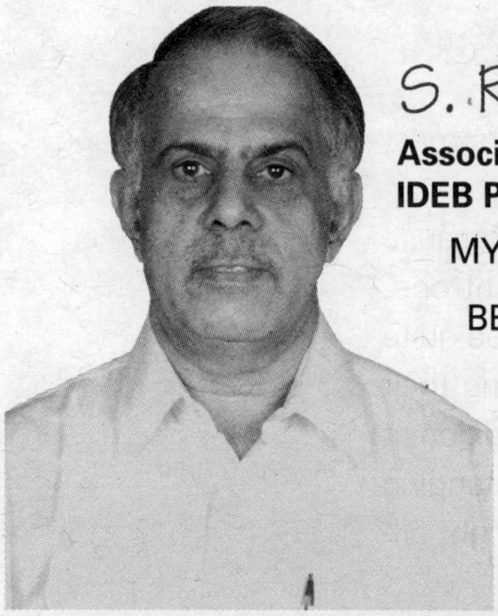

S. R. Bhat

Associate Vice President and Head, HR and Admin.
IDEB Projects Private Limited

MY JOB: Strategic HR management and change management.

BEST PART OF MY JOB: Qualitative improvement in human capital and manpower mix.

WORST PART OF MY JOB: Rationalization.

BEST MANAGEMENT ADVICE EVER RECEIVED: Balanced application of human capital management practices with effective systems approach toward resourcing, performance, reward, and manage change without losing sight of the long-term organizational objectives.

You'll be hearing more from this real manager throughout the chapter.

Chapter 13

Understanding Individual Behavior

Have you ever wondered why the people around you are behaving the way they are? People differ in their behaviors, and even the same person can behave one way one day and a completely different way another day. Managers need to understand individual behavior. Focus on the following learning outcomes as you read and study this chapter.

LEARNING OUTCOMES

13.1	Identify the focus and goals of organizational behavior.	page 286
13.2	Explain the role that attitudes play in job performance.	page 288
13.3	Describe different personality theories.	page 294
13.4	Describe perception and factors that influence it.	page 301
13.5	Discuss learning theories and their relevance in shaping behavior.	page 304
13.6	Discuss contemporary issues in organizational behavior.	page 307

A Manager's Dilemma

Like many other revolts, it began with something simple.[1] At Microsoft, it was the vanishing towels. For employees who biked to work through the often-drizzly weather in Seattle, the provided towels had become an entitlement. However, one day when employees came to work, the towels were gone . . . pulled without notice from the locker rooms in the company's underground garage. The company's human resources manager thought removing the towels, which had been done as a cost-saving measure, "wouldn't even be a blip." But it was. Irate employees waged war on company message boards and blogs. One post fumed, "It is a dark and dreary day at One Microsoft Way. Do yourself a favor and stay away." The intensity of the comments shocked senior executives. The towel fiasco, in conjunction with a languishing stock price and a little bit of "Google envy," suggested a serious morale problem. Lisa Brummel, a successful Microsoft product development manager with no HR experience, was tapped to become the new HR chief. Her mandate: Improve the mood around here. Put yourself in her position . . .

What would you do?

Lisa Brummel had a bit of a people challenge ahead of her! Like her, most other managers want to attract and retain employees with the right attitudes and personality. They want people who show up and work hard, get along with coworkers and customers, have good attitudes, and exhibit good work behaviors in other ways. But as you're probably already aware, people don't always behave like that "ideal" employee. They post critical comments in blogs. They complain about missing towels. People differ in their behaviors, and even the same person can behave one way one day and a completely different way another day. For instance, haven't you seen family members, friends, or coworkers behave in ways that prompted you to wonder: Why did they do that?

LEARNING OUTCOME 13.1 ▷ FOCUS AND GOALS OF ORGANIZATIONAL BEHAVIOR

The material in this chapter and the next three chapters draws heavily on the field of study known as *organizational behavior (OB)*. Although it's concerned with the subject of **behavior**—that is, the actions of people—**organizational behavior (OB)** is the study of the actions of people at work.

One of the challenges in understanding organizational behavior is that it addresses issues that aren't obvious. Like an iceberg, OB has a small visible dimension and a much larger hidden portion (see Exhibit 13–1). What we see when we look at an organization is its visible aspects: strategies, goals, policies and procedures, structure, technology, formal authority relationships, and chain of command. But under the surface are other elements that managers need to understand—elements that also influence how employees behave at work. As we'll show, OB provides managers with considerable insights into

Exhibit 13–1

Organization as Iceberg

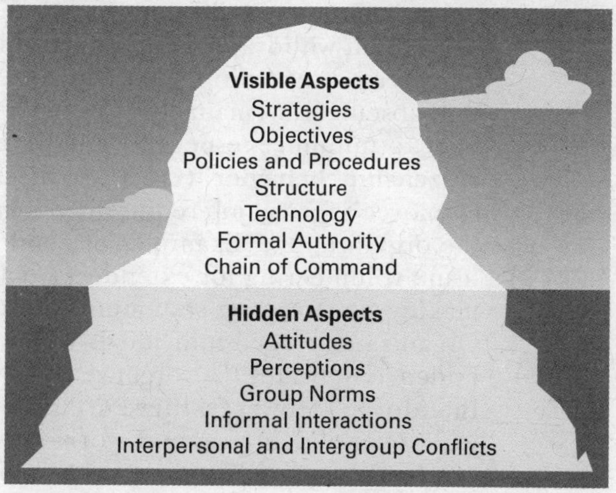

these important, but hidden, aspects of the organization. For instance, in November 2008, Commerce and Industries Minister of West Bengal, Nirupam Sen, decided to move his office in Kolkata from the Writers Building to a posh and nicer office in Camac Street. On the day of the appointment, only the minister and some of his staff reported to the new office. All other department members did not go saying that it would make their work more difficult because of the distance from other offices and also access to the new office was more difficult than the old office. The Minister then realized that he could not have made such a major decision without getting inputs from his staff.[2]

FOCUS OF ORGANIZATIONAL BEHAVIOR

Organizational behavior focuses on three major areas. First, OB looks at *individual behavior*. Based predominantly on contributions from psychologists, this area includes such topics as attitudes, personality, perception, learning, and motivation. Second, OB is concerned with *group behavior*, which includes norms, roles, team building, leadership, and conflict. Our knowledge about groups comes basically from the work of sociologists and social psychologists. Finally, OB also looks at *organizational* aspects, including structure, culture, and human resource policies and practices. We've addressed group and organizational aspects in previous chapters. In this chapter, we'll look at individual behavior.

GOALS OF ORGANIZATIONAL BEHAVIOR

The goals of OB are to *explain*, *predict*, and *influence* behavior. Managers need to be able to *explain* why employees engage in some behaviors rather than others, *predict* how employees will respond to various actions and decisions, and *influence* how employees behave.

What employee behaviors are we specifically concerned with explaining, predicting, and influencing? Six important ones have been identified: employee productivity, absenteeism, turnover, organizational citizenship behavior (OCB), job satisfaction, and workplace misbehavior. **Employee productivity** is a performance measure of both efficiency and effectiveness. Managers want to know what factors will influence the efficiency and effectiveness of employees. **Absenteeism** is the failure to show up for work. It's difficult for work to get done if employees don't show up. In a 2004–2005 survey of Indian industries, an average of 8 percent man-days were lost due to absenteeism.[3] A UNESCO report has shown that in primary schools 25 percent teachers are missing from their classes and it is estimated that this has cost 22.5 percent of the salary budget earmarked for the teachers in India.[4] Although absenteeism can't be totally eliminated, excessive levels have a direct and immediate

behavior
The actions of people.

organizational behavior (OB)
The study of the actions of people at work.

employee productivity
A performance measure of both efficiency and effectiveness.

absenteeism
The failure to show up for work.

impact on the organization's functioning. **Turnover** is the voluntary or involuntary permanent withdrawal from an organization. It can be a problem because it leads to increased recruiting, selection, and training costs as well as work disruptions. As with absenteeism, managers can never eliminate turnover, but it is something they want to minimize, especially among high-performing employees. **Organizational citizenship behavior (OCB)** is discretionary behavior that's not part of an employee's formal job requirements but that promotes the effective functioning of the organization.[5] Examples of good OCB include helping others on one's work team, volunteering for extended job activities, avoiding unnecessary conflicts, and making constructive statements about one's work group and the organization. Organizations need individuals who will do more than their usual job duties, and evidence indicates that organizations that have such employees outperform those that don't.[6] However, there are drawbacks to OCB: Employees may experience work overload, stress, and work–family conflicts.[7] **Job satisfaction** refers to an employee's general attitude toward his or her job. Although job satisfaction is an attitude rather than a behavior, it's an outcome that concerns many managers because satisfied employees are more likely than dissatisfied employees to show up for work, have higher levels of performance, and stay with an organization. **Workplace misbehavior** is any intentional employee behavior that is potentially harmful to the organization or individuals within the organization. Workplace misbehavior shows up in organizations in four ways: deviance, aggression, antisocial behavior, and violence.[8] Such behaviors can range from playing loud music just to irritate coworkers to verbal aggression to sabotaging work, all of which can create havoc in any organization. In the following pages, we'll address how an understanding of four psychological factors—employee attitudes, personality, perception, and learning—can help predict and explain these employee behaviors.

QUICK LEARNING REVIEW:

LEARNING OUTCOME 13.1

- Explain why the concept of an organization as an iceberg is important.
- Describe the focus and goals of OB.
- Define the six important employee behaviors.

Go to page 311 and see how well you know this material.

LEARNING **OUTCOME 13.2** ▷

ATTITUDES AND JOB PERFORMANCE

Attitudes are evaluative statements—favorable or unfavorable—concerning objects, people, or events. They reflect how an individual feels about something. When a person says, "I like my job," he or she is expressing an attitude about work.

An attitude is made up of three components: cognition, affect, and behavior.[9] The **cognitive component** of an attitude refers to the beliefs, opinions, knowledge, or information held by a person—for instance, the belief that "discrimination is wrong." The **affective component** of an attitude is the emotional or feeling part of an attitude. Using our example, this component would be reflected by the statement "I don't like Vikas because he discriminates against Muslims and Christians." Finally, affect can lead to behavioral outcomes. The **behavioral component** of an attitude refers to an intention to behave in a certain way toward someone or something. To continue our example, I might choose to avoid Vikas because of my feelings about him. Understanding that attitudes are made up of three components helps show their complexity. But keep in mind that the term *attitude* usually refers only to the affective component.

Naturally, managers aren't interested in every attitude an employee has. They're especially interested in job-related attitudes. The three most widely known are job satisfaction, job involvement, and organizational commitment. Also, a new concept that's generating widespread interest is employee engagement.[10]

JOB SATISFACTION

According to our earlier definition, job satisfaction refers to a person's general attitude toward his or her job. A person with a high level of job satisfaction has a positive attitude toward his or her job. A person who is dissatisfied has a negative attitude. When people speak of employee attitudes, they are usually referring to job satisfaction.

How Satisfied Are Employees? A survey of Indian employees by Watson Wyatt shows that 71 percent of them were satisfied with their work. Indians were found to be more satisfied as compared to respondents in other Asia-Pacific countries. Among the top five reasons for staying in an organization, "opportunity to utilize skills" emerged as one of them. However, only 42 percent were satisfied with the training provided to them. Indians were also high on their satisfaction of communication, teamwork, and work environment.[11] Although job satisfaction tends to increase as income increases, only 52 percent of individuals earning more than $50,000 are satisfied with their jobs. For individuals earning less than $15,000, about 45 percent of workers say they are satisfied with their jobs.[12] While it's possible that higher pay translates into higher job satisfaction, an alternative explanation for the difference in satisfaction levels is that higher pay reflects different types of jobs. Higher-paying jobs generally require more advanced skills, give jobholders greater responsibilities, are more stimulating and provide more challenges, and allow workers more control. It's likely that the reports of higher satisfaction among higher income levels reflect those factors rather than the pay.

What about job satisfaction levels in other countries? Surveys of European workers show regional variations.[13] For instance, only 68 percent of Scandinavian workers, 67 percent of Italian workers, and 53 percent of Swiss workers report being satisfied with their jobs. Only 50 percent of US workers are satisfied with their jobs.[14] Other numbers from Europe are somewhat higher: 80 percent of workers in France, 73 percent of German workers, and 72 percent of workers in Great Britain say they're satisfied with their jobs.[15] On the other hand, 60 percent of Canadian workers report being satisfied with their jobs, while 61 percent of Asia-Pacific employees are.[16]

Satisfaction and Productivity. After the Hawthorne Studies (discussed in Chapter 2), managers believed that happy workers were productive workers. Because it's not been easy to determine whether job satisfaction caused job productivity or vice versa, some management researchers felt that belief was generally wrong. However, we can say with some certainty that the correlation between satisfaction and productivity is fairly strong.[17] Also, organizations with more satisfied employees tend to be more effective than organizations with fewer satisfied employees.[18]

Satisfaction and Absenteeism. Although research shows that satisfied employees have lower levels of absenteeism than do dissatisfied employees, the correlation isn't strong.[19] While it certainly makes sense that dissatisfied employees are more likely to miss work, other factors affect the relationship. For instance, organizations that provide liberal sick leave benefits are encouraging all their employees—including those who are highly satisfied—to take "sick" days. Assuming that your job has some variety in it, you can find work satisfying and yet still take a "sick" day to enjoy a three-day weekend or to golf on a warm spring day if taking such days results in no penalties.

turnover
Voluntary or involuntary permanent withdrawal from an organization.

organizational citizenship behavior (OCB)
Discretionary behavior that is not part of an employee's formal job requirements but that promotes the effective functioning of the organization.

job satisfaction
An employee's general attitude toward his or her job.

workplace misbehavior
An intentional employee behavior that is potentially damaging to the organization or to individuals within the organization.

attitudes
Evaluative statements, either favorable or unfavorable, concerning objects, people, or events.

cognitive component
The part of an attitude that's made up of the beliefs, opinions, knowledge, or information held by a person.

affective component
The part of an attitude that's the emotional or feeling part.

behavioral component
The part of an attitude that refers to an intention to behave in a certain way toward someone or something.

Best Buy has its own staff of customer service representatives, the employees of a formerly independent company known as the Geek Squad, which Best Buy purchased in 2002. The teams of high-tech troubleshooters operate in all U.S. and Canadian stores to help customers solve their PC and other electronics-related problems. And they also make house calls. In addition to Best Buy's standard new-employee training, which focuses a great deal on customer service, the Geek Squad members go through extra technical training and additional testing. All of these things are intended to improve the job satisfaction of the techies, which research has shown can have a positive effect on customer satisfaction.

Satisfaction and Turnover. Research on the relationship between satisfaction and turnover is much stronger than the other satisfaction research mentioned so far. Satisfied employees have lower levels of turnover, while dissatisfied employees have higher levels of turnover.[20] Yet factors such as labor-market conditions, expectations about alternative job opportunities, and length of employment with the organization also affect an employee's decision to leave.[21] Research suggests that the level of satisfaction is less important in predicting turnover for superior performers because the organization typically does everything it can to keep them (for example, pay raises, praise, increased promotion opportunities).[22]

Job Satisfaction and Customer Satisfaction. Is job satisfaction related to positive customer outcomes? For frontline employees who have regular contact with customers, the answer is "yes." Satisfied employees increase customer satisfaction and loyalty.[23] Why? In service organizations, customer retention and defection are highly dependent on how frontline employees deal with customers. Satisfied employees are more likely to be friendly, upbeat, and responsive, which customers appreciate. And because satisfied employees are less likely to leave their jobs, customers are more likely to encounter familiar faces and receive experienced service. These qualities help build customer satisfaction and loyalty. However, the relationship also seems to work in reverse: Dissatisfied customers can increase an employee's job dissatisfaction. Employees who have regular contact with customers report that rude, thoughtless, or unreasonably demanding customers adversely affect their job satisfaction.[24]

A number of companies appear to understand this connection. Service-oriented businesses such as L.L.Bean, Southwest Airlines, and Johnson & Johnson work towards pleasing their customers. They also focus on building employee satisfaction, recognizing that satisfied employees will go a long way toward contributing to their goal of having happy customers. These firms seek to hire upbeat and friendly employees, they train employees in customer service, they reward customer service, they provide positive work climates, and they regularly track employee satisfaction through attitude surveys.

Job Satisfaction and OCB. It seems logical to assume that job satisfaction should be a major determinant of an employee's OCB.[25] Satisfied employees would seem to be more likely to talk positively about the organization, help others, and go above and beyond normal job expectations. Research suggests that there is a modest overall relationship between job satisfaction and OCB.[26] But that relationship is tempered by perceptions of fairness.[27] Basically, if you don't feel as though your supervisor, organizational procedures, or pay policies are fair, your job satisfaction is likely to suffer significantly. However, when you perceive that these things are fair, you have more trust in your employer and are more willing to voluntarily engage in behaviors that go beyond your formal job requirements. Another factor that influences individual OCB is the type of citizenship behavior a person's work group exhibits. In a work group with low group-level OCB, any individual in that group who engaged in OCB had higher-than-average job performance ratings. One possible explanation may be that the person was trying to find some way to "stand out" from the crowd.[28] No matter why it happens, the point is that OCB can have positive benefits for organizations.

Job Satisfaction and Workplace Misbehavior. When employees are dissatisfied with their jobs, they respond in some way. But it's not easy to predict *how* they'll respond. One person might quit. Another might respond by using work time to play computer games. And another might verbally abuse a coworker. If managers want to control the undesirable consequences of job dissatisfaction, they should attack the problem—job dissatisfaction—rather than try to control the different employee responses.

Three other job-related attitudes we need to look at include job involvement, organizational commitment, and employee engagement.

JOB INVOLVEMENT AND ORGANIZATIONAL COMMITMENT

Job involvement is the degree to which an employee identifies with his or her job, actively participates in it, and considers his or her job performance to be important to his or her self-worth.[29] Employees with a high level of job involvement strongly identify with and really care about the kind of work they do. Their positive attitude leads them to contribute to their work in positive ways. High levels of job involvement have been found to be related to fewer absences, lower resignation rates, and higher employee engagement with work.[30]

Organizational commitment is the degree to which an employee identifies with a particular organization and its goals and wishes to maintain membership in that organization.[31] Whereas job involvement is identifying with your job, organizational commitment is identifying with your employing organization. Research suggests that organizational commitment also leads to lower levels of both absenteeism and turnover and, in fact, is a better indicator of turnover than job satisfaction.[32] Why? Probably because it's a more global and enduring response to the organization than is satisfaction with a particular job.[33] However, organizational commitment is less important as a work-related attitude than it once was. Employees don't generally stay with a single organization for most of their career, and the relationship they have with their employer has changed considerably.[34] Although the commitment of *an employee to an organization* may not be as important as it once was, research about **perceived organizational support**—employees' general belief that their organization values their contribution and cares about their well-being—shows that the commitment of *the organization to an employee* can be beneficial. High levels of perceived organizational support lead to increased job satisfaction and lower turnover.[35]

EMPLOYEE ENGAGEMENT

A low-level trader employed by Société Générale, a giant French bank, lost billions of dollars through dishonest trades, and no one reported suspicious behavior. An internal investigation uncovered evidence that many back office employees failed to alert their supervisors about the suspicious trades.[36] Employee indifference can have serious consequences.

Managers want their employees to be connected to, satisfied with, and enthusiastic about their jobs. This concept is known as **employee engagement**.[37] Highly engaged

job involvement
The degree to which an employee identifies with his or her job, actively participates in it, and considers his or her job performance to be important to self-worth.

organizational commitment
The degree to which an employee identifies with a particular organization and its goals and wishes to maintain membership in that organization.

perceived organizational support
Employees' general belief that their organization values their contribution and cares about their well-being.

employee engagement
Employees being connected to, satisfied with, and enthusiastic about their jobs.

Exhibit 13–2 Key Employee Engagement Factors

Globally, respect ranks as the no. 1 factor contributing to employee engagement.

	GLOBAL	CHINA	FRANCE	GERMANY	INDIA	JAPAN	UK	U.S.
Respect	125	121	133	129	104	90	144	122
Type of Work	112	75	138	113	116	107	122	112
Work–Life Balance	112	98	133	106	97	119	119	111
Provide Good Service to Customers	108	108	110	108	103	79	122	107
Base Pay	108	113	110	105	103	140	117	114
People You Work With	107	96	105	131	98	107	120	104
Benefits	94	127	81	110	94	75	76	112
Long-Term Career Potential	92	91	89	77	108	94	88	92
Learning and Development	91	83	67	80	98	86	85	82
Flexible Working	87	85	77	92	80	88	83	88
Promotion Opportunities	85	92	79	83	113	92	68	80
Variable Pay/Bonus	80	111	77	65	86	123	56	75

Source: Mercer

Note: Scores near 100 are middle importance, scores below 100 are less important, scores above 100 are more important.

Source: Mercer; *IndustryWeek*, April 2008, p. 24

employees are passionate about and deeply connected to their work. Disengaged employees have essentially "checked out" and don't care. They show up for work but have no energy or passion for it. Exhibit 13–2 lists the key engagement factors found in a global study of more than 12,000 employees.

There are benefits to having highly engaged employees. First, highly engaged employees are 2½ times more likely to be top performers than their less-engaged coworkers. In addition, companies with highly engaged employees have higher retention rates, which helps keep recruiting and training costs low. And both of these outcomes—higher performance and lower costs—contribute to superior financial performance.[38]

ATTITUDES AND CONSISTENCY

Have you ever noticed that people change what they say so it doesn't contradict what they do? Perhaps a friend of yours has repeatedly said that she thinks joining a sorority is an important part of college life. But then she goes through rush and doesn't get accepted. All of a sudden, she's saying that sorority life isn't all that great.

Research has generally concluded that people seek consistency among their attitudes *and* between their attitudes and behavior.[39] This means that individuals try to reconcile differing attitudes and align their attitudes and behavior so they appear rational and consistent. When there is an inconsistency, individuals will do something to make it consistent by altering the attitudes, altering the behavior, or rationalizing the inconsistency.

For example, a campus recruiter for R&S Information Services who visits college campuses and sells them on the advantages of R&S as a good place to work would experience inconsistency if he personally believed that R&S had poor working conditions and few opportunities for promotion. This recruiter could, over time, find his attitudes toward R&S becoming more positive. He might actually convince himself by continually articulating the merits of working for the company. Another alternative is that the recruiter could become openly negative about R&S and the opportunities within the company for prospective applicants. The original enthusiasm that the recruiter might have had would dwindle and might be replaced by outright cynicism toward the company. Finally, the recruiter might acknowledge that R&S is an undesirable place to work, but as a professional, he might realize that his obligation is to present the positive aspects of working for the company. He might further rationalize that no workplace is perfect and that his job is to present a favorable picture of the company, not to present both sides.

COGNITIVE DISSONANCE THEORY

Can we assume from the consistency principle that an individual's behavior can always be predicted if we know his or her attitude on a subject? The answer isn't a simple "yes" or "no." Why? Cognitive dissonance theory.

Cognitive dissonance theory sought to explain the relationship between attitudes and behavior.[40] **Cognitive dissonance** is any incompatibility or inconsistency between attitudes or between behavior and attitudes. The theory argued that inconsistency is uncomfortable and that individuals will try to reduce the discomfort and, thus, the dissonance.

Of course, no one can avoid dissonance. You probably know you should floss your teeth every day but don't do it. There's an inconsistency between attitude and behavior. How do people cope with cognitive dissonance? The theory proposed that how hard we try to reduce dissonance is determined by three factors: (1) the *importance* of the factors creating the dissonance, (2) the degree of *influence* the individual believes he or she has over those factors, and (3) the *rewards* that may be involved in dissonance.

If the factors creating the dissonance are relatively unimportant, the pressure to correct the inconsistency will be low. However, if those factors are important, individuals may change their behavior, conclude that the dissonant behavior isn't so important, change their attitude, or identify compatible factors that outweigh the dissonant ones.

How much influence individuals believe they have over the factors also affects their reaction to the dissonance. If they perceive that the dissonance is something about which they have no choice, they won't be receptive to changing their attitude and won't feel a need to do so. If, for example, the dissonance-producing behavior was required as a result of a manager's order, the pressure to reduce dissonance would be less than if the behavior had been performed voluntarily. Although dissonance exists, it can be rationalized and justified by the need to follow the manager's orders—that is, the person had no choice or control.

Finally, rewards influence the degree to which individuals are motivated to reduce dissonance. Coupling high dissonance with high rewards tends to reduce the discomfort by motivating the individual to believe that there is consistency.

Let'S Get Real: F2F

ARE EMPLOYEE SURVEYS HELPFUL?

Yes—very much. It has a value to guide and support the decision-making process. It is important to ensure that the appropriate methodology is adopted and is followed effectively during the survey process.

ATTITUDE SURVEYS

Many organizations regularly survey their employees about their attitudes.[41] Exhibit 13–3 shows an example of an attitude survey. Typically, **attitude surveys** present employees with a set of statements or questions asking how they feel about their jobs, work groups, supervisors, or the organization. Ideally, the items are designed to obtain the specific information that managers desire. An attitude score is achieved by summing up responses to individual questionnaire items. These scores can then be averaged

Exhibit 13–3

Sample Employee Survey

To measure employee attitudes, some KFC and Long John Silver's restaurants ask employees to react to statements such as:

- My restaurant is a great place to work.
- People on my team help out, even if it is not their job.
- I am told whether I am doing good work or not.
- I understand the employee benefits that are available to me.

Source: Yum Brands, Inc., based on E. White, "How Surveying Workers Can Pay Off," *Wall Street Journal*, June 18, 2007, p. B3.

cognitive dissonance
Any incompatibility or inconsistency between attitudes or between behavior and attitudes.

attitude surveys
Surveys that elicit responses from employees through questions about how they feel about their jobs, work groups, supervisors, or the organization.

for work groups, departments, divisions, or the organization as a whole. For instance, the Tennessee Valley Authority, the largest U.S. government-run energy company, came up with a "Cultural Health Index" to measure employee attitudes. The organization found that business units that scored high on the attitude surveys also were the ones whose performance was high. For poorly performing business units, early signs of potential trouble had shown up in the attitude surveys.[42]

Regularly surveying employee attitudes provides managers with valuable feedback on how employees perceive their working conditions. Policies and practices that managers view as objective and fair may not be seen that way by employees. The use of regular attitude surveys can alert managers to potential problems and employees' intentions early so that action can be taken to prevent repercussions.[43]

IMPLICATIONS FOR MANAGERS

Managers should be interested in their employees' attitudes because they influence behavior. Satisfied and committed employees, for instance, have lower-than-average rates of turnover and absenteeism. If managers want to keep resignations and absences down—especially among their more productive employees—they'll want to do things that generate positive job attitudes.

Satisfied employees also perform better on the job. Managers should focus on factors that have been shown to be conducive to high levels of employee job satisfaction: making work challenging and interesting, providing equitable rewards, and creating supportive working conditions and supportive colleagues.[44] These factors are likely to help employees be more productive.

Managers should also survey employees about their attitudes. As one study put it, "A sound measurement of overall job attitude is one of the most useful pieces of information an organization can have about its employees."[45]

Finally, managers should know that employees will try to reduce dissonance. If employees are required to do things that appear inconsistent to them or that are at odds with their attitudes, managers should remember that pressure to reduce the dissonance is not as strong when the employee perceives that the dissonance is externally imposed and uncontrollable. It's also decreased if rewards are significant enough to offset the dissonance. So the manager might point to external forces such as competitors, customers, or other factors when explaining the need to perform some work that the individual may have some dissonance about. Or the manager can provide rewards that an individual desires.

QUICK LEARNING REVIEW:

LEARNING OUTCOME 13.2

- Describe the three components of an attitude.
- Explain the four job-related attitudes.
- Describe the impact job satisfaction has on employee behavior.
- Discuss how individuals reconcile inconsistencies between attitudes and behavior.

Go to page 311 and see how well you know this material.

LEARNING OUTCOME 13.3 ▷ PERSONALITY

"Let's face it, dating is a drag. There was a time when we thought the computer was going to make it all better. . . . But most of us learned the hard way that finding someone who shares our love of film noir and obscure garage bands does not a perfect match make."[46] Using in-depth personality assessment and profiling, Chemistry.com is trying to do something about making the whole dating process better.

Personality. We each have one. Some of us are quiet and passive; others are loud and aggressive. When we describe people using terms such as *quiet, passive, loud, aggressive, ambitious, extroverted, loyal, tense,* or *sociable,* we're describing their personalities.

An individual's **personality** is a unique combination of emotional, thought, and behavioral patterns that affect how a person reacts to situations and interacts with others. Personality is most often described in terms of measurable traits that a person exhibits. We're interested in looking at personality because, just like attitudes, it also affects how and why people behave the way they do.

Over the years, researchers have attempted to identify the traits that best describe personality. The two most well-known approaches are the Myers-Briggs Type Indicator (MBTI®) and the Big Five model.

MBTI®

One popular approach to classifying personality traits is the MBTI®. This personality assessment consists of more than 100 questions that ask people how they usually act or feel in different situations.[47] The way you respond to these questions puts you at one end or another of four dimensions:

1. ***Social interaction: extrovert or introvert (E or I)***—An extrovert is someone who is outgoing, dominant, and often aggressive and who wants to change the world. Extroverts need a work environment that's varied and action oriented, that lets them be with others, and that gives them a variety of experiences. An individual who's shy and withdrawn and focuses on understanding the world is described as an introvert. Introverts prefer a work environment that is quiet and concentrated, that lets them be alone, and that gives them a chance to explore in depth a limited set of experiences.
2. ***Preference for gathering data: sensing or intuitive (S or N)***—Sensing types dislike new problems unless there are standard ways to solve them, prefer an established routine, have a high need for closure, show patience with routine details, and tend to be good at precise work. On the other hand, intuitive types are individuals who like solving new problems, dislike doing the same thing over and over again, jump to conclusions, are impatient with routine details, and dislike taking time for precision.
3. ***Preference for decision making: feeling or thinking (F or T)***—Individuals who are feeling types are aware of other people and their feelings, like harmony, need occasional praise, dislike telling people unpleasant things, tend to be sympathetic, and relate well to most people. Thinking types are unemotional and uninterested in people's feelings, like analysis and putting things into logical order, are able to reprimand people and fire them when necessary, may seem hard-hearted, and tend to relate well only to other thinking types.
4. ***Style of making decisions: perceptive or judgmental (P or J)***—Perceptive types are curious, spontaneous, flexible, adaptable, and tolerant. They focus on starting a task, postpone decisions, and want to find out all about the task before starting it. Judgmental types are good planners, decisive, purposeful, and exacting. They focus on completing a task, make decisions quickly, and want only the information necessary to get a task done.

Combining these preferences provides descriptions of 16 personality types. Exhibit 13–4 summarizes a few of them.

More than two million people take the MBTI® each year in the United States alone. Some organizations that have used the MBTI include Adani Exports Ltd, Akruti Nirman Ltd, Kotak Mahindra Bank, Vodafone, and Zee Telefilms. Hospitals, educational institutions, and the Armed Forces all over the world use MBTI". There's no hard evidence that the MBTI® is a valid measure of personality, but that doesn't seem to deter its widespread use.

personality
The unique combination of emotional, thought, and behavioral patterns that affect how a person reacts to situations and interacts with others.

Exhibit 13–4

Examples of MBTI® Personality Types

Type	Description
INFJ (introvert, intuitive, feeling, judgmental)	Quietly forceful, conscientious, and concerned for others. Such people succeed through perseverance, originality, and the desire to do whatever is needed or wanted. They are often highly respected for their uncompromising principles.
ESTP (extrovert, sensing, thinking, perceptive)	Blunt and sometimes insensitive. Such people are matter-of-fact and do not worry or hurry. They enjoy whatever comes along. They work best with real things that can be assembled or disassembled.
ISFP (introvert, sensing, feeling, perceptive)	Sensitive, kind, modest, shy, and quietly friendly. Such people strongly dislike disagreements and will avoid them. They are loyal followers and quite often are relaxed about getting things done.
ENTJ (extrovert, intuitive, thinking, judgmental)	Warm, friendly, candid, and decisive; also usually skilled in anything that requires reasoning and intelligent talk, but may sometimes overestimate what they are capable of doing.

Source: Based on I. Briggs-Myers, *Introduction to Type* (Palo Alto, CA: Consulting Psychologists Press, 1980). pp 7–8.

Let'S Get Real: F2F

PERSONALITY "PROBLEMS" I'VE DEALT WITH:

Dealing with senior executives is challenging with respect to attitude, professional objectives, and organizational goals, more so in the changing business and social environment, where professionals have come from varied and different backgrounds. Hence, conflict resolutions played a key role in resolving issues and managing such personality differences.

How could the MBTI® help managers? Proponents believe that it's important to know these personality types because they influence the way people interact and solve problems. For instance, if your boss is an intuitive type and you're a sensing type, you'll gather information in different ways. An intuitive type prefers gut reactions, whereas a sensor prefers facts. To work well with your boss, you would have to present more than just facts about a situation; you'd have to also talk out how you feel about it. Also, the MBTI® has been used to help managers better match employees to certain types of jobs.

THE BIG FIVE MODEL

In recent years, research has shown that five basic personality dimensions underlie all others and encompass most of the significant variation in human personality.[48] The five personality traits in the **Big Five model** are:

1. ***Extraversion***—The degree to which someone is sociable, talkative, and assertive
2. ***Agreeableness***—The degree to which someone is good-natured, cooperative, and trusting
3. ***Conscientiousness***—The degree to which someone is responsible, dependable, persistent, and achievement oriented
4. ***Emotional stability***—The degree to which someone is calm, enthusiastic, and secure (positive) or tense, nervous, depressed, and insecure (negative)
5. ***Openness to experience***—The degree to which someone is imaginative, artistically sensitive, and intellectual

The Big Five model provides more than just a personality framework. Research has shown that important relationships exist between these personality dimensions and job performance. For example, one study examined five categories of occupations: *professionals* (such as engineers, architects, and attorneys), *police, managers, salespeople*, and *semiskilled and skilled employees*.[49] The results showed that conscientiousness predicted job performance for all five occupational groups. Predictions for the other personality dimensions depended on the situation and the occupational group. For example, extraversion predicted performance in managerial and sales positions—occupations in which high social interaction is necessary. Openness to experience was found to be important in predicting training competency. Ironically, emotional security wasn't positively related to job performance in any of the occupations.

ADDITIONAL PERSONALITY INSIGHTS

Although the traits in the Big Five model are highly relevant to understanding behavior, they aren't the only personality traits that can describe someone's personality. Five other personality traits are powerful predictors of behavior in organizations:

1. ***Locus of control***—Some people believe that they control their own fate. Others see themselves as pawns, believing that what happens to them in their lives is due to luck or chance. The **locus of control** in the first case is *internal*; these people believe that they control their own destiny. The locus of control in the second case is *external*; these people believe that their lives are controlled by outside forces.[50] Research indicates that employees who are externals are less satisfied with their jobs, more alienated from the work setting, and less involved in their jobs than are those who rate high on internality.[51] A manager might also expect externals to blame a poor performance evaluation on their boss's prejudice, their coworkers, or other events outside their control; internals would explain the same evaluation in terms of their own actions.
2. ***Machiavellianism***—The second characteristic is called **Machiavellianism (Mach)**, named after Niccolo Machiavelli, who wrote in the sixteenth century on how to gain and manipulate power. An individual who is high in Machiavellianism is pragmatic, maintains emotional distance, and believes that ends can justify means.[52] "If it works, use it" is consistent with a high Mach perspective. Do high Machs make good employees? That depends on the type of job and whether you consider ethical factors in evaluating performance. In jobs that require bargaining skills (such as a purchasing manager) or that have substantial rewards for excelling (such as a salesperson working on commission), high Machs are productive.
3. ***Self-esteem***—People differ in the degree to which they like or dislike themselves, a trait called **self-esteem.**[53] Research on self-esteem offers some interesting behavioral insights. For example, self-esteem is directly related to expectations for success. Those with high self-esteem believe that they possess the ability they need in order to succeed at work. Individuals with high self-esteem tend to take more risks in job selection and are more likely to choose unconventional jobs than are people with low self-esteem.

   The most common finding on self-esteem is that people with low self-esteem are more susceptible to external influence than are people with high self-esteem. Those with low self-esteem depend on receiving positive evaluations from others. As a result, they're more likely to seek approval from others and are more prone to conform to the beliefs and behaviors of those they respect than are individuals with high self-esteem. In managerial positions, those with low self-esteem tend to be concerned with pleasing others and, therefore, are less likely to take unpopular stands than are those with high self-esteem. Finally, self-esteem has been found to be related to job satisfaction. A number of studies confirm those with high self-esteem are more satisfied with their jobs than are those with low self-esteem.
4. ***Self-monitoring***—**Self-monitoring** refers to a person's ability to adjust behavior to external, situational factors.[54] Individuals high in self-monitoring show considerable adaptability in adjusting their behavior. They're highly sensitive to external cues and can behave differently in different situations. High self-monitors are capable of presenting striking contradictions between their public persona and their private selves. Low self-monitors can't adjust their behavior. They tend to display

Big Five model
A personality trait model that examines extraversion, agreeableness, conscientiousness, emotional stability, and openness to experience.

locus of control
The degree to which people believe they are masters of their own fate.

Machiavellianism (Mach)
A measure of the degree to which people are pragmatic, maintain emotional distance, and believe that ends justify means.

self-esteem
An individual's degree of like or dislike for himself or herself.

self-monitoring
A personality trait that measures the ability to adjust behavior to external situational factors.

Naguib Sawiris is a risk taker who makes decisions quickly. He's the chairman and CEO of Orascom Telecom Holding, one of the largest and most diversified telecommunications firms in the world, with operations mostly in the Middle East, Africa, and South Asia. Based in Cairo, the company frequently targets "difficult and primitive" areas to develop its mobile phone business, like Algeria, Tunisia, Pakistan, Congo, Zimbabwe, Bangladesh, and Iraq. Says Sawiris, "We do not fear difficult missions." Neither does Sawiris himself.

their true dispositions and attitudes in every situation, and there's high behavioral consistency between who they are and what they do.

Research on self-monitoring suggests that high self-monitors pay closer attention to the behavior of others and are more flexible than are low self-monitors.[55] In addition, high self-monitoring managers tend to be more mobile in their careers, receive more promotions (both internal and cross-organizational), and are more likely to occupy central positions in an organization.[56] A high self-monitor is capable of putting on different "faces" for different audiences, an important trait for managers who must play multiple, or even contradicting, roles.

5. ***Risk taking***—People differ in their willingness to take chances. Differences in the propensity to assume or to avoid risk have been shown to affect how long it takes managers to make a decision and how much information they require before making their choice. For instance, in one study in which managers worked on simulated exercises that required them to make hiring decisions, high risk-taking managers took less time to make decisions and used less information in making their choices than did low risk-taking managers.[57] Interestingly, the decision accuracy was the same for the two groups. To maximize organizational effectiveness, managers should try to align employee risk-taking propensity with specific job demands.

Other Personality Traits. A couple other personality traits deserve mention. One is the **Type A personality**, which describes someone who is continually and aggressively struggling to achieve more and more in less and less time.[58] In the corporate culture, the Type A personality is highly valued. Type A's subject themselves to continual time pressure and deadlines and have moderate to high levels of stress. These individuals emphasize quantity over quality. On the other hand, Type B's aren't harried by the desire to achieve more and more. They don't suffer from a sense of time urgency and are able to relax without guilt.

Another interesting trait that's been studied extensively is **proactive personality**, which describes people who identify opportunities, show initiative, take action, and persevere until meaningful change occurs. Not surprisingly, research has shown that proactives have many desirable behaviors that organizations want.[59] For instance, they are more likely than others to be seen as leaders and more likely to act as change agents in organizations; they're more likely to challenge the status quo; they have entrepreneurial abilities; and they're more likely to achieve career success.

PERSONALITY TYPES IN DIFFERENT CULTURES

Do personality frameworks, such as the Big Five model, transfer across cultures? Are dimensions such as locus of control relevant in all cultures? Let's try to answer these questions.

The five personality factors studied in the Big Five model appear in almost all cross-cultural studies.[60] This includes a wide variety of diverse cultures, such as India, China, Israel, Germany, Japan, Spain, Nigeria, Norway, Pakistan, and the United States. Differences are found in the emphasis on dimensions. Chinese people, for example, use the category of conscientiousness more often and use the category of agreeableness less often than do Americans. But there is a surprisingly high amount of agreement, especially among individuals from developed countries. As a case in point, a comprehensive review of studies covering people from the European Community found that conscientiousness was a valid predictor of performance across jobs and occupational groups.[61]

We know that there are certainly no common personality types for a given country. You can, for instance, find high risk takers and low risk takers in almost any culture. Yet a country's culture influences the *dominant* personality characteristics of its people. We can see this effect of national culture by looking at one of the personality traits we just discussed: locus of control.

National cultures differ in terms of the degree to which people believe they control their environment. For instance, North Americans believe that they can dominate their environment; other societies, such as those in Middle Eastern countries, believe that life is essentially predetermined. Notice how closely this distinction parallels the concept of internal and external locus of control. On the basis of this particular cultural characteristic, we should expect a larger proportion of internals in the U.S. and Canadian workforces than in the workforces of Saudi Arabia and Iran.

As we have seen throughout this section, personality traits influence employees' behavior. For global managers, understanding how personality traits differ takes on added significance when looked at from the perspective of national culture.

EMOTIONS AND EMOTIONAL INTELLIGENCE

Airhosts or airhostesses have to be courteous to anxious and stressed air travelers who can be sometimes rude and uncourteous. Similarly, call center executives have to be empathetic to the customers, irrespective of what they feel themselves about it. A nurse has to continuously smile and sound positive even when facing the pessimistic patients and their families. To be able to cope with such emotionally laborious situations, service personnel need help and support from their companies.

We can't leave the topic of personality without looking at another important behavioral aspect—emotions—especially because how we respond emotionally and how we deal with our emotions can be functions of our personality. **Emotions** are intense feelings that are directed at someone or something. They're object specific; that is, emotions are reactions to an object.[62] For instance, when a work colleague criticizes you for the way you spoke to a client, you might become angry at him. That is, you show emotion (anger) toward a specific object (your colleague). Because employees bring an emotional component with them to work every day, managers need to understand the role that emotions play in employee behavior.[63]

How many emotions are there? Although you could probably name several dozen, research has identified six universal emotions: anger, fear, sadness, happiness, disgust, and surprise.[64] Do these emotions surface in the workplace? Absolutely! I get *angry* after receiving a poor performance appraisal. I *fear* that I could be laid off as a result of a company cutback. I'm *sad* about one of my coworkers leaving to take a new job in another city. I'm *happy* after being selected as employee of the month. I'm *disgusted* with the way my supervisor treats women on our team. And I'm *surprised* to find out that management plans a complete restructuring of the company's retirement program.

People respond differently to identical emotion-provoking stimuli. In some cases, this can be attributed to a person's personality because people vary in their ability to express emotions. For instance, you undoubtedly know people who almost never show

Type A personality
Someone who is continually and aggressively struggling to achieve more and more in less and less time.

proactive personality
People who identify opportunities, show initiative, take action, and persevere until meaningful change occurs.

emotions
Intense feelings that are directed at someone or something.

their feelings. They rarely get angry or show rage. In contrast, you probably also know people who seem to be on an emotional roller coaster. When they're happy, they're ecstatic; when they're sad, they're deeply depressed. And two people can be in exactly the same situation, but one shows excitement and joy, and the other remains calm.

However, at other times how people respond emotionally is a result of job requirements. Jobs make different demands in terms of what types and how much emotion needs to be displayed. For instance, air traffic controllers, ER nurses, and trial judges are expected to be calm and controlled, even in stressful situations. On the other hand, public-address announcers at sporting events and lawyers in a courtroom must be able to alter their emotional intensity as the need arises.

One area of emotions research with interesting insights into personality is **emotional intelligence (EI)**, which is the ability to notice and to manage emotional cues and information.[65] It's composed of five dimensions:

1. ***Self-awareness***—The ability to be aware of what you're feeling
2. ***Self-management***—The ability to manage your own emotions and impulses
3. ***Self-motivation***—The ability to persist in the face of setbacks and failures
4. ***Empathy***—The ability to sense how others are feeling
5. ***Social skills***—The ability to handle the emotions of others

EI has been shown to be positively related to job performance at all levels. For instance, one study looked at the characteristics of engineers at Lucent Technologies whom their peers had rated as stars. The researchers concluded that stars were better at relating to others. That is, it was EI, not academic intelligence, that characterized high performers. A study of Air Force recruiters generated similar findings. Top-performing recruiters exhibited high levels of EI. Despite these findings, EI has been a controversial topic in OB. Supporters say that EI has intuitive appeal and predicts important behavior.[66] Critics say that EI is vague, can't be measured, and has questionable validity.[67] One thing we can conclude is that EI appears to be relevant to success in jobs that demand a high degree of social interaction.

IMPLICATIONS FOR MANAGERS

More than 62 percent of companies are using personality tests when recruiting and hiring.[68] And that's where the major value in understanding personality differences probably lies. Managers are likely to have higher-performing and more satisfied employees if consideration is given to matching personalities with jobs. The best-documented personality–job fit theory was developed by psychologist John Holland.[69] His theory states that an employee's satisfaction with his or her job, as well as his or her likelihood of leaving that job, depends on the degree to which the individual's personality matches the job environment. Holland identified six basic personality types, as shown in Exhibit 13–5.

Holland's theory proposes that satisfaction is highest and turnover lowest when personality and occupation are compatible. Social individuals should be in "people" type jobs, and so forth. The key points of this theory are that (1) there appear to be intrinsic differences in personality among individuals; (2) there are different types of jobs; and (3) people in job environments compatible with their personality types should be more satisfied and less likely to resign voluntarily than should people in incongruent jobs.

In addition, there are other benefits to understanding personality. By recognizing that people approach problem solving, decision making, and job interactions differently, a manager can better understand why an employee is uncomfortable with making quick decisions or why another employee insists on gathering as much information as possible before addressing a problem. Or, for instance, managers can expect that individuals with an external locus of control may be less satisfied with their jobs than internals and also that they may be less willing to accept responsibility for their actions.

Exhibit 13–5

Holland's Personality–Job Fit

Type	Personality Characteristics	Sample Occupations
Realistic. Prefers physical activities that require skill, strength, and coordination.	Shy, genuine, persistent, stable, conforming, practical	Mechanic, drill press operator, assembly-line worker, farmer
Investigative. Prefers activities involving thinking, organizing, and understanding.	Analytical, original, curious, independent	Biologist, economist, mathematician, news reporter
Social. Prefers activities that involve helping and developing others.	Sociable, friendly, cooperative, understanding	Social worker, teacher, counselor, clinical psychologist
Conventional. Prefers rule-reguated, orderly, and unambiguous activities.	Conforming, efficient, practical, unimaginative, inflexible	Accountant, corporate manager, bank teller, file clerk
Enterprising. Prefers verbal activities in which there are opportunities to influence others and attain power.	Self-confident, ambitious, energetic, domineering	Lawyer, real estate agent, public relations specialist, small business manager
Artistic. Prefers ambiguous and unsystematic activities that allow creative expression.	Imaginative, disorderly, idealistic, emotional, impractical	Painter, musician, writer, interior decorator

Source: Based on J. L. Holland, *Making Vocational Choices: A Theory of Vocational Personalities and Work Environments* (Odessa, FL: Psychological Assessment Resources, 1997).

Finally, being a successful manager and accomplishing goals means working well with others both inside and outside the organization. In order to work effectively together, people need to understand each other. This understanding comes, at least in part, from an appreciation of personality traits and emotions.

QUICK LEARNING REVIEW:

LEARNING OUTCOME 13.3

- Contrast the MBTI® and the Big Five model.
- Describe five other personality traits that help explain individual behavior in organizations.
- Explain how emotions and emotional intelligence affect behavior.

Go to page 311 and see how well you know this material.

LEARNING **OUTCOME 13.4** ▷

PERCEPTION

Nadia Aman, Mirza Baig, M. Yusuf Mohamed, and Amar Baig have three things in common: they all are young, Muslim, and live in Gujarat. Their lives have changed, mostly due to stereotypes that their neighbors and coworkers have of Muslims.

Perception is a process by which we give meaning to our environment by organizing and interpreting sensory impressions. Research on perception consistently demonstrates that individuals may look at the same thing yet perceive it differently. One manager, for instance, may interpret the fact that her assistant regularly takes several days to make important decisions as evidence that the assistant is slow, disorganized, and afraid to make decisions. Another manager with the same assistant might interpret the same tendency as evidence that the assistant is thoughtful, thorough, and deliberate. The first manager would probably evaluate her assistant negatively; the second manager would probably evaluate the person positively. The point is that none

emotional intelligence (EI)
The ability to notice and to manage emotional cues and information.

perception
A process by which we give meaning to our environment by organizing and interpreting sensory impressions.

of us sees reality. We interpret what we see and call it reality. And, of course, as the example shows, we behave according to our perceptions.

FACTORS THAT INFLUENCE PERCEPTION

How do we explain the fact that different people can perceive the same thing differently? A number of factors act to shape and sometimes distort perception. These factors are in the *perceiver*, in the *target* being perceived, or in the *situation* in which the perception occurs.

When a person looks at a target and attempts to interpret what he or she sees, the individual's personal characteristics heavily influence the interpretation. These personal characteristics include attitudes, personality, motives, interests, experiences, or expectations.

The characteristics of the target being observed can also affect what's perceived. Loud people are more likely than quiet people to be noticed in a group, as are extremely attractive or unattractive individuals. The relationship of a target to its background also influences perception, as does our tendency to group close things and similar things together. You can experience these tendencies by looking at the visual perception examples shown in Exhibit 13–6. Notice how what you see changes as you look differently at each one.

Finally, the context in which we see objects or events is also important. The time at which an object or event is seen can influence perception, as can location, light, heat, color, and a number of other situational factors.

ATTRIBUTION THEORY

Much of the research on perception is directed at inanimate objects. Managers, however, are concerned with people. Our perceptions of people differ from our perceptions of inanimate objects because we make inferences about the behaviors of people that we don't make about objects. Objects don't have beliefs, motives, or intentions; people do. The result is that when we observe an individual's behavior, we try to develop explanations of why they behave in certain ways. Our perception and judgment of a person's actions are significantly influenced by the assumptions we make about the person.

Attribution theory was developed to explain how we judge people differently, depending on what meaning we attribute to a given behavior.[70] Basically, the theory suggests that when we observe an individual's behavior, we attempt to determine whether it was internally or externally caused. Internally caused behaviors are those that are believed to be under the personal control of the individual. Externally caused behavior results from outside factors; that is, the person is forced into the behavior by the situation. That determination, however, depends on three factors: distinctiveness, consensus, and consistency.

Distinctiveness refers to whether an individual displays different behaviors in different situations. Is the employee who arrived late today the same person that some

Exhibit 13–6

What do you see?

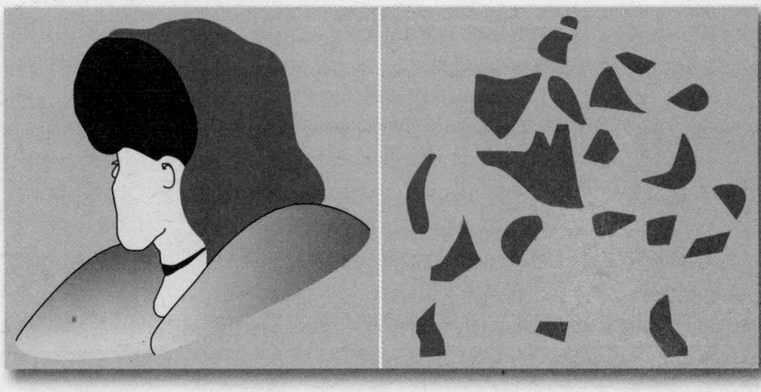

Old woman or young woman? A knight on a horse?

employees are complaining is a "goof-off"? What we want to know is whether this behavior is unusual. If it's unusual, the observer is likely to attribute the behavior to external forces, something beyond the control of the person. However, if the behavior isn't unusual, it will probably be judged as internal.

If everyone who's faced with a similar situation responds in the same way, we can say the behavior shows *consensus*. A tardy employee's behavior would meet this criterion if all employees who took the same route to work were also late. From an attribution perspective, if consensus is high, you're likely to give an external attribution to the employee's tardiness; that is, some outside factor—maybe road construction or a traffic accident—caused the behavior. However, if other employees who come the same way to work made it on time, you would conclude that the cause of the late behavior was internal.

Finally, an observer looks for *consistency* in a person's actions. Does the person engage in the behaviors regularly and consistently? Does the person respond the same way over time? Coming in 10 minutes late for work isn't perceived in the same way if, for one employee, it represents an unusual case (she hasn't been late in months), while for another employee, it's part of a routine pattern (she's late two or three times every week). The more consistent the behavior, the more the observer is inclined to attribute it to internal causes. Exhibit 13–7 summarizes the key elements of attribution theory.

One interesting finding from attribution theory is that errors or biases distort our attributions. For instance, there's substantial evidence to support the fact that when we make judgments about the behavior of other people, we have a tendency to *under*estimate the influence of external factors and to *over*estimate the influence of internal or personal factors.[71] This tendency is called the **fundamental attribution error** and can explain why a sales manager may attribute the poor performance of her sales representative to laziness rather than to the innovative product line introduced by a competitor. There's also a tendency for us to attribute our own successes to internal factors such as ability or effort while putting the blame for personal failure on external factors such as luck. This tendency is called the **self-serving bias** and suggests that feedback provided to employees in performance reviews will be distorted by them, depending on whether it's positive or negative.

Are these errors or biases that distort attributions universal across different cultures? We can't say for sure, but preliminary evidence indicates cultural differences.[72] For instance, a study of Korean managers found that, contrary to the self-serving bias, they tended to accept responsibility for group failure "because I was not a capable

Exhibit 13–7

Attribution Theory

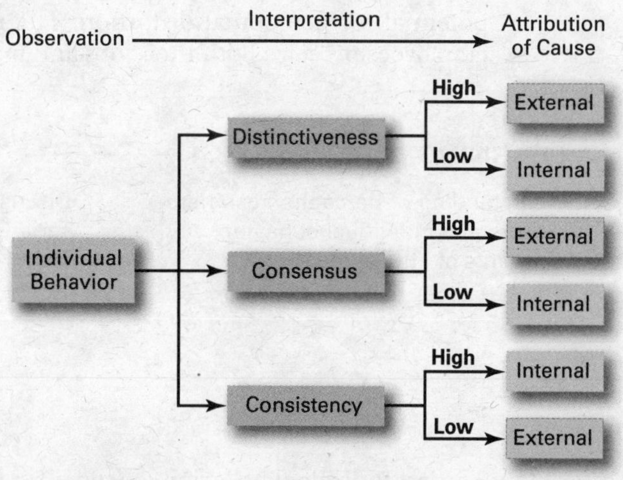

attribution theory
A theory used to explain how we judge people differently, depending on what meaning we attribute to a given behavior.

fundamental attribution error
The tendency to underestimate the influence of external factors and overestimate the influence of internal factors when making judgments about the behavior of others.

self-serving bias
The tendency for individuals to attribute their successes to internal factors while putting the blame for failures on external factors.

leader" instead of attributing it to group members.[73] Attribution theory was developed largely based on experiments with Americans and western Europeans. But the Korean study suggests caution in making attribution theory predictions in non-Western societies, especially in countries with strong collectivist traditions.

SHORTCUTS USED IN JUDGING OTHERS

Perceiving and interpreting people's behavior is a lot of work, so we use shortcuts to make the task more manageable. These techniques can be valuable because they let us make accurate perceptions quickly and provide valid data for making predictions. However, they aren't perfect. They can and do get us into trouble.

It's easy to judge others if we assume that they're similar to us. In **assumed similarity**, or the "like me" effect, the observer's perception of others is influenced more by the observer's own characteristics than by those of the person observed. For example, if you want challenges and responsibility in your job, you'll assume that others want the same. People who assume that others are like them can, of course, be right, but not always.

Let's Get Real: F2F

BIGGEST EMPLOYEE BEHAVIORAL CHALLENGE? Performance-oriented teamwork.

When we judge someone on the basis of our perception of a group to which he or she belongs we're using the shortcut called **stereotyping**. For instance, "Married persons are more stable employees than single persons" is an example of stereotyping. To the degree that a stereotype is based on fact, it may produce accurate judgments. However, many stereotypes aren't factual and distort our judgment.[74]

When we form a general impression about a person on the basis of a single characteristic, such as intelligence, sociability, or appearance, we're being influenced by the **halo effect**. This effect frequently occurs when students evaluate their classroom instructor. Students may isolate a single trait, such as enthusiasm, and allow their entire evaluation to be slanted by the perception of that one trait. An instructor may be quiet, assured, knowledgeable, and highly qualified, but if his classroom teaching style lacks enthusiasm, he might be rated lower on a number of other characteristics.

IMPLICATIONS FOR MANAGERS

Managers need to recognize that employees react to perceptions, not to reality. So whether a manager's appraisal of an employee's performance is actually objective and unbiased or whether the organization's wage levels are among the highest in the community is less relevant than what employees perceive them to be. If individuals perceive appraisals to be biased or wage levels to be low, they'll behave as if those conditions actually exist. Employees organize and interpret what they see, so there is always the potential for perceptual distortion. The message is clear: Pay close attention to how employees perceive both their jobs and management actions.

QUICK LEARNING REVIEW:

LEARNING OUTCOME 13.4

- Explain how an understanding of perception can help managers better understand individual behavior.
- Describe the key elements of attribution theory.
- Discuss the fundamental attribution error and self-serving bias.
- Name three shortcuts used in judging others.

Go to page 312 and see how well you know this material.

LEARNING **OUTCOME 13.5** ▷

LEARNING

The last individual behavior concept we're going to look at is learning. It's included for the obvious reason that almost all behavior is learned. If we want to explain, predict, and influence behavior, we need to understand how people learn.

The psychologists' definition of *learning* is considerably broader than the average person's view that "it's what we do in school." Learning occurs all the time, as we continuously learn from our experiences. A workable definition of **learning** is any relatively permanent change in behavior that occurs as a result of experience. Two learning theories help us understand how and why individual behavior occurs: operant conditioning and social learning.

OPERANT CONDITIONING

Operant conditioning argues that behavior is a function of its consequences. People learn to behave to get something they want or to avoid something they don't want. Operant behavior is voluntary or learned behavior, not reflexive or unlearned behavior. The tendency to repeat learned behavior is influenced by reinforcement or lack of reinforcement that happens as a result of the behavior. Reinforcement strengthens a behavior and increases the likelihood that it will be repeated. Lack of reinforcement weakens a behavior and reduces the likelihood that it will be repeated.

B. F. Skinner's research widely expanded our knowledge of operant conditioning.[75] Behavior is assumed to be determined from without—that is, *learned*—rather than from within—reflexive or unlearned. Skinner argued that people will most likely engage in desired behaviors if they are positively reinforced for doing so, and rewards are most effective if they immediately follow the desired response. In addition, behavior that isn't rewarded or that is punished is less likely to be repeated.

You see examples of operant conditioning everywhere. Any situation in which it's either explicitly stated or implicitly suggested that reinforcement (rewards) are contingent on some action on your part is an example of operant conditioning. Your instructor says that if you want a high grade in this course, you must perform well on tests by giving correct answers. A salesperson working on commission knows that earning a sizable income is contingent on generating high sales in his or her territory. Of course, the linkage between behavior and reinforcement can also work to teach an individual to behave in ways that work against the best interests of the organization. Assume that your boss tells you that if you'll work overtime during the next three-week busy season, you'll be compensated for it at the next performance appraisal. Then, when performance appraisal time comes, you are given no positive reinforcements (such as being praised for pitching in and helping out when needed). What will you do the next time your boss asks you to work overtime? You'll probably refuse. Your behavior can be explained by operant conditioning: If a behavior isn't positively reinforced, the probability that the behavior will be repeated declines.

SOCIAL LEARNING

It is common for family-run businesses in India to encourage their children to participate in meetings, discussions at the board level while they are young so that they can learn by observing their parents and elders of the family. In Essar, fast-track MBA's are asked to follow senior leadership within the company so that they can learn the ropes of the business and be ready to lead on their own. Individuals can also learn by observing what happens to other people and by being told about something as well as by direct experiences. Much of what we have learned comes from watching others (models)—parents, teachers, peers, television and movie actors, managers, and so forth. This view that we can learn both through observation and direct experience is called **social learning theory**.

The influence of others is central to the social learning viewpoint. The amount of influence that these models have on an individual is determined by four processes:

1. ***Attentional processes***—People learn from a model when they recognize and pay attention to its critical features. People are most influenced by models who are attractive, repeatedly available, thought to be important, or seen as similar to us.

assumed similarity
The assumption that others are like oneself.

stereotyping
Judging a person on the basis of one's perception of a group to which he or she belongs.

halo effect
A general impression of an individual that is influenced by a single characteristic.

learning
A relatively permanent change in behavior that occurs as a result of experience.

operant conditioning
A theory of learning that says behavior is a function of its consequences.

social learning theory
A theory of learning that says people can learn through observation and direct experience.

2. ***Retention processes***—A model's influence depends on how well the individual remembers the model's action, even after the model is no longer readily available.
3. ***Motor reproduction processes***—After a person has seen a new behavior by observing a model, the watching must become doing. This process then demonstrates that the individual can actually do the modeled activities.
4. ***Reinforcement processes***—Individuals will be motivated to exhibit the modeled behavior if positive incentives or rewards are provided. Behaviors that are reinforced will be given more attention, learned better, and performed more often.

SHAPING: A MANAGERIAL TOOL

Because learning takes place on the job as well as prior to it, managers are concerned with how they can teach employees to behave in ways that most benefit the organization. Thus, managers often attempt to "mold" individuals by guiding their learning in graduated steps. This is called **shaping behavior**.

Consider a situation in which an employee's behavior is significantly different from that sought by a manager. If the manager reinforced the individual only when he or she showed desirable responses, there might be very little reinforcement taking place. Shaping offers a logical approach to achieving the desired behavior. We shape behavior by systematically reinforcing each successive step that moves the individual closer to the desired behavior. If an employee who has chronically been a half-hour late for work comes in only 20 minutes late, we can reinforce the improvement. Reinforcement would increase as an employee gets closer to the desired behavior.

There are four ways to shape behavior: positive reinforcement, negative reinforcement, punishment, and extinction. A behavior followed by something pleasant, such as praising an employee for a job well done, is called *positive reinforcement.* Positive reinforcement increases the likelihood that the desired behavior will be repeated. Rewarding a response by eliminating or withdrawing something unpleasant is *negative reinforcement.* A manager who says "I won't dock your pay if you start getting to work on time" is using negative reinforcement. The desired behavior (getting to work on time) is being encouraged by the withdrawal of something unpleasant (the employee's pay being docked). On the other hand, *punishment* penalizes undesirable behavior and eliminates it. Suspending an employee for two days without pay for habitually coming to work late is an example of punishment. Finally, eliminating any reinforcement that's maintaining a behavior is called *extinction.* When a behavior isn't reinforced, it gradually disappears. In meetings, managers who wish to discourage employees from continually asking irrelevant or distracting questions can eliminate this behavior by ignoring those employees when they raise their hands to speak. Soon this behavior should disappear.

Let's Get Real: F2F

HOW I SHAPE MY EMPLOYEES' BEHAVIOR:

Understanding the employee, their role, and the key drivers and make them realize the need for change through effective mentoring, case study, education backed by change in system, and processes.

Both positive and negative reinforcement result in learning. They strengthen a desired behavior and increase the probability that the desired behavior will be repeated. Both punishment and extinction also result in learning, but they do so by weakening an undesired behavior and decreasing its frequency.

IMPLICATIONS FOR MANAGERS

Employees are going to learn on the job. Will managers manage employees' learning through the rewards they allocate and the examples they set, or will they allow it to occur haphazardly? If marginal employees are rewarded with pay raises and promotions, they will have little reason to change their behavior. In fact, productive employees who see marginal performance rewarded might change their behavior. If managers want behavior A but reward behavior B, they shouldn't be surprised to find employees' learning to engage in behavior B. Similarly, managers should expect that employees will look to them as models. Managers who are consistently late to work, or take two hours for lunch, or help themselves to company office supplies for personal use should expect employees to read the message they are sending and model their behavior accordingly.

QUICK LEARNING REVIEW:
LEARNING OUTCOME 13.5

- Explain how operant conditioning helps managers.
- Describe the implications of social learning theory.
- Discuss how managers can shape behavior.

Go to page 312 and see how well you know this material.

LEARNING **OUTCOME 13.6** ▷

CONTEMPORARY OB ISSUES

By this point, you're probably well aware of why managers need to understand how and why employees behave the way they do. We conclude this chapter by looking at two OB issues that have a major influence on managers' jobs today.

MANAGING GENERATIONAL DIFFERENCES

They're young, smart, brash. They wear flip-flops to the office and listen to iPods at their desk. They want to work but don't want work to be their life. This is Generation Y, some 70 million of them, many who are embarking on their careers, taking their place in an increasingly multigenerational workplace.[76]

Just Who Is Gen Y? There's no consensus about the exact time span that Gen Y comprises, but most definitions include individuals born from about 1982 to 1997. One thing is for sure: They're bringing new attitudes with them to the workplace. Gen Yers have grown up with an amazing array of experiences and opportunities. And they want their work life to provide that as well, as shown in Exhibit 13–8. For instance, Stella Kenyi, who is passionately interested in international development, was sent by her employer, the National Rural Electric Cooperative Association, to Yai, Sudan, to survey energy use.[77] At Best Buy's corporate offices, Beth Trippie, a senior scheduling specialist, feels that as long as the results are there, why should it matter how it gets done? She says, "I'm constantly playing video games, on a call, doing work, and the thing is, all of it gets done, and it gets done well."[78] And Katie Patterson, an assistant account executive in Atlanta, says, "We are willing and not afraid to challenge the status quo. An environment where creativity and independent thinking are looked upon as a positive is appealing to people my age. We're very independent and tech savvy."[79]

For instance, every year at IIMA, out of the batch of the graduate students, there are some who are taking risks and starting businesses in unusual fields, restaurants, schools, lingerie, juices, fresh vegetables, social development, music, etc. It is very common to meet a young programmer who is constantly playing video games, listening to music, texting to friends, e-chatting with colleagues, and doing work. The generation of qualified, metro-bred youth is able to question authority and want to rise and do something that is close to their heart rather than a conventional occupation. For example, Mr Gaurav Dagaonkar (2004 Batch) is a successful singer, music composer, and performer.

Courtesy: Gaurav Dagaonkar

shaping behavior
The process of guiding learning in graduated steps, using reinforcement or lack of reinforcement.

Exhibit 13–8

Gen Y Workers

Gen Y Workers

High Expectations of Self
They aim to work faster and better than other workers.

High Expectations of Employers
They want fair and direct managers who are highly engaged in their professional development.

Ongoing Learning
They seek out creative challenges and view colleagues as vast resources from whom to gain knowledge.

Immediate Responsibility
They want to make an important impact on Day 1.

Goal Oriented
They want small goals with tight deadlines so they can build up ownership of tasks.

Source: Bruce Tulgan of Rainmaker Thinking. Used with permission.

Dealing with the Managerial Challenges. Managing Gen Y workers presents some unique challenges. Conflicts and resentment can arise over issues such as appearance, technology, and management style.

How flexible must an organization be in terms of "appropriate" office attire? It may depend on the type of work being done and the size of the organization. In many organizations, jeans, T-shirts, and flip-flops are acceptable. However, in other settings, employees are expected to dress more conventionally. But even in those more conservative organizations, one possible solution to accommodate the more casual attire preferred by Gen Y is to be more flexible in what's acceptable. For instance, the guideline might be that when the person is not interacting with someone outside the organizations, more casual wear (with some restrictions) can be worn.

What about technology? This is a generation that has lived much of their lives with ATMs, DVDs, cell phones, e-mail, texting, laptops, and the Internet. When they don't have information they need, they just enter a few keystrokes to get it. Having grown up with technology, Gen Yers tend to be totally comfortable with it. They're quite content to meet virtually to solve problems, while bewildered baby boomers expect important problems to be solved with in-person meetings. Baby boomers complain about Gen Y's inability to focus on one task, while Gen Yers see nothing wrong with multitasking. Again, flexibility from both is the key.

Finally, what about managing Gen Yers? Like the old car advertisement that said, "this isn't your father's car," we can say that "this isn't your father's or mother's way of managing." Gen Y employees want bosses who are open minded; experts in their field, even if they aren't tech savvy; organized; teachers, trainers, and mentors; not authoritarian or paternalistic; respectful of their generation; understanding of their need for work–life balance; providing constant feedback; communicating in vivid and compelling ways; and providing stimulating and novel learning experiences.[80]

Gen Y employees have a lot to offer organizations in terms of their knowledge, passion, and abilities. Managers, however, have to recognize and understand the behaviors of this group in order to create an environment in which work can be accomplished efficiently, effectively, and without disruptive conflict.

thinking critically about **Ethics**

New scientific evidence suggests that an individual's genetics may be responsible, at least in part, for obesity, addictions, and even risky behaviors. Some people may feel liberated by the idea that their undesirable traits aren't entirely their fault. Others may feel depressed at the idea that their accomplishments may be because of genes rather than effort. If some bad behaviors are due to genetics, what role does personal responsibility play? What are the ethical implications of these findings? What impact might these findings ultimately have on organizations that want to control negative workplace behavior?

MANAGING NEGATIVE BEHAVIOR IN THE WORKPLACE

Raji notices that the saline in the patient's bottle in the next bed is low, but it is neither her patient nor her job so she does not attend to it or inform the concerned nurse. Subhash goes to see a counselor secretly because he cannot put up with the abusive language of his boss. Raghav is loud on the phone and shouts using abusive language that makes most of his coworkers uncomfortable. Rudeness, hostility, aggression, and other forms of workplace negativity have become all too common in today's organizations. There is no data available on the extent of workplace incivility in India. In a survey of U.S. employees, 10 percent said they witnessed rudeness daily within their workplaces, and 20 percent said that they personally were direct targets of incivility at work at least once a week. In a survey of Canadian workers, 25 percent reported seeing incivility daily, and 50 percent said they were the direct targets at least once per week.[81] And it's been estimated that negativity costs the U.S. economy some \$300 billion per year.[82] Only in 2007, psychological harassment has been mentioned in the Bureau of Indian Standards—Organizational Accountability at the Workplace.[83] What can managers do to manage negative behavior in the workplace?

The main thing is to recognize that it's there. Pretending that negative behavior doesn't exist or ignoring such misbehavior will only confuse employees about what is expected and acceptable behavior. Although there's some debate among researchers about the preventive or responsive actions to negative behaviors, in reality, both are needed.[84] Preventing negative behaviors by carefully screening potential employees for certain personality traits and responding immediately and decisively to unacceptable negative behaviors can go a long way toward managing negative workplace behaviors. But it's also important to pay attention to employee attitudes because negativity will show up there as well. As we said earlier, when employees are dissatisfied with their jobs, they *will* respond somehow.

QUICK LEARNING REVIEW:

LEARNING OUTCOME 13.6

- Describe the challenges managers face in managing Gen Y workers.
- Explain what managers can do to deal with workplace misbehavior.

Go to page 312 and see how well you know this material.

Let's Get Real:
My Turn

S. R. Bhat

Associate Vice President and Head, HR and Admin.
IDEB Projects Private Limited

Having decided to affect without prior communication, it is now my responsibility to handle the situation.

- *First Alternative:* This was clear sign of untimed action and an unintended response from the employees. Call on opinion leaders and communicate the need for such a decision. Take their suggestions and give confidence that I will implement workable suggestions, and also assure that there will be no such uncommunicated actions from the HR and reinstate towels with a minimal cost being shared by the employees so that the message also goes that . . . towels was not simply an entitlement.
- *Second Alternative:* Allow the noise from the employees to continue for few days; ignore them for a couple of days but apprize management effectively through proper lobbying at appropriate levels and allow this hue and cry to mellow down on its own. The moment too much of attention is given to an issue of this nature, it will encourage employees to rake up issues and stall the functioning and loose focus which will set precedence. Create an intentional diversion by recasting the project schedules so that employees become engaged, and allow the matter to settle down and mean while reinstate the towels. Simultaneously, spread of examples where credibility has been demonstrated in the organization earlier to make them feel that their present behavior is frivolous.

LEARNING OUTCOMES SUMMARY

13.1 ▷ FOCUS AND GOALS OF ORGANIZATIONAL BEHAVIOR

- Explain why the concept of an organization as an iceberg is important.
- Describe the focus and goals of OB.
- Define the six important employee behaviors.

As with an iceberg, a number of hidden organizational elements (attitudes, perceptions, norms, etc.) make understanding individual behavior challenging.

OB focuses on three areas: individual behavior, group behavior, and organizational aspects. The goals of OB are to explain, predict, and influence behavior.

Employee productivity is a performance measure of both efficiency and effectiveness. Absenteeism is the failure to report to work. Turnover is voluntary or involuntary permanent withdrawal from an organization. Organizational citizenship behavior (OCB) is discretionary behavior that's not part of an employee's formal job requirements but that promotes the effective functioning of an organization. Job satisfaction is an individual's general attitude toward his or her job. Workplace misbehavior is intentional employee behavior that is potentially harmful to the organization or individuals within the organization.

13.2 ▷ ATTITUDES AND JOB PERFORMANCE

- Describe the three components of an attitude.
- Explain the four job-related attitudes.
- Describe the impact job satisfaction has on employee behavior.
- Discuss how individuals reconcile inconsistencies between attitudes and behavior.

The cognitive component of an attitude refers to the beliefs, opinions, knowledge, or information held by a person. The affective component is the emotional or feeling part of an attitude. The behavioral component refers to an intention to behave in a certain way toward someone or something.

Job satisfaction refers to a person's general attitude toward his or her job. Job involvement is the degree to which an employee identifies with his or her job, actively participates in it, and considers his or her job performance to be important to his or her self-worth. Organizational commitment is the degree to which an employee identifies with a particular organization and its goals and wishes to maintain membership in that organization. Employee engagement means that employees are connected to, satisfied with, and enthusiastic about their jobs.

Job satisfaction positively influences productivity, reduces absenteeism levels, reduces turnover rates, promotes positive customer satisfaction, moderately promotes OCB, and helps minimize workplace misbehavior.

Individuals try to reconcile attitude and behavior inconsistencies by altering their attitudes, altering their behavior, or rationalizing the inconsistency.

13.3 ▷ PERSONALITY

- Contrast the MBTI® and the Big Five model.
- Describe five other personality traits that help explain individual behavior in organizations.
- Explain how emotions and emotional intelligence affect behavior.

The MBTI® measures four dimensions: social interaction, preference for gathering data, preference for decision making, and style of making decisions. The Big Five model consists of five personality traits: extraversion, agreeableness, conscientiousness, emotional stability, and openness to experience.

Five other personality traits that help explain individual behavior in organizations are locus of control, Machiavellianism, self-esteem, self-monitoring, and risk taking.

How a person responds emotionally and how the person deals with his or her emotions is a function of personality. A person who is emotionally intelligent has the ability to notice and to manage emotional cues and information.

13.4 ▷ PERCEPTION

- Explain how an understanding of perception can help managers better understand individual behavior.
- Describe the key elements of attribution theory.
- Discuss the fundamental attribution error and self-serving bias.
- Name three shortcuts used in judging others.

Perception is how we give meaning to our environment by organizing and interpreting sensory impressions. Managers need to understand perception because people behave according to their perceptions.

Attribution theory depends on three factors. Distinctiveness is whether an individual displays different behaviors in different situations (i.e., whether the behavior is unusual). Consensus is whether others facing a similar situation respond in the same way. Consistency is a person engaging in behaviors regularly and consistently. These three factors help managers determine whether employee behavior is attributed to external or internal causes.

The fundamental attribution error is the tendency to underestimate the influence of external factors and overestimate the influence of internal factors. The self-serving bias is the tendency to attribute our own successes to internal factors and to put the blame for personal failure on external factors.

Three shortcuts used in judging others are assumed similarity, stereotyping, and the halo effect.

13.5 ▷ LEARNING

- Explain how operant conditioning helps managers.
- Describe the implications of social learning theory.
- Discuss how managers can shape behavior.

Operant conditioning argues that behavior is a function of its consequences. Managers can use it to explain, predict, and influence behavior.

Social learning theory says that individuals learn by observing what happens to other people and by directly experiencing something.

Managers can shape behavior by using positive reinforcement (reinforcing a desired behavior by giving something pleasant), negative reinforcement (reinforcing a desired response by withdrawing something unpleasant), punishment (eliminating undesirable behavior by applying penalties), or extinction (not reinforcing a behavior to eliminate it).

13.6 ▷ CONTEMPORARY OB ISSUES

- Describe the challenges managers face in managing Gen Y workers.
- Explain what managers can do to deal with workplace misbehavior.

Gen Y workers bring new attitudes to the workplace. The main challenges of managing them have to do with issues such as appearance, technology, and management style.

Workplace misbehavior can be dealt with by recognizing that it's there, carefully screening potential employees for possible negative tendencies, and, most importantly, paying attention to employee attitudes through surveys about job satisfaction and dissatisfaction.

THINKING ABOUT MANAGEMENT ISSUES

1. Does the importance of knowledge of OB differ based on a manager's level in the organization? If so, how? If not, why not? Be specific.
2. "Instead of worrying about job satisfaction, companies should be trying to create environments where performance is enabled." What do you think this statement means? Explain. What's your reaction to this statement? Do you agree? Disagree? Why?
3. "A growing number of companies are now convinced that people's ability to understand and to manage their emotions improves their performance, their collaboration with peers, and their interaction with customers." What are the implications of this statement for managers?
4. What behavioral predictions might you make if you knew that an employee had (a) an external locus of control, (b) a low Mach score, (c) low self-esteem, or (d) high self-monitoring tendencies?

5. "Managers should never use discipline with a problem employee." Do you agree or disagree? Discuss.
6. A Gallup Organization survey shows that most workers rate having a caring boss even higher than they value money or fringe benefits. How should managers interpret this information? What are the implications?
7. Surveys indicate that during an employee's first year of employment, his or her level of satisfaction with the employer is about 69 percent. However, for employees with two to five years' experience, that employer satisfaction level falls to about 53 percent.[85] Why do you think this number drops? What, if anything, could managers do to keep the level of satisfaction high?

YOUR TURN to be a Manager

- For one week, pay close attention to how people around you behave, especially people who are close to you (roommates, siblings, significant others, coworkers, etc.). Use what you've learned about attitudes, personality, perception, and learning to understand and explain how and why they're behaving the ways they do. Write your observations and your explanations in a journal.
- Write down three attitudes you have. Identify the cognitive, affective, and behavioral components of those attitudes.
- Survey 15 employees (at your place of work or at some campus office). Be sure to obtain permission before doing this survey. Ask the employees what rude or negative behaviors they've seen at work. Compile your findings in a report and be prepared to discuss this in class. If you were the manager in this workplace, how would you handle this behavior?
- If you've never taken a personality or career compatibility test, contact your school's testing center to see if you can take one. When you get your results, evaluate what they mean for your career choice. Have you chosen a career that "fits" your personality? What are the implications?
- Complete the skill-building module Mentoring found in mymanagementlab. Your professor will tell you what to do with it.
- Have you ever heard of the "waiter rule"? A lot of businesspeople think that how you treat service workers says a lot about your character and attitudes. What do you think this means? Do you agree with this idea? Why or why not? How would you be evaluated on the "waiter rule"?
- Like it or not, each of us is continually shaping the behavior of those around us. For one week, keep track of how many times you use positive reinforcement, negative reinforcement, punishment, or extinction to shape behaviors. At the end of the week, look at your results. Which one did you tend to use most? What were you trying to do; that is, what behaviors were you trying to shape? Were your attempts successful? Evaluate. What could you have done differently if you were trying to change someone's behavior?
- Create a job satisfaction survey for a business you're familiar with.
- Now do a Web search for sample job satisfaction surveys. Find one or two samples. Write a report describing, comparing, and evaluating the examples you found and the survey you created.
- Steve's and Mary's recommended readings: Yoav Vardi and Ely Weitz, *Misbehavior in Organizations* (Lawrence Erlbaum Associates, 2004); Murray R. Barrick and A. M. Ryan (eds.), *Personality and Work* (Jossey-Bass, 2003); Daniel Goleman, *Destructive Emotions: How Can We Overcome Them?* (Bantam, 2003); L. Thomson, *Personality Type: An*

Owner's Manual (Shambhala, 1998); and Daniel Goleman, *Working with Emotional Intelligence* (Bantam, 1998).

- Survey 10 Gen Yers. Ask them three questions: (1) What do you think appropriate office attire is? (2) How comfortable are you with using technology, and what types of technology do you rely on most? (3) What do you think the "ideal" boss would be like? Compile your results into a paper that reports your data and summarizes your findings in a bulleted list format.

- In your own words, write down three things you learned in this chapter about being a good manager.

- Self-knowledge can be a powerful learning tool. Go to mymanagementlab and complete these self-assessment exercises: What's My Basic Personality? What's My Jungian 16 Type Personality? (Note that this is a miniature version of the MBTI®.) Am I a Type A? How Involved Am I in My Job? How Satisfied Am I with My Job? What's My Emotional Intelligence Score? How Committed Am I to My Organization? Using the results of your assessments, identify personal strengths and weaknesses. What will you do to reinforce your strengths and improve your weaknesses?

CASE APPLICATION

Odd Couples

A 29-year-old and a 68-year-old. How much could they possibly have in common? And what could they learn from each other? At Randstad USA's Manhattan office, such employee pairings are common. One such pair of colleagues sits inches apart, facing each other. "They hear every call the other makes. They read every e-mail the other sends or receives. Sometimes they finish each other's sentences."

Randstad, a Dutch company, has been using this pairing idea since its founding over 40 years ago. The founder's motto was "Nobody should be alone." The original intent was to boost productivity by having sales agents share one job and trade off job responsibilities. Today, these partners in the home office have an arrangement where one is in the office one week while the other one is out making sales calls, then the next week, they switch. The company brought its partner arrangement to the United States in the late 1990s, but it wasn't until 2005 that the company began recruiting new employees, the vast majority of whom were in their 20s. "Knowing that these Gen Yers need lots of attention in the workplace, Randstad executives figured that if they shared a job with someone whose own success depended on theirs, they were certain to get all the nurturing they required."

Randstad, a temporary employment agency.

Randstad doesn't simply pair up people and hope it works. It looks for people who will work well with others by conducting extensive interviews and requiring job applicants to shadow a sales agent for half a day. "One question Randstad asks is: What's your most memorable moment while being on a team? If they respond: When I scored the winning touchdown, that's a deal killer. Everything about our organization is based on the team and group." When a new hire is paired with an experienced agent, both individuals have some adjusting. One of the most interesting elements of Randstad's program is that neither person is "the boss." And both are expected to teach each other.

Discussion Questions

1. What do you think about Randstad's pairing-up idea? Would you be comfortable with such an arrangement? Why or why not?
2. What personality traits would be most needed for this type of work arrangement? Why?
3. What types of issues might a Gen Y employee and an older, more-experienced employee face when working closely together? How could two people in such a close-knit work arrangement deal with those issues?
4. Design an employee attitude survey for Randstad's employees.

Source: S. Berfield, "Bridging the Generation Gap," *BusinessWeek*, September 17, 2007, pp. 60–61.

Let's Get Real:

Meet The Manager

Subrata Beura

Sr Manager, Business Planning and Strategy
Hewlett Packard

MY JOB: To provide business planning formulation and support the business unit on the areas of labor, overheads and project, and manage key projects of the business globally.

BEST PART OF MY JOB: Gives me an understanding of global business.

WORST PART OF MY JOB: There is none that I can think of now.

BEST MANAGEMENT ADVICE EVER RECEIVED: Never sit on a problem.

You'll be hearing more from this real manager throughout the chapter.

Chapter 14

Managers and Communication

Without communication, nothing would ever get done in organizations. Managers are concerned with two types of communication: interpersonal and organizational. We look at both in this chapter and the role they play in a manager's ability to be efficient and effective. Focus on the following learning outcomes as you read and study this chapter.

LEARNING OUTCOMES

A Manager's Dilemma

Courtesy of U.S. Cellular

During 2008, each corporate user of e-mail sent or received more than 150 messages per day.[1] Estimates suggest that by 2011, that number will be well over 225. Another study found that one-third of e-mail users feel stressed by heavy e-mail volume. Once imagined to be a time-saver, has the inbox become a burden? U.S. Cellular's CEO Jay Ellison thought so and did something about it. He imposed a "no e-mail Friday" rule, a move that a growing number of companies are taking. Although most bans typically allow e-mailing clients and customers or responding to urgent matters, the intent is to slow down the routine internal e-mails that take up time and clog the system. The limits also aim to encourage more face-to-face and phone contact with coworkers and customers. Ellison also hoped that the ban would give his employees a small respite from the e-mail onslaught. What he got, however, was a rebellion. One employee confronted him, saying that Ellison didn't understand how much work had to get done and how much easier it was when using e-mail. Put yourself in Ellison's position. Should the ban stay or go?

What would you do?

Jay Ellison understands both the importance and the drawbacks of communication. Communication between managers and employees is important because it provides the information necessary to get work done in organizations. Thus, there's no doubt that communication is fundamentally linked to managerial performance.[2]

LEARNING
OUTCOME 14.1 ▷

THE NATURE AND FUNCTION OF COMMUNICATION

Unlike the character Bill Murray plays in *Groundhog Day*, Neal L. Patterson, CEO of Cerner Corporation, a health care software development company based in Kansas City, probably wishes he *could* repeat one day. Upset with the fact that employees didn't seem to be putting in enough hours, he sent an angry and emotional e-mail to about 400 company managers that said, in part:

> We are getting less than 40 hours of work from a large number of our K.C.-based EMPLOYEES. The parking lot is sparsely used at 8 A.M.; likewise at 5 P.M. As managers, you either do not know what your EMPLOYEES are doing, or you do not CARE. You have created expectations on the work effort which allowed this to happen inside Cerner, creating a very unhealthy environment. In either case, you have a problem and you will fix it or I will replace you . . . I will hold you accountable. You have allowed things to get to this state. You have two weeks. Tick, tock.[3]

Although the e-mail was meant only for the company's managers, it was leaked and posted on an Internet discussion site. The tone of the e-mail surprised industry analysts, investors, and, of course, Cerner's managers and employees. The company's stock price dropped 22 percent over the next three days. Patterson apologized to his employees and acknowledged, "I lit a match and started a firestorm." This is a good example of why it's important for managers to understand the impact of communication.

The importance of effective communication for managers can't be overemphasized for one specific reason: Everything a manager does involves communicating. Not *some* things, but everything! A manager can't make a decision without information. That information has to be communicated. Once a decision is made, communication must again take place. Otherwise, no one would know that a decision was made. The best idea, the most creative suggestion, the best plan, or the most effective job redesign can't take shape without communication.

WHAT IS COMMUNICATION?

Communication is the transfer and understanding of meaning. Note the emphasis on the *transfer* of meaning; this means that if information or ideas have not been conveyed, communication hasn't taken place. The speaker who isn't heard or the writer whose materials aren't read hasn't communicated. More importantly, however, communication involves the *understanding* of meaning. For communication to be successful, the meaning must be imparted and understood. A letter written in Spanish addressed to a person who doesn't read Spanish can't be considered communication until it's translated into a language the person does read and understand. Perfect communication, if such a thing existed, would occur if a transmitted thought or idea was received and understood by the receiver exactly as it was envisioned by the sender.

Another point to keep in mind is that *good* communication is often erroneously defined by the communicator as *agreement* with the message instead of clear understanding of the message.[4] If someone disagrees with us, we assume that the person just didn't fully understand our position. In other words, many of us define good communication as having someone accept our views. But you can clearly understand what someone means and just *not* agree with what the person says.

The final point we want to make about communication is that it encompasses both **interpersonal communication**—communication between two or more people—and **organizational communication**—all the patterns, networks, and systems of communication within an organization. Both types are important to managers.

FUNCTIONS OF COMMUNICATION

J. W. (Bill) Marriott, chairman and CEO of Marriott International, was awarded the 2008 Excellence in Communication Leadership (EXCEL) Award by the International Association of Business Communicators. According to the chair, "The Award Committee

Good communication is characterized by the understanding of the sender's meaning not necessarily by agreement between the parties. From checking in passengers on a particular flight, having all of them pass through security check, board the bus, and then the plane, the pilot communicating with the air-traffic controller, the ground staff, airport personnel, aircraft staff, and the air-traffic controller have to be in constant touch with each other no matter where they are.

communication
The transfer and understanding of meaning.

interpersonal communication
Communication between two or more people.

organizational communication
All the patterns, networks, and systems of communication within an organization.

was very impressed with Bill Marriott and his personal commitment to communications in his namesake company. Marriott International has a respected, well-documented communications program, and Bill demonstrates his personal commitment every day, from traditional employee engagement while inspecting 250 Marriott properties each year, to adopting new communications tools such as blogging."[5]

Throughout Marriott and many other organizations, communication serves four major functions: control, motivation, emotional expression, and information.[6] Each function is equally important.

Communication acts to *control* employee behavior in several ways. As we know from Chapter 9, organizations have authority hierarchies and formal guidelines that employees are expected to follow. For instance, when employees are required to communicate any job-related grievance to their immediate manager, to follow their job description, or to comply with company policies, communication is being used to control. Informal communication also controls behavior. When a work group teases a member who's ignoring the norms by working too hard, those individuals are informally controlling the member's behavior.

Next, communication *motivates* by clarifying to employees what is to be done, how well they're doing, and what can be done to improve performance if it's not up to par. As employees set specific goals, work toward those goals, and receive feedback on progress toward goals, communication is required.

For many employees, their work group is a primary source of social interaction. The communication that takes place within the group is a fundamental mechanism by which members share frustrations and feelings of satisfaction. Communication, therefore, provides a release for *emotional expression* of feelings and for fulfillment of social needs.

Finally, individuals and groups need information to get things done in organizations. Communication provides that *information.*

QUICK LEARNING REVIEW:

LEARNING OUTCOME 14.1

- Define communication, interpersonal communication, and organizational communication.
- Discuss the functions of communication.

Go to page 338 and see how well you know this material.

LEARNING OUTCOME 14.2 ▷

METHODS OF INTERPERSONAL COMMUNICATION

Before communication can take place, a purpose, expressed as a **message** to be conveyed, must exist. It passes between a source (the sender) and a receiver. The message is converted to symbolic form (called **encoding**) and passed by way of some medium (**channel**) to the receiver, who retranslates the sender's message (called **decoding**). The result is the transfer of meaning from one person to another.[7]

Exhibit 14–1 illustrates the seven elements of the **interpersonal communication process**. Note that the entire process is susceptible to **noise**—disturbances that interfere with the transmission, receipt, or feedback of a message. Typical examples of noise include illegible print, phone static, inattention by the receiver, and background sounds of machinery or coworkers. However, anything that interferes with understanding can be noise, and noise can create distortion at any point in the communication process.

Exhibit 14–1

The Interpersonal Communication Process

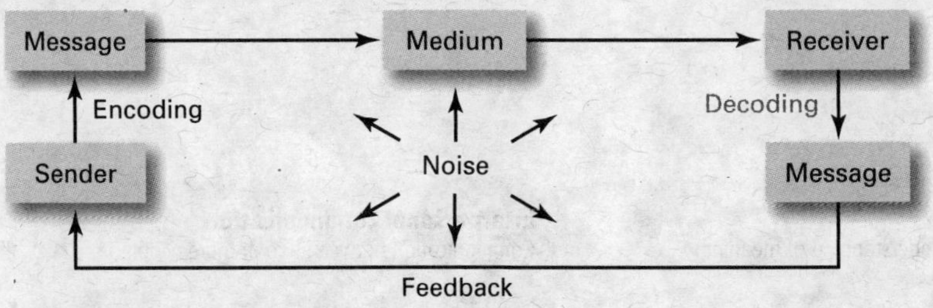

METHODS OF COMMUNICATING INTERPERSONALLY

Say that you need to communicate to your employees the organization's new policy on sexual harassment; or you want to compliment one of your workers on the extra hours she's put in to help your team complete a customer's order; or you must tell one of your employees about changes to his job; or you would like to get employees' feedback on your proposed budget for next year. In each of these instances, how would you communicate? Managers have a wide variety of communication methods from which to choose and can use 12 questions to help evaluate these methods:[8]

1. ***Feedback***—How quickly can the receiver respond to the message?
2. ***Complexity capacity***—Can the method effectively process complex messages?
3. ***Breadth potential***—How many different messages can be transmitted using this method?
4. ***Confidentiality***—Can communicators be reasonably sure their messages are received only by the intended audience?
5. ***Encoding ease***—Can senders easily and quickly use this channel?
6. ***Decoding ease***—Can receivers easily and quickly decode messages?
7. ***Time–space constraint***—Do senders and receivers need to communicate at the same time and in the same space?
8. ***Cost***—How much does it cost to use this method?
9. ***Interpersonal warmth***—How well does this method convey interpersonal warmth?
10. ***Formality***—Does this method have the needed amount of formality?
11. ***Scanability***—Does this method allow the message to be easily browsed or scanned for relevant information?
12. ***Time of consumption***—Does the sender or receiver exercise the most control over when the message is dealt with?

Exhibit 14–2 provides a comparison of various communication methods. Which method a manager ultimately chooses should reflect the needs of the sender, the attributes of the message, the attributes of the channel, and the needs of the receiver. For instance, if you need to communicate to an employee the changes being made in her job, face-to-face communication would be a better choice than a memo because you want to be able to address immediately any questions and concerns that she might have.

An important part of interpersonal communication is **nonverbal communication**—that is, communication transmitted without words. Some of the most meaningful communications are neither spoken nor written. When a college instructor is teaching a class, she doesn't need words to tell her that students are tuned out when they begin to read a newspaper in the middle of class. Similarly, when students start putting their book, papers, and notebooks away, the message is clear: Class time is about over. The size of a person's office or the clothes he or she wears also convey messages to others. Although these are all forms of nonverbal communication, the best-known types are body language and verbal intonation.

Body language refers to gestures, facial expressions, and other body movements that convey meaning. A person frowning "says" something different than a person smiling. Hand motions, facial expressions, and other gestures can communicate emotions or temperaments such as aggression, fear, shyness, arrogance, joy, and anger.

message
A purpose to be conveyed.

encoding
Converting a message into symbols.

channel
The medium along which a message travels.

decoding
Retranslating a sender's message.

interpersonal communication process
The seven elements involved in transferring meaning from one person to another.

noise
Any disturbances that interfere with the transmission, receipt, or feedback of a message.

nonverbal communication
Communication transmitted without words.

body language
Gestures, facial configurations, and other body movements that convey meaning.

Exhibit 14–2 Comparison of Communication Methods

	Criteria											
Channel	**Feedback Potential**	**Complexity Capacity**	**Breadth Potential**	**Confidentiality**	**Encoding Ease**	**Time-Decoding Ease**	**Space Constraint**	**Cost**	**Personal Warmth**	**Formality**	**Scannability**	**Consumption Time**
Face-to-face	1	1	1	1	1	1	1	2	1	4	4	S/R
Telephone	1	4	2	2	1	1	3	3	2	4	4	S/R
Group meetings	2	2	2	4	2	2	1	1	2	3	4	S/R
Formal presentations	4	2	2	4	3	2	1	1	3	3	5	Sender
Memos	4	4	2	3	4	3	5	3	5	2	1	Receiver
Postal mail	5	3	3	2	4	3	5	3	4	1	1	Receiver
Fax	3	4	2	4	3	3	5	3	3	3	1	Receiver
Publications	5	4	2	5	5	3	5	2	4	1	1	Receiver
Bulletin boards	4	5	1	5	3	2	2	4	5	3	1	Receiver
Audio/videotapes	4	4	3	5	4	2	3	2	3	3	5	Receiver
Hotlines	2	5	2	2	3	1	4	2	3	3	4	Receiver
E-mail	3	4	1	2	3	2	4	2	4	3	4	Receiver
Computer conference	1	2	2	4	3	2	3	2	3	3	4	S/R
Voice mail	2	4	2	1	2	1	5	3	2	4	4	Receiver
Teleconference	2	3	2	5	2	2	2	2	3	3	5	S/R
Videoconference	3	3	2	4	2	2	2	1	2	3	5	S/R

Note: Ratings are on a 1–5 scale where 1 = high and 5 = low. Consumption time refers to who controls the reception of communication. S/R means the sender and receiver share control.

Source: P. G. Clampitt, *Communicating for Managerial Effectiveness* (Newbury Park, CA: Sage Publications, 1991), p. 136.

Knowing the meaning behind someone's body movements and learning how to put forth your best body language can help you personally and professionally.[9]

Verbal intonation refers to the emphasis someone gives to words or phrases that conveys meaning. To illustrate how intonations can change the meaning of a message, suppose that a student asks an instructor a question. The instructor replies, "What do you mean by that?" The student's reaction will vary, depending on the tone of the instructor's response. A soft, smooth vocal tone conveys interest and creates a different meaning from one that is abrasive and puts a strong emphasis on saying the last word. Most of us would view the first intonation as coming from someone sincerely interested in clarifying the student's concern, whereas the second suggests that the person is defensive or aggressive.

Managers need to remember that as they communicate, the nonverbal component usually carries the greatest impact. It's not *what* you say but *how* you say it.

QUICK LEARNING REVIEW:

LEARNING OUTCOME 14.2

- Explain the components in the communication process.
- Discuss the criteria that managers can use to evaluate the various communication methods.
- List the communication methods managers can use.

Go to page 338 and see how well you know this material.

LEARNING **OUTCOME 14.3** ▷

EFFECTIVE INTERPERSONAL COMMUNICATION

Managers face barriers that can distort the interpersonal communication process. Let's look at these barriers to effective communication.

BARRIERS TO COMMUNICATION

Filtering. Filtering is the deliberate manipulation of information to make it appear more favorable to the receiver. For example, when a person tells his or her manager what the manager wants to hear, information is being filtered. Or if information being communicated up through organizational levels is condensed by senders, that's filtering.

Sahil Parikh, the creator of DeskAway, an online software to help virtual teams collaborate and track people, projects, and processes, started with the vision to provide a simple, powerful, and affordable collaborative online service to virtual teams in small businesses or teams.[10] It has been highly recommended by reviewers and users.[11]

Courtesy: Deskaway/Synage

verbal intonation
An emphasis given to words or phrases that conveys meaning.

filtering
The deliberate manipulation of information to make it appear more favorable to the receiver.

How much filtering takes place tends to be a function of the number of hierarchical levels in the organization and the organizational culture; more levels mean more opportunities for filtering. As organizations use more collaborative, cooperative work arrangements, information filtering may become less of a problem. In addition, e-mail reduces filtering because communication is more direct. Finally, an organization's culture encourages or discourages filtering based on the type of behavior it rewards. If organizational rewards emphasize style and appearance, managers may be motivated to filter communications in their favor.

Emotions. How a receiver feels when he or she receives a message influences how he or she interprets it. Extreme emotions are most likely to hinder effective communication. In such instances, we often disregard our rational and objective thinking processes and substitute emotional judgments.

Information Overload. A marketing manager goes on a week-long sales trip to Spain, where he doesn't have access to his e-mail, and he faces 1,000 messages on his return. It's not possible to fully read and respond to each message without facing **information overload**, which is when information exceeds a person's processing

managing workforce Diversity

The Communication Styles of Men and Women

"You don't understand what I'm saying, and you never listen!" "You're making a big deal out of nothing." Have you said statements like these to friends of the opposite sex? Most of us probably have! Research shows, as does personal experience, that men and women communicate differently.[12]

Deborah Tannen has studied the ways that men and women communicate and reports some interesting differences. The essence of her research is that men use talk to emphasize status, while women use it to create connection. She states that communication between the sexes can be a continual balancing act of juggling our conflicting needs for intimacy, which suggests closeness and commonality, and independence, which emphasizes separateness and differences. It's no wonder, then, that communication problems arise! Women speak and hear a language of connection and intimacy. Men hear and speak a language of status and independence. For many men, conversations are merely a way to preserve independence and maintain status in a hierarchical social order. Yet for many women, conversations are negotiations for closeness and are used to seek out support and confirmation. Let's look at a few examples of what Tannen described.

Men frequently complain that women talk on and on about their problems. Women, however, criticize men for not listening. What's happening is that when a man hears a woman talking about a problem, he frequently asserts his desire for independence and control by offering solutions. Many women, in contrast, view conversing about a problem as a way to promote closeness. The woman talks about a problem to gain support and connection, not to get the male's advice.

Here's another example: Men are often more direct than women in conversation. A man might say, "I think you're wrong on that point." A woman might say, "Have you looked at the marketing department's research report on that issue?" The implication in the woman's comment is that the report will point out the error. Men frequently misread women's indirectness as "covert" or "sneaky," but women aren't as concerned as men with the status and one-upmanship that directness often creates.

Finally, men often criticize women for seeming to apologize all the time. Men tend to see the phrase "I'm sorry" as a sign of weakness because they interpret the phrase to mean the woman is accepting blame, when he may know she's not to blame. The woman also knows she's not at fault. Yet she's typically using "I'm sorry" to express regret: "I know you must feel bad about this, and I do, too."

How can these differences in communication styles be managed? Keeping gender differences from becoming persistent barriers to effective communication requires acceptance, understanding, and a commitment to communicate adaptively with each other. Both men and women need to acknowledge that there are differences in communication styles, that one style isn't better than the other, and that it takes real effort to "talk" with each other successfully.

capacity. Today's employees frequently complain of information overload. Statistics show that the average business e-mail user devotes 107 minutes per day to e-mail—about 25 percent of the workday. Other statistics show that employees send and receive an average of 150 e-mail messages every day. And the number of worldwide e-mail messages sent daily is a staggering 97.3 billion.[13] The demands of keeping up with e-mail, text messages, phone calls, faxes, meetings, and professional reading create an onslaught of data. What happens when individuals have more information than they can process? They tend to ignore, pass over, forget, or selectively choose information. Or, they may stop communicating. In any case, the result is lost information and ineffective communication.

Defensiveness. When people feel that they're being threatened, they tend to react in ways that hinder effective communication and reduce their ability to achieve mutual understanding. They become defensive—verbally attacking others, making sarcastic remarks, being overly judgmental, or questioning others' motives.[14]

Language. Authors like Ruskin Bond, Khushwant Singh and young anchors on MTV speak English, but the language they use is different. Words mean different things to different people. Age, education, and cultural background are three of the most obvious variables that influence the language a person uses and the definitions he or she gives to words.

Let's Get Real: F2F

MOST FREQUENT COMMUNICATION PROBLEM:

To manage nonverbal communication and cues on the phone.

In an organization, employees come from diverse backgrounds and have different patterns of speech. Even employees who work for the same organization but in different departments often have different **jargon**—specialized terminology or technical language that members of the group use to communicate among themselves.

National Culture. For technological and cultural reasons, Chinese people dislike voice mail.[15] This illustrates how communication differences can arise from national culture as well as different languages. For example, let's compare countries that value individualism (such as the United States) with countries that emphasize collectivism (such as India).[16]

In an individualistic country such as the United States, communication is relatively formal and is clearly spelled out. Managers rely heavily on reports, memos, and other formal forms of communication. In a collectivist country such as India, there's more interpersonal contact, and face-to-face communication is encouraged. An Indian manager extensively consults with colleagues and boss about an issue and then draws up a formal document to outline the agreement that was made.

OVERCOMING BARRIERS TO COMMUNICATION

On average, an individual must hear new information seven times before he or she truly understands.[17] In light of this fact and the communication barriers just described, what can managers do to be more effective communicators?

Using Feedback. Many communication problems can be directly attributed to misunderstanding and inaccuracies. These problems are less likely to occur if the manager gets feedback, both verbal and nonverbal.

A manager can ask questions about a message to determine whether it was received and understood as intended. Or the manager can ask the receiver to restate the message

information overload
A situation in which information exceeds a person's processing capacity.

jargon
Specialized terminology or technical language that members of a group use to communicate among themselves.

in his or her own words. If the manager hears what was intended, understanding and accuracy should improve. Feedback can also be more subtle; general comments can give a manager a sense of the receiver's reaction to a message.

Feedback doesn't have to be verbal. Say that a sales manager e-mails information about a new monthly sales report that all sales representatives will need to complete, and some of them don't turn it in; the sales manager has received feedback, which suggests that the sales manager needs to clarify the initial communication. Similarly, managers can look for nonverbal cues to tell whether someone's getting the message.

Simplifying Language. Because language can be a barrier, managers should consider the audience to whom the message is directed and tailor the language to those individuals. Remember that effective communication is achieved when a message is both received and *understood.* This means, for example, that a hospital administrator should always try to communicate in clear, easily understood terms and to use language tailored to different employee groups. Messages to the surgical staff should be purposefully different from those used with office employees. Jargon can facilitate understanding if it's used within a group that knows what it means, but it can cause problems when used with people who don't understand it.

Listening Actively. When someone talks, we hear. But often we don't listen. Listening is an active search for meaning, whereas hearing is passive. In listening, the receiver is also putting effort into the communication.

Many of us are poor listeners. Why? Because it's difficult, and most of us would rather do the talking. Listening, in fact, is often more tiring than talking. Unlike hearing, **active listening**, which is listening for full meaning without making premature judgments or interpretations, demands total concentration. The average person normally speaks at a rate of about 125 to 200 words per minute. However, the average listener can comprehend up to 400 words per minute.[18] The difference leaves lots of idle brain time and opportunities for the mind to wander.

Active listening is enhanced by developing empathy with the sender—that is, by putting yourself in the sender's position. Because senders differ in attitudes, interests, needs, and expectations, empathy makes it easier to understand the actual content of a message. An empathetic listener reserves judgment on the message's content and carefully listens to what is being said. The goal is to improve one's ability to get the full meaning of a communication without distorting it through premature judgments or interpretations. Other specific behaviors that active listeners demonstrate are listed in Exhibit 14–3.

Exhibit 14–3

Active Listening Behaviors

Source: Based on P. L. Hunsaker, *Training in Management Skills* (Upper Saddle River, NJ: Prentice Hall, 2001).

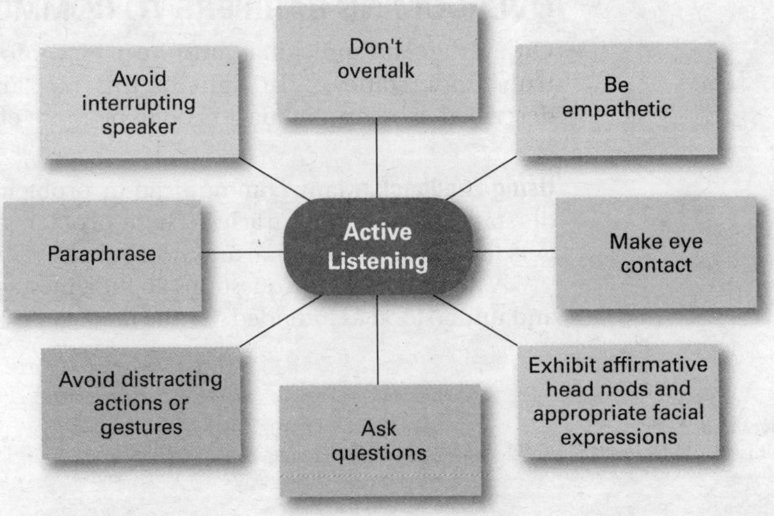

Constraining Emotions. It would be naïve to assume that managers always communicate in a rational manner. We know that emotions can cloud and distort communication. A manager who's upset about an issue is more likely to misconstrue incoming messages and fail to communicate his or her outgoing messages clearly and accurately. What to do? The simplest answer is to calm down and get emotions under control before communicating.

Watching Nonverbal Cues. Actions speak louder than words, so it's important to make sure your actions align with and reinforce the words that go along with them. An effective communicator watches his or her nonverbal cues to ensure that they convey the desired message.

QUICK LEARNING REVIEW:

LEARNING OUTCOME 14.3

- Explain the barriers to effective interpersonal communication.
- Discuss ways to overcome the barriers to effective interpersonal communication.

Go to page 338 and see how well you know this material.

LEARNING OUTCOME 14.4 ▷

ORGANIZATIONAL COMMUNICATION

An understanding of managerial communication isn't possible without looking at organizational communication. In this section, we look at several important aspects of organizational communication, including formal versus informal communication, the flow patterns of communication, and formal and informal communication networks.

FORMAL VERSUS INFORMAL COMMUNICATION

Communication within an organization is described as formal or informal. **Formal communication** refers to communication that takes place within prescribed organizational work arrangements. For example, when a manager asks an employee to complete a task, that's formal communication, as is an employee communicating a problem to his or her manager.

Informal communication is organizational communication not defined by the organization's structural hierarchy. When employees talk with each other in the lunch room, as they pass in hallways, or as they're exercising at the company wellness facility, that's informal communication. Employees form friendships and communicate with each other. The informal communication system fulfills two purposes in organizations: (1) It permits employees to satisfy their need for social interaction, and (2) it can improve an organization's performance by creating alternative, and frequently faster and more efficient, channels of communication.

DIRECTION OF COMMUNICATION FLOW

Let's look at the ways that organizational communication can flow: downward, upward, laterally, or diagonally.

Downward Communication. Every morning and often several times a day, managers at UPS package delivery facilities gather workers for mandatory meetings that last precisely 3 minutes each. During those 180 seconds, managers relay company announcements and go over local information such as traffic conditions or customer complaints. Then, each meeting ends with a safety tip. The 3-minute meetings have proved so

active listening
Listening for full meaning without making premature judgments or interpretations.

formal communication
Communication that takes place within prescribed organizational work arrangements.

informal communication
Communication that is not defined by an organization's structural hierarchy.

successful that many of the company's office workers are using the idea.[19] This is **downward communication**, which is any communication that flows from a manager to employees. It's used to inform, direct, coordinate, and evaluate employees. When managers assign goals to their employees, they're using downward communication. They're also using downward communication when they provide employees with job descriptions, inform them of organizational policies and procedures, point out problems that need attention, or evaluate employees' performance. Downward communication can take place through any of the communication methods described earlier.

Upward Communication. Managers rely on their employees for information. For instance, reports are given to managers to inform them of progress toward goals or to report problems. **Upward communication** is communication that flows from employees to managers. It keeps managers aware of how employees feel about their jobs, their coworkers, and the organization in general. Managers also rely on upward communication for ideas on how things can be improved. Some examples of upward communication include performance reports prepared by employees, employee messages in suggestion boxes, employee attitude surveys, grievance procedures, manager–employee discussions, and informal group sessions in which employees have the opportunity to discuss problems with their manager or representatives of top-level management.

How much upward communication is used depends on the organizational culture. If managers have created a climate of trust and respect and use participative decision making or empowerment, there will be considerable upward communication as employees provide input to decisions. In a more highly structured and authoritarian environment, upward communication still takes place, but it is limited.

Lateral Communication. Communication that takes place among employees on the same organizational level is called **lateral communication**. In today's dynamic environment, horizontal communications are frequently needed to save time and facilitate coordination. Cross-functional teams, for instance, rely heavily on this form of communication interaction. However, conflicts can arise if employees don't keep their managers informed about decisions they've made or actions they've taken.

Diagonal Communication. Diagonal Communication is communication that crosses both work areas and organizational levels. When a credit analyst communicates directly with a regional marketing manager about a customer's problem—note the different department and different organizational level—that's diagonal communication. Because of its efficiency and speed, diagonal communication can be beneficial. Increased e-mail use facilitates diagonal communication. In many organizations, any employee can communicate by e-mail with any other employee, regardless of organizational work area or level, even with upper-level managers. In many organizations, CEOs have adopted an "open inbox" e-mail policy. For example, William H. Swanson, head of defense contractor Raytheon Company, figures that he has received and answered more than 150,000 employee e-mails. And Henry McKinnell, Jr., former CEO of Pfizer, says that the approximately 75 internal e-mails he received every day were "an avenue of communication I didn't otherwise have."[20] When Basix, a social enterprise, started operations in Hyderabad, it maintained a policy that all e-mails received by Basix would be in an inbox for all members of the top management to see and react to. Pramod Bhasin, President and CEO of Gurgaon-based BPO firm Genpact Ltd, sent out a request to all of his 36,000 employees for ideas on cost cutting. Bhasin received 150 e-mails regarding suggestions within a day.[21] However, diagonal communication also has the potential to create problems if employees don't keep their managers informed.

ORGANIZATIONAL COMMUNICATION NETWORKS

The vertical and horizontal flows of organizational communication can be combined into a variety of patterns called **communication networks**. Exhibit 14–4 illustrates three common communication networks.

Exhibit 14–4

Organizational Communication Networks

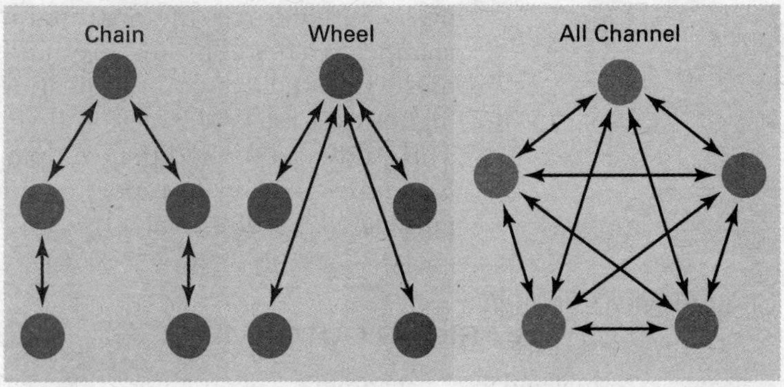

Criteria	Chain	Wheel	All Channel
Speed	Moderate	Fast	Fast
Accuracy	High	High	Moderate
Emergence of leader	Moderate	High	None
Member satisfaction	Moderate	Low	High

Types of Communication Networks. In a *chain* network, communication flows according to the formal chain of command, both downward and upward. A *wheel* network represents communication flowing between a clearly identifiable and strong leader and others in a work group or team. The leader serves as the hub through whom all communication passes. Finally, in an *all-channel* network, communication flows freely among all members of a work team.

Which type of network you should use depends on your goal. Exhibit 14–4 also summarizes each network's effectiveness according to four criteria: speed, accuracy, the probability that a leader will emerge, and the importance of member satisfaction. One observation is immediately apparent: No single network is best for all situations.

Let's Get Real: F2F

MY ADVICE FOR MANAGING THE GRAPEVINE:

Within the team communicate preemptively and sensitize team members on issues affecting the business.

The Grapevine. We can't leave our discussion of communication networks without discussing the **grapevine**—the informal organizational communication network. The grapevine is active in almost every organization. Is it an important source of information? You bet! One survey reported that 63 percent of employees say that they hear about important matters first through rumors or gossip on the grapevine.[22]

Certainly, the grapevine is an important part of any communication network and well worth understanding.[23] Acting as both a filter and a feedback mechanism, it pinpoints those bewildering issues that employees consider important. More importantly, from a managerial point of view, it *is* possible to analyze what is happening on the grapevine—what information is being passed, how information seems to flow, and what individuals seem to be key information conduits. By staying aware of the grapevine's flow and patterns, managers can identify issues that concern employees, and, in turn, use the grapevine to disseminate important information. Because the grapevine can't be eliminated, managers should "manage" it as an important information network.

Rumors that flow along the grapevine also can never be eliminated entirely. However, managers can minimize the negative consequences of rumors. How? By communicating openly, fully, and honestly with employees, particularly in situations in which employees may not like proposed or actual managerial decisions. Open and honest communication has positive benefits for an organization. A study by Watson Wyatt Worldwide concluded that effective communication "connects employees to the

downward communication
Communication that flows downward from a manager to employees.

upward communication
Communication that flows upward from employees to managers.

lateral communication
Communication that takes place among any employees on the same organizational level.

diagonal communication
Communication that cuts across work areas and organizational levels.

communication networks
The variety of patterns of vertical and horizontal flows of organizational communication.

grapevine
The informal organizational communication network.

business, reinforces the organization's vision, fosters process improvement, facilitates change, and drives business results by changing employee behavior." For those companies with effective communication, total returns to shareholders were 91 percent higher over a five-year period than for companies with less effective communication. This study also showed that companies that were highly effective communicators were four times as likely to report high levels of employee engagement as firms that communicated less effectively.[24]

QUICK LEARNING REVIEW:

LEARNING OUTCOME 14.4

- Contrast formal and informal communication.
- Explain communication flow in an organization.
- Describe the three common communication networks.
- Discuss how managers should handle the grapevine.

Go to page 338 and see how well you know this material.

LEARNING **OUTCOME 14.5** ▷

INFORMATION TECHNOLOGY AND COMMUNICATION

Technology is changing the way we live and work. Consider the following four examples: Japanese employees, managers, housewives, and teens use wireless interactive Web phones to send e-mail, surf the Web, swap photos, and play computer games. At DreamWorks Animation, a sophisticated videoconferencing system allows animators in three different locations to collaboratively edit films. Almost all sales personnel at Airtel use only cell phones and no landlines to do their work. A recent survey of employees showed that 93 percent of those polled use the Internet at work.[25]

The world of communication isn't what it used to be! Although changing technology has been a significant source of the environmental uncertainty facing organizations, these same technological changes have enabled managers to coordinate employees' work efforts in more efficient and effective ways. Information technology (IT) now touches every aspect of almost every company's business. The implications for the ways managers communicate are profound.

HOW TECHNOLOGY AFFECTS MANAGERIAL COMMUNICATION

IT has radically changed the way organizational members communicate. For example, it has significantly improved a manager's ability to monitor individual and team performance, has allowed employees to have more complete information to make faster decisions, and has provided employees more opportunities to collaborate and share information. In addition, IT has made it possible for people in organizations to be fully accessible, any time, regardless of where they are. Employees don't have to be at their desk with their computer running to communicate with others in the organization. Two IT developments that are especially significant for managerial communication are networked systems and wireless capabilities.

Networked Systems. In a networked system, an organization's computers are linked. Organizational members can communicate with each other and tap into information whether they're down the hall, across town, or halfway around the world. In this chapter, we're not looking at the mechanics of how a network system works but at its communication applications, including e-mail, instant messaging, blogs and wikis, voice mail, fax, teleconferencing and videoconferencing, and intranets.

Wireless Capabilities. At Airtel, even vendors are given cell phones so that it is possible for the project manager to keep in touch with the vendor or servicc provider to do their work better. The MERU taxi provides all drivers with a cell phone so that they can be reached with no difficulty and redirected to a client in the shortest possible time. Leonia

thinking critically about **Ethics**

According to a survey by Harris Interactive, 69 percent of employees spend time at non-work-related Web sites, and 55 percent send and receive personal e-mails at work. And younger employees are even more likely to use employer-furnished computers for personal reasons: 77 percent of workers aged 18 to 34 access the Internet at work for personal reasons, and 72 percent check their personal e-mail accounts at work. A survey by Salary.com and AOL found that personal Web surfing was the top method of goofing off at work. In addition, funny stories, jokes, and pictures make their way from one employee's e-mail inbox to another, to another, and so forth. An elf bowling game sent by e-mail was a favorite diversion during the holiday season.

Although these may seem like fun and harmless activities, it's estimated that such technological distractions cost businesses more than $54 billion annually. There's a high dollar cost associated with using the Internet at work for other than business reasons, but is there a psychological benefit to be gained by letting employees do something to relieve the stress of pressure-packed jobs? What ethical issues are associated with widely available Internet access at work for both employees and organizations?[26]

Holistic Resorts in Hyderabad invested $1.2 million in a Unified Communication (UC) solution to enable complete connectivity within the resort spread over 350 acres. The new system is designed to satisfy its guests' demands and expectations and assist employees in doing it efficiently. With the new system, the location of the guests and employees can be pinpointed with ease.[27] As this example shows, wireless communication technology has the ability to improve work for managers and employees. Even Internet access is available through Wi-Fi and WiMAX hot spots, whiμch are locations where users gain wireless access. The number of hot spot locations continues to grow. Surveys show that airports are the number-one spot where people use public Wi-Fi, hotels are second, and coffee shops/cafés are third. Also, London leads the way in total number of Wi-Fi sessions, but Amsterdam was the first city in Europe with a mobile WiMAX network.[28] Padmasree Warrior, CTO, Cisco, says, "there are four babies born every minute but 30 phones sold in the same time."[29] Because there are more than 50 million "mobile" workers in the United States, smartphones, notebook computers, and other pocket communication devices have spawned a whole new way for managers to keep in touch. And the number of mobile communication users keeps increasing.[30] Employees don't have to be at their desks to communicate with others in the organization. As wireless technology continues to improve, we'll see more organizational members using it as a way to collaborate and share information.

HOW INFORMATION TECHNOLOGY AFFECTS ORGANIZATIONS

Monsanto Company wanted to raise the visibility of some projects and to make a stronger argument for bioengineered crops. Using a YouTube approach, the company sent camera crews to the Philippines, Australia, and other countries to film testimonials from farmers using Monsanto products to grow bioengineered crops. The clips were posted on a company Web site, which now attracts more than 15,000 visitors per month. The PR manager in charge of the project said, "When the people involved relate how their life has changed and you actually see it, it's more compelling."[31] That's the power of IT at work. Employees—working in teams or as individuals—need information to make decisions and to do their work. It's clear that technology *can* significantly affect the way that organizational members communicate, share information, and do their work.

Communication and the exchange of information among organizational members are no longer constrained by geography or time. Collaborative work efforts among widely dispersed individuals and teams, sharing of information, and integration of decisions and work throughout an entire organization have the potential to increase organizational efficiency and effectiveness. And while the economic benefits of IT are obvious, managers must not forget the psychological drawbacks.[32] For instance, what is the psychological cost of an employee always being accessible? Will there be increased pressure for

Technology need not always reduce face-to-face communication. To make contact easier between employees at its call center and in its information systems department, ASB Bank, a New Zealand bank, adopted an open layout encompassing five areas on three different floors. There's a landscaped park area in the center, a café, a minigolf green, a TV room, and a barbecue area, all of which help bring people together. Since moving into the new design, bank managers have noticed that the volume of interdepartmental emails has dropped, indicating that people are communicating in person more frequently.

employees to "check in" even during their off hours? How important is it for employees to separate their work and personal lives? There are no easy answers to these questions, and managers will have to face these issues, many of which we've addressed in our "Managing in a Virtual World" boxes throughout the book.

QUICK LEARNING REVIEW:

LEARNING OUTCOME 14.5

- **Describe how technology affects managerial communication.**
- **Explain how information technology affects organizations.**

Go to page 339 and see how well you know this material.

LEARNING **OUTCOME 14.6** ▷

COMMUNICATION ISSUES IN TODAY'S ORGANIZATIONS

Let's Get Real: F2F

MAJOR COMMUNICATION ISSUE FACING MANAGERS:

Across cultures and countries with a global business words and actions do not mean the same thing and in virtual teams even the nonverbal cues are missing on the phone or e-mail.

"Pulse lunches." That's what managers at Citibank's offices throughout Malaysia used to address pressing problems of declining customer loyalty and staff morale and increased employee turnover. By connecting with employees and listening to their concerns—that is, taking their "pulse"—in informal lunch settings, managers were able to make changes that boosted both customer loyalty and employee morale by over 50 percent and reduced employee turnover to nearly zero.[33] Atul Singh, CEO of Coca-Cola India Ltd, moved to India from China in 2005 with the mandate to pull it out of the quandary of recurring controversies that had seriously influenced the image and business of the company. He began with organizing regular meetings with his employees. He would regularly update the employees on the steps being taken to meet the challenges faced by the company. This led to building of confidence in the company and improvement in business results. Atul continues to meet and share important financial details and strategies of the company with his employees in India.[34]

Being an effective communicator in today's organizations means being connected—most importantly to employees and customers, but to any of the organization's stakeholders. In this section, we examine four communication issues of particular significance to today's managers: managing communication in an Internet world, managing the organization's knowledge resources, communicating with customers, and using politically correct communication.

MANAGING COMMUNICATION IN AN INTERNET WORLD

As our chapter-opening "A Manager's Dilemma" points out, e-mail can consume employees, but it's not always easy for employees to let go of it. But e-mail is only one communication challenge in this Internet world. A recent survey found that 20 percent of employees at large companies say they contribute regularly to blogs, social networks, wikis, and other Web services.[35] Managers are learning, the hard way sometimes, that all this new technology has created special communication challenges. The two main ones are (1) legal and security issues and (2) lack of personal interaction.

Legal and Security Issues. Chevron paid $2.2 million to settle a sexual-harassment lawsuit stemming from inappropriate jokes being sent by employees over company e-mail. UK firm Norwich Union had to pay £450,000 in an out-of-court settlement after an employee sent an e-mail stating that company competitor Western Provident Association was in financial difficulties. Whole Foods Market was investigated by federal regulators and its board after CEO John P. Mackey used a pseudonym to post comments on a blog, attacking the company's rival Wild Oats Markets.[36] An investment banker, after his firm received a grand jury subpoena, instructed his staffs to delete e-mails and instant messages. He was convicted and sentenced to prison for obstructing information.[37] In 2006, the chief executive of an auction Website was arrested because a sexually explicit MMS was posted on the site. The police of Uttar Pradesh have been questioned and berated for making public the e-mails of a 14-year-old girl Arushi Talwar, who was found dead inside her home in Noida, while the case was still being investigated.[38]

Although e-mail, blogs, and other forms of online communication are quick and easy ways to communicate, managers need to be aware of potential legal problems that could result from inappropriate usage. Electronic information is potentially admissible in court. For instance, during the Enron trial, prosecutors entered into evidence e-mails and other documents they say showed that the defendants defrauded investors. Says one expert, "Today, e-mail and instant messaging are the electronic equivalent of DNA evidence."[39]

Legal problems aren't the only issue; security concerns are important as well. A survey addressing outbound e-mail and content security found that 26 percent of the companies surveyed saw their businesses affected by the exposure of sensitive or embarrassing information.[40] Managers need to ensure that confidential information is kept confidential. Employee e-mails and blogs should not communicate—inadvertently or purposely—proprietary information. Corporate computer and e-mail systems should be protected against hackers (people who try to gain unauthorized access to computer systems) and spam (electronic junk mail). These are serious issues that must be addressed if the benefits of communication technology are to be realized.

Personal Interaction. Another communication challenge posed by the Internet age is the lack of personal interaction.[41] Even when two people are communicating face-to-face, understanding is not always achieved. However, it can be especially challenging to achieve understanding and collaborate on getting work done when communication takes place in a virtual environment. In response, some companies have banned e-mail on certain days, as we saw earlier. Others have simply encouraged employees to do more in-person collaboration. Yet, there are situations and times when personal interaction isn't physically possible, such as when your colleagues work across the continent or even on the other side of the globe. In such instances, using real-time collaboration software (such as private workplace wikis, blogs, instant messaging systems, and other types of groupware) may be a better communication choice than sending an e-mail and waiting for a response.[42] Instead of fighting it, some companies are encouraging employees to utilize the power of social networks to collaborate on work and to build strong connections. This is especially appealing to younger workers who are comfortable with this communication medium. Some companies have gone so far as to create their own in-house social networks. For instance, employees at Starcom MediaVest Group tap into SMG Connected to find colleague profiles that outline their jobs, list the brands they admire, and describe their values. A company vice president says, "Giving our employees a way to connect over the Internet around the world made sense because they were doing it anyway."[43]

MANAGING AN ORGANIZATION'S KNOWLEDGE RESOURCES

Kara Johnson is a materials expert at product design firm IDEO. To make finding the right materials easier, she's building a master library of samples linked to a database that explains their properties and manufacturing processes.[44] What Johnson is doing is managing knowledge and making it easier for others at IDEO to learn and benefit from her knowledge. That's what today's managers need to do with the organization's knowledge resources—make it easy for employees to communicate and share their knowledge so they can learn from each other ways to do their jobs more effectively and efficiently. One way organizations can do this is to build online information databases that employees can access. For example, at Patni, several forums based on learning, sharing, and collaboration have been set up. In addition, many activities are conducted on an ongoing basis, to capture and manage the "tacit" knowledge of the organization at different levels. Some of the modules that are used for knowledge management and collaboration are—Treasure Chest where artifacts and documents from various contributors across Patni are stored and users rate it and cross-reference it to make it more usable; Maven–Grid, an expert (maven) in his or her field which supports users across the organization in answering subject-specific queries and users can rate the expert depending on the answers she or he provides them; PatniPedia, a Wiki-publishing tool which is peer moderated, helps building and refining content by users across the organization; and TeamTalk which enables "Discussion Forums" to quickly examine various points of views on a subject of mutual interest and create action points.[45]

In addition to online information databases for sharing knowledge, some knowledge management experts suggest that organizations create **communities of practice**, which are "groups of people who share a concern, a set of problems, or a passion about a topic, and who deepen their knowledge and expertise in that area by interacting on an ongoing basis."[46] The keys to making such communities work are that the group must actually meet in some fashion on a regular basis and also use its information exchanges to improve in some way. For example, repair technicians at Xerox tell "war stories" to communicate their experiences and to help others solve difficult problems with repairing machines.[47] To make these communities of practice work, it's important to maintain strong human interactions through communication; interactive Web sites, e-mail, and videoconferencing are essential tools. In addition, these groups face the same communication problems that individuals face—filtering, emotions, defensiveness, over-documentation, and so forth. However, groups can resolve these issues by focusing on the suggestions discussed earlier.

THE ROLE OF COMMUNICATION IN CUSTOMER SERVICE

You've been a customer many times; in fact, you probably find yourself in customer service encounters several times a day. So what does this have to do with communication? As it turns out, a lot! *What* communication takes place and *how* it takes place can have a significant impact on a customer's satisfaction with the service and the likelihood of being a repeat customer. Managers in service organizations need to make sure that employees who interact with customers are communicating appropriately and effectively with those customers. How? By first recognizing the three components in any service delivery process: the customer, the service organization, and the individual service provider.[48] Each plays a role in whether communication is working. Obviously, managers don't have a lot of control over what or how the customer communicates, but they can influence the other two.

An organization with a strong service culture already values taking care of customers—finding out what their needs are, meeting those needs, and following up to make sure that their needs were met satisfactorily. Each of these activities involves communication, whether face-to-face, by phone or e-mail, or through other channels. In addition, communication is part of the specific customer service strategies the organization pursues. One strategy that many service organizations use is personalization. For instance, at Ritz-Carlton Hotels, customers are provided with more than a clean bed and room. Customers who have stayed at a location previously and indicated that certain items are important to them—such as extra pillows, hot chocolate, or a certain brand of shampoo—will find those items waiting in their room at arrival. The hotel's database allows service to be personalized to customers' expectations. In addition, all employees are asked to communicate information related to service provision. For instance, if a room attendant overhears guests talking about celebrating an anniversary, he or she is supposed to relay the information so something special can be done.[49] Communication plays an important role in the hotel's customer personalization strategy.

Communication is also important to individual service providers or contact employees. The quality of the interpersonal interaction between a customer and a contact employee influences customer satisfaction.[50] This is especially true when the service encounter isn't up to expectations. People on the frontline involved with such "critical service encounters" are often the first to hear about or notice service failures or breakdowns. They must decide *how* and *what* to communicate during these instances. Their ability to listen actively and communicate appropriately with customers has a big impact on whether a situation is resolved to the customer's satisfaction or spirals out of control. Another important communication concern for an individual service provider is making sure that he or she has the information needed to deal with customers efficiently and effectively. If a service provider doesn't personally have the information, there should be some way to get the information easily and promptly.[51]

"POLITICALLY CORRECT" COMMUNICATION

Sears tells its employees to use phrases such as "person with a disability" instead of "disabled person" when writing or speaking about people with disabilities. The company also suggests that when talking with a customer in a wheelchair for more than a few minutes, an employee should place himself or herself at the customer's eye level by sitting down to make a more comfortable atmosphere for everyone.[52] These suggestions, provided in an employee brochure that discusses assisting customers with disabilities, reflect the importance of politically correct communication. How you communicate with someone who isn't like you, what terms you use in addressing a female customer, or what words you use to describe a colleague who is wheelchair-bound can mean the difference between losing a client, an employee, a lawsuit, a harassment claim, or a job.[53]

Most of us are acutely aware of how our vocabulary has been modified to reflect political correctness. For instance, most of us refrain from using words such as *handicapped, blind,* and *elderly* and use instead terms such as *physically challenged, visually impaired,* or *senior.* We must be sensitive to others' feelings. Certain words do stereotype, intimidate, and insult individuals. With an increasingly diverse workforce, we must be sensitive to how words might offend others. While it's complicating our vocabulary and making it more difficult for people to communicate, it is something managers can't ignore.

communities of practice
Groups of people who share a concern, a set of problems, or a passion about a topic and who deepen their knowledge and expertise in that area by interacting on an ongoing basis.

Words are the primary means by which people communicate. When we eliminate words from use because they're politically incorrect, we reduce our options for conveying messages in the clearest and most accurate form. For the most part, the larger the vocabulary used by a sender and a receiver, the greater the opportunity to accurately transmit messages. By removing certain words from our vocabulary, we make it more difficult to communicate accurately. When we replace these words with new ones whose meanings are less well understand, we reduce the likelihood that our messages will be received as we had intended them.

We must be sensitive to how our choice of words might offend others. But we need to acknowledge that politically correct language restricts communication clarity. Nothing suggests that this increased communication ambiguity is likely to be reduced anytime soon. This is just another communication challenge for managers.

QUICK LEARNING REVIEW:

LEARNING OUTCOME 14.6

- Discuss the challenges of managing communication in an Internet world.
- Explain how organizations can manage knowledge.
- Describe why communicating with customers is an important managerial issue.
- Explain how political correctness is affecting communication.

Go to page 339 and see how well you know this material.

Let's Get Real:

My Turn

Subrata Beura

Sr Manager, Business Planning and Strategy
Hewlett Packard

The ban should stay for 1 day a week, and I will encourage my employees to use a day in a week to increase face-to-face meetings to sort out issues and work on areas where you need to provide attention without being disturbed by e-mails.

To gain acceptance of this proposal, I would run a contest in innovative ways to know how my employees have used the no e-mail day to foster interaction within employees.

LEARNING OUTCOMES SUMMARY

14.1 ▷ THE NATURE AND FUNCTION OF COMMUNICATION

- Define *communication, interpersonal communication,* and *organizational communication.*
- Discuss the functions of communication.

Communication is the transfer and understanding of meaning. Interpersonal communication is communication between two or more people. Organizational communication is all the patterns, networks, and systems of communication within an organization.

The functions of communication include controlling employee behavior, motivating employees, providing a release for emotional expression of feelings and fulfillment of social needs, and providing information.

14.2 ▷ METHODS OF INTERPERSONAL COMMUNICATION

- Explain the components in the communication process.
- Discuss the criteria that managers can use to evaluate the various communication methods.
- List the communication methods managers can use.

There are seven elements in the communication process. First there is a sender who has a message. A message is a purpose to be conveyed. Encoding is converting a message into symbols. A channel is the medium a message travels along. Decoding is when the receiver retranslates a sender's message. Finally, there is feedback.

Managers can evaluate the various communication methods according to their feedback, complexity capacity, breadth potential, confidentiality, encoding ease, decoding ease, time–space constraint, cost, interpersonal warmth, formality, scannability, and time of consumption.

Communication methods include face-to-face communication, telephone communication, group meetings, formal presentations, memos, traditional mail, faxes, employee publications, bulletin boards, other company publications, audio- and videotapes, hotlines, e-mail, computer conference, voice mail, teleconferences, and videoconferences.

14.3 ▷ EFFECTIVE INTERPERSONAL COMMUNICATION

- Explain the barriers to effective interpersonal communication.
- Discuss ways to overcome the barriers to effective interpersonal communication.

Barriers to effective communication include filtering, emotions, information overload, defensiveness, language, and national culture.

Managers can overcome these barriers by using feedback, simplifying language, listening actively, constraining emotions, and watching for nonverbal clues.

14.4 ▷ ORGANIZATIONAL COMMUNICATION

- Contrast formal and informal communication.
- Explain communication flow in an organization.
- Describe the three common communication networks.
- Discuss how managers should handle the grapevine.

Formal communication is communication that takes place within prescribed organizational work arrangements. Informal communication is not defined by the organization's structural hierarchy.

Communication in an organization can flow downward, upward, laterally, and diagonally.

The three types of communication networks are the chain, in which communication flows according to the formal chain of command; the wheel, in which communication flows between a clearly identifiable and strong leader and others in a work team; and the all-channel, in which communication flows freely among all members of a work team.

Managers should manage the grapevine as an important information network. They can minimize the negative consequences of rumors by communicating with employees openly, fully, and honestly.

14.5 ▷ INFORMATION TECHNOLOGY AND COMMUNICATION

- Describe how technology affects managerial communication.
- Explain how information technology affects organizations.

Technology has radically changed the way organizational members communicate. It improves a manager's ability to monitor performance; it gives employees more complete information to make decisions more quickly; it has provided employees more opportunities to collaborate and share information; and it has made it possible for people to be fully accessible, anytime, anywhere.

IT affects organizations by affecting the way that organizational members communicate, share information, and do their work.

14.6 ▷ COMMUNICATION ISSUES IN TODAY'S ORGANIZATIONS

- Discuss the challenges of managing communication in an Internet world.
- Explain how organizations can manage knowledge.
- Describe why communicating with customers is an important managerial issue.
- Explain how political correctness is affecting communication.

The two main challenges of managing communication in an Internet world are the legal and security issues and the lack of personal interaction.

Organizations can manage knowledge by making it easy for employees to communicate and share their knowledge so they can learn from each other ways to do their jobs more effectively and efficiently. One way is through online information databases and another way is through creating communities of practice.

Communicating with customers is an important managerial issue because *what* communication takes place and *how* it takes place can significantly affect a customer's satisfaction with the service and the likelihood of being a repeat customer.

Political correctness affects communication in that it sometimes restricts communication clarity. However, managers must be sensitive as to how their choice of words might offend others.

THINKING ABOUT MANAGEMENT ISSUES

1. Why isn't effective communication synonymous with *agreement*?
2. Which do you think is more important for a manager: speaking accurately or listening actively? Why?
3. "Ineffective communication is the fault of the sender." Do you agree or disagree with this statement? Discuss.
4. How might managers use the grapevine for their benefit?
5. Is information technology helping managers be more effective and efficient? Explain your answer.
6. A recent study showed that 28 percent of an information worker's day is spent on interruptions from things that aren't urgent or important, such as unnecessary e-mail, and the time it takes to get back on track.[54] Does this statistic surprise you? What are the managerial implications?

YOUR TURN to be a Manager

- Research the characteristics of a good communicator. What are they? Now practice being a good communicator—as both a sender and a listener.
- For one day, keep track of the types of communication you use (see Exhibit 14–2 for a list of various types). Which do you use most? Least? Were your choices of communication methods effective? Why or why not? Could they have been improved? How?
- For one day, track nonverbal communication that you notice in others. What types did you observe? Was the nonverbal communication always consistent with the verbal communication taking place? Describe.

- Research new types of IT devices. Write a report describing these devices (at least three) and their applicability to employees and organizations. Be sure to look at both the positive and negative aspects.
- Complete the skill-building module Active Listening found in mymanagementlab. Your professor will tell you what to do with it.
- Survey five different managers for their advice on being a good communicator. Put this information in a bulleted list format and be prepared to present it in class.
- Steve's and Mary's recommended readings: Phillip G. Clampitt, *Communicating for Managerial Effectiveness*, 3rd edition (Sage Publications, 2005); John Baldoni, *Great Communication Secrets of Great Leaders* (McGraw-Hill, 2003); Robert Mai and Alan Akerson, *The Leader as Communicator* (AMACOM, 2003); Boyd Clarke, *The Leader's Voice: How Communication Can Inspire Action and Get Results!* (Select Books, 2002); and Jo-Ellan Dimitrius and Mark Mazzarella, *Reading People* (Random House, 1998).
- Survey 10 office workers. Ask them (1) how many e-mail messages they receive daily, on average; (2) how many times in one day they check their e-mail; and (3) whether they think a ban on e-mail messages one day a week would be a good idea and why or why not. Compile this information into a report.
- Choose one of the four topics addressed in the section "Communication Issues in Today's Organizations" and do some additional research. Put your findings in a bulleted list and be prepared to discuss in class. Be sure to cite your sources!
- In your own words, write down three things you learned in this chapter about being a good manager.
- Self-knowledge can be a powerful learning tool. Go to mymanagementlab and complete these self-assessment exercises: What's My Face-to-Face Communication Style? How Good Are My Listening Skills? How Good Am I at Giving Performance Feedback? Using the results of your assessments, identify personal strengths and weaknesses. What will you do to reinforce your strengths and improve your weaknesses?

CASE APPLICATION

Gossip Girls

How many times a day do you gossip . . . either as a sender or a receiver? Although you may think gossip is harmless, it can have some pretty serious consequences. It did for four former employees of the town of Hooksett, New Hampshire, who were fired by the city council for gossiping about their boss. They learned the hard way that gossip can cost you your job.

Former Hooksett, NH town workers, from left Sandra Piper, Joann Drewniak, Jessica Skorupski, and Michelle Bonsteel.

The longtime employees were fired because one of the women had used derogatory terms to describe the town administrator and because all of them had discussed a rumor that he was having an affair with a female subordinate. All four of the women acknowledged feeling resentment toward the woman, who worked in a specially created position and was paid more than two of the employees, despite having less experience and seniority.

Despite an appeal of their dismissal by the four employees, the Hooksett council didn't budge and stated, "These employees do not represent the best interests of the town of Hooksett and the false rumors, gossip and derogatory statements have contributed to a negative working environment and malcontent among their fellow employees." Despite national media attention and a petition signed by 419 residents, asking for the women to be reinstated, the city council hasn't wavered on its decision. An attorney for the four women said that his clients were, "legitimately questioning the conduct of their supervisor, and whether the female subordinate was getting preferential treatment. It almost cheapens it to call it gossip. It might have been idle, not particularly thoughtful, talk. But there was no harm intended."

Discussion Questions

1. What do you think of this situation? Do you agree with the council's decision about the firing and refusal to reinstate the employees?
2. In a recent survey, 60 percent of respondents indicated that the biggest pet peeve they have about their jobs is workplace gossip. Research the topic of office gossip/office rumors. Is office gossip always harmful? Discuss. Could it ever be useful to managers?
3. In retrospect, what could these four women have done differently?
4. What implications from this story can you see for managers and communication?

Sources: J. McGregor, "Mining the Office Chatter," *BusinessWeek*, May 19, 2008, p. 54; E. Zimmerman, "Gossip Is Information by Another Name," *New York Times* online, www.nytimes.com, February 3, 2008; A. Fisher, "Harmless Office Chitchat or Poisonous Gossip?" *CNNMoney.com*, November 12, 2007; S. Armour, "Did You Hear the Story About Office Gossip?" *USA Today*, September 10, 2007, pp. 1B+; "Women Lose Jobs over Office Scuttlebutt," *AARP Bulletin*, July–August 2007, p. 11; G. Cuyler, "Hooksett 4 to Seek Judge's Aid in Getting Jobs Back," *New Hampshire Union Leader*, June 25, 2007; and P. B. Erickson, "Drawing the Line Between Gossip, Watercooler Chat," *NewsOK.com*, June 15, 2007.

Let's Get Real:

Meet The Manager

Priya Jha

AVP, Business Analytics
CPA Global

MY JOB: As a Group Manager for Data Analytics, I am responsible for project delivery and client and team management. I am also responsible for pitching data analytics service offerings to clients and providing thought leadership on key projects.

BEST PART OF MY JOB: The work content: each project and dataset is unique and presents intellectual challenges requiring me to use my technical and business skills to deliver value on the project.

WORST PART OF MY JOB: Following up with team members on work assigned to them. I find it frustrating when it happens several times.

BEST MANAGEMENT ADVICE EVER RECEIVED: There is no substitute for hard work and experience.

You'll be hearing more from this real manager throughout the chapter.

Chapter 15

Motivating Employees

Motivating and rewarding employees is one of the most important and challenging activities that managers do. To get employees to put forth maximum work effort, managers need to know how and why they're motivated. That's what we discuss in this chapter. Focus on the following learning outcomes as you read and study this chapter.

LEARNING OUTCOMES

A Manager's Dilemma

Courtesy: Teach for India

The scale of illiteracy and problems with Indian education are enormous to say the least. Around 13.4 million will never attend school.[1] There are 1.12 million primary schools in India. According to a 2007 report, 9,503 primary schools across India have no teacher and another 1,22,355 have only one teacher each.[2] Fifty-two percent of children cannot read grade 2 (7 years old)[3] level material, 25 percent of all school children cannot write a dictated sentence by class 5 (10 years old)[4], and 124 million need remedial help in arithmetic.

Not getting overwhelmed by these numbers and depressed by the situation, Akanksha an organization that has been working in the area of educating slum children for the last 20 years has launched a TeachforIndia program where young, bright graduates from leading schools and companies are being asked to devote 2 years of their life in teaching in a classroom in a municipal school.[5] The program is being modeled after the success of TeachforAmerica program. Shaheen Mistry, the CEO of this program, wishes to attract 3,000 applications to be able to select 100 fellows for the first year to start teaching in a classroom in June 2009. She also needs to work with companies to have them support the program by accepting these students for placement after they finish their fellowship. In January 2009, the program had no brand and recognition. She and her team visited many colleges to promote this program and tried to reach out to their network to generate interest and resources.

If you were Shaheen Mistry, what would you do?

What would you do?

Successful managers, including Shaheen Mistry, understand that what motivates them personally may have little or no effect on others. Just because *you're* motivated by being part of a cohesive work team, don't assume that everyone else will be. Or just because *you're* motivated by your job doesn't mean that everyone is. Effective managers who get employees to put forth maximum effort know how and why those employees are motivated and tailor motivational practices to satisfy their needs and wants.

LEARNING **OUTCOME 15.1** ▷

WHAT IS MOTIVATION?

Neil Lebovits, president of Ajilon, a staffing firm based in New Jersey, had some serious employee problems.[6] Turnover was high, and morale was low. The severity of the situation hit home when he hosted an after-work party and only 5 out of 50 employees bothered to show up. Lebovits wanted to improve employees' spirits, but like many other managers, he didn't have the resources to give out big raises. So he tried some different things that wouldn't cost a lot of money. He started in-house training programs on various topics in which employees had expressed interest. He initiated monthly conference calls with every employee to discuss management decisions point by point. He set up an e-mail address employees could use to propose ideas, and he responded to every single one. And he gave every employee three "YDOs," or Your Days Off, a year with no questions asked. After he implemented these changes, staff morale skyrocketed. Company employees even sent notes to Lebovits, enthusing about how they felt reenergized.

Neil Lebovits is a good motivator. Like him, all other managers need to be able to motivate their employees. That requires understanding what motivation is. Let's begin by pointing out what motivation is not. Why? Because many people incorrectly view motivation as a personal trait; that is, they think some people are motivated and others aren't. Our knowledge of motivation tells us that we can't label people that way because individuals differ in motivational drive, and their overall motivation varies from situation to situation. For instance, you're probably more motivated in some classes than in others.

Let's Get Real: F2F

MOTIVATING EMPLOYEES IS IMPORTANT BECAUSE:
Only motivated employees function to their full potential—otherwise, people tend to work mechanically and get by with doing the bare minimum. Motivated employees demonstrate independent and innovative thinking when faced with challenging work situations.

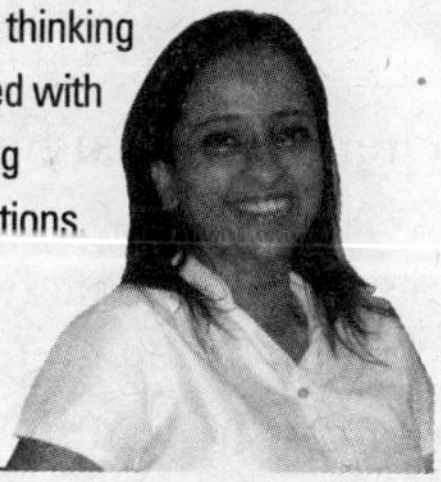

Motivation refers to the process by which a person's efforts are energized, directed, and sustained toward attaining a goal.[7] This definition has three key elements: energy, direction, and persistence.[8]

The *energy* element is a measure of intensity or drive. A motivated person puts forth effort and works hard. However, the quality of the effort must also be considered. High levels of effort don't necessarily lead to favorable job performance unless the effort is channeled in a *direction* that benefits the organization. Effort that's directed toward, and consistent with, organizational goals is the kind of effort we want from employees. Finally, motivation includes a *persistence* dimension. We want employees to persist in putting forth effort to achieve those goals.

Motivating high levels of employee performance is an important organizational concern, and managers keep looking for answers. For instance, a recent Gallup poll found that a large majority of U.S. employees—some 73 percent—are not excited about their work. As the researchers stated, "These employees are essentially 'checked out.' They're sleepwalking through their workday, putting time, but not energy or passion, into their work."[9] It's no wonder then that both managers and academics want to understand and explain employee motivation.

QUICK LEARNING REVIEW:

LEARNING OUTCOME 15.1

- Define motivation.
- Explain the three key elements of motivation.

Go to page 368 and see how well you know this material.

LEARNING **OUTCOME 15.2** ▷

EARLY THEORIES OF MOTIVATION

We begin by looking at four early motivation theories: *Maslow's hierarchy of needs theory, McGregor's Theory X and Theory Y, Herzberg's two-factor theory,* and *McClelland's three needs theory.* Although more valid explanations of motivation have been developed, these early theories are important because they represent the foundation from which contemporary motivation theories were developed and because many practicing managers still use them.

MASLOW'S HIERARCHY OF NEEDS THEORY

Having a car to get to work is a necessity for many workers. When two crucial employees of Vurv Technology in Jacksonville, Florida, had trouble getting to work, owner Derek Mercer decided to buy two inexpensive used cars for the employees. He said, "I felt that they were good employees and a valuable asset to the company." One of the employees who got one of the cars said, "It wasn't the nicest car. It wasn't the prettiest car. But boy did my overwhelming feeling of dread go from that to enlightenment. The 80-hour weeks we worked after that never meant anything. It was give and take. I was giving and the company was definitely giving back."[10] Derek Mercer understands employee needs and their impact on motivation. The first motivation theory we're going to look at addresses employee needs.

motivation
The process by which a person's efforts are energized, directed, and sustained toward attaining a goal.

Exhibit 15–1

Maslow's Hierarchy of Needs

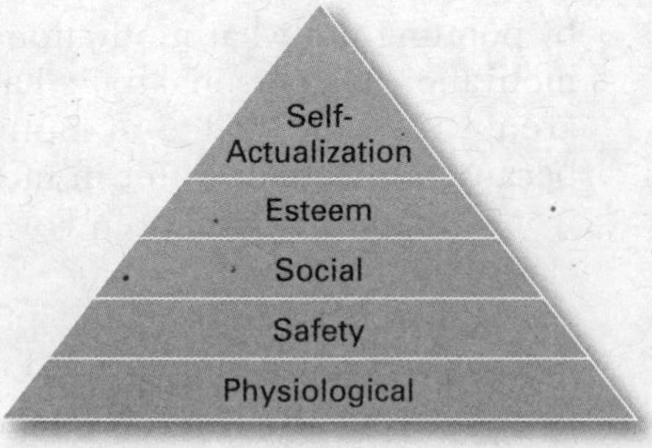

The best-known theory of motivation is probably Abraham Maslow's **hierarchy of needs theory**.[11] Maslow was a psychologist who proposed that within every person is a hierarchy of five needs:

1. **Physiological needs**—A person's needs for food, drink, shelter, sex, and other physical requirements.
2. **Safety needs**—A person's needs for security and protection from physical and emotional harm, as well as assurance that physical needs will continue to be met.
3. **Social needs**—A person's needs for affection, belongingness, acceptance, and friendship.
4. **Esteem needs**—A person's needs for internal esteem factors, such as self-respect, autonomy, and achievement, and external esteem factors, such as status, recognition, and attention.
5. **Self-actualization needs**—A person's needs for growth, achieving one's potential, and self-fulfillment; the drive to become what one is capable of becoming.

Maslow argued that each level in the needs hierarchy must be substantially satisfied before the next need becomes dominant. An individual moves up the needs hierarchy from one level to the next. (See Exhibit 15–1.) In addition, Maslow separated the five needs into higher and lower levels. He considered physiological and safety needs *lower-order needs* and he considered social, esteem, and self-actualization needs *higher-order needs.* Lower-order needs are predominantly satisfied externally, while higher-order needs are satisfied internally.

How does Maslow's theory explain motivation? Managers using Maslow's hierarchy to motivate employees do things to satisfy employees' needs. But the theory also says that once a need is substantially satisfied, an individual isn't motivated to satisfy that need. Therefore, to motivate someone, you need to understand what need level that person is on in the hierarchy and focus on satisfying needs at or above that level.

Maslow's need theory was widely recognized during the 1960s and 1970s, especially among practicing managers, probably because it was intuitively logical and easy to understand. But Maslow provided no empirical support for his theory, and several studies that sought to validate it could not.[12]

McGREGOR'S THEORY X AND THEORY Y

Douglas McGregor is best known for proposing two assumptions about human nature: Theory X and Theory Y.[13] Very simply, **Theory X** is a negative view of people which assumes that workers have little ambition, dislike work, want to avoid responsibility, and need to be closely controlled to work effectively. **Theory Y** is a positive view which assumes that employees enjoy work, seek out and accept responsibility, and exercise self-direction. McGregor believed that Theory Y assumptions should guide management practice and proposed that participation in decision making, responsible and challenging jobs, and good group relations would maximize employee motivation.

Unfortunately, there's no evidence to confirm that either set of assumptions is valid or that being a Theory Y manager is the only way to motivate employees. For instance, Jen-Hsun Huang, founder of Nvidia Corporation, an innovative and successful microchip manufacturer, has been known to use both reassuring hugs and tough love in motivating employees. But he has little tolerance for screw-ups. "In one legendary meeting, he's said to have ripped into a project team for its tendency to repeat mistakes. 'Do you suck?' he asked the stunned employees. 'Because if you suck, just get up and say you suck.'"[14] His message, delivered in classic Theory X style, was that if you need help, ask for it. It's a harsh approach, but in this case, it worked.

HERZBERG'S TWO-FACTOR THEORY

Frederick Herzberg's **two-factor theory** (also called motivation-hygiene theory) proposes that intrinsic factors are related to job satisfaction, while extrinsic factors are associated with job dissatisfaction.[15] Herzberg wanted to know when people felt exceptionally good (satisfied) or bad (dissatisfied) about their jobs. (These findings are shown in Exhibit 15–2.) He concluded that the replies people gave when they felt good about their jobs were significantly different from the replies they gave when they felt badly. Certain characteristics were consistently related to job satisfaction (factors on the left side of the exhibit), and others were related to job dissatisfaction (factors on the right side). When people felt good about their work, they tended to cite intrinsic factors arising from the job itself, such as achievement, recognition, and responsibility. On the other hand, when they were dissatisfied, they tended to cite extrinsic factors arising from the job context, such as company policy and administration, supervision, interpersonal relationships, and working conditions.

In addition, Herzberg believed that the data suggested that the opposite of satisfaction was not dissatisfaction, as traditionally had been believed. Removing dissatisfying characteristics from a job would not necessarily make that job more satisfying (or motivating). As shown in Exhibit 15–3, Herzberg proposed that a dual continuum existed: The opposite of "satisfaction" is "no satisfaction," and the opposite of "dissatisfaction" is "no dissatisfaction."

Again, Herzberg believed that the factors that led to job satisfaction were separate and distinct from those that led to job dissatisfaction. Therefore, managers who sought to eliminate factors that created job dissatisfaction could keep people from being dissatisfied but not necessarily motivate them. He called the extrinsic factors

Exhibit 15–2

Herzberg's Two-Factor Theory

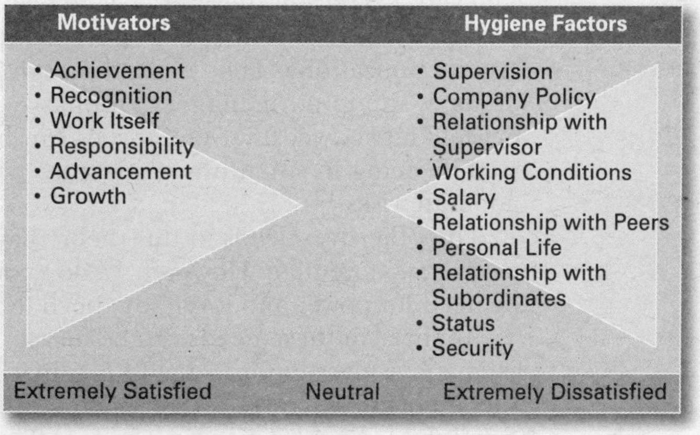

hierarchy of needs theory
Maslow's theory that there is a hierarchy of five human needs: physiological, safety, social, esteem, and self-actualization.

physiological needs
A person's needs for food, drink, shelter, sexual satisfaction, and other physical needs.

safety needs
A person's needs for security and protection from physical and emotional harm.

social needs
A person's needs for affection, belongingness, acceptance, and friendship.

esteem needs
A person's needs for internal factors, such as self-respect, autonomy, and achievement, and external factors, such as status, recognition, and attention.

self-actualization needs
A person's need to become what he or she is capable of becoming.

Theory X
The assumption that employees dislike work, are lazy, avoid responsibility, and must be coerced to perform.

Theory Y
The assumption that employees are creative, enjoy work, seek responsibility, and can exercise self-direction.

two-factor theory
Herzberg's motivation theory, which proposes that intrinsic factors are related to job satisfaction and motivation, whereas extrinsic factors are associated with job dissatisfaction.

Exhibit 15–3

Contrasting Views of Satisfaction–Dissatisfaction

Traditional View

Satisfied	Dissatisfied

Herzberg's View

Motivators		Hygiene Factors	
Satisfaction	No Satisfaction	No Dissatisfaction	Dissatisfaction

that create job dissatisfaction **hygiene factors**. When these factors are adequate, people won't be dissatisfied, but they won't be satisfied (or motivated) either. To motivate people, Herzberg suggested emphasizing **motivators**, intrinsic factors having to do with the job itself.

Two-factor theory enjoyed wide popularity from the mid-1960s to the early 1980s, despite criticisms of Herzberg's procedures and methodology. Although some critics said this theory was too simplistic, it has influenced how we currently design jobs.

McCLELLAND'S THREE-NEEDS THEORY

David McClelland and his associates proposed the **three-needs theory**, which says there are three acquired (not innate) needs that are major motivators in work.[16] These three needs are the **need for achievement (nAch)**, which is the drive to succeed and excel in relation to a set of standards; the **need for power (nPow)**, which is the need to make others behave in a way that they would not have behaved otherwise; and the **need for affiliation (nAff)**, which is the desire for friendly and close interpersonal relationships. Of these three needs, the need for achievement has been researched the most.

People with a high need for achievement strive for personal achievement rather than for the trappings and rewards of success. They have a desire to do something better or more efficiently than it's been done before.[17] They prefer jobs that offer personal responsibility for finding solutions to problems, in which they can receive rapid and unambiguous feedback on their performance in order to tell whether they're improving, and in which they can set moderately challenging goals. High achievers avoid what they perceive to be very easy or very difficult tasks. Also, a high need to achieve doesn't necessarily lead to being a good manager, especially in large organizations. That's because high achievers focus on their *own* accomplishments, while good managers emphasize helping *others* accomplish their goals.[18] McClelland showed that employees can be trained to stimulate their achievement need by being in situations where they have personal responsibility, feedback, and moderate risks.[19]

The other two needs in this theory haven't been researched as extensively as the need for achievement. However, we do know that the best managers tend to be high in the need for power and low in the need for affiliation.[20]

All three of these needs can be measured by using a projective test (known as the Thematic Apperception Test [TAT]), in which respondents react to a set of pictures. After a person briefly sees each picture, he or she writes a story based on the picture. (See Exhibit 15–4 for some examples.) Trained interpreters then determine the individual's levels of nAch, nPow, and nAff from the stories written.

QUICK LEARNING REVIEW:

LEARNING OUTCOME 15.2

- Describe how Maslow's hierarchy of needs can be used to motivate.
- Discuss how Theory X and Theory Y managers approach motivation.
- Explain Herzberg's two-factor theory.
- Describe the three-needs theory.

Go to page 368 and see how well you know this material.

Exhibit 15–4 TAT Pictures

nAch: Indicated by someone in the story wanting to perform or do something better.
nAff: Indicated by someone in the story wanting to be with someone else and enjoy mutual friendship.
nPow: Indicated by someone in the story desiring to have an impact or make an impression on others in the story.

LEARNING
OUTCOME 15.3 ▷

CONTEMPORARY THEORIES OF MOTIVATION

The theories we look at in this section represent current explanations of employee motivation. Although these theories may not be as well known as those we just discussed, they are supported by research.[21] These contemporary motivation approaches are goal-setting theory, reinforcement theory, job design theory, equity theory, and expectancy theory.

GOAL-SETTING THEORY

At Wyeth's research division, Executive Vice President Robert Ruffolo established challenging new product quotas for the company's scientists in an attempt to bring more efficiency to the innovation process. And he made bonuses contingent on meeting those goals.[22] Before a big assignment or major class project presentation, has a teacher ever encouraged you to "Just do your best"? What does that vague statement "do your best" mean? Would your performance on a class project have been higher had that teacher said you needed to score a 93 percent to keep your A in the class? Research on goal-setting theory addresses these issues; the findings, as you'll see, are impressive in terms of the effect that goal specificity, challenge, and feedback have on performance.[23]

There is substantial research support for **goal-setting theory**, which says that specific goals increase performance and that difficult goals, when accepted, result in higher performance than do easy goals. What does goal-setting theory tell us?

hygiene factors
Factors that eliminate job dissatisfaction but don't motivate.

motivators
Factors that increase job satisfaction and motivation.

three-needs theory
McClelland's motivation theory, which says that three acquired (not innate) needs—achievement, power, and affiliation—are major motives in work.

need for achievement (nAch)
The drive to succeed and excel in relation to a set of standards.

need for power (nPow)
The need to make others behave in a way that they would not have behaved otherwise.

need for affiliation (nAff)
The desire for friendly and close interpersonal relationships.

goal-setting theory
The proposition that specific goals increase performance and that difficult goals, when accepted, result in higher performance than do easy goals.

First, working toward a goal is a major source of job motivation. Studies on goal setting have demonstrated that specific and challenging goals are superior motivating forces.[24] Such goals produce a higher output than does the generalized goal "do your best." The specificity of the goal itself acts as an internal stimulus. For instance, when a sales rep commits to making eight sales calls daily, this intention gives him a specific goal to try to attain.

It's not a contradiction that goal-setting theory says that motivation is maximized by *difficult* goals, whereas achievement motivation (from three-needs theory) is stimulated by *moderately challenging* goals.[25] First, goal-setting theory deals with people in general, while the conclusions on achievement motivation are based on people who have a high nAch. Given that the achievement motivation has separate factor loadings for men and women and not everyone has high nAch (infact, it is estimated to be high in only 10–15 percent of the population).[26] Second, the conclusions of goal-setting theory apply to those who accept and are committed to the goals. Difficult goals will lead to higher performance *only* if they are accepted.

Next, will employees try harder if they have the opportunity to participate in the setting of goals? Not always. In some cases, participatively set goals elicit superior performance; in other cases, individuals perform best when their manager assign goals. However, participation is probably preferable to assigning goals when employees might resist accepting difficult challenges.[27]

Finally, we know that people will do better if they get feedback on how well they're progressing toward their goals because feedback helps identify discrepancies between what they have done and what they want to do. But all feedback isn't equally effective. Self-generated feedback—where an employee monitors his or her own progress—has been shown to be a more powerful motivator than feedback coming from someone else.[28]

Three other contingencies besides feedback influence the goal–performance relationship: goal commitment, adequate self-efficacy, and national culture.

First, goal-setting theory assumes that an individual is committed to a goal. Commitment is most likely when goals are made public, when an individual has an internal locus of control, and when the goals are self-set rather than assigned.[29]

Next, **self-efficacy** refers to an individual's belief that he or she is capable of performing a task.[30] The higher your self-efficacy, the more confidence you have in your ability to succeed in a task. So, in difficult situations, we find that people with low self-efficacy are likely to reduce their effort or give up altogether, whereas those with high self-efficacy will try harder to master the challenge.[31] In addition, individuals with high self-efficacy seem to respond to negative feedback with increased

At Ashok Leyland, in 2005, organization-wide surveys showed that the young employees were feeling strangled. The top management took these results seriously and in January 2006, it launched "Mission Young Executives"—or "Mission Yes"—to enable, engage, and empower its young executives. Cross-functional teams of youngsters were formed to work on specific projects, funds were allocated, and top management gave them inputs as and when they sought it. One visible outcome of this effort was The "Ibus" conceptualized, designed, and developed by a cross-functional team of 25, executives less than 30 years of age, in 11 months.[32]

Courtesy: Ashok Leyland

effort and motivation, whereas those with low self-efficacy are likely to reduce their effort when given negative feedback.[33]

Finally, the value of goal-setting theory depends on the national culture. It's well adapted to North American countries because its main ideas align reasonably well with those cultures. It assumes that subordinates will be reasonably independent (not a high score on power distance), that people will seek challenging goals (low in uncertainty avoidance), and that performance is considered important by both managers and subordinates (high in assertiveness). Don't expect goal setting to lead to higher employee performance in countries where the cultural characteristics aren't like this.

Exhibit 15–5 summarizes the relationships among goals, motivation, and performance. Our overall conclusion is that the intention to work toward hard and specific goals is a powerful motivating force. Under the proper conditions, it can lead to higher performance. However, there is no evidence that such goals are associated with increased job satisfaction.[34]

REINFORCEMENT THEORY

Reinforcement theory says that behavior is a function of its consequences. Consequences that immediately follow a behavior and increase the probability that the behavior will be repeated are called **reinforcers**.

Reinforcement theory ignores factors such as goals, expectations, and needs. Instead, it focuses solely on what happens to a person when he or she does something. For instance, at Flex Industries Limited, a 23-year old company, a generous incentive program has helped increase productivity by 30–40 percent as well as minimize wastage by 2–3 percent. Incentives are offered to all employees in addition to being given a chance to work in the UAE unit on a rotational basis. Such programs have helped reduce absenteeism and attrition levels from 12 percent to 8 percent.[35]

In Chapter 13 we showed how managers use reinforcers to shape behavior, but the concept is also widely believed to explain motivation. According to B. F. Skinner, people will most likely engage in desired behaviors if they are rewarded for doing so. These rewards are most effective if they immediately follow a desired behavior; and behavior that isn't rewarded or that is punished is less likely to be repeated.[36]

Using reinforcement theory, managers can influence employees' behavior by using positive reinforcers for actions that help the organization achieve its goals. And managers should ignore, not punish, undesirable behavior. Although punishment eliminates undesired behavior faster than nonreinforcement does, its effect is often

Exhibit 15–5

Goal-Setting Theory

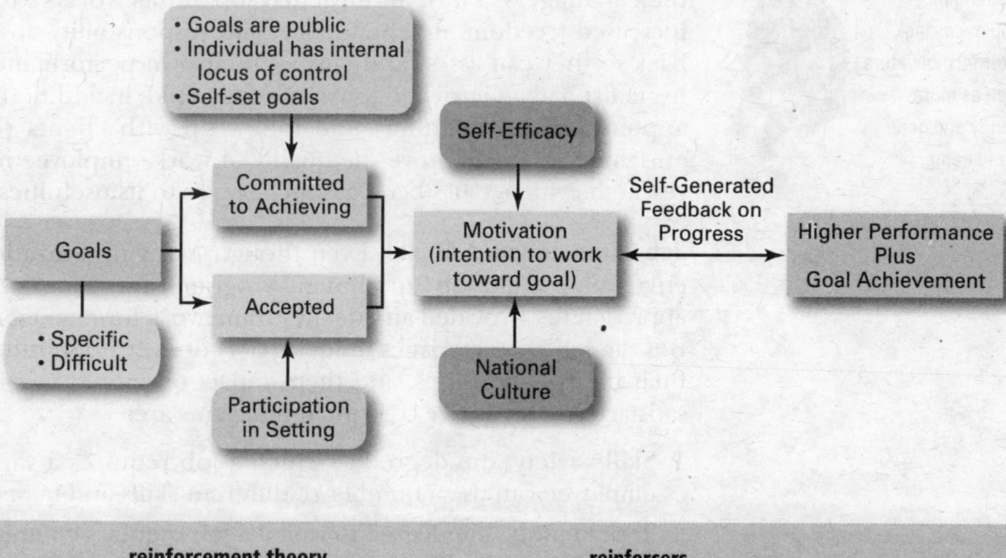

self-efficacy
An individual's belief that he or she is capable of performing a task.

reinforcement theory
The theory that behavior is a function of its consequences.

reinforcers
Consequences immediately following a behavior that increase the probability that the behavior will be repeated.

temporary, and it may have unpleasant side effects, including dysfunctional behavior such as workplace conflicts, absenteeism, and turnover. Although reinforcement is an important influence on work behavior, it isn't the only explanation for differences in employee motivation.[37]

JOB DESIGN THEORY

Because managers want to motivate individuals on the job, we need to look at ways to design motivating jobs. If you look closely at what an organization is and how it works, you'll find that it's composed of thousands of tasks. These tasks are, in turn, aggregated into jobs. We use the term **job design** to refer to the way tasks are combined to form complete jobs. The jobs that people perform in an organization should not evolve by chance. Managers should design jobs deliberately and thoughtfully to reflect the demands of the changing environment, the organization's technology, and employees' skills, abilities, and preferences.[38] When jobs are designed like that, employees are motivated to work hard. Let's look at some ways that managers can design motivating jobs.[39]

Job Enlargement. As we saw in Chapters 2 and 9, job design historically has been used to make jobs smaller and more specialized. It's difficult to motivate employees when jobs are like this. An early effort at overcoming the drawbacks of job specialization involved horizontally expanding a job through increasing **job scope**—the number of different tasks required in a job and the frequency with which those tasks are repeated. For instance, a dental hygienist's job could be enlarged so that in addition to cleaning teeth, he or she is pulling patients' files, re-filing them when finished, and sanitizing and storing instruments. This type of job design option is called **job enlargement**.

Most job enlargement efforts that focused solely on increasing the number of tasks don't seem to work. As one employee who experienced such a job redesign said, "Before, I had one lousy job. Now, thanks to job enlargement, I have three lousy jobs!" However, research has shown that *knowledge* enlargement activities (expanding the scope of knowledge used in a job) lead to more job satisfaction, enhanced customer service, and fewer errors.[40]

Let's Get Real: F2F

HARDEST PART OF MOTIVATING EMPLOYEES:

Very often work satisfaction is not a criterion on which employees are motivated. Even if work is challenging, employees look for external motivators such as more money and good brand name.

Job Enrichment. Another approach to job design is the vertical expansion of a job by adding planning and evaluating responsibilities—**job enrichment**. Job enrichment increases **job depth**, which is the degree of control employees have over their work. In other words, employees are empowered to assume some of the tasks typically done by their managers. Thus, an enriched job allows workers to do an entire activity with increased freedom, independence, and responsibility. In addition, workers get feedback so they can assess and correct their own performance. For instance, if a dental hygienist had an enriched job, he or she could, in addition to cleaning teeth, schedule appointments (planning) and follow up with clients (evaluating). Although job enrichment may improve the quality of work, employee motivation, and satisfaction, research evidence has been inconclusive as to its usefulness.[41]

Job Characteristics Model. Even though many organizations have implemented job enlargement and job enrichment programs and experienced mixed results, neither approach has provided an effective framework for managers to design motivating jobs. But the **job characteristics model (JCM)** does.[42] It identifies five core job dimensions, their interrelationships, and their impact on employee productivity, motivation, and satisfaction. These five core job dimensions are:

1. **Skill variety**, the degree to which a job requires a variety of activities so that an employee can use a number of different skills and talents.
2. **Task identity**, the degree to which a job requires completion of a whole and identifiable piece of work.
3. **Task significance**, the degree to which a job has a substantial impact on the lives or work of other people.

Exhibit 15–6

Job Characteristics Model

Source: J. R. Hackman and J. L. Suttle (eds,). *Improving Life at Work* (Glenview, IL: Scott, Foresman, 1977). With permission of the authors.

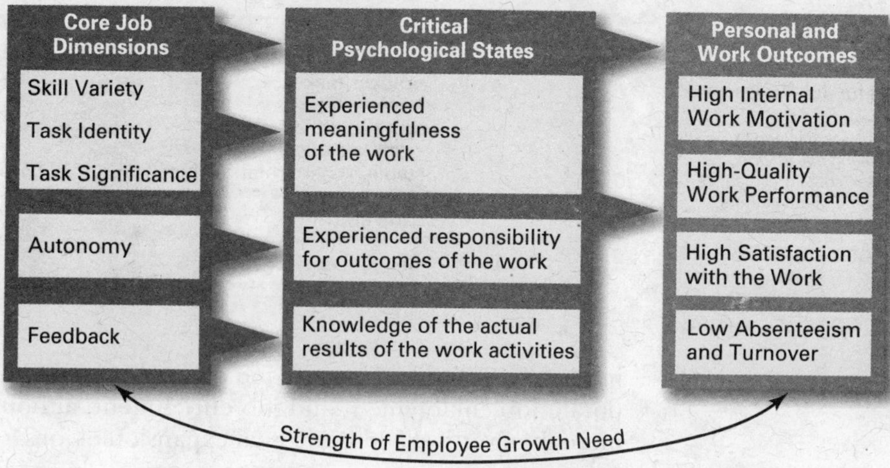

4. **Autonomy**, the degree to which a job provides substantial freedom, independence, and discretion to an individual in scheduling work and determining the procedures to be used in carrying it out.
5. **Feedback**, the degree to which doing work activities required by a job results in an individual obtaining direct and clear information about the effectiveness of his or her performance.

The JCM is shown in Exhibit 15–6. Notice how the first three dimensions—skill variety, task identity, and task significance—combine to create meaningful work. In other words, if these three characteristics exist in a job, we can predict that the person will view his or her job as being important, valuable, and worthwhile. Notice, too, that jobs that possess autonomy give the jobholder a feeling of personal responsibility for the results and that if a job provides feedback, the employee will know how effectively he or she is performing.

The JCM suggests that employees are likely to be motivated when they *learn* (knowledge of results through feedback) that they *personally* (experienced responsibility through autonomy of work) performed well on tasks that they *care about* (experienced meaningfulness through skill variety, task identity, or task significance).[43] The more a job is designed around these three elements, the greater the employee's motivation, performance, and satisfaction and the lower his or her absenteeism and likelihood of resigning. As the model shows, the links between the job dimensions and the outcomes are moderated by the strength of the individual's growth need (the person's desire for self-esteem and self-actualization). Individuals with a high growth need are more likely to experience the critical psychological states and respond positively when their jobs include the core dimensions than are individuals with a low growth need. This may explain the mixed results with job enrichment: Individuals with low growth need aren't likely to achieve high performance or satisfaction by having their jobs enriched.

The JCM provides specific guidance to managers for job design. (See Exhibit 15–7.) These suggestions specify the types of changes that are most likely to lead to

job design
The way tasks are combined to form complete jobs.

job scope
The number of different tasks required in a job and the frequency with which those tasks are repeated.

job enlargement
The horizontal expansion of a job by increasing job scope.

job enrichment
The vertical expansion of a job by adding planning and evaluating responsibilities.

job depth
The degree of control employees have over their work.

job characteristics model (JCM)
A framework for analyzing and designing jobs that identifies five primary core job dimensions, their interrelationships, and their impact on outcomes.

skill variety
The degree to which a job requires a variety of activities so that an employee can use a number of different skills and talents.

task identity
The degree to which a job requires completion of a whole and identifiable piece of work.

task significance
The degree to which a job has a substantial impact on the lives or work of other people.

autonomy
The degree to which a job provides substantial freedom, independence, and discretion to an individual in scheduling work and determining the procedures to be used in carrying it out.

feedback
The degree to which carrying out work activities required by a job results in an individual's obtaining direct and clear information about his or her performance effectiveness.

Exhibit 15–7

Guidelines for Job Redesign

Source: J. R. Hackman and I.L. Suttle (eds.). *Improving Life at work* (Glenview, IL: Scott, Foresman, 1977). With permission of the authors.

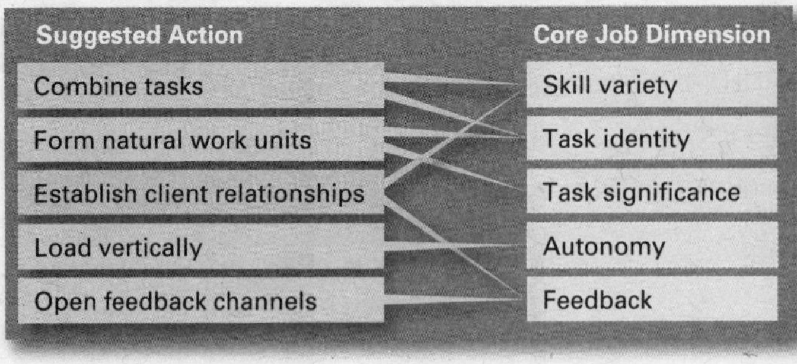

improvement in the five core job dimensions. You'll notice that two suggestions incorporate job enlargement and job enrichment, although the other suggestions involve more than vertical and horizontal expansion of jobs.

1. ***Combine tasks***—Put fragmented tasks back together to form a new, larger work module (job enlargement) to increase skill variety and task identity.
2. ***Create natural work units***—Design tasks that form an identifiable and meaningful whole to increase employee "ownership" of the work. Encourage employees to view their work as meaningful and important rather than as irrelevant and boring.
3. ***Establish client (external or internal) relationships***—Whenever possible, establish direct relationships between workers and their clients to increase skill variety, autonomy, and feedback.
4. ***Expand jobs vertically***—Vertical expansion gives employees responsibilities and controls that were formerly reserved for managers, which can increase employee autonomy.
5. ***Open feedback channels***—Direct feedback lets employees know how well they're performing their jobs and whether their performance is improving.

EQUITY THEORY

Do you ever wonder what kind of grade the person sitting next to you in class makes on a test or on a major class assignment? Most of us do! Being human, we tend to compare ourselves with others. If someone offered you Rs 8,00,000 per year on your first job after graduating from college, you'd probably jump at the offer and report to work enthusiastic, ready to tackle whatever needed to be done, and certainly satisfied with your pay. How would you react, though, if you found out a month into the job that a coworker—another recent graduate, your age, with comparable grades from a comparable school, and with comparable work experience—was getting Rs 9,00,000 a year? You'd probably be upset! Even though in absolute terms, Rs 8,00,000 is a lot of money for a new graduate to make (and you know it!), that suddenly isn't the issue. Now you see the issue as what you believe is *fair*—what is *equitable.* The term *equity* is related to the concept of fairness and equitable treatment compared with others who behave in similar ways. There's considerable evidence that employees compare themselves to others and that inequities influence how much effort employees exert.[44]

Equity theory, developed by J. Stacey Adams, proposes that employees compare what they get from a job (outcomes) in relation to what they put into it (inputs) and then compare their inputs:outcomes ratio with the inputs:outcomes ratios of relevant others (Exhibit 15–8). If an employee perceives her ratio to be equitable in comparison to those of relevant others, there's no problem. However, if the ratio is inequitable, she views herself as underrewarded or overrewarded. When inequities occur, employees attempt to do something about it.[45] The result might be lower or higher productivity, improved or reduced quality of output, increased absenteeism, or voluntary resignation.

The **referent**—the other person, system, or self an individual compares himself or herself against in order to assess equity—is an important variable in equity theory.[46] Each of the three referent categories is important. The "persons" category includes other individuals with similar jobs in the same organization but also includes friends,

Exhibit 15–8

Equity Theory

Perceived Ratio Comparison[a]	Employee's Assessment
$\frac{\text{Outcomes A}}{\text{Inputs A}} < \frac{\text{Outcomes B}}{\text{Inputs B}}$	Inequity (underrewarded)
$\frac{\text{Outcomes A}}{\text{Inputs A}} = \frac{\text{Outcomes B}}{\text{Inputs B}}$	Equity
$\frac{\text{Outcomes A}}{\text{Inputs A}} > \frac{\text{Outcomes B}}{\text{Inputs B}}$	Inequity (overrewarded)

[a]Person A is the employee, and person B is a relevant other or referent.

neighbors, or professional associates. Based on what they hear at work or read about in newspapers or trade journals, employees compare their pay with that of others. The "system" category includes organizational pay policies, procedures, and allocation. The "self" category refers to inputs:outcomes ratios that are unique to the individual. It reflects past personal experiences and contacts and is influenced by criteria such as past jobs or family commitments.

Originally, equity theory focused on **distributive justice**, which is the perceived fairness of the amount and allocation of rewards among individuals. More recent research has focused on looking at issues of **procedural justice**, which is the perceived fairness of the process used to determine the distribution of rewards. This research shows that distributive justice has a greater influence on employee satisfaction than procedural justice, while procedural justice tends to affect an employee's organizational commitment, trust in his or her boss, and intention to quit.[47] What are the implications for managers? They should consider openly sharing information on how allocation decisions are made, follow consistent and unbiased procedures, and engage in similar practices to increase the perception of procedural justice. Employees who have an increased perception of procedural justice are likely to view their bosses and the organization as positive even if they're dissatisfied with pay, promotions, and other personal outcomes.

EXPECTANCY THEORY

The most comprehensive explanation of how employees are motivated is Victor Vroom's **expectancy theory**.[48] Although the theory has critics,[49] most research evidence supports it.[50]

Expectancy theory states that an individual tends to act in a certain way based on the expectation that the act will be followed by a given outcome and on the attractiveness of that outcome to the individual. It includes three variables, or relationships (see Exhibit 15–9):

1. *Expectancy*, or *effort–performance linkage*, is the probability perceived by an individual that exerting a given amount of effort will lead to a certain level of performance.
2. *Instrumentality*, or *performance–reward linkage*, is the degree to which an individual believes that performing at a particular level is instrumental in attaining the desired outcome.
3. *Valence*, or *attractiveness of reward*, is the importance that an individual places on the potential outcome or reward that can be achieved on the job. Valence considers both the goals and needs of the individual.

equity theory
The theory that an employee compares his or her job's input:outcomes ratio with that of relevant others and then corrects any inequity.

referents
The persons, systems, or selves against which individuals compare themselves to assess equity.

distributive justice
Perceived fairness of the amount and allocation of rewards among individuals.

procedural justice
Perceived fairness of the process used to determine the distribution of rewards.

expectancy theory
The theory that an individual tends to act in a certain way, based on the expectation that the act will be followed by a given outcome and on the attractiveness of that outcome to the individual.

Exhibit 15–9

Expectancy Model

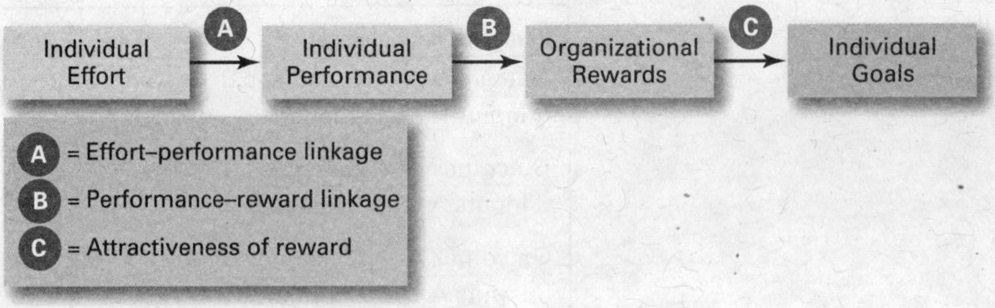

This explanation of motivation might sound complicated, but it really isn't. It can be summed up in the questions: How hard do I have to work to achieve a certain level of performance, and can I actually achieve that level? What reward will performing at that level of performance get me? How attractive is the reward to me, and does it help me achieve my own personal goals? Whether you are motivated to put forth effort (that is, to work hard) at any given time depends on your goals and your perception of whether a certain level of performance is necessary to attain those goals. Let's look at an example. Your second author had a student many years ago who went to work for IBM as a sales rep. Her favorite work "reward" was having an IBM corporate jet fly into Springfield, Missouri, to pick up her best customers and her and take them for a weekend of golfing at some fun location. But to get that particular "reward," she had to achieve at a certain level of performance, which involved exceeding her sales goals by a specified percentage. How hard she was willing to work (that is, how motivated she was to put forth effort) was dependent on the level of performance that had to be met and the likelihood that if she achieved at that level of performance she would receive that reward. Because she "valued" that reward, she always worked hard to exceed her sales goals. And the performance–reward linkage was clear because her hard work and performance achievements were always rewarded by the company with the reward she valued (access to the corporate jet).

The key to expectancy theory is understanding an individual's goal and the linkage between effort and performance, between performance and rewards, and between rewards and individual goal satisfaction. It emphasizes payoffs, or rewards. As a result, we have to believe that the rewards an organization is offering align with what the individual wants. Expectancy theory recognizes that there is no universal principle for explaining what motivates individuals and thus stresses that managers understand why employees view certain outcomes as attractive or unattractive. After all, we want to reward individuals with the things they value positively. Also, expectancy theory emphasizes expected behaviors. Do employees know what is expected of them and how they'll be evaluated? Finally, the theory is concerned with perceptions. Reality is irrelevant. An individual's own perceptions of performance, reward, and goal outcomes, not the outcomes themselves, will determine his or her motivation (level of effort).

INTEGRATING CONTEMPORARY THEORIES OF MOTIVATION

Many of the ideas underlying the contemporary motivation theories are complementary, and you'll understand better how to motivate people if you see how the theories fit together.[51] Exhibit 15–10 presents a model that integrates much of what we know about motivation. Its basic foundation is the expectancy model. Let's work through the model, starting on the left.

The individual effort box has an arrow leading into it. This arrow flows from the individual's goals. Consistent with goal-setting theory, this goals–effort link is meant to illustrate that goals direct behavior. Expectancy theory predicts that an employee will exert a high level of effort if he or she perceives that there is a strong relationship between effort and performance, performance and rewards, and rewards and satisfaction of personal goals. Each of these relationships is, in turn, influenced by certain factors. You can see from the model that the level of individual performance is determined not only by the level of individual effort but also by the individual's ability to perform and by whether

Exhibit 15–10

Integrating Contemporary Theories of Motivation

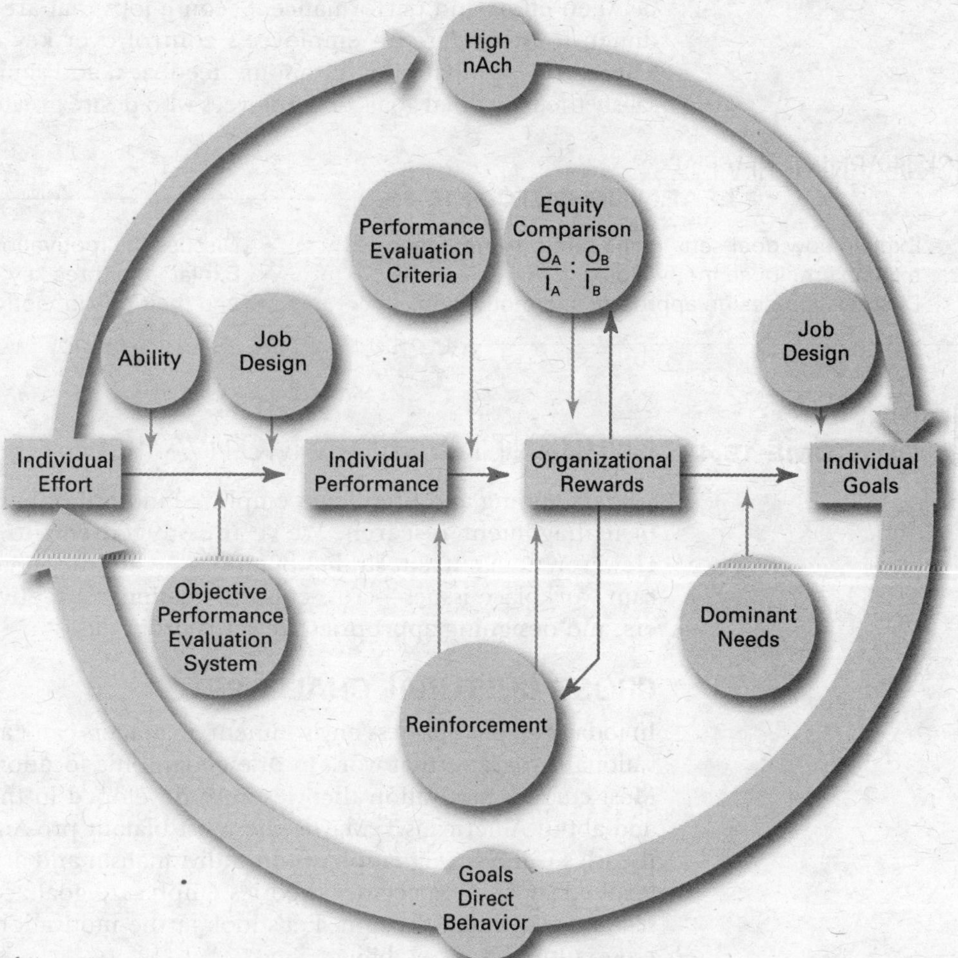

the organization has a fair and objective performance evaluation system. The performance–reward relationship will be strong if the individual perceives that it is performance (rather than seniority, personal favorites, or some other criterion) that is rewarded. The final link in expectancy theory is the rewards–goals relationship. The traditional need theories come into play at this point. Motivation would be high to the degree that the rewards an individual received for his or her high performance satisfied the dominant needs consistent with his or her individual goals.

A closer look at the model also shows that it considers the achievement-need, reinforcement, equity, and JCM theories. The high achiever isn't motivated by the organization's assessment of his or her performance or organizational rewards; hence the jump from effort to individual goals for those with a high nAch. Remember that high achievers are internally driven as long as the jobs they're doing provide them with personal responsibility, feedback, and moderate risks. They're not concerned with the effort–performance, performance–reward, or rewards–goals linkages.

Reinforcement theory is seen in the model by recognizing that the organization's rewards reinforce the individual's performance. If managers have designed a reward system that is seen by employees as "paying off" for good performance, the rewards will reinforce and encourage continued good performance. Rewards also play a key part in equity theory. Individuals will compare the rewards (outcomes) they have received from the inputs or efforts they made with the inputs:outcomes ratio of relevant others. If inequities exist, the effort expended may be influenced.

Finally, the JCM is seen in this integrative model. Task characteristics (job design) influence job motivation at two places. First, jobs that are designed around the five job dimensions are likely to lead to higher actual job performance because the individual's motivation will be stimulated by the job itself—that is, these jobs will increase the linkage between effort and performance. Second, jobs that are designed around the five job dimensions increase an employee's control over key elements in his or her work. Therefore, jobs that offer autonomy, feedback, and similar task characteristics help to satisfy the individual goals of employees who desire greater control over their work.

QUICK LEARNING REVIEW:

LEARNING OUTCOME 15.3

- Explain how goal-setting and reinforcement theories explain employee motivation.
- Describe job design approaches to motivation.
- Discuss the motivation implications of equity theory.
- Explain the three key linkages in expectancy theory and their role in motivation.

Go to page 368 and see how well you know this material.

LEARNING **OUTCOME 15.4** ▷

CURRENT ISSUES IN MOTIVATION

Understanding and predicting employee motivation is one of the most popular areas in management research. We've introduced you to several motivation theories. However, even current studies of employee motivation are influenced by some significant workplace issues—cross-cultural challenges, motivating unique groups of workers, and designing appropriate rewards programs.

CROSS-CULTURAL CHALLENGES

In today's global business environment, managers can't automatically assume that motivational programs that work in one geographic location are going to work in others. Most current motivation theories were developed in the United States by Americans and about Americans.[52] Maybe the most blatant pro-American characteristic in these theories is the strong emphasis on individualism and achievement. For instance, both goal-setting and expectancy theories emphasize goal accomplishment as well as rational and individual thought. Let's look at the motivation theories to see if there's any cross-cultural transferability.

Maslow's needs hierarchy argues that people start at the physiological level and then move progressively up the hierarchy in order. This hierarchy, if it has any application at all, aligns with American culture. In countries such as Japan, Greece, and Mexico, where uncertainty avoidance characteristics are strong, security needs would be on top of the needs hierarchy. Countries that score high on nurturing characteristics—Denmark, Sweden, Norway, the Netherlands, and Finland—would have social needs on top.[53] However, for India, the need to self-actualize remains at top given its long spiritual and philosophical history of search for salvation and purpose. We would predict, for instance, that group work will be more motivating when the country's culture scores high on the nurturing criterion.

Kensaku Konishi, the 55-year-old President and CEO of Canon India, has been managing 700 young Indian sales force as part of his job. He feels that his Indian employees are young and prone to getting excited. They are good speakers but not so good listeners. The company has started a Crucial Conversations Program to improve on the listening skills of its employees.[54]

Courtesy: Canon India

Another motivation concept that clearly has an American bias is the achievement need. The view that a high achievement need acts as an internal motivator presupposes two cultural characteristics—a willingness to accept a moderate degree of risk (which excludes countries with strong uncertainty avoidance characteristics) and a concern with performance (which applies almost singularly to countries with strong achievement characteristics). This combination is found in Anglo-American countries such as the United States, Canada, and Great Britain.[55] On the other hand, these characteristics are relatively absent in countries such as Chile and Portugal. However, the equation is not so straightforward. In India, one can find people with high need for achievement and high external locus of control.[56]

Equity theory has a relatively strong following in the United States. That's not surprising, given that U.S.-style reward systems are based on the assumption that workers are highly sensitive to equity in reward allocations. In the United States, equity is meant to closely link pay to performance. However, recent evidence suggests that in collectivist cultures, especially in the former socialist countries of central and eastern Europe, employees expect rewards to reflect their individual needs as well as their performance.[57] Moreover, consistent with a legacy of communism and centrally planned economics, employees in collectivist cultures exhibit a greater "entitlement" attitude—that is, they expect outcomes to be greater than their inputs.[58] In India, employees would be sensitive to both equity and entitlement depending on the situations.[59] These findings suggest that pay practices may need to be modified in some countries in order for employees to perceive them as fair.

Despite these cross-cultural differences in motivation, there are some cross-cultural consistencies. For instance, the desire for interesting work seems important to almost all workers, regardless of their national culture. In a study of seven countries, employees in Belgium, Britain, Israel, and the United States ranked "interesting work" first among 11 work goals. The survey by Watson Wyat found this same to be true for Indian respondents.[60] It was ranked either second or third in Japan, the Netherlands, and Germany.[61] Similarly, in a study comparing job-preference outcomes among graduate students in the United States, Canada, Australia, and Singapore, growth, achievement, and responsibility were tied for first.[62] Both studies suggest some universality to the importance of intrinsic factors identified by Herzberg in his two-factor theory. Another recent study examining workplace motivation trends in Japan also seems to indicate that Herzberg's model is applicable to Japanese employees.[63]

MOTIVATING UNIQUE GROUPS OF WORKERS

Motivating employees has never been easy! Employees come into organizations with different needs, personalities, skills, abilities, interests, and aptitudes. They have different expectations of their employers and different views of what they think their employer has a right to expect of them. And they vary widely in what they want from their jobs. For instance, some employees get more satisfaction out of their personal interests and pursuits and only want a weekly paycheck—nothing more. They're not interested in making their work more challenging or interesting or in "winning" performance contests. Others derive a great deal of satisfaction in their jobs and are motivated to exert high levels of effort. Given these differences, how can managers do an effective job of motivating the

thinking critically about Ethics

You're getting a new boss. That's scary enough in itself, but what if that new boss had an outside passion that employees were expected to support? For instance, what if your new boss was an ardent supporter of Rashtriya Swayamsevak Sangh (RSS) and expected you to contribute to RSS? What if he kept track of who did and did not contribute? Or what if your new boss had an intense love of chocolate and told you, "If you want to stay in this department, you're going to have to learn to love chocolate"? Or the obsession could be golf or tennis or opera. What ethical issues might there be in such situations for managers and employees? How should such situations be handled?

unique groups of employees in today's workforce? One thing is to understand the motivational requirements of these groups, including diverse employees, professionals, contingent workers, and low-skilled minimum-wage employees.

Motivating a Diverse Workforce. To maximize motivation among today's workforce, managers need to think in terms of *flexibility*. For instance, studies tell us that men place more importance on having autonomy in their jobs than do women. In contrast, the opportunity to learn, convenient and flexible work hours, and good interpersonal relations are more important to women.[64] Having the opportunity to be independent and to be exposed to different experiences is important to Gen Y employees, whereas older workers may be more interested in highly structured work opportunities.[65] Managers need to recognize that what motivates a single mother with two dependent children who's working full time to support her family may be very different from the needs of a single part-time employee or an older employee who is working only to supplement his or her retirement income. A diverse array of rewards is needed to motivate employees with such diverse needs. Many of the work–life balance programs (see Chapter 10) that organizations have implemented are a response to the varied needs of a diverse workforce. In addition, many organizations have developed flexible work arrangements that recognize different needs. These types of programs may become even more popular as employers look for ways to help employees cope with high fuel prices. For instance, a **compressed workweek** is a workweek in which employees work longer hours per day but fewer days per week. The most common arrangement is four 10-hour days (a 4-40 workweek). However, organizations could design whatever schedules they wanted to fit employees' needs. Another alternative is **flexible work hours** (also known as **flextime**), which is a scheduling system in which employees are required to work a specific number of hours each week but are free to vary those hours within certain limits. In a flextime schedule, there are certain common core hours when all employees are required to be on the job, but starting, ending, and lunch times are flexible. According to a survey by Hewitt Associates, 75 percent of large companies now offer flextime benefits. Another survey by Watson Wyatt of midsized and large-sized companies found that flexible work schedules was the most commonly offered benefit.[66]

In Great Britain, McDonald's is experimenting with an unusual program—dubbed the Family Contract—to reduce absenteeism and turnover at some of its restaurants. Under the Family Contract, employees from the same immediate family can fill in for one another for any work shift without having to clear it first with their manager.[67] This type of job scheduling, which can be effective in motivating a diverse workforce, is called **job sharing**—the practice of having two or more people split a full-time job. Although something like McDonald's Family Contract may be appropriate for a low-skilled job, other organizations might offer job sharing to professionals who want to work but don't want the demands and hassles of a full-time position. For instance, at Ernst & Young, employees in many of the company's locations can choose from a variety of flexible work arrangements, including job sharing.

Another alternative made possible by information technology is **telecommuting**. Here, employees work at home and are linked to the workplace by computer and modem. For example, around 40 percent of IBM's workforce has no physical office space. The number is even higher for Sun Microsystems, where nearly 50 percent of employees work offsite.[68] Canon India does not expect its sales force to come to office at all. They are given a maintenance allowance for their home office. Because many jobs can be done at offsite locations, this approach might be close to the ideal job for many people as there is no commuting, the hours are flexible, there's freedom to dress as you please, and there are little or no interruptions from colleagues. However, keep in mind that not all employees embrace the idea of telecommuting. Some workers relish the informal interactions at work that satisfy their social needs as well as being a source of new ideas.

Do flexible work arrangements motivate employees? Although such arrangements might seem highly motivational, both positive and negative relationships have been found. For instance, a recent study looking at the impact of telecommuting on

Let's Get Real: F2F

MY BIGGEST MOTIVATION CHALLENGE:

Making employees realize that if they focus on their current assignment or job, more interesting work or more money will come to them over course of time.

job satisfaction found that job satisfaction initially increased as the extent of telecommuting increased, but as the number of hours spent telecommuting increased, job satisfaction started to level off, decreased slightly, and then stabilized.[69]

Motivating Professionals. In contrast to a generation ago, the typical employee today is more likely to be a professional with a college degree than a blue-collar factory worker. What special concerns should managers be aware of when trying to motivate a team of engineers at Intel's India Development Center, software designers at SAS Institute in North Carolina, or a group of consultants at Accenture in Singapore?

Professionals are different from nonprofessionals.[71] They have a strong and long-term commitment to their field of expertise. To keep current in their field, they need to regularly update their knowledge, and because of their commitment to their profession, they rarely define their workweek as 8 A.M. to 5 P.M. five days per week.

What motivates professionals? Money and promotions typically are low on their priority list. Why? Professionals tend to be well paid and enjoy what they do. Rather than prioritize money and promotions, professionals tend to rank job challenge high. They like to tackle problems and find solutions. Their chief reward is the work itself. Professionals also value support. They want others to think that what they are working on is important. That may be true for all employees, but professionals tend to be focused on their work as their central life interest, whereas nonprofessionals typically have other interests outside work that can compensate for needs not met on the job.

managing workforce Diversity

Developing Employee Potential: The Bottom Line of Diversity

One of a manager's most important goals is helping employees develop their potential.[70] This is particularly important in managing talented diverse employees who can bring new perspectives and ideas to the business but who may find that the workplace environment is not as conducive as it could be to accepting and embracing these different perspectives. For instance, managers at Bell Labs have worked hard to develop an environment in which the ideas of diverse employees are encouraged openly.

What can managers do to ensure that their diverse employees have the opportunity to develop their potential? One thing they can do is make sure that there are diverse role models in leadership positions so that others see that there are opportunities to grow and advance. Giving motivated, talented, hard-working, and enthusiastic diverse employees opportunities to excel in decision-making roles can be a powerful motivator to other diverse employees to work hard to develop their own potential. A mentoring program in which diverse employees are given the opportunity to work closely with organizational leaders can be a powerful tool. At Silicon Graphics, for instance, new employees become part of a mentoring group called "Horizons." Through this mentoring group, diverse employees have an opportunity to observe and learn from key company decision makers.

Another way for managers to develop the potential of their diverse employees is to offer developmental work assignments that provide a variety of learning experiences in different organizational areas. Employees who are provided the opportunity to learn new processes and new technology are more likely to excel at their work and to stay with the company. These types of developmental opportunities are particularly important for diverse employees because they empower employees with tools that are critical to professional development.

compressed workweek
A workweek in which employees work longer hours per day but fewer days per week.

flexible work hours (flextime)
A scheduling system in which employees are required to work a certain number of hours per week but are free, within limits, to vary the hours of work.

job sharing
The practice of having two or more people split a full-time job.

telecommuting
A job approach in which employees work at home and are linked to the workplace by computer and modem.

Motivating Contingent Workers. As full-time jobs have been eliminated through downsizing and other organizational restructurings, the number of openings for part-time, contract, and other forms of temporary work have increased. Contingent workers don't have the security or stability that permanent employees have, and they don't identify with the organization or display the commitment that other employees do. Temporary workers also typically get little or no benefits such as health care or pensions.[72]

There's no simple solution for motivating contingent employees. For that small set of individuals who prefer the freedom of their temporary status, the lack of stability may not be an issue. In addition, temporariness might be preferred by highly compensated physicians, engineers, accountants, or financial planners who don't want the demands of a full-time job. But these are the exceptions. For the most part, temporary employees are not temporary by choice.

What will motivate involuntarily temporary employees? An obvious answer is the opportunity to become a permanent employee. In cases in which permanent employees are selected from a pool of temps, the temps will often work hard in hopes of becoming permanent. A less obvious answer is the opportunity for training. The ability of a temporary employee to find a new job is largely dependent on his or her skills. If an employee sees that the job he or she is doing can help develop marketable skills, then motivation is increased. From an equity standpoint, when temps work alongside permanent employees who earn more and get benefits too for doing the same job, the performance of temps is likely to suffer. Separating such employees or perhaps minimizing interdependence between them might help managers counteract potential problems.[73]

Motivating Low-Skilled, Minimum-Wage Employees. Suppose that in your first managerial position after graduating, you're responsible for managing a work group of low-skilled, minimum-wage employees. Offering more pay to these employees for high levels of performance is out of the question: Your company just can't afford it. In addition, these employees have limited education and skills. What are your motivational options at this point?

One trap we often fall into is thinking that people are motivated only by money. Although money is an important motivator, it's not the only reward that people seek and that managers can use. In motivating minimum-wage employees, managers might look at employee recognition programs. Many managers also recognize the power of praise, although these "pats on the back" must be sincere and given for the right reasons.

DESIGNING APPROPRIATE REWARDS PROGRAMS

Many senior people at UTI AMC had resigned because of the listless work culture, but when initiatives were taken to make the company performance driven and performance linked, incentive plans were announced out and some of the senior people who had left came back to join the UTI AMC.[74] FedEx's drivers are motivated by a pay system that rewards them for timeliness and how much they deliver.[75] Employee rewards programs play a powerful role in motivating appropriate employee behavior.

Open-Book Management. At Max New York Life, all employees are rewarded based on dual factors review of goals set by the employers at the beginning of the year and assessment of individuals on competencies specified for the position. On the basis of the transparent review, the best employees are rewarded.[76] Many organizations of various sizes involve their employees in workplace decisions by opening up the financial statements (the "books"). They share that information so that employees will be motivated to make better decisions about their work and better able to understand the implications of what they do, how they do it, and the ultimate impact on the bottom line. This approach is called **open-book management**, and many organizations are using it.[77] At Intel Technology in India, the president of the company meets all the employees face to face in small groups and has a direct communication program running. The CEO also blogs regularly to share his thoughts. Highly qualified employees are offered alternative groups to work in.[78] At McAfee India office, financial data, such as earnings, growth, revenue, and balance sheets, are shared on a quarterly basis.[79]

The goal of open-book management is to get employees to think like owners by seeing the impact their decisions have on financial results. Because many employees don't have the knowledge or background to understand the financials, they have to be taught how to read and understand the organization's financial statements. Once employees have this knowledge, however, managers need to regularly share the numbers with them. When they share this information, employees begin to see the link between their efforts, the level of performance, and operational results.

Employee Recognition Programs. **Employee recognition programs** consist of personal attention and expressing interest, approval, and appreciation for a job well done.[80] They can take numerous forms. At BPL mobile communication and Synplicity India, employees are recognized for doing anything that is beyond the call of their duty and are given a spot award. At Essar telecommunication, employees who reach their targets are given the "reach the star" award. Shoppers Stop has an employee of the month award given to those who provide excellent service to customers. At Paramount, there are employee recognition programs and letters of commendation sent to those who excel at their tasks. This has led to zero-percent attrition.[81] Apart from monetary benefits, Communicate 2's reward and recognition program includes giving the employees Ipods, mobile phones, and laptops. At GE's John F. Welch Technology Centre also, there is a mix of monetary and nonmonetary rewards; top patent holders and publication authors at the center are invited to an exclusive dinner with the leadership at the center. At Texas Instruments India, rewards such as differentiated basic pay; differentiated performance-based bonus programs, long-term incentive plans like use of equity, awards, and appreciation program, spot rewards, patent incentive, long service awards, technical ladder award, team awards, holiday packages, and famous author program, are used. Datamatics Laureate is awarded to those employees who through their unstilted efforts and total dedication make a very significant contribution to the growth and development of one or more businesses of the Datamatics Group. Arteria has the practice of giving gold coins to create awareness of importance of savings among employees; it also offers travel by airplane to employees who have not traveled so, earlier, and gifting music systems, home theatres, and TV sets according to employees' need and interest as a reward.

A recent survey of organizations found that 84 percent had some type of program to recognize worker achievements.[82] And do employees think these programs are important? You bet! In a survey conducted a few years ago, a wide range of employees was asked what they considered the most powerful workplace motivator. Their response? Recognition, recognition, and more recognition![83]

Consistent with reinforcement theory, rewarding a behavior with recognition immediately following that behavior is likely to encourage its repetition. And recognition can take many forms. You can personally congratulate an employee in private for a good job. You can send a handwritten note or an e-mail message acknowledging something positive that the employee has done. For employees with a strong need for social acceptance, you can publicly recognize accomplishments. To enhance group cohesiveness and motivation, you can celebrate team successes. For instance, you can do something as simple as throw a pizza party to celebrate a team's accomplishments. Some of these things may seem simple, but they can go a long way in showing employees they're valued.

Let's Get Real: F2F

GUIDELINES FOR BEING A GOOD MOTIVATOR:

- Empathy, i.e. being able to put oneself in the employees' shoes even while trying to explain a contrary or different point of view.
- Manager should be a good role model by being successful or good at his or her job and organizationally strong.

Pay-for-Performance Programs. Here's a survey statistic that may surprise you: 40 percent of employees see no clear link between performance and pay.[84] So what are the companies where these employees work paying for? They're obviously not clearly communicating performance expectations.[85] **Pay-for-performance programs** are variable compensation plans that pay employees on the basis of some performance measure.[86] Piece-rate pay

open-book management
A motivational approach in which an organization's financial statements (the "books") are shared with all employees.

employee recognition programs
Programs that consist of personal attention and expressing interest, approval, and appreciation for a job well done.

pay-for-performance programs
Variable compensation plans that pay employees on the basis of some performance measure.

plans, wage incentive plans, profit-sharing, and lump-sum bonuses are examples. What differentiates these forms of pay from more traditional compensation plans is that instead of paying a person for time on the job, pay is adjusted to reflect some performance measure. These performance measures might include such things as individual productivity, team or work group productivity, departmental productivity, or the overall organization's profit performance.

Pay-for-performance is probably most compatible with expectancy theory. For motivation to be maximized, individuals should perceive a strong relationship between their performance and the rewards they receive. If rewards are allocated only on nonperformance factors—such as seniority, job title, or across-the-board pay raises—then employees are likely to reduce their efforts. From a motivation perspective, making some or all of an employee's pay conditional on some performance measure focuses his or her attention and effort toward that measure and then reinforces the continuation of the effort with a reward. If the employee, team, or organization's performance declines, so does the reward. Thus, there's an incentive to keep efforts and motivation strong.

Pay-for-performance programs are popular. Studies show that there is no clear data available for extent of variable pay in Indian companies but most companies other than manufacturing sector have variable pay in some form or the other. Some 80 percent of large U.S. companies have some form of variable pay plan.[87] About 30 percent of Canadian companies and 22 percent of Japanese companies have companywide pay-for-performance plans.[88]

Do pay-for-performance programs work? For the most part, studies seem to indicate that they do. For instance, one study found that companies that used pay-for-performance programs performed better financially than those that did not.[89] Another study showed that pay-for-performance programs with outcome-based incentives had a positive impact on sales, customer satisfaction, and profits.[90] "The reason why this (pay for performance) has been adopted is because companies need to keep their top 20 percent happy with salaries that are higher than the market and this is not possible unless they have a variable component in the salary. Only then organizations will be able to reward people based on their performance and market dynamics, and not on the years of experience that they have," explains Gautam Sinha, CEO of TVA InfoTech.[91] If an organization uses work teams, managers should consider group-based performance incentives that will reinforce team effort and commitment. But whether these programs are individual based or team based, managers need to ensure that they're specific about the relationship between an individual's pay and his or her expected level of appropriate performance. Employees must clearly understand exactly how performance—theirs and the organization's—translates into dollars on their paychecks.[92]

Stock Option Programs. Alarmed by the excessive compensation package being paid to Countrywide Financial CEO Angelo Mozilo—equal to 6 percent of Countrywide's net income—shareholders challenged the company's board of directors to do something. However, that was only part of the problems facing Countrywide. The struggles in the U.S. housing market were creating a lot of uncertainty. Then, the company was sued by at least five states for using misleading marketing practices and is "under fire from housing advocates, members of Congress, and shareholders for making aggressive loans to borrowers even as top management made tens of millions in profits from stock sales."[93] All these factors caused the company's financial results to weaken and led to a vote by shareholders approving the sale of the company to Bank of America. When executives receive compensation in the millions of dollars even while their companies' performance is suffering, it's no wonder that executive bonus and stock option programs have come under fire. Such situations fly in the face of the belief that pay should align with organizational performance.

Stock options are financial instruments that give employees the right to purchase shares of stock at a set price. The original idea was to turn employees into owners and give them strong incentives to work hard to make the company successful.[94] If the company was successful, the value of the stock went up, making the stock options valuable. In other words, there was a link between performance and reward. The popularity of stock options as a motivation and compensation tool skyrocketed during the dot-com boom in the late 1990s. Because many dot-coms couldn't afford to pay employees the going market-rate salaries, stock options were offered as performance incentives. However, the shakeout among dot-com stocks in 2000 and 2001 illustrated one of the inherent risks of offering stock options. As long as the market was rising, employees were willing to give up a large salary in exchange for stock options. However, when stock prices tanked, many individuals who had joined and stayed with a dot-com for the opportunity to get rich through stock options found that their stock options had become worthless. And the declining stock market became a powerful demotivator.

Despite the risk of potential lost value and the widespread abuse of stock options, they should be considered as part of an overall motivational program. An appropriately designed stock option program can be a powerful motivational tool for employees, but with any appropriate and effective rewards system, there needs to be a clear link between performance and reward.[95]

QUICK LEARNING REVIEW:

LEARNING OUTCOME 15.4

- Describe the cross-cultural challenges of motivation.
- Discuss the challenges managers face in motivating today's workforce.
- Describe open-book management, employee recognition, pay-for-performance, and stock option programs.

Go to page 369 and see how well you know this material.

FROM THEORY TO PRACTICE: SUGGESTIONS FOR MOTIVATING EMPLOYEES

We've covered a lot of information about motivation. If you're a manager concerned with effectively and efficiently motivating your employees, what specific recommendations can you draw from the theories presented in this chapter? Although there's no simple, all-encompassing set of rules, the following suggestions draw on what we know about motivating employees.

Recognize Individual Differences. Almost every contemporary motivation theory recognizes that employees aren't identical. They have different needs, attitudes, personality, and other important individual variables.

Match People to Jobs. A great deal of evidence shows the motivational benefits of carefully matching people to jobs. For example, high achievers should have jobs that allow them to participate in setting moderately challenging goals and that involve autonomy and feedback. Also, keep in mind that not everybody is motivated by jobs that are high in autonomy, variety, and responsibility.

Use Goals. The literature on goal-setting theory suggests that managers should ensure that employees have hard, specific goals and feedback on how well they're doing in achieving those goals. Should the goals be assigned by the manager or should employees participate in setting them? The answer depends on your perception of goal acceptance and the organization's culture. If you expect resistance to goals, participation should increase acceptance. If participation is inconsistent with the culture, use assigned goals.

Ensure That Goals Are Perceived as Attainable. Regardless of whether goals are actually attainable, employees who see goals as unattainable will reduce their effort because they'll wonder why they should bother. Managers must be sure, therefore, that employees feel confident that increased efforts *can* lead to achieving performance goals.

stock options
Financial instruments that give employees the right to purchase shares of stock at a set price.

Individualize Rewards. Because employees have different needs, what acts as a reinforcer for one may not for another. Managers should use their knowledge of employee differences to individualize the rewards they control, such as pay, promotions, recognition, desirable work assignments, autonomy, and participation.

Link Rewards to Performance. Managers need to make rewards contingent on performance. Rewarding factors other than performance will only reinforce those other factors. Important rewards such as pay increases and promotions should be given for the attainment of specific goals. Managers should also look for ways to increase the visibility of rewards, making them potentially more motivating.

Check the System for Equity. Employees should perceive that rewards or outcomes are equal to the inputs. On a simple level, experience, ability, effort, and other obvious inputs should explain differences in pay, responsibility, and other obvious outcomes. And remember that one person's equity is another's inequity, so an ideal reward system should probably weigh inputs differently in arriving at the proper rewards for each job.

Use Recognition. Recognize the power of recognition. In a stagnant economy where cost-cutting is widespread, using recognition is a low-cost means to reward employees. And it's a reward that most employees consider valuable.

Show Care and Concern for Your Employees. Employees perform better for managers who care about them. Research done by Gallup with millions of employees and tens of thousands of managers consistently shows this simple truth: The best organizations create "caring" work environments.[96] When managers care about employees, performance results typically follow.

Don't Ignore Money. It's easy to get so caught up in setting goals, creating interesting jobs, and providing opportunities for participation that you forget that money is the major reason most people work. Thus, the allocation of performance-based wage increases, piecework bonuses, and other pay incentives is important in determining employee motivation. We're not saying that managers should focus solely on money as a motivational tool. Rather, we're simply stating the obvious: If money is removed as an incentive, people aren't going to show up for work. The same can't be said for removing goals, enriched work, or participation.

Let's Get Real:

My Turn

Priya Jha

AVP, Business Analytics
CPA Global

Link this program to the Corporate Social Responsibility (CSR) of three to four companies that are large and well known. The graduates can be placed in these companies after they finish their fellowship. These companies' endorsement can then be further used to promote this program in other companies. It would be important to convince young graduates that how this program will help thom to develop their leadership potential and how in the future they would be able to handle any challenge. Shaheen must use the voice of several TeachforAmerica fellows to share their experiences to motivate young graduates to apply.

LEARNING OUTCOMES SUMMARY

15.1 ▷ WHAT IS MOTIVATION?

- Define motivation.
- Explain the three key elements of motivation.

Motivation is the process by which a person's efforts are energized, directed, and sustained toward attaining a goal.

The energy element is a measure of intensity or drive. The high level of effort needs to be directed in ways that help the organization achieve its goals. Employees must persist in putting forth effort to achieve those goals.

15.2 ▷ EARLY THEORIES OF MOTIVATION

- Describe how Maslow's hierarchy of needs can be used to motivate.
- Discuss how Theory X and Theory Y managers approach motivation.
- Explain Herzberg's two-factor theory.
- Describe the three-needs theory.

Individuals move up the hierarchy of five needs (physiological, safety, social, esteem, and self-actualization) as needs are substantially satisfied. A need that's substantially satisfied no longer motivates.

A Theory X manager believes that people don't like to work or won't seek out responsibility so they have to be threatened and coerced to work. A Theory Y manager assumes that people like to work and seek out responsibility, so they will exercise self-motivation and self-direction.

Herzberg's two-factor theory proposed that intrinsic factors associated with job satisfaction were what motivated people. In addition, he proposed that extrinsic factors associated with job dissatisfaction simply kept people from being dissatisfied.

The three-needs theory proposed three acquired needs that are major motives in work: need for achievement, need for affiliation, and need for power.

15.3 ▷ CONTEMPORARY THEORIES OF MOTIVATION

- Explain how goal-setting and reinforcement theories explain employee motivation.
- Describe job design approaches to motivation.
- Discuss the motivation implications of equity theory.
- Explain the three key linkages in expectancy theory and their role in motivation.

Goal-setting theory says that specific goals increase performance and that difficult goals, when accepted, result in higher performance than do easy goals. Goal-setting theory says that intention to work toward a goal is a major source of job motivation; specific hard goals produce higher levels of output than generalized goals; participation in setting goals is probably preferable to assigning goals, but not always; feedback guides and motivates behavior, especially self-generated feedback; and contingencies that affect goal setting include goal commitment, self-efficacy, and national culture. Reinforcement theory says that behavior is a function of its consequences. To motivate, use positive reinforcers to reinforce desirable behaviors. Ignore undesirable behavior rather than punishing it.

Job enlargement involves horizontally expanding job scope by adding more tasks or increasing how many times the tasks are done. Job enrichment involves vertically expanding job depth by giving employees more control over their work. The job characteristics model says that five core job dimensions (skill variety, task identity, task significance, autonomy, and feedback) are used to design motivating jobs.

Equity theory focuses on how employees compare their inputs:outcomes ratios to relevant others' ratios. A perception of inequity will cause an employee to do something about it. Procedural justice has a greater influence on employee satisfaction than does distributive justice.

Expectancy theory says that an individual tends to act in a certain way based on the expectation that the act will be followed by a desired outcome. Expectancy is the effort–performance linkage (How much effort do I need to exert to achieve a certain level of performance?); instrumentality is the performance–reward linkage (Achieving at a certain level of performance will get me what reward?); and valence is the attractiveness of the reward (Is the reward something I want?).

15.4 ▷ CURRENT ISSUES IN MOTIVATION

- Describe the cross-cultural challenges of motivation.
- Discuss the challenges managers face in motivating today's workforce.
- Describe open-book management, employee recognition, pay-for-performance, and stock option programs.

Most motivational theories were developed in the United States and have a North American bias. Some theories (Maslow's needs hierarchy, achievement need, and equity theory) don't work well for other cultures. However, the desire for interesting work seems important to all workers, and Herzberg's motivator (intrinsic) factors may be universal.

Managers face challenges in motivating unique groups of workers. A diverse workforce is looking for flexibility. Professionals want job challenge and support, and they are motivated by the work itself. Contingent workers want the opportunity to become permanent or to receive skills training. Recognition programs and sincere appreciation for work done can be used to motivate low-skilled, minimum-wage workers.

With open-book management, financial statements (the "books") are shared with employees who have been taught what they mean. Employee recognition programs consist of personal attention, approval, and appreciation for a job well done. Pay-for-performance programs are variable compensation plans that pay employees on the basis of some performance measure. Stock option programs can be set up to reward employees for performance.

THINKING ABOUT MANAGEMENT ISSUES

1. Most of us have to work for a living, and a job is a central part of our lives. So why do managers have to worry so much about employee motivation issues?
2. Describe a task you have done recently for which you exerted a high level of effort. Explain your behavior, using any three of the motivation approaches described in this chapter.
3. If you had to develop an incentive system for a small company that makes tortillas, which elements from which motivation approaches or theories would you use? Why? Would your choice be the same if it was a software design firm?
4. Could managers use any of the motivation theories or approaches to encourage and support workforce diversity efforts? Explain.
5. Many job design experts who have studied the changing nature of work say that people do their best work when they're motivated by a sense of purpose rather than by the pursuit of money. Do you agree? Explain your position.
6. "Too many managers today have forgotten that work should be inspiring and fun and are too out of touch with what makes people productive." How would you respond to this assertion?
7. Can an individual be too motivated? Discuss.
8. "Motivation simply means taking care of your people." How would you respond to this statement?

YOUR TURN to be a Manager

- A good habit to get into if you don't already do it is goal setting. Set goals for yourself using the suggestions from goal-setting theory. Write them down and keep them in a notebook. Track your progress toward achieving these goals.
- Pay attention to times when you're highly motivated and times when you're not as motivated. Write down a description of these times. What accounts for the difference in your level of motivation?
- Interview three managers about how they motivate their employees. What have they found that works best? Write up your findings in a report and be prepared to present it in class.
- Using the job characteristics model, redesign the following jobs to be more motivating: retail store sales associate, utility company meter reader, and check-out cashier at a discount

store. In a written report, describe for each job at least two specific actions you would take for each of the five core job dimensions.

- Do some serious thinking about what you want from the job you get after graduation. Make a list of what's important to you. Are you looking for a pleasant work environment, challenging work, flexible work hours, fun coworkers, or something else? Discuss how you will discover whether a particular job will help you get those things.
- Steve's and Mary's recommended readings: Terry R. Bacon, *What People Want* (Davies-Black Publishing, 2006); Dennis W. Bakke, *Joy at Work* (PVG, 2005); Leon Martel, *High Performers* (Jossey-Bass, 2002); Jon R. Katzenbach, *Peak Performance* (Harvard Business School Press, 2000); and Steven Kerr (ed.), *Ultimate Rewards: What Really Motivates People to Achieve* (Harvard Business School Press, 1997).
- Find five different examples of employee recognition programs. These could be organizations with which you're familiar, or they could come from articles that you find. Write a report describing your examples and evaluating what you think about the various approaches.
- Go to the Web site of the Great Place to Work Institute, www.greatplacetowork.com. What does it say about what makes an organization a great place to work? Next, locate the site's lists of the best companies to work for. Choose one company from each of the international lists. Research each company and describe what it does that makes it a great place to work.
- In your own words, write down three things you learned in this chapter about being a good manager.
- Self-knowledge can be a powerful learning tool. Go to mymanagementlab and complete these self-assessment exercises: What Motivates Me? What Are My Dominant Needs? What Rewards Do I Value Most? What's My View on the Nature of People? How Sensitive Am I to Equity Differences? What's My Job's Motivating Potential? Do I Want an Enriched Job? What Are My Course Performance Goals? How Confident Am I in My Ability to Succeed? What's My Attitude Toward Achievement? Using the results of your assessments, identify personal strengths and weaknesses. What will you do to reinforce your strengths and improve your weaknesses?

CASE APPLICATION

Paradise Lost . . . Or Gained?

A massage every other week, on-site laundry, swimming pool and spa, free delicious all-you-can-eat gourmet meals. What more could an employee want? Sounds like an ideal job, doesn't it? However, at Google, many people are demonstrating by their decisions to leave the company that all those perks (and these are just a few) aren't enough to keep them there. As one analyst said, "Yes, Google's making gobs of money. Yes, it's full of smart people. Yes, it's a wonderful place to work. So why are so many people leaving?"

Google employees inspect the company's maps program.

Google has been named the "best company to work for" by *Fortune* magazine for two years running, but make no mistake: Google's executives made the decision to offer all those fabulous perks for several reasons: to attract the best knowledge workers it can in an intensely competitive, cutthroat market; to help employees work long hours and not have to deal with time-consuming personal chores outside work; to show employees they are valued; and to have employees remain Googlers (the name used for employees) for many years. But a number of Googlers have jumped ship and given up these fantastic benefits to go out on their own.

For instance, Sean Knapp and two colleagues, brothers Bismarck and Belsasar Lepe, came up with an idea about how to handle Web video. They left Google in April 2007, or as one person put it, "expelled themselves from paradise to start their own company." When the threesome left the company, Google wanted badly for them and their project to stay. Google offered them a "blank check." But they realized they would do all the hard work, and Google would own the product. So off they went, for the excitement of a start-up.

If this were an isolated occurrence, it would be easy to write off. But it's not. Other talented Google employees have done the same thing. In fact, so many of them have left that they've formed an informal alumni club of ex-Googlers turned entrepreneurs.

Discussion Questions

1. What's it like to work at Google? (Hint: Go to Google's Web site and click on About Google. Find the section Jobs at Google.) What's your assessment of the company's work environment?
2. Google is doing a lot for its employees, but obviously it's not done enough to retain several of its talented employees. Using what you've learned from studying the various motivation theories, what does this situation tell you about employee motivation?
3. What do you think is Google's biggest challenge in keeping employees motivated?
4. If you were managing a team of Google employees, how would you keep them motivated?

Sources: A. Lashinsky, "Where Does Google Go Next?" *CNNMoney.com*, May 12, 2008; K. Hafner, "Google Options Make Masseuse a Multimillionaire," *New York Times* online, www.nytimes.com, November 12, 2007; Q. Hardy, "Close to the Vest," *Forbes*, July 2, 2007, pp. 40–42; K. J. Delaney, "Start-ups Make Inroads with Google's Work Force," *Wall Street Journal* online, www.wsj.com, June 28, 2007; and "Perk Place: The Benefits Offered by Google and Others May Be Grand, but They're All Business," *Knowledge @ Wharton,* knowledge.wharton.upenn.edu, March 21, 2007.

Let's Get Real:

Meet the Manager

Nimish Vohra

Country Head, India
Regalix

MY JOB: Keep clients and employees happy. Get new business from new and existing customers. Address day-to-day fires, keep employees motivated, and run a tight ship to maintain profitability.

BEST PART OF MY JOB: New challenges that I face every day—could be financial, legal, and delivery or employee related. Working with people to find solutions.

WORST PART OF MY JOB: When I am not able to understand other people or when I think I have been let down as a friend. And the worse part is that the other person might be thinking the same!

BEST MANAGEMENT ADVICE EVER RECEIVED: If it is brickbats, you get to take them on your own self and if it is bouquet shower them on your team.

You'll be hearing more from this real manager throughout the chapter.

Chapter 16

Managers as Leaders

Leaders in organizations make things happen. But what makes leaders different from nonleaders? What's the most appropriate style of leadership? What can you do to be seen as a leader? Those are just a few of the questions we'll try to answer in this chapter. Focus on the following learning outcomes as you read and study this chapter.

LEARNING OUTCOMES

A Manager's Dilemma

Courtesy of Citigate Cunningham for HCL

HCL Technologies is headquartered in the world's largest democracy, so it's quite fitting that the New Delhi–based company is attempting a radical experiment in workplace democracy.[1] CEO Vineet Nayar is committed to creating a company where the job of company leaders is to enable people to find their own destiny by gravitating to their strengths. Although he believes that the command-and-control dictatorship approach is the easiest management style, he also thinks it's not the most productive. In his corporate democracy, employees can write a "trouble ticket" on anyone in the company. Anyone with trouble tickets has to respond, just as if he or she were dealing with a customer who had problems and needed some response. Nayar also believes that leaders should be open to criticism. He volunteered to share the weaknesses from his 360-degree feedback for all employees to see. Although a lot of people said he was crazy to communicate his weaknesses, Nayar believed that it was a good way to increase his accountability as a leader to his employees. Such an environment requires a lot of trust between leaders and followers. How can Nayar continue to build that trust?

What would you do?

Vineet Nayar is a good example of what it takes to be a good leader in today's organizations. He has created an environment in which employees feel like they're heard and trusted. However, it's important that he continue to nurture this culture *and* be seen as an effective leader. Why is leadership so important? Because the leaders in organizations make things happen.

LEARNING **OUTCOME 16.1** ▷

WHO ARE LEADERS, AND WHAT IS LEADERSHIP?

Let's begin by clarifying who leaders are and what leadership is. Our definition of a **leader** is someone who can influence others and who has managerial authority. **Leadership** is what leaders do. It's a process of leading a group and influencing that group to achieve its goals.

Are all managers leaders? Because leading is one of the four management functions, ideally, all managers *should* be leaders. Thus, we're going to study leaders and leadership from a managerial perspective.[2] However, even though we're looking at these from a managerial perspective, we're aware that groups often have informal leaders who emerge. Although these informal leaders may be able to influence others, they have not been the focus of most leadership research and are not the types of leaders we're studying in this chapter.

Leaders and leadership, like motivation, are organizational behavior topics that have been researched a lot. Most of that research has been aimed at answering the question "What is an effective leader?" We'll begin our study of leadership by looking at some early leadership theories that attempted to answer that question.

QUICK LEARNING REVIEW:

LEARNING OUTCOME 16.1

- Define leader and leadership.
- Explain why managers should be leaders.

Go to page 395 and see how well you know this material.

LEARNING **OUTCOME 16.2** ▷

EARLY LEADERSHIP THEORIES

People have been interested in leadership since they started coming together in groups to accomplish goals. However, it wasn't until the early part of the twentieth century that researchers actually began to study leadership. These early leadership theories focused on the *leader* (trait theories) and how the *leader interacted* with his or her group members (behavioral theories).

TRAIT THEORIES

Even before the 2008 presidential election, Barack Obama had captured the attention of political analysts and the public.[3] He had been compared to popular historical persons such as Abraham Lincoln and Martin Luther King, Jr. Many are saying that he has what it takes to be a leading political figure—characteristics such as self-awareness, clarity of speech, keen intellect, and an ability to relate to people. Is Obama a leader? The trait theories of leadership would answer that by focusing on his traits.

Leadership research in the 1920s and 1930s focused on isolating leader traits—that is, characteristics—that would differentiate leaders from nonleaders. Some of the traits studied included physical stature, appearance, social class, emotional stability, fluency of speech, and sociability. Despite the best efforts of researchers, it proved impossible to identify a set of traits that would *always* differentiate a leader (the person) from a nonleader. Maybe it was a bit optimistic to think that there could be consistent and unique traits that would apply universally to all effective leaders, no matter whether they were in charge of Toyota Motor Corporation, the Moscow Ballet, the country of France, a local collegiate chapter of Spic Macay, Aggarwal Sweet Shop, or Delhi University. However, later attempts to identify traits consistently associated with *leadership* (the process, not the person) were more successful. The seven traits shown to be associated with effective leadership are described briefly in Exhibit 16–1.[4]

Let's Get Real: F2F

TRAITS LEADERS SHOULD HAVE:

- Ability to grasp of the landscape in which they are operating.
- Ability to understand people.
- Ability to trust others.
- Ability to take decision fast and to back them.

Researchers eventually recognized that traits alone were not sufficient for identifying effective leaders because explanations based solely on traits ignored the interactions of leaders and their group members as well as situational factors. Possessing the appropriate traits only made it more likely that an individual would be an effective leader. Therefore, leadership research from the late 1940s to the mid-1960s concentrated on the preferred behavioral styles that leaders demonstrated. Researchers wondered whether there was something unique in what effective leaders *did*—in other words, in their *behavior.*

BEHAVIORAL THEORIES

Entrepreneur and Member of Parliament—Rajya Sabha, Rajeev Chandrashekhar founded BPL Mobile played a significant role in the development of Indian telecom sector. He founded Jupiter Capital in 2005 to venture into futuristic businesses. Rajeev Chandrashekar is known to be very autocratic, micromanaging, and demanding. In contrast, Ratan Tata, Chairman of the Tata Group, is known to be able to ignite passion and ownership in people who work with him. How? By encouraging their participation and letting them figure out how best to do things.[5] These are two successful companies whose leaders, as you can see, behave in two very different ways. What do

leader
A person who can influence others and who has managerial authority.

leadership
A process of influencing a group to achieve goals.

Exhibit 16–1

Seven Traits Associated with Leadership

1. *Drive.* Leaders exhibit a high effort level. They have a relatively high desire for achievement, they are ambitious, they have a lot of energy, they are tirelessly persistent in their activities, and they show initiative.
2. Desire to *lead.* Leaders have a strong desire to influence and lead others. They demonstrate the willingness to take responsibility.
3. *Honesty and integrity.* Leaders build trusting relationships with followers by being truthful or nondeceitful and by showing high consistency between word and deed.
4. *Self-confidence.* Followers look to leaders for an absence of self-doubt. Leaders, therefore, need to show self-confidence in order to convince followers of the rightness of their goals and decisions.
5. *Intelligence.* Leaders need to be intelligent enough to gather, synthesize, and interpret large amounts of information, and they need to be able to create visions, solve problems, and make correct decisions.
6. *Job-relevant knowledge.* Effective leaders have a high degree of knowledge about the company, industry, and technical matters. In-depth knowledge allows leaders to make well-informed decisions and to understand the implications of those decisions.
7. *Extraversion.* Leaders are energetic, lively people. They are sociable, assertive, and rarely silent or withdrawn.

Sources: S. A. Kirkpatrick and E. A. Locke, "Leadership: Do Traits Really Matter?" *Academy of Management Executive,* May 1991, pp. 48–60; and T. A. Judge, J. E. Bono, R. Ilies, and M. W. Gerhardt, "Personality and Leadership: A Qualitative and Quantitative Review," *Journal of Applied Psychology,* August 2002, pp. 765–780.

we know about leader behavior and how can it help us in our understanding of what an effective leader is?

Researchers hoped that the **behavioral theories** approach would provide more definitive answers about the nature of leadership than did the trait theories. The four main leader behavior studies are summarized in Exhibit 16–2.

University of Iowa Studies. The University of Iowa studies explored three leadership styles to find which was the most effective.[6] The **autocratic style** described a leader who dictated work methods, made unilateral decisions, and limited employee participation. The **democratic style** described a leader who involved employees in decision making, delegated authority, and used feedback as an opportunity for coaching employees. Finally, the **laissez-faire style** described a leader who let the group make decisions and complete the work in whatever way it saw fit. The researchers' results seemed to indicate that the democratic style contributed to both good quantity and quality of work. Had the answer to the question of the most effective leadership style been found? Unfortunately, it wasn't that simple. Later studies of the autocratic and democratic styles showed mixed results. For instance, the democratic style sometimes produced higher performance levels than the autocratic style, but at other times, it didn't. However, more consistent results were found when a measure of employee satisfaction was used. Group members were more satisfied under a democratic leader than under an autocratic one.[7]

Now leaders had a dilemma! Should they focus on achieving higher performance or on achieving higher member satisfaction? This recognition of the dual nature of a leader's behavior—that is, focus on the task and focus on the people—was also a key characteristic of the other behavioral studies.

The Ohio State Studies. The Ohio State studies identified two important dimensions of leader behavior.[8] Beginning with a list of more than 1,000 behavioral dimensions, the researchers eventually narrowed it down to just 2 that accounted for most of the leadership behavior described by group members. The first dimension, called **initiating structure**, referred to the extent to which a leader defined his or her role and the roles of group members in attaining goals. It included behaviors that involved attempts to organize work, work relationships, and goals. The second

Exhibit 16–2

Behavioral Theories of Leadership

	Behavioral Dimension	**Conclusion**
University of Iowa	*Democratic style:* involving subordinates, delegating authority, and encouraging participation *Autocratic style:* dictating work methods, centralizing decision making, and limiting participation *Laissez-faire style:* giving group freedom to make decisions and complete work	Democratic style of leadership was most effective, although later studies showed mixed results.
Ohio State	*Consideration:* being considerate of followers' ideas and feelings *Initiating structure:* structuring work and work relationships to meet job goals	High–high leader (high in consideration and high in initiating structure) achieved high subordinate performance and satisfaction, but not in all situations
University of Michigan	*Employee oriented:* emphasized interpersonal relationships and taking care of employees' needs *Production oriented:* emphasized technical or task aspects of job	Employee-oriented leaders were associated with high group productivity and higher job satisfaction.
Managerial Grid	*Concern for people:* measured leader's concern for subordinates on a scale of 1 to 9 (low to high) *Concern for production:* measured leader's concern for getting job done on a scale 1 to 9 (low to high)	Leaders performed best with a 9.9 style (high concern for production and high concern for people).

dimension, called **consideration**, was defined as the extent to which a leader had work relationships characterized by mutual trust and respect for group members' ideas and feelings. A leader who was high in consideration helped group members with personal problems, was friendly and approachable, and treated all group members as equals. He or she showed concern for (was considerate of) his or her followers' comfort, well-being, status, and satisfaction. Research found that a leader who was high in both initiating structure and consideration (a **high-high leader**) sometimes achieved high group task performance and high group member satisfaction, but not always.

University of Michigan Studies. Leadership studies conducted at the University of Michigan at about the same time as those being done at Ohio State also hoped to identify behavioral characteristics of leaders that were related to performance effectiveness. The Michigan group also came up with two dimensions of leadership

behavioral theories
Leadership theories that identify behaviors that differentiate effective leaders from ineffective leaders.

autocratic style
A leader who dictates work methods, makes unilateral decisions, and limits employee participation.

democratic style
A leader who involves employees in decision making, delegates authority, and uses feedback as an opportunity for coaching employees.

laissez-faire style
A leader who lets the group make decisions and complete the work in whatever way it sees fit.

initiating structure
The extent to which a leader defines his or her role and the roles of group members in attaining goals.

consideration
The extent to which a leader has work relationships characterized by mutual trust and respect for group members' ideas and feelings.

high-high leader
A leader high in both initiating structure and consideration behaviors.

behavior, which they labeled *employee oriented* and *production oriented*.[9] Leaders who were employee oriented were described as emphasizing interpersonal relationships. The production-oriented leaders, in contrast, tended to emphasize the task aspects of the job. Unlike the other studies, the Michigan studies concluded that leaders who were employee oriented were able to get high group productivity and high group member satisfaction.

The Managerial Grid. The behavioral dimensions from the early leadership studies provided the basis for the development of a two-dimensional grid for appraising leadership styles. This **managerial grid** used the behavioral dimensions "concern for people" and "concern for production" and evaluated a leader's use of these behaviors, ranking them on a scale from 1 (low) to 9 (high).[10] Although the grid (shown in Exhibit 16–3) had 81 potential categories into which a leader's behavioral style might fall, only 5 styles were named: impoverished management (1,1), task management (9,1), middle-of-the-road management (5,5), country club management (1,9), and team management (9,9). Of these five styles, the researchers concluded that managers performed best when using a 9,9 style. Unfortunately, the grid offered no explanations about what made a manager an effective leader; it only provided a framework for conceptualizing leadership style. In fact, there's little substantive evidence to support the conclusion that a 9,9 style is most effective in all situations.[11]

Exhibit 16–3 The Managerial Grid

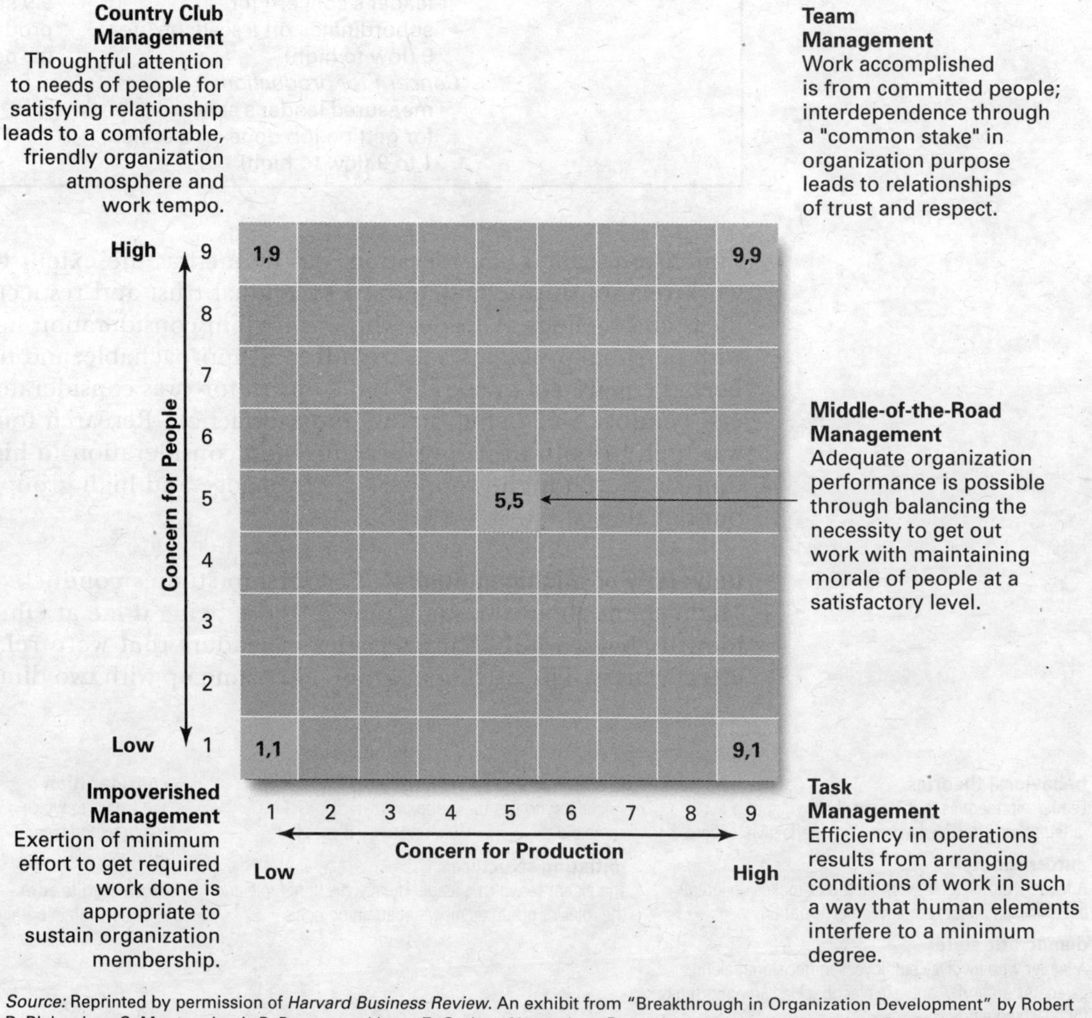

Source: Reprinted by permission of *Harvard Business Review*. An exhibit from "Breakthrough in Organization Development" by Robert R. Blake, Jane S. Mouton, Louis B. Barnes, and Larry E. Greiner, November– December 1964, p. 136.

Leadership researchers were discovering that predicting leadership success involved something more complex than isolating a few leader traits or preferable behaviors. They began looking at situational influences. Specifically, which leadership styles might be suitable in different situations, and what were these different situations?

QUICK LEARNING REVIEW:

LEARNING OUTCOME 16.2

- **Discuss what research has shown about leadership traits.**
- **Contrast the findings of the four behavioral leadership theories.**
- **Explain the dual nature of a leader's behavior.**

Go to page 395 and see how well you know this material.

LEARNING **OUTCOME 16.3** ▷

CONTINGENCY THEORIES OF LEADERSHIP

"The corporate world is filled with stories of leaders who failed to achieve greatness because they failed to understand the context they were working in."[12] In this section, we examine three contingency theories: the Fiedler model, Hersey and Blanchard's situational leadership theory, and path–goal theory. Each of these theories looks at defining leadership style and the situation, and it attempts to answer the *if–then* contingencies (that is, *if* this is the context or situation, *then* this is the best leadership style to use).

THE FIEDLER MODEL

The first comprehensive contingency model for leadership was developed by Fred Fiedler.[13] The **Fiedler contingency model** proposed that effective group performance depended on properly matching the leader's style and the amount of control and influence in the situation. The model was based on the premise that a certain leadership style would be most effective in different types of situations. The keys were to (1) define those leadership styles and the different types of situations and then (2) identify the appropriate combinations of style and situation.

Fiedler proposed that a key factor in leadership success was an individual's basic leadership style, either task oriented or relationship oriented. To measure a leader's style, Fiedler developed the **least-preferred coworker (LPC) questionnaire**. This questionnaire contained 18 pairs of contrasting adjectives—for example, pleasant–unpleasant, cold–warm, boring–interesting, and friendly–unfriendly. Respondents were asked to think of all the coworkers they had ever had and to describe that one person they *least enjoyed* working with by rating him or her on a scale of 1 to 8 for each of the 18 sets of adjectives. (The 8 always described the positive adjective out of the pair, and the 1 always described the negative adjective out of the pair.)

If the leader described the least preferred coworker in relatively positive terms (in other words, a "high" LPC score—a score of 64 or above), then the respondent was primarily interested in good personal relations with coworkers, and the style would be described as *relationship oriented.* In contrast, if the leader saw the least preferred coworker in relatively unfavorable terms (a low LPC score—a score of 57 or below), he or she was primarily interested in productivity and getting the job done; thus, the individual's style would be labeled *task oriented.* Fiedler acknowledged that a small number of people might fall in between these two extremes and not have a cut-and-dried leadership style. One other important point is that Fiedler assumed that a person's leadership style was fixed, regardless of the situation. In other words, if you were a

managerial grid
A two-dimensional grid for appraising leadership styles.

Fiedler contingency model
A leadership theory which proposed that effective group performance depended on the proper match between a leader's style and the degree to which the situation allowed the leader to control and influence.

least-preferred coworker (LPC) questionnaire
A questionnaire that measured whether a leader was task or relationship oriented.

relationship-oriented leader, you'd always be one, and if you were a task-oriented leader, you'd always be one.

After an individual's leadership style had been assessed through the LPC, it was time to evaluate the situation in order to be able to match the leader with the situation. Fiedler's research uncovered three contingency dimensions that defined the key situational factors in leader effectiveness:

- **Leader–member relations**—The degree of confidence, trust, and respect employees had for their leader; rated as either good or poor.
- **Task structure**—The degree to which job assignments were formalized and structured; rated as either high or low.
- **Position power**—The degree of influence a leader had over activities such as hiring, firing, discipline, promotions, and salary increases; rated as either strong or weak.

Each leadership situation was evaluated in terms of these three contingency variables, which when combined produced eight possible situations that were either favorable or unfavorable for the leader. (See the bottom of Exhibit 16–4.) Situations I, II, and III were classified as highly favorable for the leader. Situations IV, V, and VI were moderately favorable for the leader. And situations VII and VIII were described as highly unfavorable for the leader.

Once Fiedler had described the leader variables and the situational variables, he had everything he needed to define the specific contingencies for leadership effectiveness. To do so, he studied 1,200 groups where he compared relationship-oriented versus task-oriented leadership styles in each of the eight situational categories. He concluded that task-oriented leaders performed better in very favorable situations and in very unfavorable situations. (See the top of Exhibit 16–4, where performance is shown on the vertical axis and situation favorableness is shown on the horizontal axis.) On the other hand, relationship-oriented leaders performed better in moderately favorable situations.

Because Fiedler treated an individual's leadership style as fixed, there were only two ways to improve leader effectiveness. First, you could bring in a new leader whose style better fit the situation. For instance, if the group situation was highly unfavorable but was led by a relationship-oriented leader, the group's performance could be improved by replacing that person with a task-oriented leader. The second alternative was to change

Exhibit 16–4 The Fiedler Model

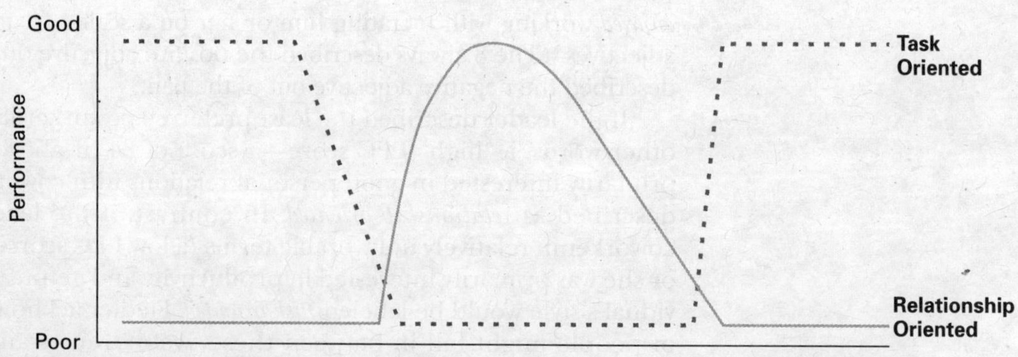

Situation Favorableness: Highly Favorable | Moderate | Highly Unfavorable

Category	I	II	III	IV	V	VI	VII	VIII
Leader–Member Relations	Good	Good	Good	Good	Poor	Poor	Poor	Poor
Task Structure	High	High	Low	Low	High	High	Low	Low
Position Power	Strong	Weak	Strong	Weak	Strong	Weak	Strong	Weak

Let's Get Real: F2F

MY LEADERSHIP STYLE WOULD BE DESCRIBED AS:

- Very much on the ground and working with the team.
- Allowing risk taking and ownership.

the situation to fit the leader. This could be done by restructuring tasks; by increasing or decreasing the power that the leader had over factors such as salary increases, promotions, and disciplinary actions; or by improving the leader–member relations.

Research testing the overall validity of Fiedler's model has shown considerable evidence in support of the model.[14] However, his theory wasn't without criticisms. The major criticism is that it's probably unrealistic to assume that a person can't change his or her leadership style to fit the situation. Effective leaders can, and do, change their styles. Another is that the LPC wasn't very practical. Finally, the situation variables were difficult to assess.[15] Despite its shortcomings, the Fiedler model showed that effective leadership style needed to reflect situational factors.

HERSEY AND BLANCHARD'S SITUATIONAL LEADERSHIP THEORY

Paul Hersey and Ken Blanchard developed a leadership theory that has gained a strong following among management development specialists.[16] This model, called **situational leadership theory (SLT)**, is a contingency theory that focuses on followers' readiness. Before we proceed, there are two points we need to clarify: Why a leadership theory focuses on the followers and what is meant by the term *readiness*.

The emphasis on the followers in leadership effectiveness reflects the reality that it *is* the followers who accept or reject the leader. Regardless of what the leader does, the group's effectiveness depends on the actions of the followers. This is an important dimension that most leadership theories have overlooked or underemphasized. **Readiness**, as defined by Hersey and Blanchard, refers to the extent to which people have the ability and willingness to accomplish a specific task.

SLT uses the same two leadership dimensions that Fiedler identified: task and relationship behaviors. However, Hersey and Blanchard go a step further by considering each as either high or low and then combining them into four specific leadership styles:

- **Telling (high task–low relationship)**—The leader defines roles and tells people what, how, when, and where to do various tasks.
- **Selling (high task–high relationship)**—The leader provides both directive and supportive behavior.
- **Participating (low task–high relationship)**—The leader and followers share in decision making; the main role of the leader is facilitating and communicating.
- **Delegating (low task–low relationship)**—The leader provides little direction or support.

The final component in the SLT model is the four stages of follower readiness:

- **R1**—People are both *unable and unwilling* to take responsibility for doing something. Followers aren't competent or confident.
- **R2**—People are *unable but willing* to do the necessary job tasks. Followers are motivated but lack the appropriate skills.
- **R3**—People are *able but unwilling* to do what the leader wants. Followers are competent but don't want to do something.
- **R4**—People are both *able and willing* to do what is asked of them.

SLT essentially views the leader–follower relationship as like that of a parent and a child. Just as a parent needs to relinquish control when a child becomes more mature and responsible, so, too, should leaders. As followers reach higher levels of readiness, the leader responds not only by decreasing control over their activities

leader–member relations
One of Fiedler's situational contingencies that described the degree of confidence, trust, and respect employees had for their leader.

task structure
One of Fiedler's situational contingencies that described the degree to which job assignments were formalized and structured.

position power
One of Fiedler's situational contingencies that described the degree of influence a leader had over activities such as hiring, firing, discipline, promotions, and salary increases.

situational leadership theory (SLT)
A leadership contingency theory that focuses on followers' readiness.

readiness
The extent to which people have the ability and willingness to accomplish a specific task.

Kristen Cardwell is an infectious diseases researcher at St. Jude's Children's Research Hospital in Memphis, Tennessee. Cardwell and other medical researchers at the hospital have a high level of follower readiness. As responsible, experienced, and mature employees, they are both able and willing to complete their tasks under leadership that gives them freedom to make and implement decisions. This leader-follower relationship is consistent with Hersey and Blanchard's situational leadership theory.

but also by decreasing relationship behaviors. The SLT says if followers are at R1 (*unable and unwilling* to do a task), the leader needs to use the telling style and give clear and specific directions; if followers are at R2 (*unable and willing*), the leader needs to use the selling style and display high task orientation to compensate for the followers' lack of ability and high relationship orientation to get followers to "buy into" the leader's desires; if followers are at R3 (*able and unwilling*), the leader needs to use the participating style to gain their support; and if employees are at R4 (both *able and willing*), the leader doesn't need to do much and should use the delegating style.

SLT has intuitive appeal. It acknowledges the importance of followers and builds on the logic that leaders can compensate for ability and motivational limitations in their followers. However, research efforts to test and support the theory have generally been disappointing.[17] Possible explanations include internal inconsistencies in the model as well as problems with research methodology. Despite its appeal and wide popularity, we have to be cautious about any enthusiastic endorsement of SLT.

PATH–GOAL THEORY

Currently, one of the most respected approaches to understanding leadership is **path–goal theory**, which states that the leader's job is to assist followers in attaining their goals and to provide direction or support needed to ensure that their goals are compatible with the goals of the group or organization. Developed by Robert House, path–goal theory takes key elements from the expectancy theory of motivation.[18] The term *path–goal* is derived from the belief that effective leaders clarify the path to help their followers get from where they are to the achievement of their work goals and make the journey along the path easier by reducing roadblocks and pitfalls.

House identified four leadership behaviors:

- **Directive leader**—The leader lets subordinates know what's expected of them, schedules work to be done, and gives specific guidance on how to accomplish tasks.
- **Supportive leader**— The leader shows concern for the needs of followers and is friendly.
- **Participative leader**—The leader consults with group members and uses their suggestions before making a decision.
- **Achievement oriented leader**—The leader sets challenging goals and expects followers to perform at their highest level.

In contrast to Fiedler's view that a leader couldn't change his or her behavior, House assumed that leaders are flexible and can display any or all of these leadership styles, depending on the situation.

Exhibit 16–5

Path–Goal Model

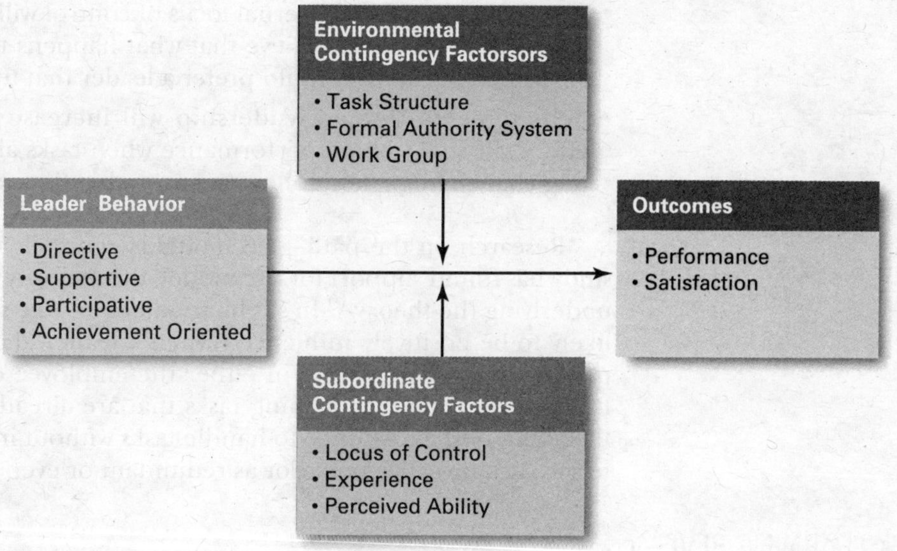

As Exhibit 16–5 illustrates, path–goal theory proposes two situational or contingency variables that moderate the leadership behavior–outcome relationship: those in the *environment* that are outside the control of the follower (factors including task structure, formal authority system, and the work group) and those that are part of the personal characteristics of the *follower* (including locus of control, experience, and perceived ability). Environmental factors determine the type of leader behavior required if subordinate outcomes are to be maximized; personal characteristics of the follower determine how the environment and leader behavior are interpreted. The theory proposes that a leader's behavior won't be effective if it's redundant with what the environmental structure is providing or is incongruent with follower characteristics. For example, the following are some predictions from path–goal theory:

- Directive leadership leads to greater satisfaction when tasks are ambiguous or stressful than when they are highly structured and well laid out. The followers aren't sure what to do, so the leader needs to give them some direction.
- Supportive leadership results in high employee performance and satisfaction when subordinates are performing structured tasks. In this situation, the leader only needs to support followers, not tell them what to do.
- Directive leadership is likely to be perceived as redundant among subordinates with high perceived ability or with considerable experience. These followers are quite capable, so they don't need a leader to tell them what to do.
- The clearer and more bureaucratic the formal authority relationships, the more leaders should exhibit supportive behavior and deemphasize directive behavior. The organizational situation has provided the structure as far as what is expected of followers, so the leader's role is simply to support.
- Directive leadership will foster higher employee satisfaction when there is substantive conflict within a work group. In this situation, the followers need a leader who will take charge.
- Subordinates with an internal locus of control will be more satisfied with a participative style. Because these followers believe that they control what happens to them, they prefer to participate in decisions.

path–goal theory
A leadership theory that says the leader's job is to assist followers in attaining their goals and to provide direction or support needed to ensure that their goals are compatible with the goals of the group or organization.

- Subordinates with an external locus of control will be more satisfied with a directive style. These followers believe that what happens to them is a result of the external environment, so they would prefer a leader that tells them what to do.
- Achievement-oriented leadership will increase subordinates' expectations that effort will lead to high performance when tasks are ambiguously structured. By setting challenging goals, followers know what the expectations are.

Research on the path–goal model is generally encouraging. Although not every study has found support for the model, the majority of the evidence supports the logic underlying the theory.[19] In summary, an employee's performance and satisfaction are likely to be positively influenced when a leader chooses a leadership style that compensates for shortcomings in either the employee or the work setting. However, if a leader spends time explaining tasks that are already clear or when an employee has the ability and experience to handle tasks without interference, the employee is likely to see such directive behavior as redundant or even insulting.

QUICK LEARNING REVIEW:

LEARNING OUTCOME 16.3

- Explain Fiedler's contingency model of leadership.
- Describe situational leadership theory.
- Discuss how path–goal theory explains leadership.

Go to page 395 and see how well you know this material.

LEARNING **OUTCOME 16.4** ▷

CONTEMPORARY VIEWS OF LEADERSHIP

What are the latest views of leadership? There are three we want to look at: transformational–transactional leadership, charismatic–visionary leadership, and team leadership.

TRANSFORMATIONAL–TRANSACTIONAL LEADERSHIP

Many early leadership theories viewed leaders as **transactional leaders**—that is, leaders who lead primarily by using social exchanges (or transactions).Transactional leaders guide or motivate followers to work toward established goals by exchanging rewards for their productivity.[20] But there's another type of leader—a **transformational leader**—who stimulates and inspires (transforms) followers to achieve extraordinary outcomes. Examples include Ashok Soola of Mindtree and B. Muthuraman of Tata Steel. They pay attention to the concerns and developmental needs of individual followers; they change followers' awareness of issues by helping those followers look at old problems in new ways; and they are able to excite, arouse, and inspire followers to exert extra effort to achieve group goals.

Let's Get Real: F2F

WHAT MAKES A LEADER SUCCESSFUL IS:

- Measurable success in his or her field!
- Ability to guide and lead people.
- Ability to understand goals and break it down into actions.

Transactional and transformational leadership shouldn't be viewed as opposing approaches to getting things done.[21] Transformational leadership develops from transactional leadership. Transformational leadership produces levels of employee effort and performance that go beyond what would occur with a transactional approach alone. Moreover, transformational leadership is more than charisma because a transformational leader attempts to instill in followers the ability to question not only established views but views held by the leader.[22]

The evidence supporting the superiority of transformational leadership over transactional leadership is overwhelmingly impressive. For instance, studies that looked at managers in different settings, including the military and business, found that transformational leaders were evaluated as more effective, higher performers, more promotable than their transactional counterparts, and more interpersonally sensitive.[23] In addition, evidence indicates that transformational leadership is strongly correlated with lower turnover rates and higher levels of productivity, employee satisfaction, creativity, goal attainment, and follower well-being.[24]

Charismatic–Visionary Leadership Jeff Bezos, founder and CEO of Amazon.com, exudes energy, enthusiasm, and drive.[25] He's fun loving (his legendary laugh has been described as a flock of Canadian geese on nitrous oxide) but has pursued his vision for Amazon with serious intensity and has demonstrated an ability to inspire his employees through the ups and downs of a rapidly growing company. Bezos is what we call a **charismatic leader**—that is, an enthusiastic, self-confident leader whose personality and actions influence people to behave in certain ways.

Several authors have attempted to identify personal characteristics of charismatic leaders.[26] The most comprehensive analysis identified five such characteristics: Charismatic leaders have a vision, ability to articulate that vision, willingness to take risks to achieve that vision, sensitivity to both environmental constraints and follower needs, and behaviors that are out of the ordinary.[27]

An increasing body of evidence shows impressive correlations between charismatic leadership and high performance and satisfaction among followers.[28] Although one study found that charismatic CEOs had no impact on subsequent organizational performance, charisma is still believed to be a desirable leadership quality.[29]

If charisma is desirable, can people learn to be charismatic leaders? Or are charismatic leaders born with their qualities? Although a small number of experts still think that charisma can't be learned, most believe that individuals can be trained to exhibit charismatic behaviors.[30] For example, researchers have succeeded in teaching undergraduate students to "be" charismatic. How? The students have been taught to articulate a far-reaching goal, communicate high performance expectations, exhibit confidence in the ability of subordinates to meet those expectations, and empathize with the needs of their subordinates; they have learned to project a powerful, confident, and dynamic presence; and they have practiced using a captivating and engaging tone of voice. The researchers have also trained student leaders to use charismatic nonverbal behaviors, including leaning toward the follower when communicating, maintaining direct eye contact, and having a relaxed posture and animated facial expressions. In groups with these "trained" charismatic leaders, members had higher task performance, higher task adjustment, and better adjustment to the leader and to the group than did group members who worked in groups led by noncharismatic leaders.

One last thing we should say about charismatic leadership is that it may not always be necessary to achieve high levels of employee performance. Charismatic leadership may be most appropriate when the follower's task has an ideological purpose or when the environment involves a high degree of stress and uncertainty.[31] This may explain why, when charismatic leaders surface, they're likely to crop up in politics, religion, or wartime; or when a business firm is starting up or facing a survival crisis. For example, Martin Luther King, Jr., used his charisma to bring about social equality through nonviolent means; and Steve Jobs achieved unwavering loyalty and commitment from Apple's technical staff in the early 1980s by articulating a vision of personal computers that would dramatically change the way people lived.

Although the term *vision* is often linked with charismatic leadership, **visionary leadership** is different because it's the ability to create and articulate a realistic, credible, and attractive vision of the future that improves on the present situation.[32] This vision, if properly selected and implemented, is so energizing that it "in effect jump-starts the future by calling forth the skills, talents, and resources to make it happen."[33]

An organization's vision should offer clear and compelling imagery that taps into people's emotions and inspires enthusiasm to pursue the organization's goals. It should be able to generate possibilities that are inspirational and unique and offer new ways of doing things that are clearly better for the organization and its members.

transactional leaders
Leaders who lead primarily by using social exchanges (or transactions).

transformational leaders
Leaders who stimulate and inspire (transform) followers to achieve extraordinary outcomes.

charismatic leaders
Enthusiastic, self-confident leaders whose personalities and actions influence people to behave in certain ways.

visionary leadership
The ability to create and articulate a realistic, credible, and attractive vision of the future that improves on the present situation.

Mallika Srinivasan, the CEO of Tractor and Farm Equipment (TAFE), runs a successful 2,500-crore company. The tractor business is highly competitive and needs a fresh outlook. In addition to acquiring Eicher Tractors to become a company with national presence, it was important to invest in R&D and find out what the farmer wanted. She role modeled and went herself directly to the farmers to learn. She believes in continuous learning and never shies away from an opportunity to learn.[34]

Courtesy: Tractor and Farm Equipment (TAFE)

Visions that are clearly articulated and have powerful imagery are easily grasped and accepted. For instance, Michael Dell of Dell Inc. created a vision of a business that sells and delivers customized PCs directly to customers in less than a week. Seven housewives living in Girgaum, a crowded residential area of Mumbai with a capital of Rs 80, took over a sick Papad-making unit in 1959. Today, Shri Mahila Griha Udyog Lijjat Papad is a household name for its high-quality product, Lijjat Papad. Sticking to its core values of business, family, and devotion for the past 50 years, Lijjat has ensured that all stakeholders are satisfied, members earn profit, agents distributed across 80,000 homes get reimbursed for their effort, consumers get the best quality at reasonable prices, and some profit is ploughed back for societal causes.[35]

Team Leadership. Because leadership is increasingly taking place within a team context and more organizations are using work teams, the role of the leader in guiding team members has become increasingly important. The role of a team leader *is* different from the traditional leadership role, as J. D. Bryant, a supervisor at the Texas Instruments Forest Lane plant in Dallas, discovered.[36] One day he was contentedly overseeing a staff of 15 circuit board assemblers. The next day he was told that the company was going to use employee teams, and he was to become a "facilitator." He said, "I'm supposed to teach the teams everything I know and then let them make their own decisions." Confused about his new role, he admitted, "There was no clear plan on what I was supposed to do." What *is* involved in being a team leader?

Many leaders are not equipped to handle the change to employee teams. As one consultant noted, "Even the most capable managers have trouble making the transition because all the command-and-control type things they were encouraged to do before are no longer appropriate. There's no reason to have any skill or sense of this."[37] This same consultant estimated that "probably 15 percent of managers are natural team leaders; another 15 percent could never lead a team because it runs counter to their personality—that is, they're unable to sublimate their dominating style for the good of the team. Then there's that huge group in the middle: Team leadership doesn't come naturally to them, but they can learn it."[38]

The challenge for many managers is learning how to become an effective team leader. They have to learn skills such as patiently sharing information, being able to trust others and give up authority, and understanding when to intervene. And effective team leaders have mastered the difficult balancing act of knowing when to leave their teams alone and when to get involved. New team leaders may try to retain too much control at a time when team members need more autonomy, or they may abandon their teams at times when the teams need support and help.[39]

One study looking at organizations that had reorganized themselves around employee teams found certain common responsibilities of all leaders. These included coaching, facilitating, handling disciplinary problems, reviewing team and individual

Exhibit 16–6

Team Leadership Roles

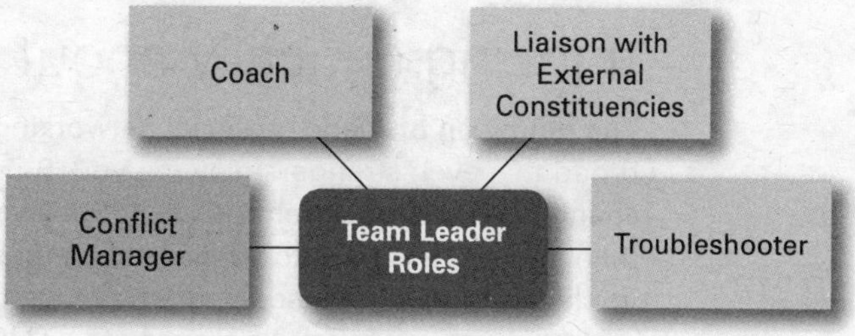

performance, training, and communication.[40] However, a more meaningful way to describe the team leader's job is to focus on two priorities: (1) managing the team's external boundary and (2) facilitating the team process.[41] These priorities entail four specific leadership roles as shown in Exhibit 16–6.

QUICK LEARNING REVIEW:

LEARNING OUTCOME 16.4

- Differentiate between transactional and transformational leaders.
- Describe charismatic and visionary leadership.
- Discuss what team leadership involves.

Go to page 395 and see how well you know this material.

LEARNING **OUTCOME 16.5** ▷

LEADERSHIP ISSUES IN THE TWENTY-FIRST CENTURY

Let's Get Real: F2F

SOMEONE I THINK IS A GOOD LEADER AND WHY:

Ronnie Screwvala of UTV. Would back new ideas on his people, even at the risk of failing. Excellent interpersonal skills. Very metrics oriented.

It's not easy being a chief information officer (CIO) today. This person, who is responsible for managing a company's information technology activities, faces a lot of external and internal pressures. Technology continues to change rapidly—almost daily, it sometimes seems. Business costs continue to rise. Rob Carter, CIO of FedEx, is on the hot seat facing such challenges.[42] He's responsible for all the computer and communication systems that provide around-the-clock and around-the-globe support for FedEx's products and services. If anything goes wrong, you know who takes the heat. However, Carter has been an effective leader in this seemingly chaotic environment.

For most leaders, leading effectively in today's environment is unlikely to involve the challenging circumstances Carter faces. However, twenty-first-century leaders do deal with some important leadership issues. In this section, we look at some of these issues: managing power, developing trust, empowering employees, leading across cultures, understanding gender differences in leadership, and becoming an effective leader.

MANAGING POWER

Where do leaders get their power—that is, their capacity to influence work actions or decisions? Five sources of leader power have been identified: legitimate, coercive, reward, expert, and referent.[43]

Legitimate power and authority are the same. Legitimate power represents the power a leader has as a result of his or her position in the organization. Although people in positions of authority are also likely to have reward and coercive power, legitimate power is broader than the power to coerce and reward.

Coercive power is the power a leader has to punish or control. Followers react to this power out of fear of the negative results that might occur if they don't comply. Managers typically have some coercive power, such as being able to suspend or demote employees or to assign them work they find unpleasant or undesirable.

legitimate power
The power a leader has as a result of his or her position in an organization.

coercive power
The power a leader has to punish or control.

thinking critically about Ethics

The definition of *friend* on social networking sites such as Facebook and MySpace is so broad that even strangers may tag you. But it doesn't feel weird because nothing really changes when a stranger does this. However, what if your boss, who isn't much older than you are, asks you to be a friend on these sites? What then? What are the implications if you refuse the offer? What are the implications if you accept? What ethical issues might arise because of this? What would you do?

Reward power is the power to give positive rewards. These can be anything that a person values, such as money, favorable performance appraisals, promotions, interesting work assignments, friendly colleagues, and preferred work shifts or sales territories.

Expert power is power that's based on expertise, special skills, or knowledge. If an employee has skills, knowledge, or expertise that's critical to a work group, that person's expert power is enhanced.

Finally, **referent power** is the power that arises because of a person's desirable resources or personal traits. If I admire you and want to be associated with you, you can exercise power over me because I want to please you. Referent power develops out of admiration of another and a desire to be like that person.

Most effective leaders rely on several different forms of power to affect the behavior and performance of their followers. For example, the captain of a ship or the project manager of Bangalore Metro employs different types of power in managing his crew or staff and equipment. He gives orders to the crew or staff (legitimate), praises them (reward), and disciplines those who commit infractions (coercive). As an effective leader, he also strives to have expert power (based on his expertise and knowledge) and referent power (based on his being admired) to influence his crew or staff.

DEVELOPING TRUST

In today's uncertain environment, an important consideration for leaders is building trust and credibility—trust that can be extremely fragile. Before we can discuss ways leaders can build trust and credibility, we have to know what trust and credibility are and why they're so important.

The main component of credibility is honesty. Surveys show that honesty is consistently singled out as the number-one characteristic of admired leaders. "Honesty is absolutely essential to leadership. If people are going to follow someone willingly, whether it be into battle or into the boardroom, they first want to assure themselves that the person is worthy of their trust."[44] In addition to being honest, credible leaders are competent and inspiring. They are personally able to effectively communicate their confidence and enthusiasm. Thus, followers judge a leader's **credibility** in terms of his or her honesty, competence, and ability to inspire.

Trust is closely entwined with the concept of credibility, and, in fact, the terms are often used interchangeably. **Trust** is defined as the belief in the integrity, character, and ability of a leader. Followers who trust a leader are willing to be vulnerable to the leader's actions because they are confident that their rights and interests will not be abused.[45] Research has identified five dimensions that make up the concept of trust:[46]

- **Integrity**—Honesty and truthfulness
- **Competence**—Technical and interpersonal knowledge and skills
- **Consistency**—Reliability, predictability, and good judgment in handling situations
- **Loyalty**—Willingness to protect a person, physically and emotionally
- **Openness**—Willingness to share ideas and information freely

Of these five dimensions, integrity seems to be the most critical when someone assesses another's trustworthiness.[47] Both integrity and competence came up in our earlier discussion of traits found to be consistently associated with leadership.

Workplace changes have reinforced why such leadership qualities are important. For instance, the trend toward empowerment (which we'll discuss shortly) and self-managed work teams has reduced many of the traditional control mechanisms used to monitor employees. If a work team is free to schedule its own work, evaluate its own performance, and even make its own hiring decisions, trust becomes critical. Employees have to trust managers to treat them fairly, and managers have to trust employees to conscientiously fulfill their responsibilities.

Also, leaders have to increasingly lead others who may not be in their immediate work group or even may be physically separated—members of cross-functional or virtual teams, individuals who work for suppliers or customers, and perhaps even people who represent other organizations through strategic alliances. These situations don't allow leaders the luxury of falling back on their formal positions for influence. Many of these relationships, in fact, are fluid and temporary. So the ability to quickly develop trust and sustain that trust is crucial to the success of the relationship.

Why is it important that followers trust their leaders? Research has shown that trust in leadership is significantly related to positive job outcomes, including job performance, organizational citizenship behavior, job satisfaction, and organizational commitment.[48] Given the importance of trust to effective leadership, how can leaders build trust? Exhibit 16–7 lists some suggestions, which are explained in the skills module on developing trust found in mymanagementlab.[49]

Now, more than ever, managerial and leadership effectiveness depends on the ability to gain the trust of followers.[50] Downsizing, corporate financial misrepresentation, and the increased use of temporary employees have undermined employees' trust in their leaders and shaken the confidence of investors, suppliers, and customers. A survey found that only 39 percent of U.S. employees and 51 percent of Canadian employees trusted their executive leaders.[51] In a recent study that included managers from India, it was found that 75 percent employees trust their immediate manager but only 60 percent trust their senior leaders.[52] Today's leaders are faced with the challenge of rebuilding and restoring trust with employees and with other important organizational stakeholders.

EMPOWERING EMPLOYEES

Employees at DuPont's facility in Uberaba, Brazil, planted trees to commemorate the site's 10th anniversary. Although they had several things to celebrate, one of the most important was the fact that since production began, the facility had had zero environmental incidents and no recordable safety violations. The primary reason for this achievement was the company's Safety Training Observation Program (STOP), a program in which empowered employees were responsible for observing one another, correcting improper procedures, and encouraging safe procedures.[53]

As we've described in different places throughout the text, managers are increasingly leading by empowering their employees. **Empowerment** involves increasing the

Exhibit 16–7

Building Trust

Practice openness.
Be fair.
Speak your feelings.
Tell the truth.
Show consistency.
Fulfill your promises.
Maintain confidences.
Demonstrate competence.

reward power
The power a leader has to give positive rewards.

expert power
Power that's based on expertise, special skills, or knowledge.

referent power
Power that arises because of a person's desirable resources or personal traits.

credibility
The degree to which followers perceive someone as honest, competent, and able to inspire.

trust
The belief in the integrity, character, and ability of a leader.

empowerment
The act of increasing the decision-making discretion of workers.

decision-making discretion of workers. Millions of individual employees and employee teams are making the key operating decisions that directly affect their work. They're developing budgets, scheduling work loads, controlling inventories, solving quality problems, and engaging in similar activities that until very recently were viewed exclusively as part of the manager's job.[54] For instance, at Airtel if a customer calls to say that he or she has found an error in his or her phone bill, the employee taking the call has been empowered to waive off the charge upto a certain amount without checking with any supervisor.

One reason more companies are empowering employees is the need for quick decisions by the people who are most knowledgeable about the issues—often those at lower organizational levels. If organizations want to successfully compete in a dynamic global economy, employees have to be able to make decisions and implement changes quickly. Another reason more companies are empowering employees is that organizational downsizings have left many managers with larger spans of control. In order to cope with the increased work demands, managers have had to empower their people. Although empowerment is not a universal answer, it can be beneficial when employees have the knowledge, skills, and experience to do their jobs competently.

LEADING ACROSS CULTURES

One general conclusion that surfaces from leadership research is that effective leaders do not use a single style. They adjust their style to the situation. Although not mentioned explicitly, national culture is certainly an important situational variable in determining which leadership style will be most effective. What works in India isn't likely to be effective in France or Canada. For instance, one study of Asian leadership styles revealed that Asian managers preferred leaders who were competent decision makers, effective communicators, and supportive of employees.[55]

National culture affects leadership style because it influences how followers will respond. Leaders can't (and shouldn't) just choose their styles randomly. They're constrained by the cultural conditions their followers have come to expect. Exhibit 16–8 provides some findings from selected examples of cross-cultural leadership studies. Because most leadership theories were developed in the United States, they have a U.S. bias. They emphasize follower responsibilities rather than rights; assume self-gratification rather than commitment to duty or altruistic motivation; assume centrality of work and democratic value orientation; and stress rationality rather than spirituality, religion, or superstition.[56] However, the GLOBE research program, first introduced in Chapter 4, is the most extensive and comprehensive cross-cultural study of leadership ever undertaken. The GLOBE study has found that there are some universal aspects to

Exhibit 16–8

Cross-Cultural Leadership

- Korean leaders are expected to be paternalistic toward employees.
- Arab leaders who show kindness or generosity without being asked to do so are seen by other Arabs as weak.
- Japanese leaders are expected to be humble and speak frequently.
- Scandinavian and Dutch leaders who single out individuals with public praise are likely to embarrass, not energize, those individuals.
- Effective leaders in Malaysia are expected to show compassion while using more of an autocratic than a participative style.
- Effective German leaders are characterized by high performance orientation, low compassion, low self-protection, low team orientation, high autonomy, and high participation.

Sources: Based on J. C. Kennedy, "Leadership in Malaysia: Traditional Values, International Outlook," *Academy of Management Executive*, August 2002, pp. 15–17; F. C. Brodbeck, M. Frese, and M. Javidan, "Leadership Made in Germany: Low on Compassion. High on Performance," *Academy of Management Executive*, February 2002, pp. 16–29; M. F. Peterson and J. G. Hunt. "International Perspectives on International Leadership," *Leadership Quarterly*, Fall 1997, pp. 203–231; R. J. House and R. N. Aditya, "The Social Scientific Study of Leadership: Quo Vadis?" *Journal of Management*, vol. 23 (3), 1997, p. 463; and R. J. House, "Leadership in the Twenty-First Century," in A. Howard (ed.), *The Changing Nature of Work* (San Francisco: Jossey-Bass, 1995), p. 442.

leadership. Specifically, a number of elements of transformational leadership appear to be associated with effective leadership, regardless of what country the leader is in.[57] These include vision, foresight, providing encouragement, trustworthiness, dynamism, positiveness, and proactiveness. The results led two members of the GLOBE team to conclude that "effective business leaders in any country are expected by their subordinates to provide a powerful and proactive vision to guide the company into the future, strong motivational skills to stimulate all employees to fulfill the vision, and excellent planning skills to assist in implementing the vision."[58] Some people suggest that the universal appeal of these transformational leader characteristics is due to the pressures toward common technologies and management practices, as a result of global competitiveness and multinational influences.

UNDERSTANDING GENDER DIFFERENCES AND LEADERSHIP

There was a time when the question "Do males and females lead differently?" could be seen as a purely academic issue—interesting, but not very relevant. That time has certainly passed! Many women now hold senior management positions, and many more around the world continue to join the management ranks. Misconceptions about the relationship between leadership and gender can adversely affect hiring, performance evaluation, promotion, and other human resource decisions for both men and women. For instance, evidence indicates that a "good" manager is still perceived as predominantly masculine.[59]

A number of studies focusing on gender and leadership style have been conducted in recent years. Their general conclusion is that males and females use different styles. Specifically, women tend to adopt a more democratic or participative style. Women are more likely to encourage participation, share power and information, and attempt to enhance followers' self-worth. They lead through inclusion and rely on their charisma, expertise, contacts, and interpersonal skills to influence others. Women tend to use transformational leadership, motivating others by transforming their self-interest into organizational goals. Men are more likely to use a directive, command-and-control style. They rely on formal position authority for their influence. Men use transactional leadership, handing out rewards for good work and punishment for bad.[60]

There is an interesting qualifier to the findings just mentioned. The tendency for female leaders to be more democratic than males declines when women are in male-dominated jobs. Apparently, group norms and male stereotypes influence women, and in some situations, women tend to act more autocratically.[61]

Although it's interesting to see how male and female leadership styles differ, a more important question is whether they differ in effectiveness. Although some researchers have shown that males and females tend to be equally effective as leaders,[62] an increasing number of studies have shown that women executives, when

Yale graduate Indra Nooyi, who played in an all-girl rock band while growing up in Chennai (India), is the savvy and irreverent chair and CEO of Pepsico Inc. Drawn to PepsiCo as chief strategist almost 15 years ago by the chance to help turnaround the company, she has helped the company double net profits to more than $5.6 billion by focusing on better nutrition and by promoting workforce diversity. "Indra can drive as deep and hard as anyone I've ever met," says former CEO Roger Enrico, "but she can do it with a sense of heart and fun." Nooyi still sings in the office and has been known to go barefoot at work.

Exhibit 16–9

Where Female Managers Do Better: A Scorecard

None of the five studies set out to find gender differences. They stumbled on them while compiling and analyzing performance evaluations.

Skill (Each check mark denotes which group scored higher on the respective studies)	**MEN**	**WOMEN**
Motivating Others		✓✓✓✓✓
Fostering Communication		✓✓✓✓✓*
Producing High-Quality Work		✓✓✓✓✓
Strategic Planning	✓✓	✓✓✓*
Listening to Others		✓✓✓✓✓
Analyzing Issues	✓✓	✓✓✓*

*In one study, women's and men's scores in these categories were statistically even.
Data: Hagberg Consulting Group, Management Research Group, Lawrence A. Pfaff, Personnel Decisions International Inc., Advanced Teamware Inc.

Source: Where Female Managers Do Better from R. Sharpe, "As Leaders, Women Rule" *Business Week*, November 20, 2000, p. 75.

rated by their peers, employees, and bosses, score higher than their male counterparts on a wide variety of measures.[63] (See Exhibit 16–9 for a summary.) Why? One possible explanation is that in today's organizations, flexibility, teamwork and partnering, trust, and information sharing are rapidly replacing rigid structures, competitive individualism, control, and secrecy. In these types of workplaces, effective managers must use more social and interpersonal behaviors. They listen, motivate, and provide support to their people. They inspire and influence rather than control. And women seem to do those things better than men.[64]

Although women seem to rate highly on the leadership skills needed to succeed in today's dynamic global environment, we don't want to fall into the same trap as the early leadership researchers who tried to find the "one best leadership style" for all situations. We know that there is no one *best* style for all situations. Instead, the most effective leadership style depends on the situation. So even if men and women differ in their leadership styles, we shouldn't assume that one is always preferable to the other.

BECOMING AN EFFECTIVE LEADER

Organizations need effective leaders. Two issues pertinent to becoming an effective leader are leader training and recognizing that sometimes being an effective leader means *not* leading. Let's take a look at these issues.

Leader Training. Organizations around the globe spend billions of dollars, yen, and euros on leadership training and development.[65] These efforts take many forms—from leadership programs offered by Indian Institute of Management, Ahmedabad to climbing experiences at iDiscoveri. Although much of the money spent on leader training may provide doubtful benefits, our review suggests that there are some things managers can do to get the maximum effect from such training.[66]

First, let's recognize the obvious: Some people don't have what it takes to be a leader. Period. For instance, evidence indicates that leadership training is more likely to be successful with individuals who are high self-monitors than with low self-monitors. Such individuals have the flexibility to change their behavior as different situations require. In addition, organizations may find that individuals with higher levels of a trait called *motivation to lead* are more receptive to leadership development opportunities.[67]

What kinds of things can individuals learn that might be related to being a more effective leader? It may be a bit optimistic to think that "vision-creation" can be taught, but implementation skills can be taught. People can be trained to develop "an understanding about content themes critical to effective visions."[68] We can also teach skills such as trust-building and mentoring. And leaders can be taught situational analysis skills. They can learn how to evaluate situations, how to modify situations to make them fit better with their style, and how to assess which leader behaviors might be most effective in given situations.

Substitutes for Leadership. Despite the belief that some leadership style will always be effective, regardless of the situation, leadership may not always be important! Research indicates that, in some situations, any behaviors a leader exhibits are irrelevant. In other words, certain individual, job, and organizational variables can act as "substitutes for leadership," negating the influence of the leader.[69]

For instance, follower characteristics such as experience, training, professional orientation, and need for independence can neutralize the effect of leadership. These characteristics can replace the employee's need for a leader's support or ability to create structure and reduce task ambiguity. Similarly, jobs that are inherently unambiguous and routine or that are intrinsically satisfying may place fewer demands on the leadership variable. Finally, such organizational characteristics as explicit formalized goals, rigid rules and procedures, and cohesive work groups can substitute for formal leadership.

QUICK LEARNING REVIEW:

LEARNING OUTCOME 16.5

- Describe the five sources of a leader's power.
- Discuss the issues today's leaders face.

Go to page 396 and see how well you know this material.

Let's Get Real: My Turn

Nimish Vohra

Country Head, India
Regalix

Nayar can build trust by being and acting as the first among *unequals* by

- working as a team with his immediate juniors (further below would not help and would create a "political situation") and
- visibly taking feedbacks of his immediate juniors and setting an example for others.

LEARNING OUTCOMES SUMMARY

16.1 ▷ WHO ARE LEADERS, AND WHAT IS LEADERSHIP?

- Define leader and leadership.
- Explain why managers should be leaders.

A leader is someone who can influence others and who has managerial authority. Leadership is a process of leading a group and influencing that group to achieve its goals.

Managers should be leaders because leading is one of the four management functions.

16.2 ▷ EARLY LEADERSHIP THEORIES

- Discuss what research has shown about leadership traits.
- Contrast the findings of the four behavioral leadership theories.
- Explain the dual nature of a leader's behavior.

Early attempts to define leadership traits were unsuccessful although later attempts found seven traits associated with leadership.

The University of Iowa studies explored three leadership styles. The only conclusion was that group members were more satisfied under a democratic leader than under an autocratic one. The Ohio State studies identified two dimensions of leader behavior: initiating structure and consideration. A leader high in both those dimensions at times achieved high group task performance and high group member satisfaction, but not always. The University of Michigan studies looked at employee-oriented leaders and production-oriented leaders. They concluded that leaders who were employee oriented could get high group productivity and high group member satisfaction. The managerial grid looked at leaders' concern for production and concern for people and identified five leader styles. Although it suggested that a leader who was high in concern for production and high in concern for people was the best, there was no substantive evidence for that conclusion.

As the behavioral studies showed, a leader's behavior has a dual nature: a focus on the task and a focus on the people.

16.3 ▷ CONTINGENCY THEORIES OF LEADERSHIP

- Explain Fiedler's contingency model of leadership.
- Describe situational leadership theory.
- Discuss how path–goal theory explains leadership.

Fiedler's model attempted to define the best style to use in particular situations. He measured leader style—relationship oriented or task oriented—using the least-preferred coworker questionnaire. Fiedler also assumed that a leader's style was fixed. He measured three contingency dimensions: leader–member relations, task structure, and position power. The model suggested that task-oriented leaders performed best in very favorable and very unfavorable situations, and relationship-oriented leaders performed best in moderately favorable situations.

Hersey and Blanchard's situational leadership theory (SLT) focused on followers' readiness. It includes four leadership styles: telling (high task–low relationship), selling (high task–high relationship), participating (low task–high relationship), and delegating (low task–low relationship). SLT also identified four stages of readiness: unable and unwilling (use telling style), unable but willing (use selling style), able but unwilling (use participative style), and able and willing (use delegating style).

The path–goal model developed by Robert House identified four leadership behaviors: directive, supportive, participative, and achievement oriented. This model assumes that a leader can and should be able to use any of these styles. The two situational contingency variables were found in the environment and in the follower. Essentially, the path–goal model says that a leader should provide direction and support as needed; that is, the leader should structure the path so the followers can achieve goals.

16.4 ▷ CONTEMPORARY VIEWS OF LEADERSHIP

- Differentiate between transactional and transformational leaders.
- Describe charismatic and visionary leadership.
- Discuss what team leadership involves.

A transactional leader exchanges rewards for productivity, whereas a transformational leader stimulates and inspires followers to achieve goals.

A charismatic leader is an enthusiastic and self-confident leader whose personality and actions influence people to behave in certain ways. People can learn to be charismatic. A visionary leader is able to create and articulate a realistic, credible, and attractive vision of the future.

A team leader has two priorities: manage the team's external boundary and facilitate the team process. Four leader roles are involved: liaison with external constituencies, troubleshooter, conflict manager, and coach.

16.5 ▷ LEADERSHIP ISSUES IN THE TWENTY-FIRST CENTURY

- Describe the five sources of a leader's power.
- Discuss the issues today's leaders face.

The five sources of a leader's power are legitimate (authority or position), coercive (punish or control), reward (give positive rewards), expert (special expertise, skills, or knowledge), and referent (desirable resources or traits).

Today's leaders face the issues of managing power, developing trust, empowering employees, leading across cultures, understanding gender differences in leadership, and becoming an effective leader.

THINKING ABOUT MANAGEMENT ISSUES

1. What types of power are available to you? Which ones do you use most? Why?
2. Do you think that most managers in real life use a contingency approach to increase their leadership effectiveness? Discuss.
3. If you ask people why a given individual is a leader, they tend to describe the person in terms such as *competent, consistent, self-assured, inspiring a shared vision,* and *enthusiastic.* How do these descriptions fit in with leadership concepts presented in this chapter?
4. What kinds of campus activities could a full-time college student do that might lead to the perception that he or she is a charismatic leader? In pursuing those activities, what might the student do to enhance this perception of being charismatic?
5. Do you think trust evolves out of an individual's personal characteristics or out of specific situations? Explain.
6. A recent study showed that CEOs of successful companies have hard-nosed personal traits, such as persistence, efficiency, attention to detail, and a tendency to set high standards, rather than softer strengths, such as teamwork, enthusiasm, and flexibility.[70] What do you think of this? Are you surprised? How would you explain this in light of the leadership theories discussed in this chapter?
7. Do followers make a difference in whether a leader is effective? Discuss.
8. How can organizations develop effective leaders?

YOUR TURN to be a Manager

- Think of the different organizations to which you belong. Note the different styles of leadership used by the leaders in those organizations. Write a paper describing these individuals' styles of leading (no names, please) and evaluate the styles being used.
- Write the names of three people you consider to be effective leaders. Make a bulleted list of the characteristics these individuals exhibit that you think make them effective leaders.
- Think about the times that you have had to lead. Describe what you think your own personal leadership style is. What could you do to improve your leadership style? Come up with an action plan of steps that you can take. Put all this information into a brief paper.
- Managers say that increasingly they must use influence to get things done. Do some research on the art of persuasion. Make a bulleted list of suggestions that you find on how to improve your skills at influencing others.

- Can leadership skills be taught in multiplayer online games? Some people think so. Select two online fantasy games and describe (1) the basics of the game—what it's based on and how it's played; (2) how players advance in the game; and (3) what leadership skills the game might help develop and how.
- Here's a list of leadership skills: building employee communities, building teams, coaching and motivating others, communicating with impact, confidence and energy, leading by example, leading change, making decisions, providing direction and focus, and valuing diversity. Choose two of these skills and develop a training exercise that will help develop or improve each of them.
- Steve's and Mary's recommended readings: Stephen M. R. Covey with Rebecca Merrell, *The Speed of Trust: The One Thing That Changes Everything* (The Free Press, 2006); Nancy S. Ahlrichs, *Manager of Choice* (Davies-Black Publishing, 2003); John H. Zenger and Joseph Folkman, *The Extraordinary Leader: Turning Good Managers into Great Leaders* (McGraw-Hill, 2002); Robert H. Rosen, *Leading People* (Viking Penguin Publishing, 1996); Margaret J. Wheatley, *Leadership and the New Science* (Berrett-Koehler Publishers, 1994); Max DePree, *Leadership Jazz* (Dell Publishing, 1992); and Max DePree, *Leadership Is an Art* (Dell Publishing, 1989).
- Select one of the topics from the section "Leadership Issues in the Twenty-First Century." Do some additional research on the topic and put your findings in a bulleted list. Be prepared to share this in class. Be sure to cite your sources.
- There are two relevant leadership skill-building modules on mymanagementlab: Choosing an Effective Leadership Style and Coaching. Complete these modules. Your professor will tell you what to do with them.
- Interview three managers about what they think it takes to be a good leader. Write up your findings in a report and be prepared to present it in class.
- In your own words, write down three things you learned in this chapter about being a good manager.
- Self-knowledge can be a powerful learning tool. Go to mymanagementlab and complete these self-assessment exercises: What's My Leadership Style? How Charismatic Am I? Do I Trust Others? Do Others See Me as Trusting? How Good Am I at Building and Leading a Team? Using the results of your assessments, identify personal strengths and weaknesses. What will you do to reinforce your strengths and improve your weaknesses?

CASE APPLICATION

Radical Leadership

Many consider Ricardo Semler, CEO of Semco Group of São Paulo, Brazil, to be a radical. He's never been the type of leader most people might expect to be in charge of a multimillion-dollar business. Why? Semler breaks all the traditional "rules" of leading. He's the ultimate hands-off leader; he doesn't even have an office at the company's headquarters. As the "leading proponent and most tireless evangelist" of participative management, Semler says his philosophy is simple: Treat people like adults, and they'll respond like adults.

Underlying Semler's participative management approach is the belief that "organizations thrive best by entrusting employees to apply their creativity and ingenuity in service of the whole enterprise, and to make important decisions close to the flow of work, conceivably including the selection and election of their bosses." And according to Semler, his approach works ... and works well. But how does it work in reality?

At Semco you won't find most of the trappings of organizations and management. There are no organizational charts, no long-term plans, no corporate values statements, no dress codes, and no written rules or policy manuals. The company's employees decide their work hours and their pay levels. Subordinates decide who their bosses will be, and they also review their boss's performance. The employees also elect the corporate leadership and decide most of the company's new strategic initiatives. Each person—including Ricardo Semler—has one vote.

Why did Semler decide that his form of radical leadership was necessary? Does it work? Semler didn't

Ricardo Semler, CEO of Semco Group of Sao Paulo, Brazil.

pursue such radical self-governance out of some altruistic ulterior motive. Instead, he felt it was the only way to build an organization that was flexible and resilient enough to flourish in chaotic and turbulent times. He maintains that this approach has enabled Semco to survive the roller-coaster nature of Brazilian politics and economy. Although the country's political leadership and economy have gone from one extreme to another, and countless Brazilian banks and companies have failed, Semco has survived. And not just survived—prospered. Semler says, "If you look at Semco's numbers, we've grown 27.5 percent a year for 14 years." And Semler attributes that fact to flexibility...of his company and, most importantly, of his employees.

Discussion Questions

1. Describe Ricardo Semler's leadership style. What do you think might be the advantages and drawbacks of his style?
2. What challenges might a radically hands-off leader face? How could those challenges be addressed?
3. How could future leaders be identified in Semco? Would leadership training be important to this organization? Discuss.
4. What could other businesses learn from Ricardo Semler's approach to leadership?

Sources: L. M. Fisher, "Ricardo Semler Won't Take Control," *Strategy and Business*, Winter 2005, pp. 78–88; R. Semler, *The Seven-Day Weekend: Changing the Way Work Works* (New York: Penguin Group, 2004); A. J. Vogl, "The Anti-CEO," *Across the Board*, May/June 2004, pp. 30–36; G. Colvin, "The Anti-Control Freak," *Fortune*, November 26, 2001, p. 22; and R. Semler, "Managing Without Managers," *Harvard Business Review*, September/October 1989, pp. 76–84.

Part Five

Controlling

Managers must establish goals and plans, organize and structure work activities, and develop programs to motivate and lead people to put forth effort to accomplish those goals. Even when managers have finished these tasks, their job is not done. Quite the opposite! Managers must then monitor activities to make sure they're being done as planned and correct any significant deviations. This process is called *controlling.* It's the final link in the management process, and although controlling happens last in the process, that doesn't make it any less important than any of the other managerial functions.

In Part Five, we look at the process of controlling. In Chapter 17, we look at the fundamental elements of controlling, which consists of a three-step process: measuring, comparing, and taking action. Chapter 18 introduces the concept of managing operations. Because an organization's operations management system controls how its products (goods or services) are produced, it's important that managers know about managing operations.

Let's Get Real:

Meet the Manager

Asha Nair

Sr Manager, IT-Flat Products
Tata Steel

MY JOB: Involves interacting with business users and designing systems that would help them in their functional areas.

BEST PART OF MY JOB: The fact that it provides a great platform to learn and keep myself updated with the current happenings in the industry.

WORST PART OF MY JOB: Striking a balance between work and a life outside work. The required long hours are also sometimes frustrating, when troubleshooting can run late into the night as well.

BEST MANAGEMENT ADVICE EVER RECEIVED: It's OK to make mistakes; one does not learn unless we make mistakes. Difficulties are meant to rouse, not discourage. Better to take the occasional knock and rise from it rather than never have the opportunity to learn. It keeps one humble and open to learning.

You'll be hearing more from this real manager throughout the chapter.

Chapter 17

Introduction to Controlling

Managers must monitor whether goals that were established as part of the planning process are being accomplished efficiently and effectively. That's what they do when they control. Appropriate controls can help managers look for specific performance gaps and areas for improvement. Focus on the following learning outcomes as you read and study this chapter.

LEARNING OUTCOMES

A Manager's Dilemma

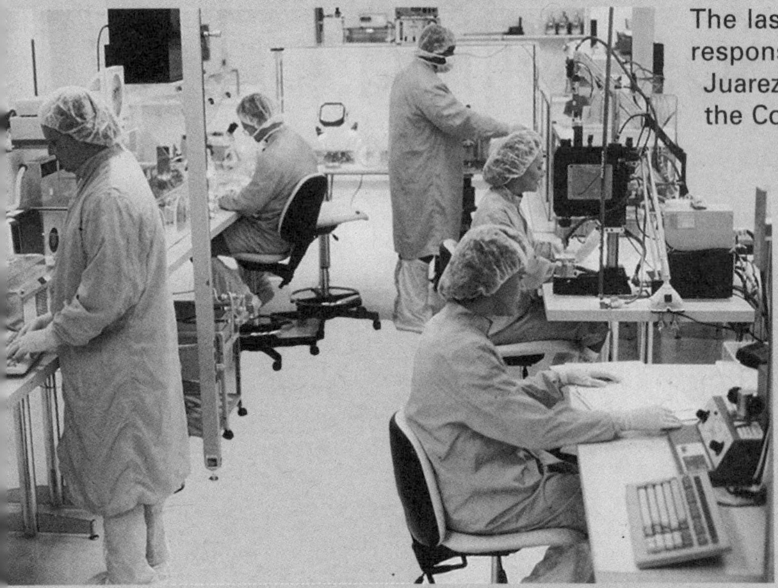

The last place you'd probably expect to see a rapid response team is at a production facility in Ciudad Juarez, Mexico.[1] But that's exactly what you'd find at the Cordis de Mexico S.A. de C.V. facility that manufactures cardiovascular and endovascular devices. Employees at this facility—which is part of Cordis Corporation, a Florida-based subsidiary of Johnson & Johnson—make catheters and stents used to treat various circulatory system problems. The rapid response team is an employee troubleshooting group that's called into action with Mozart's *Symphony No. 40* blaring through the overhead speakers. Whenever a production-line employee activates the music to signal a problem on the line, the team races toward the work area to find out what's happened and how it can be corrected quickly. Because customer demand is so great and the product manufacture so exacting, such work stoppages must be addressed promptly. Cordis plant managers want to make sure that the rapid response team gets the information it needs to do its job as quickly as possible.

What would you do?

The situation at the Cordis plant illustrates how important controls are to managers. The control "alert" in this case, Mozart's *Symphony No. 40*, has a unique way of making everyone in the plant aware that a situation needs to be addressed. Cordis managers have found something that works for them. And that's what all managers are looking for: appropriate controls that can help pinpoint specific performance gaps and areas for improvement.

LEARNING **OUTCOME 17.1** ▷

WHAT IS CONTROLLING, AND WHY IS IT IMPORTANT?

A press operator at the Denver Mint noticed a flaw—an extra up leaf or an extra down leaf—on Wisconsin state quarters being pressed at one of his five press machines. He stopped the machine and left for a meal break. When he returned, he saw the machine running and assumed that someone had changed the die in the machine. However, after a routine inspection, the machine operator realized that the die had not been changed. The faulty press had likely been running for over an hour, and thousands of the flawed coins were now comingled with unblemished quarters. As many as 50,000 of the faulty coins entered circulation, setting off a coin collector buying frenzy.[2] Can you see why controlling is such an important managerial function?

What is **controlling**? It's the process of monitoring, comparing, and correcting work performance. All managers should control, even if they think their units are performing as planned; they can't really know how units are performing unless they've evaluated what activities have been done and compared actual performance against the desired standard.[3] Effective controls ensure that activities are completed in ways that lead to the attainment of goals. Whether controls are effective, then, is determined by how well they help employees and managers achieve their goals.[4]

Why is control so important? Planning can be done, an organizational structure can be created to facilitate efficient achievement of goals, and employees can be motivated through effective leadership. But there's no assurance that activities are going as planned and that the goals employees and managers are working toward are, in fact, being attained. Control is important, therefore, because it helps managers know whether organizational goals are being met and, if they're not being met, the reasons why. The value of the control function can be seen in three specific areas: planning, empowering employees, and protecting the workplace.

In Chapter 7, we described goals, which provide specific direction to employees and managers, as the foundation of planning. However, just stating goals or having employees accept goals doesn't guarantee that the necessary actions to accomplish those goals have been taken. As the old saying goes, "The best-laid plans often go awry." An effective manager follows up to ensure that what employees are supposed to do is, in fact, being done and that goals are being achieved. As the final step in the management process, controlling provides the critical link back to planning (see Exhibit 17–1). If managers don't control, they have no way of knowing whether their goals and plans are being achieved and what future actions to take.

The second reason controlling is important is employee empowerment. Many managers are reluctant to empower their employees because they fear something will go wrong for which they will be held responsible. But an effective control system can provide information and feedback on employee performance and minimize the chance of potential problems.

The final reason that managers control is to protect the organization and its assets.[5] Today's environment brings heightened threats from natural disasters, financial scandals, workplace violence, supply chain disruptions, security breaches, and even possible terrorist attacks. Managers must protect organizational assets in the event that any of these happen. Comprehensive controls and backup plans will help assure minimal work disruptions.

Let's Get Real: F2F

CONTROLS ARE IMPORTANT BECAUSE:

Without the control there would be no focus or direction. They are essential for creating order and also bringing synergy by providing a required framework for action and execution. Preventive controls attempt to identify and preempt deviations in standards before they occur. Thus, controls are also important to ensure consistency with quality standards.

Exhibit 17–1

Planning–Controlling Link

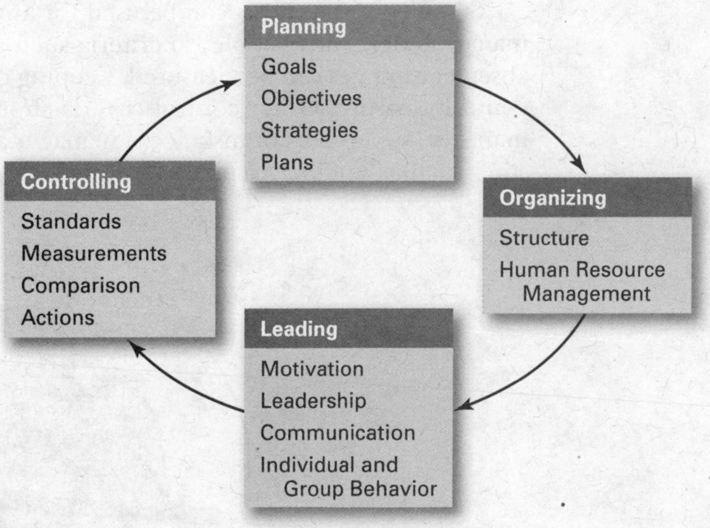

controlling
The process of monitoring, comparing, and correcting work performance.

QUICK LEARNING REVIEW:

LEARNING OUTCOME 17.1

- Define controlling.
- Explain the planning–controlling link.
- Discuss the reasons control is important.

Go to page 424 and see how well you know this material.

LEARNING **OUTCOME 17.2** ▷

THE CONTROL PROCESS

When Kirit Shah joined a chemical factory in Vadodara, Gujarat, as HR Manager, he knew that his top priority was finding ways to reduce employee injuries. At his factory, the top three causes of injuries were because of falling, stepping, striking and being struck against, and fall of objects. Kirit relied on control processes to turn the situation around.

The **control process** is a three-step process of measuring actual performance, comparing actual performance against a standard, and taking managerial action to correct deviations or to address inadequate standards (see Exhibit 17–2). The control process assumes that performance standards already exist, and they do. They're the specific goals created during the planning process.

STEP 1: MEASURING

To determine what actual performance is, a manager must first get information about it. Thus, the first step in control is measuring.

How We Measure. Four approaches managers use to measure and report actual performance are personal observations, statistical reports, oral reports, and written reports. Exhibit 17–3 summarizes the advantages and drawbacks of each approach. Most managers use a combination of these approaches.

What We Measure. What is measured is probably more critical to the control process than how it's measured. Why? Because selecting the wrong criteria can create serious problems. Besides, *what* is measured often determines what employees will do.[6] What control criteria might managers use?

Some control criteria can be used for any management situation. For instance, all managers deal with people, so criteria such as employee satisfaction or turnover and absenteeism rates can be measured. Keeping costs within budget is also a fairly common control measure. Other control criteria should recognize the different activities that managers supervise. For instance, a manager at a pizza delivery location might use measures such as number of pizzas delivered per day, average delivery time, or number of

Exhibit 17–2

The Control Process

Exhibit 17–3

Sources of Information for Measuring Performance

	Advantages	Drawbacks
Personal Observations	• Get firsthand knowledge • Information isn't filtered • Intensive coverage of work activities	• Subject to personal biases • Time-consuming • Obtrusive
Statistical Reports	• Easy to visualize • Effective for showing relationships	• Provide limited information • Ignore subjective factors
Oral Reports	• Fast way to get information • Allow for verbal and nonverbal feedback	• Information is filtered • Information can't be documented
Written Reports	• Comprehensive • Formal • Easy to file and retrieve	• Take more time to prepare

coupons redeemed. A manager in a governmental agency might use applications typed per day, client requests completed per hour, or average time to process paperwork.

Most work activities can be expressed in quantifiable terms, but when they can't, managers should use subjective measures. Although such measures may have limitations, having them is better than having no standards at all and doing no controlling.

STEP 2: COMPARING

The comparing step determines the variation between actual performance and a standard. Although some variation in performance can be expected in all activities, it's critical to determine an acceptable **range of variation** (see Exhibit 17–4). Deviations outside this range need attention. Let's work through an example.

Rahul Sharma is a sales manager for Green Earth Gardening Supply, a distributor of specialty plants and seeds in the entire North India. Rahul prepares a report during the first week of each month that describes sales for the previous month, classified by product line. Exhibit 17–5 displays both the sales goals (standard) and actual sales figures for the month of June. After looking at the numbers, should Rahul be concerned? Sales were a bit higher than originally targeted, but does that mean there were no significant deviations? That depends on what Rahul thinks is *significant*—that is, outside the acceptable

Exhibit 17–4

Acceptable Range of Variation

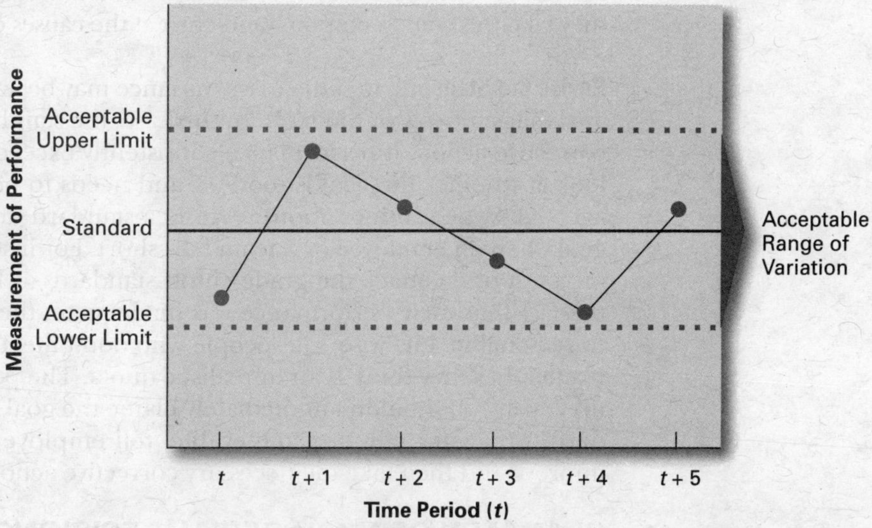

control process
A three-step process of measuring actual performance, comparing actual performance against a standard, and taking managerial action to correct deviations or inadequate standards.

range of variation
The acceptable parameters of variance between actual performance and a standard.

Exhibit 17–5

Example of Determining Significant Variation

Green Earth Gardening Supply—*June Sales*

Product	Standard	Actual	Over (Under)
Vegetable plants	1,075	913	(162)
Perennial flowers	630	634	4
Annual flowers	800	912	112
Herbs	160	140	(20)
Flowering bulbs	170	286	116
Flowering bushes	225	220	(5)
Heirloom seeds	540	672	132
Total	3,600	3,777	177

range of variation. Even though overall performance was generally quite favorable, some product lines need closer scrutiny. For instance, if sales of heirloom seeds, flowering bulbs, and annual flowers continue to be over what was expected, Rahul might need to order more product from nurseries to meet customer demand. Because sales of vegetable plants were 15 percent below goal, Rahul may need to run a special on them. As this example shows, both overvariance and undervariance may require managerial attention, which is the third step in the control process.

STEP 3: TAKING MANAGERIAL ACTION

Managers can choose among three possible courses of action: do nothing, correct the actual performance, or revise the standard. "Do nothing" is self-explanatory, so let's look at the other two.

Correct Actual Performance. Depending on what the problem is, a manager could take different corrective actions. For instance, if unsatisfactory work is the reason for performance variations, the manager could correct it by implementing training programs, taking disciplinary action, making changes in compensation practices, and so forth. One decision that a manager must make is whether to take **immediate corrective action**, which corrects problems at once to get performance back on track, or to use **basic corrective action**, which looks at how and why performance deviated before correcting the source of deviation. It's not unusual for managers to rationalize that they don't have time to find the source of a problem (basic corrective action) and continue to perpetually "put out fires" with immediate corrective action. Effective managers analyze deviations, and if the benefits justify it, they take the time to pinpoint and correct the causes of variance.

Revise the Standard. In some cases, variance may be a result of an unrealistic standard—a goal that's too low or too high. In this case, the standard—not the performance—needs corrective action. If performance consistently exceeds the goal, then a manager should look at whether the goal is too easy and needs to be raised. On the other hand, managers must be cautious about revising a standard downward. It's natural to blame the goal when an employee or a team falls short. For instance, students who get a low score on a test often attack the grade cutoff standards as being too high; rather than accept the fact that their performance was inadequate, they will argue that the standards are unreasonable. Likewise, salespeople who don't meet their monthly quota often want to blame what they think is an unrealistic quota. The point is that when performance isn't up to par, you shouldn't immediately blame the goal or standard. If you believe the standard is realistic, fair, and achievable, tell employees that you expect future work to improve and then take the necessary corrective action to help make that happen.

SUMMARY OF MANAGERIAL DECISIONS

Exhibit 17–6 summarizes the decisions a manager makes in controlling. The standards are goals that were developed during the planning process. These goals provide the basis for the control process, which involves measuring actual performance and comparing it against the standard. Depending on the results, a manager's decision is to do nothing, correct the performance, or revise the standard.

Exhibit 17–6

Managerial Decisions in the Control Process

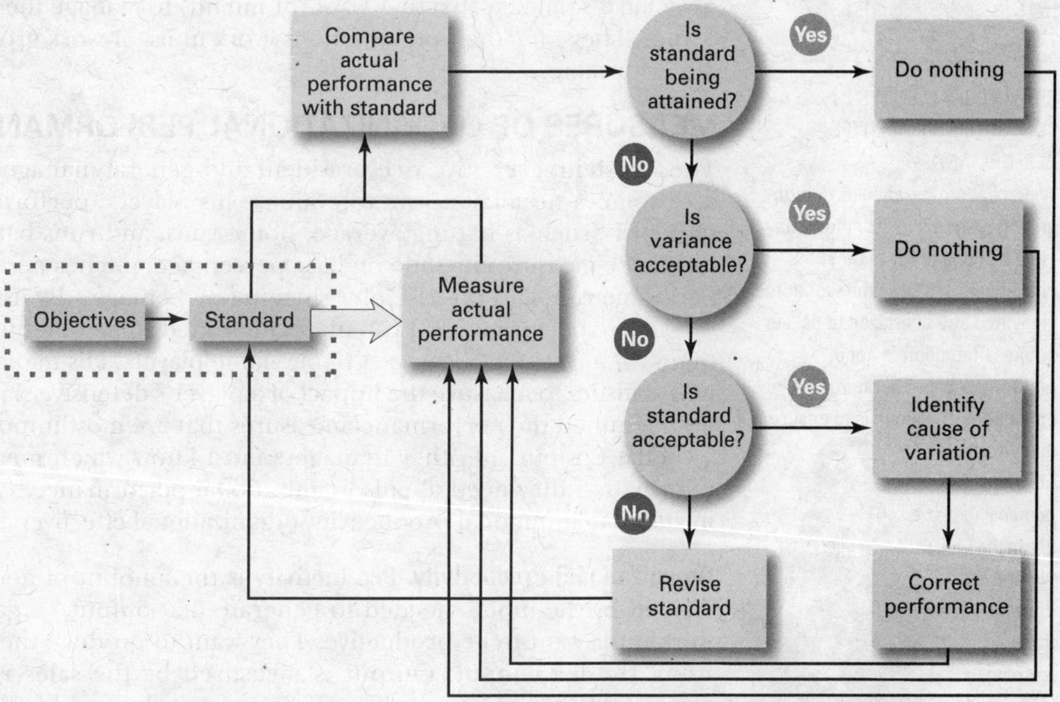

QUICK LEARNING REVIEW:

LEARNING OUTCOME 17.2

- Describe the three steps in the control process.
- Explain the three courses of action managers can take in controlling.

Go to page 424 and see how well you know this material.

LEARNING **OUTCOME 17.3** ▷

CONTROLLING FOR ORGANIZATIONAL PERFORMANCE

Cost-efficiency. The length of time customers are kept on hold. Customer satisfaction with service provided. These are just a few of the important performance indicators that executives in the intensely competitive call-center service industry measure. To make good decisions, managers in this industry want and need this type of information so they can manage organizational performance. Managers in all types of businesses are responsible for managing organizational performance.

WHAT IS ORGANIZATIONAL PERFORMANCE?

When you hear the word *performance,* what do you think of? A summer evening concert by a local community orchestra? An Olympic athlete striving for the finish line in a close race? A Indigo Airlines ramp agent in Jaipur, loading passengers as efficiently as possible in order to meet the company's 20-minute gate turnaround goal? **Performance** is all of these. It's the end result of an activity. And whether that activity is hours of intense practice before a concert or race or carrying out job responsibilities as efficiently and effectively as possible, performance is what results from that activity.

Managers are concerned with **organizational performance**—the accumulated results of all the work activities in the organization. It's a multifaceted concept, but managers need to understand the factors that contribute to organizational performance.

immediate corrective action
Corrective action that addresses problems at once to get performance back on track.

basic corrective action
Corrective action that looks at how and why performance deviated before correcting the source of deviation.

performance
The end result of an activity.

organizational performance
The accumulated results of an organization's work activities.

Let's Get Real: F2F

BIGGEST CHALLENGES IN CONTROLLING EMPLOYEE PERFORMANCE:
With a flat organization structure, growth paths are restricted. Today, I feel that one of the biggest challenges is to keep the workforce motivated and energized to deliver outputs. Retention of good performers is another hurdle. A continuous challenge is to ensure that we can achieve compelling clarity about the value we are delivering through learning.

After all, it's unlikely that they want (or intend) to manage their way to mediocre performance. They *want* their organizations, work units, or work groups to achieve high levels of performance.

MEASURES OF ORGANIZATIONAL PERFORMANCE

Theo Epstein, executive vice president and general manager of the Boston Red Sox, uses some unusual statistics to evaluate his players' performance instead of the old standards, such as batting average, home runs, and runs batted in. These "new" performance measures include on-base percentage, pitches per plate appearance, at-bats per home run, and on-base plus slugging percentage.[7] By using these statistics to predict future performance, Epstein has identified some potential star players and signed them for a fraction of the cost of big-name players. His management team is defining new statistics to measure the impact of a player's defensive skills. As a manager, Epstein has identified the performance measures that are most important to his decisions.

Like Epstein, all other managers must know which measures will give them the information they need about organizational performance. Commonly used measures include organizational productivity, organizational effectiveness, and industry rankings.

Organizational Productivity. **Productivity** is the amount of goods or services produced divided by the inputs needed to generate that output. Organizations and individual work units want to be productive. They want to produce the most goods and services using the least input. Output is measured by the sales revenue an organization receives when goods are sold (selling price multiplied by the number sold). Input is measured by the costs of acquiring and transforming resources into outputs.

Management wants to increase the ratio of output to input. Of course, the easiest way to do this is to raise prices of the outputs. But in today's competitive environment, that may not be an option. The only other option, then, is to decrease the inputs side. How? By being more efficient in performing work and thus decreasing the organization's expenses.

Organizational Effectiveness. **Organizational effectiveness** is a measure of the appropriateness of organizational goals and how well those goals are being met. That's the bottom line for managers, and it's what guides managerial decisions in designing strategies and work activities and in coordinating the work of employees.

managing workforce Diversity

Diversity Champions

U.S. companies are making progress in diversity management. Although many still have a long way to go, some companies are doing their best to bring in diverse employees as full and active participants in their businesses.[8] Every year, *DiversityInc* identifies the 50 most diverse companies. Companies that make the list demonstrate consistent strengths in the four areas the survey measures: CEO commitment, human capital, corporate and organizational communications, and supplier diversity. Each has made a strong commitment to diversity at every organizational level and in every aspect—from new hires to suppliers, and even to the charitable causes supported. Who are some of these diversity champions, and what are they doing? Let's look at a few examples.

Verizon Communications has been number 1 on the list for two consecutive years. Its diversity strengths include a strong commitment to diversity by its CEO, who meets regularly with the company's employee-resource groups. He also personally reviews the company's quarterly diversity scorecard on 17 key diversity metrics. Also, 39 percent of Verizon's managers are black, Asian, Latino, or Native American.

The Coca-Cola Company is number 2 on the list. The company's management is 33 percent black, Latino, Native American, and Asian. Even more impressive is the fact that 40 percent of managers receiving promotions come from these groups.

Finally, PricewaterhouseCoopers (PWC) is number 4 on the list. It leads the way in finding cutting-edge and valuable ways to retain and promote top talent. Some of its programs include a work–life resource and referral service, family sick days, firmwide holiday shutdowns, and reimbursement for emergency child care.

Exhibit 17–7

Popular Industry and Company Rankings

***Fortune* (www.fortune.com)** *Fortune* 500 25 Top MBA Employers Most Admired Companies 100 Best Companies to Work For 101 Dumbest Moments in Business Global 500 Top Companies for Leaders 100 Fastest-Growing Companies	***IndustryWeek* (www.industryweek.com)** *IndustryWeek* 1000 *IndustryWeek* U.S. 500 50 Best Manufacturing Companies *IndustryWeek* Best Plants
***BusinessWeek* (www.businessweek.com)** World's Most Innovative Companies *BusinessWeek 50* Top MBA Programs Customer Service Champs	**Customer Satisfaction Indexes** American Customer Satisfaction Index—University of Michigan Business School Customer Satisfaction Measurement Association
Forbes (www.forbes.com) *Forbes* 500 200 Best Small Companies 400 Best Big Companies Largest Private Companies World's 2,000 Largest Companies Global High Performers	

Industry and Company Rankings. Examining rankings is a popular way for managers to measure organizational performance. And there's not a shortage of rankings, as Exhibit 17–7 shows. Rankings are determined by specific performance measures, which are different for each list. For instance, *Fortune* chooses which companies to put on its "Best Companies to Work For" list by examining answers given by thousands of randomly selected employees on a questionnaire called "The Great Place to Work Trust Index"; materials filled out by thousands of company managers, including a corporate culture audit created by the Great Place to Work Institute; and a human resources questionnaire designed by Hewitt Associates. Such rankings give managers (and others) an indicator of how well a company is performing in comparison to others.

QUICK LEARNING REVIEW:

LEARNING OUTCOME 17.3

- Define organizational performance.
- Describe three frequently used organizational performance measures.

Go to page 424 and see how well you know this material.

LEARNING **OUTCOME 17.4** ▷

TOOLS FOR MEASURING ORGANIZATIONAL PERFORMANCE

Managers at Applebee's Neighborhood Grill & Bar restaurant chain play by their own rules. Their philosophy: Faster is better, get into a neighborhood before the competition, and keep things moving by giving customers a convenient experience.[9] Given its approach to business, what kinds of tools would Applebee's managers need for monitoring and measuring performance?

All managers need appropriate tools for monitoring and measuring organizational performance. Before describing some specific types of control tools, let's look at the concepts of feedforward, concurrent, and feedback control.

productivity
The amount of goods or services produced divided by the inputs needed to generate the output.

organizational effectiveness
A measure of the appropriateness of organizational goals and how well those goals are being met.

Exhibit 17–8

Types of Control

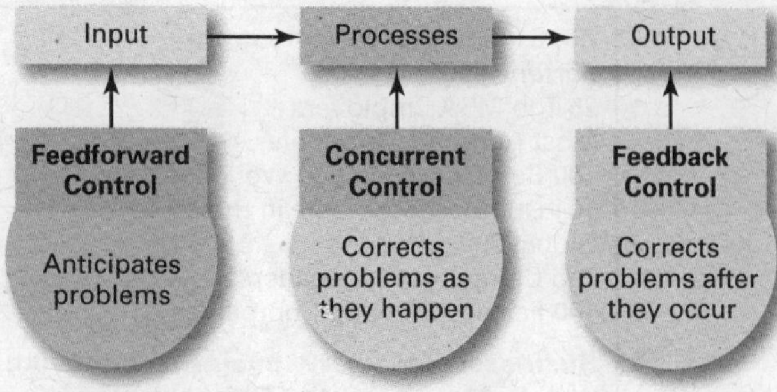

FEEDFORWARD, CONCURRENT, AND FEEDBACK CONTROL

Managers can implement controls *before* an activity begins, *during* the time an activity is going on, and *after* an activity has been completed. The first type is called feedforward control; the second, concurrent control; and the last, feedback control (see Exhibit 17–8).

Feedforward Control. The most desirable type of control—**feedforward control**—prevents problems because it takes place before the actual activity.[10] For instance, when McDonald's opened its first restaurant in Moscow, it sent company quality control experts to help Russian farmers learn techniques for growing high-quality potatoes and to help bakers learn processes for baking high-quality breads. Why? McDonald's demands consistent product quality no matter the geographical location. The company wants a cheeseburger in Moscow to taste like one in Omaha. Still another example of feedforward control is the scheduled preventive maintenance programs on aircraft done by the major airlines. These are designed to detect and hopefully prevent structural damage that might lead to an accident.

The key to feedforward control is taking managerial action *before* a problem occurs. That way, problems can be prevented rather than requiring correction after any damage—poor-quality products, lost customers, lost revenue, and so on—has already been done. However, feedforward control requires timely and accurate information that isn't always easy to get. Thus, managers frequently end up using the other two types of control.

Concurrent Control. Concurrent control, as its name implies, takes place while a work activity is in progress. For instance, Nicholas Fox is director of business product management at Google. He and his team keep a watchful eye on one of Google's most profitable businesses—online ads. They watch "the number of searches and clicks, the rate at which users click on ads, the revenue this generates—everything is tracked hour by hour, compared with the data from a week earlier and charted."[11] If they see something that's not working particularly well, they fine-tune it.

The best-known form of concurrent control is direct supervision. Another term for this is **management by walking around**, which describes a manager being in the work area, interacting directly with employees. For example, Nvidia's CEO Jen-Hsun Huang tore down his cubicle and replaced it with a conference table so he's available to employees at all times to discuss what's going on.[12] Even GE's CEO Jeff Immelt spends 60 percent of his workweek on the road talking to employees and visiting the company's numerous locations.[13] All managers can benefit from using concurrent control because it can help them correct problems before they become too costly.

Feedback Control. The most popular type of control relies on feedback. In **feedback control**, the control takes place *after* the activity is done. For instance, the Denver Mint discovered the flawed Wisconsin quarters using feedback control. The damage had already occurred, even though the organization corrected the problem when it was discovered. And that's the major problem with this type of control. By the time a manager has the information, the problems have already occurred, leading to waste or damage. However, in many work areas (for example, financial areas), feedback is the only viable type of control.

Feedback control does have two advantages.[14] First, feedback gives managers meaningful information about how effective their planning efforts have been. Feedback that shows little variance between standard and actual performance indicates that the planning was generally on target. If the deviation is significant, a manager can use that information to formulate new plans. Second, feedback can enhance motivation. People want to know how well they're doing, and feedback provides that information.

Now, let's look at some specific control tools that managers can use.

FINANCIAL CONTROLS

Every business wants to earn a profit. To achieve this goal, managers need financial controls. For instance, they might analyze quarterly income statements for excessive expenses. And they might calculate financial ratios to ensure that sufficient cash is available to pay ongoing expenses, that debt levels haven't become too high, or that assets are being used productively.

Traditional Financial Control Measures. Traditional financial measures managers might use include ratio analysis and budget analysis. Exhibit 17–9 summarizes some of the most popular financial ratios. Liquidity ratios measure an organization's ability to meet its current debt obligations. Leverage ratios examine the organization's use of debt to finance its assets and whether the organization is able to meet the interest payments on

Exhibit 17–9

Popular Financial Ratios

Objective	Radio	Calculation	Meaning
Liquidity	Current ratio	$\frac{\text{Current assets}}{\text{Current liabilities}}$	Tests the organization's ability to meet short-term obligations
	Acid test	$\frac{\text{Current assets less inventories}}{\text{Current liabilities}}$	Tests liquidity more accurately when inventories turn over slowly or are difficult to sell
Leverage	Debt to assets	$\frac{\text{Total debt}}{\text{Total assets}}$	The higher the ratio, the more leveraged the organization
	Times interest earned	$\frac{\text{Profits before interest and taxes}}{\text{Total interest charges}}$	Measures how many times the organization is able to meet its interest expenses
Activity	Inventory turnover	$\frac{\text{Sales}}{\text{Inventory}}$	The higher the ratio, the more efficiently inventory assets are being used
	Total asset turnover	$\frac{\text{Sales}}{\text{Total assets}}$	The fewer assets used to achieve a given level of sales, the more efficiently management is using the organization's total assets
Profitability	Profit margin on sales	$\frac{\text{Net profit after taxes}}{\text{Total sales}}$	Identifies the profits that are being generated
	Return on investment	$\frac{\text{Net profit after taxes}}{\text{Total assets}}$	Measures the efficiency of assets to generate profits

feedforward control
Control that takes place before a work activity is done.

concurrent control
Control that takes place while a work activity is in progress.

management by walking around
A term used to describe a manager being out in the work area, interacting directly with employees.

feedback control
Control that takes place after a work activity is done.

the debt. Activity ratios assess how efficiently a company is using its assets. Finally, profitability ratios measure how efficiently and effectively the company is using its assets to generate profits. These are calculated using selected information from the organization's two primary financial statements (the balance sheet and the income statement), which are then expressed as a percentage or ratio. (Because you've probably studied these ratios in other accounting or finance courses, or will in the near future, we don't elaborate here on how they're calculated. We mention them here to remind you that managers use such ratios as internal control tools.)

Budgets are planning and controlling tools. (See Appendix B for more information on budgeting.) When a budget is formulated, it's a planning tool because it indicates which work activities are important and what and how much resources should be allocated to those activities. But budgets are also used for controlling because they provide managers with quantitative standards against which to measure and compare resource consumption. If deviations are significant enough to require action, a manager examines what has happened and tries to uncover why. With this information, necessary action can be taken. For example, say that you use a personal budget for monitoring and controlling your monthly expenses. If you find that one month your miscellaneous expenses were higher than you had budgeted for, you might cut back spending in another area or work extra hours to get more income.

Managing Earnings. A practice that has come under increased scrutiny is managing earnings. When organizations "manage" earnings, they "time" income and expenses to enhance current financial results, which gives an unrealistic picture of the organization's financial performance. For instance, many organizations have used deferred compensation programs for their top executives. Because deferred compensation doesn't have to be counted as a current expense—although there's usually a short reference about it buried in the financial statement footnotes—earnings look better in the present. The problem is that these can add up to a huge future financial liability. For example, at Wyeth Pharmaceuticals, company executives were able to participate in a retirement program that allowed them to set aside, pretax, as much as 100 percent of their cash compensation. Wyeth guaranteed these executives a 10 percent return on this deferred pay.[15] Needless to say, this program raised serious ethical concerns and created financial uncertainty. New laws and regulations now require companies to clarify their financial information, reducing the temptation to manage earnings.

THE BALANCED SCORECARD APPROACH

Managers can use the **balanced scorecard** approach to evaluate organizational performance from more than just a financial perspective.[16] A balanced scorecard typically looks at four areas that contribute to a company's performance: finances, customers, internal processes, and people/innovation/growth assets. According to this approach, managers should develop goals in each of the four areas and then measure whether the goals are being met.

Although a balanced scorecard makes sense, managers tend to focus on areas that drive their organization's success and use scorecards that reflect those strategies.[17] For example, if strategies are customer centered, then the customer area is likely to get more attention than the other three areas. Yet, you can't focus on measuring only one performance area because others are affected as well. For instance, at IBM Global Services in Houston, managers developed a scorecard around an overriding strategy of customer satisfaction. However, the other areas (finances, internal processes, and people/innovation/growth) support that central strategy. The division manager described it as follows: "The internal processes part of our business is directly related to responding to our customers in a timely manner, and the learning and innovation aspect is critical for us since what we're selling our customers above all is our expertise. Of course, how successful we are with those things will affect our financial component."[18]

Let'S Get Real: F2F

TYPES OF CONTROL I USE:
I use more of a participatory type of control. Acknowledging my coworkers and involving them in decision-making process turns out to be more motivating than controlling their behavior and constant monitoring. I prefer to use a review mechanism that focuses more on the deliverables and incorporates feedback from multiple sources.

INFORMATION CONTROLS

A computer containing personal information (Social Security numbers, birth dates, etc.) on about some 26.5 million military veterans was stolen from the residence of a Department of Veteran Affairs employee who had taken the computer home without authorization. Although the computer was eventually recovered with no loss of personal information, the situation could have been damaging to a large number of people.[19] Talk about the need for information controls! Managers deal with information controls in two ways: (1) as tools to help them control other organizational activities and (2) as organizational areas they need to control.

How Is Information Used in Controlling? Managers need the right information at the right time and in the right amount to monitor and measure organizational activities and performance.

In measuring actual performance, managers need information about what is happening within their area of responsibility and about the standards in order to be able to compare actual performance with the standard. They also rely on information to help determine whether deviations are acceptable. Finally, they rely on information to help develop appropriate courses of action. Information *is* important! Most of the information tools that managers use come from the organization's management information system.

A **management information system (MIS)** is a system used to provide managers with needed information on a regular basis. In theory, an MIS can be manual or computer based, although most organizations have moved to computer-supported applications. The term *system* in MIS implies order, arrangement, and purpose. Further, an MIS focuses specifically on providing managers with *information* (processed and analyzed data), not merely *data* (raw, unanalyzed facts). A library provides a good analogy: Although it can contain millions of volumes, a library doesn't do you any good if you can't find what you want quickly. That's why librarians spend a great deal of time cataloging a library's collections and ensuring that materials are returned to their proper locations. Organizations today are like well-stocked libraries. There's no lack of data. There is, however, an inability to process that data so that the right information is available to the right person when he or she needs it. An MIS collects data and turns them into relevant information for managers to use.

Controlling Information. It seems that every week, there's another news story about information security breaches. A recent survey found that 85 percent of privacy and security professionals acknowledged that a reportable data breach had occurred within their organizations within the past year.[20] Because information is critically important to everything an organization does, managers must have comprehensive and secure controls in place to protect information. Such controls can range from data encryption to system firewalls to data backups, as well as other techniques.[22] Problems can lurk in places that

thinking critically about Ethics

Duplicating software for coworkers and friends is a widespread practice, but software is protected by copyright laws. Illegally copying is a cognized civil and criminal offence. It is claimed that solving piracy can create 44,000 jobs in India.[21]

Is reproducing copyrighted software ever an acceptable practice? Explain. Is it wrong for employees of a business to pirate software but permissible for struggling college students who can't afford to buy their own software? As a manager, what types of ethical guidelines could you establish for software use? What if you were a manager in another country where software piracy commonly occurs?

balanced scorecard
A performance measurement tool that looks at more than just the financial perspective.

management information system (MIS)
A system used to provide management with needed information on a regular basis.

an organization might not even have considered, such as search engines. Sensitive, defamatory, confidential, or embarrassing organizational information has found its way into search engine results. For instance, detailed monthly expenses and employee salaries on the National Speleological Society's Web site turned up in a Google search.[23] Equipment such as laptop computers and even RFID (radio-frequency identification) tags are vulnerable to viruses and hacking. Needless to say, information controls should be monitored regularly to ensure that all possible precautions are in place to protect important information.

BENCHMARKING OF BEST PRACTICES

Managers in such diverse industries as health care, education, and financial services are discovering what manufacturers have long recognized—the benefits of **benchmarking**, which is the search for best practices among competitors or noncompetitors that lead to their superior performance. The goal of benchmarking is to identify various **benchmarks**, which are the standards of excellence against which to measure and compare. For instance, in 2005, the Punjab Health System Corporation (PHSC) has set performance benchmarks for specialist doctors. This has been done to increase utilization of government health facilities by patients.[24] N. R. Narayan Murthy said, "We have benchmarked ourselves against different companies in different dimensions. For example, we need to create the corporate brand of GE and Microsoft; GE has shown growth in earnings per share for over the past hundred quarters and we have shown for probably only 30 quarters. In the area of making a difference to the context, I think HP is a great company. Similarly, in India, in the area of longevity, Hindustan Lever is a great company. In the area of quality, TVS group is a great group."[25] At its most basic, benchmarking means learning from others. As a tool for monitoring and measuring organizational performance, benchmarking can be used to identify specific performance gaps and potential areas of improvement.

Best practices aren't just found externally. Sometimes best practices can be found inside the organization and just need to be shared. One fertile area for finding good performance improvement ideas is employee suggestion boxes: When an employee has an idea about a new or improved way of doing something, he or she can put it in the box. However, suggestion boxes have long been ridiculed as a waste of time. Cartoons have lambasted the futility of employees putting ideas in the suggestion box because suggestions often sit there until someone decides to empty the box. But it doesn't have to be that way. Research shows that best practices frequently already exist within an organization but usually go unidentified and unnoticed.[26] In today's environment, organizations seeking high performance levels can't afford to ignore such potentially valuable information. Some companies are catching on. For example, Toyota Motor Corporation developed a suggestion screening system to prioritize best practices based on potential impact, benefits, and difficulty of implementation. TVS Sundaram managers used internal benchmarking to help identify performance gaps and opportunities.[27] Exhibit 17–10 provides some suggestions for internal benchmarking.

Titan has over 100 million satisfied consumers and dominates the market with a 60 percent share of the organized sector market. Titan is able to satisfy customers because of the quality and its sales and service network. In reaching these levels, Titan has constantly worked for keeping the organization current and nimble and has tackled difficult problems. At one time, Titan identified that its wage bill was very high. The company addressed this by inviting employees to avail a voluntary retirement scheme and by negotiating a new wage agreement with its trade union. Mr Bhat, the MD of Titan, in an interview said, "We went through a lot of pain, but the entire leadership team felt we had to crack these problems in order to secure our future." Titan has tried to continuously communicate, be transparent and trustworthy, and treat stakeholders as a team working towards a common goal.[28]

Courtesy: Titan Industries/Tata

Exhibit 17–10

Suggestions for Internal Benchmarking

1. *Connect best practices to strategies and goals.* The organization's strategies and goals should dictate what types of best practices might be most valuable to others in the organization.
2. *Identify best practices throughout the organization.* Organizations must have a way to find out what practices have been successful in different work areas and units.
3. *Develop best practices reward and recognition systems.* Individuals must be given an incentive to share their knowledge. The reward system should be built into the organization's culture.
4. *Communicate best practices throughout the organization.* Once best practices have been identified, that information needs to be shared with others in the organization.
5. *Create a best practices knowledge-sharing system.* There needs to be a formal mechanism for organizational members to continue sharing their ideas and best practices.
6. *Nurture best practices on an ongoing basis.* Create an organizational culture that reinforces a "we can learn from everyone" attitude an emphasizes sharing information.

Source: Based on T. Leahy, "Extracting Diamonds in the Rough," *Business Finance*, August 2000, pp. 33–37.

QUICK LEARNING REVIEW:

LEARNING OUTCOME 17.4

- Contrast feedforward, concurrent, and feedback control.
- Explain the types of financial and information controls managers can use.
- Describe how balanced scorecards and benchmarking are used in controlling.

Go to page 424 and see how well you know this material.

LEARNING **OUTCOME 17.5** ▷

CONTEMPORARY ISSUES IN CONTROL

The employees of Integrated Information Systems Inc. didn't think twice about exchanging digital music over a dedicated office server they had set up. Like office betting on college and pro sports, it was technically illegal but harmless—or so the employees thought. But after the company had to pay a $1 million settlement to the Recording Industry Association of America, managers wished they had controlled the situation better.[29] Control is an important managerial function. We're going to look at four control issues that managers face today: cross-cultural differences, workplace concerns, customer interactions, and corporate governance.

ADJUSTING CONTROLS FOR CROSS-CULTURAL DIFFERENCES

The concepts of control that we've been discussing are appropriate for an organization whose work units are not geographically separated or culturally distinct. But control techniques can be quite different in different countries. The differences are primarily in the measurement and corrective action steps of the control process. In a global corporation, managers of foreign operations tend to be less controlled by the home office, if for no other reason than distance keeps managers from being able to observe work directly. Because distance creates a tendency to formalize controls, global organizations often rely on extensive formal reports for control, most of which are communicated electronically.

Technology's impact on control is also seen when comparing technologically advanced nations with less technologically advanced countries. Managers in countries where technology is more advanced often use indirect control devices such as

benchmarking
The search for the best practices among competitors or noncompetitors that lead to their superior performance.

benchmark
The standard of excellence against which to measure and compare.

Let's Get Real:
F2F

CONTROL IS IMPORTANT TODAY BECAUSE:
It is the key to growth, adaptation, and survival. Through monitoring and measurement, they signal an alarm for correction, thus facilitating adaptation and ultimately survival in a highly competitive business environment. An organization without necessary controls is like a rudderless boat.

computer-generated reports and analyses in addition to standardized rules and direct supervision to ensure that work activities are going as planned. In less technologically advanced countries, however, managers tend to use more direct supervision and highly centralized decision making for control.

Managers in foreign countries also need to be aware of constraints on corrective actions they can take. Some countries' laws prohibit closing facilities, laying off employees, taking money out of the country, or bringing in a new management team from outside the country.

Finally, another challenge for global managers in collecting data for measurement and comparison is comparability. For instance, a company that manufactures apparel in Cambodia might also produce the same products at a facility in Scotland. However, the Cambodian facility might be more labor intensive than its Scottish counterpart to take advantage of lower labor costs in Cambodia. This makes it hard to compare, for instance, labor costs per unit.

WORKPLACE CONCERNS

Today's workplaces present considerable control challenges for managers. From monitoring employees' computer usage at work to protecting the workplace against disgruntled employees intent on doing harm, managers need controls to ensure that work can be done efficiently and effectively as planned.

Workplace Privacy. Do you think you have a right to privacy at your job? What can your employer find out about you and your work? You might be surprised at the answers! Employers can (and do), among other things, read your e-mail messages (even those marked "personal" or "confidential"), tap your telephone, monitor your work by computer, store and review computer files, monitor you in an employee bathroom or dressing room, and track your whereabouts in a company vehicle. And these actions aren't uncommon. In fact, some 26 percent of companies have fired workers for misusing the Internet, another 25 percent have terminated workers for e-mail misuse, and 6 percent have fired employees for misusing office phones.[30]

Why do managers feel they need to monitor what employees are doing? A big reason is that employees are hired to work, not to surf the Web, checking stock prices, watching online videos, getting cricket updates, or shopping for presents for family members or friends. Recreational on-the-job Web surfing is thought to cost billions of dollars in lost work productivity annually. In fact, a survey of U.S. employers showed that 87 percent of employees look at non-work-related Web sites while at work and more than half engage in personal Web site surfing every day.[31] Watching online videos has become an increasingly serious problem not only because of the time being wasted by employees but because it clogs already strained corporate computer networks.[32] (See Exhibit 17–11 for a list of the top 10 video Internet sites viewed at work.) All this non-work adds up to significant costs to busnesses.

Another reason that managers monitor employee e-mail and computer usage is that they don't want to risk being sued for creating a hostile workplace environment because of offensive messages or an inappropriate image displayed on a coworker's computer screen. Concerns about racial or sexual harassment are one reason companies might want to monitor or keep backup copies of all e-mail. Electronic records can help establish what actually happened in a situation so managers can react quickly.[33]

Finally, managers want to ensure that company secrets aren't being leaked.[34] In addition to typical e-mail and computer usage, companies are monitoring instant messaging and banning camera phones in the office. Managers need to be certain that employees are not, even inadvertently, passing information on to others who could use that information to harm the company.

Because of the potentially serious costs and given the fact that many jobs now entail computers, many companies have workplace monitoring policies. Such policies should control employee behavior in a non-demeaning way, and employees should be informed about those policies.

Exhibit 17–11

Top Internet Video Sites Viewed at Work

Source: Bobby White, "The New Workplace Rules: No Video Watching," *Wall Street Journal,* March 4, 2008, p. B3.

Top 10 Internet video brands viewed in the U.S. while at work for January 2008, in millions of streams

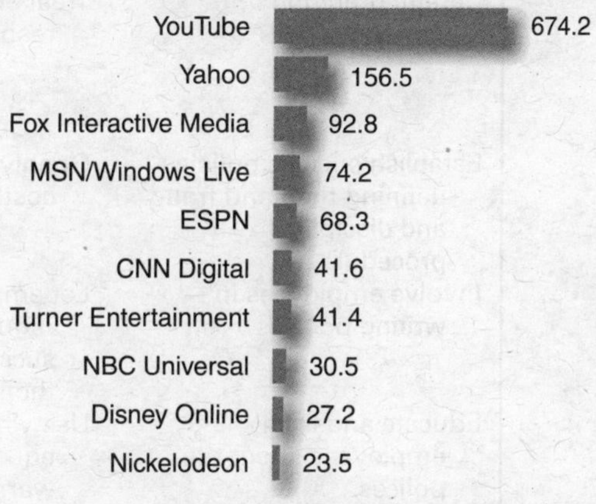

Employee Theft. You might be surprised by the fact that up to 85 percent of all organizational theft and fraud is committed by employees, not outsiders.[35] In retail, according to the Global Retail Theft Barometer 2008 Survey Report, the total loss in retail business in India in 2008 is $2.543 billion. Employee theft was responsible for 23.7 percent of the loss. The rest of the losses were contributed by consumer theft, supplier or vendor lapses, and administrative errors.[36]

Employee theft is defined as any unauthorized taking of company property by employees for their personal use.[37] It can range from embezzlement to fraudulent filing of expense reports to removal of equipment, parts, software, or office supplies from company premises. Although retail businesses have long faced serious potential losses from employee theft, loose financial controls at start-ups and small companies and the ready availability of information technology have made employee theft an escalating problem in all kinds and sizes of organizations. It's a control issue that managers need to educate themselves about and be prepared to deal with it.[38]

Why do employees steal? The answer depends on whom you ask.[39] Experts in various fields—such as industrial security, criminology, and clinical psychology—have different perspectives. The industrial security people propose that people steal because the opportunity presents itself through lax controls and favorable circumstances. Criminologists say that it's because people have financial-based pressures (such as personal financial problems) or vice-based pressures (such as gambling debts). And clinical psychologists suggest that people steal because they can rationalize whatever they're doing as being correct and appropriate behavior ("everyone does it," "they had it coming," "this company makes enough money, and they'll never miss anything this small," "I deserve this for all that I put up with," and so forth).[40] Although each approach provides compelling insight into employee theft and has been instrumental in attempts to deter it, unfortunately, employees continue to steal. What can managers do?

The concept of feedforward, concurrent, and feedback control is useful for identifying measures to deter or reduce employee theft.[41] Exhibit 17–12 summarizes several possible managerial actions.

Workplace Violence. L. K. Chaudhury, Chief Executive of Italian auto partmaker Graziano Transmissions India Pvt. Ltd, was attacked and beaten to death in September 2008 by an angry mob of sacked workers of the company.[42] In November 2005, Manjunath, an employee of the Indian Oil Corporation, was killed by goons hired by the owners of a petrol pump. Manjunath after repeated warnings had closed down the pump for continually delivering adulterated petrol.[43] On July 21, 1998, Weston, the Managing Director, was severely assaulted by four workers at the company's factory at

employee theft
Any unauthorized taking of company property by employees for their personal use.

Exhibit 17–12

Controlling Employee Theft

Feedforward	Concurrent	Feedback
Careful prehiring screening.	Treat employees with respect and dignity.	Make sure employees know when theft or fraud has occurred—not naming names but letting people know this is not acceptable.
Establish specific policies defining theft and fraud and discipline procedures.	Openly communicate the costs of stealing.	Use the services of professional investigators.
Involve employees in writing policies.	Let employees know on a regular basis about their successes in preventing theft and fraud.	Redesign control measures.
Educate and train employees about the polices.	Use video surveillance equipment if conditions warrant.	Evaluate your organization's culture and the relationships of managers and employees.
Have a professional review of your internal security controls.	Install "lock-out" options on computers, telephones, and e-mail.	
	Use corporate hotlines for reporting incidences.	
	Set a good example.	

Sources: Based on A. H. Bell and D. M. Smith, "Protecting the Company Against Theft and Fraud," *Workforce* online (www.workforce.com), December 3, 2000; J. D. Hansen, "To Catch a Thief," *Journal of Accountancy*, March 2000, pp. 43–46; and J. Greenberg, "The Cognitive Geometry of Employee Theft," in S. B. Bacharach, A. O'Leary-Kelly, J. M. Collins, and R. W. Griffin (eds.), *Dysfunctional Behavior in Organizations: Nonviolent and Deviant Behavior* (Stamford, CT: JAI Press, 1998), pp. 147–193.

Batanagar, while he was attending a business meet. The incident occurred after a member of Bata Mazdoor Union, Arup Dutta, met Weston to discuss the issue of the suspended employees.[44] Is workplace violence really an issue for managers? Yes. Despite these examples, thankfully, the number of workplace homicides has decreased with the decrease in union activism.[45] However, the U.S. National Institute of Occupational Safety and Health says that each year, some 2 million American workers are victims of some form of workplace violence. In an average week, 1 employee is killed and at least 25 are seriously injured in violent assaults by current or former coworkers. And in a Department of Labor survey, 58 percent of firms reported that managers received verbal threats from workers.[46]

Exhibit 17–13 describes the results of a survey of workers and their experiences with office rage. Anger, rage, and violence in the workplace are intimidating to coworkers and adversely affect their productivity. The annual cost to U.S. businesses is estimated to be between $20 and $35 billion.[47] Though there is no data available for Indian Corporates, it can be said with confidence that office rage isn't a uniquely American problem. In a survey of aggressive behavior in Britain's workplaces, 18 percent of managers said that they had personally experienced harassment or verbal bullying, and 9 percent said that they had experienced physical attacks.[48]

Exhibit 17–13

Workplace Violence

• Witnessed yelling or other verbal abuse	42%
• Yelled at coworkers themselves	29%
• Cried over work-related issues	23%
• Seen someone purposely damage machines or furniture	14%
• Seen physical violence in the workplace	10%
• Struck a coworker	2%

Source: Integra Realty Resources, October–November, Survey of Adults 18 and Over, in,"Desk Rage," *BusinessWeek*, November 20, 2000, p. 12.

What factors are believed to contribute to workplace violence? Undoubtedly, employee feel stress due to steep fuel prices, job uncertainties, declining value of retirement accounts, long hours, information overload, other daily interruptions, unrealistic deadlines, and uncaring managers. Even office layout designs with small cubicles where employees work amid noise and commotion from those around them have been cited as contributing to the problem.[49] Other experts have described dangerously dysfunctional work environments characterized by the following as primary contributors to the problem:[50]

- Employee work driven by time, numbers, and crises.
- Rapid and unpredictable change, where instability and uncertainty plague employees.
- Destructive communication style, where managers communicate in excessively aggressive, condescending, explosive, or passive-aggressive styles; excessive workplace teasing or scapegoating.
- Authoritarian leadership with a rigid, militaristic mind-set of managers versus employees, where employees aren't allowed to challenge ideas, participate in decision making, or engage in team-building efforts.
- Defensive attitude, where little or no performance feedback is given; only numbers count; and yelling, intimidation, and avoidance are the preferred ways of handling conflict.
- Double standards in terms of policies, procedures, and training opportunities for managers and employees.
- Unresolved grievances because there are no mechanisms or only adversarial ones in place for resolving them and because dysfunctional individuals may be protected or ignored because of long-standing rules, union contract provisions, or reluctance to take care of problems.
- Emotionally troubled employees and no attempt by managers to get help for these people.
- Repetitive, boring work, with no chance for doing something else or for new people coming in.
- Faulty or unsafe equipment or deficient training, which keeps employees from being able to work efficiently or effectively.
- Hazardous work environment in terms of temperature, air quality, repetitive motions, overcrowded spaces, noise levels, excessive overtime, and so forth. To minimize costs, no additional employees are hired when work load becomes excessive, leading to potentially dangerous work expectations and conditions.
- Culture of violence and a history of individual violence or abuse, violent or explosive role models, or tolerance of on-the-job alcohol or drug abuse.

Reading through this list, you surely hope that workplaces where you'll spend your professional life won't be like this. However, the competitive demands of succeeding in a 24/7 global economy put pressure on organizations and employees in many ways.

What can managers do to deter or reduce possible workplace violence? Once again, the concept of feedforward, concurrent, and feedback control can help identify actions that managers can take.[51] Exhibit 17–14 summarizes several suggestions.

CONTROLLING CUSTOMER INTERACTIONS

Every month, every local branch of Enterprise Rent-a-Car conducts telephone surveys with customers.[52] Each branch earns a ranking based on the percentage of its customers who say they were "completely satisfied" with their last Enterprise experience—a level of satisfaction referred to as "top box." Top box performance is important to Enterprise because completely satisfied customers are far more likely to be repeat customers. And by using this service quality index measure, employees' careers and financial aspirations are linked with the organizational goal of providing consistently superior service to each customer. Managers at Enterprise Rent-a-Car understand the

Exhibit 17–14

Controlling Workplace Violence

Feedforward	Concurrent	Feedback
MBWA (managing by walking around) to identify potential problems; observe how employees treat and interact with each other.	Ensure management commitment to functional, not dysfunctional, work environments.	Communicate openly about incidences and what's being done.
Provide employee assistance programs (EAPs) to help employees with behavioral problems.	Allow employees or work groups to "grieve" during periods of major organizational change.	Investigate incidences and take appropriate action.
Enforce organizational policy that any workplace rage, aggression, or violence will not be tolerated.	Be a good role model in how you treat others.	Review company policies and change, if necessary.
Use careful prehiring screening.	Use corporate hotlines or some other mechanism for reporting and investigating incidences.	
Never ignore threats.	Use quick and decisive intervention.	
Train employees about how to avoid danger if situation arises.	Get expert professional assistance if violence erupts.	
Clearly communicate policies to employees.	Provide necessary equipment or procedures for dealing with violent situations (cell phones, alarm system, code names or phrases, and so forth).	

Sources: Based on M. Gorkin, "Five Strategies and Structures for Reducing Workplace Violence," *Workforce* online (www.workforce.com), December 3, 2000; "Investigating Workplace Violence: Where Do You Start?" *Workforce* online (www.forceforce.com), December 3, 2000; "Ten Tips on Recognizing and Minimizing Violence," *Workforce* online (www.workforce.com), December 3, 2000; and "Points to Cover in a Workplace Violence Policy," *Workforce* online (www.workforce.com), December 3, 2000.

connection between employees and customers and the importance of controlling employee–customer interactions.

There's probably no better area to see the link between planning and controlling than in customer service. A company that proclaims customer service as one of its goals can quickly and clearly see whether it's achieving that goal by examining how satisfied customers are with their service. How can managers control the interactions between the goal and the outcome when it comes to customers? The concept of a service profit chain can help.[53]

A **service profit chain** is the service sequence from employees to customers to profit. According to this concept, a company's strategy and service delivery system influence how employees deal with customers—that is, how productive they are in providing service and the quality of that service. The level of employee service productivity and service quality influences customer perceptions of service value. When service value is high, it has a positive impact on customer satisfaction, which leads to customer loyalty. And customer loyalty improves organizational revenue growth and profitability.

What does the service profit chain concept mean for managers? Managers who want to control customer interactions should work to create long-term and mutually beneficial relationships among the company, employees, and customers. How? By creating a work environment that enables employees to deliver high levels of quality service and that makes them feel they're capable of delivering top-quality service. In such a service climate, employees are motivated to deliver superior service. Employee efforts to satisfy customers, coupled with the service value provided by the organization, improve customer satisfaction. And when customers receive high service value, they're loyal, which ultimately improves the company's growth and profitability.

J. D. Power has ranked Maruti Suzuki the highest in customer satisfaction for the ninth year. Satisfaction was measured among vehicle owners who visited authorized dealership service center for maintenance or repair work during the first 12–18 months of ownership. Seven parameters of quality of service were measured. Nearly one of seven customers reported that they did not receive their vehicle back within 3 hours and, thus, has become the leader in providing a quick service turnaround. Maruti Suzuki has effectively implemented simple procedures to improve satisfaction with the value of services provided and perceptions of the fairness and honesty of the dealer. Extensive access, intensive training, empowerment, and a relationship of trust has been established between the company and the dealers so that they provide seamless service to buyers of Maruti cars.[55]

BPL Mobile is ranked no. 1 in customer care since 2005. BPL has been consistently delivering the highest customer satisfaction in Mumbai. According to the Report on the Indian Telecom Services Performance Indicators, 2007, BPL Mobile is one of the very few cellular companies that has met and exceeded TRAI benchmarks. To ensure customer centricity, workshops on customer centricity are run for each of the departments. Also internal customers within the company are treated as clients to ensure top-class delivery to its external customers.[54]

CORPORATE GOVERNANCE

Although Andrew Fastow—Enron's former chief financial officer, who pled guilty to wire and securities fraud—had an engaging and persuasive personality, that didn't explain why Enron's board of directors failed to raise even minimal concerns about management's questionable accounting practices. The board even allowed Fastow to set up off-balance-sheet partnerships for his own profit at the expense of Enron's shareholders.

Corporate governance, the system used to govern a corporation so that the interests of corporate owners are protected, failed abysmally at Enron, as it did at many other companies caught in financial scandals. In the aftermath of these scandals, corporate governance has been reformed. Two areas where reform has taken place are the role of boards of directors and financial reporting. It is mandated that boards of directors take a more active role.

The Role of Boards of Directors. The original purpose of a board of directors was to have a group, independent from management, looking out for the interests of shareholders, who were not involved in the day-to-day management of the organization. However, it didn't always work that way. Board members often enjoyed a cozy relationship with managers in which each took care of the other. This type of "quid pro quo" arrangement has changed. Securities and Exchange Board of India (SEBI) established the regulatory framework for listed companies, specifically for corporate governance, in February 2000, based on the recommendations of Kumarmangalam Birla Committee Report. It was enshrined as Clause 49 of the Listing Agreement. The Companies Act, 1956 and Companies Bill, 2004, are two major legislations for regulating the entire corporate structure. These laws have been amended, as per

service profit chain
The service sequence from employees to customers to profit.

corporate governance
The system used to govern a corporation so that the interests of corporate owners are protected.

the requirement, to bring more transparency and accountability in the provisions of corporate governance and provide a framework for facilitating faster economic growth.[56] To protect the interests of investors in the securities markets, the Securities Contracts (Regulation) Act, 1956, Securities and Exchange Board of India Act, 1992, and Depositories Act, 1996, have been introduced by SEBI.[57]

Financial Reporting and the Audit Committee. In addition to expanding the role of boards of directors, the Sarbanes-Oxley Act also called for more disclosure and transparency of corporate financial information. In fact, senior managers in the United States are now required to certify their companies' financial results. Such changes have led to better information—that is, information that is more accurate and reflective of a company's financial condition. The Ministry of Corporate Affairs, Government of India, has set up National Foundation for Corporate Governance (NFCG) in partnership with Confederation of Indian Industry (CII), Institute of Company Secretaries of India (ICSI), and Institute of Chartered Accountants of India (ICAI) to promote better corporate governance practices in the country. The NFCG hosts an online library of principles, codes, rules, and regulations related to the corporate governance in the Indian context. Managers would be well advised to make themselves familiar with such resources.

QUICK LEARNING REVIEW:

LEARNING OUTCOME 17.5

- Describe how managers may have to adjust controls for cross-cultural differences.
- Discuss workplace concerns and how they might be controlled.
- Explain why control is important to customer interactions.
- Define corporate governance.

Go to page 425 and see how well you know this material.

Let's Get Real:

My Turn

Asha Nair

Sr Manager, IT-Flat Products
Tata Steel

In order to make the best use of the rapid response team, there has to be a seamless integration of the rapid response team with the production-line employees, since they have to work as a team to solve the issue or alternatively allow for a quick handoff. To facilitate this, work instructions, job codes, and safety norms must be clear and this should be displayed at all workstations. Chain of command for critical handoffs should also be clear, i.e., based on the nature of the job and expertise of the individual; someone from the rapid response team would have to assume leadership to steer the team toward problem resolution. Roles and responsibilities of all involved must be known and full support from the line people is also important. Proper training has to be provided to both parties to handle such situations. Some prior simulations would be helpful. Constant evaluation and upgradation of skills would be required so that the people in the rapid action team are equipped with the necessary skillset. I would also get them to share their learning's from the handling of various situations within and outside the group and document them for future reference. It is very critical for them to work as a team and the feeling of team spirit needs to be fostered. Cycle time for uptime after stoppage would need to be monitored and a reward or recognition to both the response team as well as the line employees for speedy resolution could be instituted.

LEARNING OUTCOMES **SUMMARY**

17.1 ▷ WHAT IS CONTROLLING, AND WHY IS IT IMPORTANT?

- Define *controlling*.
- Explain the planning–controlling link.
- Discuss the reasons control is important.

Controlling is the process of monitoring, comparing, and correcting work performance.

As the final step in the management process, controlling provides the link back to planning. If managers didn't control, they'd have no way of knowing whether goals were being met.

Control is important because (1) it helps a manager know whether goals are being met and if not, why; (2) it provides information and feedback so managers feel comfortable empowering employees; and (3) it helps protect an organization and its assets.

17.2 ▷ THE CONTROL PROCESS

- Describe the three steps in the control process.
- Explain the three courses of action managers can take in controlling.

The three steps in the control process are measuring, comparing, and taking action. Measuring involves deciding how to measure actual performance and what to measure. Comparing involves looking at the variation between actual performance and the standard (goal). Deviations outside an acceptable range of variation need attention.

Taking action can involve doing nothing, correcting performance, or revising the standards. Doing nothing is self-explanatory. Correcting performance can involve different corrective actions, which can either be immediate or basic. Revising standards can involve either raising or lowering them.

17.3 ▷ CONTROLLING FOR ORGANIZATIONAL PERFORMANCE

- Define organizational performance.
- Describe three frequently used organizational performance measures.

Organizational performance is the accumulated results of an organization's work activities.

Three frequently used organizational performance measures are (1) productivity, which is the output of goods or services produced divided by the inputs needed to generate that output; (2) effectiveness, which is a measure of the appropriateness of organizational goals and how well those goals are being met; and (3) industry and company rankings compiled by various business publications.

17.4 ▷ TOOLS FOR MEASURING ORGANIZATIONAL PERFORMANCE

- Contrast feedforward, concurrent, and feedback control.
- Explain the types of financial and information controls managers can use.
- Describe how balanced scorecards and benchmarking are used in controlling.

Feedforward controls take place before a work activity is done. Concurrent controls take place while a work activity is being done. Feedback controls take place after a work activity is done.

Financial controls that managers can use include financial ratios (for example, liquidity, leverage, activity, profitability) and budgets. One information control managers can use is a management information system (MIS), which provides managers with needed information on a regular basis. Other information controls are comprehensive and secure controls, such as data encryption, system firewalls, data backups, and so forth that protect the organization's information.

Balanced scorecards provide a way to evaluate an organization's performance in four different areas rather than just from the financial perspective. Benchmarking provides control by finding best practices among competitors or noncompetitors and from inside the organization itself.

17.5 ▷ CONTEMPORARY ISSUES IN CONTROL

- Describe how managers may have to adjust controls for cross-cultural differences.
- Discuss workplace concerns and how they might be controlled.
- Explain why control is important to customer interactions.
- Define corporate governance.

Adjusting controls for cross-cultural differences may be necessary primarily in the areas of measuring and taking corrective actions.

Workplace concerns include workplace privacy, employee theft, and workplace violence. For each of these, managers need to have policies in place to control inappropriate actions and ensure that work is getting done efficiently and effectively.

Control is important to customer interactions because employee service productivity and service quality influence customer perceptions of service value. Organizations want long-term and mutually beneficial relationships among their employees and customers.

Corporate governance is the system used to govern a corporation so that the interests of corporate owners are protected.

THINKING ABOUT MANAGEMENT ISSUES

1. In Chapter 12 we discussed the white-water rapids view of change. Do you think it's possible to establish and maintain effective standards and controls in this type of environment? Discuss.
2. How could you use the concept of control in your personal life? Be specific. (Think in terms of feedforward, concurrent, and feedback controls as well as specific controls for the different aspects of your life—school, work, family relationships, friends, hobbies, etc.)
3. When do electronic surveillance devices such as computers, video cameras, and telephone monitoring devices step over the line from being effective management controls to being intrusions on employee rights?
4. "Every individual employee in an organization plays a role in controlling work activities." Do you agree with this statement, or do you think control is something that only managers are responsible for? Explain.
5. What are some work activities in which the acceptable range of variation might be higher than average? What about lower than average? (Hint: Think in terms of the output from the work activities, who it might affect, and how it might affect them.)

YOUR TURN to be a Manager

- You have a major class project due in a month. Identify some performance measures that you could use to help determine whether the project is going as planned and will be completed efficiently (on time) and effectively (high quality).
- Survey 30 people about whether they have experienced the violent actions listed in Exhibit 17–13. Compile your findings in a table. Are you surprised at the results? Be prepared to present these in class.
- Pretend you're the manager of a customer call center for timeshare vacations. What types of control measures would you use to see how efficient and effective an employee is? How about measures for evaluating the entire call center?
- Disciplining employees is one of the least favorite tasks of managers, but it is something that all managers have to do. Survey three managers about their experiences with employee discipline. What types of employee actions have caused the need for disciplinary action? What disciplinary actions have these managers used? What do they think is the most difficult thing to do when disciplining employees? What suggestions do they have for disciplining employees?
- Steve's and Mary's recommended readings: Marcus Buckingham, *Go Put Your Strengths to Work* (The Free Press, 2007); W. Steven Brown, *13 Fatal Errors Managers Make and How You Can Avoid Them* (Berkley Business, 1987); and Peter F. Drucker, *Management: Tasks, Responsibilities, Practices* (Harper Business, 1974).

- Research "The Great Package Race." Write a paper describing what it is and how it's a good example of organizational control.
- Find the latest governmental statistics on workplace injuries, illnesses, and fatalities. Research ways that organizations can control the number of incidents of injuries and fatalities. Compile this information into a report.
- In your own words, write down three things you learned in this chapter about being a good manager.
- Self-knowledge can be a powerful learning tool. Go to mymanagementlab and complete these self-assessment exercises: How Good Am I at Disciplining Others? How Willing Am I to Delegate? What Time of Day Am I Most Productive? How Good Am I at Giving Performance Feedback? Using the results of your assessments, identify personal strengths and weaknesses. What will you do to reinforce your strengths and improve your weaknesses?

CASE APPLICATION

Baggage Blunders

Terminal 5, built by British Airways for $8.6 billion, is Heathrow Airport's newest state-of-the-art facility. Made of glass, concrete, and steel, it's the largest free-standing building in the United Kingdom and has over 10 miles of belts for moving luggage. At the terminal's unveiling on March 15, 2008, Queen Elizabeth II called it a "twenty-first-century gateway to Britain." Alas . . . the accolades didn't last long! After two decades of planning and 100 million hours of labor, opening day didn't work out as planned. Endless lines and severe baggage handling delays led to numerous flight cancellations, stranding many irate passengers. Airport operators said the problems were triggered by glitches in the terminal's high-tech baggage-handling system.

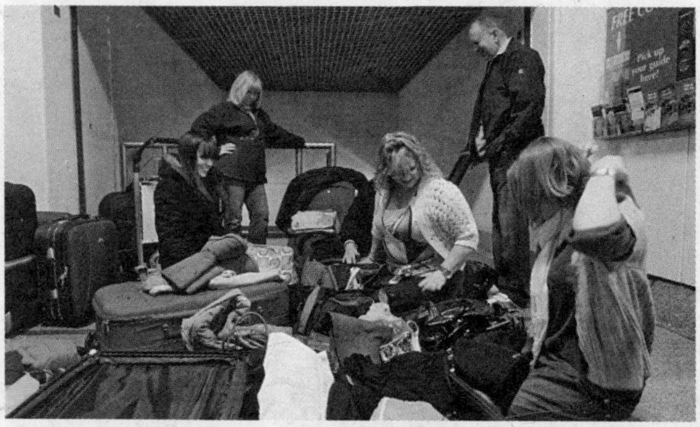

Steve Parsons/PA Wire /AP Wide World

Passengers sort through their luggage after being delayed by a baggage system failure at Heathrow Airport.

With its massive automation features, Terminal 5 was planned to ease congestion at Heathrow and improve the flying experience for the 30 million passengers expected to pass through it annually. With 96 self-service check-in kiosks, more than 90 check-in fast bag drops, 54 standard check-in desks, and over 10 miles of suitcase-moving belts that were supposed to be able to process 12,000 bags per hour, the facility's design didn't seem to support those goals.

Within the first few hours of the terminal's operation, problems developed. Baggage workers, presumably understaffed, were unable to clear incoming luggage fast enough. Many arriving passengers had to wait more than an hour to get their bags. There were problems for departing passengers, as well, as many tried in vain to check in for flights. Flights were allowed to leave with empty cargo holds. At one point that first day, the airline had no choice but to check in only those with no luggage. And it didn't help matters that the moving belt system jammed at one point. Lesser problems also became apparent: a few broken escalators, some hand dryers that didn't work, a gate that wouldn't function at the new Underground station, and inexperienced ticket sellers who didn't know the fares between Heathrow and various stations on the Piccadilly line. By the end of the first full day of operation, Britain's Department of Transportation released a statement calling for British Airways and the airport operator BAA to "work hard to resolve these issues and limit disruptions to passengers."

You might be tempted to think that all this could have been prevented if British Airways had only tested the system. But thorough runs of all systems "from toilets to check in and seating" took place six months before opening, including four full-scale test runs using 16,000 volunteers.

Discussion Questions

1. What type of control—feedforward, concurrent, or feedback—do you think would be most important in this situation? Explain your choice.
2. How might immediate corrective action have been used in this situation? How about basic corrective action?
3. Could British Airways's controls have been more effective? How?
4. What role would information controls play in this situation? Customer interaction controls? Benchmarking?

Sources: K. Capell, "British Airways Hit by Heathrow Fiasco," *BusinessWeek*, April 3, 2008, p. 6; The Associated Press, "Problems Continue at Heathrow's Terminal 5," *International Herald Tribune* online, www.iht.com, March 31, 2008; M. Scott, "New Heathrow Hub: Slick, but No Savior," *BusinessWeek*, March 28, 2008, p. 11; and G. Katz, "Flights Are Canceled, Baggage Stranded, as London's New Heathrow Terminal Opens," *The Seattle Times* online, seattletimes.nwsource.com, March 27, 2008.

Let's Get Real:

Meet the Managers

Arun Pratap Singh

Vice President and SBU Head (Auto)
M/s Electrotherm (India) Ltd

MY JOB: I am expected to make the business grow profitably while adding other lines of businesses.

BEST PART OF MY JOB: Setting up of new processes while getting the buy-in of my team and seeing the benefits of the new process for the organization.

WORST PART OF MY JOB: Explaining to the top team the reason for moving slower than expected.

BEST MANAGEMENT ADVICE EVER RECEIVED: Always reflect on strengths and weaknesses that affect the business under my stewardship.

Basudev Mohanty

Senior Manager
NTPC Limited

MY JOB: Inspection of goods, machines, raw materials, spares and consumables at suppliers place before dispatch to project or consumption site as per the procedure agreed at the time of contract finalization.

BEST PART OF MY JOB: Even if there is a written procedure and standards, a lot of common sense and technical intelligence go into the job.

WORST PART OF MY JOB: Pressure to finish the job under stringent timeline.

BEST MANAGEMENT ADVICE EVER RECEIVED: People are everything in all the processes; be sensitive to people-related issues.

You'll be hearing more from these real managers throughout the chapter.

Chapter

Managing Operations

Every organization "produces" something, whether it's a good or a service. This chapter focuses on how organizations do that through a process called *operations management*. We also look at the important role that managers play in managing those operations. Focus on the following learning outcomes as you read and study this chapter.

LEARNING OUTCOMES

A Manager's Dilemma

In Almere, Netherlands, you'll find a store that's taken a giant leap into the twenty-first century to sell a product that's been around for thousands of years.[1] The Selexyz bookstore there uses an RFID (radio-frequency identification) inventory management system that places tags on every single item on its shelves. Thanks to the RFID wireless technology, employees no longer have to take inventory by hand, there are no more lost books, and customers can easily pinpoint within seconds the exact location of any book in the store. Since implementing the system, the store's sales have been 25 percent higher than those of the average Selexyz bookstore. And the store is more efficient as well. Profits were expected to jump by 40 percent. The company's IT director said that prior to using RFID, "people had to check each box manually, and it would take up to five or six minutes to check one box. Now, the time is reduced through this technology to less than 10 seconds." Now, Boekhandels Groep Nederland, the Dutch retailer that opened the Selexyz stores, wants to implement the system throughout all 42 Selexyz outlets. Suppose that you were in charge of doing this.

What would you do?

You've probably never given much thought to how organizations "produce" the goods and services that you buy or use. But it's an important process. Without it, you wouldn't have a car to drive or McDonald's fries to snack on, or even a hiking trail in a local park to enjoy. Organizations need to have well-thought-out and well-designed operating systems, organizational control systems, and quality programs to survive in today's increasingly competitive global environment. And it's a manager's job to manage those things.

LEARNING **OUTCOME 18.1** ▷

THE ROLE OF OPERATIONS MANAGEMENT

Inside Intel's factory in New Mexico, employee Trish Roughgarden is known as a "seed"—an unofficial title for technicians who transfer manufacturing know-how from one Intel facility to another.[2] Her job is to make sure that this factory works just like an identical one that opened earlier in Oregon. When another plant opened in Ireland, several hundred other seeds copied the same techniques. The company's facility in Arizona also benefited from "seeding." What the seeds do is part of a strategy known as "Copy Exactly," which Intel implemented after frustrating variations between factories hurt productivity and product quality. In the intensely competitive chip-making industry, Intel knows that how it manages operations will determine whether it succeeds.

What is **operations management**? The term refers to the transformation process that converts resources into finished goods and services. Exhibit 18–1 portrays this process in a very simplified fashion. The system takes in inputs—people, technology, capital, equipment, materials, and information—and transforms them through various processes, procedures, work activities, and so forth into finished goods and

Exhibit 18–1

The Operations System

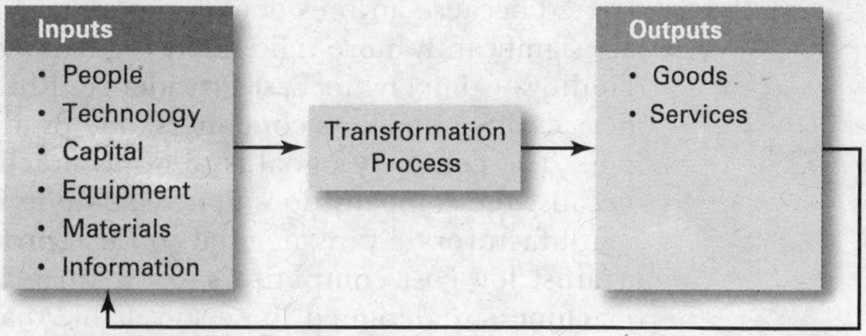

Let'S Get Real: F2F

HOW OUR "PRODUCT" IS PRODUCED:

Our product is electricity. It is produced through a closed process where heat from coal is transmitted to water making it high-temperature and high-pressure steam, and the steam is released in a controlled manner to a rotating device called turbine where the steam loses the pressure and temperature to rotate it. The turbine is coupled to a device called generator that produces electricity. The electricity thus produced is transmitted to the user by a network of wires.

Let'S Get Real: F2F

HOW OUR "PRODUCT" IS PRODUCED:

We are manufacturers of electric two wheelers. We have a manufacturing facility which produces indigenously designed electric bikes at Kutch. We have a trained staffs of 250 people to produce these vehicles.

services. Because every unit in an organization produces something, managers need to be familiar with operations management concepts in order to achieve goals efficiently and effectively.

Operations management is important to organizations and managers for three reasons: It encompasses both services and manufacturing, it's important in effectively and efficiently managing productivity, and it plays a strategic role in an organization's competitive success. Let's look at each.

SERVICES AND MANUFACTURING

With a menu that offers over 200 items, At Sundram Fasteners Limited, Chennai, over 2,000 different fasteners (nuts and bolts, special and standard) are manufactured and supplied to over 200 customers in right quality and quantity, at the right time.

Every organization produces something. Unfortunately, this fact is often overlooked except in obvious cases, such as in the manufacturing of cars, cell phones, or lawnmowers. After all, **manufacturing organizations** produce physical goods. It's easy to see the operations management (transformation) process at work in these types of organizations because raw materials are turned into recognizable physical products. But that transformation process isn't as readily evident in **service organizations** because they produce nonphysical outputs in the form of services. For instance, hospitals provide medical and health care services that help people manage their personal health, airlines provide transportation services that move people from one location to another, a cruise line provides vacation and entertainment services, military forces provide defense capabilities, and the list goes on and on. Service organizations also transform inputs into outputs, although the transformation process isn't as easily recognizable as it is in manufacturing organizations. Take a university, for example. University administrators bring together inputs—professors, books, academic journals, technology materials, computers, classrooms, and similar resources—to transform "unenlightened" students into educated and skilled individuals who are capable of making contributions to society.

The reason we're making this point is that in India, service sector of Indian economy contributes to around 55 percent of India's GDP,[3] in the United States, for instance, over 78 percent of all economic activity is services, and in the European Union, it's nearly 71 percent.[4] In lesser-developed countries, the services sector is less important. For instance, in Nigeria, it accounts for only 20 percent of economic activity; in Laos, only 26.5 percent; and in Vietnam, 38.2 percent.[5]

MANAGING PRODUCTIVITY

A jetliner has some 4 million parts. Efficiently assembling such a finely engineered product requires intense focus. Boeing and Airbus, the two major global manufacturers, have copied techniques from Toyota. However, not every technique can be

operations management
The transformation process that converts resources into finished goods and services.

manufacturing organizations
Organizations that produce physical goods.

service organizations
Organizations that produce nonphysical products in the form of services.

copied because airlines demand more customization than do car buyers, and there are significantly more rigid safety regulations for jetliners than for cars.[6] At the Evans Findings Company in East Providence, Rhode Island, which makes the tiny cutting devices on dental-floss containers, one production shift each day is run without people.[7] The company's goal is to do as much as possible with no labor. And it's not because the company doesn't care about its employees. Instead, like many other U.S. manufacturers, Evans needed to raise productivity in order to survive, especially against low-cost competitors. So it turned to "lights-out" manufacturing, where machines are designed to be so reliable that they make flawless parts on their own, without people operating them.

Although most organizations don't make products that have 4 million parts and most organizations can't function without people, improving productivity has become a major goal in virtually every organization. For countries, high productivity can lead to economic growth and development. Employees can receive higher wages, and company profits can increase without causing inflation. For individual organizations, increased productivity provides a more competitive cost structure and the ability to offer more competitive prices.

Over the past decade, Indian businesses have made dramatic improvements to increase their efficiency. For example, Sundram Fasteners Limited has won Total Productive Maintenance (TPM) awards from JIPM, Japan, for transformation of operational efficiencies. Indian Railways posted 25,000 crore profit in 2008. This has been made by appropriate strategy and increasing operational efficiencies. Revenue has been gained from increasing length, payload, seating capacity, and occupancy levels. Dynamic pricing and commercial policy across routes, regions, and periods has been implemented. Investments have been made to remove capacity bottlenecks on the high-density networks.[8] TVS Motor Company Limited has won two coveted IT Awards, bagging the SAP ACE 2008 Awards [for digitization of New Product Development process in SAP and by implementing PLM (Product Life Management)] and the 2008 Symantec South Asia Visionary Award, for its exemplary secured and managed systems and information.[9] ICICI Bank has increased its operational efficiency by investing heavily in IT from the day it started its operations.[10]

Organizations that hope to succeed globally are looking for ways to improve productivity. For example, McDonald's Corporation drastically reduced the time it takes to cook its french fries—65 seconds as compared to the 210 seconds it once took—saving time and other resources.[11] The Canadian Imperial Bank of Commerce, based in Toronto, automated its purchasing function, saving several million dollars annually.[12] And Skoda, the Czech car company owned by Germany's Volkswagen AG, improved its productivity through an intensive restructuring of its manufacturing process.[13]

Productivity is a composite of people and operations variables. To improve productivity, managers must focus on both. The late W. Edwards Deming, a renowned quality expert, believed that managers, not workers, were the primary source of increased productivity. He outlined 14 points for improving management's productivity (see Exhibit 18–2). A close look at these suggestions reveals Deming's understanding of the interplay between people and operations. High productivity can't come solely from good "people management." A truly effective organization will maximize productivity by successfully integrating people into the overall operations system. For instance, at Simplex Nails Manufacturing in Americus, Georgia, employees were an integral part of the company's much-needed turnaround effort.[14] Some production workers were redeployed on a plantwide clean-up and organization effort, which freed up floor space. The company's sales force was retrained and refocused to sell what customers wanted rather than what was in inventory. The results were dramatic: Inventory was reduced by more than 50 percent, the plant had 20 percent more floor space, orders were more consistent, and employee morale improved. Here's a company that recognized the important interplay between people and the operations system.

Exhibit 18–2

Deming's Suggestions for Improving Productivity

1. Plan for the long-term future.
2. Never be complacent concerning the quality of your product.
3. Establish statistical control over your production processes and require your suppliers to do so as well.
4. Deal with the best and fewest number of suppliers.
5. Find out whether your problems are confined to particular parts of the production process or stem from the overall process itself.
6. Train workers for the job that you are asking them to perform.
7. Raise the quality of your line supervisors.
8. Drive out fear.
9. Encourage departments to work closely together rather than to concentrate on departmental or divisional distinctions.
10. Do not adopt strictly numerical goals.
11. Require your workers to do quality work.
12. Train your employees to understand statistical methods.
13. Train your employees in new skills as the need arises.
14. Make top managers responsible for implementing these principles.

Source: W. E. Deming, "Improvement of Quality and Productivity Through Action by Management," National Productivity Review, Winter 1981–1982, pp. 12–22. With permission. Copyright 1981 by Executive Enterprises, Inc. 22 West 21st St., New York, NY 10010-6904. All rights reserved.

Let's Get Real: F2F

WHY UNDERSTANDING OPERATIONS MANAGEMENT IS IMPORTANT TO ME:

Operation management improves system efficiency, proper linkage, and synergy among the processes to produce the desired effect in a sustainable manner.

STRATEGIC ROLE OF OPERATIONS MANAGEMENT

Modern manufacturing originated over 100 years ago in the United States, primarily in Detroit's automobile factories. The success that U.S. manufacturers experienced during World War II led manufacturing executives to believe that troublesome production problems had been conquered. These executives focused, instead, on improving other functional areas, such as finance and marketing, and paid little attention to manufacturing.

Managers in Japan, Germany, and other countries have developed modern, computer-based, and technologically advanced facilities that fully integrate manufacturing operations into strategic planning decisions. These efforts lead to advances in world manufacturing. In India, once the protection regime was over in early 1990s and competition came from China, Indian manufacturers discovered that foreign goods being made were not only less expensive but also were of better quality. Indian industry recognized that they were facing a true crisis and responded. They invested heavily in improving manufacturing technology, increased the corporate authority and visibility of manufacturing executives, and began incorporating existing and future production requirements into the organization's overall strategic plan. Today, successful organizations recognize the crucial role that operations management plays as part of the overall organizational strategy to establish and maintain global leadership.[15]

The strategic role that operations management plays in successful organizational performance can be seen clearly as more organizations move toward managing their operations from a value chain perspective, which we discuss next.

QUICK LEARNING REVIEW:

LEARNING OUTCOME 18.1

- Define operations management.
- Contrast manufacturing and service organizations.
- Describe the manager's role in improving productivity.
- Discuss the strategic role of operations management.

Go to page 446 and see how well you know this material.

LEARNING **OUTCOME 18.2** ▷

WHAT IS VALUE CHAIN MANAGEMENT, AND WHY IS IT IMPORTANT?

It's 11 P.M., and you're reading a text message from your parents, saying they want to buy a laptop for you for your birthday and that you should order it. You log on to Dell's Web site and configure your dream machine. You hit the order button, and within three or four days, your dream computer is delivered to your front door, built to your exact specifications, ready to set up and use immediately to write the management assignment that's due tomorrow. Or consider that Siemens AG's computed tomography manufacturing plant in Forcheim, Germany, has established partnerships with about 30 suppliers. These suppliers are partners in the truest sense as they share responsibility with the plant for overall process performance. This arrangement has allowed Siemens to eliminate all inventory warehousing and has reduced the number of times paper changes hands when parts are ordered from 18 to 1. At the Timken plant in Canton, Ohio, electronic purchase orders are sent across the street to an adjacent "Supplier City," where many of its key suppliers have set up shop. The process takes milliseconds and costs less than 50 cents per purchase order. And when Black & Decker extended its line of handheld tools to include a glue gun, it outsourced the entire design and production to the leading glue gun manufacturer. Why? Because they understood that glue guns don't require motors, which was what Black & Decker did best.[16]

Let's Get Real: F2F

WHY UNDERSTANDING OPERATIONS MANAGEMENT IS IMPORTANT TO ME:

To effectively and efficiently utilize the resources and to fulfill the needs of the customer in the best possible way.

As these examples show, closely integrated work activities among many different players are possible. How? The answer lies in value chain management. The concepts of value chain management have transformed operations management strategies and turned organizations around the world into finely tuned models of efficiency and effectiveness, strategically positioned to exploit competitive opportunities.

WHAT IS VALUE CHAIN MANAGEMENT?

Every organization needs customers if it's going to survive and prosper. Even a not-for-profit organization must have "customers" who use its services or purchase its products. Customers want some type of value from the goods and services they purchase or use, and these customers decide what has value. Organizations must provide that value to attract and keep customers. **Value** is defined as the performance characteristics, features and attributes, and any other aspects of goods and services for which customers are willing to give up resources (usually money). For example, when you purchase a new VCD at Crossword, a new pair of Levis Jeans, a Dell laptop at the company's Web site, a hot foaming cup of cappuccino from Barista, or a haircut from your local hair salon, you are exchanging (giving up) money in return for the value you need or desire from these products—providing entertainment, keeping yourself fashionably attired, alleviating your headache since your next class starts in 15 minutes, or looking professionally groomed for the job interview you are going in the next week.

How *is* value provided to customers? Through transforming raw materials and other resources into some product or service that end users need or desire when, where, and how they want it. However, that seemingly simple act of turning varied resources into something that customers value and are willing to pay for involves a vastarray of interrelated work activities performed by different participants (suppliers, manufacturers, and even customers)—that is, it involves the value chain. The **value chain** is the entire series of organizational work activities that add value at each step, from raw materials to finished product. In its entirety, the value chain can encompass the supplier's suppliers to the customer's customers.[17]

Value chain management is the process of managing the sequence of activities and information along the entire value chain. In contrast to supply chain management, which is *internally* oriented and focuses on efficient flow of incoming materials (resources) to the organization, value chain management is *externally* oriented and focuses on both incoming materials and outgoing products and services. Whereas supply chain management is efficiency oriented (its goal is to reduce costs and make the organization more productive), value chain management is effectiveness oriented and aims to create the highest value for customers.[18]

GOAL OF VALUE CHAIN MANAGEMENT

Who has the power in the value chain? Is it the suppliers providing needed resources and materials? After all, they have the ability to dictate prices and quality. Is it the manufacturer who assembles those resources into a valuable product or service? Their contributions in creating a product or service are quite obvious. Is it the distributor that makes sure the product or service is available where and when the customer needs it? Actually, it's none of these! In value chain management, ultimately customers are the ones with power.[19] They're the ones who define what value is and how it's created and provided. Using value chain management, managers hope to find that unique combination where customers are offered solutions that truly meet their unique needs incredibly fast and at a price that competitors can't match.

The goal of value chain management is therefore to create a value chain strategy that meets and exceeds customers' needs and desires and allows for full and seamless integration among all members of the chain. A good value chain is one in which a sequence of participants work together as a team, each adding some component of value—such as faster assembly, more accurate information, better customer response and service, and so forth—to the overall process.[20] The better the collaboration among the various chain participants, the better the customer solutions. When value is created for customers and their needs and desires are satisfied, everyone along the chain benefits. For example, at Johnson Controls Inc., managing the value chain started first with improved relationships with internal suppliers and then expanded to external suppliers and customers. As the company's experience with value chain management improved, so did its connection with its customers, and this will ultimately pay off for all its value chain partners.[21]

BENEFITS OF VALUE CHAIN MANAGEMENT

Collaborating with external and internal partners in creating and managing a successful value chain strategy requires significant investments in time, energy, and other resources, as well as a serious commitment by all chain partners. So why would managers ever choose to implement value chain management? A survey of manufacturers noted four primary benefits of value chain management: improved procurement, improved logistics, improved product development, and enhanced customer order management.[22]

QUICK LEARNING REVIEW:

LEARNING OUTCOME 18.2

- Define value chain and value chain management.
- Describe the goal of value chain management.
- Describe the benefits of value chain management.

Go to page 446 and see how well you know this material.

LEARNING OUTCOME 18.3 ▷ MANAGING OPERATIONS BY USING VALUE CHAIN MANAGEMENT

Managing an organization from a value chain perspective isn't easy. Approaches to giving customers what they want that may have worked in the past are likely no longer efficient or effective. Today's dynamic competitive environment demands new solutions from global organizations. Understanding how and why value is determined by the marketplace has led some organizations to experiment with a new business model, a concept we introduced in Chapter 8. For example, IKEA transformed itself from a

value
The performance characteristics, features, and attributes, as well as any other aspects of goods and services for which customers are willing to give up resources.

value chain
The entire series of organizational work activities that add value at each step from raw materials to finished product.

value chain management
The process of managing the sequence of activities and information along the entire value chain.

small Swedish mail-order furniture operation into one of the world's largest furniture retailers by reinventing the value chain in that industry. The company offers customers well-designed products at substantially lower-than-typical prices in return for their willingness to take on certain key tasks traditionally done by manufacturers and retailers, such as getting furniture home and assembling it.[23] The company's creation of a new business model and willingness to abandon old methods and processes has worked well.

REQUIREMENTS OF VALUE CHAIN MANAGEMENT

Exhibit 18–3 shows the six main requirements of a successful value chain strategy: coordination and collaboration, technology investment, organizational processes, leadership, employees, and organizational culture and attitudes.

Coordination and Collaboration. For the value chain to achieve its goal of meeting and exceeding customers' needs and desires, collaborative relationships must exist among all chain participants.[24] Each partner must identify things they may not value but that customers do. And sharing information and being flexible in terms of who in the value chain does what are important steps in building coordination and collaboration. This sharing of information and analysis requires open communication among the various value chain partners. For example, Kraft Foods believes that better communication with customers and with suppliers has facilitated timely delivery of goods and services.[25]

Technology Investment. Successful value chain management isn't possible without a significant investment in information technology. The payoff from this investment, however, is that information technology can be used to restructure the value chain to better serve end users. For example, at American Standard's Trane facilities, a comprehensive IT strategy throughout its value chain, which extends globally, has helped it achieve significant work process improvements.[26]

Organizational Processes. Value chain management radically changes **organizational processes**—that is, the ways that organizational work is done. When managers decide to manage operations using value chain management, old processes are no longer appropriate. All organizational processes must be critically evaluated, from beginning to end, to see where value is being added. Non-value-adding activities should be eliminated. Questions such as "Where can internal knowledge be leveraged to improve the flow of material and information?" "How can we better configure our product to satisfy both customers and suppliers?" "How can the flow of material and information be improved?" and "How can we improve customer service?" should be answered for each and every process. For example, when managers at Deere and Company implemented value chain management, a thorough process evaluation revealed that work activities needed to be better synchronized and interrelationships

Exhibit 18–3

Value Chain Strategy Requirements

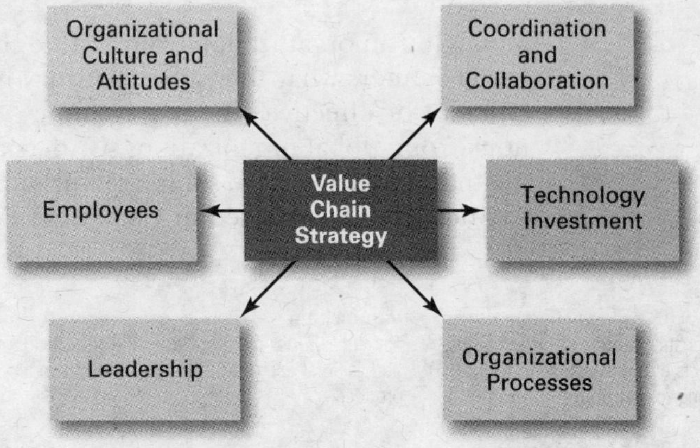

between multiple links in the value chain needed to be better managed. They changed numerous work processes divisionwide in order to do this.[27] Three important conclusions can be made about organizational processes. First, better demand forecasting is necessary *and* possible because of closer ties with customers and suppliers. For example, in an effort to make sure that Listerine was on the store shelves when customers wanted it (known in the retail industry as *product replenishment rates*), Wal-Mart and Pfizer's Consumer Healthcare Group collaborated on improving product demand forecast information. Through their mutual efforts, the partners boosted Wal-Mart's sales of Listerine, an excellent outcome for both supplier and retailer. Customers also benefited because they were able to purchase the product when and where they wanted it.

Second, selected functions may need to be done collaboratively with other partners in the value chain. This collaboration may even extend to sharing employees. For instance, Saint-Gobain Performance Plastics places its own employees in customer sites and brings in employees of suppliers and customers to work on its premises.[28]

Let's Get Real: F2F

IMPACT OF TECHNOLOGY ON MY ORGANIZATION:
Technology is a great leveler. Its impact has been spread into the way people behave, the way logistics operates in the whole gamut of production. In fact, technology has changed the advent of new products or services and the way the business is being carried out.

Finally, new measures are needed for evaluating performance of various activities along the value chain. Because the goal in value chain management is meeting and exceeding customers' needs and desires, managers need a better picture of how well this value is being created and delivered to customers. For example, when Nestlé USA implemented value chain management, it redesigned its metrics system to focus on one consistent set of measurements—including, for instance, accuracy of demand forecasts and production plans, on-time delivery, and customer service levels—that allowed the company to more quickly identify problem areas and take actions to resolve them.[29]

Leadership Successful value chain management isn't possible without strong and committed leadership. From top organizational levels to lower levels, managers must support, facilitate, and promote the implementation and ongoing practice of value chain management. Managers must seriously commit to identifying what value is, how that value can best be provided, and how successful those efforts have been. A culture where all efforts are focused on delivering superb customer value isn't possible without serious commitment on the part of the organization's leaders.

Peter Tan was the Marketing Director of UPS's Beijing Olympics Project. Among his tasks was making sure the company got every package and piece of luggage to the right place on time . . . no mean feat when you consider that the Chinese government rebuilt much of the city just for the Olympics. Tan was prepared. "Our job was not simply moving things from point A to point B," he said. "We had to plot the entire city—every point, every hill, every traffic light—to determine route times. Everything had to be synchronized down to the second. That was our job." To help motivate his Chinese staff, Tan asked them to choose a team they wanted to sponsor; they chose the Chinese women's national volleyball team.

organizational processes
The ways that organizational work is done.

Let's Get Real: F2F

IMPACT OF TECHNOLOGY ON MY ORGANIZATION:

Technology has a huge impact on our organization because ours is a new technology product for the Indian Market. There is a little learning available in the market on this technology. This leads to learning on daily basis in testing and validation of the product and also in setting up the quality-assurance processes.

Also, it's important that managers outline expectations for what's involved in the organization's pursuit of value chain management. Ideally, this starts with a vision or mission statement that expresses the organization's commitment to identifying, capturing, and providing the highest possible value to customers. For instance, when American Standard began using value chain management, the CEO held dozens of meetings across the United States to explain the new competitive environment and why the company needed to create better working relationships with its value chain partners in order to better serve the needs of its customers.[30]

Then, managers should clarify expectations regarding each employee's role in the value chain. But clear expectations aren't important only for internal partners. Being clear about expectations also extends to external partners. For example, managers at American Standard identified clear requirements for suppliers and were prepared to drop any suppliers that couldn't meet the requirements, and they did so. The upside was that the suppliers that met the expectations benefited from more business, and American Standard had partners willing to work with them in delivering better value to customers.

Employees/Human Resources. We know from our discussions of management theories throughout this textbook that employees are the organization's most important resource. Without employees, there would be no products produced or services delivered—in fact, there would be no organized efforts in the pursuit of common goals. Not surprisingly, employees play an important role in value chain management. The three main human resource requirements for value chain management are flexible approaches to job design, effective hiring process, and ongoing training.

managing in a Virtual World

IT's Role in Managing the Value Chain

Because value chain management requires such intensive collaboration among partners, getting and sharing information is critical.[31] One type of IT that many value chain partners are finding particularly relevant is RFID (radio-frequency identification). In fact, Wal-Mart has been phasing in an RFID program for its suppliers, who must comply by the company's deadlines.

What Is RFID? RFID is an automatic identification method in which information can be stored and remotely retrieved. It's similar to, but more sophisticated than, the old familiar bar code. Information is stored on and retrieved from RFID tags (sometimes called *chips*), which are like "little radio towers or transponders that send out information to a reader." An *active* RFID tag has a tiny battery in it that powers the internal circuits that store and send information at a pretty good distance. A *passive* tag has no internal power supply and must be "awakened" by a tag reader in order to send information. A complete RFID system usually requires tags, tag readers, computer servers, and software.

What Are the Benefits of RFID? RFID technology has several benefits. First, it has the potential to streamline the supply chain, eliminate theft and waste, and solve logistics problems. Another benefit is that, unlike bar codes, RFID tags don't have to be in the line of sight in order to be read. (Think about a grocery store checkout, where the products have to be directly individually scanned by a laser.) RFID tags can be read at a distance, even through crates or other packing materials. In addition, RFID tags can be attached to each item in a shipment so that manufacturers, distributors, transportation companies, retailers, and marketers can track individual units across every step of the value chain.

What Are the Drawbacks of RFID? The main drawbacks of RFID technology are the cost of the chips, the lack of standardization of chips and the machines that read them, the challenge of analyzing the vast amounts of data that RFID produces, and the privacy concerns of customers.

How Are Organizations Using RFID? Many hospitals are experimenting with RFID in patient bracelets that hold medical information and in tracking doctors and nurses so they can be located quickly in an emergency. Law firms, libraries, and research centers are using RFID to track the movement of documents, files, and books. One of the most unusual applications is probably at the University of California, where RFID tags have been inserted into cadavers used for research to thwart the illegal selling of the corpses for profit.

Flexibility is the key to job design in value chain management. Traditional functional job roles—such as marketing, sales, accounts payable, customer service, and so forth—don't work with value chain management. Instead, jobs must be designed around work processes that create and provide value to customers. It takes flexible jobs and flexible employees.

In a value chain organization, employees may be assigned to work teams that tackle particular processes and may be asked to do different things on different days, depending on need. In such an environment—where customer value is best delivered through collaborative relationships that may change as customer needs change and where there are no standardized processes or job descriptions—an employee's ability to be flexible is critical. Therefore, the organization's hiring process must be designed to identify employees who have the ability to learn and adapt.

Finally, the need for flexibility also requires that there be a significant investment in continual and ongoing employee training. Whether that training involves learning how to use information technology software, how to improve the flow of materials throughout the chain, how to identify activities that add value, how to make better decisions faster, or how to improve any number of other potential work activities, managers must see to it that employees have the knowledge and tools they need to do their jobs efficiently and effectively.

Organizational Culture and Attitudes The last requirement for value chain management is having a supportive organizational culture and attitudes. From our extensive description of value chain management, you could probably guess the type of organizational culture that's going to support its successful implementation! Those cultural attitudes include sharing, collaborating, openness, flexibility, mutual respect, and trust. And these attitudes encompass not only the internal partners in the value chain but extend to external partners as well.

OBSTACLES TO VALUE CHAIN MANAGEMENT

As desirable as the benefits of value chain management may be, managers must tackle several obstacles in managing the value chain, including organizational barriers, cultural attitudes, required capabilities, and people (see Exhibit 18–4).

Organizational Barriers Organizational barriers are among the most difficult obstacles to handle. These barriers include refusal or reluctance to share information, reluctance to shake up the status quo, and security issues. Without shared information, close coordination and collaboration is impossible. And the reluctance or refusal of employees to shake up the status quo can impede efforts toward value chain management and prevent its successful implementation. Finally, because value chain management relies heavily on a substantial information technology infrastructure, system security and Internet security breaches are issues that need to be addressed.

Cultural Attitudes. Unsupportive cultural attitudes—especially trust and control—can be obstacles to value chain management. The trust issue—both lack of trust and too much trust—is a critical one. To be effective, partners in a value chain must trust each other. There must be a mutual respect for, and honesty about, each partner's activities all along the chain. When that trust doesn't exist, the partners will be reluctant to share information, capabilities, and processes. But too much trust can also be a problem. Just about any organization is vulnerable to theft of **intellectual property**—that is, proprietary information that's critical to an organization's efficient and effective functioning and competitiveness. You need to be able to trust your value chain partners so your

RFID
An automatic identification method in which information can be stored and remotely retrieved.

intellectual property
Proprietary information that's critical to an organization's efficient and effective functioning and competitiveness.

Exhibit 18–4

Obstacles to Value Chain Management

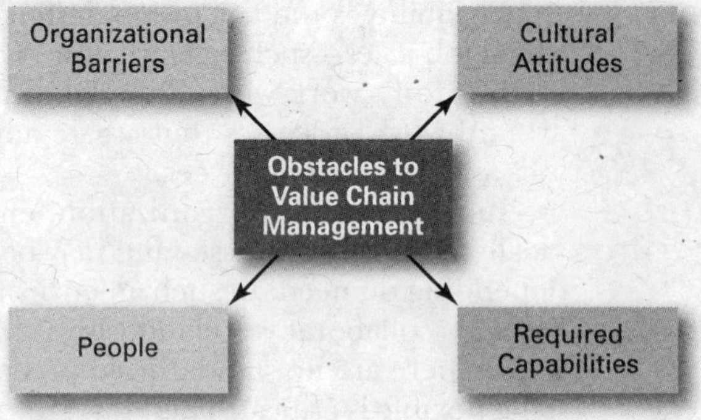

organization's valuable assets aren't compromised.[32] Another cultural attitude that can be an obstacle is the belief that when an organization collaborates with external and internal partners, it no longer controls its own destiny. However, this just isn't the case. Even with the intense collaboration that's important to value chain management, organizations still control critical decisions such as what customers value, how much value they desire, and what distribution channels are important.[33]

Required Capabilities. We know from our earlier discussion of requirements for the successful implementation of value chain management that value chain partners need numerous capabilities. Several of these—coordination and collaboration, the ability to configure products to satisfy customers and suppliers, and the ability to educate internal and external partners—aren't easy. But they're essential to capturing and exploiting the value chain. Many of the companies we've described throughout this section endured critical, and oftentimes difficult, self-evaluations of their capabilities and processes in order to become more effective and efficient at managing their value chains.

People. The final obstacles to successful value chain management can be an organization's people. Without their unwavering commitment to do whatever it takes, value chain management won't be successful. If employees refuse to be flexible in their work—how and with whom they work—collaboration and cooperation throughout the value chain will be difficult to achieve.

In addition, value chain management takes an incredible amount of time and energy on the part of an organization's employees. Managers must motivate those high levels of effort from employees, which is not an easy thing to do.

Finally, a major human resource problem is a lack of experienced managers who can lead value chain management initiatives. Value chain management isn't very widespread, so there aren't a lot of managers who've done it successfully. However, this hasn't prevented progressive organizations from pursuing the benefits to be gained from value chain management.

QUICK LEARNING REVIEW:

LEARNING OUTCOME 18.3

- Discuss the requirements for successful value chain management.
- Explain the obstacles to value chain management.

Go to page 446 and see how well you know this material.

LEARNING **OUTCOME 18.4** ▷

CURRENT ISSUES IN MANAGING OPERATIONS

Rowe Furniture has an audacious goal: Make a sofa in 10 days. It wants to "become as efficient at making furniture as Toyota is at making cars." Reaching that goal, however, requires revamping its operations management process to exploit technology

Let's Get Real: F2F

BIGGEST CHALLENGES IN "PRODUCING" OUR SERVICE:
Are in creating new capacities and sustaining the supply chain. Capital availability for new capacity additions is not easy.

and maintaining quality.[34] Rowe's actions illustrate three of today's most important operations management issues: technology, quality initiatives and goals, and mass customization.

TECHNOLOGY'S ROLE IN OPERATIONS MANAGEMENT

As we know from our previous discussion of value chain management, today's competitive marketplace has put tremendous pressure on organizations to deliver, in a timely manner, products and services that customers value. Smart companies are looking at ways to harness technology to improve operations management. Many fast-food companies are competing to see who can provide faster and better service to drive-through customers. With drive-through now representing a huge portion of sales, faster and better delivery can be a significant competitive edge. For instance, Subway has added awnings to some of its menu boards and replaced some of the text with pictures. Others use confirmation screens, a technology that helped McDonald's boost accuracy by more than 11 percent. And technology used by two national chains tells managers how much food they need to prepare by counting vehicles in the drive-through line and factoring in demand for current promotional and popular staple items.[35]

Although an organization's production activities are driven by the recognition that the customer is king, managers still need to be more responsive. For instance, operations managers need systems that can reveal available capacity, status of orders, and product quality while products are in the process of being manufactured, not just after the fact. To connect more closely with customers, production must be synchronized across the enterprise. To avoid bottlenecks and slowdowns, the production function must be a full partner in the entire business system.

Let's Get Real: F2F

BIGGEST CHALLENGES IN "PRODUCING" OUR SERVICE:
The biggest challenge in production is setting up of the quality-assurance processes for the unique items required for the product.

Technology is making such extensive collaboration possible. Technology is also allowing organizations to control costs, particularly in the areas of predictive maintenance, remote diagnostics, and utility cost savings. For instance, new Internet-compatible equipment contains embedded Web servers that can communicate proactively—for example, if a piece of equipment breaks or reaches certain preset parameters indicating that it's about to break, it asks for help. Technology can do more than sound an alarm or light up an indicator button. For instance, some devices have the ability to initiate e-mail or signal a pager of a supplier, the maintenance department, or a contractor, describing the specific problem and requesting parts and service. How much is such e-enabled maintenance control worth? It can be worth quite a lot if it prevents equipment breakdowns and subsequent production downtime.

Managers who understand the power of technology to contribute to more effective and efficient performance know that managing operations is more than the traditional view of simply producing the product. Instead, the emphasis is on working together with all the organization's business functions to find solutions to customers' business problems.

QUALITY INITIATIVES

Quality problems are expensive. For example, even though Apple has had phenomenal success with its iPod, the batteries in the first three versions died after 4 hours instead of lasting up to 12 hours, as buyers expected. Apple's settlement with consumers cost close to \$100 million. In July 2007, Honda Siel Cars India (HSCI) recalled 2,310 units of Accord Sedans to replace a relay in the fuel pump of the car and in June 2007, it recalled 4,459 units of its premium SUV model CR-V for a free fuel tank replacement.[36] Perhaps in the largest product recall in India, mobile giant Nokia recalled 46 million batteries pursuant to customer complaints across the globe. On August 15, 2007, Nokia issued a "product advisory" and said it would replace BL-5C batteries manufactured between December 2005 and November 2006 in India.[37] The help center opened to facilitate this process was inundated with 20,000 calls and received 1.45 lakh SMSes on the first day itself.[38]

Many experts believe that organizations unable to produce high-quality products won't be able to compete successfully in the global marketplace. What is quality? When you consider a product or service to have quality, what does that mean? Does it mean that the product doesn't break or quit working—that is, that it's reliable? Does it mean that the service is delivered in a way that you intended? Does it mean that the product does what it's supposed to do? Or does quality mean something else? Exhibit 18–5 provides a description of several quality dimensions. In this case, we define **quality** as the ability of a product or service to reliably do what it's supposed to do and to satisfy customer expectations.

How is quality achieved? That's an issue managers must address. A good way to look at quality initiatives is with the management functions—planning, organizing and leading, and controlling—that need to take place.

Planning for Quality. Managers must have quality improvement goals and strategies and plans to achieve those goals. Goals can help focus everyone's attention on some objective quality standard. For instance, Caterpillar has a goal of applying quality improvement techniques to help cut costs.[39] Although this goal is specific and challenging, managers and employees are partnering together to pursue well-designed strategies to achieve the goals, and they are confident they can do so.

Organizing and Leading for Quality. Because quality improvement initiatives are carried out by organizational employees, it's important for managers to look at how they can best organize and lead them. For instance, at the Atul Group of Companies unit located in Agra, involved in manufacturing pumps, every employee goes through continuous training. This training has resulted in constant research, innovations, and developments, in order to establish this company as a leader in this field.[40] Engineers at Metal Inox in Mumbai (engaged in exporting and supplying of various stainless steel industrial products) believe in ongoing improvements and training to sharpen the quality-assurance process.[41] At Tavant Software, from a small team of six involved in quality assurance, the size of the quality team has grown to 120 over the years. The best practices laid down by industry were broken down and implemented at the modular level. This has enabled them to ensure production of zero-defect software.[42]

Organizations with extensive and successful quality improvement programs tend to rely on two important people approaches: cross-functional work teams and self-directed,

Exhibit 18–5

Quality Dimensions of Goods and Services

Product Quality Dimensions

1. Performance—Operating characteristics
2. Features—Important special characteristics
3. Flexibility—Meeting operating specifications over some period of time
4. Durability—Amount of use before performance deteriorates
5. Conformance—Match with preestablished standards
6. Serviceability—Ease and speed of repair or normal service
7. Aesthetics—How a product looks and feels
8. Perceived quality—Subjective assessment of characteristics (product image)

Service Quality Dimensions

1. Timeliness—Performed in promised period of time
2. Courtesy—Performed cheerfully
3. Consistency—Giving all customers similar experiences each time
4. Convenience—Accessibility to customers
5. Completeness—Fully serviced, as required
6. Accuracy—Performed correctly each time

Sources: Adapted from J. W. Dean, Jr., and J. R. Evans, *Total Quality: Management, Organization and Society* (St. Paul, MN: West Publishing Company, 1994); H. V. Roberts and B. F. Sergesketter, *Quality is Personal* (New York: The Free Press, 1993); D. Garvin, *Managed Quality: The Strategic and Competitive Edge* (New York: The Free Press, 1988); and M. A. Hitt, R. D. Ireland, and R. E. Hoskisson, *Strategic Management*, 4th ed. (Cincinnati, OH: South-Western Publishing, 2001), p. 211.

Let's Get Real: F2F

HOW WE CONTROL QUALITY: Quality control is inbuilt in the processes from input to output. We have quality checks, quality plans, and automated hold points in every stage of the operations. Any deviation has to be documented and, if required, to bypass under exceptional requirement, the same has to be approved at a designated authority.

Let's Get Real: F2F

HOW WE CONTROL QUALITY: Quality is controlled through inspection.

or empowered, work teams. Because all employees, from upper to lower levels, must participate in achieving product quality, it's not surprising that quality-driven organizations rely on well-trained, flexible, and empowered employees.

Controlling for Quality. Quality improvement initiatives aren't possible without a means of monitoring and evaluating their progress. Whether it involves standards for inventory control, defect rate, raw materials procurement, or other operations management areas, controlling for quality is important. For instance, Indian Herbs, manufacturer and exporter of herbal animal feed supplements, focuses on quality assurance by continuous research by trained scientists, installation, and use of modern sophisticated instruments, such as the High Pressure Liquid Chromatograph with PDA (HPlC), High Performance Thin Layer Chromatography (HPTLC), procurement of raw materials only from authenticated sources and extensive testing of the procured material using sophisticated tests, and preserving and maintaining a specimen database of all plant species used by the company in their formulations.[43] In 2005, the Tata Group created the Tata Business Excellence Model, or TBEM, by using the Malcolm Baldrige National Quality Award model of the United States as a base. TBEM has been used to turn around loss-making uints (e.g., Tinplate) and spur excellence in all Tata companies. The TBEM itself has been revised to meet new initiatives and internationalization.[44]

Success stories of quality improvement are found all around the world. For example, at a Delphi assembly plant in Matamoros, Mexico, employees worked hard to improve quality and made significant strides. For instance, the customer reject rate on shipped products is now 10 ppm (parts per million), down from 3,000 ppm—an improvement of almost 300 percent.[45] Quality initiatives at several Australian companies, including Alcoa of Australia, Wormald Security, and Carlton and United Breweries, have led to significant quality improvements.[46] And at Valeo Klimasystemme GmbH of Bad Rodach, Germany, assembly teams build different climate-control systems for high-end German cars, including Mercedes and BMW. Quality initiatives by Valeo's employee teams have led to significant improvements in various quality standards.[47]

QUALITY GOALS

To publicly demonstrate their quality commitment, many organizations worldwide have pursued challenging quality goals, the two best known of which are ISO 9000 and Six Sigma.

ISO 9000. ISO 9000 is a series of international quality management standards established by the International Organization for Standardization (www.iso.org), which sets uniform guidelines for processes to ensure that products conform to customer requirements. These standards cover everything from contract review to product design to product delivery. The ISO 9000 standards have become the internationally recognized standard for evaluating and comparing companies in the global marketplace. In fact, this type of certification can be a prerequisite for doing business globally. Achieving ISO 9000 certification provides proof that a quality operations system is in place.

A recent survey of ISO 9000 certificates—awarded in 170 countries—showed that the number of registered sites worldwide was almost 900,000, an increase of 16 percent over the previous year.[48]

Six Sigma. Motorola popularized the use of stringent quality standards more than 30 years ago, through a trademarked quality improvement program called Six Sigma.[49] Very simply, **Six Sigma** is a quality standard that establishes a goal of no more than 3.4 defects per million units or procedures. What does the name mean? *Sigma* is the Greek letter that statisticians use to define a standard deviation from a bell curve. The higher the sigma, the

quality
The ability of a product or service to reliably do what it's supposed to do and to satisfy customer expectations.

ISO 9000
A series of international quality management standards that set uniform guidelines for processes to ensure that products conform to customer requirements.

Six Sigma
A quality standard that establishes a goal of no more than 3.4 defects per million units or procedures.

fewer the deviations from the norm—that is, the fewer the defects. At One Sigma, two-thirds of whatever is being measured falls within the curve. Two Sigma covers about 95 percent. At Six Sigma, you're about as close to defect free as you can get.[50] It's an ambitious quality goal! Although Six Sigma is an extremely high standard to achieve, many quality-driven businesses are using it and benefiting from it. For instance, General Electric company executives estimate that the company has saved billions in costs since 1995.[51] Other well-known companies pursuing Six Sigma include Tata Consultancies Services, Schenectidy Hedillia Chemical Ltd, Wipro, ICICI Bank, Aditya Birla Group, Airtel, Mahindra & Mahindra, Indian Oil, and Daksh (BPO).[52] Six Sigma users include manufacturer and service companies such as financial institutions, retailers, and health care organizations. What impact can Six Sigma have? Let's look at an example.

It used to take Wellmark Blue Cross & Blue Shield, a managed-care health care company, 65 days or more to add a new doctor to its medical plans. Thanks to Six Sigma, the company discovered that half the processes they were using were redundant. With those unnecessary steps gone, the job now gets done in 30 days or less, and with reduced staff. The company has also been able to reduce its administrative expenses by $3 million per year, an amount passed on to consumers through lower health insurance premiums.[53]

Quality Goals Summary. Although it's important for managers to recognize that many positive benefits come from obtaining ISO 9000 certification or Six Sigma, the key benefit comes from the quality improvement journey itself. In other words, the goal of quality certification should be having work processes and an operations system in place that enable organizations to meet customers' needs and employees to perform their jobs in a consistently high-quality way.

MASS CUSTOMIZATION

The term *mass customization* seems like an oxymoron. However, the design-to-order concept is becoming an important operations management issue for today's managers. **Mass customization** provides consumers with a product when, where, and how they want it.[54] Companies as diverse as BMW, Ford, Levi Strauss, Wells Fargo, Mattel, and Dell Computer are adopting mass customization to maintain or attain a competitive advantage. Mass customization requires flexible manufacturing techniques and continual customer dialogue.[55] Technology plays an important role in both.

With flexible manufacturing, companies have the ability to quickly readjust assembly lines to make products to order. Using technology such as computer-controlled factory equipment, intranets, industrial robots, bar-code scanners, digital printers, and logistics software, companies can manufacture, assemble, and ship customized products with customized packaging to customers in incredibly short time frames. Dell is a good example of a company that uses flexible manufacturing techniques and technology to custom-build computers to customers' specifications.

Technology is also important in the continual dialogue with customers. Using extensive databases, companies can keep track of customers' likes and dislikes. And the Internet has made it possible for companies to have ongoing dialogues with customers to learn about and respond to their exact preferences. For instance, on Amazon's Web site, customers are greeted by name and can get personalized recommendations of books and other products. The ability to customize products to a customer's desires and specifications starts an important relationship between the organization and the customer. If the customer likes the product and believes the customization provides value, he or she is more likely to be a repeat customer.

QUICK LEARNING REVIEW:

LEARNING OUTCOME 18.4

- Discuss technology's role in manufacturing.
- Explain ISO 9000 and Six Sigma.
- Describe mass customization and how operations management contributes to it.

Go to page 446 and see how well you know this material.

Let's Get Real:

Our Turn

Arun Pratap Singh

Vice President and SBU Head (Auto)
M/s Electrotherm (India) Ltd

Before making the comment on my action plan, I would like to understand as to what is the value proposition of our store; are we the seller of the standard books or special books and why does customer comes to our store, i.e.; he comes to our store because we just sell books or does he find value interacting with our experienced sales staff. If there is a value that our staff provides to our customers then I would like to see long term performance of the store wherein RFID was implemented. But, the new technology is improving efficiency; hence, I would like to use the new technology, while ensuring that my value proposition is not lost. This can be done by reducing the staff in each store to a level the customer requirement is fulfilled and the excess manpower being shifted to other new stores that the company might have in its plan.

Basudev Mohanty

Senior Manager
NTPC Limited

Well, the implementation of RFID technology at all other outlets will definitely enhance the operation of those stores. But, a cautious approach, rather than aggressiveness, is needed for the expansion of the implementation of RFID technology. The cost effectiveness has to be ascertained and the pitfall at the first store is to be looked into. A continuous learning approach will be taken, so that every mistake will be a learning for the next shop.

mass customization
Providing customers with a product when, where, and how they want it.

LEARNING OUTCOMES SUMMARY

18.1 ▷ THE ROLE OF OPERATIONS MANAGEMENT

- Define operations management.
- Contrast manufacturing and service organizations.
- Describe the manager's role in improving productivity.
- Discuss the strategic role of operations management.

Operations management is the transformation process that converts resources into finished goods and services.

Manufacturing organizations produce physical goods. Service organizations produce nonphysical outputs in the form of services.

Productivity is a composite of people and operations variables. A manager should look for ways to successfully integrate people into the overall operations systems.

Organizations must recognize the crucial role that operations management plays as part of their overall strategy in achieving successful performance.

18.2 ▷ WHAT IS VALUE CHAIN MANAGEMENT, AND WHY IS IT IMPORTANT?

- Define *value chain* and *value chain management.*
- Describe the goal of value chain management.
- Describe the benefits of value chain management.

The value chain is the sequence of organizational work activities that add value at each step from raw materials to finished product. Value chain management is the process of managing the sequence of activities and information along the entire product chain.

The goal of value chain management is to create a value chain strategy that meets or exceeds customers' needs and desires and allows for full and seamless integration among all members of the chain.

Value chain management has four benefits: improved procurement, improved logistics, improved product development, and enhanced customer order management.

18.3 ▷ MANAGING OPERATIONS BY USING VALUE CHAIN MANAGEMENT

- Discuss the requirements for successful value chain management.
- Explain the obstacles to value chain management.

The six main requirements for successful value chain management are coordination and collaboration, investment in technology, organizational processes, leadership, employees or human resources, and organizational culture and attitudes.

The obstacles to value chain management include organizational barriers (refusal to share information, reluctance to shake up the status quo, or security issues), unsupportive cultural attitudes, lack of required capabilities, and employee unwillingness or inability to do it.

18.4 ▷ CURRENT ISSUES IN MANAGING OPERATIONS

- Discuss technology's role in manufacturing.
- Explain ISO 9000 and Six Sigma.
- Describe mass customization and how operations management contributes to it.

Companies are looking at ways to harness technology to improve their operations management through extensive collaboration and cost control.

ISO 9000 is a series of international quality management standards that set uniform guidelines for processes to ensure that products conform to customer requirements. Six Sigma is a quality standard that establishes a goal of no more than 3.4 defects per million units or procedures.

Mass customization provides customers with a product when, where, and how they want it. It requires flexible manufacturing techniques and continual customer dialogue.

THINKING ABOUT MANAGEMENT ISSUES

1. Do you think that manufacturing organizations or service organizations have the greater need for operations management? Explain.
2. How might operations management apply to other managerial functions besides control?
3. How could you use value chain management concepts in your everyday life?
4. Which is more critical for success in organizations: continuous improvement or quality control? Support your position.
5. Choose a large organization that you're interested in. Research this company to find out what types of operations management strategies it uses. Focus on describing what it's doing that's unusual or effective or both.

YOUR TURN to be a Manager

- Select a company with which you're familiar and describe its value chain. Be as specific as possible in your description. Evaluate how it "uses" the value chain to create value.
- Find three examples of mass customization products. Describe them. Then try describing what you think has to take place "behind the scenes" to create these products. Focus on the operations management aspects.
- Go to the International Organization for Standardization Web site, at www.iso.org, and look for information on the ISO 9000 standards. Write a report describing the types of things organizations need to do to achieve ISO certification.
- Choose two tasks that you do every week (for example, shop for groceries, host a poker party, clean your house/apartment, do laundry). For each one, identify how you could (1) be more productive in doing that task and (2) have higher-quality output from that task.
- Interview two different managers, preferably one from a manufacturing organization and one from a service organization. Ask them how they manage operations, particularly from the aspects of being productive and producing quality outputs.
- Steve's and Mary's suggested readings: Sydney Finkelstein, *Why Smart Executives Fail and What You Can Learn from Their Mistakes* (Penguin Portfolio, 2003); James L. Heskett, W. Earl Sasser, Jr., and Leonard A. Schlesinger, *The Value Profit Chain* (The Free Press, 2003); William Joyce, Nitin Nohria, and Bruce Roberson, *What Really Works* (Harper Business, 2003); and Harvard Business Review's *Managing the Value Chain* (Harvard Business School Press, 2000).
- Go to the *IndustryWeek* Web site, at www.industryweek.com, and find *IndustryWeek*'s rankings for the best plants and the 50 best manufacturing companies. What criteria are used to select companies on these lists? Choose two companies from each of these lists and describe what they're doing. Put this information into a written report.
- In your own words, write down three things that you learned in this chapter about being a good manager.
- Self-knowledge can be a powerful learning tool. Go to mymanagementlab and complete the self-assessment exercise What Time of Day Am I Most Productive? Using the results of your assessment, identify personal strengths and weaknesses. What will you do to reinforce your strengths and improve your weaknesses?

CASE APPLICATION

Smooth Ride

Big yellow school buses. They're a common sight at the beginning and end of the school day in many communities. One company that manufactures those school buses is Blue Bird North Georgia. School buses are a product in which quality is paramount. After all, that product is carrying precious cargo! However, achieving an organizational culture that's dedicated to quality and to efficient manufacturing isn't an easy thing to do.

Blue Bird's plant in Lafayette, Georgia, started on its "lean" journey—that is, having a lean, efficient operation system—in 2003. The manager of engineering said, "Quality was at an all-time low. The plant was lacking in strategic systems and procedures to control quality, materials, production, finance and human resources." In 2003, under new management, Blue Bird got serious about tackling its quality issues and implemented specific programs, including a material review board, a quality control lab equipped with a computerized maintenance management system, an employee suggestion system, weekly management roundtable meetings, and a safety incentive program. One key contributor to the company's success is measurement. Blue Bird is determined to measure everything. The production manager said, "If you don't measure something, you don't know how well you're doing." Just how effective have these programs been?

The customer reject rate is now basically zero, with on-time delivery at 100 percent. The director of quality and risk management said, "Doing it right the first time takes a lot less time." Safety has also improved. The company's recordable injury rate was down 65 percent and time lost due to injuries was down 87 percent. After four years of hard work, the company achieved initial ISO 9001-2000 certification in March 2007. And it was named one of *IndustryWeek's* best plants in 2007.

A brightly painted Blue Bird bus in Ahuachapan bus station in El Salvador, Central America.

Discussion Questions

1. What has made Blue Bird's quality initiative successful? Discuss how you think each of the programs the company implemented contributed to what the company is today.
2. How might value chain management concepts help Blue Bird become even more productive? What would the company have to do to benefit from managing its value chain?
3. Research the concept of *lean manufacturing*. What does it mean? What benefits does "lean" offer?

Sources: D. Blanchard, "Lean .In for a Smooth Ride," *IndustryWeek*, January 2008, p. 38; and D. Blanchard, "Blue Bird North Georgia: IW Best Plants Profile 2007," www.industryweek.com, January 1, 2008.

Appendix A

Managing Entrepreneurial Ventures

THE CONTEXT OF ENTREPRENEURSHIP

Sriram Raghavan, Co-founder and President of Comat Technologies, is a successful social entrepreneur who has built a business model that transforms lives of millions of rural Indians. Comat technologies provides e-governance solutions and government and private information services to citizens across rural India. After Sriram sold his company incorporated in the United States, he traveled extensively across rural India and realized that there was an opportunity to provide information to rural Indians. Bhoomi was Comat's most remarkable project in Karnataka that involved digitization of over 20 million land records benefiting nearly 7 million farmers. Comat also benefits the youth of rural India by setting up of rural business centers in various Indian states. Sriram was nominated for the Social Entrepreneur of the Year Award by the World Technology Network in the year 2005 and also for the Super Achiever of the Year in 2006.[1]

In this appendix, we're going to look at the activities of entrepreneurs like Sriram Raghavan. We'll start by looking at the context of entrepreneurship and then examine entrepreneurship from the perspective of the four managerial functions: planning, organizing, leading, and controlling.

What Is Entrepreneurship?

Entrepreneurship is the process of starting new businesses, generally in response to opportunities. Entrepreneurs pursue opportunities by changing, revolutionizing, transforming, or introducing new products or services. For example, Captain Gopinath realized that the small towns had several moneyed people who would be willing to pay for air connectivity from their towns to metros for business and pleasure trips. Also, Lok Sabha and Rajya Sabha members from distant towns needed to travel to neighboring large cities and Delhi and would be open to do the paperwork and to provide approvals for air connectivity with small towns. The full service flights that only served large cities were comfortable but too expensive. Thus, Captain Gopinath started Air Deccan that was the first low-cost airlines in the country and flew to various never heard of destinations, such as Jamshedpur, Bellary, and Hubli. Air Deccan's success lead to meeting a latent demand of passengers all across the country. Its success can be gauged by the number of other similar airlines entering the market.

Many people think that entrepreneurial ventures and small businesses are one and the same, but they're not. There are some key differences between the two. Entrepreneurs create **entrepreneurial ventures**—organizations that pursue opportunities, are characterized by innovative practices, and have growth and profitability as their main goals. On the other hand, a **small business** is a business that is independently owned, operated, and financed; has fewer than 100 employees; doesn't necessarily engage in any new or innovative practices; and has relatively little impact on its

entrepreneurship
The process of starting new businesses, generally in response to opportunities.

entrepreneurial ventures
Organizations that are pursuing opportunities, are characterized by innovative practices, and have growth and profitability as their main goals.

small business
An organization that is independently owned, operated, and financed; has fewer than 100 employees; doesn't necessarily engage in any new or innovative practices; and has relatively little impact on its industry.

industry.[2] A small business isn't necessarily entrepreneurial just because it's small. To be entrepreneurial means that the business must be innovative, seeking out new opportunities. Even though entrepreneurial ventures may start small, they pursue growth. Some new small firms may grow, but many remain small businesses, by choice or by default.

Why Is Entrepreneurship Important?

Entrepreneurship is important to every industry sector in the United States and in most other advanced countries.[3] Its importance in the United States can be shown in three areas: innovation, number of new start-ups, and job creation.

Innovation. Innovating is a process of changing, experimenting, transforming, and revolutionizing, and it's a key aspect of entrepreneurial activity. The "creative destruction" process that characterizes innovation leads to technological changes and employment growth. Entrepreneurial firms act as "agents of change" by providing an essential source of new and unique ideas that may otherwise go untapped.[4] Statistics back this up. New small organizations generate 24 times more innovations per research and development dollar spent than do *Fortune* 500 organizations, and they account for more than 95 percent of new and "radical" product developments.[5] In addition, the U.S. Small Business Administration's Office of Advocacy reports that small entrepreneurial firms produce 13 to 14 times more patents per employee than large patenting firms.[6] This is not true in India but NIF has found that remarkable innovation is present among grass-root enterprises. This is further proof of how important small business is to innovation.

Number of New Start-ups. Because all businesses—whether they fit the definition of entrepreneurial ventures or not—were new start-ups at some point, the most suitable measure we have of the important role of entrepreneurship is to look at the number of new firms over a period of time. Data collected by the U.S. Small Business Administration show that the number of new start-ups has increased every year since 2002. An estimated 649,700 new businesses were created in 2006.[7]

Job Creation. Job creation is important to the overall long-term economic health of communities, regions, and nations. The latest figures show that small businesses accounted for most of the net new jobs.[8] Small organizations have been creating jobs at a fast pace even as many of the world's largest and well-known global corporations have continued to downsize. These facts reflect the importance of entrepreneurial firms as job creators.

Global Entrepreneurship. What about entrepreneurial activity outside the United States? What kind of impact has it had? An annual assessment of global entrepreneurship called the Global Entrepreneurship Monitor (GEM) studies the impact of entrepreneurial activity on economic growth in various countries. The 2007 GEM report covered 42 countries that were divided into two clusters: high-income countries and middle- and low-income countries. Researchers found that in the high-income group, the highest levels of early-stage entrepreneurial activity were found in Iceland, Hong Kong, and the United States. In the middle- and low-income group, the highest levels of entrepreneurial activity were found in Thailand, Peru, and Colombia. The GEM report concluded that "the importance of entrepreneurship for economic development is widely acknowledged."[9]

The Entrepreneurial Process

Entrepreneurs must address four key steps as they start and manage their entrepreneurial ventures.

The first is *exploring the entrepreneurial context.* The context includes the realities of today's economic, political/legal, social, and work environments. It's important to look at each of these aspects of the entrepreneurial context because they determine the "rules" of the game and which decisions and actions are likely to be successful. Also, it's through exploring the context that entrepreneurs confront the next critically important step in the entrepreneurial process: *identifying opportunities and possible competitive advantages.* We know from our definition of entrepreneurship that the pursuit of opportunities is an important aspect.

Once entrepreneurs have explored the entrepreneurial context and identified opportunities and possible competitive advantages, they must look at the issues involved with actually bringing an entrepreneurial venture to life. Therefore, the third step in the entrepreneurial process is *starting the venture.* Included in this phase are researching the feasibility of the venture, planning the venture, organizing the venture, and launching the venture.

Finally, when the entrepreneurial venture is up and running, the fourth and final step in the entrepreneurial process is *managing the venture,* which an entrepreneur does by managing processes, managing people, and managing growth. We can explain these important steps in the entrepreneurial process by looking at what entrepreneurs do.

What Do Entrepreneurs Do?

Describing what entrepreneurs do isn't an easy or simple task. No two entrepreneurs' work activities are exactly alike. In a general sense, entrepreneurs create something new, something different. They search for change, respond to it, and exploit it.[10]

Initially, an entrepreneur is engaged in assessing the potential for the entrepreneurial venture and then dealing with start-up issues. In exploring the entrepreneurial context, an entrepreneur gathers information, identifies potential opportunities, and pinpoints possible competitive advantage(s). Then, armed with that information, the entrepreneur researches the venture's feasibility—uncovering business ideas, looking at competitors, and exploring financing options.

After looking at the potential of a proposed venture and assessing the likelihood of pursuing it successfully, an entrepreneur proceeds to plan the venture. This includes such activities as developing a viable organizational mission, exploring organizational culture issues, and creating a well-thought-out business plan. When these planning issues have been resolved, the entrepreneur must look at organizing the venture, which involves choosing a legal form of business organization, addressing other legal issues such as patent or copyright searches, and coming up with an appropriate organizational design for structuring how work is going to be done.

Only after these start-up activities have been completed is the entrepreneur ready to actually launch the venture. This involves setting goals and strategies, as well as establishing the technology operations methods, marketing plans, information systems, financial accounting systems, and cash flow management systems.

When the entrepreneurial venture is up and running, the entrepreneur's attention switches to managing it. An important activity of managing the entrepreneurial venture is managing the various processes that are part of every business: making decisions, establishing action plans, analyzing external and internal environments, measuring and evaluating performance, and making needed changes. Also, the entrepreneur must perform activities associated with managing people, including selecting and hiring, appraising and training, motivating, managing conflict, delegating tasks, and being an effective leader. Finally, the entrepreneur must manage the venture's growth, including such activities as developing and designing growth strategies, dealing with crises, exploring various avenues for financing growth, placing a value on the venture, and perhaps even eventually exiting the venture.

Social Responsibility and Ethics Issues Facing Entrepreneurs

As they launch and manage their ventures, entrepreneurs are faced with the often-difficult issues of social responsibility and ethics. Just how important are these issues to entrepreneurs? An overwhelming majority of respondents (95 percent) in a study of small companies believed that developing a positive reputation and relationship in communities where they do business is important for achieving business goals.[11] However, despite the importance these individuals placed on corporate citizenship, more than half lacked formal programs for connecting with their communities. In fact, some 70 percent of the respondents admitted that they failed to consider community goals in their business plans.

Yet, there are some entrepreneurs who take their social responsibilities seriously. For example, Chandaben Shroff had come as part of a charity mission to help people of Kutch through one of their worst droughts in 1969. She noticed that the women of Kutch were dressed in beautiful embroidered clothes. Instead of doing charity, she thought of making the skill a source of livelihood for the women. She took loans from her husband, commitments from her friends in Mumbai, and inspired some of the women to embroider patterns in sarees in nontraditional colors (supplied by Chandaben) and traditional patterns. The first sale of sarees delivered by the rural women in Kutch was a run-away success. Chandaben ploughed the profit back into the project and Shrujan grew. Currently, Shrujan employs some 3,000 women in 114 villages in Kutch. It has a design center in Bhuj with more than 1,000 panels–some have taken over a year to embroider–by master craftswomen.[12]

Other entrepreneurs have pursued opportunities with products and services that protect the global environment. For example, Sidhraj Shah, 29, and his partners, Kiran Mehta (Technology Director), Devendra Shah (Finance and Admin. Director), Ruchir Arora (Sales), and Bhakti Mehta (Marketing) were very keen to contribute toward saving the environment. Thus, they started 360 Degree Automation, an alternate energy specialist. They have patented the Circuitry that drives the LED Light to emit lumens matching halogens at lower consumption from the source. They set up the first ever solar-powered hoarding in Mumbai and they claim that it can bring about savings of 1,825 units in a year. This is significantly given that Mumbai spends 25 kW on hoardings everyday and a 10 percent reduction can address power shortages in Mumbai. They really believe that "We don't highlight environmental problems; we are a part of the solution." They won the TATA NEN Hottest Startup Award for 2006.[13]

Ethical considerations also play a role in decisions and actions of entrepreneurs. Entrepreneurs need to be aware of the ethical consequences of what they do. The example they set—particularly for other employees—can be profoundly significant in influencing behavior.

If ethics are important, how do entrepreneurs stack up? Unfortunately, not too well! In a survey of employees from different sizes of businesses who were asked whether they thought their organization was highly ethical, 20 percent of employees at companies with 99 or fewer employees disagreed.[14]

QUICK LEARNING REVIEW

- Differentiate between entrepreneurial ventures and small businesses.
- Explain why entrepreneurship is important in the United States and globally.
- Describe the four key steps in the entrepreneurial process.
- Explain what entrepreneurs do.
- Discuss why social responsibility and ethics are important considerations for entrepreneurs.

START-UP AND PLANNING ISSUES FOR AN ENTREPRENEURIAL VENTURE

Although pouring a bowl of cereal might seem like a simple task, even the most awake and alert morning person has probably ended up with cereal on the floor at some point. Philippe Meert, a product designer based in Erpe-Mere, Belgium, has come up with a better way. Meert sensed an opportunity to correct the innate design flaw of cereal boxes and developed the Cerealtop, a plastic cover that snaps onto a cereal box and channels the cereal into a bowl.[15]

The first thing that entrepreneurs such as Philippe Meert must do is to identify opportunities and possible competitive advantages. After they've done this, they're ready to start the venture by researching its feasibility and planning for its launch. In this section, we'll look at these start-up and planning issues.

Identifying Environmental Opportunities and Competitive Advantage

How important is the ability to identify environmental opportunities? C. K. Ranganathan, Chairman and Managing Director of CavinKare, had Rs. 15,000 in his pocket and knew how to manufacture shampoo and raise pets. He had left the family business after a fight with his brothers. He thought hard and decided to take his father's idea about packaging in sachets forward. He realized that more than 70 percent of the population of India did not have the cash to buy large packages and buying a large package restricted the ability of a poor person to experiment with the new product. Most villagers would not mind spending one rupee to test out shampoo but would not want to spend Rs. 70 for the same. Thus, he launched Chik Shampoo in Rs 1 sachets in rural Tamil Nadu. With a start point of Rs 15,000, his business today is worth Rs. 700 crore (approximate number). It has diversified but continues to reach out to low-income populations all over the country with innovative products and packaging.[16]

In 1994, when Jeff Bezos first saw that Internet usage was increasing by 2,300 percent per month, he knew that something dramatic was happening. "I hadn't seen growth that fast outside of a Petri dish," he said. Bezos was determined to be a part of it. He quit his successful career as a stock market researcher and hedge fund manager on Wall Street and pursued his vision for online retailing, now Amazon.com.[17]

What would you have done had you seen that type of number somewhere? Ignored it? Written it off as a fluke? The skyrocketing Internet usage that Bezos observed and the recognition of the needs to rural Indians by C. R. Ranganathan are prime examples of identifying environmental opportunities. Remember the discussion in Chapter 8 that opportunities are positive trends in external environmental factors. These trends provide unique and distinct possibilities for innovating and creating value. Entrepreneurs need to be able to pinpoint the pockets of opportunities that a changing context provides. After all, "organizations do not see opportunities, individuals do."[18] And they need to do so quickly, especially in dynamic environments, before those opportunities disappear or are exploited by others.[19]

The late Peter Drucker, a well-known management author, identified seven potential sources of opportunity that entrepreneurs might look for in the external context:[20]

1. The unexpected—When situations and events are unanticipated, opportunities can be found. An event may be an unexpected success (positive news) or an unexpected failure (bad news). Either way, there can be opportunities for entrepreneurs to pursue. For instance, the dramatic increase in fuel prices has proved to be a bonanza for companies offering alternatives to petrol-driven cars. For instance, Reva, India's first electric car, which was producing 6,000 cars annually, has increased its production capacity to 36,000 to meet the increasing demand from the United Kingdom and other countries where people want to increasingly opt for cars that are environmentally friendly. Moreover, it attracts government incentives such as electric vehicles that get free parking in London.[21] Kanwal Mookhey realized that networks are becoming part of life for most businesses, their security would be a concern and, thus, at the age of 21, he established NII Consulting (Network Intelligence (I) Pvt. Ltd) in July 2001. They now serve very prestigious clients such as the United Nations WFP, Atos Origin, Abu Dhabi Stock Exchange, Dubai Financial Market, Capgemini, and Saudi Telecom.[22]
2. *The incongruous*—When something is incongruous, there are inconsistencies and incompatibilities in the way it appears. Things "ought to be" a certain way but aren't. When conventional wisdom about the way things should be no longer holds true, for whatever reason, there are opportunities to capture. Entrepreneurs who are willing to "think outside the box"—that is, to think beyond the traditional and conventional approaches—may find pockets of potential profitability. Large steel plants were the order of the day for over 60 years. It was felt that innovations in the process were necessary to produce special products. The Jindal's were among the first to set up a mini steel plant and used the furnace to produce steel out of scrap steel and sponge iron.[23] This led to a revolution of sorts and now there are more than 200 mini steel plants that produce one-third of the steel in India.[24] Another example of how the incongruous can be a potential source of entrepreneurial opportunity is Fred Smith, founder of FedEx, who recognized in the early 1970s the inefficiencies in the delivery of packages and documents. His approach was: Why

not? Who says that overnight delivery isn't possible? Smith's recognition of the incongruous led to the creation of FedEx, now a multibillion-dollar corporation.

3. *The process need*—What happens when technology doesn't immediately come up with the "big discovery" that's going to fundamentally change the very nature of some product or service? What happens is that there can be pockets of entrepreneurial opportunity in the various stages of the process as researchers and technicians continue to work for the monumental breakthrough. Because the full leap hasn't been possible, opportunities abound in the tiny steps. Take the medical products industry, for example. Although researchers haven't yet discovered a cure for cancer, there have been many successful entrepreneurial biotechnology ventures created as knowledge about a possible cure has continued to grow. The "big breakthrough" hasn't yet happened, but there have been numerous entrepreneurial opportunities throughout the process of discovery.

4. *Industry and market structures*—When changes in technology change the structure of an industry and market, existing firms can become obsolete if they're not attuned to the changes or are unwilling to change. Even changes in social values and consumer tastes can shift the structures of industries and markets. These markets and industries become open targets for nimble and smart entrepreneurs. For instance, while working part time at an auto body shop while finishing his engineering graduate degree, Joe Born wondered if the industrial paint buffer used to smooth out a car's paint job could be used to smooth out scratches on CDs. He tried it out on his favorite Clint Black CD that had been ruined, and the newly polished CD played flawlessly. After this experience, Born spent almost four years perfecting his disk repair kit invention, the SkipDr.[25] Then, there's the whole Internet area, which provides several good examples of existing industries and markets being challenged by upstart entrepreneurial ventures. For instance, eBay has prospered as an online go-between for buyers and sellers. eBay's CEO says that the company's job is connecting people, not selling them things. And connect them, it does! The online auction firm has more than 275 million registered users.[26]

5. *Demographics*—The characteristics of the world population are changing. These changes influence industries and markets by altering the types and quantities of products and services desired and customers' buying power. Although many of these changes are fairly predictable if you stay alert to demographic trends, others aren't as obvious. Either way, there can be significant entrepreneurial opportunities in anticipating and meeting the changing needs of the population. For example, Thay Thida was one of three partners in Khmer Internet Development Services (KIDS) in Phnom Penh, Cambodia. She and her co-founders saw the opportunities in bringing Internet service to Cambodians and profited from their entrepreneurial venture.[27]

6. *Changes in perception*—Perception is one's view of reality. When changes in perception take place, the facts do not vary, but their meanings do. Changes in perception get at the heart of people's psychographic profiles—what they value, what they believe in, and what they care about. Changes in these attitudes and values create potential market opportunities for alert entrepreneurs. For example, think about your perception of healthy foods. As our perception of whether certain food groups are good for us has changed, there have been product and service opportunities for entrepreneurs to recognize and capture. For example, Kishore Biyani popularized the concept of retail in India among lower middle class by introducing the Big Bazaar format. The format was not like the organized large retail formats (Safeway or Target) but like the chaos of a bazaar in India. A store in Trivandrum even had a handcart (lorry) placed inside the store to create ambience of a bazaar. Having started in 2001, within 7 years of their opening, they now have 101 stores and most stores have had record number of footfalls on the very day of opening.[28] Namdhari Fresh started its first store in Bangalore to sell the different vegetables and fruits being cultivated by Namdhari Fresh Farms at various productions centers across the country.

Namdhari Fresh is committed to producing and providing safe, hygienic and fresh, vegetables and fruits. Their farmers only use eco-friendly bioagents and follow integrated pest management practices. Having started in 2000 with its first store, there are 23 retail outlets in Bangalore itself. It is a certified organic producer and has the unyielding support of its customers.[29] Kolkata-based Emami Industries were the first to bring out the latent need in men to look good. In 2004, they launched the Fair and Handsome cream. Emami today has 65 percent market share in the men's fairness segment, which is worth about Rs 100 crore.[30]

7. *New knowledge*—New knowledge is a significant source of entrepreneurial opportunity. Although not all knowledge-based innovations are significant, new knowledge ranks pretty high on the list of sources of entrepreneurial opportunity. It takes more than just having new knowledge, though. Entrepreneurs must be able to do something with that knowledge and need to protect important proprietary information from competitors. For example, French scientists are using new knowledge about textiles to develop a wide array of innovative products to keep wearers healthy and smelling good. Neyret, the Parisian lingerie maker, innovated lingerie products woven with tiny perfume microcapsules that stay in the fabric through about 10 washings. Another French company, Francital, developed a fabric treated with chemicals to absorb perspiration and odors.[31]

Being alert to entrepreneurial opportunities is only part of an entrepreneur's initial efforts. He or she must also understand competitive advantage. As discussed in Chapter 8, when an organization has a competitive advantage, it has something that other competitors don't, does something better than other organizations, or does something that others can't. Competitive advantage is a necessary ingredient for an entrepreneurial venture's long-term success and survival. Getting and keeping a competitive advantage is tough. However, it is something that entrepreneurs must consider as they begin researching a venture's feasibility.

Researching a Venture's Feasibility—Generating and Evaluating Ideas

Jumbo King started in 1990 to provide *vada pav* that was hygienic and not very expensive and the most selling item in the earlier venture of Dheeraj Gupta. The idea of a branded vada pav, a favorite food for Mumbai dwellers, at a reasonable price was the brainchild of Dheeraj Gupta who was inspired by the McDonalds story.[32] In less than 18 years, Jumbo King has 120 outlets and plans to open 500 by 2010. Jumbo King has realized over the years that rather than targeting the train traveler it is the college-going youth who are hygiene conscious and have money to spend on a vada pav (the American equivalent of a burger) are their customers.[33] The store has a loyal following and customers from all walks of life are seen walking into the store.

It's important for entrepreneurs to research a venture's feasibility by generating and evaluating business ideas. Entrepreneurial ventures thrive on ideas. Generating ideas is an innovative, creative process. It's also one that takes time, not only in the beginning stages of the entrepreneurial venture but throughout the life of the business. Where do ideas come from?

Generating Ideas. Studies of entrepreneurs have shown that the sources of their ideas are unique and varied. One survey found that "working in the same industry" was the major source of ideas for an entrepreneurial venture (60 percent of respondents).[34] Other sources included personal interests or hobbies, familiar and unfamiliar products and services, and opportunities in external environmental sectors (technological, sociocultural, demographics, economic, or legal–political).

What should entrepreneurs look for as they explore the sources of ideas? They should look for limitations of what's currently available, new and different approaches, advances and breakthroughs, unfilled niches, or trends and changes. For example, John C. Diebel, founder of Meade Instruments Corporation, an Irvine, California, telescope maker, came up with the idea of putting computerized attachments on the company's inexpensive consumer models so that amateur astronomers could enter on a keypad the coordinates of planets or stars they wanted to see. The telescope would then automatically locate and focus on the desired planetary bodies. It took the company's engineers two years to figure out how to do it, but Meade now controls more than half the amateur astronomy market.[35]

Evaluating Ideas. Evaluating entrepreneurial ideas involves personal and marketplace considerations. Each of these assessments will provide an entrepreneur with key information about the idea's potential. Exhibit A–1 describes some questions that entrepreneurs might ask as they evaluate potential ideas.

A more structured evaluation approach that an entrepreneur might want to use is a **feasibility study**—an analysis of the various aspects of a proposed entrepreneurial venture, designed to determine its feasibility. Not only is a well-prepared feasibility study an effective evaluation tool to determine whether an entrepreneurial idea is a potentially successful one, it can serve as a basis for the all-important business plan.

A feasibility study should give descriptions of the most important elements of the entrepreneurial venture and the entrepreneur's analysis of the viability of these elements. Exhibit A–2 provides an outline of a possible approach to a feasibility study. Yes, it covers a lot of territory and takes a significant amount of time, energy, and effort to prepare this study, but an entrepreneur's potential future success is worth that investment.

Researching a Venture's Feasibility—Researching Competitors

Part of researching a venture's feasibility is looking at the competitors. What would entrepreneurs like to know about their potential competitors? Here are some possible questions:

What types of products or services are competitors offering?
What are the major characteristics of these products or services?
What are the strengths and weaknesses of competitors' products?
How do competitors handle marketing, pricing, and distributing?
What do competitors attempt to do differently from other companies?
Do they appear to be successful at it? Why or why not?
What are they good at?
What competitive advantage(s) do they appear to have?
What are they not so good at?
What competitive disadvantage(s) do they appear to have?
How large and profitable are these competitors?

For instance, although there are thousands of private schools in India, Mr Satya Narayanan of Career Launcher fame decided to open schools in smaller cities in India

Exhibit A–1

Evaluating Potential Ideas

Personal Considerations:	Marketplace Considerations:
• Do you have the capabilities to do what you've selected? • Are you ready to be an entrepreneur? • Are you prepared emotionally to deal with the stresses and challenges of being an entrepreneur? • Are you prepared to deal with rejection and failure? • Are you ready to work hard? • Do you have a realistic picture of the venture's potential? • Have you educated yourself about financing issues? • Are you willing and prepared to do continual financial and other types of analyses?	• Who are the potential customers for your idea: who, where, how many? • What similar or unique product features does your proposed idea have compared to what's currently on the market? • How and where will potential customers purchase your product? • Have you considered pricing issues and whether the price you'll be able to charge will allow your venture to survive and prosper? • Have you considered how you will need to promote and advertise your proposed entrepreneurial venture?

Exhibit A–2

Feasibility Study

A. Introduction, historical background, description of product or service:
1. Brief description of proposed entrepreneurial venture
2. Brief history of the industry
3. Information about the economy and important trends
4. Current status of the product or service
5. How you intend to produce the product or service
6. Complete list of goods or services to be provided
7. Strengths and weaknesses of the business
8. Ease of entry into the industry, including competitor analysis

B. Accounting considerations:
1. Pro forma balance sheet
2. Pro forma profit and loss statement
3. Projected cash flow analysis

C. Management considerations:
1. Personal expertise—strengths and weaknesses
2. Proposed organizational design
3. Potential staffing requirements
4. Inventory management methods
5. Production and operations management issues
6. Equipment needs

D. Marketing considerations:
1. Detailed product description
2. Identify target market (who, where, how many)
3. Describe place product will be distributed (location, traffic, size, channels, etc.)
4. Price determination (competition, price lists, etc.)
5. Promotion plans (role of personal selling, advertising, sales promotion, etc.)

E. Financial considerations:
1. Start-up costs
2. Working capital requirements
3. Equity requirements
4. Loans—amounts, type, conditions
5. Breakeven analysis
6. Collateral
7. Credit references
8. Equipment and building financing—costs and methods

F. Legal considerations:
1. Proposed business structure (type; conditions, terms, liability, responsibility; insurance needs; buyout and succession issues)
2. Contracts, licenses, and other legal documents

G. Tax considerations: sales/property/employee; federal, state, and local

H. Appendix: charts/graphs, diagrams, layouts, resumés, etc.

because he believes that his school will be a place where children will love to go. With the curriculum, the teaching methodology, and the facilities that Satya thinks he will provide, ensures that he will be able to go ahead of the current schools.[36]

When an entrepreneur has information on competitors, he or she should assess how the proposed entrepreneurial venture is going to fit into its competitive arena. Will the entrepreneurial venture be able to compete successfully? This type of competitor analysis becomes an important part of the feasibility study and the business plan. If, after all this analysis, the situation looks promising, the final part of researching the venture's feasibility is to look at the various financing options. This isn't the final determination of how much funding the venture will need or where this funding will come from but is simply gathering information about various financing alternatives.

feasibility study
An analysis of the various aspects of a proposed entrepreneurial venture that is designed to determine the feasibility of the venture.

Exhibit A–3

Possible Financing Options

- Entrepreneur's personal resources (personal savings, home equity, personal loans, credit cards, etc.)
- Financial institutions (banks, savings and loan institutions, government-guaranteed loan, credit unions, etc.)
- **Venture capitalists**—external equity financing provided by professionally-managed pools of investor money
- **Angel investors**—a private investor (or group of private investors) who offers financial backing to an entrepreneurial venture in return for equity in the venture
- **Initial public offering (IPO)**—the first public registration and sale of a company's stock
- National, state, and local governmental business development programs
- Unusual sources (television shows, judged competitions, etc.)

Researching a Venture's Feasibility—Researching Financing

Getting financing isn't always easy. For instance, when Girish Sarpah, Professor of IIT Bombay, launched Vegayan Systems Pvt. Ltd in December 2007, a telecom networking product company, and tried to get venture funding for its product patented and test marketed in the United States, he did not get any funds. He and his group tried several VCs for funding.[37]

Because most ventures will need funds in order to get started, an entrepreneur must research the various financing options. Possible financing options available to entrepreneurs are shown in Exhibit A–3.

Planning a Venture—Developing a Business Plan

Planning is important to entrepreneurial ventures. Once a venture's feasibility has been thoroughly researched, the entrepreneur must look at planning the venture. The most important thing that an entrepreneur does in planning a venture is develop a **business plan**—a written document that summarizes a business opportunity and defines and articulates how the identified opportunity is to be seized and exploited.

For many would-be entrepreneurs, developing and writing a business plan seems like a daunting task. However, a good business plan is valuable. It pulls together all the elements of the entrepreneur's vision into a single coherent document. The business plan requires careful planning and creative thinking. If done well, it can be a convincing document that serves many functions. It serves as a blueprint and road map for operating the business. And the business plan is a "living" document that guides organizational decisions and actions throughout the life of the business, not just in the start-up stage.

If an entrepreneur has completed a feasibility study, much of the information included in it becomes the basis for the business plan. A good business plan has six major sections: executive summary, analysis of opportunity, analysis of the context, description of the business, financial data and projections, and supporting documentation.

Executive Summary. The executive summary summarizes the key points that an entrepreneur wants to make about a proposed entrepreneurial venture. These might include a brief mission statement; primary goals; a brief history of the entrepreneurial venture, maybe in the form of a timeline; key people involved in the venture; the nature of the business; concise product or service descriptions; brief explanations of market niche, competitors, and competitive advantage; proposed strategies; and selected key financial information.

Analysis of Opportunity. In this section of the business plan, an entrepreneur presents the details of the perceived opportunity. Essentially, this means (1) sizing up the market by describing the demographics of the target market, (2) describing and evaluating industry trends, and (3) identifying and evaluating competitors.

Analysis of the Context. Whereas the opportunity analysis focuses on the opportunity in a specific industry and market, the context analysis takes a much broader perspective.

Here, the entrepreneur describes the broad external changes and trends taking place in the economic, political–legal, technological, and global environments.

Description of the Business. In this section, an entrepreneur describes how the entrepreneurial venture is going to be organized, launched, and managed. It includes a thorough description of the mission statement; a description of the desired organizational culture; marketing plans, including overall marketing strategy, pricing, sales tactics, service warranty policies, and advertising and promotion tactics; product development plans, such as an explanation of development status, tasks, difficulties and risks, and anticipated costs; operational plans, including a description of proposed geographic location, facilities and needed improvements, equipment, and work flow; human resource plans, including a description of key management persons, composition of the board of directors, including their background experience and skills, current and future staffing needs, compensation and benefits, and training needs; and an overall schedule and timetable of events.

Financial Data and Projections. Every effective business plan contains financial data and projections. Although the calculations and interpretation may be difficult, they are absolutely critical. No business plan is complete without financial information. Financial plans should cover at least three years and contain projected income statements, pro forma cash flow analysis (monthly for the first year and quarterly for the next two), pro forma balance sheets, break-even analysis, and cost controls. If major equipment or other capital purchases are expected, the items, costs, and available collateral should be listed. All financial projections and analyses should include explanatory notes, especially where the data seem contradictory or questionable.

Supporting Documentation. Supporting documentation is an important component of an effective business plan. An entrepreneur should back up his or her descriptions with charts, graphs, tables, photographs, or other visual tools. In addition, it might be important to include information (personal and work related) about the key participants in the entrepreneurial venture.

Just as the idea for an entrepreneurial venture takes time, so does the writing of a good business plan. It's important for an entrepreneur to put serious thought and consideration into the plan. The plan is not easy to create, but the resulting document should be valuable to the entrepreneur in current and future planning efforts.

QUICK LEARNING REVIEW

- Discuss how opportunities are important to entrepreneurial ventures.
- Describe each of the seven sources of potential opportunity.
- Explain why it's important for entrepreneurs to understand competitive advantage.
- List possible financing options for entrepreneurs.
- Describe the six major sections of a business plan.

ISSUES IN ORGANIZING AN ENTREPRENEURIAL VENTURE

Torrent Pharmaceuticals in Ahmedabad, Gujrat, redesigned his organization's structure by transforming it into an employee-empowered company. The company's management wanted to drive authority down through the organization so employees were responsible for their own efforts. One way they did this was by creating employee teams to handle specific projects. Employees with less experience were teamed with veteran employees.

Once the start-up and planning issues for an entrepreneurial venture have been addressed, the entrepreneur is ready to begin organizing the entrepreneurial venture. There are five organizing issues an entrepreneur must address: the legal forms of

venture capitalists
Individuals who provide external equity financing through professionally managed pools of investor money.

angel investors
A private investor (or group of private investors) who offers financial backing to an entrepreneurial venture in return for equity in the venture.

initial public offering (IPO)
The first public registration and sale of a company's stock.

business plan
A written document that summarizes a business opportunity and defines and articulates how the identified opportunity is to be seized and exploited.

organization, organizational design and structure, human resource management, how to simulate and make changes, and the continuing importance of innovation.

Legal Forms of Organization

The first organizing decision that an entrepreneur must make is a critical one: determining the form of legal ownership for the venture. The two primary factors affecting this decision are taxes and legal liability. An entrepreneur wants to minimize the impact of both of these factors. The right choice can protect the entrepreneur from legal liability as well as save tax dollars, in both the short run and the long run.

What alternatives are available? There are three basic ways to organize an entrepreneurial venture: sole proprietorship, partnership, and corporation. However, when you include the variations of these basic organizational alternatives, you end up with six possible choices, each with unique tax consequences, liability issues, and pros and cons: sole proprietorship, general partnership, limited liability partnership (LLP), C corporation, S corporation, and limited liability company (LLC). Let's briefly look at each one and its advantages and drawbacks. (Exhibit A–4 summarizes the basic information about each organizational alternative.)

Sole Proprietorship. A **sole proprietorship** is a form of legal organization in which the owner maintains sole and complete control over the business and is personally liable for business debts. There are no legal requirements for establishing a sole proprietorship other than obtaining the necessary local business licenses and permits. In a sole proprietorship, income and losses "pass through" to the owner and are taxed at the owner's personal income tax rate. The biggest drawback, however, is the unlimited personal liability for any and all debts of the business.

General Partnership. A **general partnership** is a form of legal organization in which two or more business owners share the management and risk of the business. Even though a partnership without a written agreement is possible, the potential and inevitable problems that arise in any partnership make a written partnership agreement drafted by legal counsel highly recommended.

Limited Liability Partnership (LLP). Limited liability partnership (LLP) is a form of legal organization in which there are general partners and limited partners. The general partners actually operate and manage the business. They are the ones who have unlimited liability. There must be at least one general partner in an LLP. However, there can be any number of limited partners. These partners are usually passive investors, although they can make management suggestions to the general partners. They also have the right to inspect the business and make copies of business records. The limited partners are entitled to a share of the business's profits, as spelled out in the partnership agreement, and their risk is limited to the amount of their investment in the LLP.

C corporation. Of the three basic types of ownership, the corporation (also known as a C corporation) is the most complex to form and operate. A **corporation** is a legal business entity that is separate from its owners and managers. Many entrepreneurial ventures are organized as **closely held corporations** which, very simply, are corporations owned by a limited number of people who do not trade the stock publicly. Whereas the sole proprietorship and partnership forms of organization do not exist separately from the entrepreneur, the corporation does. A corporation functions as a distinct legal entity and, as such, can make contracts, engage in business activities, own property, sue and be sued, and, of course, pay taxes. A corporation must operate in accordance with its charter and the laws of the state in which it operates.

S Corporation. An **S corporation**(also called a subchapter S corporation) is a specialized type of corporation that has the regular characteristics of a corporation but is unique in that the owners are taxed as a partnership, as long as certain criteria are met. The S corporation has been the classic organizing approach for getting the limited liability of a corporate structure without incurring corporate tax. However, this form of legal organization must meet strict criteria. If any of these criteria are violated, a venture's S status is automatically terminated.

Exhibit A–4 Legal Forms of Business Organization

Structure	Ownership Requirements	Tax Treatment	Liability	Advantages	Drawbacks
Sole proprietorship	One owner	Income and losses "pass through" to owner and are taxed at personal rate	Unlimited personal liability	*Low start-up costs* Freedom from most regulations *Owner has direct control* All profits go to owner *Easy to exit business*	Unlimited personal liability *Personal finances at risk* Miss out on many business tax deductions *Total responsibility* May be more difficult to raise financing
General partnership	Two or more owners	Income and losses "pass through" to partners and are taxed at personal rate; *flexibility in profit-loss allocations to partners*	Unlimited personal liability	*Ease of formation* Pooled talent *Pooled resources* Somewhat easier access to financing *Some tax benefits*	Unlimited personal liability *Divided authority and decisions* Potential for conflict *Continuity of transfer of ownership*
Limited liability partnership (LLP)	Two or more owners	Income and losses "pass through" to partner and are taxed at personal rate; *flexibility in profit-loss allocations to partners*	Limited, although one partner must retain unlimited liability	*Good way to acquire capital from limited partners*	Cost and complexity of forming can be high *Limited partners cannot participate in management of business without losing liability protection*
C corporation	Unlimited number shareholders; *no limits on types of stock or voting arrangements*	Dividend income is taxed at corporate and personal shareholder levels; *losses and deductions are corporate*	Limited	*Limited liability* Transferable ownership *Continuous existence* Easier access to resources	Expensive to set up *Closely regulated* Double taxation *Extensive record keeping* Charter restrictions
S corporation	Up to 75 shareholders; *no limits on types of stock or voting arrangements*	Income and losses "pass through" to partners and are taxed at personal rate; *flexibility in profit-loss allocation to partners*	Limited	*Easy to set up* Enjoy limited liability protection and tax benefits of partnership *Can have a tax-exempt entity as a shareholder*	Must meet certain requirements *May limit future financing options*
Limited liability company (LLC)	Unlimited number of "members"; *flexible membership arrangements for voting rights and income*	Income and losses "pass through" to partners and are taxed at personal rate; *flexibility in profit-loss allocations to partners*	Limited	*Greater flexibility* Not constrained by regulations on C and S corporations *Taxed as partnership, not as corporation*	Cost of switching from one form to this can be high *Need legal and financial advice in forming operating agreement*

Limited Liability Company (LLC). **Limited liability company (LLC)** is a relatively new form of business organization that's a hybrid between a partnership and a corporation. The LLC offers the liability protection of a corporation, the tax benefits of a partnership, and fewer restrictions than an S corporation. However, the main drawback of this approach is that it's quite complex and expensive to set up. Legal and financial advice is an absolute necessity in forming an LLC's **operating agreement**, which is a document that outlines the provisions governing the way the LLC will conduct business.

Summary of Legal Forms of Organization. The organizing decision regarding the legal form of organization for a venture is an important one because it can have significant tax and liability consequences. Although the legal form of organization can be changed, making this kind of change is not easy. In choosing the best form of organization, an entrepreneur needs to think carefully about what's important, especially in the areas of flexibility, taxes, and amount of personal liability.

Organizational Design and Structure

The choice of an appropriate organizational structure is an important decision when organizing an entrepreneurial venture. At some point, successful entrepreneurs find that they can't do everything alone. More people are needed. The entrepreneur must then decide on the most appropriate structural arrangement for effectively and efficiently carrying out the organization's activities. Without some suitable type of organizational structure, the entrepreneurial venture may soon find itself in a chaotic situation.

In many small firms, the organizational structure tends to evolve with very little intentional and deliberate planning by the entrepreneur. For the most part, the structure may be very simple—one person does whatever is needed. As the entrepreneurial venture grows and the entrepreneur finds it increasingly difficult to go it alone, employees are brought on board to perform certain functions or duties that the entrepreneur can't handle. These individuals tend to perform those same functions as the company grows. As the entrepreneurial venture continues to grow, each of these functional areas may require managers and employees.

With the evolution to a more deliberate structure, an entrepreneur faces a whole new set of challenges. All of a sudden, he or she must share decision making and operating responsibilities. This is typically one of the most difficult things for an entrepreneur to do—let go and allow someone else to make decisions. *After all,* he or she reasons, *how can anyone know this business as well as I do?* Also, what might have been a fairly informal, loose, and flexible atmosphere that worked well when the organization was small may no longer be effective. Many entrepreneurs are greatly concerned about keeping that "small company" atmosphere alive even as the venture grows and evolves into a more structured arrangement. But having a structured organization doesn't necessarily mean giving up flexibility, adaptability, and freedom. In fact, the structural design may be as fluid as the entrepreneur feels comfortable with and yet still have the rigidity it needs to operate efficiently.

Organizational design decisions in entrepreneurial ventures revolve around the six key elements of organizational structure discussed in Chapter 9: work specialization, departmentalization, chain of command, span of control, amount of centralization or decentralization, and amount of formalization. Decisions about these six elements will determine whether an entrepreneur designs a more mechanistic or organic organizational structure (concepts also discussed in Chapter 9). When would each be preferable? A mechanistic structure would be preferable when cost-efficiencies are critical to the venture's competitive advantage, when more control over employees' work activities is important, if the venture produces standardized products in a routine fashion, and when the external environment is relatively stable and certain. An organic structure would be most appropriate when innovation is critical to the organization's competitive advantage, for smaller organizations where rigid approaches to dividing and coordinating work aren't necessary, if the organization produces customized products in a flexible setting, and where the external environment is dynamic, complex, and uncertain.

Human Resource Management Issues in Entrepreneurial Ventures

As an entrepreneurial venture grows, additional employees will need to be hired to perform the increased work load. As employees are brought on board, an entrepreneur faces certain human resource management (HRM) issues. Two HRM issues of particular importance to entrepreneurs are employee recruitment and employee retention.

Employee Recruitment. An entrepreneur wants to ensure that the venture has the people it needs to do the required work. Recruiting new employees is one of the biggest challenges that entrepreneurs face. In fact, the ability of small firms to successfully recruit appropriate employees is consistently rated as one of the most important factors influencing organizational success.[38]

Entrepreneurs, particularly, are looking for high-potential people who can perform multiple roles during various stages of venture growth. They look for individuals who buy into the venture's entrepreneurial culture—individuals who have a passion for the business.[39] Unlike their corporate counterparts, who often focus on filling a job by matching a person to the job requirements, entrepreneurs look to fill in critical skills gaps. They look for people who are exceptionally capable and self-motivated, flexible, and multi-skilled and who can help grow the entrepreneurial venture. Whereas corporate managers tend to focus on using traditional HRM practices and techniques, entrepreneurs are more concerned with matching characteristics of the person to the values and culture of the organization; that is, they focus on matching the person to the organization.[40]

Employee Retention. Getting competent and qualified people into a venture is just the first step in effectively managing the human resources. An entrepreneur wants to keep the people he or she has hired and trained. Anil Sachdeva, president of Grow Talent, based in New Delhi, understands the importance of having good people on board and keeping them. In the rough-and-tumble, intensely competitive public relations industry, Anil knows that the loss of talented employees can harm client services. To combat this, he offers employees a wide array of desirable benefits, such as raises each year, profit sharing, trust funds for employees' children, paid sabbaticals, personal development funds, and so forth. But more importantly, Grow Talent recognizes that employees have a life outside the office and treats them accordingly. This type of HRM approach has kept their employees loyal and productive.[41]

A unique and important employee retention issue entrepreneurs must deal with is compensation. Whereas traditional organizations are more likely to view compensation from the perspective of monetary rewards (base pay, benefits, and incentives), smaller entrepreneurial firms are more likely to view compensation from a total rewards perspective. For these firms, compensation encompasses psychological rewards, learning opportunities, and recognition in addition to monetary rewards (base pay and incentives).[42]

How to Stimulate and Make Changes

We know that entrepreneurs face dynamic change. Both external and internal forces (see Chapter 12) may bring about the need for making changes in the entrepreneurial venture. Entrepreneurs need to be alert to problems and opportunities that may create

sole proprietorship
A form of legal organization in which the owner maintains sole and complete control over the business and is personally liable for business debts.

general partnership
A form of legal organization in which two or more business owners share the management and risk of the business.

limited liability partnership (LLP)
A form of legal organization that involves one or more general partners and one or more limited liability partners.

corporation
A legal business entity that is separate from its owners and managers.

closely held corporation
A corporation owned by a limited number of people who do not trade the stock publicly.

S corporation
A specialized type of corporation that has the regular characteristics of a C corporation but is unique in that the owners are taxed as a partnership as long as certain criteria are met.

limited liability company (LLC)
A form of legal organization that's a hybrid between a partnership and a corporation.

operating agreement
A document that outlines the provisions governing the way an LLC will conduct business.

the need for change. In fact, of the many hats an entrepreneur wears, that of change agent may be one of the most important.[43] If changes are needed in an entrepreneurial venture, often it is the entrepreneur who first recognizes the need for change and acts as the catalyst, coach, cheerleader, and chief change consultant. Change isn't easy in any organization, but it can be particularly challenging for entrepreneurial ventures. Even if a person is comfortable with taking risks—as entrepreneurs usually are—change can be difficult. That's why it's important for an entrepreneur to recognize the critical roles he or she plays in stimulating and implementing change. For instance, Nimish Vohra, Country Head of Regalik, Inc., is well aware of the important role he plays in stimulating and implementing changes. As the leading company in digital marketing, Nimish had to continually look for ways to keep his company competitive. One change was the partnership with Silicon India.

During any type of organizational change, an entrepreneur may also have to act as chief coach and cheerleader. Because organizational change of any type can be disruptive and scary, an entrepreneur must explain a change to employees and encourage change efforts by supporting employees, getting them excited about the change, building them up, and motivating them to put forth their best efforts.

Finally, an entrepreneur may have to guide the actual change process as changes in strategy, technology, products, structure, or people are implemented. In this role, the entrepreneur answers questions, makes suggestions, gets needed resources, facilitates conflict, and does whatever else is necessary to get the change(s) implemented.

The Continuing Importance of Innovation

In today's dynamically chaotic world of global competition, organizations must continually innovate new products and services if they want to compete successfully. Innovation is a key characteristic of entrepreneurial ventures and, in fact, is what makes an entrepreneurial venture "entrepreneurial."

What must an entrepreneur do to encourage innovation in the venture? Having an innovation-supportive culture is crucial. What does such a culture look like?[44] It's one in which employees perceive that supervisory support and organizational reward systems are consistent with a commitment to innovation. It's also important in this type of culture that employees not perceive their work load pressures as being excessive or unreasonable. And research has shown that firms with cultures supportive of innovation tend to be smaller, have fewer formalized human resource practices, and have less abundant resources.[45]

QUICK LEARNING REVIEW

- Contrast the six different forms of legal organization.
- Describe the organizational design issues entrepreneurs face as a venture grows.
- Discuss the unique HRM issues entrepreneurs face.
- Describe what an innovation-supportive culture looks like.

ISSUES IN LEADING AN ENTREPRENEURIAL VENTURE

The employees at Neelam Chhibber's company have to be flexible. Many don't have job descriptions, and everyone is expected to contribute ideas and pitch in with tasks in all departments. It is not common to hear or say 'That's not my job' doesn't belong here." In return, Neelam is a supportive leader who gives her employees considerable latitude.[46]

Leading is an important function of entrepreneurs. As an entrepreneurial venture grows and people are brought on board, an entrepreneur takes on a new role—that of a leader. In this section, we'll look at what's involved with the leading function. First, we'll look at the unique personality characteristics of entrepreneurs. Then we'll discuss the important role entrepreneurs play in motivating employees through empowerment and leading the venture and employee teams.

Personality Characteristics of Entrepreneurs

Think of someone you know who is an entrepreneur. Maybe it's someone you personally know or maybe it's someone like Bill Gates of Microsoft. How would you describe this person's personality? One of the most researched areas of entrepreneurship has been the search to determine what—if any—psychological characteristics entrepreneurs have in common, what types of personality traits entrepreneurs have that might distinguish them from non-entrepreneurs, and what traits entrepreneurs have that might predict who will be a successful entrepreneur.

Is there a classic "entrepreneurial personality"? Although trying to pinpoint specific personality characteristics that all entrepreneurs share suffers from the same problem as identifying the trait theories of leadership—that is, being able to identify specific personality traits that *all* entrepreneurs share—this hasn't stopped entrepreneurship researchers from listing common traits.[47] For instance, one list of personality characteristics includes the following: high level of motivation, abundance of self-confidence, ability to be involved for the long term, high energy level, persistent problem solving, high degree of initiative, ability to set goals, and moderate risk-taking.[48] Another list of characteristics of "successful" entrepreneurs includes high energy level, great persistence, resourcefulness, desire and ability to be self-directed, and relatively high need for autonomy.

Another development in defining entrepreneurial personality characteristics was using the proactive personality scale to predict an individual's likelihood of pursuing entrepreneurial ventures. We introduced the **proactive personality** trait in Chapter 13. Recall that it's a personality trait that describes individuals who are more prone than others to take actions to influence their environment—that is, they're more proactive. Obviously, an entrepreneur is likely to exhibit proactivity as he or she searches for opportunities and acts to take advantage of those opportunities.[49] Various items on the proactive personality scale have been found to be good indicators of a person's likelihood of becoming an entrepreneur, including being male, education, having an entrepreneurial parent, and possessing a proactive personality. In addition, studies have shown that entrepreneurs have greater risk propensity than do managers.[50] However, this propensity is moderated by the entrepreneur's primary goal. Risk propensity is greater for entrepreneurs whose primary goal is growth than for those whose focus is on producing family income.

Motivating Employees Through Empowerment

At Enercon Systems India, Hema Hattangady, MD, recognized that employee motivation was vitally important to their company's ultimate success.[51] She designed her organization so that cross-functional and cross-hierarchical teams were formed to solve problems. This has led to faster turnaround of projects. Hema has worked hard to change the culture from being top-driven to participative.

When you're motivated to do something, don't you find yourself energized and willing to work hard at doing something you're excited about? Wouldn't it be great if all of a venture's employees were energized, excited, and willing to work hard at their jobs? Having motivated employees is an important goal for any entrepreneur, and employee empowerment is an important motivational tool entrepreneurs can use.

Although it's not easy for entrepreneurs to do, employee empowerment—giving employees power to make decisions and take actions on their own—is an important motivational approach. Why? Because successful entrepreneurial ventures must be quick and nimble, ready to pursue opportunities and go off in new directions. Empowered employees can provide that flexibility and speed. When employees are empowered, they often display stronger work motivation, better work quality, higher job satisfaction, and lower turnover.

proactive personality
A personality trait that describes individuals who are prone to take actions to influence their environments.

For example, employees at Butler International, Inc., a technology consulting services firm based in Montvale, New Jersey, work at client locations. President and CEO Ed Kopko recognized that employees had to be empowered to do their jobs if they were to be successful.[52] Another entrepreneurial venture that found employee empowerment to be a strong motivational approach is Stryker Instruments in Kalamazoo, Michigan, a division of Stryker Corporation. Each of the company's production units is responsible for its operating budget, cost reduction goals, customer service levels, inventory management, training, production planning and forecasting, purchasing, human resource management, safety, and problem solving. In addition, unit members work closely with marketing, sales, and R&D during new product introductions and continuous improvement projects. Says one team supervisor, "Stryker lets me do what I do best and rewards me for that privilege."[53]

Empowerment is a philosophical concept that entrepreneurs have to buy into, and it doesn't come easily. In fact, it's difficult for many entrepreneurs to do. Their life is tied up in the business. They've built it from the ground up. But continuing to grow the entrepreneurial venture eventually requires handing over more responsibilities to employees. How can entrepreneurs empower employees? For many entrepreneurs, it's a gradual process.

Entrepreneurs can begin by using participative decision making in which employees provide input into decisions. Although getting employees to participate in decisions isn't quite taking the full plunge into employee empowerment, at least it's a way to begin tapping into the collective array of employees' talents, skills, knowledge, and abilities.

Another way to empower employees is through delegation—the process of assigning certain decisions or specific job duties to employees. By delegating decisions and duties, the entrepreneur is turning over the responsibility for carrying them out.

When an entrepreneur is comfortable with the idea of employee empowerment, fully empowering employees means redesigning their jobs so they have discretion over the way they do their work. It's allowing employees to do their work effectively and efficiently by using their creativity, imagination, knowledge, and skills.

If an entrepreneur implements employee empowerment properly—that is, with complete and total commitment to the program and with appropriate employee training—results can be impressive for the entrepreneurial venture and for the empowered employees. The business can enjoy significant productivity gains, quality improvements, more satisfied customers, increased employee motivation, and improved morale. Employees can enjoy the opportunities to do a greater variety of work that is more interesting and challenging.

In addition, employees are encouraged to take the initiative in identifying and solving problems and doing their work. For example, at Mindtree in Bangalore, employees are empowered to change their work processes in order to meet the organization's challenging quality improvement goals. However, the company's commitment to having an empowered workforce has resulted in increased profitability.

The Entrepreneur as Leader

The last topic we'll discuss in this section is the role of the entrepreneur as a leader. In this role, the entrepreneur has certain leadership responsibilities in leading the venture and in leading employee work teams.

Leading the Venture. Today's successful entrepreneur must be like the leader of a jazz ensemble known for its improvisation, innovation, and creativity. Max DePree, former head of Herman Miller, Inc., a leading office furniture manufacturer known for its innovative leadership approaches, said it best in his book *Leadership Jazz*: "Jazz band leaders must choose the music, find the right musicians, and perform—in public. But the effect of the performance depends on so many things—the environment, the volunteers playing the band, the need for everybody to perform as individuals and as a group, the absolute dependence of the leader on the members of the band, the need for the followers to play well .... The leader of the jazz band has the beautiful opportunity to draw the best out of the other musicians. We have much to learn from jazz band leaders, for jazz, like leadership, combines the unpredictability of the future with the gifts of individuals."[54]

The way an entrepreneur leads the venture should be much like the jazz band leader's job—drawing the best out of other individuals, even given the unpredictability of the situation. One way an entrepreneur does this is through the vision he or she creates for the organization. In fact, the driving force through the early stages of the entrepreneurial venture is often the visionary leadership of the entrepreneur. The entrepreneur's ability to articulate a coherent, inspiring, and attractive vision of the future is a key test of his or her leadership. But if an entrepreneur can do this, the results can be worthwhile. A study contrasting visionary and nonvisionary companies showed that visionary companies outperformed the nonvisionary ones 6 times over on standard financial criteria, and their stocks outperformed the general market 15 times over.[55]

Leading Employee Work Teams. As we know from Chapter 11, many organizations—entrepreneurial and otherwise—are using employee work teams to perform organizational tasks, create new ideas, and resolve problems.

Employee work teams tend to be popular in entrepreneurial ventures. An *IndustryWeek* census of manufacturers showed that nearly 68 percent of survey respondents used teams to varying degrees.[56] The three most common types respondents said they used were empowered teams (teams that have the authority to plan and implement process improvements), self-directed teams (teams that are nearly autonomous and responsible for many managerial activities), and cross-functional teams (work teams composed of individuals from various specialties who work together on various tasks).

These entrepreneurs also said that developing and using teams is necessary because technology and market demands are forcing them to make their products faster, cheaper, and better. Tapping into the collective wisdom of the venture's employees and empowering them to make decisions may be one of the best ways to adapt to change. In addition, a team culture can improve the overall workplace environment and morale.

For team efforts to work, however, entrepreneurs must shift from the traditional command-and-control style to a coach-and-collaboration style. (Refer to the discussion of team leadership in Chapter 16.) They must recognize that individual employees can understand the business and can innovate just as effectively as they can. For example, at Kotak Bank of Mumbai, CEO Uday Kotak recognized that he wasn't the smartest guy in the company in terms of operational problems, but he was smart enough to recognize that if he wanted his company to expand its market share, new levels of productivity needed to be reached. He gave his talented team the freedom to work together and the authority to resolve the constraints.[57]

QUICK LEARNING REVIEW

- Explain what personality research shows about entrepreneurs
- Discuss how entrepreneurs can empower employees.
- Explain how entrepreneurs can be effective at leading employee work teams.

ISSUES IN CONTROLLING AN ENTREPRENEURIAL VENTURE

Philip McCaleb still gets a kick out of riding the scooters his Chicago-based company, Genuine Scooter Co., makes. However, in building his business, McCaleb has had to acknowledge his own limitations. As a self-described "idea" guy, he knew that he would need someone else to come in and ensure that the end product was *what* it was supposed to be, *where* it was supposed to be, and *when* it was supposed to be there.[58]

Entrepreneurs must look at controlling the venture's operations in order to survive and prosper in both the short run and long run. Those unique control issues that face entrepreneurs include managing growth, managing downturns, exiting the venture, and managing personal life choices and challenges.

Managing Growth

Growth is a natural and desirable outcome for entrepreneurial ventures. Growth is what distinguishes an entrepreneurial venture. Entrepreneurial ventures pursue growth.[59] Growing slowly can be successful, but so can rapid growth.

Growing successfully doesn't occur randomly or by luck. Successfully pursuing growth typically requires an entrepreneur to manage all the challenges associated with growing. This entails planning, organizing, and controlling for growth.

Planning for Growth. Although it may seem we've reverted back to discussing planning issues instead of controlling issues, controlling is actually closely tied to planning, as we know from our discussion in Chapter 17 (see Exhibit 17–1). And the best growth strategy is a well-planned one.[60] Ideally, the decision to grow doesn't come about spontaneously but instead is part of a venture's overall business goals and plan. Rapid growth without planning can be disastrous. Entrepreneurs need to address growth.

G. P. Das is one of the partners of Oricom Systems Pvt. Ltd, a computer-related business in Bhubaneswar, Orissa. As the company grew in the first 3 years, the partners realized that they needed more systems in place to be able to grow.

Having enough capital is a major challenge facing growing entrepreneurial ventures. The money issue never seems to go away, does it? It takes capital to expand. The processes of finding capital to fund growth are much those used for the initial financing of the venture. Hopefully, at this time the venture has a successful track record to back up the request. If it doesn't, it may be extremely difficult to acquire the necessary capital. That's why we said earlier that the best growth strategy is a planned one.

Part of that planning should be how growth will be financed. For example, The Boston Beer Company, America's largest microbrewer and producer of Samuel Adams beer, grew rapidly by focusing almost exclusively on increasing its top-selling product line. However, the company was so focused on increasing market share that it had few financial controls and an inadequate financial infrastructure. During periods of growth, cash flow difficulties would force company chairman and brewmaster Jim Koch to tap into a pool of unused venture capital funding. However, when a chief financial officer joined the company, he developed a financial structure that

managing in a Virtual World

IT for Entrepreneurs

IT presents a number of challenges—and opportunities—for entrepreneurs. According to a survey of business owners, 53 percent cited integrating different applications and different software systems as one of the primary IT challenges. These IT tools (software and hardware) were often purchased separately as funds became available or as needs changed. However, it's difficult for employees to share information or to get work done when the information isn't available because it's sitting in separate and different applications or on incompatible hardware. Another similar IT challenge facing small businesses is integrating Web sites with applications software. This was noted by 47 percent of the survey respondents. The remaining IT challenges noted by the survey respondents were outgrowing the system (45 percent), insufficient IT staff (42 percent), and outdated applications (34 percent).

Despite these challenges, there are many IT tools that can support an entrepreneur's business. One of the primary tools is e-mail marketing, which can be a great way for smaller firms to maintain contact with current and potential customers. And despite the potential integration problems, there are other important IT tools, such as the various types of business applications software available. This software can make planning, organizing, leading, and controlling an entrepreneurial venture more efficient and more effective.

Source: R. Breeden, "Owners Want Software Programs Integrated," Wall Street Journal*, April 4, 2006, p. A16.*

enabled the company to manage its growth more efficiently and effectively by setting up a plan for funding growth.[61]

Another important issue that a growing entrepreneurial venture needs to address is finding people. If a venture is growing quickly, this challenge may be intensified because of time constraints. It's important to plan the numbers and types of employees needed as much as possible in order to support the increasing work load of the growing venture. It may also be necessary to provide additional training and support to employees to help them handle the increased pressures associated with the growing organization.

Finally, when a venture is growing, it's important to create a positive, growth-oriented culture that enhances the opportunities to achieve success, both organizationally and individually. This can sometimes be difficult to do, particularly when changes are happening rapidly. However, the values, attitudes, and beliefs that are established and reinforced during these times are critical to the entrepreneurial venture's continued and future success. Exhibit A–5 lists some suggestions that entrepreneurs might use to ensure that their venture's culture is one that embraces and supports a climate in which organizational growth is viewed as desirable and important. Keeping employees focused and committed to what the venture is doing is critical to the ultimate success of its growth strategies. If employees don't buy into the direction the entrepreneurial venture is headed, the growth strategies are unlikely to be successful.

Controlling for Growth. Another challenge that growing entrepreneurial ventures face is reinforcing already established organizational controls. Maintaining good financial records and financial controls over cash flow, inventory, customer data, sales orders, receivables, payables, and costs should be a priority of every entrepreneur—whether pursuing growth or not. However, it's particularly important to reinforce these controls when the entrepreneurial venture is expanding. It's all too easy to let things "get away" or to put them off when there's an unrelenting urgency to get things done. Rapid growth—or even slow growth—does not excuse the need to have effective controls in place. In fact, it's particularly important to have established procedures, protocols, and processes and to use them. Even though mistakes and inefficiencies can never be eliminated entirely, an entrepreneur should at least ensure that every effort is being made to achieve high levels of productivity and organizational effectiveness. For example, at Green Gear Cycling, co-founder Alan Scholz recognized the importance of controlling for growth. How? By following a "Customers for Life" strategy, which meant continually monitoring customer relationships and orienting organizational work decisions around their possible impacts on customers. Through this type of strategy, Green Gear hopes to keep customers for life. That's significant because the company figured that, if it could keep a customer for life, the value would range from $10,000 to $25,000 per lifetime customer.[62]

Exhibit A–5

Achieving a Supportive Growth-Oriented Culture

- Keep the lines of communication open—inform employees about major issues.
- Establish trust by being honest, open, and forthright about the challenges and rewards of being a growing organization.
- Be a good listener—find out what employees are thinking and facing.
- Be willing to delegate duties.
- Be flexible—be willing to change your plans if necessary.
- Provide consistent and regular feedback by letting employees know the outcomes—good and bad.
- Reinforce the contributions of each person by recognizing employees' efforts.
- Continually train employees to enhance their capabilities and skills.
- Maintain the focus on the venture's mission even as it grows.
- Establish and reinforce a "we" spirit since a successful growing venture takes the coordinated efforts of all the employees.

Managing Downturns

Organizational growth is a desirable and important goal for entrepreneurial ventures, but what happens when things don't go as planned—when the growth strategies don't result in the intended outcomes and, in fact, result in a decline in performance? There are challenges, as well, in managing the downturns.

Nobody likes to fail, especially entrepreneurs. However, when an entrepreneurial venture faces times of trouble, what can be done? How can downturns be managed successfully? The first step is recognizing that a crisis is brewing.

Recognizing Crisis Situations. An entrepreneur should be alert to the warning signs of a business in trouble. Some signals of potential performance decline include inadequate or negative cash flow, excess number of employees, unnecessary and cumbersome administrative procedures, fear of conflict and taking risks, tolerance of work incompetence, lack of a clear mission or goals, and ineffective or poor communication within the organization.[63]

Another perspective on recognizing performance declines is the **"boiled frog" phenomenon**, which involves recognizing subtly declining situations.[64] The "boiled frog" is a classic psychological response experiment. In one case, a live frog that's dropped into a boiling pan of water reacts instantaneously and jumps out of the pan. But, in the second case, a live frog that's dropped into a pan of mild water that is gradually heated to the boiling point fails to react and dies. A small firm may be particularly vulnerable to the boiled frog phenomenon because the entrepreneur may not recognize the "water heating up"—that is, the subtly declining situation. When changes in performance are gradual, a serious response may never be triggered or may be done too late for anything to be done about the situation.

So what does the boiled frog phenomenon teach us? It teaches us that entrepreneurs need to be alert to signals that the venture's performance may be worsening. Don't wait until the water has reached the boiling point before you react.

Dealing with Downturns, Declines, and Crises. Although an entrepreneur hopes to never have to deal with organizational downturns, declines, or crises, there may come a time when he or she must do just that. After all, nobody likes to think about things going badly or taking a turn for the worse. But that's exactly what the entrepreneur should do—think about it *before* it happens. (Remember feedforward control from Chapter 17.)[65] It's important to have an up-to-date plan for covering crises, like mapping exit routes from your home in case of a fire. An entrepreneur wants to be prepared before an emergency hits. This plan should focus on providing specific details for controlling the most fundamental and critical aspects of running the venture—cash flow, accounts receivable, costs, and debt. Beyond having a plan for controlling the venture's critical inflows and outflows, other actions would involve identifying specific strategies for cutting costs and restructuring the venture.

Exiting the Venture

Getting out of an entrepreneurial venture may seem to be a strange thing for entrepreneurs to do. However, there may come a point when the entrepreneur decides it's time to move on. That decision may be based on the fact that the entrepreneur hopes to capitalize financially on the investment in the venture—called **harvesting**—or on the fact that the entrepreneur is facing serious organizational performance problems and wants to get out, or even on the entrepreneur's desire to focus on other pursuits (personal or business). The issues involved with exiting the venture include choosing a proper business valuation method and knowing what's involved in the process of selling a business.

Business Valuation Methods. Valuation techniques generally fall into three categories: (1) asset valuations, (2) earnings valuations, and (3) cash flow valuations.[66] Setting a value on a business can be a little tricky. In many cases, the entrepreneur has sacrificed much for the business and sees it as his or her "baby." Calculating the value of the baby based on objective standards such as cash flow or some multiple of net profits can sometimes be a shock. That's why it's important for an entrepreneur who wishes to exit a venture to get a comprehensive business valuation prepared by professionals.

Other Important Considerations in Exiting a Venture. Although the hardest part of preparing to exit a venture is valuing it, other factors should also be considered.[67] These include being prepared, deciding who will sell the business, considering the tax implications, screening potential buyers, and deciding whether to tell employees before or after the sale. The process of exiting an entrepreneurial venture should be approached as carefully as the process of launching it. If the entrepreneur is selling the venture on a positive note, he or she wants to realize the value built up in the business. If the venture is being exited because of declining performance, the entrepreneur wants to maximize the potential return.

Managing Personal Life Choices and Challenges

Being an entrepreneur is extremely exciting and fulfilling yet extremely demanding. There are long hours, difficult demands, and high stress. Yet there are many rewards to being an entrepreneur as well. In this section, we'll look at how entrepreneurs can make it work—that is, how they can be successful and effectively balance the demands of their work and personal lives.[68]

Entrepreneurs are a special group. They are focused, persistent, hardworking, and intelligent. Because they put so much of themselves into launching and growing their entrepreneurial ventures, many may neglect their personal lives. Entrepreneurs often have to make sacrifices to pursue their entrepreneurial dreams. However, they can make it work. They can balance their work and personal lives. But how?

One of the most important things an entrepreneur can do is *become a good time manager.* They can prioritize what needs to be done and use a planner (daily, weekly, monthly) to help schedule priorities. Some entrepreneurs don't like taking the time to plan or prioritize, or they think it's a ridiculous waste of time. Yet identifying the important duties and distinguishing them from those that aren't so important makes an entrepreneur more efficient and effective. In addition, part of being a good time manager is delegating to trusted employees the decisions and actions the entrepreneur doesn't have to be personally involved in. Although it may be difficult to let go of some of the things they've always done, entrepreneurs who delegate effectively will see their personal productivity levels rise.

Another suggestion for finding that balance is to *seek professional advice* in areas of business where it's needed. Although entrepreneurs may be reluctant to spend scarce cash, the time, energy, and potential problems saved in the long run are well worth the investment. Competent professional advisers can provide entrepreneurs with information to make more intelligent decisions. Also, it's important to *deal with conflicts* as they arise. This includes both workplace and family conflicts. If an entrepreneur doesn't deal with conflicts, negative feelings are likely to crop up and lead to communication breakdowns. When communication falls apart, vital information may get lost, and people (both employees and family members) may start to assume the worst. It can turn into a nightmare situation that feeds on itself. The best strategy is to deal with conflicts as they come up. An entrepreneur should talk, discuss, and argue (if necessary), but he or she shouldn't avoid the conflict or pretend it doesn't exist.

Another suggestion for achieving that balance between work and personal life is to *develop a network of trusted friends and peers.* Having a group of people to talk with is a good way for an entrepreneur to think through problems and issues. The support and encouragement offered by these people can be an invaluable source of strength for an entrepreneur.

"boiled frog" phenomenon
A perspective on recognizing performance declines that suggests watching out for subtly declining situations.

harvesting
Exiting a venture when an entrepreneur hopes to capitalize financially on the investment in the venture.

Finally, an entrepreneur needs to *recognize when his or her stress levels are too high.* Entrepreneurs *are* achievers. They like to make things happen. They thrive on working hard. But too much stress can lead to significant physical and emotional problems (as discussed in Chapter 12). Entrepreneurs have to learn when stress is overwhelming them and to do something about it. After all, what's the point of growing and building a thriving entrepreneurial venture if you're not around to enjoy it?

QUICK LEARNING REVIEW

- Describe how entrepreneurs should plan, organize, and control growth.
- Describe the boiled frog phenomenon and why it's useful for entrepreneurs.
- Discuss the issues an entrepreneur needs to consider when deciding whether to exit an entrepreneurial venture.

Thinking About Entrepreneurship Issues

1. What do you think would be the most difficult thing about being an entrepreneur? What do you think would be the most fun thing?
2. How does the concept of social entrepreneurship (see Chapter 5) relate to entrepreneurs and entrepreneurial ventures?
3. Would a good manager be a good entrepreneur? Discuss.
4. Why do you think many entrepreneurs find it difficult to step aside and let others manage their business?
5. Do you think a person can be taught to be an entrepreneur? Why or why not?
6. What do you think it means to be a successful entrepreneurial venture? How about a successful entrepreneur?

Appendix B

Planning Tools and Techniques

We'll discuss three categories of basic planning tools and techniques in this appendix: techniques for assessing the environment, techniques for allocating resources, and contemporary planning techniques.

TECHNIQUES FOR ASSESSING THE ENVIRONMENT

Leigh Knopf, former senior manager for strategic planning at the American Institute of Certified Public Accountants (AICPA), says that many larger accounting firms have set up external analysis departments to "study the wider environment in which they, and their clients, operate." These organizations have recognized that, "what happens in India in today's environment may have an impact on an American accounting firm in North Dakota."[1] In our description of the strategic management process in Chapter 8, we discussed the importance of assessing an organization's environment. Three techniques help managers do that: environmental scanning, forecasting, and benchmarking.

Environmental Scanning

How important is environmental scanning? While looking around on competitor Google's company Web site, Bill Gates found a help-wanted page with descriptions of all the open jobs. What piqued his interest was that many of these posted job qualifications were identical to Microsoft's job requirements. He began to wonder why Google—a Web search company—would be posting job openings for software engineers with backgrounds that "had nothing to do with Web searches and everything to do with Microsoft's core business of operating-system design, compiler optimization, and distributed-systems architecture." Gates e-mailed an urgent message to some of his top executives, saying that Microsoft had better be on its toes because it sure looked like Google was preparing to move into being more of a software company.[2]

How can managers become aware of significant environmental changes, such as Tata Motors responding to the need to have a low cost and safe transport for a family of four also decided to make small cars; a new law in Germany permitting shopping for "tourist items" on Sunday; the increased trend of counterfeit consumer products in South Africa; the precipitous decline in the working-age populations in Japan, Germany, Italy, and Russia; or the decrease in family size in Mexico? Managers in both small and large organizations use **environmental scanning,** which is the screening of large amounts of information to anticipate and interpret changes in the environment. Extensive environmental scanning is likely to reveal issues and concerns that could affect an organization's current or planned activities. Research has shown that companies that use environmental scanning have higher performance.[3] Organizations that don't keep on top of environmental changes are likely to experience the opposite!

Competitor Intelligence. A fast-growing area of environmental scanning is **competitor intelligence.**[4] It's a process by which organizations gather information about their competitors and get answers to questions such as Who are they? What are they doing? How will what they're doing affect us? Let's look at an example of how one organization used competitor intelligence in its planning. Dun & Bradstreet (D&B), a leading

environmental scanning
The screening of large amounts of information to anticipate and interpret changes in the environment.

competitor intelligence
An environmental scanning activity by which organizations gather information about competitors.

provider of business information, has an active business intelligence division. The division manager received a call from an assistant vice president for sales in one of the company's geographic territories. This person had been on a sales call with a major customer, and the customer mentioned in passing that another company had visited and made a major presentation about its services. What was interesting was that, although D&B had plenty of competitors, this particular company wasn't one of them. The manager gathered together a team that sifted through dozens of sources (research services, Internet, personal contacts, and other external sources) and quickly became convinced that there was something to this, that this company was "aiming its guns right at us." Managers at D&B jumped into action to develop plans to counteract this competitive attack.[5]

Competitor intelligence experts suggest that 80 percent of what managers need to know about competitors can be found out from their own employees, suppliers, and customers.[6] Competitor intelligence doesn't have to involve spying. Advertisements, promotional materials, press releases, reports filed with government agencies, annual reports, want ads, newspaper reports, and industry studies are examples of readily accessible sources of information. Attending trade shows and debriefing the company's sales force can be other good sources of competitor information. Many firms regularly buy competitors' products and have their own engineers study them (through a process called *reverse engineering*) to learn about new technical innovations. In addition, the Internet has opened up vast sources of competitor intelligence, as many corporate Web pages include new product information and other press releases.

Managers need to be careful about the way competitor information is gathered to prevent any concerns about whether it's legal or ethical. For instance, at Procter & Gamble, executives hired competitive intelligence firms to spy on its competitors in the hair-care business. At least one of these firms misrepresented themselves to competitor Unilever's employees, trespassed at Unilever's hair-care headquarters in Chicago, and went through trash dumpsters to gain information. When P&G's CEO found out, he immediately fired the individuals responsible and apologized to Unilever.[7] Competitor intelligence becomes illegal corporate spying when it involves the theft of proprietary materials or trade secrets by any means. The Economic Espionage Act makes it a crime in the United States to engage in economic espionage or to steal a trade secret.[8] Difficult decisions about competitive intelligence arise because often there's a fine line between what's considered *legal and ethical* and what's considered *legal but unethical.* Although the top manager at one competitive intelligence firm contends that 99.9 percent of intelligence gathering is legal, there's no question that some people or companies will go to any lengths—some unethical—to get information about competitors.[9]

Global Scanning. One type of environmental scanning that's particularly important is global scanning. Because world markets are complex and dynamic, managers have expanded the scope of their scanning efforts to gain vital information on global forces

thinking critically about Ethics

Here are some techniques that have been suggested for gathering competitor information: (1) Get copies of lawsuits and civil suits that may have been filed against competitors. These court proceedings are public records and can expose surprising details. (2) Call the Better Business Bureau and ask if competitors have had complaints filed against them because of fraudulent product claims or questionable business practices. (3) Pretend to be a journalist and call competitors to ask questions. (4) Get copies of your competitors' in-house newsletters and read them. (5) Buy a single share of competitors' stock so you get the annual report and other information the company sends out. (6) Send someone from your organization to apply for a job at a competitor and have that person ask specific questions. (7) Dig through a competitor's trash.

Which, if any, of these are unethical? Defend your choices. What ethical guidelines would you suggest for competitor intelligence activities?

that might affect their organizations.[10] The value of global scanning to managers, of course, is largely dependent on the extent of the organization's global activities. For a company that has significant global interests, global scanning can be quite valuable. For instance, Sealed Air Corporation of Saddle Brook, New Jersey—you've probably seen Bubble Wrap, its most popular product—tracks global demographic changes. It found that as countries move from agriculture-based societies to industrial ones, the population tends to eat out more and favor prepackaged foods, which means more sales of its food-packaging products.[11]

Because the sources that managers use for scanning the domestic environment are too limited for global scanning, managers must globalize their perspectives. For instance, they can subscribe to information-clipping services that review world newspapers and business periodicals and provide summaries of desired information. Also, numerous electronic services provide topic searches and automatic updates in global areas of special interest to managers.

Forecasting

The second technique managers can use to assess the environment is forecasting. Forecasting is an important part of planning, and managers need forecasts that will help them to predict future events effectively and in a timely manner. Environmental scanning establishes the basis for **forecasts**, which are predictions of outcomes. Virtually any component in an organization's environment can be forecasted. Let's look at how managers forecast and how effective forecasts are.

Forecasting Techniques. Forecasting techniques fall into two categories: quantitative and qualitative. **Quantitative forecasting** applies a set of mathematical rules to a series of past data to predict outcomes. These techniques are preferred when managers have sufficient hard data that can be used. **Qualitative forecasting**, in contrast, uses the judgment and opinions of knowledgeable individuals to predict outcomes. Qualitative techniques typically are used when precise data are limited or difficult to obtain. Exhibit B–1 describes some popular forecasting techniques.

Today, many organizations collaborate on forecasts, using an approach known as CPFR, which stands for collaborative planning, forecasting, and replenishment.[12] CPFR provides a framework for the flow of information, goods, and services between retailers and manufacturers. Each organization relies on its own data to calculate a demand forecast for a particular product. If their respective forecasts differ by a certain amount (say 10 percent), the retailer and manufacturer exchange data and written comments until they arrive at a more accurate forecast. Such collaborative forecasting helps both organizations do a better job of planning.

Forecasting Effectiveness. The goal of forecasting is to provide managers with information that will facilitate decision making. Despite forecasting's importance to planning, managers have had mixed success with it.[13] For instance, prior to a holiday weekend at the Procter & Gamble factory in Lima, Ohio, managers were preparing to shut down the facility early so they wouldn't have to pay employees for just sitting around and to give them some extra time off. The move seemed to make sense because an analysis of purchase orders and historical sales trends indicated that the factory had already produced enough cases of Liquid Tide detergent to meet laundry demand over the holiday. However, managers got a real surprise. One of the company's largest retail customers placed a sizable—and unforeseen—order. They had to reopen the plant, pay the workers overtime, and schedule emergency shipments to meet the retailer's request.[14] As this example shows, managers' forecasts aren't always accurate. In a survey of financial managers in the United States, the United Kingdom, France, and Germany, 84 percent of the respondents said their financial forecasts were inaccurate by 5 percent or more;

forecasts
Predictions of outcomes.

quantitative forecasting
Forecasting that applies a set of mathematical rules to a series of past data to predict outcomes.

qualitative forecasting
Forecasting that uses the judgment and opinions of knowledgeable individuals to predict outcomes.

Exhibit B–1

Forecasting Techniques

Technique	Description	Application
Quantitative		
Time series analysis	Fits a trend line to a mathematical equation and projects into the future by means of this equation	Predicting next quarter's sales on the basis of 4 years of previous sales data
Regression models	Predicts one variable on the basis of known or assumed other variables	Seeking factors that will predict a certain level of sales (for example, price, advertising expenditures)
Econometric models	Uses a set of regression equations to simulate segments of the economy	Predicting change in car sales as a result of changes in tax laws
Economic indicators	Uses one or more economic indicators to predict a future state of the economy	Using change in GNP to predict discretionary income
Substitution effect	Uses a mathematical formula to predict how, when, and under what circumstances a new product or technology will replace an existing one	Predicting the effect of DVD players on the sale of VHS players
Qualitative		
Jury of opinion	Combines and averages the opinions of experts	Polling the company's human resource managers to predict next year's college recruitment needs
Sales force composition	Combines estimates from field sales personnel of customers' expected purchases	Predicting next year's sales of industrial lasers
Customer evaluation	Combines estimates from established customers' purchases	Surveying major car dealers by a car manufacturer to determine types and quantities of products desired

54 percent of the respondents reported inaccuracy of 10 percent or more.[15] But it is important to try to make forecasting as effective as possible because research shows that a company's forecasting ability can be a distinctive competence.[16] Here are some suggestions for making forecasting more effective:[17]

- Understand that forecasting techniques are most accurate when the environment is not rapidly changing. The more dynamic the environment, the more likely managers are to forecast ineffectively. Also, forecasting is relatively ineffective in predicting non-seasonal events such as recessions, unusual occurrences, discontinued operations, and the actions and reactions of competitors.
- Use simple forecasting methods. They tend to do as well as, and often better than, complex methods that may mistakenly confuse random data for meaningful information. For instance, at St. Louis–based Emerson Electric, chairman emeritus Chuck Knight found that forecasts developed as part of the company's planning process indicated that the competition wasn't just domestic anymore, but global. He didn't use any complex mathematical techniques to come to this conclusion but instead relied on the information already collected as part of his company's planning process.
- Look at involving more people in the process. At *Fortune* 100 companies, it's not unusual to have 1,000 to 5,000 managers providing forecasting input. These businesses are finding that the more people that are involved in the process, the more they can improve the reliability of the outcomes.[18]
- Compare every forecast with "no change." A no change forecast is accurate approximately half the time.

- Use *rolling* forecasts that look 12 to 18 months ahead instead of using a single, static forecast. These types of forecasts can help managers spot trends better and help their organizations be more adaptive in changing environments.[19]
- Don't rely on a single forecasting method. Make forecasts with several models and average them, especially when making longer-range forecasts.
- Don't assume that you can accurately identify turning points in a trend. What is typically perceived as a significant turning point often turns out to be simply a random event.
- Remember that forecasting *is* a managerial skill and can therefore be practiced and improved. Forecasting software has made the task somewhat less mathematically challenging, although the "number crunching" is only a small part of the activity. Interpreting the forecast and incorporating that information into planning decisions is the challenge facing managers.

Benchmarking

Suppose that you're a talented pianist or gymnast. To improve, you want to learn from the best, so you watch outstanding musicians or athletes for motions and techniques they use as they perform. That's what is involved in **benchmarking**, the search for the best practices among competitors or noncompetitors that lead to their superior performance.[20] Does benchmarking work? Studies show that users have achieved 69 percent faster growth and 45 percent greater productivity.[21]

The basic idea behind benchmarking is that managers can improve performance by analyzing and then copying the methods of the leaders in various fields. Organizations such as Nissan, Ranbaxy, ACC, TVS Motors, Tata Steel, and Volvo Construction Equipment have used benchmarking as a tool in improving performance. In fact, some companies have chosen some pretty unusual benchmarking partners! IBM studied Las Vegas casinos for ways to discourage employee theft. Many hospitals have benchmarked their admissions processes against registration processes of Marriott Hotels. And Giordano Holdings Ltd., a Hong Kong–based manufacturer and retailer of mass-market casual wear, borrowed its "good quality, good value" concept from Marks & Spencer, used Limited Brands to benchmark its point-of-sales computerized information system, and modeled its simplified product offerings on the McDonald's menu.[22]

What does benchmarking involve? Exhibit B–2 illustrates the four steps typically used in benchmarking.

Exhibit B–2

Steps in Benchmarking

Source: Based on Y. K. Shetty, "Aiming High: Competitive Benchmarking for Superior Performance," *Long Range Planning*, February 1993, p. 42.

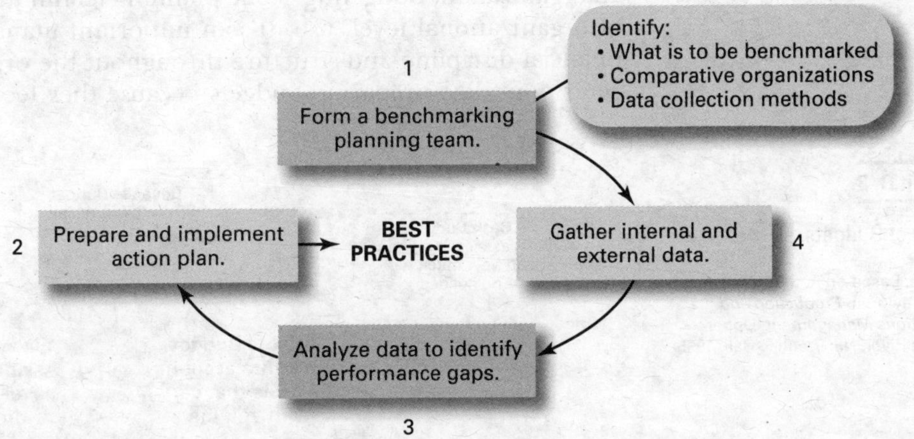

benchmarking
The search for the best practices among competitors or noncompetitors that lead to their superior performance.

QUICK LEARNING REVIEW

- List the different approaches to assessing the environment.
- Explain what competitor intelligence is and ways that managers can do it legally and ethically.
- Describe how managers can improve the effectiveness of forecasting.
- List the steps in the benchmarking process.

TECHNIQUES FOR ALLOCATING RESOURCES

Once an organization's goals have been established, it's important to determine how those goals are going to be accomplished. Before managers can organize and lead as goals are implemented, they must have **resources**, which are the assets of the organization (financial, physical, human, and intangible). How can managers allocate these resources effectively and efficiently so that organizational goals are met? Although managers can choose from a number of techniques for allocating resources (many of which are covered in courses on accounting, finance, and operations management), we'll discuss four techniques here: budgeting, scheduling, break-even analysis, and linear programming.

Budgeting

Most of us have had at least some experience with budgets. You probably learned at a very early age that unless you allocated your "revenues" carefully, your weekly allowance was spent on "expenses" before the week was half over.

A **budget** is a numerical plan for allocating resources to specific activities. Managers typically prepare budgets for revenues, expenses, and large capital expenditures such as equipment. It's not unusual, though, for budgets to be used for improving time, space, and use of material resources. These types of budgets substitute non-dollar numbers for dollar amounts. Such items as person-hours, capacity utilization, or units of production can be budgeted for daily, weekly, or monthly activities. Exhibit B–3 describes the different types of budgets that managers might use.

Why are budgets so popular? Probably because they're applicable to a wide variety of organizations and work activities within organizations. We live in a world in which almost everything is expressed in monetary units. Dollars, rupees, pesos, euros, yuan, yen, and the like are used as common measuring units within a country. That's why monetary budgets are a useful tool for allocating resources and guiding work in such diverse departments as manufacturing and information systems or at various levels in an organization. Budgeting is one planning technique that most managers, regardless of organizational level, use. It's an important managerial activity because it forces financial discipline and structure throughout the organization. However, many managers don't like preparing budgets because they feel the process is time-consuming,

Exhibit B–3

Types of Budgets

Source: Based on R. S. Russell and B. W. Taylor III, *Production and Operations Management* (Upper Saddle River, NJ: Prentice Hall, 1995), p. 287.

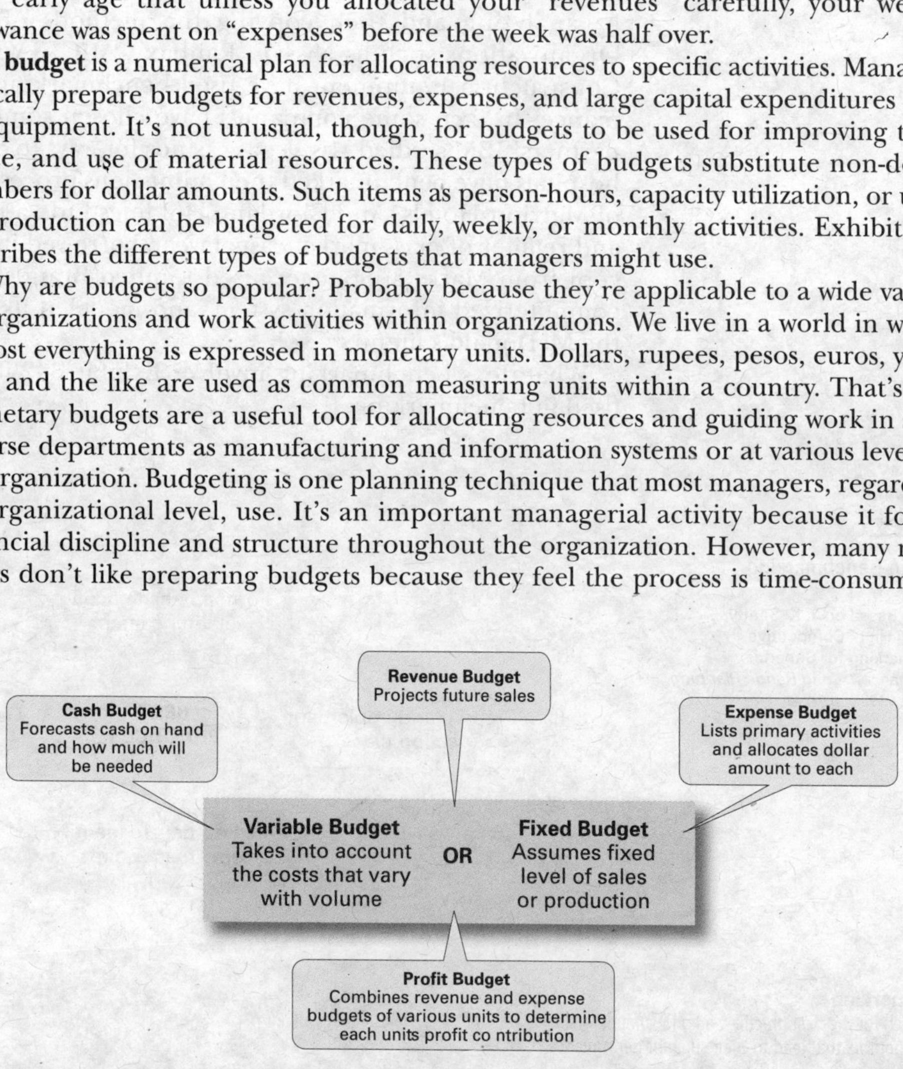

Exhibit B–4

How to Improve Budgeting

- Collaborate and communicate.
- Be flexible.
- Goals should drive budgets—budgets should not determine goals.
- Coordinate budgeting throughout the organization.
- Use budgeting/planning software when appropriate.
- Remember that budgets are tools.
- Remember that profits result from smart management, not because you budgeted for them.

inflexible, inefficient, and ineffective.[23] How can the budgeting process be improved? Exhibit B–4 provides some suggestions. Organizations such as Texas Instruments, IKEA, Max New York Life Insurance, Volvo, and Samsonite have incorporated several of these suggestions as they've revamped their budgeting processes.

Scheduling

Rohini is a manager at a retail store in Gurgaon. Every week, she determines employees' work hours and the store area where each employee will be working. If you observed any other group of supervisors or department managers for a few days, you would see them doing much the same as Rohini—allocating resources by detailing what activities have to be done, the order in which they are to be completed, who is to do each, and when they are to be completed. These managers are **scheduling**. In this section, we'll review some useful scheduling devices, including Gantt charts, load charts, and PERT network analysis.

Gantt Charts. The **Gantt chart** was developed in the early 1900s by Henry Gantt, an associate of Frederick Taylor, the scientific management expert. The idea behind a Gantt chart is simple: It's essentially a bar graph with time on the horizontal axis and the activities to be scheduled on the vertical axis. The bars show output, both planned and actual, over a period of time. A Gantt chart visually shows when tasks are supposed to be done and compares that with the actual progress on each. It's a simple but important device that lets managers easily detail what has yet to be done to complete a job or project and to assess whether an activity is ahead of, behind, or on schedule.

Exhibit B–5 shows a simplified Gantt chart for book production developed by a manager in a publishing company. Time is expressed in months across the top of the

Exhibit B–5

A Gantt Chart

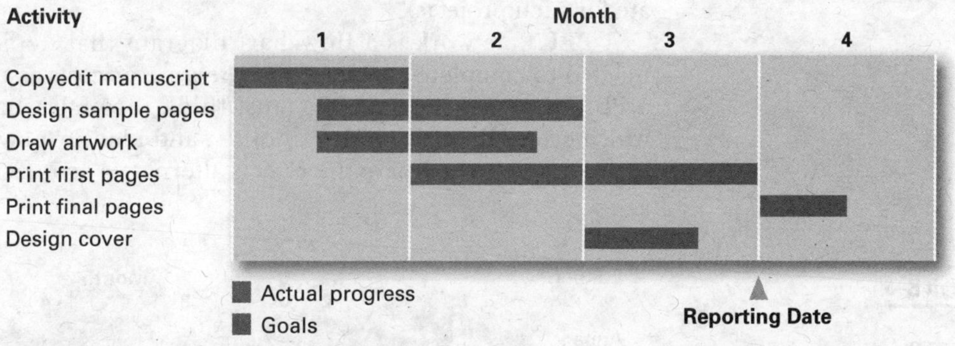

resources
The assets of an organization, including financial, physical, human, intangible, and structural/cultural assets.

budget
A numerical plan for allocating resources to specific activities.

scheduling
Detailing what activities have to be done, the order in which they are to be completed, who is to do each, and when they are to be completed.

Gannt chart
A scheduling chart developed by Henry Gantt that shows actual and planned output over a period of time.

chart. The major work activities are listed on the left side. Planning involves deciding what activities need to be done to get the book finished, the order in which those activities need to be completed, and the time that should be allocated to each activity. Where a box sits within a time frame reflects its planned sequence. The shading represents actual progress. The chart also serves as a control tool because the manager can see deviations from the plan. In this example, both the design of the cover and the printing of first pages are running behind schedule. Cover design is about three weeks behind (note that there has been no actual progress—shown by the blue line—as of the reporting date), and galley proof printing is about two weeks behind schedule (note that as of the report date, actual progress—shown by the blue line—is about six weeks, out of a goal of completing in two months). Given this information, the manager might need to take some action to either make up for the two lost weeks or to ensure that no further delays will occur. At this point, the manager can expect that the book will be published at least two weeks later than planned if no action is taken.

Load Charts. A **load chart** is a modified Gantt chart. Instead of listing activities on the vertical axis, a load chart lists either entire departments or specific resources. This arrangement allows managers to plan and control capacity utilization. In other words, load charts schedule capacity by work areas.

For example, Exhibit B–6 shows a load chart for six production editors at the same publishing company. Each editor supervises the production and design of several books. By reviewing a load chart, the executive editor, who supervises the six production editors, can see who is free to take on a new book. If everyone is fully scheduled, the executive editor might decide not to accept any new projects, to accept new projects and delay others, to make the editors work overtime, or to employ more production editors. As this exhibit shows, only Anurag and Malti are completely scheduled for the next six months. The other editors have some unassigned time and might be able to accept new projects or be available to help other editors who get behind.

PERT Network Analysis. Gantt and load charts are useful as long as the activities being scheduled are few in number and independent of each other. But what if a manager had to plan a large project, such as a departmental reorganization, the implementation of a cost-reduction program, or the development of a new product that required coordinating inputs from marketing, manufacturing, and product design? Such projects require coordinating hundreds and even thousands of activities, some of which must be done simultaneously and some of which can't begin until preceding activities have been completed. If you're constructing a building, you obviously can't start putting up the walls until the foundation is laid. How can managers schedule such a complex project? The Program Evaluation and Review Technique (PERT) is appropriate for such projects.

A **PERT network** is a flowchart diagram that depicts the sequence of activities needed to complete a project and the time or costs associated with each activity. With a PERT network, a manager must think through what has to be done, determine which events depend on one another, and identify potential trouble spots. PERT also makes it easy to compare the effects alternative actions might have on scheduling and

Exhibit B–6

A Load Chart

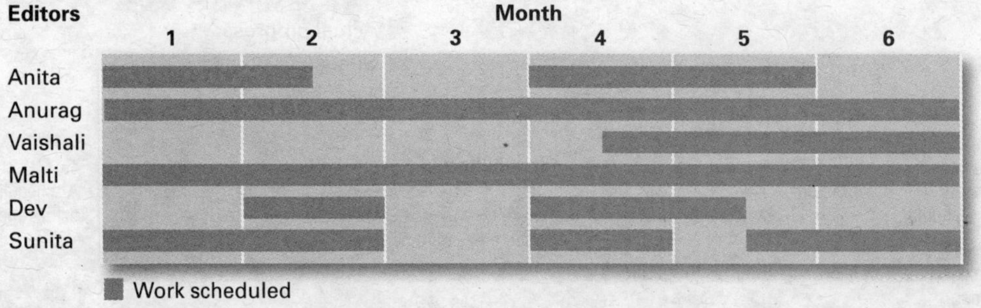

Exhibit B–7

Steps in Developing a PERT Network

1. *Identify every significant activity that must be achieved for a project to be completed.* The accomplishment of each activity results in a set of events or outcomes.
2. *Determine the order in which these events must be completed.*
3. *Diagram the flow of activities from start to finish, identifying each activity and its relationship to all other activities.* Use circles to indicate events and arrows to represent activities. This results in a flowchart diagram called a PERT network.
4. *Compute a time estimate for completing each activity.* This is done with a weighted average that uses an *optimistic* time estimate (t_o) of how long the activity would take under ideal conditions, a *most likely* estimate (t_m) of the time the activity normally should take, and a *pessimistic* estimate (t_p) that represents the time that an activity should take under the worst possible conditions. The formula for calculating the expected time (t_e) is then

$$t_e = \frac{t_o + 4t_m + t_p}{6}$$

5. *Using the network diagram that contains time estimates for each activity, determine a schedule for the start and finish dates of each activity and for the entire project.* Any delays that occur along the critical path require the most attention because they can delay the whole project.

costs. Thus, PERT allows managers to monitor a project's progress, identify possible bottlenecks, and shift resources as necessary to keep a project on schedule.

To understand how to construct a PERT network, you need to know four terms. **Events** are end points that represent the completion of major activities. **Activities** represent the time or resources required to progress from one event to another. **Slack time** is the amount of time an individual activity can be delayed without delaying the whole project. The **critical path** is the longest or most time-consuming sequence of events and activities in a PERT network. Any delay in completing events on the critical path would delay completion of the entire project. In other words, activities on the critical path have zero slack time.

Developing a PERT network requires that a manager identify all key activities needed to complete a project, rank them in order of occurrence, and estimate each activity's completion time. Exhibit B–7 explains the steps in this process.

Most PERT projects are complicated and include numerous activities. Such complicated computations can be done with specialized PERT software. However, let's work through a simple example. Assume that you're the superintendent at a construction company and have been assigned to oversee the construction of an office building. Because

Exhibit B–8

Events and Activities in Constructing an Office Building

Event	Description	Expected Time (in weeks)	Preceding Event
A	Approve design and get permits.	10	None
B	Dig subterranean garage.	6	A
C	Erect frame and siding.	14	B
D	Construct floor.	6	C
E	Install windows.	3	C
F	Put on roof.	3	C
G	Install internal wiring.	5	D, E, F
H	Install elevator.	5	G
I	Put in floor covering and paneling.	4	D
J	Put in doors and interior decorative trim.	3	I, H
K	Turn over to building management group.	1	J

load chart
A modified Gantt chart that schedules capacity by entire departments or specific resources.

PERT network
A flowchart diagram showing the sequence of activities needed to complete a project and the time or cost associated with each.

events
End points that represent the completion of major activities in a PERT network.

activities
The time or resources needed to progress from one event to another in a PERT network.

slack time
The amount of time an individual activity can be delayed without delaying the whole project.

critical path
The longest sequence of activities in a PERT network.

Exhibit B–9

A PERT Network for Constructing an Office Building

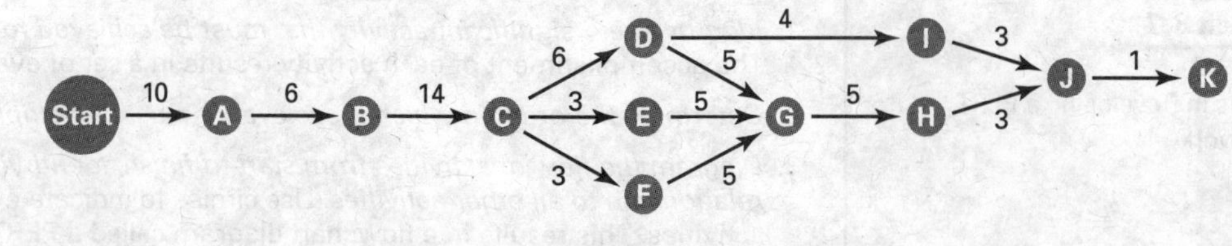

time really is money in your business, you must determine how long it will take to get the building completed. You've determined the specific activities and events. Exhibit B–8 outlines the major events in the construction project and your estimate of the expected time to complete each. Exhibit B–9 shows the actual PERT network based on the data in Exhibit B–8. You've also calculated the length of time that each path of activities will take:

A–B–C–D–I–J–K (44 weeks)
A–B–C–D–G–H–J–K (50 weeks)
A–B–C–E–G–H–J–K (47 weeks)
A–B–C–F–G–H–J–K (47 weeks)

Your PERT network shows that if everything goes as planned, the total project completion time will be 50 weeks. This is calculated by tracing the project's critical path (the longest sequence of activities)—A–B–C–D–G–H–J–K—and adding up the times. You know that any delay in completing the events on this path would delay the completion of the entire project. Taking six weeks instead of four to put in the floor covering and paneling (Event I) would have no effect on the final completion date. Why? Because that event isn't on the critical path. However, taking seven weeks instead of six to dig the subterranean garage (Event B) would likely delay the total project. A manager who needed to get back on schedule or to cut the 50-week completion time would want to concentrate on the activities along the critical path that could be completed faster. How might the manager do this? He or she could determine whether any of the other activities *not* on the critical path had slack time in which resources could be transferred to activities that *were* on the critical path.

Break-Even Analysis

Managers at Dabur want to know how many units of their newly packed Chyawanprash must be sold in order to break even—that is, the point at which total revenue is just sufficient to cover total costs. **Break-even analysis** is a widely used resource allocation technique that helps managers determine break-even point.[24]

Break-even analysis is a simple calculation, yet it's valuable to managers because it points out the relationship between revenues, costs, and profits. To compute the break-even point (*BE*), a manager needs to know the unit price of the product being sold (*P*), the variable cost per unit (*VC*), and total fixed costs (*TFC*). An organization breaks even when its total revenue is just enough to equal its total costs. But total cost has two parts: fixed and variable. *Fixed costs* are expenses that do not change regardless of volume. Examples include insurance premiums, rent, and property taxes. *Variable costs* change in proportion to output and include raw materials, labor costs, and energy costs.

Break-even point can be computed graphically or by using the following formula:

$$BE = \frac{TFC}{P - VC}$$

This formula tells us that (1) total revenue will equal total cost when we sell enough units at a price that covers all variable unit costs and (2) the difference between price and variable costs, when multiplied by the number of units sold, equals the fixed costs. Let's work through an example.

Assume that Pinky Xerox charges 75 paisa per photocopy. If fixed costs are Rs 40,000 per year and variable costs are 30 paisa per copy, Mr Shah can compute his break-even point as follows: 40,000 ÷ (0.75 − 0.35) = 1,00,000 copies, or when

Exhibit B–10

Breakeven Analysis

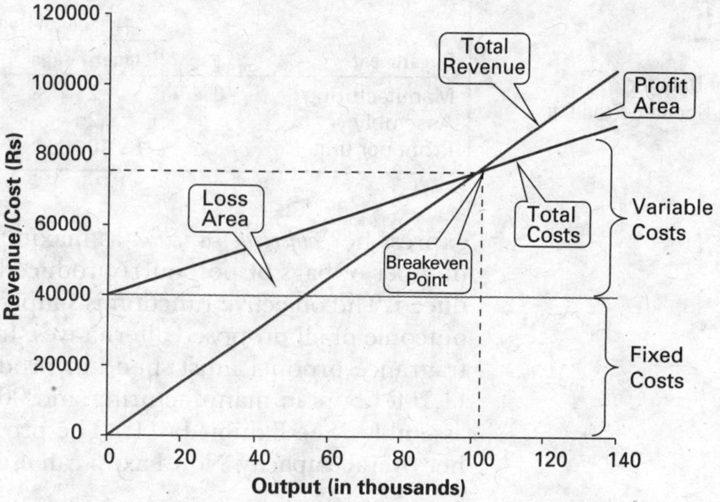

annual revenues are Rs 75,000 (1,00,000 × 0.75). This same relationship is shown graphically in Exhibit B–10.

As a planning tool, break-even analysis could help Mr Shah set his sales goal. For example, he could determine his profit goal and then calculate what sales level is needed to reach that goal. Break-even analysis could also tell Mr Shah how much volume has to increase to break even if he's currently operating at a loss or how much volume he can afford to lose and still break even.

Linear Programming

Laxmi Murty manages a manufacturing plant that produces two kinds of cinnamon-scented home fragrance products: wax candles and a woodchip potpourri sold in bags. Business is good, and she can sell all the products she can produce. This is her problem: Given that the bags of potpourri and the wax candles are manufactured in the same facility, how many of each product should she produce to maximize profits? Laxmi can use **linear programming** to solve her resource allocation problem.

Although linear programming can be used here, it can't be applied to all resource allocation problems because it requires that there be limited resources, that the goal be outcome optimization, that there be alternative ways of combining resources to produce a number of output mixes, and that there be a linear relationship between variables (a change in one variable must be accompanied by an exactly proportional change in the other).[25] For Laxmi's business, that last condition would be met if it took exactly twice the amount of raw materials and hours of labor to produce two of a given home fragrance product as it took to produce one.

What kinds of problems can be solved with linear programming? Some applications include selecting transportation routes that minimize shipping costs, allocating a limited advertising budget among various product brands, making the optimal assignment of people among projects, and determining how much of each product to make with a limited number of resources. Let's return to Laxmi's problem and see how linear programming could help her solve it. Fortunately, her problem is relatively simple, so we can solve it rather quickly. For complex linear programming problems, managers can use computer software programs designed specifically to help develop optimizing solutions.

First, we need to establish some facts about Laxmi's business. She has computed the profit margins on her home fragrance products at Rs 10 for a bag of potpourri and Rs 18 for a scented candle. These numbers establish the basis for Laxmi to be able to

break-even analysis
A technique for identifying the point at which total revenue is just sufficient to cover total costs.

linear programming
A mathematical technique that solves resource allocation problems.

Exhibit B–11

Production Data for Cinnamon-Scented Products

Department	Number of Hours Required (per unit)		Monthly Production Capacity (in hours)
	Potpourri Bags	Scented Candles	
Manufacturing	2	4	1,200
Assembly	2	2	900
Profit per unit	Rs 10	Rs 18	

express her *objective function* as maximum profit = Rs 10P + Rs 18S, where P is the number of bags of potpourri produced and S is the number of scented candles produced. The objective function is simply a mathematical equation that can predict the outcome of all proposed alternatives. In addition, Laxmi knows how much time each fragrance product must spend in production and the monthly production capacity (1,200 hours in manufacturing and 900 hours in assembly) for manufacturing and assembly. (See Exhibit B–11.) The production capacity numbers act as *constraints* on her overall capacity. Now Laxmi can establish her constraint equations:

$$2P + 4S \leq 1{,}200$$

$$2P + 2S \leq 900$$

Of course, Laxmi can also state that $P \geq 0$ and $S \geq 0$ because neither fragrance product can be produced in a volume less than zero.

Laxmi has graphed her solution in Exhibit B–12. The shaded area represents the options that don't exceed the capacity of either department. What does this mean? Well, let's look first at the manufacturing constraint line, BE. We know that total manufacturing capacity is 1,200 hours, so if Laxmi decides to produce all potpourri bags, the maximum she can produce is 600 (1,200 hours ÷ 2 hours required to produce a bag of potpourri). If she decides to produce all scented candles, the maximum she can produce is 300 (1,200 hours ÷ 4 hours required to produce a scented candle). The other constraint Laxmi faces is assembly, shown by line DF. If Laxmi decides to produce all potpourri bags, the maximum she can assemble is 450 (900 hours production capacity ÷ 2 hours required to assemble). Likewise, if Laxmi decides to produce all scented candles, the maximum she can assemble is also 450 because the scented candles also take 2 hours to assemble. The constraints imposed by these capacity limits establish Laxmi's *feasibility region.* Her optimal resource allocation will be defined at one of the corners within this feasibility region. Point C provides the maximum profits within the constraints stated. How do we know? At point A, profits would be 0 (no production of either potpourri bags or scented candles). At point B, profits would be Rs 5,400 (300 scented candles × Rs 18 profit and 0 potpourri bags produced = Rs 5,400). At point D, profits would be Rs 4,500 (450 potpourri bags produced × Rs 10 profit and 0 scented candles produced = Rs 4,500). At point C, however, profits would be Rs 5,700 (150 scented candles produced × Rs 18 profit and 300 potpourri bags produced × Rs 10 profit = Rs 5,700).

Exhibit B–12

Graphical Solution to Linear Programming Problem

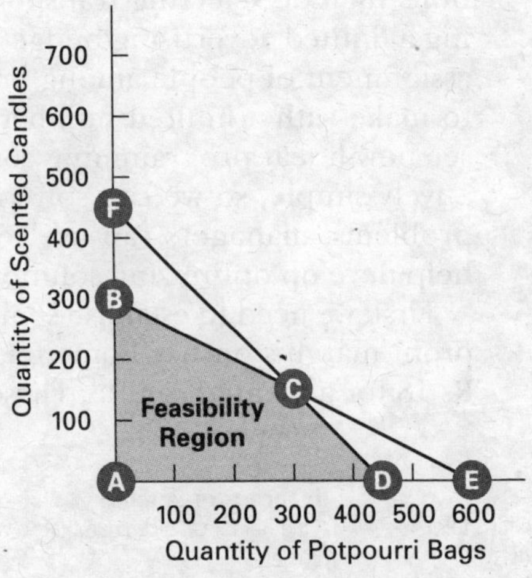

QUICK LEARNING REVIEW

- List the four techniques for allocating resources.
- Describe the different types of budgets.
- Tell what a Gantt chart and a load chart do.
- Describe how PERT network analysis works.
- Understand how to compute a break-even point.
- Describe how managers can use linear programming.

CONTEMPORARY PLANNING TECHNIQUES

Wi-Fi and WiMAX applications. Record oil prices. Bird flu pandemic. Chemical/biological attacks. Recession/inflation worries. Category 4 or 5 hurricanes. Changing competition. Today's managers face the challenges of planning in an environment that's both dynamic and complex. Two planning techniques that are appropriate for this type of environment are project management and scenarios. Both techniques emphasize *flexibility*, something that's important to making planning more effective and efficient in this type of organizational environment.

Project Management

Different types of organizations, from manufacturers such as Coleman and Boeing to software design firms such as SAS and Microsoft, use projects. A **project** is a one-time-only set of activities that has definite beginning and end points in time.[26] Projects vary in size and scope—from Bangalore's Metro Project to picnic party of friends. **Project management** is the task of getting a project's activities done on time, within budget, and according to specifications.[27]

More and more organizations are using project management because the approach fits well with the need for flexibility and rapid response to perceived market opportunities. When organizations undertake projects that are unique, have specific deadlines, contain complex interrelated tasks requiring specialized skills, and are temporary in nature, these projects often do not fit into the standardized planning procedures that guide an organization's other routine work activities. Instead, managers can use project management techniques to effectively and efficiently accomplish the project's goals. What does the project management process involve?

Project Management Process. In a typical project, work is done by a project team whose members are assigned from their respective work areas to the project and who report to a project manager. The project manager coordinates the project's activities with other departments. When the project team accomplishes its goals, it disbands, and members move on to other projects or back to their permanent work area.

The essential features of the project planning process are shown in Exhibit B–13. The process begins by clearly defining the project's goals. This step is necessary because the manager and the team members need to know what's expected. All activities in the pro-

Exhibit B–13 Project Planning Process

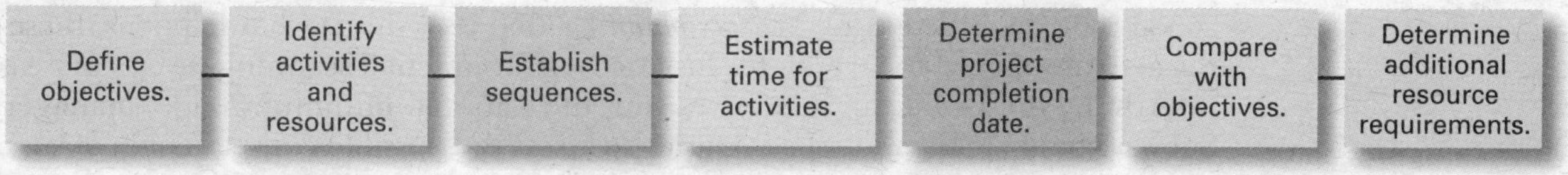

Source: Based on R. S. Russell and B. W. Taylor III, *Production and Operations Management* (Upper Saddle River, NJ: Prentice Hall, 1995), p. 287.

project
A one-time-only set of activities that has definite beginning and end points in time.

project management
The task of getting a project's activities done on time, within budget, and according to specifications.

ject and the resources needed to do them must then be identified. What materials and labor are needed to complete the project? This step may be time-consuming and complex, particularly if the project is unique and the managers have no history or experience with similar projects. Once the activities have been identified, the sequence of completion needs to be determined. What activities must be completed before others can begin? Which can be done simultaneously? This step often uses flowchart diagrams such as a Gantt chart, a load chart, or a PERT network. Next, the project activities need to be scheduled. Time estimates for each activity are created, and these estimates are used to develop an overall project schedule and completion date. Then the project schedule is compared to the goals, and any necessary adjustments are made. If the project completion time is too long, the manager might assign more resources to critical activities so they can be completed faster.

Today, the project management process can take place online as a number of Web-based software packages are available. These packages cover aspects from project accounting and estimating to project scheduling and bug and defect tracking.[28]

The Role of the Project Manager. The temporary nature of projects makes managing them different from, say, overseeing a production line or preparing a weekly tally of costs on an ongoing basis. The one-shot nature of the work makes project managers the organizational equivalent of a hired gun. There's a job to be done. It has to be defined—in detail. And the project manager is responsible for how it's done. At J.B. Hunt Transportation Services, the head of project management trains project managers on both technical and interpersonal skills so that they know how to "run a project effectively."[29]

Even with the availability of sophisticated computerized and online scheduling programs and other project management tools, the role of project manager remains difficult because he or she is managing people who typically are still assigned to their permanent work areas. The only real influence project managers have are their communication skills and their power of persuasion. To make matters worse, team members seldom work on just one project. They're usually assigned to two or three at any given time. So project managers end up competing with each other to focus a worker's attention on his or her particular project.

Scenario Planning

During the 1990s, business was so good at Colgate-Palmolive that Chairman Reuben Mark worried about what "might go wrong." He installed an "early-warning system to flag problems before they blew up into company-wrecking crises." For instance, a red-flag report alerted Mark "that officials in Baddi, India, had questions about how a plant treated wastewater." Mark's response was to quickly assign an engineering team to check it out and prevent potential problems.[30]

We already know how important it is that today's managers do what Reuben Mark was doing—monitor and assess the external environment for trends and changes. As they assess the environment, issues and concerns that could affect their organization's current or planned operations are likely to be revealed. All these won't be equally important, so it's usually necessary to focus on a limited set that are most important and to develop scenarios based on each.

A **scenario** is a consistent view of what the future is likely to be. Developing scenarios can also be described as *contingency planning*; that is, if this is what happens, these are the actions we need to take. If, for instance, environmental scanning reveals increasing interest by the U.S. Congress for raising the national minimum wage, managers at Subway could create multiple scenarios to assess the possible consequences of such an action. What would be the implications for its labor costs if the minimum wage were raised to Rs 9 per hour? How about Rs 10 per hour? What effect would these changes have on the chain's bottom line? How might competitors respond? Different assumptions lead to different outcomes. The intent of scenario planning is not to try to predict the future but to reduce uncertainty by playing out potential situations under different specified conditions.[31] Subway could, for example, develop a set of scenarios ranging from optimistic to pessimistic in terms of the minimum wage issue. It would then be prepared to implement new strategies to get and keep a competitive advantage. An expert in scenario planning said, "Just the process of doing scenarios causes executives to rethink and clarify the essence of the business environment in ways they almost certainly have never done before."[32]

Although scenario planning is useful in anticipating events that *can be* anticipated, it's difficult to forecast random events—the major surprises and aberrations that can't be foreseen. For instance, flood in Bihar in August, 2008, was a scenario that could be anticipated. The disaster recovery planning that took place after the flood was effective because this type of scenario had been experienced before. A response had already been planned, and people knew what to do. But the planning challenge comes from those totally random and unexpected events. For instance, the terrorist attacks in Mumbai on November 28, were random, unexpected, and a total shock to many organizations. Scenario planning was of little use because no one could have envisioned this scenario. As difficult as it might be for managers to anticipate and deal with these random events, they're not totally vulnerable to the consequences. One suggestion that has been identified by risk experts as particularly important is to have an early warning system in place. (A similar idea is the tsunami warning systems in the Pacific and in Alaska, which alert officials to potentially dangerous tsunamis and give them time to take action.) Early warning indicators for organizations can give managers advance notice of potential problems and changes—as in the case of Reuben Mark at Colgate-Palmolive—so managers can take action. Then, managers need to have appropriate responses (plans) in place if these unexpected events occur.

Managers can use planning tools and techniques to prepare confidently for the future. But they should remember that all those described in this chapter are just that—tools—and will never replace the manager's skills and capabilities in using the information gained to be more efficient and effective.

QUICK LEARNING REVIEW

- Explain why flexibility is important to today's planning techniques.
- Describe project management.
- List the steps in the project planning process.
- Discuss why scenario planning is an important planning tool.

Thinking About Management Issues

1. "It's a waste of time and other resources to develop a set of sophisticated scenarios for situations that may never occur." Do you agree or disagree with this statement? Support your position.
2. Do intuition and creativity have any relevance in quantitative planning tools and techniques? Explain.
3. The *Wall Street Journal* and other business periodicals often carry reports of companies that have not met their sales or profit forecasts. What are some reasons a company might not meet its forecast? What suggestions could you make for improving the effectiveness of forecasting?
4. In what ways is managing a project different from managing a department or another structured work area? In what ways are they the same?
5. "People can use statistics to prove whatever they want to prove." What do you think? What are the implications for managers and how they plan?
6. Predicting the future is very difficult, but that hasn't stopped companies from trying. How could managers make it less difficult? Or can they even do so? Explain.
7. What might be some early warning signs of (a) a new competitor coming into your market, (b) an employee work stoppage, or (c) a new technology that could change demand for your product?

scenario
A consistent view of what the future is likely to be.

Endnotes

Chapter 1

1. http://www.archives.digitaltoday.in/businesstoday/20060924/cover1.html
2. http://www.financialexpress.com/news/just-50-women-ceos-in-15-states/142279/
3. http://www.thehindubusinessline.com/2007/03/28/stories/2007032802251000.htm
4. K. A. Tucker and V. Allman, "Don't Be a Cat-and-Mouse Manager," The Gallup Organization, http://gmj.gallup.com, September 9, 2004.
5. "WorkUSA 2004/2005: Effective Employees Drive Financial Results," Watson Wyatt Worldwide, Washington, DC.
6. D. J. Campbell, "The Proactive Employee: Managing Workplace Initiative," *Academy of Management Executive*, August 2000, pp. 52–66.
7. http://www.indiancoffeehouse.com/history.php
8. http://www.theregister.co.uk/2007/12/28/google_privacy_mission_impossible/
9. http://www.herohonda.com/
10. H. Fayol, *Industrial and General Administration* (Paris: Dunod, 1916).
11. For a comprehensive review of this question, see C. P. Hales, "What Do Managers Do? A Critical Review of the Evidence," *Journal of Management*, January 1986, pp. 88–115.
12. J. T. Straub, "Put on Your Manager's Hat," *USA Today* online, www.usatoday.com, October 29, 2002; and H. Mintzberg, *The Nature of Managerial Work* (New York: Harper & Row, 1973).
13. H. Mintzberg and J. Gosling, "Educating Managers Beyond Borders," *Academy of Management Learning and Education*, September 2002, pp. 64–76.
14. See, for example, M. J. Martinko and W. L. Gardner, "Structured Observation of Managerial Work: A Replication and Synthesis," *Journal of Management Studies*, May 1990, pp. 330–357; A. I. Kraut, P. R. Pedigo, D. D. McKenna, and M. D. Dunnette, "The Role of the Manager: What's Really Important in Different Management Jobs," *Academy of Management Executive*, November 1989, pp. 286–293; and C. M. Pavett and A. W. Lau, "Managerial Work: The Influence of Hierarchical Level and Functional Specialty," *Academy of Management Journal*, March 1983, pp. 170–177.
15. Pavett and Lau, "Managerial Work."
16. S. J. Carroll and D. A. Gillen, "Are the Classical Management Functions Useful in Describing Managerial Work?" *Academy of Management Review*, January 1987, p. 48.
17. E. White, "Firms Step Up Training for Front-Line Managers," *Wall Street Journal*, August 27, 2007, p. B3.
18. R. L. Katz, "Skills of an Effective Administrator," *Harvard Business Review*, September–October 1974, pp. 90–102.
19. http://www.knowledge.wharton.upenn.edu/article.cfm?articleid=1439.
20. http://www.financialexpress.com/news/pay-for-performance-companies-now-set-to-roll-it-acrossthe-board/50307/1
21. http://www.rediff.com/money/2007/oct/09pay.htm
22. http://www.infotech.indiatimes.com/articleshow/47665038.cms
23. F. F. Reichheld, "Lead for Loyalty," *Harvard Business Review*, July–August 2001, p. 76.
24. http://www.hclstore.in/
25. Data from *The World Factbook 2008*, www.cia.gov/library/publications/the-world-factbook/index.html, February 22, 2008; and K. A. Eddleston, D. L. Kidder, and B. E. Litzky, "Who's the Boss? Contending with Competing Expectations from Customers and Management," *Academy of Management Executive*, November 2002, pp. 85–95.
26. See, for instance, B. A. Gutek, M. Groth, and B. Cherry, "Achieving Service Success Through Relationships and Enhanced Encounters," *Academy of Management Executive*, November 2002, pp. 132–144; Eddleston, Kidder, and Litzky, "Who's the Boss? Contending with Competing Expectations from Customers and Management"; S. D. Pugh, J. Dietz, J. W. Wiley, and S. M. Brooks, "Driving Service Effectiveness Through Employee–Customer Linkages," *Academy of Management Executive*, November 2002, pp. 73–84; S. D. Pugh, "Service with a Smile: Emotional Contagion in the Service Encounter," *Academy of Management Journal*, October 2001, pp. 1018–1027; W. C. Tsai, "Determinants and Consequences of Employee Displayed Positive Emotions," *Journal of Management*, vol. 27 (4), 2001, pp. 497–512; Naumann and Jackson, Jr., "One More Time: How Do You Satisfy Customers?"; and M. D. Hartline and O. C. Ferrell, "The Management of Customer-Contact Service Employees: An Empirical Investigation," *Journal of Marketing*, October 1996, pp. 52–70.
27. R. A. Hattori and J. Wycoff, "Innovation DNA," *Training and Development*, January 2002, p. 24.
28. R. A. Hattori, "Sometimes Innovation Starts with a Relationship," found on the Innovation Network, www.thinksmart.com, March 14, 2003.
29. R. Wagner, "One Store, One Team at Best Buy," *Gallup Brain*, http://gmj.gallup.com, August 12, 2004.
30. Q. Hardy, "Google Thinks Small," *Forbes*, November 14, 2005, pp. 198–202.
31. J. Sandberg, "Down Over Moving Up: Some New Bosses Find They Hate Their Jobs," *Wall Street Journal*, July 27, 2005, p. B1.

Chapter 2

1. http://www.ethicalcorp.com/content.asp?ContentID=5858&ContTypeID=38
2. C. S. George, Jr., *The History of Management Thought*, 2d ed. (Upper Saddle River, NJ: Prentice Hall, 1972), p. 4.
3. Ibid., pp. 35–41.
4. F. W. Taylor, *Principles of Scientific Management* (New York: Harper, 1911), p. 44. For other information on Taylor,

see S. Wagner-Tsukamoto, "An Institutional Economic Reconstruction of Scientific Management: On the Lost Theoretical Logic of Taylorism," *Academy of Management Review*, January 2007, pp. 105–117; R. Kanigel, *The One Best Way: Frederick Winslow Taylor and the Enigma of Efficiency* (New York: Viking, 1997); and M. Banta, *Taylored Lives: Narrative Productions in the Age of Taylor, Veblen, and Ford* (Chicago: University of Chicago Press, 1993).

5. See for example, F. B. Gilbreth, *Motion Study* (New York: Van Nostrand, 1911); and F. B. Gilbreth and L. M. Gilbreth, *Fatigue Study* (New York: Sturgis and Walton, 1916).

6. H. Fayol, *Industrial and General Administration* (Paris: Dunod, 1916).

7. M. Weber, *The Theory of Social and Economic Organizations*, ed. T. Parsons, trans. A. M. Henderson and T. Parsons (New York: The Free Press, 1947); and M. Lounsbury and E. J. Carberry, "From King to Court Jester? Weber's Fall from Grace in Organizational Theory," *Organization Studies*, vol. 26 (4), 2005, pp. 501–525.

8. N. Zamiska, "Plane Geometry: Scientists Help Speed Boarding of Aircraft," *Wall Street Journal*, November 2, 2005, pp. A1+.

9. http://www.alternativeperspective.blogspot.com/

10. "Face Recognition Software Raises Ethical Dilemma," *Ethics Newsline*, May 14, 2007; and "New Airport Security Scanner Peers Underneath Clothes," *Ethics Newsline*, February 26, 2007.

11. http://www.telegraphindia.com/1060115/asp/nation/story_5722902.asp, retrieved in December 2008.

12. M. Barbaro, "A Long Line for a Shorter Wait at the Supermarket," *New York Times* online, www.nytimes.com, June 23, 2007.

13. http://www.dellbatteryprogram.com, retrieved on December 2008.

14. http://www.speedtex.in/about_us.htm retrieved on December 2008.

15. http://www.bird.in/press_info4.asp

16. E. Mayo, *The Human Problems of an Industrial Civilization* (New York: Macmillan, 1933); and F. J. Roethlisberger and W. J. Dickson, *Management and the Worker* (Cambridge, MA: Harvard University Press, 1939).

17. See for example, G. W. Yunker, "An Explanation of Positive and Negative Hawthorne Effects: Evidence from the Relay Assembly Test Room and Bank Wiring Observation Room Studies," paper presented at the Academy of Management Annual Meeting, August 1993, Atlanta; S. R. Jones, "Was There a Hawthorne Effect?" *American Sociological Review*, November 1992, pp. 451–468; S. R. G. Jones, "Worker Interdependence and Output: The Hawthorne Studies Reevaluated," *American Sociological Review*, April 1990, pp. 176–190; J. A. Sonnenfeld, "Shedding Light on the Hawthorne Studies," *Journal of Occupational Behavior*, April 1985, pp. 111–130; B. Rice, "The Hawthorne Defect: Persistence of a Flawed Theory," *Psychology Today*, February 1982, pp. 70–74; R. H. Franke and J. Kaul, "The Hawthorne Experiments: First Statistical Interpretations," *American Sociological Review*, October 1978, pp. 623–643; and A. Carey, "The Hawthorne Studies: A Radical Criticism," *American Sociological Review*, June 1967, pp. 403–416.

18. K. B. DeGreene, *Sociotechnical Systems: Factors in Analysis, Design, and Management* (Upper Saddle River, NJ: Prentice Hall, 1973), p. 13.

19. Information for this box came from K. Fuson, "iPods Now Double as Study Aids," *USA Today*, March 15, 2006, p. 4D; P. Tyre, "Professor in Your Pocket," *Newsweek*, November 28, 2005; S. Hamm, "Motivating the Troops," *Business Week*, November 21, 2005, pp. 88–103; P. Davidson, "Gadgets Rule on College Campuses," *USA Today*, March 29, 2005, pp. 1B+; and M. J. Tippins and R. S. Sohi, "IT Competency and Firm Performance: Is Organizational Learning a Missing Link?" *Strategic Management Journal*, August 2003, pp. 745–761.

20. http://www.cxotoday.com/cxo/jsp/article.jsp?print=1&article_id=89565&cat_id=1004

Chapter 3

1. http://www.blonnet.com/2008/08/08/stories/2008080851800500.htm; http://www.business-standard.com/india/storypage.php?autono=329490

2. http://www.cricket.zeenews.com/fullstory.aspx?nid=9957; http://www.cricweed.com/blogs/mallya-speaks-blames-dravid-and-charu

3. P. Rozenzweig, "The Halo Effect and Other Managerial Delusions," *The McKinsey Quarterly* online, www.mckinseyquarterly.com, March 9, 2007.

4. For insights into the symbolic view, see "Why CEO Churn Is Healthy," *BusinessWeek*, November 13, 2000, p. 230; S. M. Puffer and J. B. Weintrop, "Corporate Performance and CEO Turnover: The Role of Performance Expectations," *Administrative Science Quarterly*, March 1991, pp. 1–19; C. R. Schwenk, "Illusions of Management Control? Effects of Self-Serving Attributions on Resource Commitments and Confidence in Management," *Human Relations*, April 1990, pp. 333–347; J. R. Meindl and S. B. Ehrlich, "The Romance of Leadership and the Evaluation of Organizational Performance," *Academy of Management Journal*, March 1987, pp. 91–109; J. A. Byrne, "The Limits of Power," *BusinessWeek*, October 23, 1987, pp. 33–35; D. C. Hambrick and S. Finkelstein, "Managerial Discretion: A Bridge Between Polar Views of Organizational Outcomes," in L. L. Cummings and B. M. Staw (eds.), *Research in Organizational Behavior*, vol. 9 (Greenwich, CT: JAI Press, 1987), pp. 369–406; and J. Pfeffer, "Management as Symbolic Action: The Creation and Maintenance of Organizational Paradigms," in L. L. Cummings and B. M. Staw (eds.), *Research in Organizational Behavior*, vol. 3 (Greenwich, CT: JAI Press, 1981), pp. 1–52.

5. T. M. Hout, "Are Managers Obsolete?" *Harvard Business Review*, March–April 1999, pp. 161–168; and Pfeffer, "Management as Symbolic Action."

6. http://www.mindtree.com/

7. http://www.itpeopleindia.com/20040719/management2.shtml (A matter of values WORK CULTURE: MindTree Consulting by Abhinav Singh, Bangalore; Issue dated July 19, 2004)

8. K. Shadur and M. A. Kienzle, "The Relationship Between Organizational Climate and Employee Perceptions of Involvement," *Group & Organization Management*, December 1999, pp. 479–503; M. J. Hatch, "The Dynamics of Organizational Culture," *Academy of Management Review*, October 1993, pp. 657–693; D. R. Denison, "What Is

the Difference Between Organizational Culture and Organizational Climate? A Native's Point of View on a Decade of Paradigm Wars," paper presented at Academy of Management Annual Meeting, Atlanta, GA, 1993; and L. Smircich, "Concepts of Culture and Organizational Analysis," *Administrative Science Quarterly*, September 1983, p. 339.

9. J. A. Chatman and K. A. Jehn, "Assessing the Relationship Between Industry Characteristics and Organizational Culture: How Different Can You Be?" *Academy of Management Journal*, June 1994, pp. 522–553; and C. A. O'Reilly III, J. Chatman, and D. F. Caldwell, "People and Organizational Culture: A Profile Comparison Approach to Assessing Person–Organization Fit," *Academy of Management Journal*, September 1991, pp. 487–516.

10. E. H. Schien, *Organizational Culture and Leadership* (San Francisco: Jossey-Bass, 1985), pp. 314–315.

11. A. E. M. Va Vianen, "Person–Organization Fit: The Match Between Newcomers' and Recruiters' Preferences for Organizational Cultures," *Personnel Psychology*, Spring 2000, pp. 113–149; K. Shadur and M. A. Kienzle, *Group & Organization Management*; P. Lok and J. Crawford, "The Relationship Between Commitment and Organizational Culture, Sub-culture, and Leadership Style," *Leadership & Organization Development Journal*, vol. 20 (6/7), 1999, pp. 365–374; C. Vandenberghe, "Organizational Culture, Person-Culture Fit, and Turnover: A Replication in the Health Care Industry," *Journal of Organizational Behavior*, March 1999, pp. 175–184; and C. Orphen, "The Effect of Organizational Cultural Norms on the Relationships between Personnel Practices and Employee Commitment," *Journal of Psychology*, September 1993, pp. 577–579.

12. See, for example, J. B. Sorensen, "The Strength of Corporate Culture and the Reliability of Firm Performance," *Administrative Science Quarterly*, vol. 47 (1), 2002, pp. 70–91; R. Goffee and G. Jones, "What Holds the Modern Company Together?" *Harvard Business Review*, November–December 1996, pp. 133–148; Collins and Porras, "Building Your Company's Vision," *Harvard Business Review*, September–October 1996, pp. 65–77; J. C. Collins and J. I. Porras, *Built to Last* (New York: HarperBusiness, 1994); G. G. Gordon and N. DiTomaso, "Predicting Corporate Performance from Organizational Culture," *Journal of Management Studies*, November 1992, pp. 793–798; J. P. Kotter and J. L. Heskett, *Corporate Culture and Performance* (New York: The Free Press, 1992), pp. 15–27; and D. R. Denison, *Corporate Culture and Organizational Effectiveness* (New York: Wiley, 1990).

13. Sorensen, "The Strength of Corporate Culture and the Reliability of Firm Performance"; and L. B. Rosenfeld, J. M. Richman, and S. K. May, "Information Adequacy, Job Satisfaction, and Organizational Culture in a Dispersed-Network Organization," *Journal of Applied Communication Research*, vol. 32, 2004, pp. 28–54.

14. S. E. Ante, "The New Blue," *BusinessWeek*, March 17, 2003, p. 82.

15. P. Kafka, "Bean Counter," *Forbes*, February 28, 2005, pp. 78–80; A. Overholt, "Listening to Starbucks," *Fast Company*, July 2004, pp. 50–56; B. Filipczak, "Trained by Starbucks," *Training*, June 1995, pp. 73–79; and S. Gruner, "Lasting Impressions," *Inc.*, July 1998, p. 126.

16. P. Guber, "The Four Truths of the Storyteller," *Harvard Business Review*, December 2007, pp. 53–59; S. Denning, "Telling Tales," *Harvard Business Review*, May 2004, pp. 122–129; T. Terez, "The Business of Storytelling," *Workforce*, May 2002, pp. 22–24; J. Forman, "When Stories Create an Organization's Future," *Strategy & Business*, Second Quarter 1999, pp. 6–9; C. H. Deutsch, "The Parables of Corporate Culture," *New York Times*, October 13, 1991, p. F25; and D. M. Boje, "The Storytelling Organization: A Study of Story Performance in an Office-Supply Firm," *Administrative Science Quarterly*, March 1991, pp. 106–126.

17. E. Ransdell, "The Nike Story? Just Tell It!" *Fast Company*, January–February 2000, pp. 44–46.

18. J. Useem, "Jim McNerney Thinks He Can Turn 3M from a Good Company into a Great One—With a Little Help from His Former Employer, General Electric," *Fortune*, August 12, 2002, pp. 127–132.

19. Denning, "Telling Tales"; and A. M. Pettigrew, "On Studying Organizational Cultures," *Administrative Science Quarterly*, December 1979, p. 576.

20. D. Drickhamer, "Straight to the Heart," *Industry Week*, October 2003, pp. 36–38.

21. http://www.adityabirla.com/careers/recognition.asp

22. E. H. Schein, "Organizational Culture," *American Psychologist*, February 1990, pp. 109–119.

23. M. Zagorski, "Here's the Drill," *Fast Company*, February 2001, p. 58.

24. "Slogans That Work," *Forbes* online, www.forbes.com, January 7, 2008.

25. A. Bryant, "The New Power Breakfast," *Newsweek*, May 15, 2000, p. 52.

26. C. Palmeri, "The Fastest Drill in the West," *BusinessWeek*, October 24, 2005, pp. 86–88.

27. St. Luke's Web site, www.stlukes.co.uk; P. LaBarre, "Success: Here's the Inside Story," *Fast Company*, November 1999, pp. 128–132; and A. Law, *Creative Company: How St. Luke's Became "The Ad Agency to End all Ad Agencies"* (New York: Wiley, 1999).

28. W. M. Bulkeley, "A Data-Storage Titan Confronts Bias Claims," *Wall Street Journal*, September 12, 2007, pp. A1+.

29. A. Raghavan, K. Kranhold, and A. Barrionuevo, "Full Speed Ahead: How Enron Bosses Created a Culture of Pushing Limits," *Wall Street Journal*, August 26, 2002, pp. A1+.

30. J. A. Byrne, et al., "How to Fix Corporate Governance," *BusinessWeek*, May 6, 2002, pp. 68–78.

31. See M. W. Dickson, D. B. Smith, M. W. Grojean, and M. Ehrhart, "An Organizational Climate Regarding Ethics: The Outcome of Leader Values and the Practices That Reflect Them," *Leadership Quarterly*, Summer 2001, pp. 197–217; L. K. Trevino, "A Cultural Perspective on Changing and Developing Organizational Ethics," in W. A. Pasmore and R. W. Woodman (eds.), *Research in Organizational Change and Development*, vol. 4 (Greenwich, CT: JAI Press, 1990); and B. Victor and J. B. Cullen, "The Organizational Bases of Ethical Work Climates," *Administrative Science Quarterly*, March 1988, pp. 101–125.

32. J. A. Byrne, "After Enron: The Ideal Corporation," *BusinessWeek*, August 26, 2002, p. 74.

33. "The World's 50 Most Innovative Companies," *Fast Company*, March 2008, p. 93; T. Kelley and J. Littman, *The Ten Faces of Innovation: IDEO's Strategies for Defeating the Devil's Advocate and Driving Creativity Throughout Your Organization* (New York: Currency, 2005); C. Fredman, "The IDEO Difference," *Hemispheres*, August 2002, pp. 52–57; and T. Kelley and J. Littman, *The Art of Innovation* (New York: Currency, 2001).

34. D. Belkin, "Talent Scouts for Cirque du Soleil Walk a Tightrope," *Wall Street Journal*, September 8/9, 2007, pp. A1+; L. Tischler, "Join the Circus," *Fast Company*, July 2005, pp. 52–58; and "Cirque du Soleil: Creating a Culture of Extraordinary Creativity," Innovation Network, www.thinksmart.com, March 14, 2003.

35. J. Yang and R. W. Ahrens, "Culture Spurs Innovation," *USA Today*, February 25, 2008, p. 1B.

36. L. Simpson, "Fostering Creativity," *Training*, December 2001, p. 56.

37. L. Gary, "Simplify and Execute: Words to Live by in Times of Turbulence," *Harvard Management Update*, January 2003, p. 12.

38. Based on J. McGregor, "Customer Service Champs," *BusinessWeek*, March 3, 2008, pp. 37–57; B. Schneider, et al., "Understanding Organization–Customer Links in Service Settings," *Academy of Management Journal*, December 2006, pp. 1017–1032; B. A. Gutek, M. Groth, and B. Cherry, "Achieving Service Success Through Relationships and Enhanced Encounters," *Academy of Management Executive*, November 2002, pp. 132–144; K. A. Eddleston, D. L. Kidder, and B. E. Litzky, "Who's the Boss? Contending with Competing Expectations from Customers and Management," *Academy of Management Executive*, November 2002, pp. 85–95; S. D. Pugh, J. Dietz, J. W. Wiley, and S. M. Brooks, "Driving Service Effectiveness Through Employee–Customer Linkages," *Academy of Management Executive*, November 2002, pp. 73–84; L. A. Bettencourt, K. P. Gwinner, and M. L. Mueter, "A Comparison of Attitude, Personality, and Knowledge Predictors of Service-Oriented Organizational Citizenship Behaviors," *Journal of Applied Psychology*, February 2001, pp. 29–41; M. D. Hartline, J. G. Maxham III, and D. O. McKee, "Corridors of Influence in the Dissemination of Customer-Oriented Strategy to Customer Contact Service Employees," *Journal of Marketing*, April 2000, pp. 35–50; M. L. Lengnick-Hall and C. A. Lengnick-Hall, "Expanding Customer Orientation in the HR Function," *Human Resource Management*, Fall 1999, pp. 201–214; M. D. Hartline and O. C. Ferrell, "The Management of Customer-Contact Service Employees: An Empirical Investigation," *Journal of Marketing*, October 1996, pp. 52–70; and M. J. Bitner, B. H. Booms, and L. A. Mohr, "Critical Service Encounters: The Employee's Viewpoint," *Journal of Marketing*, October 1994, pp. 95–106.

39. This box is based on Y. Cole, "Holding Managers Accountable for Diversity Success," *DiversityInc.* Special Issue 2006, pp. 14–19; "Diversity Is Important to the Bottom Line," *HR Powerhouse*, www.hrpowerhouse.com, January 21, 2006; P. Rosinski, *Coaching Across Cultures: New Tools for Leveraging National, Corporate, and Professional Differences* (London: Nicholas Brealey Publishing), 2003; "Diversity at the Forefront," *BusinessWeek*, November 4, 2002, pp. 27–38; "Talking to Diversity Experts: Where Do We Go from Here?" *Fortune*, September 30, 2002, pp. 157–172; "Keeping Your Edge: Managing a Diverse Corporate Culture," *Fortune*, June 11, 2001, pp. S1–S18; "Diversity Today," *Fortune*, June 12, 2000, pp. S1–S24; O. C. Richard, "Racial Diversity, Business Strategy, and Firm Performance: A Resource-Based View," *Academy of Management Journal*, April 2000, pp. 164–177; A. Markels, "How One Hotel Manages Staff's Diversity," *Wall Street Journal*, November 20, 1996, pp. B1+; C. A. Deutsch, "Corporate Diversity in Practice," *New York Times*, November 20, 1996, pp. C1+; and D. A. Thomas and R. J. Ely, "Making Differences Matter: A New Paradigm for Managing Diversity," *Harvard Business Review*, September–October 1996, pp. 79–90.

40. R. A. Giacalone and C. L. Jurkiewicz (eds.), *Handbook of Workplace Spirituality and Organizational Performance* (New York: M. E. Sharp), 2003.

41. M. B. Marklein, "Study: College Students Seeking Meaning of Life," *USA Today*, December 22, 2007, p. 6C.

42. This section is based on D. Grant, "What Should a Science of Workplace Spirituality Study? The Case for a Relational Approach," *Academy of Management Proceedings* Best Conference Paper, August 2005; C. D. Pielstick, "Teaching Spirituality Synchronicity in a Business Leadership Class," *Journal of Management Education*, February 2005, pp. 153–168; H. Ashar and M. Lane-Maher, "Success and Spirituality in the New Business Paradigm," *Journal of Management Inquiry*, June 2004, pp. 249–260; G. A. Gull and J. Doh, "The 'Transmutation' of the Organization: Toward a More Spiritual Workplace," *Journal of Management Inquiry*, June 2004, pp. 128–139; K. C. Cash and G. R. Gray, "A Framework for Accommodating Religion and Spirituality in the Workplace," *Academy of Management Executive*, August 2000, pp. 124–133; F. Wagnor-Marsh and J. Conley, "The Fourth Wave: The Spiritually-Based Firm," *Journal of Organizational Change Management*, vol. 12 (3), 1999, pp. 292–302; E. H. Burack, "Spirituality in the Workplace," *Journal of Organizational Change Management*, vol. 12 (3), 1999, pp. 280–291; J. Milliman, J. Ferguson, D. Trickett, and B. Condemi, "Spirit and Community at Southwest Airlines: An Investigation of a Spiritual Values-Based Model," *Journal of Organizational Change Management*, vol. 12 (3), 1999, pp. 221–233; and I. A. Mitroff and E. A. Denton, *A Spiritual Audit of Corporate America: A Hard Look at Spirituality, Religion, and Values in the Workplace* (San Francisco: Jossey-Bass, 1999).

43. J. Reingold, "Walking the Walk," *Fast Company*, November 2005, p. 82.

44. http://www.indiabizclub.com/qxzpmd/?q=;0tw30t7geonp00aki

45. P. Paul, "A Holier Holiday Season," *American Demographics*, December 2001, pp. 41–45; and M. Conlin, "Religion in the Workplace: The Growing Presence of Spirituality in Corporate America," *BusinessWeek*, November 1, 1999, pp. 151–58.

46. Cited in M. Conlin, "Religion in the Workplace," p. 153.

47. C. P. Neck and J. F. Milliman, "Thought Self-Leadership: Finding Spiritual Fulfillment in Organizational Life," *Journal of Managerial Psychology*, vol. 9 (8), 1994, p. 9.

48. A. K. Miles, S. Sledge, and S. Coppage, "Linking Spirituality to Workplace Performance: A Qualitative Study of the Brazilian Candomble," *Academy of Management Proceedings* Best Conference Paper, August 2005; J. Millman, A. Czaplewski, and J. Ferguson, "An Exploratory Empirical Assessment of the Relationship Between Spirituality and Employee Work Attitudes," paper presented at Academy of Management, Washington, DC, August 2001; P. H. Mirvis, "Soul Work in Organizations," *Organization Science*, vol. 8 (2), 1997, p. 193; P. Leigh, "The New Spirit at Work," *Training and Development*, vol. 51 (3), 1997, p. 26; E. Brandt, "Corporate Pioneers Explore Spiritual Peace," *HRMagazine*, vol. 41 (4), 1996, p. 82; and D. W. McCormick, "Spirituality and Management," *Journal of Managerial Psychology*, vol. 9 (6), 1994, p. 5.

49. http://www.infochangeindia.org/200807107215/Environment/News/India-unveils-national-climate-change-action-plan.html

50. D. Streitfield, "A Global Need for Grain That Farms Can't Fill," *New York Times* online, www.nytimes.com, March 8, 2008; E. Weise, "The Shape of Lights to Come? Not Everyone's Buying It," *USA Today*, February 28, 2008, p. 1A+; and J. M. Manyika, R. P. Roberts, and K. L. Sprague, "Eight Business Technology Trends to Watch," *The McKinsey Quarterly* online, www.mckinseyquarterly.com, December 2007.
51. A. Martin, "Is It Healthy? Food Rating Systems Battle It Out," *New York Times* online, www.nytimes.com, December 1, 2007.
52. http://www.socialjustice.nic.in/disabled/act.htm
53. See, for instance, A.S. Hayes, "Layoffs Take Careful Planning to Avoid Losing the Suits That Are Apt to Follow," *Wall Street Journal*, November 2, 1990, p. B1.
54. A good source of information about political/legal factors in other countries is Cornell University's Legal Information Institute, www.law.cornell.edu/world.
55. http://www.businessweek.com/1999/99_41/b3650015.htm; http://www.forbes.com/2007/09/22/india-happiness-survey-face-cx_rd_0921autofacescan02.html
56. F. Hansen, "Mega Shifts Remake Marketing," *Business Finance*, March 2003, p. 9.
57. J. P. Walsh, "Book Review Essay: Taking Stock of Stakeholder Management," *Academy of Management Review*, April 2005, pp. 426–438; R. E. Freeman, A. C. Wicks, and B. Parmar, "Stakeholder Theory and 'The Corporate Objective Revisited,'" *Organization Science*, 15, 2004, pp. 364–369; T. Donaldson and L. E. Preston, "The Stakeholder Theory of the Corporation: Concepts, Evidence, and Implications," *Academy of Management Review*, January 1995, pp. 65–91; and R. E. Freeman, *Strategic Management: A Stakeholder Approach* (Boston: Pitman/Ballinger), 1984.
58. J. S. Harrison and C. H. St. John, "Managing and Partnering with External Stakeholders," *Academy of Management Executive*, May 1996, pp. 46–60.
59. S. L. Berman, R. A. Phillips, and A. C. Wicks, "Resource Dependence, Managerial Discretion, and Stakeholder Performance," *Academy of Management Proceedings* Best Conference Paper, August 2005; A. J. Hillman and G. D. Keim, "Shareholder Value, Stakeholder Management, and Social Issues: What's the Bottom Line?" *Strategic Management Journal*, March 2001, pp. 125–139; J. S. Harrison and R. E. Freeman, "Stakeholders, Social Responsibility, and Performance: Empirical Evidence and Theoretical Perspectives," *Academy of Management Journal*, July 1999, pp. 479–487; and J. Kotter and J. Heskett, *Corporate Culture and Performance* (New York: The Free Press, 1992).
60. Harrison and St. John, "Managing and Partnering with External Stakeholders."

Chapter 4

1. J. Katz, "Worlds of Difference," *IW*, December 2007, pp. 39–41.
2. G. Koretz, "Things Go Better with Multinationals—Except Jobs," *BusinessWeek*, May 2, 1994, p. 20.
3. The idea for this quiz was adapted from R. M. Hodgetts and F. Luthans, *International Management*, 2d ed. (New York: McGraw-Hill, 1994).
4. Reuters Limited, *USA Today* online, www.usatoday.com, February 21, 2006; and "Learning the Lingo," *USA Today*, January 26, 2006, p. 1A.
5. Ibid.
6. N. Adler, *International Dimensions of Organizational Behavior*, 5th ed. (Cincinnati: South-Western Publishing, 2008).
7. M. R. F. Kets De Vries and E. Florent-Treacy, "Global Leadership from A to Z: Creating High Commitment Organizations," *Organizational Dynamics*, Spring 2002, pp. 295–309; P. R. Harris and R. T. Moran, *Managing Cultural Differences*, 4th ed. (Houston: Gulf Publishing Co., 1996); R. T. Moran, P. R. Harris, and W. G. Stripp, *Developing the Global Organization: Strategies for Human Resource Professionals* (Houston: Gulf Publishing Co., 1993); Y. Wind, S. P. Douglas, and H. V. Perlmutter, "Guidelines for Developing International Marketing Strategies," *Journal of Marketing*, April 1973, pp. 14–23; and H. V. Perlmutter, "The Tortuous Evolution of the Multinational Corporation," *Columbia Journal of World Business*, January–February 1969, pp. 9–18.
8. A. K. Gupta and V. Govindarajan, "Cultivating a Global Mindset," *Academy of Management Executive*, February 2002, pp. 117–118.
9. S. Kotkin, "The World as an Imperfect Globe," New York Times online, www.nytimes.com, December 2, 2007.
10. http://www.sdc.gov.in/AboutSAARC/aboutsaarc.htm; http://en.wikipedia.org/wiki/South_Asian_Association_for_Regional_Cooperation
11. SAARC Web site, www.saarc-sec.org; and N. George, "South Asia trade Zone in Words," Springfield, Missouri, News-Leader, January 4, 2004, p. 164.
12. Europa Web site, http://europa.eu/index_en.htm.
13. Ibid.
14. "Treaty of Lisbon," Europa, http://europa.eu/Lisbon_treaty/index_en.htm; N. Knox, "Leaders of Embattled EU Head to Washington," *USA Today*, June 20, 2005, p. A8; and N. Knox, "European Union Struggles with Constitution Rejection," *USA Today*, May 31, 2005, p. A10.
15. Information from ASEAN Web site, www.aseansec.org.
16. "NAFTA Facts," Office of the United States Trade Representative, www.ustr.gov, October 2007.
17. J. Lyons, "Costa Rica CAFTA Vote Bolsters U.S. Policy," *Wall Street Journal*, October 9, 2007, p. A2.
18. J. Forero, "U.S. and Colombia Reach Trade Deal After 2 Years of Talks," *New York Times* online, www.nytimes.com, February 28, 2006.
19. "Ministerial Declaration," Free Trade Area of the Americas, www.ftaa-alca.org, January 23, 2006; and M. Moffett and J. D. McKinnon, "Failed Summit Casts Shadow on Global Trade Talks," *Wall Street Journal*, November 7, 2005, pp. A1+.
20. "2004–2007 Strategic Plan," Commission of the African Union, www.africa-union.org; and D. Kraft, "Leaders Question, Praise African Union," *Springfield, Missouri, News-Leader*, July 10, 2002, p. 8A.
21. This section is based on materials from the World Trade Organization Web site, www.wto.org; and D. A. Irwin, "GATT Turns 60," *Wall Street Journal*, April 9, 2007, p. A13.
22. C. A. Barlett and S. Ghoshal, *Managing Across Borders: The Transnational Solution* 2d ed. (Boston: Harvard Business School Press), 2002; and N. J. Adler, *International Dimensions of Organizational Behavior*, 4th ed. (Cincinnati, OH: South-Western Publishing, 2002), pp. 9–11.
23. P. F. Drucker, "The Global Economy and the Nation-State," *Foreign Affairs*, September-October, 1997, pp. 159–171.

24. P. Dvorak, "Why Multiple Headquarters Multiply," *Wall Street Journal*, November 19, 2007, pp. B1+.

25. D. A. Aaker, *Developing Business Strategies*, 5th ed. (New York: John Wiley & Sons, 1998); and J. A. Byrne, et al., "Borderless Management," *BusinessWeek*, May 23, 1994, pp. 24–26.

26. B. Davis, "Migration of Skilled Jobs Abroad Unsettles Global-Economy Fans," *Wall Street Journal*, January 26, 2004, p. A1.

27. This box based on information from K. Bahadur, D. Desmet, and E. van Bommel, "Smart IT Spending: Insights from European Banks," *The McKinsey Quarterly* online, www.mckinseyquarterly.com, January 2006; and S. Hamm, "Motivating the Troops," *BusinessWeek*, November 21, 2005, pp. 88–103.

28. T. Pincus, "Globalization vs. Political Risk," *Chicago Sun-Times*, February 25, 2008, p. 41.

29. "Emerging Economies Are Following the Global Trend of Disinflation," *The Economist*, October 19, 2002, p. 36.

30. Central Intelligence Agency, *World Factbook* (Washington, DC: Potomac Books), 2008, www.cia.gov/library/publications/the-world-factbook/index.html.

31. J. McGregor and S. Hamm, "Managing the Global Workforce," *BusinessWeek*, January 28, 2008, pp. 34–51.

32. These examples taken from L. Khosla, "You Say Tomato," *Forbes*, May 21, 2001, p. 36; and T. Raphael, "Savvy Companies Build Bonds with Hispanic Employees," *Workforce*, September 2001, p. 19.

33. See G. Hofstede, *Culture's Consequences: International Differences in Work-Related Values*, 2nd ed., (Thousand Oaks, CA: Sage Publications, 2001), pp. 9–15.

34. S. Bhaskaran and N. Sukumaran, "National Culture, Business Culture and Management Practices: Consequential Relationships?" *Cross Cultural Management: An International Journal*, vol. 14 (7), 2007, pp. 54–67; G. Hofstede, *Culture's Consequences*; and G. Hofstede, "The Cultural Relativity of Organizational Practices and Theories," *Journal of International Business Studies*, Fall 1983, pp. 75–89.

35. J. S. Chhokar, F. C. Brodbeck, and R. J. House, *Culture and Leadership Across the World: The GLOBE Book of In-Depth Studies of 25 Societies* (Philadelphia: Lawrence Erlbaum Associates), 2007; and R. J. House, et al., *Culture, Leadership, and Organizations: The GLOBE Study of 62 Societies* (Thousand Oaks, CA: Sage Publications), 2004.

36. These examples taken from J. M. Olsen, "Toy Maker Lego Moves Production to Czech Republic," *USA Today* online, www.usatoday.com, September 1, 2005; M. Gunther, "Cops of the Global Village," *Fortune*, June 27, 2005, pp. 158–166; J. Sapsford, "Nissan to Sell in China Minivans Made in the U.S.," *Wall Street Journal*, March 17, 2005, p. A14; D. Michaels, "Sukhoi Has the World in Its Sights," *Wall Street Journal*, August 7, 2003, p. A9; and J. Slater, "GE Takes Advantage of India's Talented Research Pool," *Wall Street Journal*, March 26, 2003, p. A10.

37. D. Yergin, "Globalization Opens Door to New Dangers," *USA Today*, May 28, 2003, p. 11A; K. Lowrey Miller, "Is It Globaloney?" *Newsweek*, December 16, 2002, pp. E4–E8; L. Gomes, "Globalization Is Now a Two-Way Street—Good News for the U.S.," *Wall Street Journal*, December 9, 2002, p. B1; J. Kurlantzick and J. T. Allen, "The Trouble with Globalism," *U.S. News & World Report*, February 11, 2002, pp. 38–41; and J. Guyon, "The American Way," *Fortune*, November 26, 2001, pp. 114–120.

38. Guyon, "The American Way," p. 114.

Chapter 5

1. http://www.expresspharmaonline.com/20060228/management03.shtml; http://www.karmayog.org/;http://www.in.kpmg.com/pdf/CSR_Whitepaper.pdf

2. M. L. Barnett, "Stakeholder Influence Capacity and the Variability of Financial Returns to Corporate Social Responsibility," *Academy of Management Review*, July 2007, pp. 794–816; A. Mackey, T. B. Mackey, and J. B. Barney, "Corporate Social Responsibility and Firm Performance: Investor Preferences and Corporate Strategies," *Academy of Management Review*, July 2007, pp. 817–835; and A. B. Carroll, "A Three-Dimensional Conceptual Model of Corporate Performance," *Academy of Management Review*, October 1979, p. 499.

3. See K. Basu and G. Palazzo, "Corporate Social Performance: A Process Model of Sensemaking," *Academy of Management Review*, January 2008, pp. 122–136; and S. P. Sethi, "A Conceptual Framework for Environmental Analysis of Social Issues and Evaluation of Business Response Patterns," *Academy of Management Review*, January 1979, pp. 68–74.

4. M. Friedman, *Capitalism and Freedom* (Chicago: University of Chicago Press, 1962); and M. Friedman, "The Social Responsibility of Business Is to Increase Profits," *New York Times Magazine*, September 13, 1970, p. 33.

5. S. Liebs, "Do Companies Do Good Well?" *CFO*, July 2007, p. 16.

6. See, for example, D. J. Wood, "Corporate Social Performance Revisited," *Academy of Management Review*, October 1991, pp. 703–708; and S. L. Wartick and P. L. Cochran, "The Evolution of the Corporate Social Performance Model," *Academy of Management Review*, October 1985, p. 763.

7. Information from "Giving Back," found on American Express Web site, www.americanexpress.com, March 28, 2008.

8. See, for example, R. A. Buccholz, *Essentials of Public Policy for Management*, 2d ed. (Upper Saddle River, NJ: Prentice Hall, 1990).

9. I. Brat, "The Extra Step," *Wall Street Journal*, March 24, 2008, p. R12.

10. http://www.teach.timesofindia.com/

11. This section is based on J. D. Margolis and J. P. Walsh, "Misery Loves Companies: Rethinking Social Initiatives by Business," *Administrative Science Quarterly*, vol. 48 (2), 2003, pp. 268–305; K. Davis and W. C. Frederick, *Business and Society: Management, Public Policy, Ethics*, 5th ed. (New York: McGraw-Hill, 1984), pp. 28–41; and R. J. Monsen, Jr., "The Social Attitudes of Management," in J. M. McGuire (ed.), *Contemporary Management: Issues and Views* (Upper Saddle River, NJ: Prentice Hall, 1974), p. 616.

12. See, for instance, R. Trudel and J. Cotte, " Does Being Ethical Pay?" *Wall Street Journal*, May 12, 2008, p. R8; J. D. Margolis and H. Anger Elfenbein, "Do Well by Doing Good? Don't Count on It," *Harvard Business Review*, January 2008, pp. 19–20; M. L. Barnett, "Stakeholder Influence Capacity and the Variability of Financial Returns to Corporate Social Responsibility," 2007; D. O. Neubaum and S. A. Zahra, "Institutional Ownership and Corporate Social Performance: The Moderating Effects of Investment Horizon, Activism, and Coordination," *Journal of Management*, February 2006, pp. 108–131; B. A. Waddock and S. B. Graves, "The Corporate Social Performance–Financial Performance Link," *Strategic Management Journal*, April

1997, pp. 303–319; J. B. McGuire, A. Sundgren, and T. Schneeweis, "Corporate Social Responsibility and Firm Financial Performance," *Academy of Management Journal*, December 1988, pp. 854–872; K. Aupperle, A. B. Carroll, and J. D. Hatfield, "An Empirical Examination of the Relationship Between Corporate Social Responsibility and Profitability," *Academy of Management Journal*, June 1985, pp. 446–463; and P. Cochran and R. A. Wood, "Corporate Social Responsibility and Financial Performance," *Academy of Management Journal*, March 1984, pp. 42–56.

13. See J. Surroca and J. A. Tribo, "The Corporate Social and Financial Performance Relationship: What's the Ultimate Determinant?" *Academy of Management Proceedings* Best Conference Paper, 2005; D. J. Wood and R. E. Jones, "Stakeholder Mismatching: A Theoretical Problem in Empirical Research on Corporate Social Performance," *International Journal of Organizational Analysis*, July 1995, pp. 229–267; R. Wolfe and K. Aupperle, "Introduction to Corporate Social Performance: Methods for Evaluating an Elusive Construct," pp. 265–268, in J. E. Post (ed.), *Research in Corporate Social Performance and Policy*, vol. 12, 1991; and A. A. Ullmann, "Data in Search of a Theory: A Critical Examination of the Relationships among Social Performance, Social Disclosure, and Economic Performance of U.S. Firms," *Academy of Management Review*, July 1985, pp. 540–557.

14. B. Seifert, S. A. Morris, and B. R. Bartkus, "Having, Giving, and Getting: Slack Resources, Corporate Philanthropy, and Firm Financial Performance," *Business & Society*, June 2004, pp. 135–161; and McGuire, Sundgren, and Schneeweis, "Corporate Social Responsibility and Firm Financial Performance."

15. A. McWilliams and D. Siegel, "Corporate Social Responsibility and Financial Performance: Correlation or Misspecification?" *Strategic Management Journal*, June 2000, pp. 603–609.

16. A. J. Hillman and G. D. Keim, "Shareholder Value, Stakeholder Management, and Social Issues: What's the Bottom Line?" *Strategic Management Journal*, vol. 22, 2001, pp. 125–139.

17. M. Orlitzky, F. L. Schmidt, and S. L. Rynes, "Corporate Social and Financial Performance," *Organization Studies*, vol. 24 (3), 2003, pp. 403–441.

18. http://www.financialexpress.com/news/socially-responsible-investing-a-primer/287728/2

19. http://www.atimes.com/atimes/South_Asia/ID05Df01.html

20. Social Investment Forum, *Socially Responsible Mutual Fund Charts: Financial Performance*, February 29, 2008, www.socialinvest.org.

21. T. Delis, "Bag Revolution," *Fortune*, May 12, 2008, pp. 18–19; and E. Royte, "Moneybags," *Fast Company*, October 2007, p. 64.

22. http://www.isesco.org.ma/ISESCO_Technology_Vision/NUM01/Pradeep%20Chaturvedi/Pradeep%20Chaturvedi.pdf

23. A. White, "The Greening of the Balance Sheet," *Harvard Business Review*, March 2006, pp. 27–28; N. Guenster, J. Derwall, R. Bauer, and K. Koedijk, "The Economic Value of Eco-Efficiency," *Academy of Management Conference*, Honolulu, Hawaii, August 2005; F. Bowen and S. Sharma, "Resourcing Corporate Environmental Strategy: Behavioral and Resource-Based Perspectives," *Academy of Management Conference*, August 2005; M. P. Sharfman, T. M. Shaft, and L. Tihanyi, "A Model of the Global and Institutional Antecedents of High-Level Corporate Environmental Performance," *Business & Society*, March 2004, pp. 6–36; S. L. Hart and M. B. Milstein, "Creating Sustainable Value," *Academy of Management Executive*, May 2003, pp. 56–67; K. Buysse and A. Verbeke, "Proactive Environmental Strategies: A Stakeholder Management Perspective," *Strategic Management Journal*, May 2003, pp. 453–470; C. Marsden, "The New Corporate Citizenship of Big Business: Part of the Solution to Sustainability?" *Business & Society Review*, Spring 2000, pp. 9–25; R. D. Klassen and D. C. Whybark, "The Impact of Environmental Technologies on Manufacturing Performance," *Academy of Management Journal*, December 1999, pp. 599–615; H. Bradbury and J. A. Clair, "Promoting Sustainable Organizations With Sweden's Natural Step," *Academy of Management Executive*, October 1999, pp. 63–73; F. L. Reinhardt, "Bringing the Environment Down to Earth," *Harvard Business Review*, July–August 1999, pp. 149–157; I. Henriques and P. Sadorsky, "The Relationship Between Environmental Commitment and Managerial Perceptions of Stakeholder Importance," *Academy of Management Journal*, February 1999, pp. 87–99; and M. A. Berry and D. A. Rondinelli, "Proactive Corporate Environmental Management: A New Industrial Revolution," *Academy of Management Executive*, May 1998, pp. 38–50.

24. The concept of shades of green can be found in R. E. Freeman, J. Pierce, and R. Dodd, *Shades of Green: Business Ethics and the Environment* (New York: Oxford University Press, 1995).

25. Information from ISO Web site, www.iso.org.

26. The Global 100 list is a collaborative effort of Corporate Knights Inc. and Innovest Strategic Value Advisors. Information from Global 100 Web site, www.global100.org.

27. C. Chandler, "Livedoor Slammed," *Fortune*, February 20, 2006, p. 25; "$64B Diamond Industry Rocked by Fraud," *CNNMoney*, cnnmoney.com, December 20, 2005; D. Searcey, S. Young, and K. Scannell, "Ebbers Is Sentenced to 25 Years for $11 Billion WorldCom Fraud," *Wall Street Journal*, July 14, 2005, p. A1+; and E. B. Smith, "Wal-Mart Sets New Policy on Ethics," *USA Today*, January 28, 2005, p. 1B.

28. This last example is based on J. F. Viega, T. D. Golden, and K. Dechant, "Why Managers Bend Company Rules," *Academy of Management Executive*, May 2004, pp. 84–90.

29. Davis and Frederick, *Business and Society*, p. 76.

30. F. D. Sturdivant, *Business and Society: A Managerial Approach*, 3d ed. (Homewood, IL: Richard D. Irwin, 1985), p. 128.

31. L. K. Treviño, G. R. Weaver, and S. J. Reynolds, "Behavioral Ethics in Organizations: A Review," *Journal of Management*, December 2006, pp. 951–990; T. Kelley, "To Do Right or Just to Be Legal," *New York Times*, February 8, 1998, p. BU12; J. W. Graham, "Leadership, Moral Development, and Citizenship Behavior," *Business Ethics Quarterly*, January 1995, pp. 43–54; L. Kohlberg, *Essays in Moral Development: The Psychology of Moral Development*, vol. 2 (New York: Harper & Row, 1984); and L. Kohlberg, *Essays in Moral Development: The Philosophy of Moral Development*, vol. 1 (New York: Harper & Row, 1981).

32. See, for example, J. Weber, "Managers' Moral Reasoning: Assessing Their Responses to Three Moral Dilemmas," *Human Relations*, July 1990, pp. 687–702.

33. W. C. Frederick and J. Weber, "The Value of Corporate Managers and Their Critics: An Empirical Description and

Normative Implications," in W. C. Frederick and L. E. Preston (eds.), *Business Ethics: Research Issues and Empirical Studies* (Greenwich, CT: JAI Press, 1990), pp. 123–144; and J. H. Barnett and M. J. Karson, "Personal Values and Business Decisions: An Exploratory Investigation," *Journal of Business Ethics*, July 1987, pp. 371–382.

34. M. E. Baehr, J. W. Jones, and A. J. Nerad, "Psychological Correlates of Business Ethics Orientation in Executives," *Journal of Business and Psychology*, Spring 1993, pp. 291–308; and L. K. Treviño and S. A. Youngblood, "Bad Apples in Bad Barrels: A Causal Analysis of Ethical Decision-Making Behavior," *Journal of Applied Psychology*, August 1990, pp. 378–385.

35. M. E. Schweitzer, L. Ordonez, and B. Douma, "Goal Setting as a Motivator of Unethical Behavior," *Academy of Management Journal*, June 2004, pp. 422–432.

36. M. C. Jensen, "Corporate Budgeting Is Broken—Let's Fix It," *Harvard Business Review*, June 2001, pp. 94–101.

37. R. L. Cardy and T. T. Selvarajan, "Assessing Ethical Behavior Revisited: The Impact of Outcomes on Judgment Bias," paper presented at the Annual Meeting of the Academy of Management, Toronto, 2000.

38. G. Weaver, "Ethics and Employees: Making the Connection," *Academy of Management Executive*, May 2004, pp. 121–125; V. Anand, B. E. Ashforth, and M. Joshi, "Business as Usual: The Acceptance and Perpetuation of Corruption in Organizations," *Academy of Management Executive*, May 2004, pp. 39–53; J. Weber, L. B. Kurke, and D. W. Pentico, "Why Do Employees Steal?" *Business & Society*, September 2003, pp. 359–380; V. Arnold and J. C. Lampe, "Understanding the Factors Underlying Ethical Organizations: Enabling Continuous Ethical Improvement," *Journal of Applied Business Research*, Summer 1999, pp. 1–19; R. R. Sims, "The Challenge of Ethical Behavior in Organizations," *Journal of Business Ethics*, July 1992, pp. 505–513; and J. B. Cullen, B. Victor, and C. Stephens, "An Ethical Weather Report: Assessing the Organization's Ethical Climate," *Organizational Dynamics*, Autumn 1989, pp. 50–62; and B. Victor and J. B. Cullen, "The Organizational Bases of Ethical Work Climates," *Adminis-trative Science Quarterly*, March 1988, pp. 101–125.

39. P. Van Lee, L. Fabish, and N. McCaw, "The Value of Corporate Values," *Strategy & Business*, Summer 2005, pp. 52–65.

40. G. Weaver, "Ethics and Employees: Making the Connection," May 2004; G. R. Weaver, L. K. Treviño, and P. L. Cochran, "Integrated and Decoupled Corporate Social Performance: Management Commitments, External Pressures, and Corporate Ethics Practices," *Academy of Management Journal*, October 1999, pp. 539–552; G. R. Weaver, L. K. Treviño, and P. L. Cochran, "Corporate Ethics Programs as Control Systems: Influences of Executive Commitment and Environmental Factors," *Academy of Management Journal*, February 1999, pp. 41–57; R. B. Morgan, "Self- and Co-Worker Perceptions of Ethics and Their Relationships to Leadership and Salary," *Academy of Management Journal*, February 1993, pp. 200–214; and B. Z. Posner and W. H. Schmidt, "Values and the American Manager: An Update," *California Management Review*, Spring 1984, pp. 202–216.

41. IBM Corporate Responsibility Report, 2007, www.ibm.com; and A. Schultz, "Integrating IBM," *CRO Newsletter*, March/April 2007, pp. 16–21.

42. T. Barnett, "Dimensions of Moral Intensity and Ethical Decision Making: An Empirical Study," *Journal of Applied Social Psychology*, May 2001, pp. 1038–1057; and T. M. Jones, "Ethical Decision Making by Individuals in Organizations: An Issue-Contingent Model," *Academy of Management Review*, April 1991, pp. 366–395.

43. W. Bailey and A. Spicer, "When Does National Identity Matter? Convergence and Divergence in International Business Ethics," *Academy of Management Journal*, December 2007, pp. 1462–1480; and R. L. Sims, "Comparing Ethical Attitudes Across Cultures," *Cross Cultural Management: An International Journal*, vol. 13 (2), 2006, pp. 101–113.

44. "Legal Review of Overseas Bribery," *BBC News* online, http://news.bbc.co.uk, November 29, 2007.

45. L. Paine, R. Deshpande, J. D. Margolis, and K. E. Bettcher, "Up to Code: Does Your Company's Conduct Meet World-Class Standards?" *Harvard Business Review*, December 2005, pp. 122–133; G. R. Simpson, "Global Heavyweights Vow 'Zero Tolerance' for Bribes," *Wall Street Journal*, January 27, 2005, pp. A2+; A. Spicer, T. W. Dunfee, and W. J. Bailey, "Does National Context Matter in Ethical Decision Making? An Empirical Test of Integrative Social Contracts Theory," *Academy of Management Journal*, August 2004, pp. 610–620; J. White and S. Taft, "Frameworks for Teaching and Learning Business Ethics Within the Global Context: Background of Ethical Theories," *Journal of Management Education*, August 2004, pp. 463–477; J. Guyon, "CEOs on Managing Globally," *Fortune*, July 26, 2004, p. 169; A. B. Carroll, "Managing Ethically with Global Stakeholders: A Present and Future Challenge," *Academy of Management Executive*, May 2004, pp. 114–120; and C. J. Robertson and W. F. Crittenden, "Mapping Moral Philosophies: Strategic Implications for Multinational Firms," *Strategic Management Journal*, April 2003, pp. 385–392.

46. "The New Social Steward," *Fortune*, Special Advertising Section, November 12, 2007, pp. 57–63; and A. Savitz and M. Choi, "The Future of the Global Compact," *CRO Newsletter*, January/ February 2007, pp. 47–48.

47. Organization for Economic Cooperation and Development, "About Bribery in International Business," www.oecd.org, March 28, 2008.

48. Enron example taken from P. M. Lencioni, "Make Your Values Mean Something," *Harvard Business Review*, July 2002, p. 113; and Sears example taken from series of posters called "Sears Ethics and Business Practices: A Century of Tradition," *Business Ethics*, May/June 1999, pp. 12–13; and B. J. Feder, "The Harder Side of Sears," *New York Times*, July 20, 1997, pp. BU1+.

49. Treviño and Youngblood, "Bad Apples in Bad Barrels," p. 384.

50. http://www.tata.com/media/Speeches/inside.aspx?artid=AjUrVgmvnJQ=

51. M. Weinstein, "Survey Says: Ethics Training Works," *Training*, November 2005, p. 15.

52. J. E. Fleming, "Codes of Ethics for Global Corporations," *Academy of Management News*, June 2005, p. 4.

53. "Global Ethics Codes Gain Importance as a Tool to Avoid Litigation and Fines," *Wall Street Journal*, August 19, 1999, p. A1; and J. Alexander, "On the Right Side," *World Business*, January/February 1997, pp. 38–41.

54. F. R. David, "An Empirical Study of Codes of Business Ethics: A Strategic Perspective," paper presented at the 48th Annual Academy of Management Conference, Anaheim, California, August 1988.

55. *National Business Ethics Survey* (Arlington, VA: Ethics Resource Center, 2007).

56. Codes of conduct information from the Center for Ethical Business Cultures Web site, www.cebcglobal.org, February 15, 2006; Paine, et al., "Up to Code: Does Your Company's Conduct Meet World-Class Standards"; and A. K. Reichert and M. S. Webb, "Corporate Support for Ethical and Environmental Policies: A Financial Management Perspective," *Journal of Business Ethics*, May 2000, pp. 53–64.

57. L. Nash, "Ethics Without the Sermon," *Harvard Business Review*, November–December 1981, p. 81.

58. V. Wessler, "Integrity and Clogged Plumbing," *Straight to the Point*, Fall 2002, pp. 1–2.

59. T. A. Gavin, "Ethics Education," *Internal Auditor*, April 1989, pp. 54–57.

60. L. Myyry and K. Helkama, "The Role of Value Priorities and Professional Ethics Training in Moral Sensitivity," *Journal of Moral Education*, vol. 31 (1), 2002, pp. 35–50; and W. Penn and B. D. Collier, "Current Research in Moral Development as a Decision Support System," *Journal of Business Ethics*, January 1985, pp. 131–136.

61. J. A. Byrne, "After Enron: The Ideal Corporation," *BusinessWeek*, August 19, 2002, pp. 68–71; D. Rice and C. Dreilinger, "Rights and Wrongs of Ethics Training," *Training & Development Journal*, May 1990, pp. 103–109; and J. Weber, "Measuring the Impact of Teaching Ethics to Future Managers: A Review, Assessment, and Recommendations," *Journal of Business Ethics*, April 1990, pp. 182–190.

62. E. White, "What Would You Do? Ethics Courses Get Context," *Wall Street Journal*, June 12, 2006, p. B3; and D. Zielinski, "The Right Direction: Can Ethics Training Save Your Company," *Training*, June 2005, pp. 27–32.

63. G. Farrell and J. O'Donnell, "Ethics Training As Taught by Ex-Cons: Crime Doesn't Pay," *USA Today*, November 16, 2005, p. 1B+.

64. J. Weber, "The New Ethics Enforcers," *Business Week*, February 13, 2006, pp. 76–77.

65. The Ethics and Compliance Officer Association Web site, www.theecoa.org; and K. Maher, "Global Companies Face Reality of Instituting Ethics Programs," *Wall Street Journal*, November 9, 2004, p. B8.

66. Ethics Newsline, "Survey Reveals How Many Workers Commit Office Taboos," www.globalethics.org, September 18, 2007.

67. H. Oh, "Biz Majors Get an F for Honesty," *Business Week*, February 6, 2006, p. 14.

68. "Students Aren't Squealers," *USA Today*, March 27, 2003, p. 1D; and J. Merritt, "You Mean Cheating Is Wrong?" *BusinessWeek*, December 9, 2002, p. 8.

69. J. Hyatt, "Unethical Behavior: Largely Unreported in Offices and Justified by Teens," *CRO Newsletter* online, www.thecro.com/enewsletter, February 13, 2008.

70. D. Lidsky, "Transparency: It's Not Just for Shrink Wrap Anymore," *Fast Company*, January 2005, p. 87.

71. W. Zellner, et al., "A Hero—And a Smoking-Gun Letter," *BusinessWeek*, January 28, 2002, pp. 34–35.

72. *National Business Ethics Survey* (Arlington, VA: Ethics Resource Center, 2007).

73. S. Armour, "More Companies Urge Workers to Blow the Whistle," *USA Today*, December 16, 2002, p. 1B.

74. J. Wiscombe, "Don't Fear Whistleblowers," *Workforce*, July 2002, pp. 26–27.

75. T. Reason, "Whistle Blowers: The Untouchables," *CFO*, March 2003, p. 18; and C. Lachnit, "Muting the Whistle-Blower?" *Workforce*, September 2002, p. 18.

76. J. Hyatt, "Corporate Whistleblowers Might Need a Monetary Nudge, Researchers Suggest," *CRO Newsletter* online, www.thecro.com/enewsletter, April 11, 2007; J. O'Donnell, "Blowing the Whistle Can Lead to Harsh Aftermath, Despite Law," *USA Today*, August 1, 2005, p. 2B; and D. Solomon, "For Financial Whistle-Blowers, New Shield Is an Imperfect One," *Wall Street Journal*, October 4, 2004, pp. A1+.

77. B. Dobbin, "Dealers Market Global Trade with Social Conscience," The Associated Press, *Springfield Missouri) News-Leader*, February 16, 2005, p. 5B.

78. This definition based on P. Tracey and N. Phillips, "The Distinctive Challenge of Educating Social Entrepreneurs: A Postscript and Rejoinder to the Special Issue on Entrepreneurship Education," *Academy of Management Learning & Education*, June 2007, pp. 264–271; Schwab Foundation for Social Entrepreneurship, www.schwabfound.org, February 20, 2006; and J. G. Dees, J. Emerson, and P. Economy, *Strategic Tools for Social Entrepreneurs* (New York: John Wiley & Sons, Inc., 2002).

79. D. Bornstein, *How to Change the World: Social Entrepreneurs and the Power of New Ideas* (New York: Oxford University Press, 2004), inside cover jacket.

80. http://www.narayanahospitals.com/

81. http://www.bangalorebest.com/cityresources/Health_and_Medicine/narayana.asp

82. http://www.aravind.org/

83. http://www.giveindia.org/

84. http://www.khemkafoundation.org/awards/sey_awards.php

85. http://www.karmayog.org/redirect/strred.asp?docld=9041

86. R. J. Bies, J. M. Bartunek, T. L. Fort, and M. N Zald, "Corporations as Social Change Agents: Individual, Interpersonal, Institutional, and Environmental Dynamics," *Academy of Management Review*, July 2007, pp. 788–793.

87. "The State of Corporate Philanthropy: A McKinsey Global Survey," *The McKinsey Quarterly* online, www.mckinseyquarterly.com, February 2008.

88. http://www.karmayog.org/

89. K. J. Delaney, "Google: From 'Don't Be Evil' to How to Do Good," *Wall Street Journal*, January 18, 2008, pp. B1+; H. Rubin, "Google Offers a Map for Its Philanthropy," *New York Times* online, www.nytimes.com, January 18, 2008; and K. Hafner, "Philanthropy Google's Way: Not the Usual," *New York Times* online, www.nytimes.com, September 14, 2006.

90. http://www.abc.net.au/worldtoday/content/2003/s926955.htm2

91. http://www.hsbc.co.in/1/2/miscellaneous/about-hsbc/corporate-sustainability/employee-volunteering

Chapter 6

1. The name of the manager and the company has been changed or not mentioned to maintain anonymity. However, the case is real.

2. http://www.prsichennai.org/html/eventscurrent.htm
3. D. A. Garvin and M. A. Roberto, "What You Don't Know About Making Decisions," *Harvard Business Review*, September 2001, pp. 108–116.
4. W. Pounds, "The Process of Problem Finding," *Industrial Management Review*, Fall 1969, pp. 1–19.
5. R. J. Volkema, "Problem Formulation: Its Portrayal in the Texts," *Organizational Behavior Teaching Review*, vol. 11 (3), 1986–1987, pp. 113–126.
6. T. A. Stewart, "Did You Ever Have to Make Up Your Mind?" *Harvard Business Review*, January 2006, p. 12; and E. Pooley, "Editor's Desk," *Fortune*, June 27, 2005, p. 16.
7. J. Pfeffer and R. I. Sutton, "Why Managing by Facts Works," *Strategy & Business*, Spring 2006, pp. 9–12.
8. See A. Langley, "In Search of Rationality: The Purposes Behind the Use of Formal Analysis in Organizations," *Administrative Science Quarterly*, December 1989, pp. 598–631; and H. A. Simon, "Rationality in Psychology and Economics," *Journal of Business*, October 1986, pp. 209–224.
9. J. G. March, "Decision-Making Perspective: Decisions in Organizations and Theories of Choice," in A. H. Van de Ven and W. F. Joyce (eds.), *Perspectives on Organization Design and Behavior* (New York: Wiley-Interscience, 1981), pp. 232–233.
10. See D. R. A. Skidd, "Revisiting Bounded Rationality," *Journal of Management Inquiry*, December 1992, pp. 343–347; B. E. Kaufman, "A New Theory of Satisficing," *Journal of Behavioral Economics*, Spring 1990, pp. 35–51; and N. M. Agnew and J. L. Brown, "Bounded Rationality: Fallible Decisions in Unbounded Decision Space," *Behavioral Science*, July 1986, pp. 148–161.
11. See, for example, G. McNamara, H. Moon, and P. Bromiley, "Banking on Commitment: Intended and Unintended Consequences of an Organization's Attempt to Attenuate Escalation of Commitment," *Academy of Management Journal*, April 2002, pp. 443–452; V. S. Rao and A. Monk, "The Effects of Individual Differences and Anonymity on Commitment to Decisions," *Journal of Social Psychology*, August 1999, pp. 496–515; C. F. Camerer and R. A. Weber, "The Econometrics and Behavioral Economics of Escalation of Commitment: A Re-examination of Staw's Theory," *Journal of Economic Behavior and Organization*, May 1999, pp. 59–82; D. R. Bobocel and J. P. Meyer, "Escalating Commitment to a Failing Course of Action: Separating the Roles of Choice and Justification," *Journal of Applied Psychology*, June 1994, pp. 360–363; and B. M. Staw, "The Escalation of Commitment to a Course of Action," *Academy of Management Review*, October 1981, pp. 577–587.
12. W. Cole, "The Stapler Wars," *Time Inside Business*, April 2005, p. A5.
13. See E. Dane and M. G. Pratt, "Exploring Intuition and Its Role in Managerial Decision Making," *Academy of Management Review*, January 2007, pp. 33–54; M. H. Bazerman and D. Chugh, "Decisions Without Blinders," *Harvard Business Review*, January 2006, pp. 88–97; C. C. Miller and R. D. Ireland, "Intuition in Strategic Decision Making: Friend or Foe in the Fast-Paced 21st Century," *Academy of Management Executive*, February 2005, pp. 19–30; E. Sadler-Smith and E. Shefy, "The Intuitive Executive: Understanding and Applying 'Gut Feel' in Decision-Making," *Academy of Management Executive*, November 2004, pp. 76–91; and L. A. Burke and M. K. Miller, "Taking the Mystery Out of Intuitive Decision Making," *Academy of Management Executive*, October 1999, pp. 91–99.
14. Miller and Ireland, "Intuition in Strategic Decision Making: Friend or Foe," p. 20.
15. E. Sadler-Smith and E. Shefy, "Developing Intuitive Awareness in Management Education," *Academy of Management Learning & Education*, June 2007, pp. 186–205.
16. M. G. Seo and L. Feldman Barrett, "Being Emotional During Decision Making—Good or Bad? An Empirical Investigation," *Academy of Management Journal*, August 2007, pp. 923–940.
17. K. R. Brousseau, M. J. Driver, G. Hourihan, and R. Larsson, "The Seasoned Executive's Decision-Making Style," *Harvard Business Review*, February 2006, pp. 111–121.
18. R. M. Kidder, "Commentary: Hunt Down a Perpetrator or Hold to a Principle: A High School's Dilemma," *Ethics Newsline* online, http://ethicsnewsline.wordpress.com, October 23, 2006
19. Information for this box came from D. Jones and A. Shaw, "Slowing Momentum: Why BPM Isn't Keeping Pace with Its Potential," *BPM Magazine*, February 2006, pp. 4–12; B. Violino, "IT Directions," *CFO*, January 2006, pp. 68–72; D. Weinberger, "Sorting Data to Suit Yourself," *Harvard Business Review*, March 2005, pp. 16–18; and C. Winkler, "Getting a Grip on Performance," *CFO-IT*, Winter 2004, pp. 38–48.
20. S. Holmes, "Inside the Coup at Nike," *Business Week*, February 6, 2006, pp. 34–37; and M. Barbaro, "Slightly Testy Nike Divorce Came Down to Data vs. Feel," *New York Times* online, www.nytimes.com, January 28, 2006.
21. C. M. Vance, K. S. Groves, Y. Paik, and H. Kindler, "Understanding and Measuring Linear-NonLinear Thinking Style for Enhanced Management Education and Professional Practice," *Academy of Management Learning & Education*, June 2007, pp. 167–185.
22. Information for this box came from N. J. Adler (ed.), *International Dimensions of Organizational Behavior*, 4th ed. (Cincinnati, OH: South-Western Publishing, 2001); B. C. McDonald and D. Hutcheson, "Dealing with Diversity Is Key to Tapping Talent," *Atlanta Business Chronicle*, December 18, 1998, pp. 45A+; and P. M. Elsass and L. M. Graves, "Demographic Diversity in Decision-Making Groups: The Experience of Women and People of Color," *Academy of Management Review*, October 1997, pp. 946–973.
23. E. Teach, "Avoiding Decision Traps," *CFO*, June 2004, pp. 97–99; and D. Kahneman and A. Tversky, "Judgment Under Uncertainty: Heuristics and Biases," *Science*, vol. 185, 1974, pp. 1124–1131.
24. Information for this section taken from S. P. Robbins, *Decide & Conquer* (Upper Saddle River, NJ: Financial Times/Prentice Hall), 2004.
25. L. Margonelli, "How Ikea Designs Its Sexy Price Tags," *Business 2.0*, October 2002, p. 108.
26. P. C. Chu, E. E. Spires, and T. Sueyoshi, "Cross-Cultural Differences in Choice Behavior and Use of Deci-sion Aids: A Comparison of Japan and the United States," *Organizational Behavior & Human Decision Processes*, vol. 77 (2), 1999, pp. 147–170.
27. J. S. Hammond, R. L. Keeney, and H. Raiffa, *Smart Choices: A Practical Guide to Making Better Decisions* (Boston, MA: Harvard Business School Press, 1999), p. 4.

28. This discussion is based on E. W. Ford, et al., "Mitigating Risks, Visible Hands, Inevitable Disasters, and Soft Variables: Management Research That Matters," *Academy of Management Executive*, November 2005, pp. 24–38; K. H. Hammonds, "5 Habits of Highly Reliable Organizations: An Interview with Karl Weick," *Fast Company*, May 2002, pp. 124–128; and K. E. Weick, "Drop Your Tools: An Allegory for Organizational Studies," *Administrative Science Quarterly*, vol. 41 (2), 1996, pp. 301–313.

Chapter 7

1. Alibaba Group Web site, www.alibaba.com, February 6, 2008; and C. Chandler, "China's Web King," *Fortune*, December 10, 2007, pp. 172–180.
2. J. L. Lunsford, "Boeing Delays Dreamliner Delivery Again," *Wall Street Journal*, April 10, 2008, p. B3; and J. Teresko, "The Boeing 787: A Matter of Materials," *IndustryWeek*, December 2007, pp. 34–38.
3. See, for example, F. Delmar and S. Shane, "Does Business Planning Facilitate the Development of New Ventures?" *Strategic Management Journal*, December 2003, pp. 1165–1185; R. M. Grant, "Strategic Planning in a Turbulent Environment: Evidence from the Oil Majors," *Strategic Management Journal*, June 2003, pp. 491–517; P. J. Brews and M. R. Hunt, "Learning to Plan and Planning to Learn: Resolving the Planning School/Learning School Debate," *Strategic Management Journal*, December 1999, pp. 889–913; C. C. Miller and L. B. Cardinal, "Strategic Planning and Firm Performance: A Synthesis of More Than Two Decades of Research," *Academy of Management Journal*, March 1994, pp. 1649–1685; N. Capon, J. U. Farley, and J. M. Hulbert, "Strategic Planning and Financial Performance: More Evidence," *Journal of Management Studies*, January 1994, pp. 22–38; D. K. Sinha, "The Contribution of Formal Planning to Decisions," *Strategic Management Journal*, October 1990, pp. 479–492; J. A. Pearce II, E. B. Freeman, and R. B. Robinson, Jr., "The Tenuous Link Between Formal Strategic Planning and Financial Performance," *Academy of Management Review*, October 1987, pp. 658–675; L. C. Rhyne, "Contrasting Planning Systems in High, Medium, and Low Performance Companies," *Journal of Management Studies*, July 1987, pp. 363–385; and J. A. Pearce II, K. K. Robbins, and R. B. Robinson, Jr., "The Impact of Grand Strategy and Planning Formality on Financial Performance," *Strategic Management Journal*, March–April 1987, pp. 125–134.
4. http://www.en.wikipedia.org/wiki/Karsanbhai_Patel
5. R. Molz, "How Leaders Use Goals," *Long Range Planning*, October 1987, p. 91.
6. C. Hymowitz, "When Meeting Targets Becomes the Strategy, CEO Is on Wrong Path," *Wall Street Journal*, March 8, 2005, p. B1.
7. http://www.inhome.rediff.com/money/2005/nov/08jrg.htm
8. http://www.in.rediff.com/money/2008/jan/21bsnl.htm
9. Annual reports from Nike (2005), Winnebago (2005), and Deutsche Bank (2004) and the EnCana Corporate Constitution 2004, www.encana.com.
10. See, for instance, J. Pfeffer, *Organizational Design* (Arlington Heights, IL: AHM Publishing, 1978), pp. 5–12; and C. K. Warriner, "The Problem of Organizational Purpose," *Sociological Quarterly*, Spring 1965, pp. 139–146.
11. J. D. Hunger and T. L. Wheelen, *Strategic Management and Business Policy*, 10th ed. (Upper Saddle River, NJ: Prentice Hall, 2006).
12. J. L. Roberts, "Signed. Sealed. Delivered?" *Newsweek*, June 20, 2005, pp. 44–46.
13. D. Drickhamer, "Braced for the Future," *IndustryWeek*, October 2004, pp. 51–52.
14. P. N. Romani, "MBO By Any Other Name Is Still MBO," *Supervision*, December 1997, pp. 6–8; and A. W. Schrader and G. T. Seward, "MBO Makes Dollar Sense," *Personnel Journal*, July 1989, pp. 32–37.
15. R. Rodgers and J. E. Hunter, "Impact of Management by Objectives on Organizational Productivity," *Journal of Applied Psychology*, April 1991, pp. 322–336.
16. G. P. Latham, "The Motivational Benefits of Goal-Setting," *Academy of Management Executive*, November 2004, pp. 126–129.
17. For additional information on goals, see, for instance, P. Drucker, *The Executive in Action* (New York: HarperCollins, 1996), pp. 207–214; and E. A. Locke and G. P. Latham, *A Theory of Goal Setting and Task Performance* (Upper Saddle River, NJ: Prentice Hall, 1990).
18. Several of these factors were suggested in R. K. Bresser and R. C. Bishop, "Dysfunctional Effects of Formal Planning: Two Theoretical Explanations," *Academy of Management Review*, October 1983, pp. 588–599; and J. S. Armstrong, "The Value of Formal Planning for Strategic Decisions: Review of Empirical Research," *Strategic Management Journal*, July–September 1982, pp. 197–211.
19. Brews and Hunt, "Learning to Plan and Planning to Learn: Resolving the Planning School/Learning School Debate."
20. S. Hamm, "It's Too Darn Hot," *BusinessWeek*, March 31, 2008, pp. 60–63; and D. Clark, "Power-Hungry Computers Put Data Centers in Bind," *Wall Street Journal*, November 14, 2005, pp. A1+.
21. C. Prystay, M. Hiebert, and K. Linebaugh, "Companies Face Ethical Issues over Tamiflu," *Wall Street Journal*, January 16, 2006, pp. B1+.
22. A. Campbell, "Tailored, Not Benchmarked: A Fresh Look at Corporate Planning," *Harvard Business Review*, March–April 1999, pp. 41–50.
23. J. H. Sheridan, "Focused on Flow," *IW*, October 18, 1999, pp. 46–51.
24. H. Mintzberg, *The Rise and Fall of Strategic Planning* (New York: The Free Press, 1994).
25. Ibid.
26. Ibid.
27. G. Hamel and C. K. Prahalad, *Competing for the Future* (Boston: Harvard Business School Press, 1994).
28. D. Miller, "The Architecture of Simplicity," *Academy of Management Review*, January 1993, pp. 116–138.
29. M. C. Mankins and R. Steele, "Stop Making Plans—Start Making Decisions," *Harvard Business Review*, January 2006, pp. 76–84; L. Bossidy and R. Charan, *Execution: The Discipline of Getting Things Done* (New York: Crown/Random House), 2002; and P. Roberts, "The Art of Getting Things Done," *Fast Company*, June 2000, p. 162.
30. Associated Press, "Dow Jones to Shrink 'Wall Street Journal,' Cut Some Data," *USA Today* online, www.usatoday.com, October 12, 2005.
31. Brews and Hunt, "Learning to Plan and Planning to Learn: Resolving the Planning School/Learning School Debate."
32. Information on Wipro Limited from Hoover's online, www.hoovers.com, March 21, 2006; R. J. Newman, "Coming and

Going," *U.S. News & World Report*, January 23, 2006, pp. 50–52; T. Atlas, "Bangalore's Big Dreams," *U.S. News & World Report*, May 2, 2005, pp. 50–52; and K. H. Hammonds, "Smart, Determined, Ambitious, Cheap: The New Face of Global Competition," *Fast Company*, February 2003, pp. 90–97.

Chapter 8

1. S. Reed, "The Master Builder of the Middle East," *BusinessWeek*, July 2, 2007, pp. 48–49.
2. The Hindu Business Line, January 3, 2009.
3. J. W. Dean, Jr., and M. P. Sharfman, "Does Decision Process Matter? A Study of Strategic Decision-Making Effectiveness," *Academy of Management Journal*, April 1996, pp. 368–396.
4. Based on A. A. Thompson, Jr., A. J. Strickland III, and J. E. Gamble, *Crafting and Executing Strategy*, 14th ed. (New York: McGraw-Hill Irwin), 2005.
5. J. Magretta, "Why Business Models Matter," *Harvard Business Review*, May 2002, pp. 86–92.
6. H. J. Cho and V. Pucik, "Relationship Between Innovativeness, Quality, Growth, Profitability, and Market Value," *Strategic Management Journal*, June 2005, pp. 555–575; W. F. Joyce, "What Really Works," *Organizational Dynamics*, May 2005, pp. 118–129; M. A. Roberto, "Strategic Decision-Making Processes," *Group & Organization Management*, December 2004, pp. 625–658; A. Carmeli and A. Tischler, "The Relationships Between Intangible Organizational Elements and Organizational Performance," *Strategic Management Journal*, December 2004, pp. 1257–1278; D. J. Ketchen, C. C. Snow, and V. L. Street, "Improving Firm Performance by Matching Strategic Decision-Making Processes to Competitive Dynamics," *Academy of Management Executive*, November 2004, pp. 29–43; E. H. Bowman and C. E. Helfat, "Does Corporate Strategy Matter?" *Strategic Management Journal*, vol. 22, 2001, pp. 1–23; P. J. Brews and M. R. Hunt, "Learning to Plan and Planning to Learn: Resolving the Planning School-Learning School Debate," *Strategic Management Journal*, vol. 20, 1999, pp. 889–913; D. J. Ketchen, Jr., J. B. Thomas, and R. R. McDaniel, Jr., "Process, Content and Context; Synergistic Effects on Performance," *Journal of Management*, vol. 22 (2), 1996, pp. 231–257; C. C. Miller and L. B. Cardinal, "Strategic Planning and Firm Performance: A Synthesis of More Than Two Decades of Research," *Academy of Management Journal*, December 1994, pp. 1649–1665; and N. Capon, J. U. Farley, and J. M. Hulbert, "Strategic Planning and Financial Performance: More Evidence," *Journal of Management Studies*, January 1994, pp. 105–110.
7. These examples came from S. Dutta and I. Mia, *The Global Information Technology Report 2007–2008*, World Economic Forum, www.weforum.org; J. D. Miller, "Which Colleges Will Make the Grade?" *Better Investing*, April 2006, p. 49; and W. Cole, S. Steptoe, and S. S. Dale, "The Multitasking Generation," *Time*, March 27, 2006, pp. 48–55.
8. C. K. Prahalad and G. Hamel, "The Core Competence of the Corporation," *Harvard Business Review*, May–June 1990, pp. 79–91.
9. http://www.reuters.com/article/rbssIndustryMaterialsUtilitiesNews/idUSBOM14412420080428
10. Information in this box from D. McGinn, "From Harvard to Las Vegas," *Newsweek*, April 18, 2005, pp. E8–E14; G. Lindsay, "Prada's High-Tech Misstep," *Business 2.0*, March 2004, pp. 72–75; G. Loveman, "Diamonds in the Data Mine," *Harvard Business Review*, May 2003, pp. 109–113; and L. Gary, "Simplify and Execute: Words to Live By in Times of Turbulence," *Harvard Management Update*, January 2003, p. 12.
11. Damodaran, H. (2008). India's New Capitalists: Cases Business and Industry in a Modern Nation.
12. http://www.godrej.com/GodrejNew/GodrejHome/About/CorporateProfile/CorporateProfile/BusinessProfile.htm
13. H. Quarls, T. Pernsteiner, and K. Rangan, "Love Your Dogs," *Strategy & Business*, Spring 2006, pp. 58–65; and P. Haspeslagh, "Portfolio Planning: Uses and Limits," *Harvard Business Review*, January–February 1982, pp. 58–73.
14. *Perspective on Experience* (Boston: Boston Consulting Group, 1970).
15. Rumelt, "Towards a Strategic Theory of the Firm," in R. Lamb (ed.), *Competitive Strategic Management* (Upper Saddle River, NJ: Prentice Hall, 1984), pp. 556–570; M. E. Porter, *Competitive Advantage: Creating and Sustaining Superior Performance* (New York: The Free Press, 1985); J. Barney, "Firm Resources and Sustained Competitive Advantage," *Journal of Management*, vol. 17 (1), 1991, pp. 99–120; M. A. Peteraf, "The Cornerstones of Competitive Advantage: A Resource-Based View," *Strategic Management Journal*, March 1993, pp. 179–191; and J. B. Barney, "Looking Inside for Competitive Advantage," *Academy of Management Executive*, November 1995, pp. 49–61.
16. N. A. Shepherd, "Competitive Advantage: Mapping Change and the Role of the Quality Manager of the Future," *Annual Quality Congress*, May 1998, pp. 53–60; T. C. Powell, "Total Quality Management as Competitive Advantage: A Review and Empirical Study," *Strategic Management Journal*, January 1995, pp. 15–37; and R. D. Spitzer, "TQM: The Only Source of Sustainable Competitive Advantage," *Quality Progress*, June 1993, pp. 59–64.
17. See R. J. Schonenberger, "Is Strategy Strategic? Impact of Total Quality Management on Strategy," *Academy of Management Executive*, August 1992, pp. 80–87; C. A. Barclay, "Quality Strategy and TQM Policies: Empirical Evidence," *Management International Review*, Special Issue 1993, pp. 87–98; R. Jacob, "TQM: More Than a Dying Fad?" *Fortune*, October 18, 1993, pp. 66–72; R. Krishnan, A. B. Shani, R. M. Grant, and R. Baer, "In Search of Quality Improvement Problems of Design and Implementation," *Academy of Management Executive*, November 1993, pp. 7–20; B. Voss, "Quality's Second Coming," *Journal of Business Strategy*, March–April 1994, pp. 42–46; and special issue of *Academy of Management Review* devoted to total quality management, July 1994, pp. 390–584.
18. See, for example, M. E. Porter, *Competitive Strategy: Techniques for Analyzing Industries and Competitors* (New York: The Free Press, 1980); Porter, *Competitive Advantage: Creating and Sustaining Superior" Performance*; G. G. Dess and P. S. Davis, "Porter's (1980) Generic Strategies as Determinants of Strategic Group Membership and Organizational Performance," *Academy of Management Journal*, September 1984, pp. 467–488; G. G. Dess and P. S. Davis, "Porter's (1980) Generic Strategies and Performance: An Empirical Examination with American Data—Part I: Testing Porter," *Organization Studies*, vol. 1, 1986, pp. 37–55; G. G. Dess and P. S. Davis, "Porter's (1980) Generic Strategies and

Performance: An Empirical Examination with American Data—Part II: Performance Implications," *Organization Studies*, vol. 3, 1986, pp. 255–261; M. E. Porter, "From Competitive Advantage to Corporate Strategy," *Harvard Business Review*, May–June 1987, pp. 43–59; A. I. Murray, "A Contingency View of Porter's 'Generic Strategies,'" *Academy of Management Review*, July 1988, pp. 390–400; C. W. L. Hill, "Differentiation Versus Low Cost or Differentiation and Low Cost: A Contingency Framework," *Academy of Management Review*, July 1988, pp. 401–412; I. Bamberger, "Developing Competitive Advantage in Small and Medium-Sized Firms," *Long Range Planning*, October 1989, pp. 80–88; D. F. Jennings and J. R. Lumpkin, "Insights Between Environmental Scanning Activities and Porter's Generic Strategies: An Empirical Analysis," *Strategic Management Journal*, vol. 18 (4), 1992, pp. 791–803; N. Argyres and A. M. McGahan, "An Interview with Michael Porter," *Academy of Management Executive*, May 2002, pp. 43–52; and A. Brandenburger, "Porter's Added Value: High Indeed!" *Academy of Management Executive*, May 2002, pp. 58–60.

19. Hill, "Differentiation Versus Low Cost or Differentiation and Low Cost"; R. E. White, "Organizing to Make Business Unit Strategies Work," in H. E. Glass (ed.), *Handbook of Business Strategy*, 2d ed. (Boston: Warren Gorham and Lamont, 1991), pp. 24.1–24.14; D. Miller, "The Generic Strategy Trap," *Journal of Business Strategy*, January–February 1991, pp. 37–41; S. Cappel, P. Wright, M. Kroll, and D. Wyld, "Competitive Strategies and Business Performance: An Empirical Study of Select Service Businesses," *International Journal of Management*, March 1992, pp. 1–11; and J. W. Bachmann, "Competitive Strategy: It's O.K. to be Different," *Academy of Management Executive*, May 2002, pp. 61–65.

20. IFPI Web site, www.ifpi.org, April 28, 2008; Recording Industry Association of America Web site, www.riaa.com, April 28, 2008; and E. Pfanner, "Music Industry Steps Up Search for Digital Revenue," *International Herald Tribune* online, www.iht.com, January 24, 2008.

21. K. Shimizu and M. A. Hitt, "Strategic Flexibility: Organizational Preparedness to Reverse Ineffective Decisions," *Academy of Management Executive*, November 2004, p. 44.

22. T. Lowry, "ESPN's Cell-Phone Fumble," *Business-Week* online, www.businessweek.com, October 30, 2006; and T. Lowry, "In the Zone," *BusinessWeek*, October 17, 2005, pp. 66–77.

23. E. Kim, D. Nam, and J. L. Stimpert, "The Applicability of Porter's Generic Strategies in the Digital Age: Assumptions, Conjectures, and Suggestions," *Journal of Management*, vol. 30 (5), 2004, pp. 569–589; and G. T. Lumpkin, S. B. Droege, and G. G. Dess, "E-Commerce Strategies: Achieving Sustainable Competitive Advantage and Avoiding Pitfalls," *Organizational Dynamics*, Spring 2002, pp. 325–340.

24. Kim, Nam, and Stimpert, "The Applicability of Porter's Generic Strategies in the Digital Age: Assumptions, Conjectures, and Suggestions."

25. J. Gaffney, "Shoe Fetish," *Business 2.0*, March 2002, pp. 98–99.

26. J. Doebele, "The Engineer," *Forbes*, January 9, 2006, pp. 122–124.

27. S. Ellison, "P&G to Unleash Dental Adult-Pet Food," *Wall Street Journal*, December 12, 2002, p. B4.

Chapter 9

1. J. Greene, "Where Designers Rule," *BusinessWeek*, November 5, 2007, pp. 46–51.

2. http://www.iloveindia.com/economy-of-india/top-50-companies/tvs.html

3. D. Hudepohl, "Finesse a Flexible Work Schedule," *Wall Street Journal*, February 19, 2008, p. B8.

4. See, for example, R. L. Daft, *Organization Theory and Design*, 9th ed. (Mason, OH: South-Western Publishing), 2007.

5. Manikuti, S., and Vohra, N. (2004), Aravind Eye Care System: Giving Them the Most Precious Gift. IIM, Ahmedabad.

6. S. E. Humphrey, J. D. Nahrgang, and F. P. Morgeson, "Integrating Motivational, Social, and Contextual Work Design Features: A Meta-Analytic Summary and Theoretical Expansion of the Work Design Literature," *Journal of Applied Psychology*, September 2007, pp. 1332–1356.

7. http://www.energymanagertraining.com/eca2004/award2004/Paper%20Pulp/ITC%20Saharanpur.pdf.

8. For a discussion of authority, see W. A. Kahn and K. E. Kram, "Authority at Work: Internal Models and Their Organizational Consequences," *Academy of Management Review*, January 1994, pp. 17–50.

9. http://www.itgo.in/index.php?option=com_content&task=view&id=9037&sectionid=1&secid=7&Itemid=1

10. R. Ashkenas, "Simplicity-Minded Management," *Harvard Business Review*, December 2007, pp. 101–109; and P. Glader, "It's Not Easy Being Lean," *Wall Street Journal*, June 19, 2006, pp. B1+.

11. R. C. Morais, "The Old Lady Is Burning Rubber," *Forbes*, November 26, 2007, pp. 146–150.

12. D. Van Fleet, "Span of Management Research and Issues," *Academy of Management Journal*, September 1983, pp. 546–552.

13. G. Anders, "Overseeing More Employees—With Fewer Managers," *Wall Street Journal*, March 24, 2008, p. B6.

14. See, for example, H. Mintzberg, *Power In and Around Organizations* (Upper Saddle River, NJ: Prentice Hall, 1983); J. Child, *Organization: A Guide to Problems and Practices* (London: Kaiser & Row, 1984).

15. http://www.resources.greatplacetowork.com/article/pdf/building_a_great_place_to_work-a_matter_of_trust.pdf

16. J. Badal, "Can a Company Be Run as a Democracy?" *Wall Street Journal*, April 23, 2007, pp. B1+.

17. http://www.siliconindia.com/magazine/prnt.php?WUO967504715;http://www.eetindia.co.in/ART_8800550077_1800001_NT_f78b6179.HTM; http://www.drivingexcellence.biz/

18. Ibid.

19. T. Burns and G. M. Stalker, *The Management of Innovation* (London: Tavistock, 1961); and D. A. Morand, "The Role of Behavioral Formality and Informality in the Enactment of Bureaucratic Versus Organic Organizations," *Academy of Management Review*, October 1995, pp. 831–872.

20. J. Whalen, "Bureaucracy Buster? Glaxo Lets Scientists Choose Its New Drugs," *Wall Street Journal*, March 27, 2006, p. B1+.

21. http://www.iciciprulife.com/public/About-us/ProfileTeam-ShikhaSharma.htm.

22. A. D. Chandler, Jr., *Strategy and Structure: Chapters in the History of the Industrial Enterprise* (Cambridge, MA: MIT Press, 1962).

23. See, for instance, L. L. Bryan and C. I. Joyce, "Better Strategy Through Organizational Design," *The McKinsey Quarterly*, no. 2, 2007, pp. 21–29; D. Jennings and S. Seaman, "High and Low Levels of Organizational Adaptation: An Empirical Analysis of Strategy, Structure, and Performance," *Strategic Management Journal*, July 1994, pp. 459–475; D. C. Galunic and K. M. Eisenhardt, "Renewing the Strategy–Structure–Performance Paradigm," in B. M. Staw and L. L. Cummings (eds.), *Research in Organizational Behavior*, vol. 16 (Greenwich, CT: JAI Press, 1994), pp. 215–255; R. Parthasarthy and S. P. Sethi, "Relating Strategy and Structure to Flexible Automation: A Test of Fit and Performance Implications," *Strategic Management Journal*, vol. 14 (6), 1993, pp. 529–549; H. A. Simon, "Strategy and Organizational Evolution," *Strategic Management Journal*, January 1993, pp. 131–142; H. L. Boschken, "Strategy and Structure: Re-conceiving the Relationship," *Journal of Management*, March 1990, pp. 135–150; D. Miller, "The Structural and Environmental Correlates of Business Strategy," *Strategic Management Journal*, January–February 1987, pp. 55–76; and R. E. Miles and C. C. Snow, *Organiza-tional Strategy, Structure, and Process* (New York: McGraw-Hill, 1978).

24. See, for instance, P. M. Blau and R. A. Schoenherr, *The Structure of Organizations* (New York: Basic Books, 1971); D. S. Pugh, "The Aston Program of Research: Retrospect and Prospect," in A. H. Van de Ven and W. F. Joyce (eds.), *Perspectives on Organization Design and Behavior* (New York: John Wiley, 1981), pp. 135–166; and R. Z. Gooding and J. A. Wagner III, "A Meta-Analytic Review of the Relationship Between Size and Performance: The Productivity and Efficiency of Organizations and Their Subunits," *Administrative Science Quarterly*, December 1985, pp. 462–481.

25. J. Woodward, *Industrial Organization: Theory and Practice* (London: Oxford University Press, 1965).

26. See, for instance, C. Perrow, "A Framework for the Comparative Analysis of Organizations," *American Sociological Review*, April 1967, pp. 194–208; J. D. Thompson, *Organizations in Action* (New York: McGraw-Hill, 1967); J. Hage and M. Aiken, "Routine Technology, Social Structure, and Organizational Goals," *Administrative Science Quarterly*, September 1969, pp. 366–377; and C. C. Miller, W. H. Glick, Y. D. Wang, and G. Huber, "Understanding Technology–Structure Relationships: Theory Development and Meta-Analytic Theory Testing," *Academy of Management Journal*, June 1991, pp. 370–399.

27. D. M. Rousseau and R. A. Cooke, "Technology and Structure: The Concrete, Abstract, and Activity Systems of Organizations," *Journal of Management*, Fall–Winter 1984, pp. 345–361; and D. Gerwin, "Relationships Between Structure and Technology," in P. C. Nystrom and W. H. Starbuck (eds.), *Handbook of Organizational Design*, vol. 2 (New York: Oxford University Press, 1981), pp. 3–38.

28. S. Rausch and J. Birkinshaw, "Organizational Ambidexterity: Antecedents, Outcomes, and Moderators," *Journal of Management*, June 2008, pp. 375–409; M. Yasai-Ardekani, "Structural Adaptations to Environments," *Academy of Management Review*, January 1986, pp. 9–21; P. Lawrence and J. W. Lorsch, *Organization and Environment: Managing Differentiation and Integration* (Boston: Harvard Business School, Division of Research, 1967); and F. E. Emery and E. Trist, "The Causal Texture of Organizational Environments," *Human Relations*, February 1965, pp. 21–32.

29. S. Reed, "He's Brave Enough to Shake Up Shell," *BusinessWeek*, July 18, 2005, p. 53.

30. H. Mintzberg, *Structure in Fives: Designing Effective Organizations* (Upper Saddle River, NJ: Prentice Hall, 1983), p. 157.

31. R. J. Williams, J. J. Hoffman, and B. T. Lamont, "The Influence of Top Management Team Characteristics on M-Form Implementation Time," *Journal of Managerial Issues*, Winter 1995, pp. 466–480.

32. See, for example, G. J. Castrogiovanni, "Organization Task Environments: Have They Changed Fundamentally Over Time?" *Journal of Management*, vol. 28 (2), 2002, pp. 129–150; D. F. Twomey, "Leadership, Organizational Design, and Competitiveness for the 21st Century," *Global Competitiveness*, 2002, pp. S31–S40; M. Hammer, "Processed Change: Michael Hammer Sees Process as 'The Clark Kent of Business Ideas'—A Concept That Has the Power to Change a Company's Organizational Design," *Journal of Business Strategy*, November–December 2001, pp. 11–15; T. Clancy, "Radical Surgery: A View from the Operating Theater," *Academy of Management Executive*, February 1994, pp. 73–78; I. I. Mitroff, R. O. Mason, and C. M. Pearson, "Radical Surgery: What Will Tomorrow's Organizations Look Like?" *Academy of Management Executive*, February 1994, pp. 11–21; and R. E. Hoskisson, C. W. L. Hill, and H. Kim, "The Multidivisional Structure: Organizational Fossil or Source of Value?" *Journal of Management*, vol. 19 (2), 1993, pp. 269–298.

33. Q. Hardy, "Google Thinks Small," *Forbes*, November 14, 2005, pp. 198–202.

34. See, for example, D. R. Denison, S. L. Hart, and J. A. Kahn, "From Chimneys to Cross-Functional Teams: Developing and Validating a Diagnostic Model," *Academy of Management Journal*, December 1996, pp. 1005–1023; D. Ray and H. Bronstein, *Teaming Up: Making the Transition to a Self-Directed Team-Based Organization* (New York: McGraw Hill, 1995); J. R. Katzenbach and D. K. Smith, *The Wisdom of Teams* (Boston: Harvard Business School Press, 1993); J. A. Byrne, "The Horizontal Corporation," *BusinessWeek*, December 20, 1993, pp. 76–81; B. Dumaine, "Payoff from the New Management," *Fortune*, December 13, 1993, pp. 103–110; and H. Rothman, "The Power of Empowerment," *Nation's Business*, June 1993, pp. 49–52.

35. P. Kaihla, "Best-Kept Secrets of the World's Best Companies," *Business 2.0*, April 2006, p. 83; C. Taylor, "School of Bright Ideas," *Time Inside Business*, April 2005, pp. A8–A12; and B. Nussbaum, "The Power of Design," *BusinessWeek*, May 17, 2004, pp. 86–94.

36. See, for example, G. G. Dess, A. M. A. Rasheed, K. J. McLaughlin, and R. L. Priem, "The New Corporate Architecture," *Academy of Management Executive*, August 1995, pp. 7–20.

37. For additional readings on boundaryless organizations, see Rausch and Birkinshaw, "Organizational Ambidexterity: Antecedents, Outcomes, and Moderators"; M. F. R. Kets de Vries, "Leadership Group Coaching in Action: The Zen of Creating High Performance Teams," *Academy of Management Executive*, February 2005, pp. 61–76; J. Child and R. G. McGrath, "Organizations Unfettered: Organizational Form in an Information-Intensive Economy," *Academy of Management Journal*, December 2001, pp. 1135–1148; M. Hammer and S. Stanton, "How Process Enterprises Really Work," *Harvard Business Review*, November–December 1999, pp. 108–118; T. Zenger and W. Hesterly, "The Disaggregation of Corporations: Selective Intervention, High-Powered Incentives, and Modular

Units," *Organization Science*, vol. 8, 1997, pp. 209–222; R. Ashkenas, D. Ulrich, T. Jick, and S. Kerr, *The Boundaryless Organization: Breaking the Chains of Organizational Structure* (San Francisco: Jossey-Bass, 1997); R. M. Hodgetts, "A Conversation with Steve Kerr," *Organizational Dynamics*, Spring 1996, pp. 68–79; and J. Gebhardt, "The Boundaryless Organization," *Sloan Management Review*, Winter 1996, pp. 117–119. For another view of boundaryless organizations, see B. Victor, "The Dark Side of the New Organizational Forms: An Editorial Essay," *Organization Science*, November 1994, pp. 479–482.

38. See, for instance, Y. Shin, "A Person–Environment Fit Model for Virtual Organizations," *Journal of Management*, December 2004, pp. 725–743; D. Lyons, "Smart and Smarter," *Forbes*, March 18, 2002, pp. 40–41; W. F. Cascio, "Managing a Virtual Workplace," *Academy of Management Executive*, August 2000, pp. 81–90; G. G. Dess, A. M. A. Rasheed, K. J. McLaughlin, and R. L. Priem, "The New Corporate Architecture"; H. Chesbrough and D. Teece, "When is Virtual Virtuous: Organizing for Innovation," *Harvard Business Review*, January–February 1996, pp. 65–73; and W. H. Davidow and M. S. Malone, *The Virtual Corporation* (New York: HarperCollins, 1992).

39. "Could Your Brand Pass the Tee Shirt Test?" *Fortune*, May 28, 2007, p. 122; M. Maddever, "The New School: An Inconvenient Truth," *Strategy Magazine*, www.strategymag.com, April 2007; K. Hugh, "Goodson Forecasts Future Shock," *ADWEEK*, www.adweek.com, March 5, 2007; J. Ewing, "Amsterdam's Red-Hot Ad Shops," *BusinessWeek*, December 18, 2006, p. 52; and T. Howard, "StrawberryFrog Hops to a Different Drummer," *USA Today*, October 10, 2005, p. 4B.

40. R. E. Miles, et al., "Organizing in the Knowledge Age: Anticipating the Cellular Form," *Academy of Management Executive*, November 1997, pp. 7–24; C. Jones, W. Hesterly, and S. Borgatti, "A General Theory of Network Governance: Exchange Conditions and Social Mechanisms," *Academy of Management Review*, October 1997, pp. 911–945; R. E. Miles and C. C. Snow, "The New Network Firm: A Spherical Structure Built on Human Investment Philosophy," *Organizational Dynamics*, Spring 1995, pp. 5–18; and R. E. Miles and C. C. Snow, "Causes of Failures in Network Organizations," *California Management Review*, vol. 34 (4), 1992, pp. 53–72.

41. G. Hoetker, "Do Modular Products Lead to Modular Organizations?" *Strategic Management Journal*, June 2006, pp. 501–518; C. H. Fine, "Are You Modular or Integral?" *Strategy & Business*, Summer 2005, pp. 44–51; D. A. Ketchen, Jr., and G. T. M. Hult, "To Be Modular or Not to Be? Some Answers to the Question," *Academy of Management Executive*, May 2002, pp. 166–167; M. A. Schilling, "The Use of Modular Organizational Forms: An Industry-Level Analysis," *Academy of Management Journal*, December 2001, pp. 1149–1168; D. Lei, M. A. Hitt, and J. D. Goldhar, "Advanced Manufacturing Technology: Organizational Design and Strategic Flexibility," *Organization Studies*, vol. 17, 1996, pp. 501–523; R. Sanchez and J. Mahoney, "Modularity Flexibility and Knowledge Management in Product and Organization Design," *Strategic Management Journal*, vol. 17, 1996, pp. 63–76; and R. Sanchez, "Strategic Flexibility in Product Competition," *Strategic Management Journal*, vol. 16, 1995, pp. 135–159.

42. C. Hymowitz, "Have Advice, Will Travel," *Wall Street Journal*, June 5, 2006, pp. B1+.

43. S. Reed, A. Reinhardt, and A. Sains, "Saving Ericsson," *BusinessWeek*, November 11, 2002, pp. 64–68.

44. P. Engardio, "The Future of Outsourcing," *BusinessWeek*, January 30, 2006, pp. 50–58.

45. C. E. Connelly and D. G. Gallagher, "Emerging Trends in Contingent Work Research," *Journal of Management*, November, 2004, pp. 959–983.

46. Information in this box from R. Yu, "Work Away from Work Gets Easier with Technology," *USA Today*, November 28, 2006, p. 8B; M. Weinstein, "Going Mobile," *Training*, September 2006, pp. 24–29; C. Cobbs, "Technology Helps Boost Multitasking," *Springfield (Missouri) News-Leader*, June 15, 2006, p. 5B; C. Edwards, "Wherever You Go, You're On the Job," *BusinessWeek*, June 20, 2005, pp. 87–90; and S. E. Ante, "The World Wide Work Space," *BusinessWeek*, June 6, 2005, pp. 106–108.

47. P. Olson, "Tesco's Landing," *Forbes*, June 4, 2007, pp. 116–118; and P. M. Senge, *The Fifth Discipline: The Art and Practice of Learning Organizations* (New York: Doubleday, 1990).

48. D. A. Garvin, A. C. Edmondson, and F. Gino, "Is Yours a Learning Organization?" *Harvard Business Review*, March 2008, pp. 109–116; A. N. K. Chen and T. M. Edgington, "Assessing Value in Organizational Knowledge Creation: Considerations for Knowledge Workers," *MIS Quarterly*, June 2005, pp. 279–309; K. G. Smith, C. J. Collins, and K. D. Clark, "Existing Knowledge, Knowledge Creation Capability, and the Rate of New Product Introduction in High-Technology Firms," *Academy of Management Journal*, April 2005, pp. 346–357; R. Cross, A. Parker, L. Prusak, and S. P. Borgati, "Supporting Knowledge Creation and Sharing in Social Networks," *Organizational Dynamics*, Fall 2001, pp. 100–120; M. Schulz, "The Uncertain Relevance of Newness: Organizational Learning and Knowledge Flows," *Academy of Management Journal*, August 2001, pp. 661–681; G. Szulanski, "Exploring Internal Stickiness: Impediments to the Transfer of Best Practice within the Firm," *Strategic Management Journal*, Winter Special Issue, 1996, pp. 27–43; and J. M. Liedtka, "Collaborating Across Lines of Business for Competitive Advantage," *Academy of Management Executive*, April 1996, pp. 20–37.

49. N. M. Adler, *International Dimensions of Organi-zational Behavior*, 5th ed. (Cincinnati, OH: South-Western Publishing), 2008, p. 62.

50. P. B. Smith and M. F. Peterson, "Demographic Effects on the Use of Vertical Sources of Guidance by Managers in Widely Differing Cultural Contexts," *International Journal of Cross Cultural Management*, April 2005, pp. 5–26.

Chapter 10

1. http://www.merinews.com/catFull.jsp?articleID=146144

2. http://www.change.godrej.com/marapr/marc_int.htm

3. A. Carmeli and J. Shaubroeck, "How Leveraging Human Resource Capital with Its Competitive Distinctiveness Enhances the Performance of Commercial and Public Organizations," *Human Resource Management*, Winter 2005, pp. 391–412; L. Bassi and D. McMurrer, "How's Your Return on People?" *Harvard Business Review*, March 2004, p. 18; C. J. Collins and K. D. Clark, "Strategic Human Resource Practices, Top Management Team Social Networks, and Firm Performance: The Role of Human Resource Practices in Creating Organizational Competitive Advantage," *Academy of Management Journal*, December 2003, pp. 740–751; J. Pfeffer, *The Human Equation* (Boston:

Harvard Business School Press, 1998); J. Pfeffer, *Competitive Advantage Through People* (Boston: Harvard Business School Press, 1994); A. A. Lado and M. C. Wilson, "Human Resource Systems and Sustained Competitive Advantage," *Academy of Management Review*, October 1994, pp. 699–727; and P. M. Wright and G. C. McMahan, "Theoretical Perspectives for Strategic Human Resource Management," *Journal of Management*, vol. 18 (1), 1992, pp. 295–320.

4. Watson Wyatt Worldwide, *Maximizing the Return on Your Human Capital Investment: The 2005 Watson Wyatt Human Capital Index® Report*, Washington, DC: Watson Wyatt Worldwide; Watson Wyatt Worldwide, *WorkAsia 2004/2005: A Study of Employee Attitudes in Asia*, Washington, DC: Watson Wyatt Worldwide; and Watson Wyatt Worldwide, *European Human Capital Index 2002*, Washington, DC: Watson Wyatt Worldwide.

5. See, for example, L. Sun, S. Aryee, and K. S. Law, "High-Performance Human Resource Practices, Citizenship Behavior, and Organizational Performance: A Relational Perspective," *Academy of Management Journal*, June 2007, pp. 558–577; Carmeli and Shaubroeck, "How Leveraging Human Resource Capital With Its Competitive Distinctiveness Enhances the Performance of Commercial and Public Organizations"; Y. Y. Kor and H. Leblebici, "How Do Interdependencies Among Human-Capital Deployment, Development, and Diversification Strategies Affect Firms' Financial Performance?" *Strategic Management Journal*, October 2005, pp. 967–985; D. E. Bowen and C. Ostroff, "Understanding HRM–Firm Performance Linkages: The Role of the 'Strength' of the HRM System," *Academy of Management Review*, April 2004, pp. 203–221; A. S. Tsui, J. L. Pearce, L. W. Porter, and A. M. Tripoli, "Alternative Approaches to the Employee-Organization Relationship: Does Investment in Employees Pay Off?" *Academy of Management Journal*, October 1997, pp. 1089–1121; M. A. Huselid, S. E. Jackson, and R. S. Schuler, "Technical and Strategic Human Resource Management Effectiveness as Determinants of Firm Performance," *Academy of Management Journal*, January 1997, pp. 171–188; J. T. Delaney and M. A. Huselid, "The Impact of Human Resource Management Practices on Perceptions of Organizational Performance," *Academy of Management Journal*, August 1996, pp. 949–969; B. Becker and B. Gerhart, "The Impact of Human Resource Management on Organizational Performance: Progress and Prospects," *Academy of Management Journal*, August 1996, pp. 779–801; M. J. Koch and R. G. McGrath, "Improving Labor Productivity: Human Resource Management Policies Do Matter," *Strategic Management Journal*, May 1996, pp. 335–354; and M. A. Huselid, "The Impact of Human Resource Management Practices on Turnover, Productivity, and Corporate Financial Performance," *Academy of Management Journal*, June 1995, pp. 635–672.

6. "Human Capital a Key to Higher Market Value," *Business Finance*, December 1999, p. 15.

7. M. Boyle, "Happy People, Happy Returns," *Fortune*, January 11, 2006, p. 100.

8. Bureau of Labor Statistics, *Union Members Summary 2007*, www.bls.gov, January 25, 2008; J. Visser, "Union Membership Statistics in 24 Countries," *Monthly Labor Review*, January 2006, pp. 38–49; T. Fuller, "Workers and Bosses: Friends or Foes?" *International Herald Tribune* online, www.iht.com, January 11, 2005; and U.S. Department of Labor, *Foreign Labor Trends—Mexico*, www.dol.gov, 2002.

9. http://labourbureau.nic.in/TU%202k2%20Chapter%202.htm

10. S. Greenhouse, "Wal-Mart Settles U.S. Suit About Overtime," *New York Times* online, www.nytimes.com, January 26, 2007.

11. P. Digh, "Religion in the Workplace," *HRMagazine*, December 1998, p. 88.

12. S. Armour, "Lawsuits Pin Target on Managers," *USA Today* online, www.usatoday.com, October 1, 2002.

13. http://www.jsk.gov.in/indias_population.asp; http://www.census.gov/ipc/prod/ib-9701.pdf.

14. J. Sullivan, "Workforce Planning: Why to Start Now," *Workforce*, September 2002, pp. 46–50.

15. N. Byrnes, "Star Search," *BusinessWeek*, October 10, 2005, pp. 68–78.

16. J. W. Boudreau and P. M. Ramstad, "Where's Your Pivotal Talent?" *Harvard Business Review*, April 2005, pp. 23–24.

17. A. S. Bargerstock and G. Swanson, "Four Ways to Build Cooperative Recruitment Alliances," *HRMagazine*, March 1991, p. 49; and T. J. Bergmann and M. S. Taylor, "College Recruitment: What Attracts Students to Organizations?" *Personnel*, May–June 1984, pp. 34–46.

18. J. R. Gordon, *Human Resource Management: A Practical Approach* (Boston: Allyn & Bacon, 1986), p. 170.

19. J. Hitt, "Are Brands Out of Hand?" *Fast Company*, November 2000, p. 52.

20. M. Helft, "In Fierce Competition, Google Finds Novel Ways to Feed Hiring Machine," *New York Times* online, www.nytimes.com, May 28, 2007.

21. K. Plourd, "Lights, Camera, Audits!" *CFO*, November 2007, p. 18.

22. S. Leibs, "Online Talent Shopping," *CFO-IT*, Fall 2005, p. 25.

23. See, for example, R. W. Griffeth, P. W. Hom, L. S. Fink, and D. J. Cohen, "Comparative Tests of Multivariate Models of Recruiting Sources Effects," *Journal of Management*, vol. 23 (1), 1997, pp. 19–36; and J. P. Kirnan, J. E. Farley, and K. F. Geisinger, "The Relationship Between Recruiting Source, Applicant Quality, and Hire Performance: An Analysis by Sex, Ethnicity, and Age," *Personnel Psychology*, Summer 1989, pp. 293–308.

24. A. M. Ryan and R. E. Ployhart, "Applicants' Perceptions of Selection Procedures and Decisions: A Critical Review and Agenda for the Future," *Journal of Management*, vol. 26 (3), 2000, pp. 565–606; C. Fernandez-Araoz, "Hiring Without Firing," *Harvard Business Review*, July–August, 1999, pp. 108–120; A. K. Korman, "The Prediction of Managerial Performance: A Review," *Personnel Psychology*, Summer 1986, pp. 295–322; G. C. Thornton, *Assessment Centers in Human Resource Management* (Reading, MA: Addison-Wesley, 1992); I. T. Robertson and R. S. Kandola, "Work Sample Tests: Validity, Adverse Impact, and Applicant Reaction," *Journal of Occupational Psychology*, vol. 55 (3), 1982, pp. 171–183; E. E. Ghiselli, "The Validity of Aptitude Tests in Personnel Selection," *Personnel Psychology*, Winter 1973, p. 475; G. Grimsley and H. F. Jarrett, "The Relation of Managerial Achievement to Test Measures Obtained in the Employment Situation: Methodology and Results," *Personnel Psychology*, Spring 1973, pp. 31–48; J. J. Asher, "The Biographical Item: Can It Be Improved?" *Personnel Psychology*, Summer 1972, p. 266; and G. W. England, *Development and Use of Weighted Application Blanks*, rev. ed. (Minneapolis: Industrial Relations Center, University of Minnesota, 1971).

25. See, for example, Y. Ganzach, A. Pazy, Y. Ohayun, and E. Brainin, "Social Exchange and Organizational Commitment: Decision-Making Training for Job Choice as an Alternative to the Realistic Job Preview," *Personnel Psychology*, Autumn 2002, pp. 613–637; B. M. Meglino, E. C. Ravlin, and A. S. DeNisi, "A Meta-Analytic Examination of Realistic Job Preview Effectiveness: A Test of Three Counterintuitive Propositions," *Human Resource Management Review*, vol. 10 (4), 2000, pp. 407–434; J. A. Breaugh and M. Starke, "Research on Employee Recruitment: So Many Studies, So Many Remaining Questions," *Journal of Management*, vol. 26 (3), 2000, pp. 405–434; and S. L. Premack and J. P. Wanous, "A Meta-Analysis of Realistic Job Preview Experiments," *Journal of Applied Psychology*, November 1985, pp. 706–720.
26. "Lawyers Warn Facebook a Risky Tool for Background," *Workforce Management*, www.workforce.com, May 6, 2008; M. Conlin, "You Are What You Post," *BusinessWeek*, March 27, 2006, pp. 52–53; and J. Kornblum and M. B. Marklein, "What You Say Online Could Haunt You," *USA Today*, March 8, 2006, pp. 1A+.
27. Narayanan, R., and Neethi, S. (2001) Building Software Engineering Professionals: TCS Experience, IEEE, 162–171.
28. D. G. Allen, "Do Organizational Socialization Tactics Influence Newcomer Embeddedness and Turn-over?" *Journal of Management*, April 2006, pp. 237–256; C. L. Cooper, "The Changing Psychological Contract at Work: Revisiting the Job Demands-Control Model," *Occupational and Environmental Medicine*, June 2002, p. 355; D. M. Rousseau and S. A. Tijoriwala, "Assessing Psychological Contracts: Issues, Alternatives and Measures," *Journal of Organizational Behavior*, vol. 19, 1998, pp. 679–695; and S. L. Robinson, M. S. Kraatz, and D. M. Rousseau, "Changing Obligations and the Psychological Contract: A Longitudinal Study," *Academy of Management Journal*, February 1994, pp. 137–152.
29. http://www.constructionupdate.com/products/constructionworld/2008/feb2008/002.html
30. http://www.thehindubusinessline.com/2007/02/10/stories/2007021007190100.html
31. Ibid.
32. K. Sulkowicz, "Straight Talk at Review Time," *BusinessWeek*, September 10, 2007, p. 16.
33. This section based on R. I. Henderson, *Compen-sation Management in a Knowledge-Based World*, 9th ed. (Upper Saddle River, NJ: Prentice Hall, 2003).
34. M. P. Brown, M. C. Sturman, and M. J. Simmering, "Compensation Policy and Organizational Performance: The Efficiency, Operational and Financial Implications of Pay Levels and Pay Structure," *Academy of Management Journal*, December 2003, pp. 752–762; J. D. Shaw, N. P. Gupta, and J. E. Delery, "Pay Dispersion and Workforce Performance: Moderating Effects of Incentives and Interdependence," *Strategic Management Journal*, June 2002, pp. 491–512; E. Montemayor, "Congruence between Pay Policy and Competitive Strategy in High-Performing Firms," *Journal of Management*, vol. 22 (6), 1996, pp. 889–908; and L. R. Gomez-Mejia, "Structure and Process of Diversification, Compensation Strategy, and Firm Performance," *Strategic Management Journal*, vol. 13, 1992, pp. 381–397.
35. http://www.timesascent.co.in/index.aspx?page=article&sectid=8&contentid=2008093020080930131456199f393b7d7In B
36. http://www.india-today.com/btoday/20010121/lifeandwork.html
37. J. D. Shaw, N. Gupta, A. Mitra, and G. E. Ledford, Jr., "Success and Survival of Skill-Based Pay Plans," *Journal of Management*, February 2005, pp. 28–49; C. Lee, K. S. Law, and P. Bobko, "The Importance of Justice Perceptions on Pay Effectiveness: A Two-Year Study of a Skill-Based Pay Plan," *Journal of Management*, vol. 26 (6), 1999, pp. 851–873; G. E. Ledford, "Paying for the Skills, Knowledge and Competencies of Knowledge Workers," *Compensation and Benefits Review*, July–August 1995, pp. 55–62; and E. E. Lawler III, G. E. Ledford, Jr., and L. Chang, "Who Uses Skill-Based Pay and Why," *Compensation and Benefits Review*, March–April 1993, p. 22.
38. Shaw, Gupta, Mitra, and Ledford, "Success and Survival of Skill-Based Pay Plans."
39. http://www.expresscomputeronline.com/20050905/technologylife01.shtml
40. Information in this box from R. E. DeRouin, B. A. Fritzsche, and E. Salas, "E-Learning in Organizations," *Journal of Management*, December 2005, pp. 920–940; K. O'Leonard, *HP Case Study: Flexible Solutions for Multi-Cultural Learners* (Oakland, CA: Bersin & Associates), 2004; S. Greengard, "The Dawn of Digital HR," *Business Finance*, October 2003, pp. 55–59; and J. Hoekstra, "Three in One," *Online Learning*, vol. 5, 2001, pp. 28–32.
41. S. Berfield, "After the Layoff, the Redesign," *BusinessWeek*, April 14, 2008, pp. 54–56; L. Uchitelle, "Retraining Laid-Off Workers, But for What?" *New York Times* online, www.nytimes.com, March 26, 2006; D. Tourish, N. Paulsen, E. Hobman, and P. Bordia, "The Downsides of Downsizing: Communication Processes and Information Needs in the Aftermath of a Workforce Reduction Strategy," *Management Communication Quarterly*, May 2004, pp. 485–516; J. Brockner, et al., "Perceived Control As an Antidote to the Negative Effects of Layoffs on Survivors' Organizational Commitment and Job Performance," *Administrative Science Quarterly*, 49, 2004, pp. 76–100; and E. Krell, "Defusing Downsizing," *Business Finance*, December 2002, pp. 55–57.
42. "Bill Gates on Rewiring the Power Structure," *Working Woman*, April 1994, p. 62; and F. Moody, "Wonder Women in the Rude Boys' Paradise," *Fast Company* online, www.fastcompany.com, April 17, 1997.
43. R. Leger, "Linked by Differences," *Springfield (Missouri) News-Leader*, December 31, 1993, pp. B6+.
44. http://www.boloji.com/wfs/wfs066.htm
45. http://nrcw.nic.in/shared/report/69.pdf
46. Stop Violence Against Women, *Effects of Sexual Harassment*, www.stopvaw.org, May 9, 2007; and V. Di Martino, H. Hoel, and C. L. Cooper, "Preventing Violence and Harassment in the Workplace," *European Foundation for the Improvement of Living and Working Conditions*, 2003, p. 39.
47. The Associated Press, "Corruption, Sexual Harassment Charges Cloud Oxford Debating Club Presidential Election," *International Herald Tribune* online, www.iht.com, February 6, 2008; G. L. Maatman, Jr., "A Global View of Sexual Harassment: Global Human Resource Strategies," *HRMagazine*, July 2000, pp. 151–156; and W. Hardman and J. Heidelberg, "When Sexual Harassment Is a Foreign Affair," *Personnel Journal*, April 1996, pp. 91–97.
48. http://ncw.nic.in/sexualharassmentatworkplacebill2005.pdf
49. Ibid.
50. A. Fisher, "After All This Time, Why Don't People Know What Sexual Harassment Means?" *Fortune*, January 12,

1998, p. 68; and A. R. Karr, "Companies Crack Down on the Increasing Sexual Harassment by E-Mail," *Wall Street Journal*, September 21, 1999, p. A1.

51. See T. S. Bland and S. S. Stalcup, "Managing Harassment," *Human Resource Management*, Spring 2001, pp. 51–61; K. A. Hess and D. R. M. Ehrens, "Sexual Harassment—Affirmative Defense to Employer Liability," *Benefits Quarterly*, 2nd Quarter 1999, p. 57; J. A. Segal, "The Catch-22s of Remedying Sexual Harassment Complaints," *HRMagazine*, October 1997, pp. 111–117; S. C. Bahls and J. E. Bahls, "Hand-off Policy," *Entrepreneur*, July 1997, pp. 74–76; J. A. Segal, "Where Are We Now?," *HRMagazine*, October 1996, pp. 69–73; B. McAfee and D. L. Deadrick, "Teach Employees to Just Say No," *HRMagazine*, February 1996, pp. 86–89; G. D. Block, "Avoiding Liability for Sexual Harassment," *HRMagazine*, April 1995, pp. 91–97; and J. A. Segal, "Stop Making Plaintiffs' Lawyers Rich," *HRMagazine*, April 1995, pp. 31–35. Also, it should be noted here that under Title VII and the Civil Rights Act of 1991, the maximum award that can be given is $300,000. However, many cases are tried under state laws that permit unlimited punitive damages, such as the $7.1 million that Rena Weeks received in her trial based on California statutes.

52. http://www.groundreport.com/Lifestyle/Romance-at-Work-place-makes-Indians-happy

53. J. Yang and V. Salazar, "Would You Date a Co-worker?" *USA Today*, February 14, 2008, p. 1B.

54. Jayson, "Workplace Romance No Longer Gets the Kiss-off."

55. R. Mano and Y. Gabriel, "Workplace Romances in Cold and Hot Organizational Climates: The Experience of Israel and Taiwan," *Human Relations*, January 2006, pp. 7–35; J. A. Segal, "Dangerous Liaisons," *HR Magazine*, December 2005, pp. 104–108; "Workplace Romance Can Create Unforeseen Issues for Employers," *HR Focus*, October 2005, p. 2; C. A. Pierce and H. Aguinis, "Legal Standards, Ethical Standards, and Responses to Social-Sexual Conduct at Work," *Journal of Organizational Behavior*, September 2005, pp. 727–732; and C. A. Pierce, B. J. Broberg, J. R. McClure, and H. Aguinis, "Responding to Sexual Harassment Complaints: Effects of a Dissolved Workplace Romance on Decision-Making Standards," *Organizational Behavior and Human Decision Processes*, September 2004, pp. 66–82.

56. J. Miller and M. Miller, "Get a Life!" *Fortune*, November 28, 2005, pp. 108–124.

57. M. Elias, "The Family-First Generation," *USA Today*, December 13, 2004, p. 5D.

58. M. Mandel, "The Real Reasons You're Working So Hard . . . and What You Can Do About It," *BusinessWeek*, October 3, 2005, pp. 60–67.

59. C. Farrell, "The Overworked, Networked Family," *BusinessWeek*, October 3, 2005, p. 68.

60. http://www.siliconindia.com/shownews/39722; http://www.thehindubusinessline.com/2008/02/22/stories/2008022252400500.htm

61. F. Hansen, "Truths and Myths about Work/Life Balance," *Workforce*, December 2002, pp. 34–39.

62. K. Palmer, "The New Mommy Track," *U.S. News & World Report*, September 3 2007, pp. 40–45; and J. H. Greenhaus and G. N. Powell, "When Work and Family Are Allies: A Theory of Work–Family Enrichment," *Academy of Management Review*, January 2006, pp. 72–92.

63. Greenhaus and Powell, "When Work and Family Are Allies: A Theory of Work–Family Enrichment," p. 73.

64. S. Shellenbarger, "What Makes a Company a Great Place to Work Today," *Wall Street Journal*, October 4, 2007, p. D1; and L. B. Hammer, et al., "A Longitudinal Study of the Effects of Dual-Earner Couples' Utilization of Family-Friendly Workplace Supports on Work and Family Outcomes," *Journal of Applied Psychology*, July 2005, pp. 799–810.

65. M. M. Arthur, "Share Price Reactions to Work–Family Initiatives: An Institutional Perspective," *Academy of Management Journal*, August 2003, pp. 497–505.

66. N. P. Rothbard, T. L. Dumas, and K. W. Phillips, *The Long Arm of the Organization: Work–Family Policies and Employee Preferences for Segmentation*, paper presented at the 61st Annual Academy of Management meeting, Washington, DC, August 2001.

67. http://www.livemint.com/2009/01/05212739/Office-Work-outs-Worklife-ba.html

68. http://www.icmrindia.org/casestudies/catalogue/Human%20Resource%20and%20Organization%20Behavior/Johnson%20and%20Johnson%20Health%20and%20Wellness%20Program.htm

69. Vishal Arora (2009) Work-Life Balance in Indian Industry for MidCareers.com. Retrieved on Jan. 6th 2009 from http://vishalarora.co.in/index.php?option=com_content&task=view&id=45&Itemid=19

70. http://wellness-programs.net/employee-wellness-program-companies-save-millions-through-employee-wellness-programs/

71. B. Pyenson and K. Fitch, "Smoking May Be Hazardous to Your Bottom Line," *Workforce* online, www.workforce.com, December 2007; and L. Cornwell, "Companies Tack on Fees on Insurance for Smokers," *Springfield (Missouri) News-Leader*, February 17, 2006, p. 5B.

72. M. Scott, "Obesity More Costly to U.S. Companies Than Smoking, Alcoholism," *Workforce Management* online, www.workforce.com, April 9, 2008.

73. "Obesity Weighs Down Production," *IndustryWeek*, March 2008, pp. 22–23.

74. J. Appleby, "Companies Step Up Wellness Efforts," *USA Today*, August 1, 2005, pp. 1A+.

75. G. Kranz, "Prognosis Positive: Companies Aim to Get Workers Healthy," *Workforce Management* online, www.workforce.com, April 15, 2008.

76. M. Conlin, "Hide the Doritos! Here Comes HR," *BusinessWeek*, April 28, 2008, pp. 94–96.

77. J. Fox, "Good Riddance to Pensions," *CNN Money*, www.cnnmoney.com, January 12, 2006.

78. M. Adams, "Broken Pension System in Crying Need of a Fix," *USA Today*, November 15, 2005, p. 1B+.

79. J. Appleby, "Traditional Pensions Are Almost Gone. Is Employer-Provided Health Insurance Next?" *USA Today*, November 13, 2007, pp.1A+; S. Kelly, "FedEx, Goodyear Make Big Pension Plan Changes," *Workforce Management* online, www.workforce.com, March 1, 2007; G. Colvin, "The End of a Dream," *Fortune* online, www.fortune.com, June 22, 2006; E. Porter and M. Williams Nash, "Benefits Go the Way of Pensions," *New York Times* online, www.nytimes.com, February 9, 2006; and Fox, "Good Riddance to Pensions."

Chapter 11

1. Google Web site, www.google.com, February 13, 2008; and S. Prasso, "Google Goes to India," *Fortune*, October 29, 2007, pp. 160–166.
2. B. Mezrich, *Bringing Down the House: The Inside Story of Six MIT Students Who Took Vegas for Millions* (New York: The Free Press, 2002). The 2008 film *21* was a fictional work based loosely on this story.
3. B. W. Tuckman and M. C. Jensen, "Stages of Small-Group Development Revisited," *Group and Organizational Studies*, December 1977, pp. 419–427; and M. F. Maples, "Group Development: Extending Tuckman's Theory," *Journal for Specialists in Group Work*, Fall 1988, pp. 17–23.
4. L. N. Jewell and H. J. Reitz, *Group Effectiveness in Organizations* (Glenview, IL: Scott Foresman, 1981); and M. Kaeter, "Repotting Mature Work Teams," *Training*, April 1994, pp. 54–56.
5. A. Sobel, "The Beatles Principles," *Strategy & Business*, Spring 2006, p. 42.
6. This model is based on the work of P. S. Goodman, E. Ravlin, and M. Schminke, "Understanding Groups in Organizations," in L. L. Cummings and B. M. Staw (eds.), *Research in Organizational Behavior*, vol. 9 (Greenwich, CT: JAI Press, 1987), pp. 124–128; J. R. Hackman, "The Design of Work Teams," in J. W. Lorsch (ed.), *Handbook of Organizational Behavior* (Upper Saddle River, NJ: Prentice Hall, 1987), pp. 315–342; G. R. Bushe and A. L. Johnson, "Contextual and Internal Variables Affecting Task Group Outcomes in Organizations," *Group and Organization Studies*, December 1989, pp. 462–482; M. A. Campion, C. J. Medsker, and A. C. Higgs, "Relations Between Work Group Characteristics and Effectiveness: Implications for Designing Effective Work Groups," *Personnel Psychology*, Winter 1993, pp. 823–850; D. E. Hyatt and T. M. Ruddy, "An Examination of the Relationship Between Work Group Characteristics, and Performance: Once More into the Breach," *Personnel Psychology*, Autumn 1997, pp. 553–585; and P. E. Tesluk and J. E. Mathieu, "Overcoming Roadblocks to Effectiveness: Incorporating Management of Performance Barriers into Models of Work Group Effectiveness," *Journal of Applied Psychology*, April 1999, pp. 200–217.
7. G. L. Stewart, "A Meta-Analytic Review of Relationships Between Team Design Features and Team Performance," *Journal of Management*, February 2006, pp. 29–54; T. Butler and J. Waldroop, "Understanding 'People' People," *Harvard Business Review*, June 2004, pp. 78–86; J. S. Bunderson, "Team Member Functional Background and Involvement in Management Teams: Direct Effects and the Moderating Role of Power Centralization," *Academy of Management Journal*, August 2003, pp. 458–474; and M. J. Stevens and M. A. Campion, "The Knowledge, Skill, and Ability Requirements for Teamwork: Implications for Human Resource Management," *Journal of Management*, Summer 1994, pp. 503–530.
8. V. U. Druskat and S. B. Wolff, "The Link Between Emotions and Team Effectiveness: How Teams Engage Members and Build Effective Task Processes," *Academy of Management Proceedings*, on CD-ROM, 1999; D. C. Kinlaw, *Developing Superior Work Teams: Building Quality and the Competitive Edge* (San Diego: Lexington, 1991); and M. E. Shaw, *Contemporary Topics in Social Psychology* (Morristown, NJ: General Learning Press, 1976), pp. 350–351.
9. Information in this box from L. Copeland, "Making the Most of Cultural Differences at the Workplace," *Personnel*, June 1988, pp. 52–60; C. R. Bantz, "Cultural Diversity and Group Cross-Cultural Team Research," *Journal of Applied Communication Research*, February 1993, pp. 1–19; L. Strach and L. Wicander, "Fitting In: Issues of Tokenism and Conformity for Minority Women," *SAM Advanced Management Journal*, Summer 1993, pp. 22–25; M. L. Maznevski, "Understanding Our Differences: Performance in Decision-Making Groups with Diverse Members," *Human Relations*, May 1994, pp. 531–552; F. Rice, "How to Make Diversity Pay," *Fortune*, August 8, 1994, pp. 78–86; J. Jusko, "Diversity Enhances Decision Making," *IndustryWeek*, April 2, 2001, p. 9; K. Lovelace, D. L. Shapiro, and L. R. Weingart, "Maximizing Cross-Functional New Product Teams' Innovativeness and Constraint Adherence: A Conflict Communications Perspective," *Academy of Management Journal*, August 2002, pp. 779–793; B. L. Kirkman, P. E. Tesluk, and B. Rosen, "The Impact of Demographic Heterogeneity and Team Leader-Team Member Demographic Fit on Team Empowerment and Effectiveness," *Group & Organization Management*, June 2004, pp. 334–368; and K. B. Dahlin, L. R. Weingart, and P. J. Hinds, "Team Diversity and Information Use," *Academy of Management Journal*, December 2005, pp. 1107–1123.
10. McMurry, Inc., "The Roles Your People Play," *Managing People at Work*, October 2005, p. 4; G. Prince, "Recognizing Genuine Teamwork," *Supervisory Management*, April 1989, pp. 25–36; R. F. Bales, *SYMOLOG Case Study Kit*, (New York: The Free Press, 1980); and K. D. Benne and P. Sheats, "Functional Roles of Group Members," *Journal of Social Issues*, vol. 4, 1948, pp. 41–49.
11. A. Erez, H. Elms, and E. Fong, "Lying, Cheating, Stealing: Groups and the Ring of Gyges," paper presented at the Academy of Management Annual meeting, Honolulu, HI, August 8, 2005.
12. S. E. Asch, "Effects of Group Pressure upon the Modification and Distortion of Judgments," in H. Guetzkow (ed.), *Groups, Leadership and Men* (Pittsburgh: Carnegie Press, 1951), pp. 177–190; and S. E. Asch, "Studies of Independence and Conformity: A Minority of One Against a Unanimous Majority," *Psychological Monographs: General and Applied*, vol. 70 (9), 1956, pp. 1–70.
13. R. Bond and P. B. Smith, "Culture and Conformity: A Meta-Analysis of Studies Using Asch's [1952, 1956] Line Judgment Task," *Psychological Bulletin*, January 1996, pp. 111–137.
14. M. E. Turner and A. R. Pratkanis, "Mitigating Groupthink by Stimulating Constructive Conflict," in C. DeDreu and E. Van deVliert (eds.), *Using Conflict in Organizations* (London: Sage, 1997), pp. 53–71.
15. A. Deutschman, "Inside the Mind of Jeff Bezos," *Fast Company*, August 2004, pp. 50–58.
16. See, for instance, E. J. Thomas and C. F. Fink, "Effects of Group Size," *Psychological Bulletin*, July 1963, pp. 371–384; and M. E. Shaw, *Group Dynamics: The Psychology of Small Group Behavior*, 3rd ed. (New York: McGraw-Hill, 1981).
17. R. C. Liden, S. J. Wayne, R. A. Jaworski, and N. Bennett, "Social Loafing: A Field Investigation," *Journal of Management*, April 2004, pp. 285–304; and D. R. Comer, "A Model of Social Loafing in Real Work Groups," *Human Relations*, June 1995, pp. 647–667.
18. S. G. Harkins and K. Szymanski, "Social Loafing and Group Evaluation," *Journal of Personality and Social Psychology*, December 1989, pp. 934–941.
19. C. R. Evans and K. L. Dion, "Group Cohesion and Performance: A Meta-Analysis," *Small Group Research*, May

1991, pp. 175–186; B. Mullen and C. Copper, "The Relation Between Group Cohesiveness and Performance: An Integration," *Psychological Bulletin*, March 1994, pp. 210–227; and P. M. Podsakoff, S. B. MacKenzie, and M. Ahearne, "Moderating Effects of Goal Acceptance on the Relationship Between Group Cohesiveness and Productivity," *Journal of Applied Psychology*, December 1997, pp. 974–983.

20. See, for example, L. Berkowitz, "Group Standards, Cohesiveness, and Productivity," *Human Relations*, November 1954, pp. 509–519; and Mullen and Copper, "The Relation Between Group Cohesiveness and Performance: An Integration."

21. S. E. Seashore, *Group Cohesiveness in the Industrial Work Group* (Ann Arbor: University of Michigan, Survey Research Center, 1954).

22. Information in this box from P. Evans, "The Wiki Factor," *BizEd*, January–February 2006, pp. 28–32; and M. McCafferty, "A Human Inventory," *CFO*, April 2005, pp. 83–85.

23. C. Shaffran, "Mind Your Meeting: How to Become the Catalyst for Culture Change," *Communication World*, February–March 2003, pp. 26–29.

24. I. L. Janis, *Victims of Groupthink* (Boston: Houghton Mifflin, 1972); R. J. Aldag and S. Riggs Fuller, "Beyond Fiasco: A Reappraisal of the Groupthink Phenomenon and a New Model of Group Decision Processes," *Psychological Bulletin*, May 1993, pp. 533–552; and T. Kameda and S. Sugimori, "Psychological Entrapment in Group Decision Making: An Assigned Decision Rule and a Groupthink Phenomenon," *Journal of Personality and Social Psychology*, August 1993, pp. 282–292.

25. See, for example, L. K. Michaelson, W. E. Watson, and R. H. Black, "A Realistic Test of Individual vs. Group Consensus Decision Making," *Journal of Applied Psychology*, vol. 74 (5), 1989, pp. 834–839; R. A. Henry, "Group Judgment Accuracy: Reliability and Validity of Postdiscussion Confidence Judgments," *Organizational Behavior and Human Decision Processes*, October 1993, pp. 11–27; P. W. Paese, M. Bieser, and M. E. Tubbs, "Framing Effects and Choice Shifts in Group Decision Making," *Organizational Behavior and Human Decision Processes*, October 1993, pp. 149–165; N. J. Castellan, Jr. (ed.), *Individual and Group Decision Making* (Hillsdale, NJ: Lawrence Erlbaum Associates, 1993); and S. G. Straus and J. E. McGrath, "Does the Medium Matter? The Interaction of Task Type and Technology on Group Performance and Member Reactions," *Journal of Applied Psychology*, February 1994, pp. 87–97.

26. E. J. Thomas and C. F. Fink, "Effects of Group Size," *Psychological Bulletin*, July 1963, pp. 371–384; F. A. Shull, A. L. Delbecq, and L. L. Cummings, *Organizational Decision Making* (New York: McGraw-Hill, 1970), p. 151; A. P. Hare, *Handbook of Small Group Research* (New York: The Free Press, 1976); M. E. Shaw, *Group Dynamics: The Psychology of Small Group Behavior*, 3rd ed. (New York: McGraw-Hill, 1981); and P. Yetton and P. Bottger, "The Relationships Among Group Size, Member Ability, Social Decision Schemes, and Performance," *Organizational Behavior and Human Performance*, October 1983, pp. 145–159.

27. This section is adapted from S. P. Robbins, *Managing Organizational Conflict: A Nontraditional Approach* (Upper Saddle River, NJ: Prentice Hall, 1974), pp. 11–14. Also see D. Wagner-Johnson, "Managing Work Team Conflict: Assessment and Preventative Strategies," Center for the Study of Work Teams, University of North Texas, www.workteams.unt.edu/reports, November 3, 2000; and M. Kennedy, "Managing Conflict in Work Teams," Center for the Study of Work Teams, University of North Texas, www.workteams.unt.edu/reports, November 3, 2000.

28. See K. A. Jehn, "A Multimethod Examination of the Benefits and Detriments of Intragroup Conflict," *Administrative Science Quarterly*, June 1995, pp. 256–282; K. A. Jehn, "A Qualitative Analysis of Conflict Type and Dimensions in Organizational Groups," *Administrative Science Quarterly*, September 1997, pp. 530–557; K. A. Jehn, "Affective and Cognitive Conflict in Work Groups: Increasing Performance Through Value-Based Intragroup Conflict," in C. DeDreu and E. Van deVliert (eds.), *Using Conflict in Organizations* (London: Sage Publications, 1997), pp. 87–100; K. A. Jehn and E. A. Mannix, "The Dynamic Nature of Conflict: A Longitudinal Study of Intragroup Conflict and Group Performance," *Academy of Management Journal*, April 2001, pp. 238–251; C. K. W. DeDreu and A. E. M. Van Vianen, "Managing Relationship Conflict and the Effectiveness of Organizational Teams," *Journal of Organizational Behavior*, May 2001, pp. 309–328; and J. Weiss and J. Hughes, "Want Collaboration? Accept—And Actively Manage—Conflict," *Harvard Business Review*, March 2005, pp. 92–101.

29. C. K. W. DeDreu, "When Too Little or Too Much Hurts: Evidence for a Curvilinear Relationship Between Task Conflict and Innovation in Teams," *Journal of Management*, February 2006, pp. 83–107.

30. K. W. Thomas, "Conflict and Negotiation Processes in Organizations," in M. D. Dunnette and L. M. Hough (eds.), *Handbook of Industrial and Organizational Psychology*, 2nd ed., vol. 3 (Palo Alto, CA: Consulting Psychologists Press, 1992), pp. 651–717.

31. See, for example, J. R. Hackman and C. G. Morris, "Group Tasks, Group Interaction Process, and Group Performance Effectiveness: A Review and Proposed Integration," in L. Berkowitz (ed.), *Advances in Experimental Social Psychology* (New York: Academic Press, 1975), pp. 45–99; R. Saavedra, P. C. Earley, and L. Van Dyne, "Complex Interdependence in Task-Performing Groups," *Journal of Applied Psychology*, February 1993, pp. 61–72; M. J. Waller, "Multiple-Task Performance in Groups," *Academy of Management Proceedings* on Disk, 1996; and K. A. Jehn, G. B. Northcraft, and M. A. Neale, "Why Differences Make a Difference: A Field Study of Diversity, Conflict, and Performance in Workgroups," *Administrative Science Quarterly*, December 1999, pp. 741–763.

32. Cited in T. Purdum, "Teaming, Take 2," *Industry-Week*, May 2005, p. 43; and C. Joinson, "Teams at Work," *HRMagazine*, May 1999, p. 30.

33. See, for example, S. A. Mohrman, S. G. Cohen, and A. M. Mohrman, Jr., *Designing Team-Based Organizations* (San Francisco: Jossey-Bass, 1995); P. MacMillan, *The Performance Factor: Unlocking the Secrets of Teamwork* (Nashville, TN: Broadman & Holman, 2001); and E. Salas, C. A. Bowers, and E. Eden (eds.), *Improving Teamwork in Organizations: Applications of Resource Management Training* (Mahwah, NJ: Lawrence Erlbaum, 2002).

34. See, for instance, E. Sunstrom, DeMeuse, and D. Futrell, "Work Teams: Applications and Effectiveness," *American Psychologist*, February 1990, pp. 120–133.

35. M. Cianni and D. Wanuck, "Individual Growth and Team Enhancement: Moving Toward a New Model of Career Development," *Academy of Management Executive*, February 1997, pp. 105–115.

36. "Teams," *Training*, October 1996, p. 69; and C. Joinson "Teams at Work," p. 30.

37. G. M. Spreitzer, S. G. Cohen, and G. E. Ledford, Jr., "Developing Effective Self-Managing Work Teams in Service Organizations," *Group & Organization Management*, September 1999, pp. 340–366.

38. "Meet the New Steel," *Fortune*, October 1, 2007, pp. 68–71.

39. http://www.indiacar.net/news/n2479.htm

40. A. Malhotra, A. Majchrzak, R. Carman, and V. Lott, "Radical Innovation without Collocation: A Case Study at Boeing-Rocketdyne," *MIS Quarterly*, June 2001, pp. 229–249.

41. A. Stuart, "Virtual Agreement," *CFO*, November 2007, p. 24.

42. A. Malhotra, A. Majchrzak, and B. Rosen, "Leading Virtual Teams," *Academy of Management Perspectives*, February 2007, pp. 60–70; B. L. Kirkman and J. E. Mathieu, "The Dimensions and Antecedents of Team Virtuality," *Journal of Management*, October 2005, pp. 700–718; J. Gordon, "Do Your Virtual Teams Deliver Only Virtual Performance?" *Training*, June 2005, pp. 20–25; L. L. Martins, L. L. Gilson, and M. T. Maynard, "Virtual Teams: What Do We Know and Where Do We Go from Here?" *Journal of Management*, December 2004, pp. 805–835; S. A. Furst, M. Reeves, B. Rosen, and R. S. Blackburn, "Managing the Life Cycle of Virtual Teams," *Academy of Management Executive*, May 2004, pp. 6–20; B. L. Kirkman, B. Rosen, P. E. Tesluk, and C. B. Gibson, "The Impact of Team Empowerment on Virtual Team Performance: The Moderating Role of Face-to-Face Interaction," *Academy of Management Journal*, April 2004, pp. 175–192; F. Keenan and S. E. Ante, "The New Teamwork," *BusinessWeek e.biz*, February 18, 2002, pp. EB12–EB16; and G. Imperato, "Real Tools for Virtual Teams" *Fast Company*, July 2000, pp. 378–387.

43. J. Mathieu, M. T. Maynard, T. Rapp, and L. Gilson, "Team Effectiveness 1997–2007: A Review of Recent Advancements and a Glimpse into the Future," *Journal of Management*, June 2008, pp. 410–476; S. W. Lester, B. W. Meglino, and M. A. Korsgaard, "The Antecedents and Consequences of Group Potency: A Longitudinal Investigation of Newly Formed Work Groups," *Academy of Management Journal*, April 2002, pp. 352–368; M. A. Marks, M. J. Sabella, C. S. Burke, and S. J. Zaccaro, "The Impact of Cross-Training on Team Effectiveness," *Journal of Applied Psychology*, February 2002, pp. 3–13; J. A. Colquitt, R. A. Noe, and C. L. Jackson, "Justice in Teams: Antecedents and Consequences of Procedural Justice Climate," *Personnel Psychology*, vol. 55, 2002, pp. 83–100; J. M. Phillips and E. A. Douthitt, "The Role of Justice in Team Member Satisfaction With the Leader and Attachment to the Team," *Journal of Applied Psychology*, April 2001, pp. 316–325; J. E. Mathieu, et al., "The Influence of Shared Mental Models on Team Process and Performance," *Journal of Applied Psychology*, April 2000, pp. 273–283; G. L. Stewart and M. R. Barrick, "Team Structure and Performance: Assessing the Mediating Role of Intrateam Process and the Moderating Role of Task Type," *Academy of Management Journal*, April 2000, pp. 135–148; J. D. Shaw, M. K. Duffy, and E. M. Stark, "Interdependence and Preference for Group Work: Main and Congruence Effects on the Satisfaction and Performance of Group Members," *Journal of Management*, vol. 26 (2), 2000, pp. 259–279; V. U. Druskat and S. B. Wolff, "The Link Between Emotions and Team Effectiveness: How Teams Engage Members and Build Effective Task Processes"; R. Forrester and A. B. Drexler, "A Model for Team-Based Organization Performance," *Academy of Management Executive*, August 1999, pp. 36–49; A. R. Jassawalla and H. C. Sashittal, "Building Collaborative Cross-Functional New Product Teams," *Academy of Management Executive*, August 1999, pp. 50–63; and G. R. Jones and G. M. George, "The Experience and Evolution of Trust: Implications for Cooperation and Teamwork," *Academy of Management Review*, July 1998, pp. 531–546.

44. B. L. Kirkman, C. B. Gibson, and D. L. Shapiro, "Exporting Teams: Enhancing the Implementation and Effectiveness of Work Teams in Global Affiliates," *Organizational Dynamics*, Summer 2001, pp. 12–29; J. W. Bing and C. M. Bing, "Helping Global Teams Compete," *Training & Development*, March 2001, pp. 70–71; C. G. Andrews, "Factors That Impact Multi-Cultural Team Performance," Center for the Study of Work Teams, University of North Texas, www.workteams.unt.edu/reports/, November 3, 2000; P. Christopher Earley and E. Mosakowski, "Creating Hybrid Team Cultures: An Empirical Test of Transnational Team Functioning," *Academy of Management Journal*, February 2000, pp. 26–49; J. Tata, "The Cultural Context of Teams: An Integrative Model of National Culture, Work Team Characteristics, and Team Effectiveness," *Academy of Management Proceedings*, on CD-ROM, 1999; D. I. Jung, K. B. Baik, and J. J. Sosik, "A Longitudinal Investigation of Group Characteristics and Work Group Performance: A Cross-Cultural Comparison," *Academy of Management Proceedings*, on CD-ROM, 1999; and C. B. Gibson, "They Do What They Believe They Can? Group-Efficacy Beliefs and Group Performance Across Tasks and Cultures," *Academy of Management Proceedings*, on CD-ROM, 1996.

45. Bond and Smith, "Culture and Conformity: A Meta-Analysis of Studies Using Asch's [1952, 1956] Line Judgment Task."

46. I. L. Janis, *Groupthink*, 2nd ed. (New York: Houghton Mifflin Company, 1982), p. 175.

47. See P. C. Earley, "Social Loafing and Collectivism: A Comparison of the United States and the People's Republic of China," *Administrative Science Quarterly*, December 1989, pp. 565–581; and P. C. Earley, "East Meets West Meets Mideast: Further Explorations of Collectivistic and Individualistic Work Groups," *Academy of Management Journal*, April 1993, pp. 319–348.

48. N. J. Adler, *International Dimensions of Organizational Behavior*, 4th ed. (Cincinnati, OH: South-Western Publishing, 2002), p. 142.

49. K. B. Dahlin, L. R. Weingart, and P. J. Hinds, "Team Diversity and Information Use," *Academy of Management Journal*, December 2005, pp. 1107–1123.

50. Adler, *International Dimensions of Organizational Behavior*, p. 142.

51. S. Paul, I. M. Samarah, P. Seetharaman, and P. P. Mykytyn, "An Empirical Investigation of Collaborative Conflict Management Style in Group Support System-Based Global Virtual Teams," *Journal of Management Information Systems*, Winter 2005, pp. 185–222.

52. S. Chang and P. Tharenou, "Competencies Needed for Managing a Multicultural Workgroup," *Asia Pacific Journal of Human Resources*, vol. 42 (1), 2004, pp. 57–74; and Adler, *International Dimensions of Organizational Behavior*, p. 153.

53. C. E. Nicholls, H. W. Lane, and M. Brehm Brechu, "Taking Self-Managed Teams to Mexico," *Academy of Management Executive*, August 1999, pp. 15–27.

54. J. Reingold and J. L. Yang, "The Hidden Workplace: What's Your OQ?" *Fortune*, July 23, 2007, pp. 98–106; and P. Balkundi and D. A. Harrison, "Ties, Leaders, and Time in Teams: Strong Inference About Network Structures' Effects on Team Viability and Performance," *Academy of Management Journal*, February 2006, pp. 49–68.

55. T. Casciaro and M. S. Lobo, "Competent Jerks, Lovable Fools, and the Formation of Social Networks," *Harvard Business Review*, June 2005, pp. 92–99.

56. Balkundi and Harrison, "Ties, Leaders, and Time in Teams: Strong Inference About Network Structures' Effects on Team Viability and Performance."

57. J. McGregor, "The Office Chart That Really Counts," *BusinessWeek*, February 27, 2006, pp. 48–49.

Chapter 12

1. http://www.moneycontrol.com/stocks/company_inf/chairmans_speech.php?sc_did=m13;http://www.iki.dk/media/bestyrelsen/ranjan_malik_and_erehwon.doc; Lafley, A. G. and Charan, R. (2008). The Game Changer: How Every Leader Can Drive Everyday Innovation, pp. 49-52. Penguin Portfolio: New Delhi.

2. G. Nadler and W. J. Chandon, "Making Changes: The FIST Approach," *Journal of Management Inquiry*, September 2004, pp. 239–246; and C. R. Leana and B. Barry, "Stability and Change as Simultaneous Experiences in Organizational Life," *Academy of Management Review*, October 2000, pp. 753–759.

3. The idea for these metaphors came from J. E. Dutton, S. J. Ashford, R. M. O'Neill, and K. A. Lawrence, "Moves That Matter: Issue Selling and Organizational Change," *Academy of Management Journal*, August 2001, pp. 716–736; B. H. Kemelgor, S. D. Johnson, and S. Srinivasan, "Forces Driving Organizational Change: A Business School Perspective," *Journal of Education for Business*, January/February 2000, pp. 133–137; G. Colvin, "When It Comes to Turbulence, CEOs Could Learn a Lot from Sailors," *Fortune*, March 29, 1999, pp. 194–196; and P. B. Vaill, *Managing as a Performing Art: New Ideas for a World of Chaotic Change* (San Francisco: Jossey-Bass, 1989).

4. K. Lewin, *Field Theory in Social Science* (New York: Harper & Row, 1951).

5. G. Hamel, "Take It Higher," *Fortune*, February 5, 2001, pp. 169–170.

6. A. Sains and S. Reed, "Electrolux Cleans Up," *BusinessWeek*, February 27, 2006, pp. 42–43.

7. http://www.bluestarindia.com/investors/speech-arc

8. Hallmark, *2008 and Beyond: Emerging and Evolving Trends*, www.hallmark.com.

9. J. Jesitus, "Change Management: Energy to the People," *IW*, September 1, 1997, pp. 37, 40.

10. D. Lavin, "European Business Rushes to Automate," *Wall Street Journal*, July 23, 1997, p. A14.

11. See, for example, B. B. Bunker, B. T. Alban, and R. J. Lewicki, "Ideas in Currency and OD Practice," *The Journal of Applied Behavioral Science*, December 2004, pp. 403–422; L. E. Greiner and T. G. Cummings, "Wanted: OD More Alive Than Dead!" *Journal of Applied Behavioral Science*, December 2004, pp. 374–391; S. Hicks, "What is Organization Development?" *Training & Development*, August 2000, p. 65; W. Nicolay, "Response to Farias and Johnson's Commentary," *Journal of Applied Behavioral Science*, September 2000, p. 380–81; G. Farias, "Organizational Development and Change Management," *Journal of Applied Behavioral Science*, September 2000, pp. 376–379; N. A. Worren, K. Ruddle, and K. Moore, "From Organizational Development to Change Management," *Journal of Applied Behavioral Science*, September 1999, pp. 273–286; W. L. French and C. H. Bell, Jr., *Organization Development: Behavioral Science Interventions for Organization Improvement*, 6th ed. (Upper Saddle River, NJ: Prentice Hall, 1998); A. H. Church, W. W. Burke, and D. F. Van Eynde, "Values, Motives, and Interventions of Organization Development Practitioners," *Group & Organization Management*, March 1994, pp. 5–50; and T. C. Head and P. F. Sorensen, "Cultural Values and Organizational Development: A Seven-Country Study," *Leadership & Organization Development Journal*, March 1993, pp. 3–7.

12. M. Javidan, P. W. Dorfman, M. S. deLuque, and R. J. House, "In the Eye of the Beholder: Cross-Cultural Lessons in Leadership from Project GLOBE," *Academy of Management Perspective*, February 2006, pp. 67–90; and E. Fagenson-Eland, E. A. Ensher, and W. W. Burke, "Organization Development and Change Interventions: A Seven-Nation Comparison," *The Journal of Applied Behavioral Science*, December 2004, pp. 432–464.

13. Fagenson-Eland, Ensher, and Burke, "Organization Development and Change Interventions: A Seven-Nation Comparison," p. 461.

14. J. Pfeffer, "Breaking Through Excuses," *Business 2.0*, May 2005, p. 76.

15. See, for example, A. Deutschman, "Making Change: Why Is It So Hard to Change Our Ways?" *Fast Company*, May 2005, pp. 52–62; S. B. Silverman, C. E. Pogson, and A. B. Cober, "When Employees at Work Don't Get It: A Model for Enhancing Individual Employee Change in Response to Performance Feedback," *Academy of Management Executive*, May 2005, pp. 135–147; C. E. Cunningham, et al., "Readiness for Organizational Change: A Longitudinal Study of Workplace, Psychological and Behavioral Correlates," *Journal of Occupational and Organizational Psychology*, December 2002, pp. 377–392; M. A. Korsgaard, H. J. Sapienza, and D. M. Schweiger, "Beaten Before Begun: The Role of Procedural Justice in Planning Change," *Journal of Management*, vol. 28 (4), 2002, pp. 497–516; R. Kegan and L. L. Lahey, "The Real Reason People Won't Change," *Harvard Business Review*, November 2001, pp. 85–92; S. K. Piderit, "Rethinking Resistance and Recognizing Ambivalence: A Multidimensional View of Attitudes Toward an Organizational Change," *Academy of Management Review*, October 2000, pp. 783–794; C. R. Wanberg and J. T. Banas, "Predictors and Outcomes of Openness to Changes in a Reorganizing Workplace," *Journal of Applied Psychology*, February 2000, pp. 132–142; A. A. Armenakis and A. G. Bedeian, "Organizational Change: A Review of Theory and Research in the 1990s," *Journal of Management*, vol. 25 (3), 1999, pp. 293–315; and B. M. Staw, "Counterforces to Change," in P. S. Goodman and Associates (eds.), *Change in Organizations* (San Francisco: Jossey-Bass, 1982), pp. 87–121.

16. A. Reichers, J. P. Wanous, and J. T. Austin, "Understanding and Managing Cynicism About Organizational Change," *Academy of Management Executive*, February 1997, pp. 48–57; P. Strebel, "Why Do Employees Resist Change?" *Harvard Business Review*, May–June 1996, pp. 86–92; and J. P. Kotter and L.A. Schlesinger, "Choosing Strategies for Change," *Harvard Business Review*, March–April 1979, pp. 107–09.

17. S. Oreg, "Resistance to Change: Developing an Individual Differences Measure," *Journal of Applied Psychology*, August 2003, pp. 680–693; J. A. LePine, J. A. Colquitt, and A. Erez, "Adaptability to Changing Task Contexts: Effects of General Cognitive Ability, Conscientiousness, and Openness to Experience," *Personnel Psychology*, Fall 2000, pp. 563–593; Piderit, "Rethinking Resistance and Recognizing Ambivalence: A Multidimensional View of Attitudes Toward An Organizational Change"; K. W. Mossholder, R. P. Settoon, A. A. Armenakis, S. G. Harris, "Emotion During Organizational Transformations," *Group & Organization Management*, September 2000, pp. 220–243; J. P. Wanous, A. E. Reichers, and J. T. Austin, "Cynicism About Organizational Change," *Group & Organization Management*, June 2000, pp. 132–153; T. A. Judge, C. J. Thoresen, V. Pucki, and T. M. Welbourne, "Managerial Coping with Organizational Change: A Dispositional Perspective," *Journal of Applied Psychology*, February 1999, pp. 107–122; and A. Sagie and M. Koslowsky, "Organizational Attitudes and Behaviors as a Function of Participation in Strategic and Tactical Change Decisions: An Application of Path-Goal Theory," *Journal of Organizational Behavior*, January 1994, pp. 37–47.

18. See P. Anthony, *Managing Culture* (Philadelphia: Open University Press, 1994); P. Bate, *Strategies for Cultural Change* (Boston: Butterworth-Heinemann, 1994); C. G. Smith and R. P. Vecchio, "Organizational Culture and Strategic Management: Issues in the Strategic Management of Change," *Journal of Managerial Issues*, Spring 1993, pp. 53–70; P. F. Drucker, "Don't Change Corporate Culture—Use It!" *Wall Street Journal*, March 28, 1991, p. A14; and T. H. Fitzgerald, "Can Change in Organizational Culture Really Be Managed?" *Organizational Dynamics*, Autumn 1988, pp. 5–15.

19. K. Maney, "Famously Gruff Gerstner Leaves IBM a Changed Man," *USA Today*, November 11, 2002, pp. 1B+; and Louis V. Gerstner, *Who Says Elephants Can't Dance: Inside IBM's Historic Turnaround* (New York: Harper Business, 2002).

20. See, for example, D. C. Hambrick and S. Finkelstein, "Managerial Discretion: A Bridge Between Polar Views of Organizational Outcomes," in L. L. Cummings and B. M. Staw (eds.), *Research in Organizational Behavior*, vol. 9 (Greenwich, CT: JAI Press, 1987), p. 384; and R. H. Kilmann, M. J. Saxton, and R. Serpa (eds.), *Gaining Control of the Corporate Culture* (San Francisco: Jossey-Bass, 1985).

21. Information in this box from C. Lindsay, "Paradoxes of Organizational Diversity: Living within the Paradoxes," in L. R. Jauch and J. L. Wall (eds.), *Proceedings of the 50th Academy of Management Conference*, San Francisco, 1990, pp. 374–378.

22. C. Daniels, "The Last Taboo," *Fortune*, October 28, 2002, pp. 137–144; J. Laabs, "Time-Starved Workers Rebel," *Workforce*, October 2000, pp. 26–28; M. A. Verespej, "Stressed Out," *IndustryWeek*, February 21, 2000, pp. 30–34; and M. A. Cavanaugh, W. R. Boswell, M. V. Roehling, and J. W. Boudreau, "An Empirical Examination of Self-Reported Work Stress Among U.S. Managers," *Journal of Applied Psychology*, February 2000, pp. 65–74.

23. American Institute of Stress, *Stress...At Work* [www.stress.org/job.htm], 2002–2003.

24. V. P. Sudhashree, K. Rohith, and K. Shrinivas, "Issues and Concerns of Health Among Call Center Employees," *Indian Journal of Occupational Environmental Medicine*, vol. 9 (3), 2005, pp. 129–132; E. Muehlchen, *An Ounce of Prevention Goes a Long Way*, www.wilsonbanwell.com, January 2004; UnionSafe, *Stressed Employees Worked to Death*, unionsafe.labor.net.au/news, August 23, 2003; O. Siu, "Occupational Stressors and Well-Being Among Chinese Employees: The Role of Organizational Commitment," *Applied Psychology: An International Review*, October 2002, pp. 527–544; O. Siu, et al., "Managerial Stress in Greater China: The Direct and Moderator Effects of Coping Strategies and Work Locus of Control," *Applied Psychology: An International Review*, October 2002, pp. 608–632; A. Oswald, *New Research Reveals Dramatic Rise in Stress Levels in Europe's Workplaces*, University of Warwick press release, www.warwick.ac.uk/news/pr, 1999; and Y. Shimizu, S. Makino, and T. Takata, "Employee Stress Status During the Past Decade [1982–1992] Based on a Nation-Wide Survey Conducted by the Ministry of Labour in Japan," Japan Industrial Safety and Health Association, July 1997, pp. 441–450.

25. G. Kranz, "Job Stress Viewed Differently by Workers, Employers," *Workforce Management* online, www.workforce.com, January 15, 2008.

26. Adapted from the UK National Work-Stress Network Web site, www.workstress.net.

27. R. S. Schuler, "Definition and Conceptualization of Stress in Organizations," *Organizational Behavior and Human Performance*, April 1980, p. 191.

28. "Jobs for Life," *The Economist* online, www.economist.com, December 19, 2007; and B. L. de Mente, *Karoshi: Death from Overwork*, www.apmforum.com, May 2002.

29. H. Benson, "Are You Working Too Hard?" *Harvard Business Review*, November 2005, pp. 53–58; B. Cryer, R. McCraty, and D. Childre, "Pull the Plug on Stress," *Harvard Business Review*, July 2003, pp. 102–107; C. Daniels, "The Last Taboo"; C.L. Cooper and S. Cartwright, "Healthy Mind, Healthy Organization—A Proactive Approach to Occupational Stress," *Human Relations*, April 1994, pp. 455–71; C. A. Heaney, et al., "Industrial Relations, Worksite Stress Reduction and Employee Well-Being: A Participatory Action Research Investigation," *Journal of Organizational Behavior*, September 1993, pp. 495–510; C. D. Fisher, "Boredom at Work: A Neglected Concept," *Human Relations*, March 1993, pp. 395–417; and S. E. Jackson, "Participation in Decision Making as a Strategy for Reducing Job-Related Strain," *Journal of Applied Psychology*, February 1983, pp. 3–19.

30. C. Mamberto, "Companies Aim to Combat Job-Related Stress," *Wall Street Journal*, August 13, 2007, p. B6.

31. D. Cole, "The Big Chill," *US News & World Report*, December 6, 2004, pp. EE2–EE5.

32. J. Goudreau, "Dispatches from the War on Stress," *BusinessWeek*, August 6, 2007, pp. 74–75.

33. P. A. McLagan, "Change Leadership Today," *T&D*, November 2002, pp. 27–31.

34. Ibid, p. 29.

35. C. Haddad, "UPS: Can It Keep Delivering?" *BusinessWeek* online, www.businessweek.com, Spring 2003.

36. W. Pietersen, "The Mark Twain Dilemma: The Theory and Practice for Change Leadership," p. 35.

37. P. A. McLagan, "The Change-Capable Organization," *T&D*, January 2003, pp. 50–58.

38. http://www.nasscom.in/Nasscom/templates/NormalPage.aspx?id=2208

39. J. E. Perry-Smith and C. E. Shalley, "The Social Side of Creativity: A Static and Dynamic Social Network Perspective," *Academy of Management Review*, January 2003, pp. 89–106; and P. K. Jagersma, "Innovate or Die: It's Not Easy, But It Is Possible to Enhance Your Organization's Ability to Innovate," *Journal of Business Strategy*, January–February 2003, pp. 25–28.
40. These definitions are based on T. M. Amabile, *Creativity in Context* (Boulder, CO: Westview Press, 1996).
41. R. W. Woodman, J. E. Sawyer, and R. W. Griffin, "Toward a Theory of Organizational Creativity," *Academy of Management Review*, April 1993, pp. 293–321.
42. T. M. Egan, "Factors Influencing Individual Creativity in the Workplace: An Examination of Quantitative Empirical Research," *Advances in Developing Human Resources*, May 2005, pp 160–181; N. Madjar, G. R. Oldham, and M. G. Pratt, "There's No Place Like Home? The Contributions of Work and Nonwork Creativity Support to Employees' Creative Performance," *Academy of Management Journal*, August 2002, pp. 757–767; T. M. Amabile, C. N. Hadley, and S. J. Kramer, "Creativity Under the Gun," *Harvard Business Review*, August 2002, pp. 52–61; J. B. Sorensen and T. E. Stuart, "Aging, Obsolescence, and Organizational Innovation," *Admin-istrative Science Quarterly*, March 2000, pp. 81–112; G. R. Oldham and A. Cummings, "Employee Creativity: Personal and Contextual Factors at Work," *Academy of Management Journal*, June 1996, pp. 607–634; and F. Damanpour, "Organizational Innovation: A Meta-Analysis of Effects of Determinants and Moderators," *Academy of Management Journal*, September 1991, pp. 555–590.
43. P. R. Monge, M. D. Cozzens, and N. S. Contractor, "Communication and Motivational Predictors of the Dynamics of Organizational Innovations," *Organization Science*, May 1992, pp. 250–274.
44. Amabile, Hadley, and Kramer, "Creativity Under the Gun."
45. Madjar, Oldham, and Pratt, "There's No Place Like Home? The Contributions of Work and Nonwork Creativity Support to Employees' Creative Performance."
46. C. Salter, "Mattel Learns to 'Throw the Bunny,'" *Fast Company*, November 2002, p. 22.
47. See, for instance, J. E. Perry-Smith, "Social Yet Creative: The Role of Social Relationships in Facilitating Individual Creativity," *Academy of Management Journal*, February 2006, pp. 85–101; C. E. Shalley, J. Zhou, and G. R. Oldham, "The Effects of Personal and Contextual Characteristics on Creativity: Where Should We Go from Here?" *Journal of Management*, vol. 30 (6), 2004, pp. 933–958; Perry-Smith and Shalley, "The Social Side of Creativity: A Static and Dynamic Social Network Perspective"; J. M. George and J. Zhou, "When Openness to Experience and Conscientiousness are Related to Creative Behavior: An Interactional Approach," *Journal of Applied Psychology*, June 2001, pp. 513–524; J. Zhou, "Feedback Valence, Feedback Style, Task Autonomy, and Achievement Orientation: Interactive Effects on Creative Behavior," *Journal of Applied Psychology*, vol. 83, 1998, pp. 261–276; T. M. Amabile, et al., "Assessing the Work Environment for Creativity," *Academy of Management Journal*, October 1996, pp. 1154–1184; S. G. Scott and R. A. Bruce, "Determinants of Innovative People: A Path Model of Individual Innovation in the Workplace," *Academy of Management Journal*, June 1994, pp. 580–607; R. Moss Kanter, "When a Thousand Flowers Bloom: Structural, Collective, and Social Conditions for Innovation in Organization," in B. M. Staw and L. L. Cummings (eds.), *Research in Organizational Behavior*, vol. 10 (Greenwich, CT: JAI Press, 1988), pp. 169–211; and Amabile, *Creativity in Context*.
48. McGregor, "The World's Most Innovative Companies," p. 70.
49. Ibid., p. 74.
50. J. Ramos, "Producing Change That Lasts," *Across the Board*, March 1994, pp. 29–33; T. Stjernberg and A. Philips, "Organizational Innovations in a Long-Term Perspective: Legitimacy and Souls-of-Fire as Critical Factors of Change and Viability," *Human Relations*, October 1993, pp. 1193–2023; and J. M. Howell and C. A. Higgins, "Champions of Change," *Business Quarterly*, Spring 1990, pp. 31–32.

Chapter 13

1. M. Conlin and J. Greene, "How to Make a Microserf Smile," *BusinessWeek*, September 10, 2007, pp. 56–59.
2. http://www.indianexpress.com/news/staff-against-ministers-plan-to-shift-office/384649/
3. http://www.labourbureau.nic.in/ASI2k4-5VolII%20table%202.1.2a.htm
4. http://www.timesofindia.indiatimes.com/articleshow/2145953.cms
5. D. W. Organ, *Organizational Citizenship Behavior: The Good Soldier Syndrome* (Lexington, MA: Lexington Books, 1988), p. 4. See also J. L. Lavell, D. E. Rupp, and J. Brockner, "Taking a Multifoci Approach to the Study of Justice, Social Exchange, and Citizenship Behavior: The Target Similarity Model," *Journal of Management*, December 2007, pp. 841–866; and J. A. LePine, A. Erez, and D. E. Johnson, "The Nature and Dimensionality of Organizational Citizenship Behavior: A Critical Review and Meta-Analysis," *Journal of Applied Psychology*, February 2002, pp. 52–65.
6. R. Ilies, B. A. Scott, and T. A. Judge, "The Interactive Effects of Personal Traits and Experienced States on Intraindividual Patterns of Citizenship Behavior," *Academy of Management Journal*, June 2006, pp. 561–575; P. Cardona, B. S. Lawrence, and P. M. Bentler, "The Influence of Social and Work Exchange Relationships on Organizational Citizenship Behavior," *Group & Organization Management*, April 2004, pp. 219–247; M. C. Bolino and W. H. Turnley, "Going the Extra Mile: Cultivating and Managing Employee Citizenship Behavior," *Academy of Management Executive*, August 2003, pp. 60–73; M. C. Bolino, W. H. Turnley, and J. J. Bloodgood, "Citizenship Behavior and the Creation of Social Capital in Organizations," *Academy of Management Review*, October 2002, pp. 505–522; and P. M. Podsakoff, S. B. MacKenzie, J. B. Paine, and D. G. Bachrach, "Organizational Citizenship Behaviors: A Critical Review of the Theoretical and Empirical Literature and Suggestions for Future Research," *Journal of Management*, vol. 26 (3), 2000, pp. 543–548.
7. M. C. Bolino and W. H. Turnley, "The Personal Costs of Citizenship Behavior: The Relationship Between Individual Initiative and Role Overload, Job Stress, and Work–Family Conflict," *Journal of Applied Psychology*, July 2005, pp. 740–748.
8. This definition adapted from R. W. Griffin and Y. P. Lopez, "Bad Behavior in Organizations: A Review and Typology for Future Research," *Journal of Management*, December 2005, pp. 988–1005.
9. S. J. Breckler, "Empirical Validation of Affect, Behavior, and Cognition as Distinct Components of Attitude," *Journal of*

Personality and Social Psychology, May 1984, pp. 1191–1205; and S. L. Crites, Jr., L. R. Fabrigar, and R. E. Petty, "Measuring the Affective and Cognitive Properties of Attitudes: Conceptual and Methodological Issues," *Personality and Social Psychology Bulletin*, December 1994, pp. 619–634.

10. D. R. May, R. L. Gilson, and L. M. Harter, "The Psychological Conditions of Meaningfulness, Safety and Availability and the Engagement of the Human Spirit at Work," *Journal of Occupational and Organizational Psychology*, March 2004, pp. 11–37; R. T. Keller, "Job Involvement and Organizational Commitment as Longitudinal Predictors of Job Performance: A Study of Scientists and Engineers," *Journal of Applied Psychology*, August 1997, pp. 539–545; W. Kahn, "Psychological Conditions of Personal Engagement and Disengagement at Work," *Academy of Management Journal*, December 1990, pp. 692–794; and P. P. Brooke, Jr., D. W. Russell, and J. L. Price, "Discriminant Validation of Measures of Job Satisfaction, Job Involvement, and Organizational Commitment," *Journal of Applied Psychology*, May 1988, pp. 139–145.

11. What Indian employees want. http://www.domain-b.com/management/hrd/20060328_employees.htm.

12. The Conference Board, *U.S. Job Satisfaction Keeps Falling, the Conference Board Reports Today*, www.conference-board.org, February 28, 2005.

13. A. B. Krueger, "Job Satisfaction Is Not Just a Matter of Dollars," *New York Times*, December 8, 2005, p. C3.

14. The Conference Board, US Job Satisfaction Declines, The Conference Board Reports, http://www.conference-board.org. Accessed February 23, 2007.

15. Harris Interactive, *Six Nation Survey Finds Satisfaction with Current Job*, www.harrisinteractive.com, October 9, 2007; and SwissInfo, *Swiss Like Work, but Not Their Salaries*, www.swissinfo.org, June 7, 2005.

16. Watson Wyatt Worldwide, *A Comparison of Attitudes Around the Globe*, www.watsonwyatt.com/research, 2006.

17. T. A. Judge, C. J. Thoresen, J. E. Bono, and G. K. Patton, "The Job Satisfaction–Job Performance Relationship: A Qualitative and Quantitative Review," *Psychological Bulletin*, May 2001, pp. 376–407.

18. J. K. Harter, F. L. Schmidt, and T. L. Hayes, "Business-Unit-Level Relationship Between Employee Satisfaction, Employee Engagement, and Business Outcomes: A Meta-Analysis," *Journal of Applied Psychology*, April 2002, pp. 268–279; A. M. Ryan, M. J. Schmit, and R. Johnson, "Attitudes and Effectiveness: Examining Relations at an Organizational Level," *Personnel Psychology*, Winter 1996, pp. 853–882; and C. Ostroff, "The Relationship Between Satisfaction, Attitudes, and Performance: An Organizational Level Analysis," *Journal of Applied Psychology*, December 1992, pp. 963–974.

19. E. A. Locke, "The Nature and Causes of Job Satisfaction," in M. D. Dunnette (ed.), *Handbook of Industrial and Organizational Psychology* (Chicago: Rand McNally, 1976), p. 1331; S. L. McShane, "Job Satisfaction and Absenteeism: A Meta-Analytic Re-examination," *Canadian Journal of Administrative Science*, June 1984, pp. 61–77; R. D. Hackett and R. M. Guion, "A Reevaluation of the Absenteeism–Job Satisfaction Relationship," *Organizational Behavior and Human Decision Processes*, June 1985, pp. 340–381; K. D. Scott and G. S. Taylor, "An Examination of Conflicting Findings on the Relationship Between Job Satisfaction and Absenteeism: A Meta-Analysis," *Academy of Management Journal*, September 1985, pp. 599–612; R. D. Hackett, *Work Attitudes and Employee Absenteeism: A Synthesis of the Literature*, paper presented at the 1988 National Academy of Management Meeting, Anaheim, CA, August 1988; and R. Steel and J. R. Rentsch, "Influence of Cumulation Strategies on the Long-Range Prediction of Absenteeism," *Academy of Management Journal*, December 1995, pp. 1616–1634.

20. P. W. Hom and R. W. Griffeth, *Employee Turnover* (Cincinnati, OH: South-Western Publishing, 1995); R. W. Griffeth, P. W. Hom, and S. Gaertner, "A Meta-Analysis of Antecedents and Correlates of Employee Turnover: Update, Moderator Tests, and Research Implications for the Next Millennium," *Journal of Management*, vol. 26(3), 2000, p. 479; and P. W. Hom and A. J. Kinicki, "Toward a Greater Understanding of How Dissatisfaction Drives Employee Turnover," *Academy of Management Journal*, October 2001, pp. 975–987.

21. See, for example, J. M. Carsten and P. E. Spector, "Unemployment, Job Satisfaction, and Employee Turnover: A Meta-Analytic Test of the Muchinsky Model," *Journal of Applied Psychology*, August 1987, pp. 374–381; and C. L. Hulin, M. Roznowski, and D. Hachiya, "Alternative Opportunities and Withdrawal Decisions: Empirical and Theoretical Discrepancies and an Integration," *Psychological Bulletin*, July 1985, pp. 233–250.

22. T. A. Wright and D. G. Bonett, "Job Satisfaction and Psychological Well-Being as Nonadditive Predictors of Workplace Turnover," *Journal of Management*, April 2007, pp. 141–160; and D. G. Spencer and R. M. Steers, "Performance as a Moderator of the Job Satisfaction–Turnover Relationship," *Journal of Applied Psychology*, August 1981, pp. 511–514.

23. See, for instance, X. Luo and C. Homburg, "Neglected Outcomes of Customer Satisfaction," *Journal of Marketing*, April 2007, pp. 133–149; P. B. Barger and A. A. Grandey, "Service with a Smile and Encounter Satisfaction: Emotional Contagion and Appraisal Mechanisms," *Academy of Management Journal*, December 2006, pp. 1229–1238; C. Homburg and R. M. Stock, "The Link Between Salespeople's Job Satisfaction and Customer Satisfaction in a Business-to-Business Context: A Dyadic Analysis," *Journal of the Academy of Marketing Science*, Spring 2004, pp. 144–158; Harter, Schmidt, and Hayes, "Business-Unit-Level Relationship Between Employee Satisfaction, Employee Engagement, and Business Outcomes: A Meta-Analysis"; J. Griffith, "Do Satisfied Employees Satisfy Customers? Support-Services Staff Morale and Satisfaction Among Public School Administrators, Students, and Parents," *Journal of Applied Social Psychology*, August 2001, pp. 1627–1658; D. J. Koys, "The Effects of Employee Satisfaction, Organizational Citizenship Behavior, and Turnover on Organizational Effectiveness: A Unit-Level, Longitudinal Study," *Personnel Psychology*, Spring 2001, pp. 101–114; E. Naumann and D. W. Jackson, Jr., "One More Time: How Do You Satisfy Customers?" *Business Horizons*, May–June 1999, pp. 71–76; W. W. Tornow and J. W. Wiley, "Service Quality and Management Practices: A Look at Employee Attitudes, Customer Satisfaction, and Bottom-Line Consequences," *Human Resource Planning*, vol. 4 (2), 1991, pp. 105–116; and B. Schneider and D. E. Bowen, "Employee and Customer Perceptions of Service in Banks: Replication and Extension," *Journal of Applied Psychology*, August 1985, pp. 423–433.

24. M. J. Bittner, B. H. Blooms, and L. A. Mohr, "Critical Service Encounters: The Employees' Viewpoint," *Journal of Marketing*, October 1994, pp. 95–106.

25. See LePine, Erez, and Johnson, "The Nature and Dimensionality of Organizational Citizenship Behavior: A Critical Review and Meta-Analysis"; P. Podsakoff, S. B. Mackenzie, J. B. Paine, and D. G. Bachrach, "Organizational Citizenship Behaviors: A Critical Review of the Theoretical and Empirical Literature and Suggestions for Future Research," *Journal of Management*, May 2000, pp. 513–563; and T. S. Bateman and D. W. Organ, "Job Satisfaction and the Good Soldier: The Relationship Between Affect and Employee 'Citizenship,'" *Academy of Management Journal*, December 1983, pp. 587–595.
26. B. J. Hoffman, C. A. Blair, J. P. Maeriac, and D. J. Woehr, "Expanding the Criterion Domain? A Quantitative Review of the OCB Literature," *Journal of Applied Psychology*, vol. 92 (2), 2007, pp. 555–566; LePine, Erez, and Johnson, "The Nature and Dimensionality of Organizational Citizenship Behavior: A Critical Review and Meta-Analysis"; and D. W. Organ and K. Ryan, "A Meta-Analytic Review of Attitudinal and Dispositional Predictors of Organizational Citizenship Behavior," *Personnel Psychology*, Winter 1995, pp. 775–802.
27. N. A. Fassina, D. A. Jones, and K. L. Uggerslev, "Relationship Clean-up Time: Using Meta-Analysis and Path Analysis to Clarify Relationships Among Job Satisfaction, Perceived Fairness, and Citizenship Behaviors," *Journal of Management*, April 2008, pp. 161–188; M. A. Konovsky and D. W. Organ, "Dispositional and Contextual Determinants of Organizational Citizenship Behavior," *Journal of Organizational Behavior*, May 1996, pp. 253–266; R. H. Moorman, "Relationship Between Organization Justice and Organizational Citizenship Behaviors: Do Fairness Perceptions Influence Employee Citizenship?" *Journal of Applied Psychology*, December 1991, pp. 845–855; and J. Fahr, P. M. Podsakoff and D. W. Organ, "Accounting for Organizational Citizenship Behavior: Leader Fairness and Task Scope versus Satisfaction," *Journal of Management*, December 1990, pp. 705–722.
28. W. H. Bommer, E. C. Dierdorff, and R. S. Rubin, "Does Prevalence Mitigate Relevance? The Moderating Effect of Group-Level OCB on Employee Performance," *Academy of Management Journal*, December 2007, pp. 1481–1494.
29. See, for example, S. Rabinowitz and D. T. Hall, "Organizational Research in Job Involvement," *Psychological Bulletin*, March 1977, pp. 265–288; G. J. Blau, "A Multiple Study Investigation of the Dimensionality of Job Involvement," *Journal of Vocational Behavior*, August 1985, pp. 19–36; and N. A. Jans, "Organizational Factors and Work Involvement," *Organizational Behavior and Human Decision Processes*, June 1985, pp. 382–396.
30. D. A. Harrison, D. A. Newman, and P. L. Roth, "How Important Are Job Attitudes? Meta-Analytic Comparisons of Integrative Behavioral Outcomes and Time Sequences," *Academy of Management Journal*, April 2006, pp. 305–325; G. J. Blau, "Job Involvement and Organizational Commitment as Interactive Predictors of Tardiness and Absenteeism," *Journal of Management*, Winter 1986, pp. 577–584; and K. Boal and R. Cidambi, *Attitudinal Correlates of Turnover and Absenteeism: A Meta-Analysis*, paper presented at the meeting of the American Psychological Association, Toronto, 1984.
31. G. J. Blau and K. Boal, "Conceptualizing How Job Involvement and Organizational Commitment Affect Turnover and Absenteeism," *Academy of Management Review*, April 1987, p. 290.
32. See, for instance, P. W. Hom, R. Katerberg, and C. L. Hulin, "Comparative Examination of Three Approaches to the Prediction of Turnover," *Journal of Applied Psychology*, June 1979, pp. 280–290; R. T. Mowday, L. W. Porter, and R. M. Steers, *Employee Organization Linkages: The Psychology of Commitment, Absenteeism, and Turnover* (New York: Academic Press, 1982); H. Angle and J. Perry, "Organizational Commitment: Individual and Organizational Influence," *Work and Occupations*, May 1983, pp. 123–145; and J. L. Pierce and R. B. Dunham, "Organizational Commitment: Pre-employment Propensity and Initial Work Experiences," *Journal of Management*, Spring 1987, pp. 163–178.
33. L. W. Porter, R. M. Steers, R. T. Mowday, and V. Boulian, "Organizational Commitment, Job Satisfaction, and Turnover among Psychiatric Technicians," *Journal of Applied Psychology*, October 1974, pp. 603–609.
34. D. M. Rousseau, "Organizational Behavior in the New Organizational Era," in J. T. Spence, J. M. Darley, and D. J. Foss (eds.), *Annual Review of Psychology*, vol. 48 (Palo Alto, CA: Annual Reviews, 1997), p. 523.
35. P. Eder and R. Eisenberger, "Perceived Organizational Support: Reducing the Negative Influence of Coworker Withdrawal Behavior," *Journal of Management*, February 2008, pp. 55–68; R. Eisenberger, et al., "Perceived Supervisor Support: Contributions to Perceived Organizational Support and Employee Retention," *Journal of Applied Psychology*, June 2002, pp. 565–573; L. Rhoades and R. Eisenberger, "Perceived Organizational Support: A Review of the Literature," *Journal of Applied Psychology*, August 2002, pp. 698–714; J. L. Kraimer and S. J. Wayne, "An Examination of Perceived Organizational Support as a Multidimensional Construct in the Context of an Expatriate Assignment," *Journal of Management*, vol. 30 (2), 2004, pp. 209–237; J. W. Bishop, K. D. Scott, J. G. Goldsby, and R. Cropanzano, "A Construct Validity Study of Commitment and Perceived Support Variables," *Group & Organization Management*, April 2005, pp. 153–180; and J. A.-M. Coyle-Shapiro and N. Conway, "Exchange Relationships: Examining Psychological Contracts and Perceived Organizational Support," *Journal of Applied Psychology*, July 2005, pp. 774–781.
36. J. Marquez, "Disengaged Employees Can Spell Trouble at Any Company," *Workforce Management* online, www.workforce.com, May 13, 2008.
37. J. Smythe, "Engaging Employees to Drive Performance," *Communication World*, May/June 2008, pp. 20–22; A. B. Bakker and W. B. Schaufeli, "Positive Organizational Behavior: Engaged Employees in Flourishing Organizations," *Journal of Organizational Behavior*, February 2008, pp. 147–154; U. Aggarwal, S. Datta, and S. Bhargava, "The Relationship Between Human Resource Practices, Psychological Contract, and Employee Engagement—Implications for Managing Talent," *IIMB Management Review*, September 2007, pp. 313–325; M. C. Christian and J. E. Slaughter, "Work Engagement: A Meta-Analytic Review and Directions for Research in an Emerging Area," *AOM Proceedings*, August 2007, pp. 1–6; C. H. Thomas, "A New Measurement Scale for Employee Engagement: Scale Development, Pilot Test, and Replication," *AOM Proceedings*, August 2007, pp. 1–6; A. M. Saks, "Antecedents and Consequences of Employee Engagement," *Journal of Managerial Psychology*, vol. 21 (7), 2006, pp. 600–619; and A. Parsley, "Road Map for Employee Engagement," *Management Services*, Spring 2006, pp. 10–11.
38. Watson Wyatt Worldwide, *Driving Employee Engage-ment in a Global Workforce*, www.watsonwyatt.com/research, 2007/2008, p. 2.

39. A. J. Elliott and P. G. Devine, "On the Motivational Nature of Cognitive Dissonance: Dissonance as Psychological Discomfort," *Journal of Personality and Social Psychology*, September 1994, pp. 382–394.

40. L. Festinger, *A Theory of Cognitive Dissonance* (Stanford, CA: Stanford University Press, 1957); and C. Crossen, "Cognitive Dissonance Became a Milestone in 1950s Psychology," *Wall Street Journal*, December 4, 2006, p. B1.

41. See, for example, S. V. Falletta, "Organizational Intelligence Surveys," *T&D*, June 2008, p. 52–58; R. Fralicx, et al., *Point of View: Using Employee Surveys to Drive Business Decisions*, Mercer Human Resource Consulting, July 1, 2004; L. Simpson, "What's Going on in Your Company? If You Don't Ask, You'll Never Know," *Training*, June 2002, pp. 30–34; and B. Fishel, "A New Perspective: How to Get the Real Story from Attitude Surveys," *Training*, February 1998, pp. 91–94.

42. A. Kover, "And the Survey Says..." *IndustryWeek*, September 2005, pp. 49–52.

43. See J. Welch and S. Welch, "Employee Polls: A Vote in Favor," *BusinessWeek*, January 28, 2008, p. 90; E. White, "How Surveying Workers Can Pay Off," *Wall Street Journal*, June 18, 2007, p. B3; Kover, "And the Survey Says..."; Fralicx, et al., *Point of View: Using Employee Surveys to Drive Business Decisions*; and S. Shellenbarger, "Companies Are Finding It Really Pays to Be Nice to Employees," *Wall Street Journal*, July 22, 1998, p. B1.

44. L. Saari and T. A. Judge, "Employee Attitudes and Job Satisfaction," *Human Resource Management*, Winter 2004, pp. 395–407; and T. A. Judge and A. H. Church, "Job Satisfaction: Research and Practice," in C. L. Cooper and E. A. Locke (eds.), *Industrial and Organizational Psychology: Linking Theory with Practice* (Oxford, UK: Blackwell, 2000).

45. Harrison, Newman, and Roth, "How Important Are Job Attitudes? Meta-Analytic Comparisons of Integrative Behavioral Outcomes and Time Sequences," pp. 320–321.

46. C. Arnst, "Better Loving Through Chemistry," *BusinessWeek*, October 24, 2005, p. 48.

47. I. Briggs-Myers, *Introduction to Type* (Palo Alto, CA: Consulting Psychologists Press, 1980); W. L. Gardner and M. J. Martinko, "Using the Myers-Briggs Type Indicator to Study Managers: A Literature Review and Research Agenda," *Journal of Management*, vol. 22 (1), 1996, pp. 45–83; and N. L. Quenk, *Essentials of Myers-Briggs Type Indicator Assessment* (New York: Wiley, 2000).

48. J. M. Digman, "Personality Structure: Emergence of the Five-Factor Model," in M. R. Rosenweig and L. W. Porter (eds.), *Annual Review of Psychology*, vol. 41 (Palo Alto, CA: Annual Review, 1990), pp. 417–440; O. P. John, "The Big Five Factor Taxonomy: Dimensions of Personality in the Natural Language and in Questionnaires," in L. A. Pervin (ed.), *Handbook of Personality Theory and Research* (New York: Guilford Press, 1990), pp. 66–100; M. K. Mount, M. R. Barrick, and J. P. Strauss, "Validity of Observer Ratings of the Big Five Personality Factors," *Journal of Applied Psychology*, April 1996, pp. 272–280; G. M. Hurtz and J. J. Donovan, "Personality and Job Performance: The Big Five Revisited," *Journal of Applied Psychology*, December 2000, pp. 869–879; T. A. Judge, D. Heller, and M. K. Mount, "Five-Factor Model of Personality and Job Satisfaction: A Meta-Analysis," *Journal of Applied Psychology*, June 2002, pp. 530–541; and C. G. DeYoung, L. C. Quilty, and J. B. Peterson, "Between Facets and Domains: 10 Aspects of the Big Five," *Journal of Personality and Social Psychology*, November 2007, pp. 880–896.

49. M. R. Barrick and M. K. Mount, "The Big Five Personality Dimensions and Job Performance: A Meta-Analysis," *Personnel Psychology*, vol. 44, 1991, pp. 1–26; A. J. Vinchur, J. S. Schippmann, F. S. Switzer III, and P. L. Roth, "A Meta-Analytic Review of Predictors of Job Performance for Salespeople," *Journal of Applied Psychology*, August 1998, pp. 586–597; G. M. Hurtz and J. J. Donovan, "Personality and Job Performance Revisited," *Journal of Applied Psychology*, December 2000, pp. 869–879; T. A. Judge and J. E. Bono, "Relationship of Core Self-Evaluations Traits—Self Esteem, Generalized Self-Efficacy, Locus of Control, and Emotional Stability—With Job Satisfaction and Job Performance: A Meta-Analysis," *Journal of Applied Psychology*, February 2001, pp. 80–92; Judge, Heller, and Mount, "Five-Factor Model of Personality and Job Satisfaction: A Meta-Analysis"; and D. M. Higgins, J. B. Peterson, R. O. Pihl, and A. G. M. Lee, "Prefrontal Cognitive Ability, Intelligence, Big Five Personality, and the Prediction of Advanced Academic and Workplace Performance," *Journal of Personality and Social Psychology*, August 2007, pp. 298–319.

50. J. B. Rotter, "Generalized Expectancies for Internal Versus External Control of Reinforcement," *Psychological Monographs*, vol. 80(609), 1966, pp. 1–28.

51. See, for instance, D. W. Organ and C. N. Greene, "Role Ambiguity, Locus of Control, and Work Satisfaction," *Journal of Applied Psychology*, February 1974, pp. 101–102; and T. R. Mitchell, C. M. Smyser, and S. E. Weed, "Locus of Control: Supervision and Work Satisfaction," *Academy of Management Journal*, September 1975, pp. 623–631.

52. R. G. Vleeming, "Machiavellianism: A Preliminary Review," *Psychological Reports*, February 1979, pp. 295–310; and S. A. Snook, "Love and Fear and the Modern Boss," *Harvard Business Review*, January 2008, pp. 16–17.

53. See J. Brockner, *Self-Esteem at Work: Research, Theory, and Practice* (Lexington, MA: Lexington Books, 1988), Chapters 1–4; and N. Branden, *Self-Esteem at Work* (San Francisco: Jossey-Bass, 1998).

54. See M. Snyder, *Public Appearances/Private Realities: The Psychology of Self-Monitoring* (New York: W. H. Freeman, 1987); and D. V. Day, D. J. Schleicher, A. L. Unckless, and N. J. Hiller, "Self-Monitoring Personality at Work: A Meta-Analytic Investigation of Construct Validity," *Journal of Applied Psychology*, April 2002, pp. 390–401.

55. Snyder, *Public Appearances/Private Realities*; and J. M. Jenkins, "Self-Monitoring and Turnover: The Impact of Personality on Intent to Leave," *Journal of Organizational Behavior*, January 1993, pp. 83–90.

56. M. Kilduff and D. V. Day, "Do Chameleons Get Ahead? The Effects of Self-Monitoring on Managerial Careers," *Academy of Management Journal*, August 1994, pp. 1047–1060; and A. Mehra, M. Kilduff, and D. J. Brass, "The Social Networks of High and Low Self-Monitors: Implications for Workplace Performance," *Administrative Science Quarterly*, March 2001, pp. 121–146.

57. N. Kogan and M. A. Wallach, "Group Risk Taking as a Function of Members' Anxiety and Defensiveness," *Journal of Personality*, March 1967, pp. 50–63; and J. M. Howell and C. A. Higgins, "Champions of Technological Innovation," *Administrative Science Quarterly*, June 1990, pp. 317–341.

58. M. Friedman and R. H. Rosenman, *Type A Behavior and Your Heart* (New York: Alfred A. Knopf, 1974).

59. J. D. Kammeyer-Mueller and C. R. Wanberg, "Unwrapping the Organizational Entry Process: Disentangling Multiple

Antecedents and Their Pathways to Adjustment," *Journal of Applied Psychology*, October 2003, pp. 779–794; S. E. Seibert, M. L. Kraimer, and J. M. Crant, "What Do Proactive People Do? A Longitudinal Model Linking Proactive Personality and Career Success," *Personnel Psychology*, Winter 2001, pp. 845–874; J. M. Crant, "Proactive Behavior in Organizations," *Journal of Management*, vol. 26 (3), 2000, pp. 435–462; J. M. Crant and T. S. Bateman, "Charismatic Leadership Viewed from Above: The Impact of Proactive Personality," *Journal of Organizational Behavior*, February 2000, pp. 63–75; S. E. Seibert, J. M. Crant, and M. L. Kraimer, "Proactive Personality and Career Success," *Journal of Applied Psychology*, June 1999, pp. 416–427; R. C. Becherer and J. G. Maurer, "The Proactive Personality Disposition and Entrepreneurial Behavior Among Small Company Presidents," *Journal of Small Business Management*, January 1999, pp. 28–36; and T. S. Bateman and J. M. Crant, "The Proactive Component of Organi-zational Behavior: A Measure and Correlates," *Journal of Organizational Behavior*, March 1993, pp. 103–118.

60. See, for instance, G. W. M. Ip and M. H. Bond, "Culture, Values, and the Spontaneous Self-Concept," *Asian Journal of Psychology*, vol. 1, 1995, pp. 30–36; J. E. Williams, et al., "Cross-Cultural Variation in the Importance of Psychological Characteristics: A Seven-Year Country Study," *International Journal of Psychology*, October 1995, pp. 529–550; V. Benet and N. G. Walker, "The Big Seven Factor Model of Personality Description: Evidence for Its Cross-Cultural Generalizability in a Spanish Sample," *Journal of Personality and Social Psychology*, October 1995, pp. 701–718; R. R. McCrae and P. T. Costa, Jr., "Personality Trait Structure as a Human Universal," *American Psychologist*, 1997, pp. 509–516; and M. J. Schmit, J. A. Kihm, and C. Robie, "Development of a Global Measure of Personality," *Personnel Psychology*, Spring 2000, pp. 153–193.

61. J. F. Salgado, "The Five Factor Model of Personality and Job Performance in the European Community," *Journal of Applied Psychology*, February 1997, pp. 30–43. Note: This study covered the 15-nation European community and did not include the 10 countries that joined in 2004.

62. N. H. Frijda, "Moods, Emotion Episodes, and Emotions," in M. Lewis and J. M. Havilland (eds.), *Handbook of Emotions* (New York: Guilford Press, 1993), pp. 381–403.

63. N. M. Ashkanasy and C. S. Daus, "Emotion in the Workplace: The New Challenge for Managers," *Academy of Management Executive*, February 2002, pp. 76–86; and N. M. Ashkanasy, C. E. J. Hartel, and C. S. Daus, "Diversity and Emotions: The New Frontiers in Organizational Behavior Research," *Journal of Management*, vol. 28 (3), 2002, pp. 307–338.

64. H. M. Weiss and R. Cropanzano, "Affective Events Theory," in B. M. Staw and L. L. Cummings (eds.), *Research in Organizational Behavior*, vol. 18 (Greenwich, CT: JAI Press, 1996), pp. 20–22.

65. This section is based on D. Goleman, *Emotional Intelligence* (New York: Bantam, 1995); M. Davies, L. Stankov, and R. D. Roberts, "Emotional Intelligence: In Search of an Elusive Construct," *Journal of Personality and Social Psychology*, October 1998, pp. 989–1015; D. Goleman, *Working with Emotional Intelligence* (New York: Bantam, 1999); R. Bar-On and J. D. A. Parker (eds.), *The Handbook of Emotional Intelligence: Theory, Development, Assessment, and Application at Home, School, and in the Workplace* (San Francisco: Jossey-Bass, 2000); and P. J. Jordan, N. M. Ashkanasy, and C. E. J. Hartel, "Emotional Intelligence as a Moderator of Emotional and Behavioral Reactions to Job Insecurity," *Academy of Management Review*, July 2002, pp. 361–372.

66. R. D. Shaffer and M. A. Shaffer, "Emotional Intelligence Abilities, Personality, and Workplace Performance," *Academy of Management Best Conference Paper—HR*, August 2005; K. S. Law, C. Wong, and L. J. Song, "The Construct and Criterion Validity of Emotional Intelligence and Its Potential Utility for Management Studies," *Journal of Applied Psychology*, August 2004, pp. 483–496; D. L. Van Rooy and C. Viswesvaran, "Emotional Intelligence: A Meta-Analytic Investigation of Predictive Validity and Nomological Net," *Journal of Vocational Behavior*, August 2004, pp. 71–95; P. J. Jordan, N. M. Ashkanasy, and C. E. J. Hartel, "The Case for Emotional Intelligence in Organizational Research," *Academy of Management Review*, April 2003, pp. 195–197; H. A. Elfenbein and N. Ambady, "Predicting Workplace Outcomes from the Ability to Eavesdrop on Feelings," *Journal of Applied Psychology*, October 2002, pp. 963–971; and C. Cherniss, *The Business Case for Emotional Intelligence*, Consortium for Research on Emotional Intelligence in Organizations, www.eiconsortium.org, 1999.

67. F. J. Landy, "Some Historical and Scientific Issues Related to Research on Emotional Intelligence," *Journal of Organizational Behavior*, June 2005, pp. 411–424; E. A. Locke, "Why Emotional Intelligence Is an Invalid Concept," *Journal of Organizational Behavior*, June 2005, pp. 425–431; J. M. Conte, "A Review and Critique of Emotional Intelligence Measures," *Journal of Organizational Behavior*, June 2005, pp. 433–440; T. Becker, "Is Emotional Intelligence a Viable Concept?" *Academy of Management Review*, April 2003, pp. 192–195; and M. Davies, L. Stankov, and R. D. Roberts, "Emotional Intelligence: In Search of an Elusive Construct," *Journal of Personality and Social Psychology*, October 1998, pp. 989–1015.

68. G. Kranz, "Organizations Look to Get Personal in '07," *Workforce Management* online, www.workforce.com, June 19, 2007.

69. J. L. Holland, *Making Vocational Choices: A Theory of Vocational Personalities and Work Environments* (Odessa, FL: Psychological Assessment Resources, 1997).

70. See, for instance, M. J. Martinko (ed.), *Attribution Theory: An Organizational Perspective* (Delray Beach, FL: St. Lucie Press, 1995); and H. H. Kelley, "Attribution in Social Interaction," in E. Jones et al. (eds.), *Attribution: Perceiving the Causes of Behavior* (Morristown, NJ: General Learning Press, 1972).

71. See A. G. Miller and T. Lawson, "The Effect of an Informational Option on the Fundamental Attribution Error," *Personality and Social Psychology Bulletin*, June 1989, pp. 194–204.

72. See, for instance, G. R. Semin, "A Gloss on Attribution Theory," *British Journal of Social and Clinical Psychology*, November 1980, pp. 291–330; and M. W. Morris and K. Peng, "Culture and Cause: American and Chinese Attributions for Social and Physical Events," *Journal of Personality and Social Psychology*, December 1994, pp. 949–971.

73. S. Nam, *Cultural and Managerial Attributions for Group Performance*, unpublished doctoral dissertation, University of Oregon. Cited in R. M. Steers, S. J. Bischoff, and L. H. Higgins, "Cross-Cultural Management Research," *Journal of Management Inquiry*, December 1992, pp. 325–326.

74. See, for example, S. T. Fiske, "Social Cognition and Social Perception," *Annual Review of Psychology*, 1993,

pp. 155–194; G. N. Powell and Y. Kido, "Managerial Stereotypes in a Global Economy: A Comparative Study of Japanese and American Business Students' Perspectives," *Psychological Reports*, February 1994, pp. 219–226; and J. L. Hilton and W. von Hippel, "Stereotypes," in J. T. Spence, J. M. Darley, and D. J. Foss (eds.), *Annual Review of Psychology*, vol. 47 (Palo Alto, CA: Annual Reviews Inc., 1996), pp. 237–271.

75. B. F. Skinner, *Contingencies of Reinforcement* (East Norwalk, CT: Appleton-Century-Crofts, 1971).
76. S. Armour, "Generation Y: They've Arrived at Work with a New Attitude," *USA Today*, November 6, 2005, pp. 1B+.
77. N. Ramachandran, "New Paths at Work," *U.S. News & World Report*, March 20, 2006, p. 47.
78. D. Sacks, "Scenes from the Culture Clash," *Fast Company*, January/February 2006, p. 75.
79. Armour, "Generation Y: They've Arrived at Work with a New Attitude," p. 2B.
80. Armour, "Generation Y: They've Arrived at Work with a New Attitude"; B. Moses, "The Challenges of Managing Gen Y," *The Globe and Mail*, March 11, 2005, p. C1; and C. A. Martin, *Managing Generation Y* (Amherst, MA: HRD Press, 2001).
81. C. M. Pearson and C. L. Porath, "On the Nature, Consequences, and Remedies of Workplace Incivility: No Time for Nice? Think Again," *Academy of Management Executive*, February 2005, pp. 7–18.
82. J. Robison, "Be Nice: It's Good for Business," *Gallup Brain*, http://brain.gallup.com, August 12, 2004.
83. http://www.tifr.res.in/~aset/full_text/Psychological%20Harassment%20Talk.ppt
84. Y. Vardi and E. Weitz, *Misbehavior in Organizations* (Mahwah, NJ: Lawrence Erlbaum Associates, 2004), pp. 246–247.
85. "Q&A," *Training*, April 2007, p. 7.

Chapter 14

1. S. Shellenbarger, "A Day Without Email Is Like..." *Wall Street Journal*, October 11, 2007, pp. D1+; M. Kessler, "Fridays Go from Casual to E-Mail-Free," *USA Today*, October 5, 2007, p. 1A; D. Beizer, "Email Is Dead," *Fast Company*, July/August 2007, p. 46; O. Malik, "Why Email Is Bankrupt," *Business 2.0*, July 2007, p. 46; and D. Brady, "*!#?@ the E-Mail. Can We Talk?" *BusinessWeek*, December 4, 2006, p. 109.
2. P. G. Clampitt, *Communicating for Managerial Effectiveness*, 3rd ed. (Thousand Oaks, CA: Sage Publications, 2005); T. Dixon, *Communication, Organization, and Performance* (Norwood, NJ: Ablex Publishing Corporation, 1996), p. 281; and L. E. Penley, E. R. Alexander, I. Edward Jernigan, and C. I. Henwood, "Communication Abilities of Managers: The Relationship to Performance," *Journal of Management*, March 1991, pp. 57–76.
3. "Electronic Invective Backfires," *Workforce*, June 2001, p. 20; and E. Wong, "A Stinging Office Memo Boomerangs," *New York Times*, April 5, 2001, pp. C1+.
4. C. O. Kursh, "The Benefits of Poor Communication," *Psychoanalytic Review*, Summer–Fall 1971, pp. 189–208.
5. International Association of Business Communicators, "Excellence in Communication Leadership (EXCEL) Award," *News Release*, news.iabc.com, April 2, 2008.
6. W. G. Scott and T. R. Mitchell, *Organization Theory: A Structural and Behavioral Analysis* (Homewood, IL: Richard D. Irwin, 1976).
7. D. K. Berlo, *The Process of Communication* (New York: Holt, Rinehart & Winston, 1960), pp. 30–32.
8. Clampitt, *Communicating for Managerial Effectiveness*.
9. A. Warfield, "Do You Speak Body Language?" *Training & Development*, April 2001, pp. 60–61; D. Zielinski, "Body Language Myths," *Presentations*, April 2001, pp. 36–42; and "Visual Cues Speak Loudly in Workplace," *Springfield (Missouri) News-Leader*, January 21, 2001, p. 8B.
10. http://www.icanhasstartup.com/interviews/interview-deskaway/
11. http://www.blog.teqedge.com/2008/08/20/deskaway-project-management-made-easy/
12. Information in this box from J. Langdon, "Differences Between Males and Females at Work," *USA Today*, www.usatoday.com, February 5, 2001; J. Manion, "He Said, She Said," *Materials Management in Health Care*, November 1998, pp. 52–62; G. Franzwa and C. Lockhart, "The Social Origins and Maintenance of Gender Communication Styles, Personality Types, and Grid-Group Theory," *Sociological Perspectives*, vol. 41 (1), 1998, pp. 185–208; and D. Tannen, *Talking From 9 to 5: Women and Men in the Workplace* (New York: Avon Books, 1995).
13. Shellenbarger, "A Day Without Email Is Like..."; Kessler, "Fridays Go From Casual to E-Mail-Free"; Beizer, "Email Is Dead"; Malik, "Why Email Is Bankrupt"; Brady, "*!#?@ the E-Mail. Can We Talk?"; and American Management Association, *E-Mail Rules, Policies, and Practices Survey*, www.amanet.org, 2003.
14. Berlo, *The Process of Communication*, p. 103.
15. R. Buckman, "Why the Chinese Hate to Use Voice Mail," *Wall Street Journal*, December 1, 2005, p. B1+.
16. A. Mehrabian, "Communication Without Words," *Psychology Today*, September 1968, pp. 53–55.
17. L. Haggerman, "Strong, Efficient Leadership Minimizes Employee Problems," *Springfield (Missouri) Business Journal*, December 9–15, 2002, p. 23.
18. See, for instance, S. P. Robbins and P. L. Hunsaker, *Training in InterPersonal Skills*, 4th ed. (Upper Saddle River, NJ: Prentice Hall, 2006); M. Young and J. E. Post, "Managing to Communicate, Communicating to Manage: How Leading Companies Communicate with Employees," *Organizational Dynamics*, Summer 1993, pp. 31–43; J. A. DeVito, *The Interpersonal Communication Book*, 6th ed. (New York: HarperCollins, 1992); and A. G. Athos and J. J. Gabarro, *Interpersonal Behavior* (Upper Saddle River, NJ: Prentice Hall, 1978).
19. O. Thomas, "Best-Kept Secrets of the World's Best Companies: The Three Minute Huddle," *Business 2.0*, April 2006, p. 94.
20. J. S. Lublin, "The 'Open Inbox,'" *Wall Street Journal*, October 10, 2005, pp. B1+.
21. http://www.livemint.com/2008/12/21222828/Crisis-Time–CEO-needs-to-be.html
22. Cited in "Shut Up and Listen," *Money*, November 2005, p. 27.
23. See, for instance, D. Sagario and L. Ballard, "Workplace Gossip Can Threaten Your Office," *Springfield (Missouri) News-Leader*, September 26, 2005, p. 5B; A. Bruzzese, "What to Do About Toxic Gossip," *USA Today* online,

www.usatoday.com, March 14, 2001; N. B. Kurland and L. H. Pelled, "Passing the Word: Toward a Model of Gossip and Power in the Workplace," *Academy of Management Review*, April 2000; pp. 428–438; N. DiFonzo, P. Bordia, and R. L. Rosnow, "Reining in Rumors," *Organizational Dynamics*, Summer 1994, pp. 47–62; M. Noon and R. Delbridge, "News from Behind My Hand: Gossip in Organizations," *Organization Studies*, vol. 14 (1), 1993, pp. 23–26; and J. G. March and G. Sevon, "Gossip, Information and Decision Making," in J. G. March (ed.), *Decisions and Organizations* (Oxford, UK: Blackwell, 1988, pp. 429–442.

24. Watson Wyatt Worldwide, *Secrets of Top Performers: How Companies with Highly Effective Employee Communication Differentiate Themselves, 2007/2008 Communication ROI Study*. Washington, DC: Watson Wyatt Worldwide.
25. These examples taken from S. Kirsner, "Being There," *Fast Company*, January/February 2006, pp. 90–91; R. Breeden, "More Employees Are Using the Web at Work," *Wall Street Journal*, May 10, 2005, p. B4; C. Woodward, "Some Offices Opt for Cellphones Only," *USA Today*, January 25, 2005, p. 1B; and J. Rohwer, "Today, Tokyo. Tomorrow, the World," *Fortune*, September 18, 2000, pp. 140–152.
26. Based on J. D. Copeland, "Personal Tasks at Work a Problem," *Springfield (Missouri) Business Journal*, June 11–17, 2007, p. 51; E. Frauenheim, "Stop Reading This Headline and Get Back to Work," *CNET*, www.news.com, July 13, 2005; and Breeden, "More Employees Are Using the Web at Work."
27. http://www.expresscomputeronline.com/20080915/management01.shtml
28. M. Broersma, "Amsterdam Switches on Europe's First Mobile WiMAX Network," *Computerworld*, www.computerworld.com, June 18, 2008; S. Dean, "Wi-Fi Hotspot Usage: The Numbers Are Up...Way Up," *Web Worker Daily*, www.webworkerdaily.com, September 19, 2007; and B. White, "Helpless, Hopeless, Wireless," *Wall Street Journal*, June 26, 2007, pp. B1+.
29. http://www.gigaom.com/2008/09/18/mobilize-padmasree-warrior-cto-cisco/
30. S. Srivastava, "Doing More on the Go," *Wall Street Journal*, June 12, 2007, p. B3.
31. B. White, "Firms Take a Cue from YouTube," *Wall Street Journal*, January 2, 2007, p. B3.
32. K. Hafner, "For the Well Connected, All the World's an Office," *New York Times*, March 30, 2000, pp. D1+.
33. S. Luh, "Pulse Lunches at Asian Citibanks Feed Workers' Morale, Lower Job Turnover," *Wall Street Journal*, May 22, 2001, p. B11.
34. http://www.livemint.com/2008/12/21222828/Crisis-Time–CEO-needs-to-be.html
35. H. Green, "The Water Cooler Is Now on the Web," *BusinessWeek*, October 1, 2007, pp. 78–79.
36. The Associated Press, "Whole Foods Chief Apologizes for Posts," *New York Times* online, www.nytimes.com, July 18, 2007; E. White, J. S. Lublin, and D. Kesmodel, "Executives Get the Blogging Bug," *Wall Street Journal*, July 13, 2007, pp. B1+; C. Alldred, "U.K. Libel Case Slows E-Mail Delivery," *Business Insurance*, August 4, 1997, pp. 51–53; and T. Lewin, "Chevron Settles Sexual Harassment Charges," *New York Times* online, www.nytimes.com, February 22, 1995.
37. http://www.siliconindia.com/magazine/articledesc.php?articleid=BZT240511918
38. http://www.ciol.com/News/News-Reports/Can-police-make-chat,-email-transcripts-public/6608106794/0/
39. J. Eckberg, "E-mail: Messages Are Evidence," *Cincinnati Enquirer*, www.enquirer.com, July 27, 2004.
40. M. Scott, "Worker E-Mail and Blog Misuse Seen as Growing Risk for Companies," *Workforce Management* online, www.workforce.com, July 20, 2007.
41. K. Byron, "Carrying Too Heavy a Load? The Communication and Miscommunication of Emotion by E-mail," *Academy of Management Review*, April 2008, pp. 309–327.
42. J. Marquez, "Virtual Work Spaces Ease Collaboration, Debate Among Scattered Employees," *Workforce Management*, May 22, 2006, p. 38; and M. Conlin, "E-Mail Is So Five Minutes Ago," *BusinessWeek*, November 28, 2005, pp. 111–112
43. H. Green, "The Water Cooler Is Now on the Web"; E. Frauenheim, "Starbucks Employees Carve Out Own Space," *Workforce Management*, October 22, 2007, p. 32; and S. H. Wildstrom, "Harnessing Social Networks," *BusinessWeek*, April 23, 2007, p. 20.
44. J. Scanlon, "Woman of Substance," *Wired*, July 2002, p. 27.
45. http://www.patni.com/about-us/knowledge-management.html
46. E. Wenger, R. McDermott, and W. Snyder, *Cultivating Communities of Practice: A Guide to Managing Knowledge* (Boston: Harvard Business School Press, 2002), p. 4.
47. Ibid., p. 39.
48. B. A. Gutek, M. Groth, and B. Cherry, "Achieving Service Success Through Relationship and Enhanced Encounters," *Academy of Management Executive*, November 2002, pp. 132–144.
49. R. C. Ford and C. P. Heaton, "Lessons From Hospitality That Can Serve Anyone," *Organizational Dynamics*, Summer 2001, pp. 30–47.
50. M. J. Bitner, B. H. Booms, and L. A. Mohr, "Critical Service Encounters: The Employee's Viewpoint," *Journal of Marketing*, October 1994, pp. 95–106.
51. S. D. Pugh, J. Dietz, J. W. Wiley, and S. M. Brooks, "Driving Service Effectiveness Through Employee–Customer Linkages," *Academy of Management Executive*, November 2002, pp. 73–84.
52. "Assisting Customers with Disabilities: A Summary of Policies and Guidelines Regarding the Assistance of Customers with Disabilities for the Sears Family of Companies," pamphlet from Sears, Roebuck and Company, obtained at Springfield, Missouri, Sears store, May 28, 2003.
53. M. L. LaGanga, "Are There Words That Neither Offend Nor Bore?" *Los Angeles Times*, May 18, 1994, pp. 11–27; and J. Leo, "Language in the Dumps," *U.S. News & World Report*, July 27, 1998, p. 16.
54. M. Richtel, "Lost in E-Mail, Tech Firms Face Self-Made Beast," *New York Times* online, www.nytimes.com, June 14, 2008.

Chapter 15

1. Times of India, February 28, 2006
2. UNI Report,/ April 17, 2007

3. Survey on education undertaken by Pratham in September--December 2005
4. Nationwide survey conducted in June and August 2004 by Pratham
5. http://www.teachforindia.org/
6. C. Taylor, "Rallying the Troops," *Smart Money*, February 2003, pp. 105–106.
7. R. M. Steers, R. T. Mowday, and D. L. Shapiro, "The Future of Work Motivation Theory," *Academy of Management Review*, July 2004, pp. 379–387.
8. N. Ellemers, D. De Gilder, and S. A. Haslam, "Motivating Individuals and Groups at Work: A Social Identity Perspective on Leadership and Group Performance," *Academy of Management Review*, July 2004, pp. 459–478.
9. J. Krueger and E. Killham, "At Work, Feeling Good Matters," *Gallup Management Journal*, http://gmj.gallup.com, December 8, 2005.
10. M. Meece, "Using the Human Touch to Solve Workplace Problems," *New York Times* online, www.nytimes.com, April 3, 2008.
11. A. Maslow, *Motivation and Personality* (New York: McGraw-Hill, 1954); A. Maslow, D. C. Stephens, and G. Heil, *Maslow on Management* (New York: Wiley, 1998); M. L. Ambrose and C. T. Kulik, "Old Friends, New Faces: Motivation Research in the 1990s," *Journal of Management*, vol. 25 (3), 1999, pp. 231–292; and "Dialogue," *Academy of Management Review*, October 2000, pp. 696–701.
12. See, for example, D. T. Hall and K. E. Nongaim, "An Examination of Maslow's Need Hierarchy in an Organizational Setting," *Organizational Behavior and Human Performance*, February 1968, pp. 12–35; E. E. Lawler III and J. L. Suttle, "A Causal Correlational Test of the Need Hierarchy Concept," *Organizational Behavior and Human Performance*, April 1972, pp. 265–287; R. M. Creech, "Employee Motivation," *Management Quarterly*, Summer 1995, pp. 33–39; J. Rowan, "Maslow Amended," *Journal of Humanistic Psychology*, Winter 1998, pp. 81–92; J. Rowan, "Ascent and Descent in Maslow's Theory," *Journal of Humanistic Psychology*, Summer 1999, pp. 125–133; and Ambrose and Kulik, "Old Friends, New Faces: Motivation Research in the 1990s."
13. D. McGregor, *The Human Side of Enterprise* (New York: McGraw-Hill, 1960). For an updated description of Theories X and Y, see an annotated edition with commentary of *The Human Side of Enterprise* (McGraw-Hill, 2006); and G. Heil, W. Bennis, and D. C. Stephens, *Douglas McGregor, Revisited: Managing the Human Side of Enterprise* (New York: Wiley, 2000).
14. J. M. O'Brien, "The Next Intel," *Wired*, July 2002, pp. 100–107.
15. F. Herzberg, B. Mausner, and B. Snyderman, *The Motivation to Work* (New York: John Wiley, 1959); F. Herzberg, *The Managerial Choice: To Be Effective or to Be Human*, rev. ed. (Salt Lake City: Olympus, 1982); Creech, "Employee Motivation"; and Ambrose and Kulik, "Old Friends, New Faces: Motivation Research in the 1990s."
16. D. C. McClelland, *The Achieving Society* (New York: Van Nostrand Reinhold, 1961); J. W. Atkinson and J. O. Raynor, *Motivation and Achievement* (Washington, DC: Winston, 1974); D. C. McClelland, *Power: The Inner Experience* (New York: Irvington, 1975); and M. J. Stahl, *Managerial and Technical Motivation: Assessing Needs for Achievement, Power, and Affiliation* (New York: Praeger, 1986).
17. McClelland, *The Achieving Society*.
18. McClelland, *Power: The Inner Experience*; D. C. McClelland and D. H. Burnham, "Power Is the Great Motivator," *Harvard Business Review*, March–April 1976, pp. 100–110.
19. D. Miron and D. C. McClelland, "The Impact of Achievement Motivation Training on Small Businesses," *California Management Review*, Summer 1979, pp. 13–28.
20. "McClelland: An Advocate of Power," *International Management*, July 1975, pp. 27–29.
21. Steers, Mowday, and Shapiro, "The Future of Work Motivation Theory"; E. A. Locke and G. P. Latham, "What Should We Do About Motivation Theory? Six Recommendations for the Twenty-First Century," *Academy of Management Review*, July 2004, pp. 388–403; and Ambrose and Kulik, "Old Friends, New Faces: Motivation Research in the 1990s."
22. A. Barrett, "Cracking the Whip at Wyeth," *BusinessWeek*, February 6, 2006, pp. 70–71.
23. Ambrose and Kulik, "Old Friends, New Faces: Motivation Research in the 1990s."
24. J. C. Naylor and D. R. Ilgen, "Goal Setting: A Theoretical Analysis of a Motivational Technique," in B. M. Staw and L. L. Cummings (eds.), *Research in Organizational Behavior*, vol. 6 (Greenwich, CT: JAI Press, 1984), pp. 95–140; A. R. Pell, "Energize Your People," *Managers Magazine*, December 1992, pp. 28–29; E. A. Locke, "Facts and Fallacies About Goal Theory: Reply to Deci," *Psychological Science*, January 1993, pp. 63–64; M. E. Tubbs, "Commitment as a Moderator of the Goal–Performance Relation: A Case for Clearer Construct Definition," *Journal of Applied Psychology*, February 1993, pp. 86–97; M. P. Collingwood, "Why Don't You Use the Research?" *Management Decision*, May 1993, pp. 48–54; M. E. Tubbs, D. M. Boehne, and J. S. Dahl, "Expectancy, Valence, and Motivational Force Functions in Goal-Setting Research: An Empirical Test," *Journal of Applied Psychology*, June 1993, pp. 361–373; E. A. Locke, "Motivation Through Conscious Goal Setting," *Applied and Preventive Psychology*, vol. 5, 1996, pp. 117–124; Ambrose and Kulik, "Old Friends, New Faces: Motivation Research in the 1990s"; E. A. Locke and G. P. Latham, "Building a Practically Useful Theory of Goal Setting and Task Motivation: A 35-Year Odyssey," *American Psychologist*, September 2002, pp. 705–717; Y. Fried and L. H. Slowik, "Enriching Goal-Setting Theory with Time: An Integrated Approach," *Academy of Management Review*, July 2004, pp. 404–422; and G. P. Latham, "The Motivational Benefits of Goal-Setting," *Academy of Management Executive*, November 2004, pp. 126–129.
25. J. B. Miner, *Theories of Organizational Behavior* (Hinsdale, IL: Dryden Press, 1980), p. 65.
26. Panda, K. C. (1989). Research in Psychology if Education. A Trend Report. *In Fourth Survey of Research in Education* 1983–88 (Vol. I). Available at http://www.education.nic.in/cd50years/home.htm accessed on January 10, 2009.
27. J. A. Wagner III, "Participation's Effects on Performance and Satisfaction: A Reconsideration of Research and Evidence," *Academy of Management Review*, April 1994, pp. 312–330; J. George-Falvey, "Effects of Task Complexity and Learning Stage on the Relationship between Participation in Goal Setting and Task Performance," *Academy of Management Proceedings*, on disk, 1996; T. D. Ludwig and E. S. Geller, "Assigned Versus Participative Goal Setting and Response Generalization: Managing Injury Control among Professional Pizza Deliverers," *Journal of Applied*

Psychology, April 1997, pp. 253–261; and S. G. Harkins and M. D. Lowe, "The Effects of Self-Set Goals on Task Performance," *Journal of Applied Social Psychology*, January 2000, pp. 1–40.

28. J. M. Ivancevich and J. T. McMahon, "The Effects of Goal Setting, External Feedback, and Self-Generated Feedback on Outcome Variables: A Field Experiment," *Academy of Management Journal*, June 1982, pp. 359–372; and Locke, "Motivation Through Conscious Goal Setting."

29. J. R. Hollenbeck, C. R. Williams, and H. J. Klein, "An Empirical Examination of the Antecedents of Commitment to Difficult Goals," *Journal of Applied Psychology*, February 1989, pp. 18–23; see also J. C. Wofford, V. L. Goodwin, and S. Premack, "Meta-Analysis of the Antecedents of Personal Goal Level and of the Antecedents and Consequences of Goal Commitment," *Journal of Management*, September 1992, pp. 595–615; Tubbs, "Commitment as a Moderator of the Goal–Performance Relation"; J. W. Smither, M. London, and R. R. Reilly, "Does Performance Improve Following Multisource Feedback? A Theoretical Model, Meta-Analysis, and Review of Empirical Findings," *Personnel Psychology*, Spring 2005, pp. 171–203.

30. M. E. Gist, "Self-Efficacy: Implications for Organizational Behavior and Human Resource Management," *Academy of Management Review*, July 1987, pp. 472–485; and A. Bandura, *Self-Efficacy: The Exercise of Control* (New York: Freeman, 1997).

31. E. A. Locke, E. Frederick, C. Lee, and P. Bobko, "Effect of Self-Efficacy, Goals, and Task Strategies on Task Performance," *Journal of Applied Psychology*, May 1984, pp. 241–251; M. E. Gist and T. R. Mitchell, "Self-Efficacy: A Theoretical Analysis of Its Determinants and Malleability," *Academy of Management Review*, April 1992, pp. 183–211; A. D. Stajkovic and F. Luthans, "Self-Efficacy and Work Related Performance: A Meta-Analysis," *Psychological Bulletin*, September 1998, pp. 240–261; and A. Bandura, "Cultivate Self-Efficacy for Personal and Organizational Effectiveness," in E. Locke (ed.), *Handbook of Principles of Organizational Behavior* (Malden, MA: Blackwell, 2004), pp.120–136.

32. http://www.businesstoday.digitaltoday.in/index.php?option=com_content&task=view&issueid=47&id=9493&sectionid=22&Itemid=1

33. A. Bandura and D. Cervone, "Differential Engagement in Self-Reactive Influences in Cognitively-Based Motivation," *Organizational Behavior and Human Decision Processes*, August 1986, pp. 92–113; and R. Ilies and T. A. Judge, "Goal Regulation Across Time: The Effects of Feedback and Affect," *Journal of Applied Psychology*, May 2005, pp. 453–467.

34. See J. C. Anderson and C. A. O'Reilly, "Effects of an Organizational Control System on Managerial Satisfaction and Performance," *Human Relations*, June 1981, pp. 491–501; and J. P. Meyer, B. Schacht-Cole, and I. R. Gellatly, "An Examination of the Cognitive Mechanisms by Which Assigned Goals Affect Task Performance and Reactions to Performance," *Journal of Applied Social Psychology*, vol. 18 (5), 1988, pp. 390–408.

35. http://www.timesjobs.com/Mailers/HRDialogue/may06/05.05.06/hr_dialogue/landing.htm

36. B. F. Skinner, *Science and Human Behavior* (New York: The Free Press, 1953); and B. F. Skinner, *Beyond Freedom and Dignity* (New York: Knopf, 1972).

37. The same data, for instance, can be interpreted in either goal-setting or reinforcement terms, as shown in E. A. Locke, "Latham vs. Komaki: A Tale of Two Paradigms," *Journal of Applied Psychology*, February 1980, pp. 16–23. Also see Ambrose and Kulik, "Old Friends, New Faces: Motivation Research in the 1990s."

38. See, for example, R. W. Griffin, "Toward an Integrated Theory of Task Design," in L. L. Cummings and B. M. Staw (eds.), *Research in Organizational Behavior*, vol. 9 (Greenwich, CT: JAI Press, 1987), pp. 79–120; and M. Campion, "Interdisciplinary Approaches to Job Design: A Constructive Replication with Extensions," *Journal of Applied Psychology*, August 1988, pp. 467–481.

39. S. Caudron, "The De-Jobbing of America," *IndustryWeek*, September 5, 1994, pp. 31–36; W. Bridges, "The End of the Job," *Fortune*, September 19, 1994, pp. 62–74; and K. H. Hammonds, K. Kelly, and K. Thurston, "Rethinking Work," *BusinessWeek*, October 12, 1994, pp. 75–87.

40. M. A. Campion and C. L. McClelland, "Follow-up and Extension of the Interdisciplinary Costs and Benefits of Enlarged Jobs," *Journal of Applied Psychology*, June 1993, pp. 339–351; and Ambrose and Kulik, "Old Friends, New Faces: Motivation Research in the 1990s."

41. See, for example, J. R. Hackman and G. R. Oldham, *Work Redesign* (Reading, MA: Addison-Wesley, 1980); and Miner, *Theories of Organizational Behavior*, pp. 231–266; R. W. Griffin, "Effects of Work Redesign on Employee Perceptions, Attitudes, and Behaviors: A Long-Term Investigation," *Academy of Management Journal*, June 1991, pp. 425–435; J. L. Cotton, *Employee Involvement* (Newbury Park, CA: Sage, 1993), pp. 141–172; and Ambrose and Kulik, "Old Friends, New Faces: Motivation Research in the 1990s."

42. J. R. Hackman and G. R. Oldham, "Development of the Job Diagnostic Survey," *Journal of Applied Psychology*, April 1975, pp. 159–170; and J. R. Hackman and G. R. Oldham, "Motivation Through the Design of Work: Test of a Theory," *Organizational Behavior and Human Performance*, August 1976, pp. 250–279.

43. J. R. Hackman, "Work Design," in J. R. Hackman and J. L. Suttle (eds.), *Improving Life at Work* (Glenview, IL: Scott, Foresman, 1977), p. 129; and Ambrose and Kulik, "Old Friends, New Faces: Motivation Research in the 1990s."

44. J. S. Adams, "Inequity in Social Exchanges," in L. Berkowitz (ed.), *Advances in Experimental Social Psychology*, vol. 2 (New York: Academic Press, 1965), pp. 267–300; and Ambrose and Kulik, "Old Friends, New Faces: Motivation Research in the 1990s."

45. See, for example, P. S. Goodman and A. Friedman, "An Examination of Adams' Theory of Inequity," *Administrative Science Quarterly*, September 1971, pp. 271–288; M. R. Carrell, "A Longitudinal Field Assessment of Employee Perceptions of Equitable Treatment," *Organizational Behavior and Human Performance*, February 1978, pp. 108–118; E. Walster, G. W. Walster, and W. G. Scott, *Equity: Theory and Research* (Boston: Allyn & Bacon, 1978); R. G. Lord and J. A. Hohenfeld, "Longitudinal Field Assessment of Equity Effects on the Performance of Major League Baseball Players," *Journal of Applied Psychology*, February 1979, pp. 19–26; J. E. Dittrich and M. R. Carrell, "Organizational Equity Perceptions, Employee Job Satisfaction, and Departmental Absence and Turnover Rates," *Organizational Behavior and Human Performance*, August 1979, pp. 29–40; and J. Greenberg, "Cognitive Reevaluation of Outcomes in Response to Underpayment Inequity," *Academy of Management Journal*, March 1989, pp. 174–184.

46. P. S. Goodman, "An Examination of Referents Used in the Evaluation of Pay," *Organizational Behavior and Human Performance*, October 1974, pp. 170–195; S. Ronen, "Equity Perception in Multiple Comparisons: A Field Study," *Human Relations*, April 1986, pp. 333–346; R. W. Scholl, E. A. Cooper, and J. F. McKenna, "Referent Selection in Determining Equity Perception: Differential Effects on Behavioral and Attitudinal Outcomes," *Personnel Psychology*, Spring 1987, pp. 113–127; and C. T. Kulik and M. L. Ambrose, "Personal and Situational Determinants of Referent Choice," *Academy of Management Review*, April 1992, pp. 212–237.

47. See, for example, R. C. Dailey and D. J. Kirk, "Distributive and Procedural Justice as Antecedents of Job Dissatisfaction and Intent to Turnover," *Human Relations*, March 1992, pp. 305–316; D. B. McFarlin and P. D. Sweeney, "Distributive and Procedural Justice as Predictors of Satisfaction with Personal and Organizational Outcomes," *Academy of Management Journal*, August 1992, pp. 626–637; M. A. Konovsky, "Understanding Procedural Justice and Its Impact on Business Organizations," *Journal of Management*, vol. 26 (3), 2000, pp. 489–511; J. A. Colquitt, "Does the Justice of One Interact with the Justice of Many? Reactions to Procedural Justice in Teams," *Journal of Applied Psychology*, August 2004, pp. 633–646; J. Brockner, "Why It's So Hard to Be Fair," *Harvard Business Review*, March 2006, pp. 122–129; and B. M. Wiesenfeld, W. B. Swann, Jr., J. Brockner, and C. A. Bartel, "Is More Fairness Always Preferred? Self-Esteem Moderates Reactions to Procedural Justice," *Academy of Management Journal*, October 2007, pp. 1235–1253.

48. V. H. Vroom, *Work and Motivation* (New York: John Wiley, 1964).

49. See, for example, H. G. Heneman III and D. P. Schwab, "Evaluation of Research on Expectancy Theory Prediction of Employee Performance," *Psychological Bulletin*, July 1972, pp. 1–9; and L. Reinharth and M. Wahba, "Expectancy Theory as a Predictor of Work Motivation, Effort Expenditure, and Job Performance," *Academy of Management Journal*, September 1975, pp. 502–537.

50. See, for example, V. H. Vroom, "Organizational Choice: A Study of Pre- and Postdecision Processes," *Organizational Behavior and Human Performance*, April 1966, pp. 212–225; L. W. Porter and E. E. Lawler III, *Managerial Attitudes and Performance* (Homewood, IL: Richard D. Irwin, 1968); W. Van Eerde and H. Thierry, "Vroom's Expectancy Models and Work-Related Criteria: A Meta-Analysis," *Journal of Applied Psychology*, October 1996, pp. 575–586; and Ambrose and Kulik, "Old Friends, New Faces: Motivation Research in the 1990s."

51. See, for instance, M. Siegall, "The Simplistic Five: An Integrative Framework for Teaching Motivation," *The Organizational Behavior Teaching Review*, vol. 12 (4), 1987–1988, pp. 141–143.

52. N. J. Adler with A. Gundersen, *International Dimensions of Organizational Behavior*, 5th ed. (Cincinnati, OH: South-Western Publishing, 2008.

53. G. Hofstede, "Motivation, Leadership and Organization: Do American Theories Apply Abroad?" *Organi-zational Dynamics*, Summer 1980, p. 55.

54. http://www.businesstoday.digitaltoday.in/index.php?option=com_content&task=view&issueid=47&id=9489&sectionid=22&Itemid=1

55. Ibid.

56. Sinha, J. B. P. (forthcoming). The Shifting Mindset of Indians. To appear in: G. Misra (Ed.), "Discipline of Psychology".

57. J. K. Giacobbe-Miller, D. J. Miller, and V. I. Victorov, "A Comparison of Russian and U.S. Pay Allocation Decisions, Distributive Justice Judgments and Productivity under Different Payment Conditions," *Personnel Psychology*, Spring 1998, pp. 137–163.

58. S. L. Mueller and L. D. Clarke, "Political–Economic Context and Sensitivity to Equity: Differences Between the United States and the Transition Economies of Central and Eastern Europe," *Academy of Management Journal*, June 1998, pp. 319–329.

59. Krishnan, L. (2004). Attitudes, Social Cognition and Justice. In J. Pandey (Ed.) Psychology in India revisited—Development in the Discipline. (Vol. 3). Applied Social and Organizational Psychology. New Delhi, Sage Publications.

60. What Indian employees want. http://www.domain-b.com/management/hrd/20060328_employees.htm

61. I. Harpaz, "The Importance of Work Goals: An International Perspective," *Journal of International Business Studies*, First Quarter 1990, pp. 75–93.

62. G. E. Popp, H. J. Davis, and T. T. Herbert, "An International Study of Intrinsic Motivation Composition," *Management International Review*, January 1986, pp. 28–35.

63. R. W. Brislin, et al., "Evolving Perceptions of Japanese Workplace Motivation: An Employee–Manager Comparison," *International Journal of Cross-Cultural Management*, April 2005, pp. 87–104.

64. J. R. Billings and D. L. Sharpe, "Factors Influencing Flextime Usage Among Employed Married Women," *Consumer Interests Annual*, 1999, pp. 89–94; and I. Harpaz, "The Importance of Work Goals: An International Perspective," *Journal of International Business Studies*, First Quarter 1990, pp. 75–93.

65. N. Ramachandran, "New Paths at Work," *U.S. News & World Report*, March 20, 2006, p. 47; S. Armour, "Generation Y: They've Arrived at Work with a New Attitude," *USA Today*, November 6, 2005, pp. B1+; and R. Kanfer and P. L. Ackerman, "Aging, Adult Development, and Work Motivation," *Academy of Management Review*, July 2004, pp. 440–458.

66. J. Sahadi, "Flex-time, Time Off—Who's Getting These Perks?" *CNNMoney.com*, June 25, 2007.

67. M. Arndt, "The Family That Flips Together . . ." *BusinessWeek*, April 17, 2006, p. 14.

68. Ibid.

69. T. D. Golden and J. F. Veiga, "The Impact of Extent of Telecommuting on Job Satisfaction: Resolving Inconsistent Findings," *Journal of Management*, April 2005, pp. 301–318.

70. Information in this box from D. Jones, "Ford, Fannie Mae Tops in Diversity," *USA Today* online, www.usatoday.com, May 7, 2003; S. N. Mehta, "What Minority Employees Really Want," *Fortune*, July 10, 2000, pp. 180–186; K. H. Hammonds, "Difference Is Power," *Fast Company*, July 2000, pp. 258–266; "Building a Competitive Workforce: Diversity, the Bottom Line," *Forbes*, April 3, 2000, pp. 181–194; and "Diversity: Developing Tomorrow's Leadership Talent Today," *BusinessWeek*, December 20, 1999, pp. 85–100.

71. See, for instance, M. Alpert, "The Care and Feeding of Engineers," *Fortune*, September 21, 1992, pp. 86–95; G. Poole, "How to Manage Your Nerds," *Forbes ASAP*,

December 1994, pp. 132–136; T. J. Allen and R. Katz, "Managing Technical Professionals and Organizations: Improving and Sustaining the Performance of Organizations, Project Teams, and Individual Contributors," *Sloan Management Review*, Summer 2002, pp. S4–S5; and S. R. Barley and G. Kunda, "Contracting: A New Form of Professional Practice," *Academy of Management Perspectives*, February 2006, pp. 45–66.

72. R. J. Bohner, Jr., and E. R. Salasko, "Beware the Legal Risks of Hiring Temps," *Workforce*, October 2002, pp. 50–57.

73. J. P. Broschak and A. Davis-Blake, "Mixing Standard Work and Nonstandard Deals: The Consequences of Heterogeneity in Employment Arrangements," *Academy of Management Journal*, April 2006, pp. 371–393; M. L. Kraimer, S. J. Wayne, R. C. Liden, and R. T. Sparrowe, "The Role of Job Security in Understanding the Relationship Between Employees' Perceptions of Temporary Workers and Employees' Performance," *Journal of Applied Psychology*, March 2005, pp. 389–398; and C. E. Connelly and D. G. Gallagher, "Emerging Trends in Contingent Work Research," *Journal of Management*, November 2004, pp. 959–983.

74. http://www.timesjobs.com/Mailers/HRDialogue/may06/05.05.06/hr_dialogue/landing.htm

75. C. Haddad, "FedEx: Gaining on the Ground," *BusinessWeek*, December 16, 2002, pp. 126–128; and L. Landro, "To Get Doctors to Do Better, Health Plans Try Cash Bonuses," *Wall Street Journal*, September 17, 2004, pp. A1+.

76. http://www.businesstoday.digitaltoday.in/index.php?option=com_content&task=view&issueid=47&id=9461&Itemid=

77. J. Case, "The Open-Book Revolution," *Inc.*, June 1995, pp. 26–50; J. P. Schuster, J. Carpenter, and M. P. Kane, *The Power of Open-Book Management* (New York: Wiley, 1996); J. Case, "Opening the Books," *Harvard Business Review*, March–April 1997, pp. 118–127; and D. Drickhamer, "Open Books to Elevate Performance," *IndustryWeek*, November 2002, p. 16.

78. http://www.203.197.126.103/imz/article.asp?code=attrition_oct_07

79. http://www.siliconindia.com/guestcontributor/guestarticle/52

80. F. Luthans and A. D. Stajkovic, "Provide Recognition for Performance Improvement," in E. A. Locke (ed.), *Principles of Organizational Behavior* (Oxford, UK: Blackwell, 2000), pp. 166–180.

81. http://www.tataelxsi.com/htmls/news_Press%20Releases_rewards.htm

82. K. J. Dunham, "Amid Sinking Workplace Morale, Employers Turn to Recognition, *Wall Street Journal*, November 19, 2002, p. B8.

83. Cited in S. Caudron, "The Top 20 Ways to Motivate Employees," *IndustryWeek*, April 3, 1995, pp. 15–16. See also B. Nelson, "Try Praise," *Inc.*, September 1996, p. 115; and J. Wiscombe," "Rewards Get Results," *Workforce*, April 2002, pp. 42–48.

84. V. M. Barret, "Fight the Jerks," *Forbes*, July 2, 2007, pp. 52–54.

85. E. White, "The Best vs. the Rest," *Wall Street Journal*, January 30, 2006, pp. B1+.

86. R. K. Abbott, "Performance-Based Flex: A Tool for Managing Total Compensation Costs," *Compensation and Benefits Review*, March–April 1993, pp. 18–21; J. R. Schuster and P. K. Zingheim, "The New Variable Pay: Key Design Issues," *Compensation and Benefits Review*, March–April 1993, pp. 27–34; C. R. Williams and L. P. Livingstone, "Another Look at the Relationship Between Performance and Voluntary Turnover," *Academy of Management Journal*, April 1994, pp. 269–298; A. M. Dickinson and K. L. Gillette, "A Comparison of the Effects of Two Individual Monetary Incentive Systems on Productivity: Piece Rate Pay Versus Base Pay Plus Incentives," *Journal of Organizational Behavior Management*, Spring 1994, pp. 3–82; and C. B. Cadsby, F. Song, and F. Tapon, "Sorting and Incentive Effects of Pay for Performance: An Experimental Investigation," *Academy of Management Journal*, April 2007, pp. 387–405.

87. E. White, "Employers Increasingly Favor Bonuses to Raises," *Wall Street Journal*, August 28, 2006, p. B3.

88. "More Than 20 Percent of Japanese Firms Use Pay Systems Based on Performance," *Manpower Argus*, May 1998, p. 7; and E. Beauchesne, "Pay Bonuses Improve Productivity, Study Shows," *Vancouver Sun*, September 13, 2002, p. D5.

89. H. Rheem, "Performance Management Programs," *Harvard Business Review*, September–October 1996, pp. 8–9; G. Sprinkle, "The Effect of Incentive Contracts on Learning and Performance," *Accounting Review*, July 2000, pp. 299–326; and "Do Incentive Awards Work?," *HRFocus*, October 2000, pp. 1–3.

90. R. D. Banker, S. Y. Lee, G. Potter, and D. Srinivasan, "Contextual Analysis of Performance Impacts on Outcome-Based Incentive Compensation," *Academy of Management Journal*, August 1996, pp. 920–948.

91. http://www.expresscomputeronline.com/20050905/technologylife01.shtml

92. T. Reason, "Why Bonus Plans Fail," *CFO*, January 2003, p. 53; and "Has Pay for Performance Had Its Day?" *The McKinsey Quarterly*, no. 4, 2002, accessed on *Forbes* Web site, www.forbes.com.

93. J. R. Hagerty and G. R. Simpson, "Countrywide CEO Helped Many Get Loans," *Wall Street Journal*, June 27, 2008, p. A3; R. Simon, "Countrywide's Pressures Mount," *Wall Street Journal*, June 26, 2007, p. A3; C. Palmeri, "At Countrywide's End, an Emotional CEO," *BusinessWeek* online, www.businessweek.com, June 25, 2008; and "Shareholders Fight For Say on Countrywide CEO Angelo Mozilo's 'Godzilla-size' Pay," esop2007.wordpress.com, June 9, 2006.

94. W. J. Duncan, "Stock Ownership and Work Motivation," *Organizational Dynamics*, Summer 2001, pp. 1–11.

95. P. Brandes, R. Dharwadkar, and G. V. Lemesis, "Effective Employee Stock Option Design: Reconciling Stakeholder, Strategic, and Motivational Factors," *Academy of Management Executive*, February 2003, pp. 77–95; and J. Blasi, D. Kruse, and A. Bernstein, *In the Company of Owners: The Truth About Stock Options* (New York: Basic Books, 2003).

96. K. A. Tucker and V. Allman, "Don't Be a Cat-and-Mouse Manager," *Gallup Brain*, brain.gallup.com, September 9, 2004.

Chapter 16

1. "Management by Democracy," *USA Today*, December 17, 2007, p. 2B; and L. A. Hill, T. Khanna, and E. A. Stecker, *HCL Technologies (A)*, Harvard Business School Case Study, August 2007.

2. Most leadership research has focused on the actions and responsibilities of managers and extrapolated the results to leaders and leadership in general.

3. P. Bacon, Jr., and M. Calabresi, "The Up-and-Comers," *Time Canada*, April 24, 2006, p. 28; P. Bacon, Jr., "The Exquisite Dilemma of Being Obama," *Time*, February 20, 2006, pp. 24–28; A. Stephen, "10 People Who Will Change the World," *New Statesman*, October 17, 2005, pp. 18–20; "Ten to Watch," *Fortune*, September 9, 2005, p. 282; P. Bacon, Jr., "Barack Obama," *Time*, April 18, 2005, pp. 60–61; and A. Ripley, D. E. Thigpen, and J. McCabe, "Obama's Ascent," *Time*, November 11, 2004, pp. 74–78.

4. See T. A. Judge, J. E. Bono, R. Ilies, and M. W. Gerhardt, "Personality and Leadership: A Qualitative and Quantitative Review," *Journal of Applied Psychology*, August 2002, pp. 765–780; and S. A. Kirkpatrick and E. A. Locke, "Leadership: Do Traits Matter?" *Academy of Management Executive*, May 1991, pp. 48–60.

5. C. Hymowitz, "Bosses Need to Learn Whether They Inspire, or Just Drive, Staffers," *Wall Street Journal*, August 14, 2001, p. B1; and P. C. Judge, "From Country Boys to Big Cheese," *Fast Company*, December 2001, pp. 38–40.

6. K. Lewin and R. Lippitt, "An Experimental Approach to the Study of Autocracy and Democracy: A Preliminary Note," *Sociometry*, vol. 1, 1938, pp. 292–300; K. Lewin, "Field Theory and Experiment in Social Psychology: Concepts and Methods," *American Journal of Sociology*, vol. 44, 1939, pp. 868–896; K. Lewin, R. Lippitt, and R. K. White, "Patterns of Aggressive Behavior in Experimentally Created Social Climates," *Journal of Social Psychology*. vol. 10, 1939, pp. 271–301; and R. Lippitt, "An Experimental Study of the Effect of Democratic and Authoritarian Group Atmospheres," *University of Iowa Studies in Child Welfare*, vol. 16, 1940, pp. 43–95.

7. B. M. Bass, *Stogdill's Handbook of Leadership* (New York: The Free Press, 1981), pp. 289–299.

8. R. M. Stogdill and A. E. Coons (eds.), *Leader Behavior: Its Description and Measurement*, Research Monograph No. 88 (Columbus: Ohio State University, Bureau of Business Research, 1951). For an updated literature review of Ohio State research, see S. Kerr, C. A. Schriesheim, C. J. Murphy, and R. M. Stogdill, "Toward a Contingency Theory of Leadership Based upon the Consideration and Initiating Structure Literature," *Organizational Behavior and Human Performance*, August 1974, pp. 62–82; and B. M. Fisher, "Consideration and Initiating Structure and Their Relationships with Leader Effectiveness: A Meta-Analysis," in F. Hoy (ed.), *Proceedings of the 48th Annual Academy of Management Conference*, Anaheim, CA, 1988, pp. 201–205.

9. R. Kahn and D. Katz, "Leadership Practices in Relation to Productivity and Morale," in D. Cartwright and A. Zander (eds.), *Group Dynamics: Research and Theory*, 2nd ed. (Elmsford, NY: Row, Paterson, 1960).

10. R. R. Blake and J. S. Mouton, *The Managerial Grid III* (Houston: Gulf Publishing, 1984).

11. L. L. Larson, J. G. Hunt, and R. N. Osborn, "The Great Hi-Hi Leader Behavior Myth: A Lesson from Occam's Razor," *Academy of Management Journal*, December 1976, pp. 628–641; and P. C. Nystrom, "Managers and the Hi-Hi Leader Myth," *Academy of Management Journal*, June 1978, pp. 325–331.

12. W. G. Bennis, "The Seven Ages of the Leader," *Harvard Business Review*, January 2004, p. 52.

13. F. E. Fiedler, *A Theory of Leadership Effectiveness* (New York: McGraw-Hill, 1967).

14. R. Ayman, M. M. Chemers, and F. Fiedler, "The Contingency Model of Leadership Effectiveness: Its Levels of Analysis," *Leadership Quarterly*, Summer 1995, pp. 147–167; C. A. Schriesheim, B. J. Tepper, and L. A. Tetrault, "Least Preferred Co-worker Score, Situational Control, and Leadership Effectiveness: A Meta-Analysis of Contingency Model Performance Predictions," *Journal of Applied Psychology*, August 1994, pp. 561–573; and L. H. Peters, D. D. Hartke, and J. T. Pholmann, "Fiedler's Contingency Theory of Leadership: An Application of the Meta-Analysis Procedures of Schmidt and Hunter," *Psychological Bulletin*, March 1985, pp. 274–285.

15. See E. H. Schein, *Organizational Psychology*, 3rd ed. (Upper Saddle River, NJ: Prentice Hall, 1980), pp. 116–117; and B. Kabanoff, "A Critique of Leader Match and Its Implications for Leadership Research," *Personnel Psychology*, Winter 1981, pp. 749–764.

16. P. Hersey and K. Blanchard, "So You Want to Know Your Leadership Style?" *Training and Development Journal*, February 1974, pp. 1–15; and P. Hersey and K. H. Blanchard, *Management of Organizational Behavior: Leading Human Resources*, 8th ed. (Upper Saddle River, NJ: Prentice Hall, 2001).

17. See, for instance, E. G. Ralph, "Developing Managers' Effectiveness: A Model with Potential," *Journal of Management Inquiry*, June 2004, pp. 152–163; C. L. Graeff, "Evolution of Situational Leadership Theory: A Critical Review," *Leadership Quarterly*, vol. 8 (2), 1997, pp. 153–170; and C. F. Fernandez and R. P. Vecchio, "Situational Leadership Theory Revisited: A Test of an Across-Jobs Perspective," *Leadership Quarterly*, vol. 8 (1), 1997, pp. 67–84.

18. R. J. House, "A Path–Goal Theory of Leader Effectiveness," *Administrative Science Quarterly*, September 1971, pp. 321–338; House and T. R. Mitchell, "Path–Goal Theory of Leadership," *Journal of Contemporary Business*, Autumn 1974, p. 86; and R. J. House, "Path–Goal Theory of Leadership: Lessons, Legacy, and a Reformulated Theory," *Leadership Quarterly*, Fall 1996, pp. 323–352.

19. J. C. Wofford and L. Z. Liska, "Path–Goal Theories of Leadership: A Meta-Analysis," *Journal of Management*, Winter 1993, pp. 857–876; and A. Sagie, and M. Koslowsky, "Organizational Attitudes and Behaviors as a Function of Participation in Strategic and Tactical Change Decisions: An Application of Path–Goal Theory," *Journal of Organizational Behavior*, January 1994, pp. 37–47.

20. B. M. Bass and R. E. Riggio, *Transformational Leadership*, 2nd ed. (Mahwah, NJ: Lawrence Erlbaum Associates, 2006), p. 3.

21. B. M. Bass, "Leadership: Good, Better, Best," *Organizational Dynamics*, Winter 1985, pp. 26–40; and J. Seltzer and B. M. Bass, "Transformational Leadership: Beyond Initiation and Consideration," *Journal of Management*, December 1990, pp. 693–703.

22. B. J. Avolio and B. M. Bass, "Transformational Leadership, Charisma, and Beyond." Working paper, School of Management, State University of New York, Binghamton, 1985, p. 14.

23. R. S. Rubin, D. C. Munz, and W. H. Bommer, "Leading from Within: The Effects of Emotion Recognition and Personality on Transformational Leadership Behavior," *Academy of Management Journal*, October 2005, pp. 845–858; T. A. Judge and J. E. Bono, "Five-Factor Model of Personality

and Transformational Leadership," *Journal of Applied Psychology*, October 2000, pp. 751–765; B. M. Bass and B. J. Avolio, "Developing Transformational Leadership: 1992 and Beyond," *Journal of European Industrial Training*, January 1990, p. 23; and J. J. Hater and B. M. Bass, "Supervisors' Evaluation and Subordinates' Perceptions of Transformational and Transactional Leadership," *Journal of Applied Psychology*, November 1988, pp. 695–702.

24. A. E. Colbert, A. L. Kristof-Brown, B. H. Bradley, and M. R. Barrick, "CEO Transformational Leadership: The Role of Goal Importance Congruence in Top Management Teams," *Academy of Management Journal*, February 2008, pp. 81–96; R. F. Piccolo and J. A. Colquitt, "Transformational Leadership and Job Behaviors: The Mediating Role of Core Job Characteristics," *Academy of Management Journal*, April 2006, pp. 327–340; O. Epitropaki and R. Martin, "From Ideal to Real: A Longitudinal Study of the Role of Implicit Leadership Theories on Leader-Member Exchanges and Employee Outcomes," *Journal of Applied Psychology*, July 2005, pp. 659–676; J. E. Bono and T. A. Judge, "Self-Concordance at Work: Toward Understanding the Motivational Effects of Transformational Leaders," *Academy of Management Journal*, October 2003, pp. 554–571; T. Dvir, D. Eden, B. J. Avolio, and B. Shamir, "Impact of Transformational Leadership on Follower Development and Performance: A Field Experiment," *Academy of Management Journal*, August 2002, pp. 735–744; N. Sivasubramaniam, W. D. Murry, B. J. Avolio, and D. I. Jung, "A Longitudinal Model of the Effects of Team Leadership and Group Potency on Group Performance," *Group and Organization Management*, March 2002, pp. 66–96; J. M. Howell and B. J. Avolio, "Transformational Leadership, Transactional Leadership, Locus of Control, and Support for Innovation: Key Predictors of Consolidated-Business-Unit Performance," *Journal of Applied Psychology*, December 1993, pp. 891–911; R. T. Keller, "Transformational Leadership and the Performance of Research and Development Project Groups," *Journal of Management*, September 1992, pp. 489–501; and Bass and Avolio, "Developing Transformational Leadership: 1992 and Beyond."

25. F. Vogelstein, "Mighty Amazon," *Fortune*, May 26, 2003, pp. 60–74.

26. J. M. Crant and T. S. Bateman, "Charismatic Leadership Viewed from Above: The Impact of Proactive Personality," *Journal of Organizational Behavior*, February 2000, pp. 63–75; G. Yukl and J. M. Howell, "Organizational and Contextual Influences on the Emergence and Effectiveness of Charismatic Leadership," *Leadership Quarterly*, Summer 1999, pp. 257–283; and J. A. Conger and R. N. Kanungo, "Behavioral Dimensions of Charismatic Leadership," in J. A. Conger, et al., *Charismatic Leadership* (San Francisco: Jossey-Bass, 1988), pp. 78–97.

27. J. A. Conger and R. N. Kanungo, *Charismatic Leadership in Organizations* (Thousand Oaks, CA: Sage, 1998).

28. K. S. Groves, "Linking Leader Skills, Follower Attitudes, and Contextual Variables via an Integrated Model of Charismatic Leadership," *Journal of Management*, April 2005, pp. 255–277; J. J. Sosik, "The Role of Personal Values in the Charismatic Leadership of Corporate Managers: A Model and Preliminary Field Study," *Leadership Quarterly*, April 2005, pp. 221–244; A. H. B. deHoogh, et al., "Leader Motives, Charismatic Leadership, and Subordinates' Work Attitudes in the Profit and Voluntary Sector," *Leadership Quarterly*, February 2005, pp. 17–38; J. M. Howell and B. Shamir, "The Role of Followers in the Charismatic Leadership Process: Relationships and Their Consequences," *Academy of Management Review*, January 2005, pp. 96–112; J. Paul, et al., "The Effects of Charismatic Leadership on Followers' Self-Concept Accessibility," *Journal of Applied Social Psychology*, September 2001, pp. 1821–1844; J. A. Conger, R. N. Kanungo, and S. T. Menon, "Charismatic Leadership and Follower Effects," *Journal of Organizational Behavior*, vol. 21, 2000, pp. 747–767; R. W. Rowden, "The Relationship Between Charismatic Leadership Behaviors and Organizational Commitment," *Leadership & Organization Development Journal*, January 2000, pp. 30–35; G. P. Shea and C. M. Howell, "Charismatic Leadership and Task Feedback: A Laboratory Study of Their Effects on Self-Efficacy," *Leadership Quarterly*, Fall 1999, pp. 375–396; S. A. Kirkpatrick and E. A. Locke, "Direct and Indirect Effects of Three Core Charismatic Leadership Components on Performance and Attitudes," *Journal of Applied Psychology*, February 1996, pp. 36–51; D. A. Waldman, B. M. Bass, and F. J. Yammarino, "Adding to Contingent–Reward Behavior: The Augmenting Effect of Charismatic Leadership," *Group & Organization Studies*, December 1990, pp. 381–394; and R. J. House, J. Woycke, and E. M. Fodor, "Charismatic and Noncharismatic Leaders: Differences in Behavior and Effectiveness," in J. A. Conger, et al., *Charismatic Leadership* (San Francisco: Jossey-Bass, 1988), pp. 103–104.

29. B. R. Agle, N. J. Nagarajan, J. A. Sonnenfeld, and D. Srinivasan, "Does CEO Charisma Matter? An Empirical Analysis of the Relationships Among Organizational Performance, Environmental Uncertainty, and Top Management Team Perceptions of CEO Charisma," *Academy of Management Journal*, February 2006, pp. 161–174.

30. R. Birchfield, "Creating Charismatic Leaders," *Management*, June 2000, pp. 30–31; S. Caudron, "Growing Charisma," *IndustryWeek*, May 4, 1998, pp. 54–55; and J. A. Conger and R. N. Kanungo, "Training Charismatic Leadership: A Risky and Critical Task," in J. A. Conger, et al., *Charismatic Leadership* (San Francisco: Jossey-Bass, 1988), pp. 309–323.

31. J. G. Hunt, K. B. Boal, and G. E. Dodge, "The Effects of Visionary and Crisis-Responsive Charisma on Followers: An Experimental Examination," *Leadership Quarterly*, Fall 1999, pp. 423–448; R. J. House and R. N. Aditya, "The Social Scientific Study of Leadership: Quo Vadis?," *Journal of Management*, vol. 23 (3), 1997, pp. 316–323; and R. J. House, "A 1976 Theory of Charismatic Leadership," in J. G. Hunt and L. Larson (eds.), *Leadership: The Cutting Edge* (Carbondale, Illinois: Southern Illinois University Press), 1977, pp. 189–207.

32. This definition is based on M. Sashkin, "The Visionary Leader," in J. A. Conger, et al., *Charismatic Leadership* (San Francisco: Jossey-Bass, 1988), pp. 124–125; B. Nanus, *Visionary Leadership* (New York: The Free Press, 1992), p. 8; N. H. Snyder and M. Graves, "Leadership and Vision," *Business Horizons*, January–February 1994, p. 1; and J. R. Lucas, "Anatomy of a Vision Statement," *Management Review*, February 1998, pp. 22–26.

33. Nanus, *Visionary Leadership*, p. 8.

34. http://www.sip-trunking.tmcnet.com/news/2006/02/27/1409727.htm

35. http://www.en.wikipedia.org/wiki/Shri_Mahila_Griha_Udyog_Lijjat_Papad

36. S. Caminiti, "What Team Leaders Need to Know," *Fortune*, February 20, 1995, pp. 93–100.

37. Ibid., p. 93.

38. Ibid., p. 100.

39. N. Steckler and N. Fondas, "Building Team Leader Effectiveness: A Diagnostic Tool," *Organizational Dynamics*, Winter 1995, p. 20.

40. R. S. Wellins, W. C. Byham, and G. R. Dixon, *Inside Teams* (San Francisco: Jossey-Bass, 1994), p. 318.

41. Steckler and Fondas, "Building Team Leader Effectiveness: A Diagnostic Tool," p. 21.

42. G. Colvin, "The FedEx Edge," *Fortune*, April 3, 2006, pp. 77–84.

43. See J. R. P. French, Jr., and B. Raven, "The Bases of Social Power," in D. Cartwright and A. F. Zander (eds.), *Group Dynamics: Research and Theory* (New York: Harper & Row, 1960), pp. 607–623; P. M. Podsakoff and C. A. Schriesheim, "Field Studies of French and Raven's Bases of Power: Critique, Reanalysis, and Suggestions for Future Research," *Psychological Bulletin*, May 1985, pp. 387–411; R. K. Shukla, "Influence of Power Bases in Organizational Decision Making: A Contingency Model," *Decision Sciences*, July 1982, pp. 450–470; D. E. Frost and A. J. Stahelski, "The Systematic Measurement of French and Raven's Bases of Social Power in Workgroups," *Journal of Applied Social Psychology*, April 1988, pp. 375–389; and T. R. Hinkin and C. A. Schriesheim, "Development and Application of New Scales to Measure the French and Raven (1959) Bases of Social Power," *Journal of Applied Psychology*, August 1989, pp. 561–567.

44. J. M. Kouzes and B. Z. Posner, *Credibility: How Leaders Gain and Lose It, and Why People Demand It* (San Francisco: Jossey-Bass, 1993), p. 14.

45. Based on F. D. Schoorman, R. C. Mayer, and J. H. Davis, "An Integrative Model of Organizational Trust: Past, Present, and Future," *Academy of Management Review*, April 2007, pp. 344–354; G. M. Spreitzer and A. K. Mishra, "Giving Up Control Without Losing Control," *Group & Organization Management*, June 1999, pp. 155–187; R. C. Mayer, J. H. Davis, and F. D. Schoorman, "An Integrative Model of Organizational Trust," *Academy of Management Review*, July 1995, p. 712; and L. T. Hosmer, "Trust: The Connecting Link between Organizational Theory and Philosophical Ethics," *Academy of Management Review*, April 1995, p. 393.

46. P. L. Schindler and C. C. Thomas, "The Structure of Interpersonal Trust in the Workplace," *Psychological Reports*, October 1993, pp. 563–573.

47. H. H. Tan and C. S. F. Tan, "Toward the Differentiation of Trust in Supervisor and Trust in Organization," *Genetic, Social, and General Psychology Monographs*, May 2000, pp. 241–260.

48. R. C. Mayer and M. B. Gavin, "Trust in Management and Performance: Who Minds the Shop While the Employees Watch the Boss?" *Academy of Management Journal*, October 2005, pp. 874–888; and K. T. Dirks and D. L. Ferrin, "Trust in Leadership: Meta-Analytic Findings and Implications for Research and Practice," *Journal of Applied Psychology*, August 2002, pp. 611–628.

49. See for example, Dirks and Ferrin, "Trust in Leadership: Meta-Analytic Findings and Implications for Research and Practice"; J. K. Butler, Jr., "Toward Understanding and Measuring Conditions of Trust: Evolution of a Conditions of Trust Inventory," *Journal of Management*, September 1991, pp. 643–663; and F. Bartolome, "Nobody Trusts the Boss Completely—Now What?" *Harvard Business Review*, March–April 1989, pp. 135–142.

50. R. Zemke, "The Confidence Crisis," *Training*, June 2004, pp. 22–30; J. A. Byrne, "Restoring Trust in Corporate America," *BusinessWeek*, June 24, 2002, pp. 30–35; S. Armour, "Employees' New Motto: Trust No One," *USA Today*, February 5, 2002, p. 1B; J. Scott, "Once Bitten, Twice Shy: A World of Eroding Trust," *New York Times*, April 21, 2002, p. WK5; J. Brockner, et al., "When Trust Matters: The Moderating Effect of Outcome Favorability," *Administrative Science Quarterly*, September 1997, p. 558; and J. Brockner, et al., "When Trust Matters: The Moderating Effect of Outcome Favorability," *Administrative Science Quarterly*, September 1997, p. 558.

51. Watson Wyatt, "Weathering the Storm: A Study of Employee Attitudes and Opinions," *WorkUSA 2002 Study*, www.watsonwyatt.com.

52. http://www.talentmgt.com/industry_news/2008/June/2911/index.php

53. T. Vinas, "DuPont: Safety Starts at the Top," *IndustryWeek*, July 2002, p. 55.

54. A. Srivastava, K. M. Bartol, and E. A. Locke, "Empowering Leadership in Management Teams: Effects on Knowledge Sharing, Efficacy, and Performance," *Academy of Management Journal*, December 2006, pp. 1239–1251; P. K. Mills and G. R. Ungson, "Reassessing the Limits of Structural Empowerment: Organizational Constitution and Trust as Controls," *Academy of Management Review*, January 2003, pp. 143–153; W. A. Rudolph and M. Sashkin, "Can Organizational Empowerment Work in Multinational Settings?" *Academy of Management Executive*, February 2002, pp. 102–115; C. Gomez and B. Rosen, "The Leader–Member Link Between Managerial Trust and Employee Empowerment," *Group & Organization Management*, March 2001, pp. 53–69; C. Robert and T. M. Probst, "Empowerment and Continuous Improvement in the United States, Mexico, Poland, and India," *Journal of Applied Psychology*, October 2000, pp. 643–658; R. C. Herrenkohl, G. T. Judson, and J. A. Heffner, "Defining and Measuring Employee Empowerment," *Journal of Applied Behavioral Science*, September 1999, p. 373; R. C. Ford and M. D. Fottler, "Empowerment: A Matter of Degree," *Academy of Management Executive*, August 1995, pp. 21–31; and W. A. Rudolph, "Navigating the Journey to Empowerment," *Organizational Dynamics*, Spring 1995, pp. 19–32.

55. F. W. Swierczek, "Leadership and Culture: Comparing Asian Managers," *Leadership & Organization Development Journal*, December 1991, pp. 3–10.

56. R. J. House, "Leadership in the Twenty-First Century," in J. S. Chhokar, F. C. Brodbeck, and R. J. House (eds), *Culture & Leadership Across the World* (Mahwah, New Jersey: Lawrence Erlbaum Associates), 2007; M. F. Peterson and J. G. Hunt, "International Perspectives on International Leadership," *Leadership Quarterly*, Fall 1997, pp. 203–231; and J. R. Schermerhorn and M. H. Bond, "Cross-Cultural Leadership in Collectivism and High Power Distance Settings," *Leadership & Organization Development Journal*, vol. 18 (4/5), 1997, pp. 187–193.

57. R. J. House, et al., "Culture Specific and Cross-Culturally Generalizable Implicit Leadership Theories: Are the Attributes of Charismatic/Transformational Leadership Universally Endorsed?" *Leadership Quarterly*, Summer 1999, pp. 219–256; and D. E. Carl and M. Javidan, "Universality of Charismatic Leadership: A Multi-Nation Study," paper presented at the National Academy of Management Conference, Washington, DC, August 2001.

58. D. E. Carl and M. Javidan, "Universality of Charismatic Leadership," p. 29.

59. G. N. Powell, D. A. Butterfield, and J. D. Parent, "Gender and Managerial Stereotypes: Have the Times Changed?" *Journal of Management*, vol. 28 (2), 2002, pp. 177–193.

60. See K. M. Bartol, D. C. Martin, and J. A. Kromkowski, "Leadership and the Glass Ceiling: Gender and Ethnic Influences on Leader Behaviors at Middle and Executive Managerial Levels," *Journal of Leadership & Organizational Studies*, Winter 2003, pp. 8–19; A. H. Eagly and S. J. Karau, "Role Congruity Theory of Prejudice Toward Female Leaders," *Psychological Review*, July 2002, pp. 573–598; J. Becker, R. A. Ayman, and K. Korabik, "Discrepancies in Self/Subordinates' Perceptions of Leadership Behavior: Leader's Gender, Organizational Context, and Leader's Self-Monitoring," *Group & Organization Management*, June 2002, pp. 226–244; N. Z. Selter, "Gender Differences in Leadership: Current Social Issues and Future Organizational Implications," *Journal of Leadership Studies*, Spring 2002, pp. 88–99; J. M. Norvilitis and H. M. Reid, "Evidence for an Association Between Gender-Role Identity and a Measure of Executive Function," *Psychological Reports*, February 2002, pp. 35–45; W. H. Decker and D. M. Rotondo, "Relationships Among Gender, Type of Humor, and Perceived Leader Effectiveness," *Journal of Managerial Issues*, Winter 2001, pp. 450–465; C. L. Ridgeway, "Gender, Status, and Leadership," *Journal of Social Issues*, Winter 2001, pp. 637–655; M. Gardiner and M. Tiggemann, "Gender Differences in Leadership Style, Job Stress and Mental Health in Male- and Female-Dominated Industries," *Journal of Occupational and Organizational Psychology*, September 1999, pp. 301–315; and F. J. Yammarino, A. J. Dubinsky, L. B. Comer, and M. A. Jolson, "Women and Transformational and Contingent Reward Leadership: A Multiple-Levels-of-Analysis Perspective," *Academy of Management Journal*, February 1997, pp. 205–222.

61. Gardiner and Tiggemann, "Gender Differences in Leadership Style, Job Stress and Mental Health in Male- and Female-Dominated Industries."

62. Norvilitis and Reid, "Evidence for an Association Between Gender-Role Identity and a Measure of Executive Function"; Decker and Rotondo, "Relationships Among Gender, Type of Humor, and Perceived Leader Effectiveness"; H. Aguinis and S. K. R. Adams, "Social-role Versus Structural Models of Gender and Influence Use in Organizations: A Strong Inference Approach," *Group & Organization Management*, December 1998, pp. 414–446; and A. H. Eagly, S. J. Karau, and M. G. Makhijani, "Gender and the Effectiveness of Leaders: A Meta-Analysis," *Psychological Bulletin*, vol. 117, 1995, pp. 125–145.

63. Bartol, Martin, and Kromkowski, "Leadership and the Glass Ceiling: Gender and Ethnic Group Influences on Leader Behaviors at Middle and Executive Managerial Levels"; and R. Sharpe, "As Leaders, Women Rule," *BusinessWeek*, November 20, 2000, pp. 74–84.

64. Bartol, Martin, and Kromkowski, "Leadership and the Glass Ceiling: Gender and Ethnic Group Influences on Leader Behaviors at Middle and Executive Managerial Levels."

65. See, for instance, R. Lofthouse, "Herding the Cats," *EuroBusiness*, February 2001, pp. 64–65; and M. Delahoussaye, "Leadership in the 21st Century," *Training*, September 2001, pp. 60–72.

66. See, for instance, A. A. Vicere, "Executive Education: The Leading Edge," *Organizational Dynamics*, Autumn 1996, pp. 67–81; J. Barling, T. Weber, and E. K. Kelloway, "Effects of Transformational Leadership Training on Attitudinal and Financial Outcomes: A Field Experiment," *Journal of Applied Psychology*, December 1996, pp. 827–832; and D. V. Day, "Leadership Development: A Review in Context," *Leadership Quarterly*, Winter 2000, pp. 581–613.

67. K. Y. Chan and F. Drasgow, "Toward a Theory of Individual Differences and Leadership: Understanding the Motivation to Lead," *Journal of Applied Psychology*, June 2001, pp. 481–498.

68. M. Sashkin, "The Visionary Leader," in J. A. Congeret al., *Charismatic Leadership* (San Francisco: Jossey-Bass, 1988), p. 150.

69. S. Kerr and J. M. Jermier, "Substitutes for Leadership: Their Meaning and Measurement," *Organiza-tional Behavior and Human Performance*, December 1978, pp. 375–403; J. P. Howell, P. W. Dorfman, and S. Kerr, "Leadership and Substitutes for Leadership," *Journal of Applied Behavioral Science*, vol. 22 (1), 1986, pp. 29–46; J. P. Howell, et al., "Substitutes for Leadership: Effective Alternatives to Ineffective Leadership," *Organizational Dynamics*, Summer 1990, pp. 21–38; and P. M. Podsakoff, B. P. Niehoff, S. B. MacKenzie, and M. L. Williams, "Do Substitutes for Leadership Really Substitute for Leadership? An Empirical Examination of Kerr and Jermier's Situational Leadership Model," *Organizational Behavior and Human Decision Processes*, February 1993, pp. 1–44.

70. G. Anders, "Tough CEOs Often Most Successful, a Study Finds," *Wall Street Journal*, November 19, 2007, p. B3.

Chapter 17

1. J. Katz, "Empowered to Drive Production," *IndustryWeek*, October 2006, p. 33.

2. B. Hagenbauh, "State Quarter's Extra Leaf Grew Out of Lunch Break," *USA Today*, January 20, 2006, p. 1B.

3. K. A. Merchant, "The Control Function of Management," *Sloan Management Review*, Summer 1982, pp. 43–55.

4. E. Flamholtz, "Organizational Control Systems Managerial Tool," *California Management Review*, Winter 1979, p. 55.

5. T. Vinas and J. Jusko, "5 Threats That Could Sink Your Company," *IndustryWeek*, September 2004, pp. 52–61; "Workplace Security: How Vulnerable Are You?" Special section in *Wall Street Journal*, September 29, 2003, pp. R1–R8; P. Magnusson, "Your Jitters Are Their Lifeblood," *BusinessWeek*, April 14, 2003, p. 41; and T. Purdum, "Preparing for the Worst," *IndustryWeek*, January 2003, pp. 53–55.

6. S. Kerr, "On the Folly of Rewarding A, While Hoping for B," *Academy of Management Journal*, December 1975, pp. 769–783.

7. M. Starr, "State-of-the-Art Stats," *Newsweek*, March 24, 2003, pp. 47–49.

8. Information in this box from Top 50 Companies for Diversity, *DiversityInc*, June 2008, pp. 48–122.

9. D. Twiddy, "Applebee's IHOP Deal Completed," *USA Today*, November 30, 2007, p. 5B; J. Adamy, "A Shift in Dining Scene Nicks a Once-Hot Chain," *Wall Street Journal*, June 29, 2007, pp. A1+; and R. Barker, "Applebee's Looks Appetizing," *BusinessWeek*, September 19, 2005, p. 30.

10. H. Koontz and R. W. Bradspies, "Managing Through Feedforward Control," *Business Horizons*, June 1972, pp. 25–36.

11. M. Helft, "The Human Hands Behind the Google Money Machine," *New York Times* online, www.nytimes.com, June 2, 2008.

12. B. Caulfield, "Shoot to Kill," *Forbes*, January 7, 2008, pp. 92–96.

13. T. Laseter and L. Laseter, "See for Yourself," *Strategy+Business*, www.strategy-business.com, November 29, 2007.

14. W. H. Newman, *Constructive Control: Design and Use of Control Systems* (Upper Saddle River, NJ: Prentice Hall, 1975), p. 33.

15. L. Pulliam Weston, "The Secret Pensions of Fat-Cat Executives," *MSN Money*, moneycentral.msn.com, May 27, 2006; M. V. Rafter, "IRS Advice to Large Companies: Hit the Books," *Workforce*, January 2005, pp. 60–612; and E. E. Schultz and T. Francis, "Buried Treasure: Well-Hidden Perk Means Big Money for Top Executives," *Wall Street Journal*, October 11, 2002, pp. A1+.

16. R. S. Kaplan and D. P. Norton, "How to Implement a New Strategy Without Disrupting Your Organization," *Harvard Business Review*, March 2006, pp. 100–109; L. Bassi and D. McMurrer, "Developing Measurement Systems for Managers in the Knowledge Era," *Organizational Dynamics*, May 2005, pp. 185–196; G. M. J. DeKoning, "Making the Balanced Scorecard Work (Part 2)," *Gallup Brain*, brain.gallup.com, August 12, 2004; G. J. J. DeKoning, "Making the Balanced Scorecard Work (Part 1)," *Gallup Brain*, brain.gallup.com, July 8, 2004; K. Graham, "Balanced Scorecard," *New Zealand Management*, March 2003, pp. 32–34; K. Ellis, "A Ticket to Ride: Balanced Scorecard," *Training*, April 2001, p. 50; and T. Leahy, "Tailoring the Balanced Scorecard," *Business Finance*, August 2000, pp. 53–56.

17. Leahy, "Tailoring the Balanced Scorecard."

18. Ibid.

19. D. Stout and T. Zeller, Jr., "Vast Data Cache About Veterans Has Been Stolen," *New York Times* online, www.nytimes.com, May 23, 2006.

20. Deloitte & Touche and the Ponemon Institute, "Research Report: Reportable and Multiple Privacy Breaches Rising at Alarming Rate," *Ethics Newsline* ethicsnewsline.wordpress.com, January 1, 2008.

21. http://www.economictimes.indiatimes.com/rssarticleshow/msid-3006451,prtpage-1.cms

22. B. Grow, K. Epstein, and C.-C. Tschang, "The New E-Spionage Threat," *BusinessWeek*, April 21, 2008, pp. 32–41; S. Leibs, "Firewall of Silence," *CFO*, April 2008, pp. 31–35; J. Pereira, "How Credit-Card Data Went Out Wireless Door," *Wall Street Journal*, May 4, 2007, pp. A1+; and B. Stone, "Firms Fret as Office E-Mail Jumps Security Walls," *New York Times* online, www.nytimes.com, January 11, 2007.

23. D. Whelan, "Google Me Not," *Forbes*, August 16, 2004, pp. 102–104.

24. http://www.cbhi-hsprod.nic.in/listdetails.asp?roid=165

25. Business Gurus Speak, p. 88 by S. N. Chary (First published, 2002: Reprinted, 2003, 2004).

26. T. Leahy, "Extracting Diamonds in the Rough," *Business Finance*, August 2000, pp. 33–37.

27. B. Bruzina, B. Jessop, R. Plourde, B. Whitlock, and L. Rubin, "Ameren Embraces Benchmarking As a Core Business Strategy," *Power Engineering*, November 2002, pp. 121–124.

28. http://www.titanworld.com/titan_stories/ http://www.domain-b.com/businessleaders/interviews/clear_prescient_ranger_bhaskar.htm

29. J. Yaukey and C. L. Romero, "Arizona Firm Pays Big for Workers' Digital Downloads," Associated Press, *Springfield (Missouri) News-Leader*, May 6, 2002, p. 6B.

30. AMA/ePolicy Institute, "2005 Electronic Monitoring & Surveillance Survey," *American Management Association*, www.amanet.org.

31. S. Armour, "Companies Keep an Eye on Workers' Internet Use," *USA Today*, February 21, 2006, p. 2B.

32. B. White, "The New Workplace Rules: No Video-Watching," *Wall Street Journal*, March 4, 2008, pp. B1+.

33. P.-W. Tam, E. White, N. Wingfield, and K. Maher, "Snooping E-Mail by Software Is Now a Workplace Norm," *Wall Street Journal*, March 9, 2005, pp. B1+; D. Hawkins, "Lawsuits Spur Rise in Employee Monitoring," *U.S. News & World Report*, August 13, 2001, p. 53; and L. Guernsey, "You've Got Inappropriate Mail," *New York Times*, April 5, 2000, pp. C1+.

34. S. Armour, "More Companies Keep Track of Workers' E-Mail," *USA Today*, June 13, 2005, p. 4B; and E. Bott, "Are You Safe? Privacy Special Report," *PC Computing*, March 2000, pp. 87–88.

35. A. M. Bell and D. M. Smith, "Theft and Fraud May Be an Inside Job," *Workforce Management* online, www.workforce.com, December 3, 2000.

36. http://www.thehindubusinessline.com/2008/12/02/stories/2008120251411800.htm

37. J. Greenberg, "The STEAL Motive: Managing the Social Determinants of Employee Theft," in R. Giacalone and J. Greenberg (eds.), *Antisocial Behavior in Organizations* (Newbury Park, CA: Sage, 1997), pp. 85–108.

38. B. E. Litzky, K. A. Eddleston, and D. L. Kidder, "The Good, the Bad, and the Misguided: How Managers Inadvertently Encourage Deviant Behaviors," *Academy of Management Perspective*, February 2006, pp. 91–103; "Crime Spree," *BusinessWeek*, September 9, 2002, p. 8; B. P. Niehoff and R. J. Paul, "Causes of Employee Theft and Strategies That HR Managers Can Use for Prevention," *Human Resource Management*, Spring 2000, pp. 51–64; and G. Winter, "Taking at the Office Reaches New Heights: Employee Larceny Is Bigger and Bolder," *New York Times*, July 12, 2000, pp. C1+.

39. This section is based on J. Greenberg, *Behavior in Organizations: Understanding and Managing the Human Side of Work*, 8th ed. (Upper Saddle River, NJ: Prentice Hall, 2003), pp. 329–330.

40. A. H. Bell and D. M. Smith, "Why Some Employees Bite the Hand That Feeds Them," *Workforce Management* online, www.workforce.com, December 3, 2000.

41. Litzky, Eddleston, and Kidder, "The Good, the Bad, and the Misguided"; A. H. Bell and D. M. Smith, "Protecting the Company Against Theft and Fraud," *Workforce Management* online, www.workforce.com, December 3, 2000; J. D. Hansen, "To Catch a Thief," *Journal of Accountancy*, March 2000, pp. 43–46; and J. Greenberg, "The Cognitive Geometry of Employee Theft," in *Dysfunctional Behavior in Organizations: Nonviolent and Deviant Behavior* (Stamford, CT: JAI Press, 1998), pp. 147–193.

42. http://news.in.msn.com/national/article.aspx?cp-documentid=1768387

43. http://www.indianexpress.com/news/manjunath-inspected-mittals-petrol-pump-the-day-he-was-killed/13170/

44. http://www.india-today.com/itoday/10081998/newsnote.html

45. Occupational Health and Safety, *BLS: Workplace Homicides Drop to Lowest Number on Record*, www.ohsonline.com, August 17, 2007.

46. J. McCafferty, "Verbal Chills," *CFO*, June 2005, p. 17; S. Armour, "Managers Not Prepared for Workplace Violence," July 15, 2004, pp. 1B+; and Occupational Safety and Health Administration, *Workplace Violence* OSHA Fact Sheet, U.S. Department of Labor, 2002.

47. "Ten Tips on Recognizing and Minimizing Violence," *Workforce Management* online, www.workforce.com, December 3, 2000.

48. "Bullying Bosses Cause Work Rage Rise," *Management-Issues*, www.management-issues.com, January 28, 2003.

49. R. McNatt, "Desk Rage," *BusinessWeek*, November 27, 2000, p. 12.

50. M. Gorkin, "Key Components of a Dangerously Dysfunctional Work Environment," *Workforce Management* online, www.workforce.com, December 3, 2000.

51. "Ten Tips on Recognizing and Minimizing Violence"; M. Gorkin, "Five Strategies and Structures for Reducing Workplace Violence," *Workforce Management* online www.workforce.com, December 3, 2000; "Investigating Workplace Violence: Where Do You Start?" *Workforce Management* online, www.workforce.com, December 3, 2000; and "Points to Cover in a Workplace Violence Policy," *Workforce Management* online www.workforce.com, December 3, 2000.

52. A. Taylor, "Enterprise Asks What Customer's Thinking and Acts," *USA Today*, May 22, 2006, p. 6B; and A. Taylor, "Driving Customer Satisfaction," *Harvard Business Review*, July 2002, pp. 24–25.

53. S. D. Pugh, J. Dietz, J. W. Wiley, and S. M. Brooks, "Driving Service Effectiveness Through Employee–Customer Linkages," *Academy of Management Executive*, November 2002, pp. 73–84; J. L. Heskett, W. E. Sasser, and L. A. Schlesinger, *The Service Profit Chain* (New York: The Free Press, 1997); and J. L. Heskett, et al., "Putting the Service Profit Chain to Work," *Harvard Business Review*, March–April 1994, pp. 164–170.

54. http://www.bplmobile.com/; http://www.telecomindiaonline.com/BPL-Mobile-brings-call-center-on-mobile-phone.html

55. http://www.finance.indiainfo.com/2008/11/11/0811112138_jd_power-maruti.html

56. http://www.business.gov.in/corporate_governance/index.php

57. http://www.business.gov.in/corporate_governance/index.php

Chapter 18

1. Case Study on Boekhandels Groep Nederland, Progress Software Company, www.progress.com/realtime/docs/case_studies/selexyz_cs.pdf, 2007; and E. Schonfeld, "Tagged for Growth," *Business 2.0*, December 2006, pp. 58–61.

2. D. Clark, "Inside Intel, It's All Copying," *Wall Street Journal*, October 28, 2002, pp. B1+.

3. http://www.business.mapsofindia.com/sectors/service.htm

4. U.S. Central Intelligence Agency, *World Factbook 2008*, www.cia.gov/library/publications.

5. Ibid.

6. D. Michaels and J. L. Lunsford, "Streamlined Plane Making," *Wall Street Journal*, April 1, 2005, pp. B1+.

7. T. Aeppel, "Workers Not Included," *Wall Street Journal*, November 19, 2002, pp. B1+.

8. http://www.budgetwithet.economictimes.indiatimes.com/Railway_Budget/Railways_turnaround_exemplary_sustainable/rlyarticleshow/2817689.cms

9. http://www.tvsmotor.in/news.asp?id=181&yr=2008

10. http://www.economictimes.indiatimes.com/News/News_By_Industry/The_new_face_of_technology_in_ICICI_Bank/articleshow/3542162.cms

11. J. Ordonez, "McDonald's to Cut the Cooking Time of Its French Fries," *Wall Street Journal*, May 19, 2000, p. B2.

12. C. Fredman, "The Devil in the Details," *Executive Edge*, April–May, 1999, pp. 36–39.

13. Information from new.skoda-auto.com/Documents/AnnualReports/skoda_auto_annual_report_2007_%20EN_FINAL.pdf, July 8, 2008; and T. Mudd, "The Last Laugh," *IndustryWeek*, September 18, 2000, pp. 38–44.

14. T. Vinas, "Little Things Mean a Lot," *IndustryWeek*, November 2002, p. 55.

15. P. Panchak, "Shaping the Future of Manufacturing," *IndustryWeek*, January 2005, pp. 38–44; M. Hammer, "Deep Change: How Operational Innovation Can Transform Your Company," *Harvard Business Review*, April 2004, pp. 84–94; S. Levy, "The Connected Company," *Newsweek*, April 28, 2003, pp. 40–48; and J. Teresko, "Plant Floor Strategy," *IndustryWeek*, July 2002, pp. 26–32.

16. T. Laseter, K. Ramdas, and D. Swerdlow, "The Supply Side of Design and Development," *Strategy & Business*, Summer 2003, p. 23; J. Jusko, "Not All Dollars and Cents," *IndustryWeek*, April 2002, p. 58; and D. Drickhamer, "Medical Marvel," *IndustryWeek*, March 2002, pp. 47–49.

17. J. H. Sheridan, "Managing the Value Chain," *IndustryWeek*, September 6, 1999, pp. 1–4.

18. Ibid, p. 3.

19. J. Teresko, "Forward, March!" *IndustryWeek*, July 2004, pp. 43–48; D. Sharma, C. Lucier, and R. Molloy, "From Solutions to Symbiosis: Blending with Your Customers," *Strategy + Business*, Second Quarter 2002, pp. 38–48; and S. Leibs, "Getting Ready: Your Suppliers," *IndustryWeek*, September 6, 1999.

20. D. Bartholomew, "The Infrastructure," *IndustryWeek*, September 6, 1999, p. 1.

21. T. Stevens, "Integrated Product Development," *IndustryWeek*, June 2002, pp. 21–28.

22. T. Vinas, "A Map of the World: IW Value-Chain Survey," *IndustryWeek*, September 2005, pp. 27–34.

23. R. Normann and R. Ramirez, "From Value Chain to Value Constellation," *Harvard Business Review on Managing the Value Chain* (Boston: Harvard Business School Press, 2000), pp. 185–219.

24. J. Teresko, "The Tough Get Going," *IndustryWeek*, March 2005, pp. 25–32; D. M. Lambert and A. M. Knemeyer, "We're in This Together," *Harvard Business Review*, December 2004, pp. 114–122; and V. G. Narayanan and A. Raman, "Aligning Incentives in Supply Chains," *Harvard Business Review*, November 2004, pp. 94–102.

25. D. Drickhamer, "Looking for Value," *IndustryWeek*, December 2002, pp. 41–43.

26. J. Teresko, "Tying IT Assets to Process Success," *IndustryWeek*, September 2005, p. 21.

27. Sheridan, "Managing the Value Chain," p. 3.

28. S. Leibs, "Getting Ready: Your Customers," *IndustryWeek*, September 6, 1999, p. 1.

29. G. Taninecz, "Forging the Chain," *IndustryWeek*, May 15, 2000, pp. 40–46.

30. Leibs, "Getting Ready: Your Customers."

31. Information in this box from J. McPartlin, "Making Waves," *CFO-IT*, Spring 2005, pp. 32–37.

32. ASIS International and Pinkerton, *Top Security Threats and Management Issues Facing Corporate America: 2003 Survey of Fortune 1000 Companies*, www.asisonline.org.

33. Sheridan, "Managing the Value Chain," p. 4.

34. R. Russell and B. W. Taylor, *Operations Management*, 5th ed. (New York: Wiley, 2005); C. Liu-Lien Tan, "U.S. Response: Speedier Delivery," *Wall Street Journal*, November 18, 2004, pp. D1+; and C. Salter, "When Couches Fly," *Fast Company*, July 2004, pp. 80–81.

35. S. Anderson, "Restaurants Gear Up for Window Wars," *Springfield (Missouri) News-Leader*, January 27, 2006, p. 5B.

36. http://www.economictimes.indiatimes.com/Honda_to_recall_2310_Accords_for_fuel_pump_snag/rssarticleshow/2228675.cms

37. http://www.domain-b.com/companies/companies_n/nokia/20070814_phone_batteries.html

38. http://www.businessstandard.com/india/storypage.php?autono=294819, Product recalls to cost MNCs dear. Leslie D'Monte in Mumbai August 17, 2007 14:01 IST

39. T. Vinas, "Six Sigma Rescue," *IndustryWeek*, March 2004, p. 12.

40. http://www.engineering.indiabizclub.com/profile/1736876~atul+group+of+companies~agra_india

41. http://www.metalinoxindia.co.in

42. http://www.tavant.com/it_services/quality_assurance_at_tavant.html

43. http://www.indianherbs.org/index.htm

44. http://www.tata.com/aboutus/articles/inside.aspx?artid=3vAdAoEGvJU=

45. W. Royal, "Spotlight Shines on Maquiladora," *IndustryWeek*, October 16, 2000, pp. 91–92.

46. See B. Whitford and R. Andrew (eds.), *The Pursuit of Quality* (Perth, UK: Beaumont Publishing, 1994).

47. D. Drickhamer, "Road to Excellence," *IndustryWeek*, October 16, 2000, pp. 117–118.

48. Information from International Organization for Standardization, *The ISO Survey—2006*, www.iso.org/iso/survey2006.pdf.

49. G. Hasek, "Merger Marries Quality Efforts," *IndustryWeek*, August 21, 2000, pp. 89–92.

50. M. Arndt, "Quality Isn't Just for Widgets," *BusinessWeek*, July 22, 2002, pp. 72–73.

51. E. White, "Rethinking the Quality Improvement Program," *Wall Street Journal*, September 19, 2005, p. B3.

52. http://www.isixsigma.com/forum/showmessage.asp?messageID=89778

53. Arndt, "Quality Isn't Just for Widgets."

54. S. McMurray, "Ford's F-150: Have It Your Way," *Business 2.0*, March 2004, pp. 53–55; "Made-to-Fit Clothes Are on the Way," *USA Today*, July 2002, pp. 8–9; and L. Elliott, "Mass Customization Comes a Step Closer," *Design News*, February 18, 2002, p. 21.

55. E. Schonfeld, "The Customized, Digitized, Have-It-Your-Way Economy," *Fortune*, October 28, 1998, pp. 114–120.

Appendix A

1. http://www.rediff.com/money/2008/feb/18nasscom.htm; http://www.comat.com/leadership_profile.html

2. J. W. Carland, F. Hoy, W. R. Boulton, and J. C. Carland, "Differentiating Entrepreneurs from Small Business Owners: A Conceptualization," *Academy of Management Review*, vol. 9 (2), 1984, pp. 354–359.

3. J. McDowell, *Small Business Continues to Drive U.S. Economy*, Office of Advocacy, U.S. Small Business Administration, October 3, 2005, www.sba.gov.

4. P. Almeida and B. Kogut, "The Exploration of Technological Diversity and Geographic Localization in Innovation: Start-up Firms inthe Semiconductor Industry," *Small Business Economics*, vol. 9 (1), 1997, pp. 21–31.

5. R. J. Arend, "Emergence of Entrepreneurs Following Exogenous Technological Change," *Strategic Management Journal*, vol. 20 (1), 1999, pp. 31–47.

6. U.S. Small Business Administration, Office of Advocacy, *Frequently Asked Questions*, www.sba.gov, April 16, 2007.

7. U.S. Small Business Administration, Office of Advocacy, *The Small Business Economy: A Report to the President*,www.sba.gov/advo/research/sb_econ2007.pdf, December 2007.

8. Ibid.

9. N. Bosma, K. Jones, E. Autio, and J. Levie, *Global Entrepreneurship Monitor: 2007 Executive Report*, www.gemconsortium.org, p. 12.

10. P. F. Drucker, *Innovation and Entrepreneurship: Practice and Principles* (New York: Harper & Row, 1985).

11. W. Royal, "Real Expectations," *IndustryWeek*, September 4, 2000, pp. 31–34.

12. http://www.ibef.org/artdisplay.aspx?cat_id=60&art_id=13853&refer=n10

13. http://www.360d.biz/; http://www.hotteststartups.in/viewandvote.do?method=fetch&businessFn=viewandvote&startupId=442

14. C. Sandlund, "Trust Is a Must," *Entrepreneur*, October 2002, pp. 70–75.

15. B. I. Koerner, "Cereal in the Bowl, Not on the Floor," *New York Times* online, www.nytimes.com, June 18, 2006.

16. http://www.rediff.com/money/2007/mar/22bspec.htm; http://www.livemint.com/2009/01/07224328/CavinKare-to-enterdairy-biz.html?d=1

17. G. B. Knight, "How Wall Street Whiz Found a Niche Selling Books on the Internet," *Wall Street Journal*, May 15, 1996, pp. A1+.

18. N. F. Krueger, Jr., "The Cognitive Infrastructure of Opportunity Emergence," *Entrepreneurship Theory and Practice*, Spring 2000, p. 6.

19. D. P. Forbes, "Managerial Determinants of Decision Speed in New Ventures," *Strategic Management Journal*, April 2005, pp. 355–366.
20. Drucker, *Innovation and Entrepreneurship*.
21. http://www.financialexpress.com/news/Reva-unveils-new-electric-car/407486/; http://www.cartradeindia.com/car-bike-news/electric-car-reva-to-have-increased-production-110536.html
22. http://www.niiconsulting.com/
23. http://www.iloveindia.com/economy-of-india/top-50-companies/jindal.html
24. http://www.indianetzone.com/24/iron_steel_industry_india.htm
25. S. Schubert, "The Ultimate Music Buff," *Business 2.0*, March 2006, p. 64.
26. Latest figures on registered users from Hoover's, www.hoovers.com, July 13, 2008; and A. Cohen, "eBay's Bid to Conquer All," *Time*, February 5, 2001, pp. 48–51.
27. S. McFarland, "Cambodia's Internet Service Is in Kids' Hands," Wall Street Journal, May 15, 2000, p. A9A.
28. http://www.moneycontrol.com/india/news/pressnews/big-bazaar-%E2%80%93-india%E2%80%99s-real-retail-story/19/37/358768
29. http://www.hindu.com/thehindu/mp/2002/12/09/stories/2002120900870300.htm; http://www.namdharifresh.com/introduction.htm#Vegetable%20Production
30. http://www.business.outlookindia.com/inner.aspx?articleid=2276&editionid=61&catgid=1&subcatgid=1027
31. A. Eisenberg, "What's Next: New Fabrics Can Keep Wearers Healthy and Smelling Good," *New York Times*, February 3, 2000, pp. D1+.
32. http://www.knowledge.wharton.upenn.edu/india/article.cfm?articleid=4321
33. http://www.domain-b.com/brand_dossier/adv_brnd/20081111_jumbo_king.html
34. S. Greco, "The Start-up Years," *Inc. 500*, October 21, 1997, p. 57.
35. T. Stevens, "Master of His Universe," *IndustryWeek*, January 15, 2001, pp. 76–80; and R. Grover, "Back from a Black Hole," *BusinessWeek*, May 29, 2000, p. 186.
36. http://www.rsatya.in/?p=80
37. http://www.livemint.com/2007/12/03003049/Indian-startups-find-angel-fu.html
38. I. O. Williamson, "Employer Legitimacy and Recruitment Success in Small Businesses," *Entrepreneurship Theory and Practice*, Fall 2000, pp. 27–42.
39. R. L. Heneman, J. W. Tansky, and S. M. Camp, "Human Resource Management Practices in Small and Medium-Sized Enterprises: Unanswered Questions and Future Research Perspectives," *Entrepreneurship Theory and Practice*, Fall 2000, pp. 11–26.
40. Ibid.
41. "Best Employer," *Working Woman*, May 1999, p. 54.
42. Heneman, Tansky, and Camp, "Human Resource Management Practices in Small and Medium-Sized Enterprises: Unanswered Questions and Future Research Perspectives."
43. Based on G. Fuchsberg, "Small Firms Struggle With Latest Management Trends," *Wall Street Journal*, August 26, 1993, p. B2; M. Barrier, "Re-engineering Your Company," *Nation's Business*, February 1994, pp. 16–22; J. Weiss,"Re-engineering the Small Business," *Small Business Reports*, May 1994, pp. 37–43; and K. D. Godsey, "Back on Track," *Success*, May 1997, pp. 52–54.
44. G. N. Chandler, C. Keller, and D. W. Lyon, "Unraveling the Determinants and Consequences of an Innovation-Supportive Organizational Culture," *Entrepreneurship Theory and Practice*, Fall 2000, pp. 59–76.
45. Ibid.
46. P. Gogoi, "Pregnant with Possibility," *BusinessWeek*, December 26, 2005, p. 50.
47. P. B. Robinson, D. V. Simpson, J. C. Huefner, and H. K. Hunt, "An Attitude Approach to the Prediction of Entrepreneurship," *Entrepreneurship Theory and Practice*, Summer 1991, pp. 13–31.
48. B. M. Davis, "Role of Venture Capital in the Economic Renaissance of an Area," in R. D. Hisrich (ed.), *Entrepreneurship, Intrapreneurship, and Venture Capital* (Lexington, MA: Lexington Books, 1986), pp. 107–18.
49. J. M. Crant, "The Proactive Personality Scale as Predictor of Entrepreneurial Intentions," *Journal of Small Business Management*, July 1996, pp. 42–49.
50. W. H. Stewart, "Risk Propensity Differences Between Entrepreneurs and Managers: A Meta-Analytic Review," *Journal of Applied Psychology*, February 2001, pp. 145–153.
51. Information from Sapient Web site, www.sapient.com, July 7, 2003; and S. Herrera, "People Power," *Forbes*, November 2, 1998, p. 212.
52. "Saluting the Global Awards Recipients of Arthur Andersen's Best Practices Awards 2000," *Fortune Online*, www.fortune.com, January 16, 2001.
53. T. Purdum, "Winning with Empowerment," *IndustryWeek*, October 16, 2000, pp. 109–110.
54. M. DePree, *Leadership Jazz* (New York: Currency Doubleday, 1992), pp. 8–9.
55. J. C. Collins and J. I. Porras, *Built to Last: Successful Habits of Visionary Companies* (New York: Harper Business, 1994).
56. P. Strozniak, "Teams at Work," *IndustryWeek*, September 18, 2000, pp. 47–50.
57. Ibid.
58. T. Siegel Bernard, "Scooter's Popularity Offers a Chance for Growth," *Wall Street Journal*, September 20, 2005, p. B3.
59. G. R. Merz, P. B. Weber, and V. B. Laetz, "Linking Small Business Management with Entrepreneurial Growth," *Journal of Small Business Management*, October 1994, pp. 48–60.
60. J. Bailey, "Growth Needs a Plan or Only Losses May Build," *Wall Street Journal*, October 29, 2002, p. B9; and L. Beresford, "Growing Up," *Entrepreneur*, July 1995, pp. 124–28.
61. J. Summer, "More, Please!" *Business Finance*, July 2000, pp. 57–61.
62. T. Stevens, "Pedal Pushers," *IndustryWeek*, July 17, 2000, pp. 46–52.
63. P. Lorange and R. T. Nelson, "How to Recognize—and Avoid—Organizational Decline," *Sloan Management Review*, Spring 1987, pp. 41–48.

64. S. D. Chowdhury and J. R. Lange, "Crisis, Decline, and Turnaround: A Test of Competing Hypotheses for Short-Term Performance Improvement in Small Firms," *Journal of Small Business Management*, October 1993, pp. 8–17.

65. C. Farrell, "How to Survive a Downturn," *BusinessWeek*, April 28, 1997, pp. ENT4-ENT6.

66. R. W. Pricer and A. C. Johnson, "The Accuracy of Valuation Methods in Predicting the Selling Price of Small Firms," *Journal of Small Business Management*, October 1997, pp. 24–35.

67. J. Bailey, "Selling the Firm and Letting Go of the Dream," *Wall Street Journal*, December 10, 2002, p. B6; P. Hernan, "Finding the Exit," *IndustryWeek*, July 17, 2000, pp. 55–61; D. Rodkin, "For Sale by Owner," *Entrepreneur*, January 1998, pp. 148–153; A. Livingston, "Avoiding Pitfalls When Selling a Business," *Nation's Business*, July 1998, pp. 25–26; and G. Gibbs Marullo, "Selling Your Business: A Preview of the Process," *Nation's Business*, August 1998, pp. 25–26.

68. K. Stringer, "Time Out," *Wall Street Journal*, March 27, 2002, p. R14; T. Stevens, "Striking a Balance," *IndustryWeek*, November 20, 2000, pp. 26–36; and S. Caudron, "Fit to Lead," *IndustryWeek*, July 17, 2000, pp. 63–68.

Appendix B

1. J. Trotsky, "The Futurists," *U.S. News & World Report*, April 19, 2004, pp. EE4–EE6.

2. F. Vogelstein, "Search and Destroy," *Fortune*, May 2, 2005, pp. 73–82.

3. S. C. Jain, "Environmental Scanning in U.S. Corporations," *Long Range Planning*, April 1984, pp. 117–128; see also L. M. Fuld, *Monitoring the Competition* (New York: John Wiley & Sons, 1988); E. H. Burack and N. J. Mathys, "Environmental Scanning Improves Strategic Planning," *Personnel Administrator*, April 1989, pp. 82–87; R. Subramanian, N. Fernandes, and E. Harper, "Environmental Scanning in U.S. Companies: Their Nature and Their Relationship to Performance," *Management International Review*, July 1993, pp. 271–286; B. K. Boyd and J. Fulk, "Executive Scanning and Perceived Uncertainty: A Multidimensional Model," *Journal of Management*, 22 (1), 1996, pp. 1–21; D. S. Elkenov, "Strategic Uncertainty and Environmental Scanning: The Case for Institutional Influences on Scanning Behavior," *Strategic Management Journal*, vol. 18, 1997, pp. 287–302; K. Kumar, R. Subramanian, and K. Strandholm, "Competitive Strategy, Environmental Scanning and Performance: A Context Specific Analysis of Their Relationship," *International Journal of Commerce and Management*, Spring 2001, pp. 1–18; C. G. Wagner, "Top 10 Reasons to Watch Trends," *The Futurist*, March–April 2002, pp. 68–69; and V. K. Garg, B. A. Walters, and R. L. Priem, "Chief Executive Scanning Emphases, Environmental Dynamism, and Manufacturing Firm Performance," *Strategic Management Journal*, August 2003, pp. 725–744.

4. B. Gilad, "The Role of Organized Competitive Intelligence in Corporate Strategy," *Columbia Journal of World Business*, Winter 1989, pp. 29–35; L. Fuld, "A Recipe for Business Intelligence," *Journal of Business Strategy*, January–February 1991, pp. 12–17; J. P. Herring, "The Role of Intelligence in Formulating Strategy," *Journal of Business Strategy*, September–October 1992, pp. 54–60; K. Western, "Ethical Spying," *Business Ethics*, September–October 1995, pp. 22–23; D. Kinard, "Raising Your Competitive IQ: The Payoff of Paying Attention to Potential Competitors," *Association Management*, February 2003, pp. 40–44; K. Girard, "Snooping on a Shoestring," *Business 2.0*, May 2003, pp. 64–66; and "Know Your Enemy," *Business 2.0*, June 2004, p. 89.

5. C. Davis, "Get Smart," *Executive Edge*, October/November 1999, pp. 46–50.

6. B. Ettore, "Managing Competitive Intelligence," *Management Review*, October 1995, pp. 15–19.

7. A. Serwer, "P&G's Covert Operation," *Fortune*, September 17, 2001, pp. 42–44.

8. B. Rosner, "HR Should Get a Clue: Corporate Spying Is Real," *Workforce*, April 2001, pp. 72–75.

9. Western, "Ethical Spying."

10. W. H. Davidson, "The Role of Global Scanning in Business Planning," *Organizational Dynamics*, Winter 1991, pp. 5–16.

11. T. Smart, "Air Supply," *U.S. News & World Report*, February 28, 2005, p. EE10.

12. "Is Supply Chain Collaboration Really Happening?" *ERI Journal*, www.eri.com, January/February 2006; L. Denend and H. Lee, "West Marine: Driving Growth Through Shipshape Supply Chain Management, A Case Study" *Stanford Graduate School of Business*, www.vics.org, April 7, 2005; N. Nix, et al., "Keys to Effective Supply Chain Collaboration: A Special Report from the Collaborative Practices Research Program," *Neeley School of Business, Texas Christian University*, www.vics.org, November 15, 2004; Collaborative, Planning, Forecasting, and Replenishment Committee Web site, www.cpfr.org, May 20, 2003; and J. W. Verity, "Clearing the Cobwebs from the Stockroom," *BusinessWeek*, October 21, 1996, p. 140.

13. See A. B. Fisher, "Is Long-Range Planning Worth It?" *Fortune*, April 23, 1990, pp. 281–284; J. A. Fraser, "On Target," *Inc.*, April 1991, pp. 113–114; P. Schwartz, *The Art of the Long View* (New York: Doubleday/Currency, 1991); G. Hamel and C. K. Prahalad, "Competing for the Future," *Harvard Business Review*, July–August 1994, pp. 122–128; F. Elikai and W. Hall, Jr., "Managing and Improving the Forecasting Process," *Journal of Business Forecasting Methods & Systems*, Spring 1999, pp. 15–19; L. Lapide, "New Developments in Business Forecasting," *Journal of Business Forecasting Methods & Systems*, Summer 1999, pp. 13–14; and T. Leahy, "Building Better Forecasts," *Business Finance*, December 1999, pp. 10–12.

14. J. Goff, "Start with Demand," *CFO*, January 2005, pp. 53–57.

15. L. Brannen, "Upfront: Global Planning Perspectives," *Business Finance*, March 2006, pp. 12+.

16. R. Durand, "Predicting a Firm's Forecasting Ability: The Roles of Organizational Illusion of Control and Organizational Attention," *Strategic Management Journal*, September 2003, pp. 821–838.

17. P. N. Pant and W. H. Starbuck, "Innocents in the Forest: Forecasting and Research Methods," *Journal of Management*, June 1990, pp. 433–460; Elikai and Hall, "Managing and Improving the Forecasting Process"; M. A. Giullian, M. D. Odom, and M. W. Totaro, "Developing Essential Skills for Success in the Business World: A Look at Forecasting," *Journal of Applied Business Research*, Summer 2000, pp. 51–65; and T. Leahy, "Turning Managers into Forecasters," *Business Finance*, August 2002, pp. 37–40.

18. Leahy, "Turning Managers into Forecasters."

19. J. Hope, "Use a Rolling Forecast to Spot Trends," *Harvard Business School Working Knowledge*, hbswk.hbs.edu, March 13, 2006.

20. This section is based on Y. K. Shetty, "Benchmarking for Superior Performance," *Long Range Planning*, vol. 1, April 1993, pp. 39–44; G. H. Watson, "How Process Benchmarking Supports Corporate Strategy," *Planning Review*, January–February 1993, pp. 12–15; S. Greengard, "Discover Best Practices," *Personnel Journal*, November 1995, pp. 62–73; J. Martin, "Are You as Good as You Think You Are?" *Fortune*, September 30, 1996, pp. 142–152; R. L. Ackoff, "The Trouble with Benchmarking," *Across the Board*, January 2000, p. 13; V. Prabhu, D. Yarrow, and G. Gordon-Hart, "Best Practice and Performance Within Northeast Manufacturing," *Total Quality Management*, January 2000, pp. 113–121; "E-Benchmarking: The Latest E-Trend," *CFO*, March 2000, p. 7; E. Krell, "Now Read This," *Business Finance*, May 2000, pp. 97–103; and H. Johnson, "All in Favor Say Benchmark!" *Training*, August 2004, pp. 30–34.

21. "Newswatch," *CFO*, July 2002, p. 26.

22. Benchmarking examples from S. Carey, "Racing to Improve," *Wall Street Journal*, March 24, 2006, pp. B1+; D. Waller, "NASCAR: The Army's Unlikely Adviser," *Time*, July 4, 2005, p. 19; A. Taylor, III, "Double Duty," *Fortune*, March 7, 2005, p. 108; P. Gogoi, "Thinking Outside the Cereal Box," *BusinessWeek*, July 28, 2003, pp. 74–75; "Benchmarkers Make Strange Bedfellows," *IndustryWeek*, November 15, 1993, p. 8; G. Fuchsberg, "Here's Help in Finding Corporate Role Models," *Wall Street Journal*, June 1, 1993, p. B1; and A. Tanzer, "Studying at the Feet of the Masters," *Forbes*, May 10, 1993, pp. 43–44.

23. E. Krell, "The Case Against Budgeting," *Business Finance*, July 2003, pp. 20–25; J. Hope and R. Fraser, "Who Needs Budgets?" *Harvard Business Review*, February 2003, pp. 108–115; T. Leahy, "The Top 10 Traps of Budgeting," *Business Finance*, November 2001, pp. 10–16; T. Leahy, "Necessary Evil," *Business Finance*, November 1999, pp. 41–45; J. Fanning, "Businesses Languishing in a Budget Comfort Zone?" *Management Accounting*, July/August 1999, p. 8; "Budgeting Processes: Inefficiency or Inadequate?" *Management Accounting*, February 1999, p. 5; A. Kennedy and D. Dugdale, "Getting the Most from Budgeting," *Management Accounting*, February 1999, pp. 22–24; G. J. Nolan, "The End of Traditional Budgeting," *Bank Accounting & Finance*, Summer 1998, pp. 29–36; and J. Mariotti, "Surviving the Dreaded Budget Process," *IndustryWeek*, August 17, 1998, p. 150.

24. See, for example, S. Stiansen, "Breaking Even," *Success*, November 1988, p. 16.

25. S. E. Barndt and D. W. Carvey, *Essentials of Operations Management* (Upper Saddle River, NJ: Prentice Hall, 1982), p. 134.

26. E. E. Adam, Jr., and R. J. Ebert, *Production and Operations Management*, 5th ed. (Upper Saddle River, NJ: Prentice Hall, 1992), p. 333.

27. See, for instance, C. Benko and F. W. McFarlan, *Connecting the Dots: Aligning Projects with Objectives in Unpredictable Times* (Boston: Harvard Business School Press, 2003); M. W. Lewis, M. A. Welsh, G. E. Dehler, and S. G. Green, "Product Development Tensions: Exploring Contrasting Styles of Project Management," *Academy of Management Journal*, June 2002, pp. 546–564; C. E. Gray and E. W. Larsen, *Project Management: The Managerial Process* (Columbus, OH: McGraw-Hill Higher Education, 2000); J. Davidson Frame, *Project Management Competence: Building Key Skills for Individuals, Teams, and Organizations* (San Francisco: Jossey-Bass, 1999).

28. For more information, see Project Management Software Directory Web site, infogoal.com/pmc/pmcswr.htm.

29. D. Zielinski, "Soft Skills, Hard Truth," *Training*, July 2005, pp. 19–23.

30. H. Collingwood, "Best Kept Secrets of the World's Best Companies: Secret 05, Bad News Folders," *Business 2.0*, April 2006, p. 84.

31. G. Colvin, "An Executive Risk Handbook," *Fortune*, October 3, 2005, pp. 69–70; A. Long and A. Weiss, "Using Scenario Planning to Manage Short-Term Uncertainty," *Outward Insights*, www.outwardinsights.com, 2005; B. Fiora, "Use Early Warning to Strengthen Scenario Planning," *Outward Insights*, www.outwardinsights.com, 2003; L. Fahey, "Scenario Learning," *Management Review*, March 2000, pp. 29–34; S. Caudron, "Frontview Mirror," *Business Finance*, December 1999, pp. 24–30; and J. R. Garber, "What if...?" *Forbes*, November 2, 1998, pp. 76–79.

32. Caudron, "Frontview Mirror," p. 30.

Glindex

P

Q

R

T

U

Photo Credits

Chapter 1: page 4: Rajeev Bhargava/Picture-it Studio; page 10: Gaurav Jain; page 13: HCL Technologies; page 21: Jeff Minton Photography

Chapter 2: page 24: Sameer Gupta; page 26: S.C. Williams Library; page 28: UPI/Corbis/Bettmann; page 31: Abhishek R. Gowda; page 41: Brian Harkin/Redux Pictures

Chapter 3: page 44: Avnish Garg; page 49: Anshul Yadav; page 54: Tony Avelar/AFP/Getty Images, Inc. AFP; page 60: Praveen Tiwari; page 67: Toshifumi Kitamura/AFP/Getty Images

Chapter 4: page 70: Andrea Greenan/W.R. Grace & Co.; page 72: Cabrera Georges/Kempinski Hotels; page 79: Nitin Valecha; page 89: Vario Images/GmbH & Co. KG/Alamy Images

Chapter 5: page 94: Jubilant Organosys; page 99: Clive Sawyer/Alamy Images; page 109: Neharika Vohra; page 118: D. Hurst/Alamy Images

Chapter 6: page 122: Vibhor Kataria; page 124: Avnish Garg; page 130: Quentin Shih aka Shi Xiaofan; page 136: Brian Smith; page 142: Jeff Haynes/AFP/ Getty Images, Inc. AFP

Chapter 7: page 146: Alibaba.com; page 147: Nirma University; page 150: AP Wide World Photos; page 161: PSL Images/Alamy Images

Chapter 8: page 164: Steve LaBadessa/Steve LaBeadessa Photography; page 170: Ara Koopelian Photography; page 173: Dave Yoder/Polaris Images; page 182: Steve Goldstein/The New York Times/Redux Pictures

Chapter 9: page 186: Nicky Bonne/Redux Pictures; page 192: Microchip Technology Inc.; page 198: Gautam Singh/AP Wide World Photos; page 204: Mark Mahaney/Mark Mahaney

Chapter 10: page 208: Varun Arora; page 212: Price Waterhouse Cooper; page 218: Lynsey Addario/CORBIS-NY; page 225: Joel Salcido/Salcido Photography; page 232: Tom Dodge, Columbus Dispatch/AP Wide World Photos

Chapter 11: page 236: Robyn Twomey/Robyn Twomey Photography; page 238: Micha Bar Am/Magnum Photos, Inc.; page 251: Rajeev Bhargava/Picture-it Studio; page 258: Thomas Strand/Thomas Strand Studio

Chapter 12: page 262: Avnish Garg; page 267: Rajan Bharucha; page 281: Helen Cathcart/Alamy Images

Chapter 13: page 286: Steve Ringman/Seattle Times/MCT/Newscom; page 290: Joshua Lutz/Redux Pictures; page 298: Thomas Hartwell/Redux Pictures; page 307: Gaurav Dagaonkar/Gaurav Online; page 314: imagebroker/Alamy Images

Chapter 14: page 318: U.S. Cellular; page 319: Rajeev Bhargava/Picture-it Studio; page 323: Sahil Parikh/Deskaway/Synage; page 332: Kim Christiansen Photography; page 340: Cheryl Senter/AP Wide World Photos

Chapter 15: page 344: Teach for India; page 350: Ashok Leyland; page 358: Canon India; page 370: Marcio Jose Sanchez/AP Wide World Photos

Chapter 16: page 374: Sandy George/Citigate Cunningham for HCL; page 382: AP Wide World Photos; page 386: Tractors and Farm Equipment (TAFE); page 391: Manish Swarup; AP Wide World Photos; page 398: Paulo Fridman

Chapter 17: page 402: Newscom; page 414: Titan Industries/Tata; page 421: Sameer Gupta; page 426: Steve Parsons/PA Wire URN: 5705238/AP Wide World Photos

Chapter 18: page 430: Frank van der Most; page 437: Tony Law/Redux Pictures; page 448: Fabienne Fossez/Alamy Images